Chapter	Exercises	Tutorials	Active Figures	Additional Resources
8 **Atomic Electron Configurations and Chemical Periodicity**	• Screen 8.6: Effective Nuclear Charge, Z*	• Screen 8.7: Atomic Electron Configurations • Screen 8.8: Electron Configuration in Ions	• 8.2: Observing and Measuring Paramagnetism • 8.4: Experimentally Determined Order of Subshell Energies • 8.7: Electron Configurations and the Periodic Table • 8.9: Examples of the Periodicity of Group 1A and Group 7A Elements • 8.11: Atomic Radii in Picometers for Main Group Elements • 8.13: First Ionization Energies of the Main Group Elements of the First Four Periods • 8.14: Electron Affinity • 8.15: Relative Sizes of Some Common Ions	• Screen 8.3: Spinning Electrons and Magnetism • Screen 8.6: Effective Nuclear Charge, Z* • Screen 8.7: Atomic Electron Configurations • Screen 8.8: Electron Configuration in Ions • Screen 8.9: Atomic Properties and Periodic Trends • Screen 8.10: Atomic Sizes • Screen 8.11: Ionization Energy • Screen 8.12: Electron Affinity • Screen 8.14: Ion Size • Screen 8.15: Chemical Reactions and Periodic Properties
9 **Bonding and Molecular Structure: Fundamental Concepts**	• Screen 9.8: Drawing Lewis Structures • Screen 9.14: Determining Molecular Shape	• Screen 9.7: Lewis Electron Dot Structures • Screen 9.8: Drawing Lewis Structures • Screen 9.9: Resonance Structures • Screen 9.10: Exceptions to the Octet Rule • Screen 9.13: Ideal Electron Repulsion Shapes • Screen 9.14: Determining Molecular Shape	• 9.3: Lattice Energy • 9.8: Various Geometries Predicted by VSEPR • 9.14: Electronegativity Values for the Elements According to Pauling • 9.16: Polarity of Triatomic Molecules, AB_2 • 9.17: Polar and Nonpolar Molecules of the Type AB_3	• Screen 9.2: Valence Electrons • Screen 9.4: Lattice Energy • Screen 9.5: Chemical Reactions and Periodic Properties • Screen 9.6: Chemical Bond Formation—Covalent Bonding • Screen 9.13: Ideal Electron Repulsion Shapes • Screen 9.16: Formal Charge • Screen 9.17: Bond Polarity and Electronegativity • Screen 9.18: Molecular Polarity • Screen 9.19: Bond Properties • Screen 9.20: Bond Energy and ΔH_{rxn}
10 **Bonding and Molecular Structure: Orbital Hybridization and Molecular Orbitals**	• Screen 10.8: Molecular Fluxionality • Screen 10.9: Molecular Orbital Theory • Screen 10.11: Homonuclear Diatomic Molecules	• Screen 10.5: Sigma Bonding • Screen 10.6: Determining Hybrid Orbitals • Screen 10.7: Multiple Bonding	• 10.1: Potential Energy Change During H—H Bond Formation • 10.5: Hybrid Orbitals for Two to Six Electron Pairs • 10.6: Bonding in the Methane (CH_4) Molecule • 10.10: The Valence Bond Model of Bonding in Ethylene, C_2H_4 • 10.13: Rotation Around Bonds • 10.22: Molecular Orbital Energy Level Diagram	• Screen 10.3: Valence Bond Theory • Screen 10.4: Hybrid Orbitals • Screen 10.10: Molecular Orbital Configurations

Media Integration Guide

Chapter	Exercises	Tutorials	Active Figures	Additional Resources
11 **Carbon: More Than Just Another Element**	• Screen 11.6: Functional Groups (1): Reactions of Alcohols	• Screen 11.4: Hydrocarbons and Addition Reactions • Screen 11.6: Functional Groups	• 11.2: Optical Isomers • 11.4: Alkanes • 11.7: Bacon Fat and Addition Reactions • 11.13: Polyethylene • 11.18: Nylon-6,6	• Screen 11.3: Hydrocarbons • Screen 11.4: Hydrocarbons and Addition Reactions • Screen 11.6: Functional Groups • Screens 11.9, 11.10: Synthetic Organic Polymers
12 **Gases & Their Properties**	• Screen 12.5: Gas Density • Screen 12.12: Application of the Kinetic-Molecular Theory: Diffusion	• Screen 12.6: Using Gas Laws: Determining Molar Mass • Screen 12.7: Gas Laws and Chemical Reactions: Stoichiometry • Screen 12.8: Gas Mixtures and Partial Pressures	• 12.4: An Experiment to Demonstrate Boyle's Law • 12.6: Charles's Law • 12.18: Gaseous Diffusion	• Screen 12.3: Gas Laws • Screen 12.4: The Ideal Gas Law • Screen 12.5: Gas Density • Screen 12.9: The Kinetic-Molecular Theory of Gases: Gases on the Molecular Scale • Screen 12.10: Gas Laws and Kinetic-Molecular Theory • Screen 12.11: Distribution of Molecular Speeds: Maxwell-Boltzmann Curves • Screen 12.12: Application of the Kinetic-Molecular Theory: Diffusion
13 **Intermolecular Forces, Liquids, and Solids**	• Screen 13.5: Intermolecular Forces (3) • Screen 13.17: Phase Changes	• Screen 13.5: Intermolecular Forces (3) • Screen 13.9: Properties of Liquids	• 13.2: Ion–Dipole Interactions • 13.8: The Boiling Points of Some Simple Hydrogen Compounds • 13.11: The Temperature Dependence of the Densities of Ice and Water • 13.17: Vapor Pressure • 13.18: Vapor Pressure Curves for Diethyl Ether [(C₂H₅)₂O], Ethanol (C₂H₅OH), and Water • 13.39: Phase Diagram for Water	• Screen 13.2: Phases of Matter • Screens 13.3, 13.4, 13.5: Intermolecular Forces • Screen 13.6: Hydrogen Bonding • Screen 13.7: The Weird Properties of Water • Screens 13.8, 13.9, 13.10, 13.11: Properties of Liquids • Screens 13.12, 13.13, 13.14, 13.15: Solid Structures • Screens 13.17: Phase Changes
14 **Solutions and Their Behavior**	• Screen 14.2: Solubility • Screen 14.5: Factors Affecting Solubility (1)—Henry's Law and Gas Pressure • Screens 14.7, 14.8: Colligative Properties	• Screens 14.5, 14.6: Factors Affecting Solubility • Screens 14.7, 14.8, 14.9: Colligative Properties	• 14.6: Solubility of Nonpolar Iodine in Polar Water and Nonpolar Carbon Tetrachloride • 14.9: Dissolving an Ionic Solid in Water	• Screen 14.3: The Solution Process: Intermolecular Forces • Screen 14.4: Energetics of Solution Formation—Dissolving Ionic Compounds • Screen 14.9: Colligative Properties
15 **Principles of Reactivity: Chemical Kinetics**	• Screen: 15.4 Concentration Dependence • Screen: 15.5 Determination of the Rate Equation (1) • Screen 15.12: Reaction Mechanisms • Screen 15.13: Reaction Mechanisms and Rate Equations • Screen 15.14: Catalysis and Reaction Rate	• Screen 15.4: Concentration Dependence • Screen 15.5: Determination of the Rate Equation (1) • Screen 15.6: Concentration–Time Relationships • Screen 15.7: Determination of Rate Equation (2) • Screen 15.8: Half-Life • Screen 15.10: Control of Reaction Rates (3)	• 15.2: A Plot of Reactant Concentration Versus Time for the Decomposition of N_2O_5 • 15.7: The Decomposition of H_2O_2 • 15.9: Half-Life of a First-Order Reaction • 15.13: Activation Energy • 15.14: Arrhenius Plot	• Screen 15.2: Rates of Chemical Reactions • Screens 15.3, 15.4, 15.10: Control of Reaction Rates • Screen 15.4: Concentration Dependence • Screen 15.5: Determination of the Rate Equation (1) • Screens 15.9, 15.10: Microscopic View of Reactions • Screen 15.14: Catalysis and Reaction Rate

Media Integration Guide

		3A (13)	4A (14)	5A (15)	6A (16)	7A (17)	8A (18)
							Helium 2 **He** 4.0026
		Boron 5 **B** 10.811	Carbon 6 **C** 12.011	Nitrogen 7 **N** 14.0067	Oxygen 8 **O** 15.9994	Fluorine 9 **F** 18.9984	Neon 10 **Ne** 20.1797
2B (12)		Aluminum 13 **Al** 26.9815	Silicon 14 **Si** 28.0855	Phosphorus 15 **P** 30.9738	Sulfur 16 **S** 32.066	Chlorine 17 **Cl** 35.4527	Argon 18 **Ar** 39.948
Zinc 30 **Zn** 65.39	Gallium 31 **Ga** 69.723	Germanium 32 **Ge** 72.61	Arsenic 33 **As** 74.9216	Selenium 34 **Se** 78.96	Bromine 35 **Br** 79.904	Krypton 36 **Kr** 83.80
Cadmium 48 **Cd** 112.411	Indium 49 **In** 114.818	Tin 50 **Sn** 118.710	Antimony 51 **Sb** 121.760	Tellurium 52 **Te** 127.60	Iodine 53 **I** 126.9045	Xenon 54 **Xe** 131.29
Mercury 80 **Hg** 200.59	Thallium 81 **Tl** 204.3833	Lead 82 **Pb** 207.2	Bismuth 83 **Bi** 208.9804	Polonium 84 **Po** (208.98)	Astatine 85 **At** (209.99)	Radon 86 **Rn** (222.02)
— 112 — Discovered 1996	— 113 — Discovered 2004	— 114 — Discovered 1999	— 115 — Discovered 2004	— 116 — Discovered 1999		

Terbium 65 **Tb** 158.9253	Dysprosium 66 **Dy** 162.50	Holmium 67 **Ho** 164.9303	Erbium 68 **Er** 167.26	Thulium 69 **Tm** 168.9342	Ytterbium 70 **Yb** 173.04	Lutetium 71 **Lu** 174.967
Berkelium 97 **Bk** (247.07)	Californium 98 **Cf** (251.08)	Einsteinium 99 **Es** (252.08)	Fermium 100 **Fm** (257.10)	Mendelevium 101 **Md** (258.10)	Nobelium 102 **No** (259.10)	Lawrencium 103 **Lr** (262.11)

GENERAL Chemistry ⚛ Now™

Completely integrated with this text!
http://now.brookscole.com/kotz6e

What do you need to learn now?

Take charge of your learning with General ChemistryNow™!

This powerful online learning companion helps you manage and make the most of your study time—and it's included with every new copy of this text! This collection of dynamic technology resources gauges your unique study needs to provide you with a *Personalized Learning Plan* that will enhance your problem-solving skills and understanding of core concepts. Designed to optimize your time investment, **General ChemistryNow™** helps you succeed by focusing your study time on the concepts you need to master.

GENERAL
Chemistry ⚛ Now™

See the General ChemistryNow CD-ROM or website:

- **Screen 6.17 Product-Favored Systems,** for an exercise on the reaction when a Gummi Bear is placed in molten potassium chlorate

Look for these references in the text, such as this one from page 270. They direct you to the corresponding media-enhanced activities on **General ChemistryNow**. The text's *Chapter Goals Revisited* sections are also reinforced through the *Homework and Goals* found on **General ChemistryNow**.

This precise page-by-page integration enables you to go beyond reading about chemistry—you'll actually experience it in action!

Easy to use

The **General ChemistryNow** system includes two powerful assessment components:

- ► **WHAT DO I KNOW?**
  This diagnostic *Exam-Prep Quiz*, based on the text's *Chapter Goals*, gives you an initial assessment of your understanding of core concepts.

- ► **WHAT DO I NEED TO LEARN?**
  A *Personalized Learning Plan* outlines key elements for review.

With a click of the mouse, **General ChemistryNow's** unique activities allow you to:

- ► Create a *Personalized Learning Plan* or review for an exam using the *Exam-Prep Quiz* web quizzes

- ► Explore chemical concepts through tutorials, simulations, and animations

- ► View *Active Figures* and interact with text illustrations. These *Active Figures* will help you master key concepts from the book. Each figure is paired with corresponding questions to help you focus on chemistry at work to ensure that you understand the concepts played out in the animations.

By providing you with a better understanding of exactly what you need to focus on, **General ChemistryNow** saves you time and brings you a step closer to success!

Make the most of your study time— log on to *General ChemistryNow* today!
http://now.brookscole.com/kotz6e

THOMSON™
BROOKS/COLE

Chapter	Exercises	Tutorials	Active Figures	Additional Resources
16 Principles of Reactivity: Chemical Equilibria		• Screen 16.6: Writing Equilibrium Expressions • Screen 16.8: Determining an Equilibrium Constant • Screen 16.9: Systems at Equilibrium • Screen 16.10: Estimating Equilibrium Concentrations • Screens 16.12, 16.13: Disturbing a Chemical Equilibrium	• 16.3: The Reaction of H_2 and I_2 Reaches Equilibrium • 16.9: Changing Concentrations	• Screen 16.2: The Principle of Microscopic Reversibility • Screen 16.3: Equilibrium State • Screen 16.4: Equilibrium Constant • Screen 16.5: The Meaning of the Equilibrium Constant • Screen 16.6: Writing Equilibrium Expressions • Screen 16.9: Systems at Equilibrium • Screens 16.11, 16.13, 16.14: Disturbing a Chemical Equilibrium
17 Principles of Reactivity: The Chemistry of Acids and Bases	• Screen 17.2: Bronsted Acids and Bases	• Screen 17.2: Bronsted Acids and Bases • Screen 17.4: The pH Scale • Screen 17.5: Strong Acids and Bases • Screen 17.8: Determining K_a and K_b Values • Screen 17.9: Estimating the pH of Weak Acid Solutions • Screen 17.11: Estimating the pH Following an Acid-Base Reaction • Screen 17.13: Lewis Acids and Bases • Screen 17.15: Neutral Lewis Acids	• 17.2: pH and pOH	• Screen 17.3: The Acid–Base Properties of Water • Screen 17.4: The pH Scale • Screen 17.6: Weak Acids and Bases • Screen 17.7: Acid–Base Reactions • Screen 17.12: Acid–Base Properties of Salts • Screen 17.14: Cationic Lewis Acids • Screen 17.16: Molecular Interpretation of Acid–Base Behavior
18 Principles of Reactivity: Other Aspects of Aqueous Equilibria		• Screen 18.3: Buffer Solutions • Screen 18.4: pH of Buffer Solutions • Screen 18.5: Preparing Buffer Solutions • Screen 18.6: Adding Reagents to a Buffer Solution • Screen 18.7: Titration Curves • Screen 18.12: Solubility Product Constant • Screen 18.13: Determining K_{sp} Experimentally • Screen 18.14: Estimating Salt Solubility: Using K_{sp} • Screen 18.15: Common Ion Effect • Screen 18.16: Solubility and pH • Screen 18.17: Can a Precipitation Reaction Occur? • Screen 18.19: Complex Ion Formation and Solubility	• 18.2: Buffer Solutions • 18.5: The Change in pH During the Titration of a Weak Acid with a Strong Base	• Screen 18.2: Common Ion Effect • Screen 18.3: Buffer Solutions • Screen 18.4: pH of Buffer Solutions • Screen 18.5: Preparing Buffer Solutions • Screen 18.7: Titration Curves • Screen 18.8: Titration of a Weak Polyprotic Acid • Screen 18.9: Titration of a Weak Base with a Strong Acid • Screen 18.10: Acid–Base Indicators • Screen 18.11: Precipitation Reactions • Screen 18.12: Solubility Product Constant • Screen 18.15: Common Ion Effect • Screen 18.16: Solubility and pH • Screen 18.17: Can a Precipitation Reaction Occur? • Screen 18.18: Simultaneous Equilibria • Screen 18.20: Using Solubility
19 Principles of Reactivity: Entropy and Free Energy		• Screen 19.5: Calculating ΔS for a Chemical Reaction • Screen 19.6: The Second Law of Thermodynamics • Screen 19.7: Gibbs Free Energy • Screen 19.8: Free Energy and Temperature • Screen 19.9: Thermodynamics and the Equilibrium Constant	• 19.12: Spontaneity ΔG^{o} with Temperature • 19.13: Free Energy Changes as a Reaction Approaches Equilibrium	• Screen 19.2: Reaction Spontaneity • Screen 19.3: Directionality of Reactions • Screen 19.4: Entropy: Matter Dispersal and Disorder • Screen 19.6: The Second Law of Thermodynamics • Screen 19.8: Free Energy and Temperature • Screen 19.9: Thermodynamics and the Equilibrium Constant

Media Integration Guide

Chapter	Exercises	Tutorials	Active Figures	Additional Resources
20 Principles of Reactivity: Electron Transfer Reactions		• Screen 20.6: Standard Potentials • Screen 20.8: Cells at Nonstandard Conditions • Screen 20.12: Coulometry Counting Electrons	• 20.13: A Voltaic Cell Using $Zn|Zn^{2+}$(aq, 1.0 M) and $H_2|H^+$(aq, 1.0 M) Half-Cells	• Screen 20.2: Redox Reactions: Electron Transfer • Screen 20.3: Balancing Equations for Redox Reactions • Screen 20.4: Electrochemical Cells • Screen 20.5: Batteries • Screen 20.5: Electrochemical Cells and Potentials • Screen 20.6: Standard Potentials • Screen 20.11: Electrolysis: Chemical Change from Electrical Energy
21 The Chemistry of the Main Group Elements	• Screen 21.4: Boron Hydrides Structures • Screen 21.5: Aluminum Compounds • Screen 21.6: Silicon-Oxygen Compounds: Formulas and Structures • Screen 21.8: Sulfur Allotropes • Screen 21.9: Structures of Sulfur Compounds	• Screen 21.2: Formation of Ionic Compounds by Main Group Elements	• 21.15: Industrial Production of Aluminum • 21.22: Compounds and Oxidation Numbers for Nitrogen • 21.32: A Membrane Cell for the Production of NaOH and Cl_2 Gas from a Saturated, Aqueous Solution of NaCl (Brine)	
22 The Chemistry of the Transition Elements	• Screen 22.2: Formulas and Oxidation Numbers in Transition Metal Complexes • Screen 22.5: Geometry of Coordination Compounds • Screen 22.6: Geometric Isomerism in Coordination Compounds		• 22.8: A Blast Furnace	• Screen 21.7: Electronic Structure in Transition Metal Complexes • Screen 21.8: Spectroscopy of Transition Metal Complexes • Screen 22.3: Periodic Trends for Transition Elements
23 Nuclear Chemistry	• Screen 23.5: Kinetics of Nuclear Decay	• Screen 23.2: Radioactive Decay • Screen 23.3: Balancing Nuclear Reaction Equations • Screen 23.4: Stability of Atomic Nuclei • Screen 23.5: Kinetics of Nuclear Decay		• Screen 23.4: Stability of Atomic Nuclei • Screen 23.6: Nuclear Fission

Media Integration Guide

Chemistry
& CHEMICAL REACTIVITY

SIXTH EDITION

John C. Kotz

SUNY Distinguished Teaching Professor
State University of New York
College at Oneonta

Paul M. Treichel

Professor of Chemistry
University of Wisconsin–Madison

Gabriela C. Weaver

Associate Professor of Chemistry
Purdue University

THOMSON

BROOKS/COLE

Australia • Canada • Mexico • Singapore • Spain • United Kingdom • United States

THOMSON

BROOKS/COLE

Publisher/Executive Editor: DAVID HARRIS
Development Editor: PETER MCGAHEY
Assistant Editor: ANNIE MAC
Editorial Assistant: CANDACE LUM
Technology Project Manager: DONNA KELLEY
Executive Marketing Manager: JULIE CONOVER
Senior Marketing Manager: AMEE MOSLEY
Marketing Communications Manager: NATHANIEL BERGSON-MICHELSON
Project Manager, Editorial Production: LISA WEBER
Creative Director: ROB HUGEL
Print Buyers: REBECCA CROSS AND JUDY INOUYE
Permissions Editor: KIELY SEXTON

Production Service: THOMPSON STEELE, INC.
Text Designers: ROB HUGEL AND JOHN WALKER DESIGN
Photo Researcher: JANE SANDERS MILLER
Copy Editor: THOMPSON STEELE, INC.
Developmental Artist: PATRICK A. HARMAN
Illustrators: ROLIN GRAPHICS AND THOMPSON STEELE, INC.
Cover Designer: JOHN WALKER DESIGN
Cover Images: MOTOHIKO MURAKAMI
Cover Printer: TRANSCONTINENTAL PRINTING/INTERGLOBE
Compositor: THOMPSON STEELE, INC.
Printer: TRANSCONTINENTAL PRINTING/INTERGLOBE

Printed in Canada
1 2 3 4 5 6 7 08 07 06 05 04

For more information about our products, contact us at:
THOMSON LEARNING ACADEMIC RESOURCE CENTER
1-800-423-0563

For permission to use material from this text or product, submit a request online at: **http://www.thomsonrights.com**

Any additional questions about permissions can be submitted by email to: **thomsonrights@thomson.com**

Library of Congress Control Number: 2004109955

Student Edition: ISBN 0-534-99766-X

Volume 1: ISBN 0-495-01013-8

Volume 2: ISBN 0-495-01014-6

Two-volume set: ISBN 0-534-40800-1

Instructor's Edition: ISBN 0-534-99848-8

International Student Edition: ISBN 0-534-39597-X
(Not for sale in the United States)

Thomson Brooks/Cole
10 Davis Drive
Belmont, CA 94002-3098
USA

Asia
Thomson Learning
5 Shenton Way #01-01
UIC Building
Singapore 068808

Australia/New Zealand
Thomson Learning
102 Dodds Street
Southbank, Victoria 3006
Australia

Canada
Nelson
1120 Birchmount Road
Toronto, Ontario M1K 5G4
Canada

Europe/Middle East/Africa
Thomson Learning
High Holborn House
50/51 Bedford Row
London WC1R 4LR
United Kingdom

About the Cover
What lies beneath the Earth's surface? The mantle of the Earth consists largely of silicon-oxygen based minerals. But about 2900 km below the surface the solid silicate rock of the mantle gives way to the liquid iron alloy core of the planet. To explore the nature of the rocks at the core-mantle boundary, scientists in Japan examined magnesium silicate ($MgSiO_3$) at a high pressure (125 gigapascals) and high temperature (2500 K). The cover image is what they saw. The solid consists of SiO_6 octahedra (blue) and magnesium ions (Mg^{2+}; yellow spheres). Each SiO_6 octahedron shares the four O atoms in opposite edges with two neighboring octahedra, thus forming a chain of octahedra. These chains are interlinked by sharing the O atoms at the "top" and "bottom" of SiO_6 octahedra in neighboring chains. The magnesium ions lie between the layers of interlinked SiO_6 chains. For more information see M. Murakami, K. Hirose, K. Kawamura, N. Sata, and Y. Ohishi, *Science,* Volume 304, page 855, May 7, 2004.

Brief Contents

Contents

This text is available in these student versions:
- Complete text ISBN 0-534-99766-X • Volume 1 (Chapters 1–12) ISBN 0-495-01013-8
- Volume 2 (Chapters 12–23) ISBN 0-495-01014-6 • Two-volume set ISBN 0-534-40800-1

Charles D. Winters

page 19

Charles D. Winters

page 25

Charles D. Winters

page 82

6   Principles of Reactivity: Energy and Chemical Reactions   232

INTERCHAPTER

The Chemistry of Fuels and Energy Sources   282

Charles D. Winters

page 145

Charles D. Winters

page 214

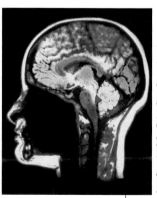

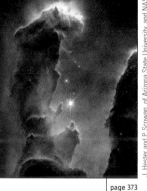

Scott Camazine & Sue Trainor/Photo Researchers, Inc.

J. Hester and P. Scowan, of Arizona State University, and NASA.

page 339   page 373

page 515   page 568

page 646   page 651

Part 4

The Control of Chemical Reactions

page 686

page 763

Charles D. Winters

Charles D. Winters

page 882   page 921

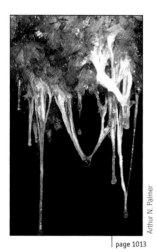

Arthur N. Palmer

© Ludovic Maisant/Corbis

| page 1013 | page 1052 |

Preface

We are gratified that *Chemistry & Chemical Reactivity* has been used by more than a million students in its first five editions. Because this is one indication our book has been successful in helping students learn chemistry, we believe the goals we set out in the first edition are still appropriate. Our principal goals have always been to provide a broad overview of the principles of chemistry, the reactivity of the chemical elements and their compounds, and the applications of chemistry. We have organized this approach around the close relation between the observations chemists make of chemical and physical changes in the laboratory and in nature and the way these changes are viewed at the atomic and molecular levels.

Charles D. Winters

page 36

Another of our goals has been to convey a sense of chemistry not only as a field that has a lively history but also as one that is highly dynamic, with important new developments occurring every year. Furthermore, we want to provide some insight into the chemical aspects of the world around us. Indeed, a major objective of this book is to provide the tools needed for you to function as a chemically literate citizen. Learning something of the chemical world is just as important as understanding some basic mathematics and biology, and as important as having an appreciation for history, music, and literature. For example, you should know which materials are important to our economy, what some of the reactions in plants and animals and in our environment are, and what role chemists play in protecting the environment.

Among the most exciting and satisfying aspects of our careers as chemists has been our ability to discover new compounds and to find new ways to apply chemical principles and explain what we observe. We hope we have conveyed that sense of enjoyment in this book as well as our awe at what is known about chemistry—and, just as important, what is not known!

Emerging Developments in Content Usage and Delivery

The use of media, presentation tools, and homework management tools has expanded significantly in the last three years. About ten years ago we incorporated electronic media into this text with the first edition of our interactive CD-ROM. It has been used by thousands of students worldwide and has been the most successful attempt to date to encourage students to interact with chemistry.

Multimedia technology has evolved over the past ten years, and so have our students. Students are not only focused on conceptual understanding, but are also keenly aware of the necessity of preparing for examinations. Our challenge as authors and educators is to use students' focus on assessment as a way to help them reach a higher level of conceptual understanding. In light of this goal, we have made major changes in our integrated media program. We have found that few students explore multimedia for its own sake. Therefore, we have redesigned the media so that students now have the opportunity to interact with media based on

Charles D. Winters

page 52

clearly stated chapter goals that are correlated to end-of-chapter questions. By using new diagnostic tools, students will be directed to specific resources based on their levels of understanding. This new program, called *General ChemistryNow*, is described in detail later. The closely related OWL homework management system has also been used by tens of thousands of students, and we are pleased to announce that selected end-of-chapter questions are now available for use within the OWL system.

Audience for the Textbook, the General ChemistryNow CD-ROM and Website, and OWL

The textbook, CD-ROM and website, and OWL are designed to serve introductory courses in chemistry for students interested in further study in science, whether that science is biology, chemistry, engineering, geology, physics, or related subjects. Our assumption is that students beginning this course have had some preparation in algebra and in general science. Although undeniably helpful, a previous exposure to chemistry is neither assumed nor required.

Philosophy and Approach of the Program

We have had three major, albeit not independent, goals since the first edition of the book. The first goal was to write a book that students would enjoy reading and that would offer, at a reasonable level of rigor, chemistry and chemical principles in a format and organization typical of college and university courses today. Second, we wanted to convey the utility and importance of chemistry by introducing the properties of the elements, their compounds, and their reactions as early as possible and by focusing the discussion as much as possible on these subjects. Finally, with the new, integrated media program, we hope to bring students to a higher level of conceptual understanding.

The American Chemical Society has been urging educators to put "chemistry" back into introductory chemistry courses. We agree with this position wholeheartedly. Therefore, we have tried to describe the elements, their compounds, and their reactions as early and as often as possible in several ways. First, numerous color photographs depict reactions occurring, the elements and common compounds, and common laboratory operations and industrial processes. Second, we have tried to bring material on the properties of elements and compounds as early as possible into the Exercises and Study Questions and to introduce new principles using realistic chemical situations. Finally, relevant highlights are given in Chapters 21 and 22 as a capstone to the principles described earlier.

Organization of the Book

Chemistry & Chemical Reactivity has two overarching themes: *chemical reactivity* and *bonding and molecular structure*. The chapters on *principles of reactivity* introduce the factors that lead chemical reactions to be successful in converting reactants to products. Under this topic you will study common types of reactions, the energy involved in reactions, and the factors that affect the speed of a reaction. One reason for the enormous advances in chemistry and molecular biology in the last several decades has been an understanding of molecular structure. Sections of the book on *principles of bonding and molecular structure* lay the groundwork for understanding these developments. Particular attention is paid to an understanding of the structural aspects of such biologically important molecules as DNA.

A glance at the introductory chemistry texts currently available shows there is a generally common order of topics used by educators. With a few minor variations, we have followed that order as well. That is not to say that the chapters in our book cannot be used in some other order. We have written it to be as flexible as possible. For example, the chapter on the behavior of gases (Chapter 12) is placed with chapters on liquids, solids, and solutions (Chapters 13 and 14) because it logically fits with these topics. It can easily be read and understood, however, after covering only the first four or five chapters of the book. Similarly, chapters on atomic and molecular structure (Chapters 7–10) could be used before the chapters on stoichiometry and common reactions (Chapters 4 and 5). Also, the chapters on chemical equilibria (Chapters 16–18) can be covered before those on solutions and kinetics (Chapters 14 and 15).

Organic chemistry (Chapter 11) is often left to one of the final chapters in chemistry textbooks. We believe that the importance of organic compounds in biochemistry and in consumer products means we should present that material earlier in the sequence of chapters. This coverage follows the chapters on structure and bonding, because organic chemistry nicely illustrates the application of models of chemical bonding and molecular structure. However, one can use the remainder of the book without including this chapter.

The order of topics in the text was also devised to introduce as early as possible the background required for the laboratory experiments usually done in General Chemistry courses. For this reason, chapters on chemical and physical properties, common reaction types, and stoichiometry begin the book. In addition, because an understanding of energy is so important in the study of chemistry, thermochemistry is introduced in Chapter 6.

In addition to the regular chapters, uses and applications of chemistry are described in more detail in interchapters on *The Chemistry of Fuels and Energy Sources, The Chemistry of Life: Biochemistry, The Chemistry of Modern Materials,* and *The Chemistry of the Environment.* These chapters, new to this edition, are described in more detail later in this Preface.

Additionally, *Chemical Perspectives* attempt to bring relevance and perspective to a study of chemistry. These features delve into such topics as nanotechnology, using isotopes, what it means to be in the "limelight," the importance of sulfuric acid in the world economy, sunscreens, and the newly recognized importance of the NO molecule. *Historical Perspectives* describe the historical development of chemical principles and the people who made the advances in our understanding of chemistry.

A *Closer Look* boxes describe ideas that form the background to material under discussion or provide another dimension of the subject. For example, in Chapter 11 on organic chemistry, the "A Closer Look" boxes are devoted to a discussion of structural aspects of important molecules, to petroleum, and to fats and oils. In other chapters we delve into molecular modeling, magnetic resonance, and mass spectrometry.

Finally, *Problem-Solving Tips* provide students with important insights into problem solving. They also identify where, from our experience, students often make mistakes and suggest alternative ways to solve problems.

The chapters of *Chemistry & Chemical Reactivity* are organized into five sections, each grouping with a common theme.

Part 1: The Basic Tools of Chemistry

There are fundamental ideas and methods that are the basis of all of chemistry, and these are introduced in Part 1. Chapter 1 defines important terms and reviews units and mathematical methods. Chapters 2 and 3 introduce basic ideas of atoms,

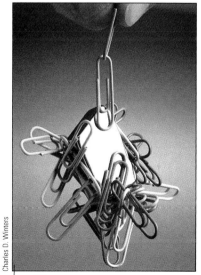

Charles D. Winters

page 368

molecules, and ions, and Chapter 2 describes the most important organizational device in chemistry, the periodic table. In Chapters 4 and 5, we begin to discuss the principles of chemical reactivity and to introduce the numerical methods used by chemists to extract quantitative information from chemical reactions. Chapter 6 introduces the energy involved in chemical processes. The interchapter *The Chemistry of Fuels and Energy Sources* follows Chapter 6 and uses many of the concepts developed in the preceding chapters.

Part 2: The Structure of Atoms and Molecules

The goal of this section is to outline the current theories of the arrangement of electrons in atoms and some of the historical developments that led to these ideas (Chapters 7 and 8). This discussion is tied closely to the arrangement of elements in the periodic table, so that these properties can be recalled and predictions made. In Chapter 9, we discuss for the first time how the electrons of atoms in a molecule participate in chemical bonding and lead to the properties of these bonds. In addition, we show how to derive the three-dimensional structure of simple molecules. Chapter 10 considers the major theories of chemical bonding in more detail.

This part of the book finishes with a discussion of organic chemistry (Chapter 11), primarily from a structural point of view. Organic chemistry is such an enormous area of chemistry that we cannot hope to cover it in detail in this book. Therefore, we have focused on compounds of particular importance, including synthetic polymers and the structures of these materials.

In this section of the book you will find the molecular modeling software on the General ChemistryNow CD-ROM and website to be especially useful.

To cap this section, the interchapter *The Chemistry of Life: Biochemistry* provides an overview of some of the most important aspects of biochemistry.

Part 3: States of Matter

The behavior of the three states of matter—gas, liquid, and solid—is described in that order in Chapters 12 and 13. The discussion of liquids and solids is tied to gases through the description of intermolecular forces, with particular attention given to liquid and solid water. In Chapter 14, we describe the properties of solutions, intimate mixtures of gases, liquids, and solids.

The interchapter *The Chemistry of Modern Materials* is placed after Chapter 13, following coverage of the solid state. Designing and making new materials with useful properties is one of the most exciting areas of modern chemistry.

Part 4: The Control of Chemical Reactions

Part 4 is wholly concerned with the principles of reactivity. Chapter 15 examines the important question of the rates of chemical processes and the factors controlling these rates. With this material on chemical kinetics in mind, we move to Chapters 16–18, which describe chemical reactions at equilibrium. After an introduction to equilibrium in Chapter 16, we highlight reactions involving acids and bases in water (Chapters 17 and 18) and reactions leading to insoluble salts (Chapter 18). To tie together the discussion of chemical equilibria, we again explore thermodynamics in Chapter 19. As a final topic in Part 4, we describe in Chapter 20 a major class of chemical reactions, those involving the transfer of electrons, and the use of these reactions in electrochemical cells.

The *Chemistry of the Environment* interchapter appears at the end of Part 4. This chapter uses ideas from kinetics and chemical equilibria in particular, as well as principles described in earlier chapters in the book.

Royalty-Free/Corbis

page 235

Part 5: The Chemistry of the Elements and Their Compounds

Although the chemistry of the various elements has been described throughout the book to this point, Part 5 considers this topic in a more systematic way. Chapter 21, which has been expanded for this edition, is devoted to the chemistry of the representative elements, whereas Chapter 22 discusses the transition elements and their compounds. Finally, Chapter 23 offers a brief discussion of nuclear chemistry.

Changes for the Sixth Edition

Colleagues and students often ask why yet another edition of the book has been prepared. We all understand, however, that even the most successful books can be improved. In addition, our experience in the classroom suggests that student interests change and that there are ever more effective ways to help our students learn chemistry. For these reasons, we made a number of changes in this book from the fifth edition. For this new, sixth edition, the material and our approach have been refined further to take students to a higher level of conceptual understanding, and several important ideas have been added.

In summary, while this sixth edition retains the overall structure and goals of the previous five editions, we have done much more than change a few words and illustrations. Significant changes have been made that we believe will aid our students in learning and understanding the important principles of chemistry and in discovering that it is an exciting and dynamic field.

Book Revisions

Readability and Clarity   A hallmark of the first five editions of *Chemistry & Chemical Reactivity* has been the book's readability. Nonetheless, each sentence and paragraph in the book has been examined with an eye toward improving clarity and shortening the material without reducing content coverage or readability. Many of the illustrations have been revised and new ones added.

Expanded Coverage   We have worked to raise the level of the text by introducing new material on, among other things, molecular orbital theory and the solid state and on biochemistry and environmental chemistry. The Clausius-Clapeyron equation has been given greater prominence, and "cumulative" and more challenging Study Questions have been added.

Accuracy   Although previous editions of the book have always been relatively free of errors, even greater effort has been made in this edition, and seven accuracy reviewers—four for the text and three for the supplemental chapters—have been brought into our team.

Supplemental Material on Mathematics

A knowledge of basic mathematics is required to be successful in general chemistry. For students unsure of their abilities, a special section (Section 1.8) has been added that reviews exponential notation, significant figures, dimensional analysis, plotting graphs, and reading graphical information.

Supplemental Interchapters

Applications of chemical principles are pervasive in our lives. Although the sixth edition describes many applications as chemical principles are developed, a number of important and interesting areas are left untouched. Therefore, four areas of chemistry are covered in interchapters in a magazine style.

- *The Chemistry of Fuels and Energy Sources* (page 282). This material explores the energy situation confronting our planet and examines such subjects as alternative energy sources, hybrid cars, fuel cells, and "the hydrogen economy."

- *The Chemistry of Life: Biochemistry* (page 530). Perhaps more chemists work in biochemistry than in any other area. This chapter delves into amino acids and proteins, nucleic acids, and metabolism.

- *The Chemistry of Modern Materials* (page 642). The past few decades have seen the development of new electronic devices (such LEDs in car and traffic lights), nanostructures, superconductors, and new adhesives. This supplemental chapter touches on some of these areas as well as others. In addition, there is a discussion of the molecular orbital approach to bonding in metals and semiconductors, material that was in Chapter 10 in the previous edition.

- *The Chemistry of the Environment* (page 998). Environmental issues such as smog, the hole in the earth's ozone layer, global warming, and water quality are regularly encountered in the news. This chapter describes how our water is treated, discusses the effect of particulate pollutants in our atmosphere, and explores the new efforts chemists are making worldwide to produce the products we all rely on in an environmentally safe manner.

Introducing General ChemistryNow Linked to Chapter Goals

Students have always been concerned about "what's on the exam." Although this is certainly a legitimate concern, our challenge as educators has been to help students come to a conceptual understanding and not have them simply learn patterns of thought and memorize equations. To that end, each chapter in the textbook is introduced by 4–6 Chapter Goals that have a conceptual underpinning and are covered in the chapter. These goals are revisited at the end of the chapter, where each goal is divided into several subtopics with which the student should be familiar. Study Questions relevant to the goals are noted in the Chapter Goals Revisited section and are marked with the ■ icon in the Study Questions.

General ChemistryNow at http://now.brookscole.com/kotz6e is a web-based program, which we also offer on a CD-ROM for students who have difficulty accessing the World Wide Web. The program, which is available with each new copy of the book, incorporates material from our original General Chemistry Interactive CD-ROM and includes *more than 400 new step-by-step tutorial modules keyed to end-of-chapter Study Questions*. The system is completely flexible, so students have access to the material through a variety of methods.

- A **Chapter Outline** screen for each chapter matches the text organization.

- A **Homework and Goals** screen is keyed to the Chapter Goals Revisited section in each chapter. Each goal is linked to Simulations, Exercises, and Tutorials and to selected end-of-chapter Study Questions taken from the book. (These questions are marked in the book with ■.) Students can attempt to answer each of the selected Study Questions any number of times, view feedback on the solution, and submit answers online to the instructor for grading.

- A **Diagnostic Exam-Prep Quiz** ("What Do I Know") provides diagnostic questions that have been carefully crafted to assess student understanding of the Chapter Goals. Upon completing a quiz, students receive feedback and a personalized Learning Plan, and, if applicable, will be directed to the relevant Chapter Goals and accompanying resources.

Study Questions

Several important changes have been made in the end-of-chapter questions:

William James Warren/Corbis

page 235

- As noted earlier, approximately 20 Study Questions in each chapter have been selected as illustrative of the chapter goals, and these questions are available in interactive form in General ChemistryNow. These questions are marked in the book with the ■ icon.

- As in previous editions, a number of Study Questions are provided that refer to a particular section of the book. These questions are paired; that is, there are two similar questions with one question (indicated with a blue number) having an answer in Appendix O and a solution in the Student Solutions Manual. The idea is that you can learn how to solve the question without an answer in the appendix by first doing the question for which an answer is provided. Furthermore, for questions on a given section or subsection of the chapter, we note which Example questions or Exercises are relevant. Also, we refer to a particular screen or screens of General ChemistryNow that may be helpful.

- After the sections containing paired questions on specific topics, General Questions integrate concepts from several parts of the chapter.

- Challenging questions are marked with the ▲ icon. The number of these challenging questions has been increased in this sixth edition.

- Some questions rely more heavily than usual on material in preceding chapters or are more conceptual. These questions, sometimes called "cumulative questions," are set out in a separate section called Summary and Conceptual Questions.

- Some questions have been added that call upon students to understand the chemistry at the molecular level.

Homework Management Options

Thousands of students around the country are successfully using the OWL program (Online Web-based Learning) developed at the University of Massachusetts–Amherst. (OWL is described in detail later.) We have heard from many chemistry instructors that they would like to be able to assign specific, parameterized (algorithmic) questions from the end-of-chapter problem set, so we are pleased to announce that *approximately 20 questions per chapter are available to assign in this new OWL format*. These are the same 20 or so questions marked in the Study Questions section as relevant to the Chapter Goals. In addition, all of the end-of-chapter problems are available in Web CT and Blackboard formats.

Book Design

A major effort was made with the fifth edition to design a book that would aid students by clearly delineating the functions of the various parts of the book. (Although seemingly simple, one of many innovations was to use different typographic fonts for text and chemical equations so that these are clearly separated. Another was to label chemicals or parts of an apparatus in photos so that the reader does not have to move continually between caption and photo to understand the photo's message.) For this new edition, we have continued to put a great deal of thought into book design for functional clarity.

Supporting Materials for the Student

> Visit http://chemistry.brookscole.com to see samples of selected student supplements or to purchase them online from Brooks/Cole. To locate products at your local retailer, provide them with the ISBN.

NEW! General ChemistryNow CD-ROM and Website by William Vining, University of Massachusetts–Amherst, and John Kotz, State University of New York–Oneonta.

General ChemistryNow at http://now.brookscole.com/kotz6e is a powerful, assessment-based online learning companion designed to help students master chapter goals by directing them to interactive resources based on their level of conceptual understanding. Incorporating material from the best-selling *General Chemistry Interactive CD-ROM*, this new media resource includes more than 400 new step-by-step tutorial modules keyed to end-of-chapter Study Questions. The system is completely flexible so students have access to the material through a variety of methods:

- A **Chapter Outline** screen matches the text organization.

- A **Homework and Goals** screen is keyed to the Chapter Goals Revisited feature from the sixth edition and provides selected end-of-chapter Study Questions. The goals are linked to simulations, exercises, and tutorials. Students can attempt each question a number of times, view feedback on the solution, and submit answers online to their instructor for grading. These questions are indicated with the ■ icon.

- An **Exam-Prep Quiz** ("What Do I Know?") provides diagnostic questions that have been carefully crafted to assess students' understanding of the chapter goals. Upon completing a quiz, students will receive feedback and a personalized Learning Plan, and, if applicable, will be directed to the relevant chapter goals screens and accompanying interactive resources.

To accommodate a variety of access methods, the CD-ROM and website duplicate much of the core content. Access to this program is included with the purchase of a new text.

Enhanced! OWL (Online Web-based Learning system),
University of Massachusetts–Amherst

Learning chemistry takes practice, and that usually means completing homework assignments. With a new, easier-to-use interface, the class-tested, Web-based OWL system at http://owl.thomsonlearning.com presents students with a series of questions—many from the text itself for this new edition—and students respond with numerical answers or with a selection from a menu of choices. Questions are generated from a database of numerical and chemical information, so each student in a course receives a different variant of the question each time he or she accesses an instructional unit. Each question has extensive, question-specific feedback keyed to a student's answer. Instructors can customize the unit by determining when questions are available, how many attempts students may make, and how many questions students must answer successfully before they are considered to have mastered the topic. Gradable reports on each attempt at the unit are provided to the instructor, who has access to course management tools such as a gradebook and report-generating functions. Students find OWL an excellent exam review and studies at the University of Massachusetts–Amherst show a positive correlation between use of the OWL system and course performance. **The end-of-chapter questions in the text that are correlated to the Chapter Goals are now fully assignable within the OWL program.**

Student Solutions Manual by Alton Banks, North Carolina State University

This ancillary contains detailed solutions to selected end-of-chapter Study Questions found in the text. Solutions match the problem-solving strategies used in the text. Sample chapters are available for review at the book's website. ISBN 0-534-99852-6

NEW! *Study Guide* by John R. Townsend, West Chester University of Pennsylvania

This completely new study guide contains learning tools explicitly linked to the goals introduced in each chapter. It includes chapter overviews, key terms and definitions, and sample tests. Emphasis is placed on the chapter goals presented in this text by means of further commentary and study tips, worked-out examples, and direct references back to the text. Sample chapters are available for review at the book's website. ISBN 0-534-99851-8

vMentor included with General ChemistryNow

vMentor is an online live tutoring service from Brooks/Cole in partnership with *Elluminate*. vMentor is included in General ChemistryNow. Whether it's one-to-one tutoring help with daily homework or exam review tutorials, vMentor lets students interact with experienced tutors right from their own computers at school or at home. All tutors have not only specialized degrees in the particular subject area (biology, chemistry, mathematics, physics, or statistics), but also extensive teaching experience. Each tutor also has a copy of the textbook the student is using in class. Students can ask as many questions as they want when they access vMentor—and they don't need to set up appointments in advance! Access is provided with vClass, an Internet-based virtual classroom featuring two-way voice, a shared whiteboard, chat, and more. For proprietary, college, and university adopters only. For additional information, consult your local Thomson representative.

NEW! *Chemistry & Chemical Reactivity, Sixth Edition in Two Hardbound Volumes (Volume 1: Chapters 1–12 and Volume 2: Chapters 12–23)*

We recognize that students are concerned about price and portability of their textbooks, and that some students take only one semester of general chemistry. Therefore, we are pleased to announce that the sixth edition is available in two volumes. Volume 1 covers Chapters 1–12 and Volume 2 covers Chapters 12–23. Note that both volumes contain Chapter 12 so as to serve differing curricula. Both volumes will include full access to all the media resources. Consult your Thomson representative for special pricing options. Volume 1 ISBN 0-495-01013-8; Volume 2 ISBN 0-495-01014-6; Two-volume set ISBN 0-534-40800-1.

Essential Algebra for Chemistry Students, Second Edition by David W. Ball, Cleveland State University

This supplement focuses on the skills needed to survive in General Chemistry, with worked examples showing how these skills translate into successful chemical problem solving. This text is an ideal tool for students lacking in confidence or competency in the essential math skills required for general chemistry. Consult your Thomson representative for special bundling pricing. ISBN 0-495-01327-7.

Survival Guide for General Chemistry with Math Review by Charles H. Atwood, University of Georgia

Designed to help students gain a better understanding of the basic problem-solving skills and concepts of General Chemistry, this guide assists students who lack confidence and/or competency in the essential skills necessary to survive general chemistry. The text can be fully customized so that you can incorporate, if you so wish, your old exams. Consult your Brooks/Cole representative for special bundling pricing. ISBN 0-534-99370-2

Supporting Materials for the Instructor

Supporting instructor materials are available to qualified adopters. Please consult your local Thomson Brooks/Cole representative for details. Visit http://chemistry.brookscole.com to:

- See samples of materials
- Locate your local representative
- Download electronic files of books, PowerPoint slides, and text art
- Request a desk copy
- Purchase a book online

Instructor's Resource Manual by Susan Young, Hartwick College

Contains worked-out solutions to *all* end-of-chapter Study Questions and features ideas for instructors on how to fully utilize resources and technology in their courses. The *Manual* provides questions for electronic response systems, suggests classroom demonstrations, and emphasizes good and innovative teaching practices. Electronic files of the *Instructor's Resource Manual* are available for download on the instructor's website. ISBN 0-534-99856-9

General ChemistryNow Website and CD-ROM

A powerful, personalized learning companion that offers your students a variety of tools with which to learn the material, test their knowledge, and identify which tools will best meet their needs. General ChemistryNow is included with every new copy of the book. (Please see the description in the "For the Student" list of ancillary materials.)

Multimedia Manager Instructor CD-ROM

The Multimedia Manager is a dual-platform digital library and presentation tool that provides art, photos, and tables from the main text in a variety of electronic formats that can be used to make transparencies and are easily exported into other software packages. This enhanced CD-ROM also contains simulations, molecular models, and QuickTime movies to supplement lectures as well as electronic files of various print supplements. In addition, instructors can customize presentations by importing personal lecture slides or other selected materials. ISBN 0-534-99855-0

OWL (Online Web-based Learning System)

An online homework, quizzing, and testing tool with course management capability. (Please see the description in the "For the Student" list of ancillary materials.)

PowerPoint Lecture Slides by John Kotz, State University of New York–Oneonta

These class-tested, fully customizable, lecture slides have been used by author John Kotz for many years and are available for instructor download at the text's website at http://chemistry.brookscole.com. Hundreds of slides cover the entire year of general chemistry. Slides use the full power of Microsoft PowerPoint and incorporate videos, animations, and other assets from General ChemistryNow. Instructors can customize their lecture presentations by adding their own slides or by deleting or changing existing slides.

Test Bank by David Treichel, Nebraska Wesleyan University

This printed test bank contains more than 1250 questions, over 90% of which are revised or newly written for this edition. Questions range in difficulty and variety and correlate directly to the chapter sections found in the main text. Numerical, open-ended, or conceptual problems are written in multiple choice, fill-in-the-blank, or short-answer formats. Both single- and multiple-step problems are presented for each chapter. Electronic files of the Test Bank are available for instructor download at the text's website at http://chemistry.brookscole.com. ISBN 0-53-499850-X

Transparencies

A collection of 150 full-color transparencies of key images selected by the authors from the text. Instructors have access on the Multimedia Manager CD-ROM to all text art and many photos to aid in preparing transparencies for material not present in this set. ISBN 0-534-99854-2

iLrn Testing

With a balance of efficiency and high performance, simplicity and versatility, iLrn Testing lets instructors test the way they teach, giving them the power to transform the learning and teaching experience. iLrn Testing is a revolutionary, Internet-ready, cross-platform text-specific testing suite that allows instructors to customize exams and track student progress in an accessible, browser-based format delivered via the Web at www.iLrn.com. Results flow automatically to instructors' gradebooks so that they are better able to assess students' understanding of the material prior to class or an actual test. iLrn offers full algorithmic generation of problems as well as free-response problems using intuitive mathematical notation. **Populated with the questions from the printed Test Bank.** ISBN 0-534-99857-7

JoinIn on TurningPoint for Response Systems

Thomson Brooks/Cole is now pleased to offer book-specific JoinIn content for Response Systems tailored to _Chemistry & Chemical Reactivity_, allowing you to transform your classroom and assess your students' progress with instant in-class quizzes and polls. Our exclusive agreement to offer TurningPoint software lets you pose book-specific questions and display students' answers seamlessly within the Microsoft PowerPoint slides of your own lecture, in conjunction with the "clicker" hardware of your choice. Enhance how your students interact with you, your lecture, and each other. Contact your local Thomson representative to learn more.

WebTutor ToolBox for WebCT and WebTutor ToolBox for Blackboard

Preloaded with content and available via a free access code when packaged with this text, WebTutor ToolBox pairs the content of this text's rich Book Companion website with sophisticated course management functionality. **The end-of-chapter Study Questions in the text are available in WebCT and Blackboard formats.** Instructors can assign materials (including online quizzes) and have the results flow automatically to their gradebooks. ToolBox is ready to use upon logging on—or instructors can customize its preloaded content by uploading images and other resources, adding weblinks, or creating their own practice materials. Students have access only to student resources on the website. Instructors can enter an access code to utilize password-protected Instructor Resources. Contact your Thomson representative for information on packaging WebTutor ToolBox with this text.

For the Laboratory

Chemical Education Resources (CER) at http://www.CERLabs.com

Allows instructors to customize laboratory manuals for their courses from a wide range of more than 300 experiments refereed by the CER board.

Brooks/Cole Laboratory Series for General Chemistry

Brooks/Cole offers a variety of printed manuals to meet all General Chemistry laboratory needs. Instructors can visit the chemistry website at http://chemistry.brookscole.com for a full listing and description of these laboratory manuals and laboratory notebooks. All Brooks/Cole lab manuals can be customized for your specific needs.

Acknowledgments

Because significant changes have been made, preparing this new edition of *Chemistry & Chemical Reactivity* took almost three years of continuous effort. However, as in our work on the first five editions, we have enjoyed the support and encouragement of our families and of some wonderful friends, colleagues, and students.

Brooks/Cole Publishing

The first four editions of this book were published by Saunders College Publishing, a part of Harcourt College Publishing. About a year before the fifth edition was published, however, the company came under new ownership, the Brooks/Cole group of Thomson Higher Education. Throughout the period during which the first five editions were developed, we had the guidance of John Vondeling as our Editor-Publisher and friend. John was responsible for much of the success the book enjoyed, but he passed away in January 2001. Angus McDonald guided us through the final stages of the publication of the fifth edition. We owe Angus a great debt of gratitude for taking over under difficult circumstances and for bringing the project to a successful conclusion.

Following the final acquisition of Harcourt by Thomson Higher Education, we were introduced to our new Editor in Chief, Michelle Julet, and our new Publisher, David Harris. Both have been invaluable in guiding this new edition, and both have become good friends. We look forward to doing future editions with them—and to more sailing with David.

Peter McGahey was the Developmental Editor for the fifth edition and again for this sixth edition. He is blessed with energy, creativity, enthusiasm, intelligence, and good humor. Peter is a trusted friend and confidant. And he cheerfully answered our many questions during almost-daily phone calls.

No book can be successful without proper marketing. Julie Conover is a whiz at marketing and a delight to work with. She is knowledgeable about the market and has worked tirelessly to bring the book to everyone's attention.

Our team at Brooks/Cole is completed with Lisa Weber, Production Manager, and Rob Hugel, Creative Director. Schedules are very demanding in textbook publishing, and Lisa has helped to keep us on schedule. We certainly appreciate her organizational skills. Rob has been involved in product and advertising design for many years, and he has brought his design skills to bear in making this a very attractive book.

People outside of publishing often do not realize the number of people involved in producing a textbook. Karla Maki and Nicole Barone of Thompson Steele, the production company, guided the book through the almost year-long production process. Jane Sanders Miller was the photo researcher for the book and was successful in filling our sometimes off-beat requests for a particular photo. Finally, Jill Hobbs did a very thorough job copyediting the manuscript, and Jay Freedman once again did a masterful job on the index.

Photography, Art, and Design

Most of the color photographs for this edition were again beautifully done by Charles D. Winters. He produced several dozen new images for this book, often under deadline pressure and always with a creative eye. Charlie's work gets better and better with each edition. We have worked with Charlie for almost 20 years and have become close friends. We listen to his jokes, both new and old—and always forget them. When we finish the book, we look forward to a kayaking trip.

When the fifth edition was being planned, we brought in Patrick Harman as a member of the team. Pat designed the first edition of the General ChemistryNow CD-ROM, and we believe its success is in no small way connected to his design skill. For the fifth edition of the book Pat went over almost every figure, and almost every word, to bring a fresh perspective to ways to communicate chemistry. Pat also worked on designing and producing new illustrations for the sixth edition, and his creativity is obvious in their clarity and beauty. As we have worked together so closely for so many years, Pat has become a good friend as well, and we share interests not only in beautiful books but also in interesting music.

Other Collaborators

We have been fortunate to have a number of other colleagues who have played valuable roles in this project.

- Bill Vining (University of Massachusetts–Amherst), the lead author of the General ChemistryNow CD-ROM and website, has been a colleague and friend for many years. Not only has he applied his considerable energy and creativity to preparing a thorough revision of the CD-ROM, but he was also a valuable advisor on the book.

- Susan Young (Hartwick College) has been a good friend and collaborator through four editions and has again prepared the *Instructor's Resource Manual*. She has always been helpful in proofreading, in answering questions on content, and in giving us good advice.

- Alton Banks (North Carolina State University) has also been involved for several editions preparing the *Student Solutions Manual*. Both Susan and Alton have been very helpful in ensuring the accuracy of the Study Questions answers in the book as well as in their respective manuals.

- John Townsend (West Chester University) prepared the *Study Guide* for this edition. This book has had a history of excellent study guides, and John's manual follows that tradition. As described later, John also contributed the supplemental chapter on biochemistry.
- Beatrice Botch (University of Massachusetts–Amherst) gave advice on parts of the text and supplied the information for Figure 13.13.

A major task is proofreading the book once it has been set in type. The book is read in its entirety by the authors and accuracy reviewers. After making corrections, the book is read a second time. Any errors remaining at this point are certainly the responsibility of the authors, and students and instructors should contact the authors by email to offer their suggestions. If this is done in a timely manner, corrections can be made when the book is reprinted.

We want to thank the following accuracy reviewers for their invaluable assistance. The book is immeasurably improved by their work.

Rodney Boyer, Ph.D., Hope College
Larry Fishel, Ph.D.
Michael Grady, Ph.D., College of the Redwoods
Frances Houle, Ph.D., IBM Almaden Research Center
Wayne E. Jones, Jr., Ph.D., Binghamton University
Kathy Mitchell, St. Petersburg College
Barbara Mowery, Ph.D., York College of Pennsylvania
David Shinn, Ph.D.

Reviewers for the Sixth Edition

Patricia Amateis, Virginia Tech
Todd L. Austell, University of North Carolina, Chapel Hill
Joseph Bularzik, Purdue University, Calumet
Stephen Carlson, Lansing Community College
Robert L. Carter, University of Massachusetts, Boston
Paul Charlesworth, Michigan Technological University
Paul Gilletti, Mesa Community College
Stan Genda, University of Nevada, Las Vegas
C. Alton Hassell, Baylor University
Margaret Kerr, Worcester State University
Jeffrey A. Mack, California State University, Sacramento
Elizabeth M. Martin, College of Charleston
Shelley D. Minteer, Saint Louis University
Jason R. Telford, University of Iowa
Wayne Tikkanen, California State University, Los Angeles
Mark A. Whitener, Montclair State University
Marcy Whitney, University of Alabama

Reviewers for the Fifth Edition

David W. Ball, Cleveland State University
Roger Barth, West Chester University
John G. Berberian, Saint Joseph's University
Don A. Berkowitz, University of Maryland
Simon Bott, University of Houston
Wendy Clevenger Cory, University of Tennessee, Chattanooga
Richard Cornelius, Lebanon Valley College
James S. Falcone, West Chester University
Martin Fossett, Tabor Academy
Michelle Fossum, Laney College
Sandro Gambarotta, University of Ottawa
Robert Garber, California State University, Long Beach
Michael D. Hampton, University of Central Florida
Paul Hunter, Michigan State University
Michael E. Lipschutz, Purdue University
Shelley D. Minteer, Saint Louis University
Jessica N. Orvis, Georgia Southern University
David Spurgeon, University of Arizona
Stephen P. Tanner, University of West Florida
John Townsend, West Chester University
John A. Weyh, Western Washington University
Marcy Whitney, University of Alabama
Sheila Woodgate, University of Auckland

Contributors

When we designed this edition, we decided to seek chemists outside of our team to author the supplemental interchapters. John Townsend prepared the chapter on The Chemistry of Life: Biochemistry, and Meredith Newman authored the chapter on The Chemistry of the Environment. We thank them for their very valuable contributions.

John R. Townsend, Associate Professor of Chemistry at West Chester University of Pennsylvania, completed his B.A. in Chemistry as well as the Approved Program for Teacher Certification in Chemistry at the University of Delaware. After a career teaching high school science and mathematics, he earned his M.S. and Ph.D. in biophysical chemistry at Cornell University. At Cornell he also performed experiments in the origins of life field and received the DuPont Teaching Award. After teaching at Bloomsburg University, Dr. Townsend joined the faculty at West Chester University, where he coordinates the chemistry education program for prospective high school teachers and the general chemistry program for science majors. He is also the co-leader of his university's local team of the Collaborative for Excellence in Teacher Preparation in Pennsylvania. His research interests lie in the fields of chemical education and biochemistry.

Meredith E. Newman is an associate professor of chemistry and geology at Hartwick College in Oneonta, New York. She received her B.S. in biology and her M.S. and Ph.D. in environmental engineering. After a postdoctoral appointment in the Department of Analytical Chemistry at the University of Geneva, Switzerland, and work at the Idaho National Environmental and Engineering Laboratory, she joined the faculty at Hartwick College. She has been a visiting scientist in the Environmental Engineering Department at Clemson University and the Institute for Alpine and Arctic Research at the University of Colorado in Boulder. Having previously been an affiliate faculty member at the University of Idaho in Idaho Falls, she is currently an affiliate faculty member at Clemson University. Her research on groundwater contaminant transport, subsurface colloid transport, and environmental education has been published in a variety of scientific journals and texts.

Advisory Board

Many decisions on topic placement, level of text, illustrations, and so on must be made when a textbook is being developed. We have benefited from the help of some wonderful colleagues who met with us on several occasions and who carried on email conversations in between. We certainly acknowledge their significant contributions.

Kevin Chambliss, Baylor University
Michael Hampton, University of Central Florida
Andy Jorgenson, University of Toledo
Laura Kibler-Herzog, Georgia State University
Cathy Middlecamp, University of Wisconsin, Madison
Norbert Pienta, University of Iowa
John Townsend, West Chester University

About the Authors

Left to right: Paul Treichel, Gabriela Weaver, and John Kotz

JOHN C. KOTZ, a State University of New York Distinguished Teaching Professor at the College at Oneonta, was educated at Washington and Lee University and Cornell University. He held National Institutes of Health postdoctoral appointments at the University of Manchester Institute for Science and Technology in England and at Indiana University.

He has coauthored three textbooks in several editions (*Inorganic Chemistry, Chemistry & Chemical Reactivity,* and *The Chemical World*) and the General ChemistryNow CD-ROM. He has also published on his research in inorganic chemistry and electrochemistry.

Dr. Kotz was a Fulbright Lecturer and Research Scholar in Portugal in 1979 and a Visiting Professor there in 1992. He was also a Visiting Professor at the Institute for Chemical Education (University of Wisconsin, 1991–1992) and at Auckland University in New Zealand (1999). He has been an invited speaker at a meeting of the South African Chemical Society and at the biennial conference for secondary school chemistry teachers in Christchurch, New Zealand. He was recently named a mentor of the U.S. Chemistry Olympiad Team.

Dr. Kotz has received several awards, among them a State University of New York Chancellor's Award (1979), a National Catalyst Award for Excellence in Teaching (1992), the Estee Lecturership in Chemical Education at the University of South Dakota (1998), the Visiting Scientist Award from the Western Connecticut Section of the American Chemical Society (1999), and the first annual Distinguished Education Award from the Binghamton (New York) Section of the American Chemical Society (2001). He may be contacted by email at kotzjc@oneonta.edu.

PAUL M. TREICHEL received his B.S. degree at the University of Wisconsin in 1958 and a Ph.D. from Harvard University in 1962. After a year of postdoctoral study in London, he assumed a faculty position at the University of Wisconsin–Madison, where he is currently Helfaer Professor of Chemistry. He served as department chair from 1986 through 1995. He has held visiting faculty positions in South Africa (1975) and in Japan (1995). Currently, he teaches courses in general chemistry, inorganic chemistry, and scientific ethics. Dr. Treichel's research in organometallic and metal cluster chemistry and in mass spectrometry, aided by 75 graduate and undergraduate students, has led to publication of more than 170 papers in scientific journals. He may be contacted by email at treichel@chem.wisc.edu.

GABRIELA C. WEAVER received her B.S. in 1989 from the California Institute of Technology and her Ph.D. in 1994 from the University of Colorado at Boulder. She served as Assistant Professor at the University of Colorado at Denver from 1994 to 2001 and as Associate Professor at Purdue University since 2001. She has been an invited speaker at more than 35 national and international meetings, including the 2001 Gordon Conference on Chemical Education Research and the DVD Summit in Dublin, Ireland. She is currently Director of the Center for Authentic Science Practice in Education at Purdue University. Her work in instructional technology development and on active learning has led to numerous publications in addition to her publications on surface physical chemistry. She may be contacted by email at gweaver@purdue.edu.

An Introduction to Chemistry

Dr. Donald Catlin, the director of the Olympic Analytical Laboratory in Los Angeles, California.

Ann Johansson/Corbis

Chemical Sleuthing

On June 13, 2003, a colorless liquid arrived at the Olympic Analytical Laboratory in Los Angeles, California. This laboratory, headed by Dr. Donald H. Catlin, annually tests about 25,000 samples for the presence of illegal drugs. Among its clients are the U.S. Olympic Committee, the National Collegiate Athletic Association, and the National Football League.

At about the time of the U.S. Outdoor Track and Field Championships in the summer of 2003, a coach in Colorado tipped off the U.S. Anti-Doping Agency (USADA) that several athletes were using a new steroid. The coach had found a syringe with an unknown substance and sent it to the USADA. The USADA dissolved the contents of the syringe in a few milliliters of an alcohol, and then sent the solution to Catlin's laboratory for analysis. That submission initiated weeks of intense work that led to the identification of a previously unknown steroid that was presumably being used by athletes. The head of the USADA later said that the story behind the discovery suggested a "conspiracy involving chemists, coaches, and certain athletes using . . . undetectable designer steroids to defraud their fellow competitors and the world public."

To identify the unknown substance, chemists at the Olympic Analytical Laboratory used a GC-MS, an instrument widely employed in forensic science work. They first passed the sample through a gas chromatograph (GC), an instrument that can separate different chemical compounds in a mixture of liquids. A GC has a very-small-diameter, coiled tube (a typical inside diameter is 0.025 mm), in which the inside surface has been specially treated so that chemicals are attracted to the surface. This

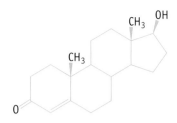

The steroid testosterone.
All steroids, including cholesterol, have the same basic four-ring structure, but they differ in detail.

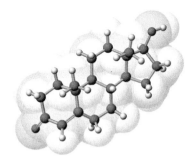

A molecular model of testosterone.

A photo of crystals of the steroid cholesterol taken with a microscope using polarized light.

tube is placed in an oven and heated to a temperature of 200 °C or higher. Different substances in a mixture are swept along the tube by a stream of helium gas. Because each component in the sample binds differently to the material on the inside surface of the tube, each component moves through the column at a different rate and exits from the end of the column at a different time. Thus, separation of the components in the mixture is achieved.

After exiting the GC, each compound is routed directly into a mass spectrometer (MS). (Scientists would describe the two instruments as being interfaced, or linked together, and operating as a single unit.) In a mass spectrometer, the compounds are bombarded with high-energy electrons, and each compound is turned into an ion, a species with a positive electric charge. These ions are then passed through a strong magnetic field, causing the ions to be deflected. The path an ion takes in the magnetic field (the extent of deflection) is related to its mass. The mass of the particle is a key piece of information that will help to identify the compound.

Such a straightforward process: separate the compounds in a GC and identify them in a MS. What can go wrong? In fact, many things can potentially go awry that require ingenuity to overcome. In this case, the unknown steroid did not survive the high temperatures of the GC. It broke apart into pieces, making it possible to study only the pieces of the original molecule. However, this analysis gave enough evidence to convince scientists that the compound was a steroid. But what steroid? According to Catlin, one hypothesis was that "the new steroid was made by people who knew it was not going to be detectable"—that is, the molecule had been designed in a way that would guarantee that it would not be detected by the standard GC-MS procedure.

Intrigued, Catlin and his colleagues set out to identify the steroid. First, they made the molecule stable during the analysis. This was done by attaching new atoms to the molecule to make what chemists call a *derivative*. A number of approaches were tested, and one gave a molecule that did not break down in the GC. MS data allowed scientists to identify the intact molecule (the derivative). Based on this identification and the chemistry used to prepare the derivative, they now knew the identity of the unknown steroid only a few weeks after they had received the sample.

The final step to solve the mystery was to try to make the compound in the laboratory and then to use the GC-MS on this sample. If the material behaved the same way as the unknown sample, then the scientists could be as certain as possible they knew what they had received from the track coach. These experiments worked, confirming the identity of the compound. It was an entirely new steroid, never seen

A GC-MS (gas chromatograph-mass spectrometer). A GC-MS is one of the major tools used in forensic chemistry. A gas chromatograph (GC) separates chemical compounds in a mixture by using differences in the ability of compounds to bind to a chemically treated surface in a thin, coiled tube. When substances emerge from the chromatograph, they are analyzed and identified by the mass spectrometer (MS). The GC-MS pictured has an automated sample changer (carousel, center). An operator will load dozens of samples into the carousel, and the instrument will then process the samples automatically, with data recorded and stored in a computer.

Varian, Inc.

before in nature or in the laboratory. Its formula is $C_{21}H_{28}O_2$, and its name is tetrahydrogestrinone or THG. THG resembled two well-known steroids: gestrinone, used to treat gynecological problems, and trenbolone, a steroid used by ranchers to beef up cattle.

There are two sequels to the story. First, a scientific problem is not solved until its solution has been verified in another laboratory. Not only was this confirmatory analysis done, but a test was soon devised to find THG in urine samples. Second, armed with the new analytical procedures, the USADA asked Catlin's lab to retest 550 urine samples—and THG was found in several.

What is the problem with athletes taking steroids? THG is one of a class of steroids called anabolic steroids. They elevate the body's natural testosterone levels and increase body mass, muscle strength, and muscle definition. They can also improve an athlete's capacity to train and compete at the highest levels. Aside from giving steroid users an illegal competitive advantage, steroids have some damaging potential side effects—liver damage, heart disease, anxiety, and rage.

A check of the Internet shows that there are hundreds of sources of steroids for athletes. The known performance-enhancing drugs can be detected and their users banned from competitive sports. But what about as-yet-unknown steroids? Catlin believes that other steroids are available on the market, made by secret labs without safety standards, a problem he calls horrifying.

Chemistry and Its Methods

Chemistry is about change. It was once only about changing one natural substance into another—wood and oil burn, grape juice turns into wine, and cinnabar (Figure 1), a red mineral from the earth, changes into shiny quicksilver. Today chemistry is still about change, but now chemists focus on the change of one pure substance, whether natural or synthetic, into another (Figure 2).

(a)

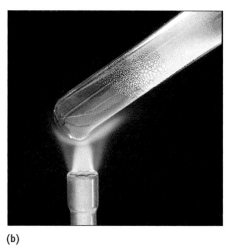

(b)

Charles D. Winters

Figure 1 Cinnabar and mercury. (a) The red crystals of cinnabar consist of the chemical compound mercury sulfide. It is heated in air to change it into orange mercury oxide (b), which, on further heating, decomposes to the elements oxygen and mercury metal. (The droplets you see on the test tube are mercury.)

Although chemistry is endlessly fascinating—at least to chemists—why should you study it? Each person probably has a different answer, but many of you may be taking this chemistry course because someone else has decided it is an important part of preparing for a particular career. Chemistry is especially useful because it is central to our understanding of disciplines as diverse as biology, geology, materials science, medicine, physics, and many branches of engineering. In addition, chemistry plays a major role in our economy; chemistry and chemicals affect our daily lives in a wide variety of ways. Furthermore, a course in chemistry can help you see how a scientist thinks about the world and how to solve problems. The knowledge and skills developed in such a course will benefit you in many career paths and will

Solid sodium, Na

+

Chlorine gas, Cl$_2$

Sodium chloride solid, NaCl

Photos: Charles D. Winters

Figure 2 Forming a chemical compound. (Sodium chloride, table salt, can be made by combining sodium metal (Na) and yellow chlorine gas (Cl$_2$). The result is a crystalline solid, common salt. (The spheres show how the atoms are arranged in the substances.)

Charles D. Winters

Figure 3   The metallic element sodium reacts vigorously with water. (*See General ChemistryNow Screen 8.15 Chemical Reactions and Periodic Properties, for a video of the reactions of lithium, sodium, and potassium with water.*)

help you become a better-informed citizen in a world that is becoming technologically more complex—and more interesting.

Hypotheses, Laws, and Theories

To begin your study of chemistry, this Preface discusses some fundamental ideas used by scientists of all kinds.

As scientists, we study questions of our own choosing or ones that someone else poses in the hope of finding an answer or of discovering some useful information. In the story of the revelation of the banned steroid, THG, Dr. Catlin and his group of chemists were handed a problem to solve, and they followed the usual methods of science to arrive at the answer.

After some preliminary tests, they recognized that the mystery substance was most likely a steroid. That is, they formed a **hypothesis,** a tentative explanation or prediction based on experimental observations.

After formulating one or more hypotheses, scientists perform experiments that are designed to give results that confirm or invalidate these hypotheses. In chemistry this usually requires that both quantitative and qualitative information be collected. **Quantitative** information is numerical data, such as the temperature at which a chemical substance melts. **Qualitative** information, in contrast, consists of nonnumerical observations, such as the color of a substance or its physical appearance.

Catlin and his colleagues assembled a great deal of qualitative and quantitative information. Based on their experience, and on experiments done in the past by chemists around the world, they became more certain they knew the identity of the substance. Their preliminary experiments led them to perform still more experiments, such as looking for a way to stabilize the molecule so that it would not decompose and attempting to make the molecule in the laboratory. Finally, to make certain they had the right molecule and knew how to detect it, their work was confirmed by scientists in other laboratories.

After scientists have performed a number of experiments, and the results have been checked to ensure they are reproducible, a pattern of behavior or results may emerge. At this point it may be possible to summarize the observations in the form of a general rule or conclusion. After making a number of experimental observations, Catlin and his associates could conclude, for example, that the unknown substance was a steroid because it had properties characteristic of many other steroids they had observed.

Finally, after numerous experiments have been conducted by many scientists over an extended period of time, the original hypothesis may become a **law**—a concise verbal or mathematical statement of a behavior or a relation that seems always to be the same under the same conditions. An example might be the law of mass conservation in chemical reactions.

We base much of what we do in science on laws because they help us predict what may occur under a new set of circumstances. For example, we know from experience that if the chemical element sodium comes in contact with water, a violent reaction will occur and new substances will be formed (Figure 3). We also know that the mass of the substances produced in the reaction is exactly the same as the mass of sodium and water used in the reaction. That is, mass is conserved. But the result of an experiment might be different from what is expected based on a general rule. When that happens, chemists get excited because experiments that do not follow the known rules of chemistry are often the most interesting. We know that understanding the exceptions almost invariably gives new insights.

Once enough reproducible experiments have been conducted, and experimental results have been generalized as a law or general rule, it may be possible to conceive a theory to explain the observation. A **theory** is a unifying principle that explains a body of facts and the laws based on them. It is capable of suggesting new hypotheses.

Sometimes nonscientists use the word "theory" to imply that someone has made a guess and that an idea is not yet substantiated. To scientists, a theory is based on carefully determined and reproducible evidence. Theories are the cornerstone of our understanding of the natural world at any given time. Remember, though, that theories are inventions of the human mind. Theories can and do change as new facts are uncovered.

Goals of Science

The sciences, including chemistry, have several goals. Two of these are prediction and control. We do experiments and seek generalities because we want to be able to predict what may occur under a given set of circumstances. We also want to know how we might control the outcome of a chemical reaction or process.

A third goal is explanation and understanding. We know, for example, that certain elements will react vigorously with water (see Figure 3). But why should this be true? And why is this extreme reactivity unique to these elements? To explain and understand this phenomenon, we turn to theories such as those developed in Chapters 9 and 10.

The Importance of Serendipity and Creativity

People who work outside of science usually have the idea that science is an intensely logical field. They picture white-coated chemists moving logically from hypothesis to experiment and then to laws and theories without human emotion or foibles. This picture is a great simplification—and quite wrong!

Often, scientific results and understanding arise quite by accident, otherwise known as **serendipity.** Creativity and insight are needed to transform a fortunate accident into useful and exciting results. The discovery of the cancer drug cisplatin by Barnett Rosenberg in 1965 or of penicillin by Alexander Fleming (1881–1955) in 1928 are wonderful examples of serendipity.

A material familiar to many of you—Teflon®—was found by a combination of serendipity and curiosity. In 1938 Dr. Roy Plunkett was a young scientist working in a DuPont laboratory on the chemistry of fluorine-containing refrigerants (which we now know by their trademark name, Freon). For one experiment, Plunkett and his assistants opened a tank of tetrafluoroethylene gas. The tank supposedly held 1000 g of gas, but only 990 g came out. What happened to the other 10 g? Curiosity is the mark of a good scientist, so they sawed open the tank. A white, waxy substance coated the inside (Figure 4). Following his curiosity further, Plunkett tested the material and found it had remarkable properties. It was more inert than sand! Strong acids and bases did not affect it, nothing could dissolve it, and it was resistant to heat. Unlike sand, it was slippery.

Were it not so expensive, the remarkable properties of this new substance should have led to an immediate search for uses in consumer products. However, Teflon found its first use in the World War II atomic bomb project as a sealant in the equipment used in the separation of uranium. The project was of such national importance that the expense of the material was of no concern. Not until the 1960s did Teflon begin to show up in consumer items. Today one of its most important uses is in medical products (Figure 5). Because it is one of the few substances the body does not reject, it can be used for hip and knee joints, heart valves, and many other body parts.

Hagley Museum and Library

Figure 4   Discovery of Teflon. In a photo taken of a reenactment of the actual event in 1938, Roy Plunkett (*right*) (1910–1994) and his assistants find a white solid coating the inside of a gas cylinder. This solid, now called Teflon, was discovered by accident.

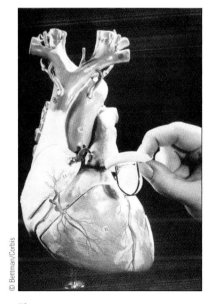

© Bettman/Corbis

Figure 5   Medical products such as heart valves use the polymer Teflon.

Dilemmas and Integrity in Science

You may think research in science is straightforward: Do an experiment, draw a conclusion. In reality, research is seldom that easy. Frustrations and disappointments are common enough, and results can be inconclusive. Complicated experiments often contain some level of uncertainty, and spurious or contradictory data can be collected. For example, suppose you perform an experiment expecting to find a direct relationship between two experimental quantities. You collect six data sets. When plotted, four of the sets lie on a straight line, but two others lie far away from the line. Should you ignore the last two points? Or should you do more experiments when you know the time they take might mean someone else could publish first and thus get the credit for discovering a new scientific principle? What if the two points not on the line indicate that your original hypothesis is wrong, so that you will have to abandon a favorite idea you have worked on for a year? Scientists have a responsibility to remain objective in these situations, but it is sometimes hard to do.

It is important to remember that scientists are human and therefore subject to the same moral pressures and dilemmas as any other person. To help ensure integrity in science, some simple principles have emerged over time that guide scientific practice:

- Experimental results should be reproducible. Furthermore, these results should be reported in sufficient detail that they can be used or reproduced by others.

- Conclusions should be reasonable and unbiased.

- Credit should be given where it is due.

Moral and ethical issues frequently arise in science. Consider the ban on using the pesticide DDT (Figure 6). This is a classic case of the law of "unintended consequences." DDT was developed during World War II and promoted as effective in controlling pests but harmless to people. In fact, it was thought to be so effective that it was used in larger and larger quantities around the world. It was especially effective in controlling mosquitoes carrying malaria. Unfortunately, it soon became evident that there were negative consequences to DDT use. In Borneo, the World Health Organization used large quantities of DDT to kill mosquitoes. The mosquito population did indeed decline, as did malaria incidence. Soon, however, the thatch roofs of people's houses fell down. A parasitic wasp, which ate thatch-eating caterpillars, had also been wiped out by the DDT. Worse still was that geckoes, small lizards, which had eaten DDT-laced caterpillars, were eaten by cats, which then died. The end result was an infestation of rats. Unintended consequences, indeed.

DDT use has been banned in many parts of the world because of its very real, but unforeseen, environmental consequences. The DDT ban occurred in the United States in 1972 because evidence accumulated that the pesticide affected the reproduction of birds such as the bald eagle. DDT is also known to accumulate slowly in human body fat.

The ban on DDT has affected the control of malaria-carrying insects, however. Several million people, primarily children in sub-Saharan Africa, die every year from malaria. The chairman of the Malaria Foundation International has said that "the malaria epidemic is like loading up seven Boeing 747 airliners each day and crashing them into Mt. Kilimanjaro." Consequently, there is a movement to return DDT to the arsenal of weapons in fighting the spread of malaria.

There are many, many moral and ethical issues for chemists. Chemistry has extended and improved the lives of millions of people. But just as certainly, chemicals can cause harm, particularly when misused. It is incumbent on all of us to understand enough science to ask pertinent questions and to evaluate sources of infor-

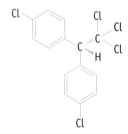

(a) The molecular structure of DDT.

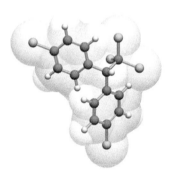

(b) A molecular model of DDT.

Martin Dohrn/Photo Researchers, inc.

(c) DDT can be used to control malaria-carrying insects such as mosquitos.

Figure 6 **The pesticide DDT, an example of the moral and ethical issues in science.**

mation sufficiently to reach reasonable conclusions regarding the health and safety of ourselves and our communities.

A Final Word to Students

Why study chemistry? The reasons are clear. Whether you want to become a biologist, a geologist, an engineer, or a physician, or pursue any of dozens of other professions, chemistry will be at the core of your discipline. It will always be useful to you, sometimes when least expected.

In addition, you will be called upon to make many decisions in your life for your own good or for the good of those in your community—whether that be your neighborhood or the world. An understanding of the nature of science in general, and of chemistry in particular, can only serve to help in these decisions.

Because the authors of this book were students once—and still are—we know chemistry can be a challenging area of study. Like anything worthwhile, it takes time and effort to reach genuine understanding. Be sure to give it time, and talk with your professors and your fellow students. We are sure you will find it as exciting, as useful, and as interesting as we do.

Readings About Science

You will find the following books about science both interesting and informative:

- Rachel Carson: *Silent Spring*, New York, Houghton Mifflin, 1962.
- John Emsley: *Molecules at an Exhibition, Portraits of Intriguing Materials in Everyday Life*, New York, Oxford University Press, 1998.
- John Emsley: *The 13th Element: The Sordid Tale of Murder, Fire, and Phosphorus*, New York, John Wiley & Sons, 2000.
- John Emsley: *Nature's Building Blocks, An A–Z Guide to the Elements*, New York, Oxford University Press, 2001.
- Richard Feynman: *What Do You Care What Other People Think?*, New York, W. W. Norton and Company, 1988; and *Surely You're Joking, Mr. Feynman*, New York, W. W. Norton and Company, 1985.
- Arthur Greenberg: *A Chemical History Tour, Picturing Chemistry from Alchemy to Modern Molecular Science*, New York, Wiley-Interscience, 2000.
- Roald Hoffmann and Vivian Torrance: *Chemistry Imagined, Reflections on Science*, Washington, D.C., Smithsonian Institution Press, 1993.
- Primo Levi: *The Periodic Table*, New York, Schocken Books, 1984. An autobiography of a chemist, a resistance fighter in World War II, and a man who survived some years in a concentration camp.
- Sharon D. McGrayne: *Nobel Prize Women in Science*, New York, Birch Lane Press, 1993.
- Royston M. Roberts: *Serendipity: Accidental Discoveries in Science*, New York, John Wiley & Sons, 1989.
- Oliver Sacks: *Uncle Tungsten, Memories of a Chemical Boyhood*, New York, Alfred Knopf, 2001.
- Lewis Thomas: *The Lives of a Cell*, New York, Penguin Books, 1978.
- J. D. Watson: *The Double Helix, A Personal Account of the Discovery of the Structure of DNA*, New York, Atheneum, 1968.

1—Matter and Measurement

Platinum resistance thermometer. This device measures temperatures over a range from about −259 °C to +962 °C.

NPL photograph © Crown copyright 1997. Reprinted with permission of the controller of HMSO.

How Hot Is It?

"It's so hot outside you could fry an egg on the sidewalk!" This is an expression we heard as children. But what does it mean to say that something is hot? We would say it has a high temperature—but what is temperature and how is it measured?

Temperature and heat are related but different concepts. Although we will discuss the difference in more detail in Chapter 6, for the moment it is easy to think of them this way: Temperature determines the direction of heat transfer. That is, heat transfers from something at a higher temperature to something at a lower temperature. If you touch your finger to a hot match, heat is transferred to your finger, and you decide the match is hot.

Early scientists learned that gases, liquids, and solids expand when heated. In his investigations of heat, Galileo Galilei (1564–1642) invented the "thermoscope," a simple device that depended on the expansion of a liquid in a tube with increasing temperature. Others developed instruments based on this principle, using liquids such as alcohol and mercury. Among them was Daniel Gabriel Fahrenheit (1686–1736). To create his scale, Fahrenheit initially assigned the freezing point of water as 7.5 °F and body temperature as 22.5 °F. He multiplied these values by 4, and then later adjusted them so that the freezing point of water was 32 °F and body temperature was 96 °F. After Fahrenheit's death a further revision of the scale established the reference temperatures at their current values, 32 °F for the freezing point of water and 212 °F for the boiling point. On the current scale, normal body temperature is 98.6 °F.

Archives of the Royal Swedish Academy of Sciences

Anders Celsius (1701–1744). Swedish astronomer and geographer.

A significant advance in temperature measurement came from Anders Celsius (1701–1744). Celsius was a Swedish geographer and astronomer who constructed the Celsius thermometer, which used liquid mercury in a glass tube. The Celsius thermometer scale originally used 0 as the boiling point of water, and 100 as the freezing point of water—reference points that were reversed after Celsius's death. His contribution to thermometry was to show experimentally that the freezing point of water is unchanged by atmospheric pressure or the latitude at which the experiment is done. Celsius also showed that, in contrast, the boiling point of water does depend on atmospheric pressure. Both of these observations were important to establishing a standard temperature scale that could be used around the world.

In modern science there is an interest in determining low and high temperatures well outside the ranges where alcohol and mercury are liquids. Scientists have created new temperature measuring devices for this purpose. The platinum resistance thermometer, for example, relies on the fact that the electrical resistance of platinum wire changes with temperature in a predictable manner. Such devices are extremely sensitive and can make measurements to within one thousandth of a degree over temperatures ranging from -259.25 °C to $+961.78$ °C (the melting point of silver).

How do you measure a very high temperature—say, a temperature high enough to boil mercury or melt glass or platinum? From watching the heater element on a stove or in a toaster, you know that heated objects emit light. It turns out that the wavelength of the emitted light can be correlated with temperature. A pyrometer, an optical device, is commonly used for this purpose.

You might have had your temperature taken with a device that is inserted in your ear. This instrument is essentially a pyrometer. Warm humans emit light, albeit at longer wavelengths than a toaster element. A sensor in the ear thermometer scans the wavelength emitted from the eardrum and reports the temperature. This is a useful measure of body temperature because the eardrum shares blood vessels with the hypothalamus, the area of the brain that regulates body temperature.

Charles D. Winters

Infrared thermometer. This device depends on the long wavelength radiation emitted by a warm object.

Thinking about matter. Is this a glass of
pure water? How can you prove it is?

Imagine a tall glass filled with a clear liquid. Sunlight from a nearby window causes the liquid to sparkle, and the glass is cool to the touch. A drink of water would certainly taste good, but should you take a sip? If the glass were sitting in your kitchen you might say yes. But what if this scene occurred in a chemical laboratory? How would you know that the glass held pure water? Or, to pose a more "chemical" question, how would you *prove* this liquid is water?

We usually think of the water we drink as being pure, but this is not strictly true. In some instances material may be suspended in it or bubbles of gases such as oxygen may be visible to the eye. Some tap water has a slight color from dissolved iron. In fact, drinking water is almost always a mixture of substances, some dissolved and some not. As with any mixture, we could ask many questions. What are the components of the mixture—dust particles, bubbles of oxygen, dissolved sodium, calcium, or iron salts—and what are their relative amounts? How can these substances be separated from one another, and how are the properties of one substance changed when it is mixed with another?

This chapter begins our discussion of how chemists think about matter. After looking at a way to classify matter, we will turn to some basic ideas about elements, atoms, compounds, and molecules and discover how chemists characterize these building blocks of matter. Finally, we will see how we can use numerical information.

1.1—Classifying Matter

A chemist looks at a glass of drinking water and sees a liquid. This liquid could be the chemical compound water. More likely, the liquid is a homogeneous mixture of water and dissolved substances—that is, a solution. It is also possible the water sample is a heterogeneous mixture, with solids being suspended in the liquid. These descriptions represent some of the ways we can classify matter (Figure 1.1).

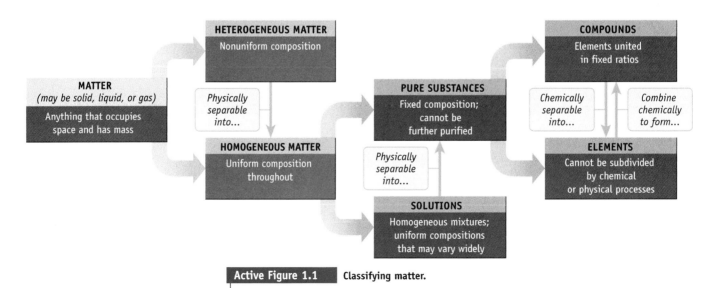

Active Figure 1.1   **Classifying matter.**

Photos: Charles D. Winters

Bromine solid and liquid   Bromine gas and liquid

Active Figure 1.2   **States of matter—solid, liquid, and gas.** Elemental bromine exists in all three states near room temperature. The tiny spheres represent bromine (Br) atoms. In elemental bromine, two Br atoms join to form a Br_2 molecule. (See Section 1.3 and Chapter 3.)

GENERAL
Chemistry . . Now™ See the General ChemistryNow CD-ROM or website to explore an interactive version of this figure accompanied by an exercise.

States of Matter and Kinetic-Molecular Theory

An easily observed property of matter is its **state**—that is, whether a substance is a solid, liquid, or gas (Figure 1.2). You recognize a solid because it has a rigid shape and a fixed volume that changes little as temperature and pressure change. Like solids, liquids have a fixed volume, but a liquid is fluid—it takes on the shape of its container and has no definite shape of its own. Gases are fluid as well, but the volume of a gas is determined by the size of its container. The volume of a gas varies more than the volume of a liquid with temperature and pressure.

At low enough temperatures, virtually all matter is found in the solid state. As the temperature is raised, solids usually melt to form liquids. Eventually, if the temperature is high enough, liquids evaporate to form gases. Volume changes typically accompany changes in state. For a given mass of material, there is usually a small increase in volume on melting—water being a significant exception—and then a large increase in volume occurs upon evaporation.

The **kinetic-molecular theory** of matter helps us interpret the properties of solids, liquids, and gases. According to this theory, all matter consists of extremely tiny particles (atoms, molecules, or ions), which are in constant motion.

- In solids these particles are packed closely together, usually in a regular array. The particles vibrate back and forth about their average positions, but seldom does a particle in a solid squeeze past its immediate neighbors to come into contact with a new set of particles.

- The atoms or molecules of liquids are arranged randomly rather than in the regular patterns found in solids. Liquids and gases are fluid because the particles are not confined to specific locations and can move past one another.

- Under normal conditions, the particles in a gas are far apart. Gas molecules move extremely rapidly because they are not constrained by their neighbors. The molecules of a gas fly about, colliding with one another and with the container walls. This random motion allows gas molecules to fill their container, so the volume of the gas sample is the volume of the container.

An important aspect of the kinetic-molecular theory is that the higher the temperature, the faster the particles move. The energy of motion of the particles (their **kinetic energy**) acts to overcome the forces of attraction between particles. A solid melts to form a liquid when the temperature of the solid is raised to the point at which the particles vibrate fast enough and far enough to push one another out of the way and move out of their regularly spaced positions. As the temperature increases even more, the particles move even faster until finally they can escape the clutches of their comrades and enter the gaseous state. Increasing temperature corresponds to faster and faster motions of atoms and molecules, a general rule you will find useful in many future discussions.

Matter at the Macroscopic and Particulate Levels

The characteristic properties of gases, liquids, and solids just described are observed by the unaided human senses. They are determined using samples of matter large enough to be seen, measured, and handled. Using such samples, we can also determine, for example, what the color of a substance is, whether it dissolves in water, or whether it conducts electricity or reacts with oxygen. Observations and manipulations generally take place in the **macroscopic** world of chemistry (Figure 1.3). This is the world of experiments and observations.

Now let us move to the level of atoms, molecules, and ions—a world of chemistry we cannot see. Take a macroscopic sample of material and divide it, again and again, past the point where the amount of sample can be seen by the naked eye, past the point where it can be seen using an optical microscope. Eventually you reach the level of individual particles that make up all matter, a level that chemists refer to as the **submicroscopic** or **particulate** world of atoms and molecules (Figures 1.2 and 1.3).

Chemists are interested in the structure of matter at the particulate level. Atoms, molecules, and ions cannot be "seen" in the same way that one views the macroscopic world, but they are no less real to chemists. Chemists imagine what atoms must look like and how they might fit together to form molecules. They create models to represent atoms and molecules (Figures 1.2 and 1.3)—where tiny spheres are used to represent atoms—and then use these models to think about chemistry and to explain the observations they have made about the macroscopic world.

It has been said that chemists carry out experiments at the macroscopic level, but they think about chemistry at the particulate level. They then write down their observations as "symbols," the letters (such as H_2O for water or Br_2 for bromine molecules) and drawings that signify the elements and compounds involved. This is a useful perspective that will help you as you study chemistry. Indeed, one of our goals is to help you make the connections in your own mind among the symbolic, particulate, and macroscopic worlds of chemistry.

Pure Substances

Let us think again about a glass of drinking water. How would you tell whether the water is pure (a single substance) or a mixture of substances? Begin by making a few simple observations. Is solid material floating in the liquid? Does the liquid have an odor or an unexpected taste or color?

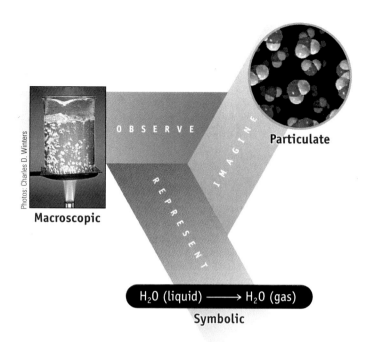

OBSERVE

IMAGINE

REPRESENT

Macroscopic

Particulate

H_2O (liquid) $\longrightarrow$ H_2O (gas)

Symbolic

Photos: Charles D. Winters

Active Figure 1.3 **Levels of matter.** We observe chemical and physical processes at the macroscopic level. To understand or illustrate these processes, scientists often try to imagine what has occurred at the particulate atomic and molecular levels and write symbols to represent these observations. A beaker of boiling water can be visualized at the particulate level as rapidly moving H_2O molecules. The process is symbolized by indicating that the liquid H_2O molecules are becoming H_2O molecules in the gaseous state.

GENERAL Chemistry Now™ See the General ChemistryNow CD-ROM or website to explore an interactive version of this figure accompanied by an exercise.

Every substance has a set of unique properties by which it can be recognized. Pure water, for example, is colorless, is odorless, and certainly does not contain suspended solids. If you wanted to identify a substance conclusively as water, you would have to examine its properties carefully and compare them against the known properties of pure water. Melting point and boiling point serve the purpose well here. If you could show that the substance melts at 0 °C and boils at 100 °C at atmospheric pressure, you can be certain it is water. No other known substance melts and boils at precisely these temperatures.

A second feature of a pure substance is that it cannot be separated into two or more different species by any physical technique such as heating in a Bunsen flame. If it could be separated, our sample would be classified as a mixture.

Mixtures: Homogeneous and Heterogeneous

A cup of noodle soup is obviously a mixture of solids and liquids (Figure 1.4a). A **mixture** in which the uneven texture of the material can be detected is called a **heterogeneous** mixture. Heterogeneous mixtures may appear completely uniform but on closer examination are not. Blood, for example, may not look heterogeneous until you examine it under a microscope and red and white blood cells are revealed (Figure 1.4b). Milk appears smooth in texture to the unaided eye, but magnification

(a)

(b)

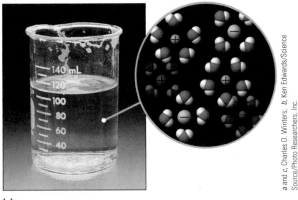

(c)

a and *c*, Charles D. Winters; *b*, Ken Edwards/Science Source/Photo Researchers, Inc.

Figure 1.4   Mixtures. (a) A cup of noodle soup is a heterogeneous mixture. (b) A sample of blood may look homogeneous, but examination with an optical microscope shows it is, in fact, a heterogeneous mixture of liquids and suspended particles (blood cells). (c) A homogeneous mixture, here consisting of salt in water. The model shows that salt consists of separate, electrically charged particles (ions) in water, but the particles cannot be seen with an optical microscope.

would reveal fat and protein globules within the liquid. In a heterogeneous mixture the properties in one region are different from those in another region.

A **homogeneous** mixture consists of two or more substances in the same phase (Figure 1.4c). No amount of optical magnification will reveal a homogeneous mixture to have different properties in different regions. Homogeneous mixtures are often called **solutions**. Common examples include air (mostly a mixture of nitrogen and oxygen gases), gasoline (a mixture of carbon- and hydrogen-containing compounds called hydrocarbons), and an unopened soft drink.

When a mixture is separated into its pure components, the components are said to be *purified* (see Figure 1.5). Efforts at separation are often not complete in a sin-

(a)

(b)

a, Charles D. Winters; *b*, Littleton, Massachusetts, Spectacle Pond Iron and Manganese Treatment Facility

Figure 1.5   Purifying water by filtration. (a) A laboratory setup. A beaker full of muddy water is passed through a paper filter, and the mud and dirt are removed. (b) A water treatment plant uses filtration to remove suspended particles from the water.

gle step, however, and repetition almost always gives an increasingly pure substance. For example, soil particles can be separated from water by filtration (Figure 1.5). When the mixture is passed through a filter, many of the particles are removed. Repeated filtrations will give water a higher and higher state of purity. This purification process uses a property of the mixture, its clarity, to measure the extent of purification. When a perfectly clear sample of water is obtained, all of the soil particles are assumed to have been removed.

GENERAL
Chemistry ⚛ Now™

See the General ChemistryNow CD-ROM or website:
- **Screen 1.5 Mixtures and Pure Substances,** for an exercise on identifying pure substances and types of mixtures
- **Screen 1.6 Separation of Mixtures,** to watch a video on heterogeneous mixtures

Homogeneous and heterogeneous mixtures. Which is homogeneous? See Exercise 1.1.

Exercise 1.1—Mixtures and Pure Substances

The photo in the margin shows two mixtures. Which is a homogeneous mixture and which is a heterogeneous mixture?

■ **Exercise Answers**
In each chapter of the book you will find a number of Exercises. Their purpose is to help you to check your knowledge of the material in that chapter. Solutions to the Exercises are found in Appendix N.

1.2—Elements and Atoms

Passing an electric current through water can decompose it to gaseous hydrogen and oxygen (Figure 1.6a). Substances like hydrogen and oxygen that are composed of only one type of atom are classified as **elements**. Currently 116 elements are known. Of these, only about 90—some of which are illustrated in Figure 1.6—are found in nature. The remainder have been created by scientists. The name and symbol for each element are listed in the tables at the front and back of this book. Carbon (C), sulfur (S), iron (Fe), copper (Cu), silver (Ag), tin (Sn), gold (Au), mercury (Hg), and lead (Pb) were known to the early Greeks and Romans and to the alchemists of ancient China, the Arab world, and medieval Europe. However, many other elements—such as aluminum (Al), silicon (Si), iodine (I), and helium (He)—were not discovered until the 18th and 19th centuries. Finally, artificial elements—those that do not exist in nature, such as technetium (Tc), plutonium (Pu), and americium (Am)—were made in the 20th century using the techniques of modern physics.

Many elements have names and symbols with Latin or Greek origins. Examples include helium (He), named from the Greek word *helios* meaning "sun," and lead, whose symbol, Pb, comes from the Latin word for "heavy," *plumbum*. More recently discovered elements have been named for their place of discovery or for a person or place of significance. Examples include americium (Am), californium (Cf), and curium (Cm).

The table inside the front cover of this book, in which the symbol and other information for the elements are enclosed in a box, is called the **periodic table**. We will describe this important tool of chemistry in more detail beginning in Chapter 2.

An **atom** is the smallest particle of an element that retains the characteristic chemical properties of that element. Modern chemistry is based on an understanding and exploration of nature at the atomic level. We will have much more to say about atoms and atomic properties in Chapters 2, 7, and 8, in particular.

■ **Writing Element Symbols**
Notice that only the first letter of an element's symbol is capitalized. For example, cobalt is Co, not CO. The notation CO represents the chemical compound carbon monoxide. Also note that the element name is not capitalized, except at the beginning of a sentence.

■ **Periodic Table**
See the periodic table at General ChemistryNow. It can be accessed from Screen 1.5 or from the Toolbox. See also the extensive information on the periodic table and the elements at the American Chemical Society website:
- www.chemistry.org/periodic_table.html
- http://pubs.acs.org/cen/80th/elements .html

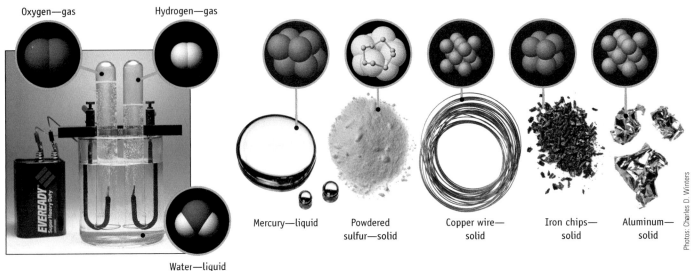

Oxygen—gas   Hydrogen—gas

Water—liquid

Mercury—liquid   Powdered sulfur—solid   Copper wire—solid   Iron chips—solid   Aluminum—solid

Photos: Charles D. Winters

(a)   **(b)**

Figure 1.6   Elements. (a) Passing an electric current through water produces the elements hydrogen (test tube on the right) and oxygen (test tube on the left). (b) Chemical elements can often be distinguished by their color and their state at room temperature.

GENERAL
Chemistry·ᴥ·Now™

See the General ChemistryNow CD-ROM or website:
• **Screen 1.7 Elements and Atoms,** and the Periodic Table tool on this screen or in the Toolbox

---

Exercise 1.2—Elements

Using the periodic table inside the front cover of this book or on the CD-ROM:
(a) Find the names of the elements having the symbols Na, Cl, and Cr.
(b) Find the symbols for the elements zinc, nickel, and potassium.

1.3—Compounds and Molecules

A pure substance like sugar, salt, or water, which is composed of two or more different elements held together by a **chemical bond**, is referred to as a **chemical compound**. Even though only 116 elements are known, there appears to be no limit to the number of compounds that can be made from those elements. More than 20 million compounds are now known, with about a half million added to the list each year.

When elements become part of a compound, their original properties, such as their color, hardness, and melting point, are replaced by the characteristic properties of the compound. Consider common table salt (sodium chloride), which is composed of two elements (Figure 1.7):

• Sodium is a shiny metal that reacts violently with water. It is composed of sodium atoms tightly packed together.

• Chlorine is a light yellow gas that has a distinctive, suffocating odor and is a powerful irritant to lungs and other tissues. The element is composed of Cl_2 units in which two chlorine atoms are tightly bound together.

Photos: Charles D. Winters

Solid sodium, Na

+

Chlorine gas, Cl₂

Sodium chloride solid, NaCl

Figure 1.7 **Forming a chemical compound.** Sodium chloride, commonly known as table salt, can be made by combining sodium metal (Na) and yellow chlorine gas (Cl₂). The result is a crystalline solid.

- Sodium chloride, or common salt, is a colorless, crystalline solid. Its properties are completely unlike those of the two elements from which it is made (Figure 1.7). Salt is composed of sodium and chlorine bound tightly together. (The meaning of chemical formulas such as NaCl is explored in Sections 3.3 and 3.4.)

It is important to distinguish between a mixture of elements and a chemical compound of two or more elements. Pure metallic iron and yellow, powdered sulfur (Figure 1.8a) can be mixed in varying proportions. In the chemical compound iron pyrite (Figure 1.8b), however, there is no variation in composition. Not only does iron pyrite exhibit properties peculiar to itself and different from those of either iron or sulfur, or a mixture of these two elements, but it also has a definite percentage composition by weight (46.55% Fe and 53.45% S). Thus, two major differences

Charles D. Winters

(a)

(b)

Figure 1.8 **Mixtures and compounds.** (a) The substance in the dish is a mixture of iron chips and sulfur. The iron can be removed easily by using a magnet. (b) Iron pyrite is a chemical compound composed of iron and sulfur. It is often found in nature as perfect, golden cubes.

Figure 1.9 **Names, formulas, and mod-els of some common molecules.** Models of molecules appear throughout this book. In such models C atoms are gray, H atoms are white, N atoms are blue, and O atoms are red.

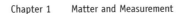

NAME	Water	Methane	Ammonia	Carbon dioxide
FORMULA	H_2O	CH_4	NH_3	CO_2
MODEL				

exist between mixtures and pure compounds: Compounds have distinctly different characteristics from their parent elements, and they have a definite percentage composition (by mass) of their combining elements.

Some compounds—such as table salt, NaCl—are composed of **ions**, which are electrically charged atoms or groups of atoms [▶ Chapter 3]. Other compounds—such as water and sugar—consist of **molecules**, the smallest discrete units that retain the composition and chemical characteristics of the compound.

The composition of any compound is represented by its **chemical formula**. In the formula for water, H_2O, for example, the symbol for hydrogen, H, is followed by a subscript "2" indicating that two atoms of hydrogen occur in a single water molecule. The symbol for oxygen appears without a subscript, indicating that one oxygen atom occurs in the molecule.

As you shall see throughout this book, molecules can be represented with models that depict their composition and structure. Figure 1.9 illustrates the names, formulas, and models of the structures of a few common molecules.

1.4—Physical Properties

You recognize your friends by their physical appearance: their height and weight and the color of their eyes and hair. The same is true of chemical substances. You can tell the difference between an ice cube and a cube of lead of the same size not only because of their appearance (one is clear and colorless, and the other is a lustrous metal) (Figure 1.10), but also because one is much heavier (lead) than the other (ice). Properties such as these, which can be observed and measured without changing the composition of a substance, are called **physical properties**. The chemical elements in Figures 1.6 and 1.7, for example, clearly differ in terms of their color, appearance, and state (solid, liquid, or gas). Physical properties allow us to classify and identify substances. Table 1.1 lists a few physical properties of matter that chemists commonly use.

Figure 1.10 **Physical properties.** An ice cube and a piece of lead can be differentiated easily by their physical properties (such as density, color, and melting point).

Exercise 1.3—Physical Properties

Identify as many physical properties in Table 1.1 as you can for the following common substances: **(a)** iron, **(b)** water, **(c)** table salt (chemical name is sodium chloride), and **(d)** oxygen.

Density

Density, the ratio of the mass of an object to its volume, is a physical property useful for identifying substances.

$$\text{Density} = \frac{\text{mass}}{\text{volume}}$$

(1.1)

Table 1.1 Some Physical Properties

Property	Using the Property to Distinguish Substances
Color	Is the substance colored or colorless? What is the color and what is its intensity?
State of matter	Is it a solid, liquid, or gas? If it is a solid, what is the shape of the particles?
Melting point	At what temperature does a solid melt?
Boiling point	At what temperature does a liquid boil?
Density	What is the substance's density (mass per unit volume)?
Solubility	What mass of substance can dissolve in a given volume of water or other solvent?
Electric conductivity	Does the substance conduct electricity?
Malleability	How easily can a solid be deformed?
Ductility	How easily can a solid be drawn into a wire?
Viscosity	How easily will a liquid flow?

Your brain unconsciously uses the density of an object you want to pick up by estimating volume visually and preparing your muscles to lift the expected mass. For example, you can readily tell the difference between an ice cube and a cube of lead of identical size (Figure 1.10). Lead has a high density, 11.35 g/cm³ (11.35 grams per cubic centimeter), whereas the density of ice is slightly less than 0.917 g/cm³. An ice cube with a volume of 16.0 cm³ has a mass of 14.7 g, whereas a cube of lead with the same volume has a mass of 182 g.

Density relates the mass and volume of a substance. If any two of three quantities—mass, volume, and density—are known for a sample of matter, the third can be calculated. For example, the mass of an object is the product of its density and volume.

$$\text{Mass (g)} = \text{volume} \times \text{density} = \text{volume (cm}^3) \times \frac{\text{mass (g)}}{\text{volume (cm}^3)}$$

You can use this approach to find the mass of 32 cm³ [or 32 mL (milliliters)] of mercury in the graduated cylinder in the photo. A handbook of information for chemistry lists the density of mercury as 13.534 g/cm³ (at 20 °C).

$$\text{Mass (g)} = 32 \text{ cm}^3 \times \frac{13.534 \text{ g}}{1 \text{ cm}^3} = 430 \text{ g}$$

Be sure to notice that the units of cm³ cancel to leave the answer in units of g as required.

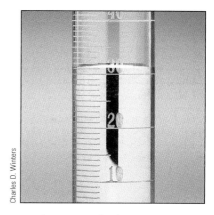

Charles D. Winters

Density, mass, and volume. What is the mass of 32 mL of mercury?

■ **Dimensional Analysis**
The approach to problem solving used in this book is often called *dimensional analysis*. The essence of this approach is to change one number (A) into another (B) using a conversion factor so that the units of A are changed to the desired unit. See Section 1.8.

See the General ChemistryNow CD-ROM or website:
• **Screen 1.10 Density,** for two step-by-step tutorials on determining density and volume

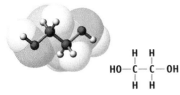

ethylene glycol, $C_2H_6O_2$
density = 1.11 g/cm³ (or 1.11 g/mL)

HO—C—C—OH (with H, H above and H, H below the carbons)

■ Units of Density

The SI unit of mass is the kilogram and the SI unit of length is the meter. Therefore, the SI unit of density is kg/m³. In chemistry the more commonly used unit is g/cm³. To convert from kg/m³ to g/cm³, divide by 1000.

Example 1.1—Using Density

Problem Ethylene glycol, $C_2H_6O_2$, is widely used in automobile antifreeze. It has a density of 1.11 g/cm³ (or 1.11 g/mL). What volume of ethylene glycol will have a mass of 1850 g?

Strategy You know the density and mass of the sample. Because density is the ratio of the mass of a sample to its volume, volume = (mass)(1/density).

Solution

$$\text{Volume (cm}^3) = 1850 \text{ g} \left(\frac{1 \text{ cm}^3}{1.11 \text{ g}} \right) = 1670 \text{ cm}^3$$

Comment Here we multiply the mass (in grams) by the conversion factor (1 cm³/1.11 g) so that units of g cancel to leave an answer in the desired unit of cm³.

Exercise 1.4—Density

The density of dry air is 1.18×10^{-3} g/cm³ (= 0.00118 g/cm³; see Section 1.8 on using scientific notation). What volume of air, in cubic centimeters, has a mass of 15.5 g?

Temperature Dependence of Physical Properties

The temperature of a sample of matter often affects the numerical values of its properties. Density is a particularly important example. Although the change in water density with temperature seems small, it affects our environment profoundly. For example, as the water in a lake cools, the density of the water increases, and the denser water sinks (Figure 1.11a). This continues until the water temperature reaches 3.98 °C, the point at which water has its maximum density (0.999973 g/cm³). If the water temperature drops further, the density decreases slightly, and the colder water floats on top of water at 3.98 °C.

If water is cooled below about 0 °C, solid ice forms. Water is unique among substances in the universe: Ice is much less dense than water, so it floats on water.

Because the density of liquids changes with temperature, it is necessary to report the temperature when you make accurate volume measurements. Laboratory glassware used to make such measurements always specifies the temperature at which it was calibrated (Figure 1.11b).

Temperature Dependence of Water Density

Temperature (°C)	Density of Water (g/cm³)
0 (ice)	0.917
0 (liq water)	0.99984
2	0.99994
4	0.99997
10	0.99970
25	0.99707
100	0.95836

Problem-Solving Tip 1.1

Finding Data

All the information you need to solve a problem in this book may not be presented in the problem. For example, we could have left out the value of the density in Example 1.1 and assumed you would (a) recognize that you needed density to convert a mass to a volume and (b) know where to find the information. The Appendices of this book contain a wealth of information, and even more is available on the General ChemistryNow CD-ROM and website. Various handbooks of information are available in most libraries; among the best are the *Handbook of Chemistry and Physics* (CRC Press) and *Lange's Handbook of Chemistry* (McGraw-Hill). The most up-to-date source of data is the National Institute for Standards and Technology (**www.nist.org**). See also the World Wide Web site Webelements (**www.webelements.com**).

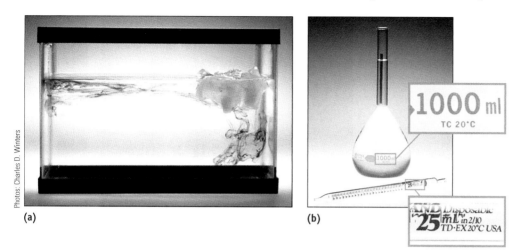

Figure 1.11 Temperature dependence of physical properties. (a) Change in density with temperature. Ice cubes were placed in the right side of the tank and blue dye in the left side. The water beneath the ice is cooler and denser than the surrounding water, so it sinks. The convection current created by this movement of water is traced by the dye movement as the denser, cooler water sinks. **(b) Temperature and calibration.** Laboratory glassware is calibrated for specific temperatures. This pipet or volumetric flask will contain the specified volume at the indicated temperature.

Exercise 1.5—Density and Temperature

The density of mercury at 0 °C is 13.595 g/cm^3, at 10 °C it is 13.570 g/cm^3, and at 20 °C it is 13.546 g/cm^3. Estimate the density of mercury at 30 °C.

Extensive and Intensive Properties

Extensive properties depend on the amount of a substance present. The mass and volume of the samples of elements in Figures 1.2 and 1.6 are extensive properties, for example. In contrast, **intensive properties** do not depend on the amount of substance. A sample of ice will melt at 0 °C, no matter whether you have an ice cube or an iceberg. Density is also an intensive property. The density of gold, for example, is the same (19.3 g/cm^3) whether you have a flake of pure gold or a solid gold ring.

1.5—Physical and Chemical Changes

Changes in physical properties are called **physical changes**. In a physical change the identity of a substance is preserved even though it may have changed its physical state or the gross size and shape of its pieces. An example of a physical change is the melting of a solid. The temperature at which this occurs (the melting point) is often so characteristic that it can be used to identify the solid (Figure 1.12).

A physical property of hydrogen gas (H$_2$) is its low density, so a balloon filled with H$_2$ floats in air (Figure 1.13). Suppose a lighted candle is brought up to the balloon. When the heat causes the skin of the balloon to rupture, the hydrogen combines with the oxygen (O$_2$) in the air, and the heat of the candle sets off a chemical reaction (Figure 1.13), producing water, H$_2$O. This reaction is an example of a **chemical change**, in which one or more substances (the **reactants**) are transformed into one or more different substances (the **products**).

Figure 1.12 A physical property used to distinguish compounds.
Aspirin and naphthalene are both white solids at 25 °C. You can tell them apart by, among other things, a difference in physical properties. At the temperature of boiling water, 100 °C, naphthalene is a liquid (*left*), whereas aspirin is a solid (*right*).

Photos: Charles D. Winters

Naphthalene is a white solid at 25 °C but has a melting point of 80.2 °C.

Aspirin is a white solid at 25 °C. It has a melting point of 135 °C.

The reaction of H_2 with O_2 is an example of a chemical property of hydrogen. A chemical property involves a change in the identity of a substance. Here the H atoms of the gaseous H_2 molecules have become incorporated into H_2O. Similarly, a chemical change occurs when gasoline burns in air in an automobile engine or an old car rusts in the air. Burning of gasoline or rusting of iron are characteristic chemical properties of these substances.

A chemical change at the particulate level is illustrated by the reaction of hydrogen and oxygen molecules to form water molecules.

$$2\ H_2(gas) + O_2(gas) \longrightarrow 2\ H_2O(gas)$$

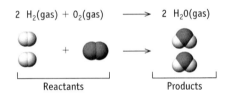

Reactants   Products

The representation of the change with chemical formulas is called a **chemical equation**. It shows that the substances on the left (the **reactants**) produce the substances on the right (the **products**). As this equation shows, there are four atoms of H and two atoms of O before *and* after the reaction, but the molecules before the reaction are different from those after the reaction.

Unlike a chemical change, a physical change does not result in a new chemical substance being produced. The substances (atoms, molecules, or ions) present before and after the change are the same, but they might be farther apart in a gas or closer together in a solid (Figure 1.2).

Finally, as described more fully in Chapter 6, physical changes and chemical changes are often accompanied by transfer of energy. The reaction of hydrogen and oxygen to give water (Figure 1.13), for example, transfers a tremendous amount of energy (in the form of heat and light) to its surroundings.

GENERAL
Chemistry ⚛ Now™

See the General ChemistryNow CD-ROM or website:
- **Screen 1.12 Chemical Changes,** for an exercise on identifying physical and chemical changes
- **Screen 1.13 Chemical Change on the Molecular Scale,** to watch a video and view an animation of the molecular changes when chlorine gas and solid phosphorus react

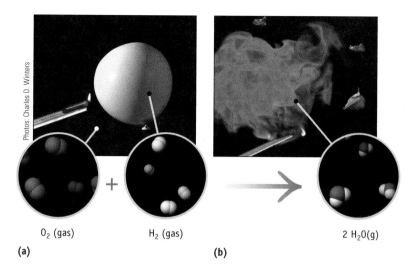

(a)

O_2 (gas)　　　H_2 (gas)

(b)

$2 H_2O(g)$

Figure 1.13 A chemical change—the reaction of hydrogen and oxygen. (a) A balloon filled with molecules of hydrogen gas, and surrounded by molecules of oxygen in the air. (The balloon floats in air because gaseous hydrogen is less dense than air.) (b) When ignited with a burning candle, H_2 and O_2 react to form water, H_2O. (*See General ChemistryNow Screen 1.11 Chemical Change, for a video of this reaction.*)

Chemical and physical changes. A pot of water has been put on a campfire. What chemical and physical changes are occurring here (Exercise 1.6)?

Exercise 1.6—Chemical Reactions and Physical Changes

When camping in the mountains, you boil a pot of water on a campfire. What physical and chemical changes take place in this process?

1.6—Units of Measurement

Doing chemistry requires observing chemical reactions and physical changes. Suppose you mix two solutions in the laboratory and see a golden yellow solid form. Because this new solid is denser than water, it drops to the bottom of the test tube (Figure 1.14). The color and appearance of the substances, and whether heat is involved, are **qualitative** observations. No measurements and numbers were involved.

To understand a chemical reaction more completely, chemists usually make **quantitative** observations. These involve numerical information. For example, if two compounds react with each other, how much product forms? How much heat, if any, is evolved?

In chemistry, quantitative measurements of time, mass, volume, and length, among other things, are common. On page 31 you can read about one of the fastest growing areas of science, nanotechnology, which involves the creation and study of matter on the nanometer scale. A nanometer (nm) is equivalent to 1×10^{-9} m

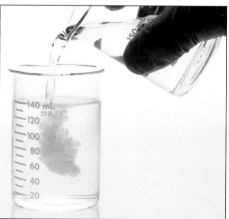

Figure 1.14 Qualitative and quantitative observations. A new substance is formed by mixing two known substances in solution. Of the substance produced we can make several observations. *Qualitative observations:* yellow, fluffy solid. *Quantitative observations:* mass of solid formed.

(meter), a common dimension in chemistry and biology. For example, a typical molecule is only about 1 nm across and a bacterium is about 1000 nm in length.

The scientific community has chosen a modified version of the **metric system** as the standard system for recording and reporting measurements. This decimal system, used internationally in science, is called the Système International d'Unités (International System of Units), abbreviated **SI**.

Table 1.2   Some SI Base Units

Measured Property	Name of Unit	Abbreviation
Mass	kilogram	kg
Length	meter	m
Time	second	s
Temperature	kelvin	K
Amount of substance	mole	mol
Electric current	ampere	A

All SI units are derived from base units, some of which are listed in Table 1.2. Larger and smaller quantities are expressed by using appropriate prefixes with the base unit (Table 1.3).

GENERAL
Chemistry·⚛·Now™

See the General ChemistryNow CD-ROM or website:
- **Screen 1.16 The Metric System,** for a step-by-step tutorial on converting metric units

Temperature Scales

Three temperature scales are commonly used: the Fahrenheit, Celsius, and Kelvin scales (Figure 1.15). The Fahrenheit scale is used in the United States to report everyday temperatures, but most other countries use the Celsius scale. The Celsius scale is generally used worldwide for measurements in the laboratory. When calculations incorporate temperature data, however, kelvin degrees must be used.

The Celsius Temperature Scale
The size of the Celsius degree is defined by assigning zero as the freezing point of pure water (0 °C) and 100 as its boiling point (100 °C) (page 10). You can readily interconvert Fahrenheit and Celsius temperatures using the equation

$$T(°C) = \frac{5\ °C}{9\ °F}[T\ (°F) - 32]$$

but it is best to "calibrate" your senses on the Celsius scale. Pure water freezes at 0 °C, a comfortable room temperature is around 20 °C, your body temperature is 37 °C, and the warmest water you could stand to immerse a finger in is probably about 60 °C.

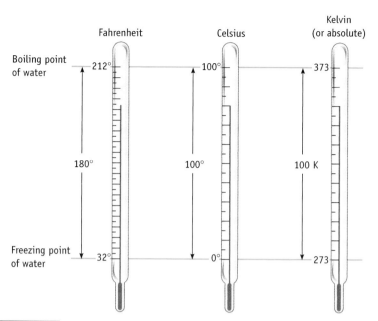

Active Figure 1.15   **A comparison of Fahrenheit, Celsius, and Kelvin scales.** The reference, or starting point, for the Kelvin scale is absolute zero (0 K = −273.15 °C), which has been shown theoretically and experimentally to be the lowest possible temperature.

GENERAL
Chemistry ⚛ Now™ See General ChemistryNow CD-ROM or website to explore an interactive version of this figure accompanied by an exercise.

Table 1.3   Selected Prefixes Used in the Metric System

Prefix	Abbreviation	Meaning	Example
mega-	M	10^6 (million)	1 megaton = 1×10^6 tons
kilo-	k	10^3 (thousand)	1 kilogram (kg) = 1×10^3 g
deci-	d	10^{-1} (tenth)	1 decimeter (dm) = 1×10^{-1} m
centi-	c	10^{-2} (one hundredth)	1 centimeter (cm) = 1×10^{-2} m
milli-	m	10^{-3} (one thousandth)	1 millimeter (mm) = 1×10^{-3} m
micro-	μ	10^{-6} (one millionth)	1 micrometer (μm) = 1×10^{-6} m
nano-	n	10^{-9} (one billionth)	1 nanometer (nm) = 1×10^{-9} m
pico-	p	10^{-12}	1 picometer (pm) = 1×10^{-12} m
femto-	f	10^{-15}	1 femtometer (fm) = 1×10^{-15} m

■ **Common Conversion Factors**
1 kg = 1000 g
1×10^9 nm = 1 m
10 mm = 1 cm
100 cm = 10 dm = 1 m
1000 m = 1 km
Conversion factors for SI units are given in Appendix C and inside the back cover of this book.

■ **Lord Kelvin**
William Thomson (1824–1907), known as Lord Kelvin, was a professor of natural philosophy at the University in Glasgow, Scotland, from 1846 to 1899. He was best known for his work on heat and work, from which came the concept of the absolute temperature scale.

E. F. Smith Collection/Van Pelt Library/University of Pennsylvania.

The Kelvin Temperature Scale

William Thomson, known as Lord Kelvin (1824–1907), first suggested the temperature scale that now bears his name. The Kelvin scale uses the same size unit as the Celsius scale, but it assigns zero as the lowest temperature that can be achieved, a point called **absolute zero**. Many experiments have found that this limiting temperature is −273.15 °C (−459.67 °F). Kelvin units and Celsius degrees are the same size. Thus, the freezing point of water is reached at 273.15 K; that is, 0 °C = 273.15 K.

The boiling point of pure water is 373.15 K. Temperatures in Celsius degrees are readily converted to kelvins, and vice versa, using the relation

$$T\ (\text{K}) = \frac{1\ \text{K}}{1\ ^\circ\text{C}}\ [T\ ^\circ\text{C} + 273.15\ ^\circ\text{C}] \qquad (1.2)$$

Thus, a common room temperature of 23.5 °C is

$$T\ (\text{K}) = \frac{1\ \text{K}}{1\ ^\circ\text{C}}\ (23.5\ ^\circ\text{C}\ +\ 273.15\ ^\circ\text{C}) = 296.7\ \text{K}$$

Finally, notice that the degree symbol (°) is not used with Kelvin temperatures. The name of the unit on this scale is the kelvin (not capitalized), and such temperatures are designated with a capital K.

■ **Temperature Conversions**
When converting 23.5 °C to kelvins, adding the two numbers gives 296.65. However, the rules of "significant figures" tell us that the sum or difference of two numbers can have no more decimal places than the number with the fewest decimal places. (See page 40.) Thus, we round the answer to 296.7 K, a number with one decimal place.

GENERAL
Chemistry•ᐧ•Now™

See the General ChemistryNow CD-ROM or website:
• **Screen 1.15 Temperature,** for a step-by-step tutorial on converting temperatures

Exercise 1.7—Temperature Scales

Liquid nitrogen boils at 77 K. What is this temperature in Celsius degrees?

Length

The meter is the standard unit of length, but objects observed in chemistry are frequently smaller than 1 meter. Measurements are often reported in units of centimeters or millimeters, and objects on the atomic and molecular scale have dimensions of nanometers (nm; 1 nm = 1.0×10^{-9} m) or picometers (pm; 1 pm = 1×10^{-12} m). Your hand, for example, is about 18 cm from the wrist to the fingertips, and the ant in the photo here is about 1 cm long. Using a special microscope—a scanning electron microscope (SEM)—scientists can zoom in on the face of an ant, then to the ant's eye, and finally to one segment of the eye (Figure 1.16).

If we could continue to zoom in on the ant's eye in Figure 1.16, we would enter the nanoscale molecular world (Figure 1.17). The DNA (deoxyribonucleic acid) in the ant's eye is a helical coil of atoms many nanometers long. The rungs of the DNA ladder are approximately 0.34 nm apart, and the helix repeats itself about every 3.4 nm. Zooming in even more, we might encounter a water molecule. Here the distance between the two hydrogen atoms on either side of the oxygen atom is 0.152 nm or 152 pm (pm; picometer, 1 pm = 1×10^{-12} m).

Ant. Your hand is about 18 centimeters long from your wrist to your fingertips. The ant here is about 1 cm in length.

Charles D. Winters

Example 1.2—Distances on the Molecular Scale

Problem The distance between an O atom and an H atom in a water molecule is 95.8 pm. What is this distance in meters (m)? In nanometers (nm)?

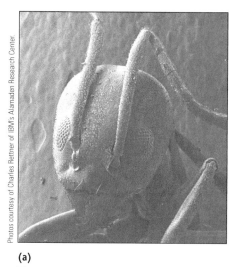

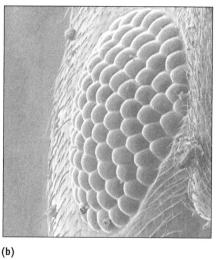

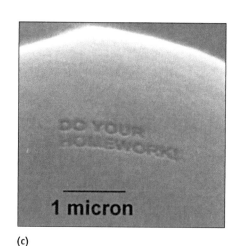

(a) **(b)** **(c)**

Figure 1.16 **Dimensions in biology.** These photos were done at the IBM Laboratories using a scanning electron microscope (SEM). The subject was a dead ant. (a) The head of the ant is about 600 micrometers (microns, μm) wide. (This is equivalent to 6×10^{-4} m or 0.6 mm.) (b) The compound eye of the ant. (c) The scientists at IBM used a special probe to write, on one lens of the ant eye, their advice to science students. The word "homework" is about 1.5 micrometers (microns, μm) long.

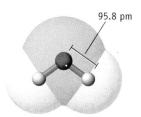

95.8 pm

Strategy You can solve this problem by knowing the conversion factor between the units in the information you are given (picometers) and the desired units (meters or nanometers). (For more about conversion factors and their use in problem solving see Section 1.8.) There is no conversion factor given in Table 1.3 to change nanometers to picometers, but relationships are listed between meters and picometers and between meters and nanometers (Table 1.3). First, we convert picometers to meters, and then we convert meters to nanometers.

$$\times \frac{m}{pm} \qquad \times \frac{nm}{m}$$

$$\text{Picometers} \longrightarrow \text{Meters} \longrightarrow \text{Nanometers}$$

Solution Using the appropriate conversion factors (1 pm = 1×10^{-12} m and 1 nm = 1×10^{-9} m), we have

$$95.8 \text{ pm} \times \frac{1 \times 10^{-12} \text{ m}}{1 \text{ pm}} = 9.58 \times 10^{-11} \text{ m}$$

$$9.58 \times 10^{-11} \text{ m} \times \frac{1 \text{ nm}}{1 \times 10^{-9} \text{ m}} = 9.58 \times 10^{-2} \text{ nm} \text{ or } 0.0958 \text{ nm}$$

Comment Notice how the units cancel to leave an answer whose unit is that of the numerator of the conversion factor. The process of using units to guide a calculation is called dimensional analysis and is discussed further on pages 41–43.

■ **Powers of Ten**
The book *Powers of Ten* explores the dimensions of our universe (Philip and Phylis Morrison, Scientific American Books, 1982). See also the following website in which the "powers of ten" is elegantly animated. http://micro.magnet.fsu.edu/primer/java/scienceopticsu/powersof10/

Figure 1.17 Dimensions in the molecular world. Objects on the molecular scale are often given in terms of nanometers (1 nm = 1×10^{-9} m) or picometers (1 pm = 1×10^{-12} m). An older non-SI unit is the angstrom unit, where 1 Å = 1.0×10^{-10} m.

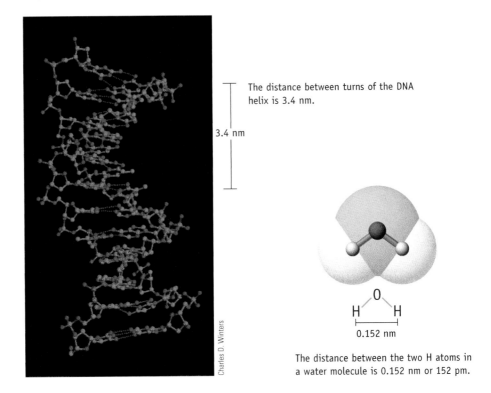

The distance between turns of the DNA helix is 3.4 nm.

3.4 nm

0.152 nm

The distance between the two H atoms in a water molecule is 0.152 nm or 152 pm.

Charles D. Winters

Exercise 1.8—Interconverting Units of Length

The pages of a typical textbook are 25.3 cm long and 21.6 cm wide. What is each dimension in meters? In millimeters? What is the area of a page in square centimeters? In square meters?

Exercise 1.9—Using Units of Length and Density

A platinum sheet is 2.50 cm square and has a mass of 1.656 g. The density of platinum is 21.45 g/cm³. What is the thickness of the platinum sheet in millimeters?

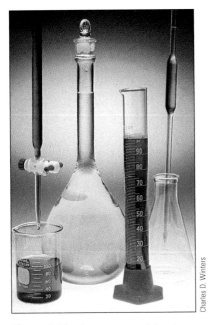

Figure 1.18 Some common laboratory glassware. Volumes are marked in units of milliliters (mL). Remember that 1 mL is equivalent to 1 cm³.

Charles D. Winters

Volume

Chemists often use glassware such as beakers, flasks, pipets, graduated cylinders, and burets, which are marked in volume units (Figure 1.18). The SI unit of volume is the cubic meter (m^3), which is too large for everyday laboratory use. Therefore, chemists usually use the **liter**, symbolized by **L**. A cube with sides equal to 10 cm (0.1 m) has a volume of 10 cm × 10 cm × 10 cm = 1000 cm³ (or 0.001 m³). This is defined as 1 liter.

$$1 \text{ liter (L)} = 1000 \text{ mL} = 1000 \text{ cm}^3$$

The liter is a convenient unit to use in the laboratory, as is the milliliter (mL). Because there are exactly 1000 mL (= 1000 cm³) in a liter, this means that

$$1 \text{ cm}^3 = 0.001 \text{ L} = 1 \text{ mL}$$

Chemical Perspectives

It's a Nanoworld!

A nanometer is one billionth of a meter, a dimension in the realm of atoms and molecules—eight oxygen atoms in a row span a distance of about 1 nanometer. Nanotechnology is one of the hottest fields in science today because the building blocks of those materials having nanoscale dimension can have unique properties.

Carbon nanotubes are excellent examples of nanomaterials. These lattices of carbon atoms form the walls of tubes

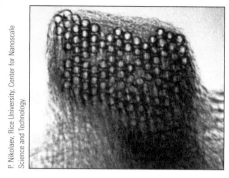

A bundle of carbon nanotubes. Each tube has a diameter of 1.4 nm, and the bundle is 10–20 nm thick.

having diameters of only a few nanometers. Carbon nanotubes are at least 100 times stronger than steel, but only one-sixth as dense. In addition, they conduct heat and electricity far better than copper. As a consequence, carbon nanotubes could be used in tiny, physically strong, conducting devices. Recently, carbon nanotubes have been filled with potassium atoms, making them even better electrical conductors. And even more recently, molecular-sized bearings have been made by sliding one nanotube inside another.

Nanomaterials are by no means new. For the last century tire companies have reinforced tires by adding nanosized particles called carbon black to rubber.

Atomic force microscopy (AFM) is an important tool in chemistry and physics to observe materials at the nanometer level. A tiny probe, often a whisker of a carbon nanotube, moves over the surface of a substance and interacts with individual molecules. Here you see an AFM image of a silicon surface about 460 nm on a side and 5 nm high.

Professor Alex Zettl of the University of California–Berkeley, holding a model of a carbon nanotube.

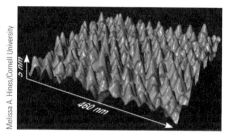

An AFM image of nanobumps on a silicon surface. The average spacing between nanobumps is 38 nm, or about 160 silicon atoms. The average nanobump width is 25 nm or 100 silicon atoms.

The units *milliliter and cubic centimeter* (or "cc") *are interchangeable.* Therefore, a flask that contains exactly 125 mL has a volume of 125 cm³.

Although not widely used in the United States, the cubic decimeter (dm³) is a common unit in the rest of the world. A length of 10 cm is called a decimeter (dm). Because a cube 10 cm on a side defines a volume of 1 liter, *a liter is equivalent to a cubic decimeter:* $1 \text{ L} = 1 \text{ dm}^3$. Products in Europe and other parts of the world are often sold by the cubic decimeter.

The *deciliter, dL,* which is exactly equivalent to 0.100 L or 100 mL, is widely used in medicine. For example, standards for amounts of environmental contaminants are often set as a certain mass per deciliter. The state of Massachusetts recommends that children with more than 10 micrograms (10×10^{-6} g) of lead per deciliter of blood undergo further testing for lead poisoning.

Example 1.3—Units of Volume

Problem A laboratory beaker has a volume of 0.6 L. What is its volume in cubic centimeters (cm³), milliliters (mL), and deciliters?

Strategy Use the information in Table 1.3 to interconvert between units, and use dimensional analysis (see *The Mathematics of Chemistry*, pages 41–43) as a guide.

Solution You should multiply 0.6 L by the conversion factor (1000 cm³/L). The units of L cancel to leave an answer with units of cm³.

$$0.6 \text{ L} \cdot \frac{1000 \text{ cm}^3}{1 \text{ L}} = 600 \text{ cm}^3$$

Because cubic centimeters and milliliters are equivalent, we can also say that the volume of the beaker is 600 mL. The deciliter is 0.100 L or 100 mL. In deciliters, the volume is

$$600 \text{ mL} \cdot \frac{1 \text{ dL}}{100 \text{ mL}} = 6 \text{ dL}$$

Exercise 1.10—Volume

(a) A standard wine bottle has a volume of 750 mL. How many liters does this represent? How many deciliters?

(b) One U.S. gallon is equivalent to 3.7865 L. How many liters are in a 2.0-quart carton of milk? (There are 4 quarts in a gallon.) How many cubic decimeters?

Mass

The mass of a body is the fundamental measure of the quantity of matter, and the SI unit of mass is the kilogram (kg). Smaller masses are expressed in grams (g) or milligrams (mg).

$$1 \text{ kg} = 1000 \text{ g}$$
$$1 \text{ g} = 1000 \text{ mg}$$

■ **Micrograms**
Very small masses are often given in micrograms. A microgram is 1/1000 of a milligram or one millionth of a gram.

Exercise 1.11—Mass

(a) A new U.S. quarter has a mass of 5.59 g. Express this mass in kilograms and milligrams.

(b) An environmental study of a river found a pesticide present to the extent of 0.02 microgram per liter of water. Express this amount in grams per liter.

■ **Accuracy**
The National Institute for Standards and Technology (NIST) is the most important resource for the standards used in science. Comparison with the NIST data is the best test of the accuracy of the measurement. See www.nist.gov.

1.7—Making Measurements: Precision, Accuracy, and Experimental Error

The **precision** of a measurement indicates how well several determinations of the same quantity agree. This is illustrated by the results of throwing darts at a target. In Figure 1.19a, the dart thrower was apparently not skillful, and the precision of the dart's placement on the target is low. In Figures 1.19b and 1.19c, the darts are clustered together, indicating much better consistency on the part of the thrower—that is, greater precision.

Accuracy is the agreement of a measurement with the accepted value of the quantity. Figure 1.19c shows that our thrower was accurate as well as precise—the average of all shots is close to the targeted position, the bull's eye.

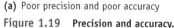

(a) Poor precision and poor accuracy   **(b)** Good precision and poor accuracy   **(c)** Good precision and good accuracy

Figure 1.19   Precision and accuracy.

Figure 1.19b shows that it is possible to be precise without being accurate—the thrower has consistently missed the bull's eye, although all the darts are clustered precisely around one point on the target. This is analogous to an experiment with some flaw (either in design or in a measuring device) that causes all results to differ from the correct value by the same amount.

The precision of a measurement is often expressed in terms of its **standard deviation**, a technique of data analysis explored in *A Closer Look: Standard Deviation.* For

---

A Closer Look

Standard Deviation

Laboratory measurements can be in error for two basic reasons. First, there may be "determinate" errors caused by faulty instruments or human errors such as incorrect record keeping. So-called "indeterminate" errors arise from uncertainties in a measurement where the cause is not known and cannot be controlled by the lab worker. One way to judge the indeterminate error in a result is to calculate the standard deviation.

The standard deviation of a series of measurements is equal to the square root of the sum of the squares of the deviations for each measurement divided by the number of measurements. It has a precise statistical significance: 68% of the values collected are expected to be within one standard deviation of the value determined. (This value assumes a large number of measurements is used to calculate the deviation.)

Consider a simple example. Suppose you carefully measured the mass of water delivered by a 10-mL pipet. For five attempts at the measurement (shown in the table, column 2), the standard deviation is found as follows: First, the average of the measurements is calculated (here, 9.984). Next, the deviation of each individual measurement from this value is determined (column 3). These values are squared, giving the values in column 4, and the sum of these values is determined. The standard deviation is then

Determination	Measured Mass, (g)	Difference between Average and Measurement (g)	Square of Difference
1	9.990	0.006	4×10^{-5}
2	9.993	0.009	8×10^{-5}
3	9.973	0.011	12×10^{-5}
4	9.980	0.004	2×10^{-5}
5	9.982	0.002	0.4×10^{-5}

calculated by dividing this number by 5 (the number of determinations) and taking the square root of the result.

$$\text{Average mass} = 9.984 \text{ g}$$
$$\text{Sum of squares of differences} = 26 \times 10^{-5}$$

$$\text{Standard deviation} = \sqrt{\frac{26 \times 10^{-5}}{5}} = \pm 0.007$$

Based on this calculation it would be appropriate to represent the measured mass as 9.984 ± 0.007 g. This would tell a reader that if this experiment were repeated, approximately 68% of the values would fall in the range of 9.977 g to 9.991 g.

example, suppose a series of measurements led to a distance of 2.965 cm, and the standard deviation was 0.006 cm. Because the uncertainty shows up in the thousandths position, the value should be reported to the nearest thousandth—that is, 2.965 cm. A standard deviation of 0.006 cm means that 68% of the random measurements we make will be within 1 standard deviation—that is, within $\pm$ 0.006 cm.

If you are measuring a quantity in the laboratory, you may be required to report the error in the result, the difference between your result and the accepted value,

$$\text{Error} = \text{experimentally determined value} - \text{accepted value}$$

or the **percent error**.

$$\text{Percent error} = \frac{\text{error in measurement}}{\text{accepted value}} \times 100\%$$

Example 1.4—Precision and Accuracy

Problem A coin has an "accepted" diameter of 28.054 mm. In an experiment, two students measure this diameter. Student A makes four measurements of the diameter of a coin using a precision tool called a micrometer. Student B measures the same coin using a simple plastic ruler. The two students report the following results:

Student A	Student B
28.246 mm	27.9 mm
28.244	28.0
28.246	27.8
28.248	28.1

What is the average diameter and percent error obtained in each case? Which student's data are more accurate? Which are more precise?

Strategy For each set of values we calculate the average of the results and then compare this average with 28.054 mm.

Solution The average for each set of data is obtained by summing the four values and dividing by 4.

Student A	Student B
28.246 mm	27.9 mm
28.244	28.0
28.246	27.8
28.248	28.1
Average = 28.246	Average = 28.0

Student A's data are all very close to the average value, so they are quite precise. Student B's data, in contrast, have a wider range and are less precise. However, student A's result is less accurate than that of student B. The average diameter for student A differs from the "accepted" value by 0.192 mm and has a percent error of 0.684%:

$$\text{Percent error} = \frac{28.246 \text{ mm} - 28.054 \text{ mm}}{28.054 \text{ mm}} \times 100\% = 0.684\%$$

Student B's measurement has an error of only about 0.2%.

Comment Possible reasons for the error in Students A's result are incorrect use of the micrometer or a flaw in the instrument.

Exercise 1.12—Error, Precision, and Accuracy

Two students measured the freezing point of an unknown liquid. Student A used an ordinary laboratory thermometer calibrated in 0.1 °C units. Student B used a thermometer certified by NIST and calibrated in 0.01 °C units. Their results were as follows:

Student A: −0.3 °C; 0.2 °C; 0.0 °C; and −0.3 °C

Student B: 273.13 K; 273.17 K; 273.15 K; 273.19 K

Calculate the average value and, knowing that the liquid was water, calculate the percent error for each student. Which student has the more precise values? Which has the smaller error?

1.8—Mathematics of Chemistry

At its core, chemistry is a quantitative science. Chemists make measurements of, among other things, size, mass, volume, time, and temperature. Scientists then manipulate that quantitative numerical information to search for relationships among properties and to provide insight into the molecular basis of matter.

This section reviews some of the mathematical skills you will need in chemical calculations. It also describes ways to perform calculations and ways to handle quantitative information. The background you should have to be successful includes the following skills:

- Ability to express and use numbers in exponential or scientific notation.
- Ability to make unit conversions (such as liters to milliliters).
- Ability to express quantitative information in an algebraic expression and solve that expression. An example would be to solve the equation $a = (b/x)c$ for x.
- Ability to read information from graphs.
- Ability to prepare a graph of numerical information. If the graph produces a straight line, find the slope and equation of the line.

Examples and Exercises using some of these skills follow, and some problems involving unit conversions and solving algebraic expressions are included in the Study Questions at the end of this chapter.

Exponential or Scientific Notation

Lake Otsego in northern New York is also called Glimmerglass, a name suggested by James Fenimore Cooper (1789–1851), the great American author and an early resident of the village now known as Cooperstown. Extensive environmental studies

Figure 1.20   Lake Otsego. This lake, with a surface area of 2.33×10^7 m², is located in northern New York. Cooperstown is a village at the base of the lake, where the Susquehanna River originates. To learn more about the environmental biology and chemistry of the lake, go to www.oneonta.edu/academics/biofld/

Charles D. Winters

Figure 1.21 Exponential numbers used in astronomy. The spiral galaxy M-83 is 3.0×10^6 parsecs away and has a diameter of 9.0×10^3 parsecs. The unit used in astronomy, the parsec (pc), is 206265 AU (astronomical units), and 1 AU is 1.496×10^8 km. Therefore, the galaxy is about 9.3×10^{19} km away from Earth.

have been done along this lake (Figure 1.20), and some quantitative information useful to chemists, biologists, and geologists is given in the following table:

Lake Otsego Characteristics	Quantitative Information
Area	2.33×10^7 m^2
Maximum depth	505 m
Dissolved solids in lake water	2×10^2 mg/L
Average rainfall in the lake basin	1.02×10^2 cm/year
Average snowfall in the lake basin	198 cm/year

All of the data collected are in metric units. However, some data are expressed in **fixed notation** (505 m, 198 cm/year), whereas other data are expressed in **exponential,** or **scientific, notation** (2.33×10^7 m^2). Scientific notation is a way of presenting very large or very small numbers in a compact and consistent form that simplifies calculations. Because of its convenience it is widely used in sciences such as chemistry, physics, engineering, and astronomy (Figure 1.21).

In scientific notation the number is expressed as a product of two numbers: $N \times 10^n$. N is the *digit* term and is a number between 1 and 9.9999. . . . The second number, 10^n, the *exponential* term, is some integer power of 10. For example, 1234 is written in scientific notation as 1.234×10^3, or 1.234 multiplied by 10 three times:

$$1.234 = 1.234 \times 10^1 \times 10^1 \times 10^1 = 1.234 \times 10^3$$

Conversely, a number less than 1, such as 0.01234, is written as 1.234×10^{-2}. This notation tells us that 1.234 should be divided twice by 10 to obtain 0.01234:

$$0.01234 = \frac{1.234}{10^1 \times 10^1} = 1.234 \times 10^{-1} \times 10^{-1} = 1.234 \times 10^{-2}$$

Some other examples of scientific notation follow:

$$
\begin{array}{ll}
10000 = 1 \times 10^4 & 12345 = 1.2345 \times 10^4 \\
1000 = 1 \times 10^3 & 1234.5 = 1.2345 \times 10^3 \\
100 = 1 \times 10^2 & 123.45 = 1.2345 \times 10^2 \\
10 = 1 \times 10^1 & 12.345 = 1.2345 \times 10^1 \\
1 = 1 \times 10^0 \text{ (any number to the zero power} = 1) \\
1/10 = 1 \times 10^{-1} & 0.12 = 1.2 \times 10^{-1} \\
1/100 = 1 \times 10^{-2} & 0.012 = 1.2 \times 10^{-2} \\
1/1000 = 1 \times 10^{-3} & 0.0012 = 1.2 \times 10^{-3} \\
1/10000 = 1 \times 10^{-4} & 0.00012 = 1.2 \times 10^{-4} \\
\end{array}
$$

When converting a number to scientific notation, notice that the exponent n is positive if the number is greater than 1 and negative if the number is less than 1. The value of n is the number of places by which the decimal is shifted to obtain the number in scientific notation:

$$1 \,2\, 3\, 4\, 5. = 1.2345 \times 10^4$$

(a) Decimal shifted four places to the left. Therefore, n is positive and equal to 4.

■ **Comparing the Earth and a Plant Cell—Powers of Ten**

Earth = 12,760,000 meters wide
    = 12.76 million meters
    = 1.276×10^7 meters
Plant cell = 0.00001276 meter wide
      = 12.76 millionths of a meter
      = 1.276×10^{-5} meters

Problem-Solving Tip 1.2

Using Your Calculator

You will be performing a number of calculations in general chemistry, most of them using a calculator. Many different types of calculators are available, but this problem-solving tip describes several of the kinds of operations you will need to perform on a typical calculator. Be sure to consult your calculator manual for specific instructions to enter scientific notation and to find powers and roots of numbers.

1. Scientific Notation

When entering a number such as 1.23 $\times$ 10^{-4} into your calculator, you first enter 1.23 and then press a key marked EE or EXP (or something similar). This enters the " $\times$ 10" portion of the notation for you. You then complete the entry by keying in the exponent of the number, -4. (To change the exponent from $+4$ to -4, press the "$+/-$" key.)

A common error made by students is to enter 1.23, press the multiply key (x), and then key in 10 before finishing by pressing EE or EXP followed by -4. This gives you an entry that is 10 times too large. Try this! Experiment with your calculator so you are sure you are entering data correctly.

2. Powers of Numbers

Electronic calculators usually offer two methods of raising a number to a power. To square a number, enter the number and then press the "x^2" key. To raise a number to any power, use the "y^x" (or similar) key. For example, to raise 1.42 $\times$ 10^2 to the fourth power:

1. Enter 1.42 $\times$ 10^2.
2. Press "y^x".
3. Enter 4 (this should appear on the display).
4. Press "=" and 4.0659 $\times$ 10^8 appears on the display.

3. Roots of Numbers

To take a square root on an electronic calculator, enter the number and then press the "$\sqrt{x}$" key. To find a higher root of a number, such as the fourth root of 5.6 $\times$ 10^{-10}:

1. Enter the number.
2. Press the "$\sqrt[x]{y}$" key. (On many calculators, the sequence you actually use is to press "2ndF" and then "=." Alternatively, you press "INV" and then "y^x".)
3. Enter the desired root, 4 in this case.
4. Press "=". The answer here is 4.8646 $\times$ 10^{-3}.

A general procedure for finding any root is to use the "y^x" key. For a square root, x is 0.5 (or 1/2), whereas it is 0.3333 (or 1/3) for a cube root, 0.25 (or 1/4) for a fourth root, and so on.

$$0.0\ 0\ 1\ 2 = 1.2 \times 10^{-3}$$

(b) Decimal shifted three places to the right. Therefore, n is negative and equal to 3.

If you wish to convert a number in scientific notation to one using fixed notation (that is, not using powers of 10), the procedure is reversed:

$$6\ .\ 2\ 7\ 3 \times 10^2 = 627.3$$

(a) Decimal point moved two places to the right because n is positive and equal to 2.

$$0\ 0\ 6.273 \times 10^{-3} = 0.006273$$

(b) Decimal point shifted three places to the left because n is negative and equal to 3.

Two final points should be made concerning scientific notation. First, be aware that calculators and computers often express a number such as 1.23×10^3 as 1.23E3 or 6.45×10^{-5} as 6.45E-5. Second, some electronic calculators can readily convert numbers in fixed notation to scientific notation. If you have such a calculator, you may be able to do this by pressing the EE or EXP key and then the "=" key (but check your calculator manual to learn how your device operates).

In chemistry you will often have to use numbers in exponential notation in mathematical operations. The following five operations are important:

- *Adding and Subtracting Numbers Expressed in Scientific Notation*

When adding or subtracting two numbers, first convert them to the same powers of 10. The digit terms are then added or subtracted as appropriate:

$$(1.234 \times 10^{-3}) + (5.623 \times 10^{-2}) = (0.1234 \times 10^{-2}) + (5.623 \times 10^{-2})$$
$$= 5.746 \times 10^{-2}$$

- *Multiplication of Numbers Expressed in Scientific Notation*

The digit terms are multiplied in the usual manner, and the exponents are added algebraically. The result is expressed with a digit term with only one nonzero digit to the left of the decimal:

$$(6.0 \times 10^{23})(2.0 \times 10^{-2}) = (6.0)(2.0) \times 10^{23-2} = 12 \times 10^{21} = 1.2 \times 10^{22}$$

- *Division of Numbers Expressed in Scientific Notation*

The digit terms are divided in the usual manner, and the exponents are subtracted algebraically. The quotient is written with one nonzero digit to the left of the decimal in the digit term:

$$\frac{7.60 \times 10^{3}}{1.23 \times 10^{2}} = \frac{7.60}{1.23} \times 10^{3-2} = 6.18 \times 10^{1}$$

- *Powers of Numbers Expressed in Scientific Notation*

When raising a number in exponential notation to a power, treat the digit term in the usual manner. The exponent is then multiplied by the number indicating the power:

$$(5.28 \times 10^{3})^{2} = (5.28)^{2} \times 10^{3 \times 2} = 27.9 \times 10^{6} = 2.79 \times 10^{7}$$

- *Roots of Numbers Expressed in Scientific Notation*

Unless you use an electronic calculator, the number must first be put into a form in which the exponent is exactly divisible by the root. For example, for a square root, the exponent should be divisible by 2. The root of the digit term is found in the usual way, and the exponent is divided by the desired root:

$$\sqrt{3.6 \times 10^{7}} = \sqrt{36 \times 10^{6}} = \sqrt{36} \times \sqrt{10^{6}} = 6.0 \times 10^{3}$$

Significant Figures

In most experiments several kinds of measurements must be made, and some can be made more precisely than others. It is common sense that a result calculated from experimental data can be no more precise than the least precise piece of information that went into the calculation. This is where the rules for significant figures come in. **Significant figures** are the digits in a measured quantity that reflect the accuracy of the measurement.

When describing standard deviation on page 33, we used the example of a measurement that was known to be 9.984 with an uncertainty of ± 0.007 cm. That is, the last number of our measurement, 0.004 cm, was uncertain to some degree. Our measurement is said to have four significant figures, the last of which is uncertain to some extent.

Suppose we want to calculate the density of a piece of metal (Figure 1.22). The mass and dimensions were determined by standard laboratory techniques. Most of these numbers have two digits to the right of the decimal, but they have different numbers of significant figures.

Measurement	Data Collected	Significant Figures
Mass of metal	13.56 g	4
Length	6.45 cm	3
Width	2.50 cm	3
Thickness	3.1 mm	2

The quantity 3.1 mm has two significant figures. That is, the 3 in 3.1 is exactly known, but the 1 is not. In general, *in a number representing a scientific measurement, the last digit to the right is taken to be inexact.* Unless stated otherwise, it is common practice to assign an uncertainty of ± 1 to the last significant digit. This means the thickness of the metal piece may have been as small as 3.0 mm or as large as 3.2 mm.

When the data on the piece of metal are combined to calculate the density, the result will be 2.7 g/cm^3, a number with two significant figures. (The complete calculation of the metal density is given on page 41). The reason for this is that a calculated result can be no more precise than the least precise data used, and here the thickness has only two significant figures.

When doing calculations using measured quantities, we follow some basic rules so that the results reflect the precision of all the measurements that go into the calculations. The *rules used for significant figures in this book* are as follows:

Rule 1. To determine the number of significant figures in a measurement, read the number from left to right and count all digits, starting with the first digit that is not zero.

Example	Number of Significant Figures
1.23	3; all nonzero digits are significant.
0.00123 g	3; the zeros to the left of the 1 (the first significant digit) simply locate the decimal point. To avoid confusion, write numbers of this type in scientific notation; thus, $0.00123 = 1.23 \times 10^{-3}$.
2.040 g	4; when a number is greater than 1, *all zeros to the right of the decimal point are significant.*
0.02040 g	4; for a number less than 1, only zeros to the right of the first nonzero digit are significant.
100 g	1; in numbers that do not contain a decimal point, "trailing" zeros may or may not be significant. *The practice followed in this book is to include a decimal point if the zeros are significant.* Thus, 100. is used to represent three significant digits, whereas 100 has only one significant digit. To avoid confusion, an alternative method is to write numbers in scientific notation because all digits are significant when written in scientific notation. Thus, 1.00×10^2 has three significant digits, whereas 1×10^2 has only one significant digit.
100 cm/m	Infinite number of significant digits. This is a *defined quantity*. Defined quantities do not limit the number of significant figures in a calculated result.
$\pi = 3.1415926$	The value of certain constants such as π is known to a greater number of significant figures than you will ever use in a calculation.

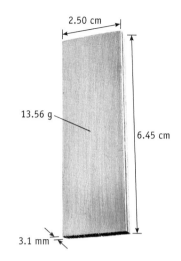

2.50 cm

13.56 g

6.45 cm

3.1 mm

Figure 1.22 Data to determine the density of a metal.

Charles D. Winters

Standard laboratory balance and significant figures. Such balances can determine the mass of an object to the nearest milligram. Thus, an object may have a mass of 13.456 g (13456 mg, five significant figures), 0.123 g (123 mg, three significant figures), or 0.072 g (72 mg, two significant figures).

Rule 2. When adding or subtracting numbers, the number of decimal places in the answer is equal to the number of decimal places in the number with the fewest digits after the decimal.

0.12	2 decimal places	2 significant figures
+ 1.9	1 decimal place	2 significant figures
+10.925	3 decimal places	5 significant figures
12.945	3 decimal places	

The sum should be reported as 12.9, a number with one decimal place, because 1.9 has only one decimal place.

Rule 3. In multiplication or division, the number of significant figures in the answer should be the same as that in the quantity with the fewest significant figures.

$$\frac{0.01208}{0.0236} = 0.512 \text{ or, in scientific notation, } 5.12 \times 10^{-1}$$

Because 0.0236 has only three significant digits and 0.01208 has four, the answer should have three significant digits.

Rule 4. When a number is rounded off, the last digit to be retained is increased by one only if the following digit is 5 or greater.

Full Number	Number Rounded to Three Significant Digits
12.696	12.7
16.349	16.3
18.35	18.4
18.351	18.4

One last word on significant figures and calculations: When working problems, you should do the calculation with all the digits allowed by your calculator and round off only at the end of the calculation. Rounding off in the middle can introduce errors.

■ **To Multiply or to Add?**
Take the number 4.68.

(a) Take the sum of 4.68 + 4.68 + 4.68. The answer is 14.04, a number with four significant figures.

(b) Multiply 4.68 times 3. The answer can have only three significant figures (14.0). You should recognize that different outcomes are possible depending on the type of mathematical operation.

■ **Who Is Right—You or the Book?**
If your answer to a problem in this book does not quite agree with the answers in Appendix N or O, the discrepancy may be the result of rounding the answer after each step and then using that rounded answer in the next step. This book follows these conventions:

(a) Final answers to numerical problems in this book result from retaining full calculator accuracy throughout the calculation and rounding only at the end.

(b) In Example problems, the answer to each step is given to the correct number of significant figures for that step, but the full calculator accuracy is carried to the next step. The number of significant figures in the final answer is dictated by the number of significant figures in the original data.

GENERAL
Chemistry ⚛ Now™

See the General ChemistryNow CD-ROM or website:

- **Screen 1.17 Using Numerical Information,** for tutorials on multiplying and dividing with significant figures, raising significant figures to a power, and taking square roots of significant figures

Example 1.5—Using Significant Figures

Problem An example of a calculation you will do later in the book (Chapter 12) is

$$\text{Volume of gas (L)} = \frac{(0.120)(0.08206)(273.15 + 23)}{(230/760.0)}$$

Calculate the final answer to the correct number of significant figures.

Strategy Let us first decide on the number of significant figures represented by each number (Rule 1), and then apply Rules 2 and 3.

Solution

Number	Number of Significant Figures	Comments
0.120	3	The trailing 0 is significant. See Rule 1.
0.08206	4	The first 0 to the immediate right of the decimal is not significant. See Rule 1.
$273.15 + 23 = 296$	3	23 has no decimal places, so the sum can have none. See Rule 2.
$230/760.0 = 0.30$	2	230 has two significant figures because the last zero is not significant. In contrast, there is a decimal point in 760.0, so there are four significant digits. The quotient may have only two significant digits. See Rules 1 and 3.

Analysis shows that one of the pieces of information is known to only two significant figures. Therefore, the volume of gas is 9.6 L, a number with two significant figures.

Exercise 1.13—Using Significant Figures

(a) How many significant figures are indicated by 2.33×10^7, by 50.5, and by 200?

(b) What are the sum and the product of 10.26 and 0.063?

(c) What is the result of the following calculation?

$$x = \frac{(110.7 - 64)}{(0.056)(0.00216)}$$

Problem Solving by Dimensional Analysis

Suppose you want to find the density of a rectangular piece of metal (Figure 1.22) in units of grams per cubic centimeter (g/cm^3). Because density is the ratio of mass to volume, you need to measure the mass and determine the volume of the piece. To find the volume of the sample in cubic centimeters, you multiply its length by its width and its thickness. First, however, all the measurements must have the same units, meaning that the thickness must be converted to centimeters. Recognizing that there are 10 mm in 1 cm, we use this relationship to get a thickness of 0.31 cm:

$$3.1 \text{ mm} \times \frac{1 \text{ cm}}{10 \text{ mm}} = 0.31 \text{ cm}$$

With all the dimensions in the same unit, the volume and then the density can be calculated:

$$\text{Length} \times \text{width} \times \text{thickness} = \text{volume}$$

$$6.45 \text{ cm} \times 2.50 \text{ cm} \times 0.31 \text{ cm} = 5.0 \text{ cm}^3$$

$$\text{Density} = \frac{13.56 \text{ g}}{5.0 \text{ cm}^3} = 2.7 \text{ g/cm}^3$$

■ **Data to Calculate Metal Density**
(See Figure 1.22)
Mass of metal = 13.56 g
Length = 6.45 cm
Width = 2.50 cm
Thickness = 3.1 mm

Dimensional analysis (sometimes called the *factor-label method*) is a general problem-solving approach that uses the dimensions or units of each value to guide you through calculations. This approach was used above to change 3.1 mm to its equivalent in centimeters. We multiplied the number we wished to convert (3.1 mm) by a **conversion factor** (1 cm/10 mm) to produce the result in the desired unit (0.31 cm). Units are handled like numbers: Because the unit "mm" was in both the numerator and the denominator, dividing one by the other leaves a quotient of 1. The units are said to "cancel out." Here this leaves the answer in centimeters, the desired unit.

A conversion factor expresses the equivalence of a measurement in two different units (1 cm $\equiv$ 10 mm; 1 g $\equiv$ 1000 mg; 12 eggs $\equiv$ 1 dozen; 12 inches $\equiv$ 1 foot). Because the numerator and the denominator describe the same quantity, the conversion factor is equivalent to the number 1. Therefore, multiplication by this factor does not change the measured quantity, only its units. A conversion factor is always written so that it has the form "new units divided by units of original number."

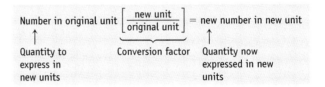

Number in original unit $\left[\dfrac{\text{new unit}}{\text{original unit}}\right]$ = new number in new unit

↑   ↑
Quantity to   Conversion factor   Quantity now
express in   expressed in new
new units   units

See the General ChemistryNow CD-ROM or website:
- **Screen 1.17 Using Numerical Information,** for a tutorial on dimensional analysis

Example 1.6—Using Conversion Factors— Density in Different Units

Problem Oceanographers often express the density of sea water in units of kilograms per cubic meter. If the density of sea water is 1.025 g/cm^3 at 15 °C, what is its density in kilograms per cubic meter?

Strategy To simplify this problem, break it into three steps. First, change grams to kilograms. Next, convert cubic centimeters to cubic meters. Finally, calculate the density by dividing the mass in kilograms by the volume in cubic meters.

Solution First convert the mass in grams to kilograms.

$$1.025 \text{ g} \times \frac{1 \text{ kg}}{1000 \text{ g}} = 1.025 \times 10^{-3} \text{ kg}$$

No conversion factor is available in one of our tables to *directly* change units of cubic centimeters to cubic meters. You can find one, however, by cubing (raising to the third power) the relation between the meter and the centimeter:

$$1 \text{ cm}^3 \times \left(\frac{1 \text{ m}}{100 \text{ cm}}\right)^3 = 1 \text{ cm}^3 \times \left(\frac{1 \text{ m}^3}{1 \times 10^6 \text{ cm}^3}\right) = 1 \times 10^{-6} \text{ m}^3$$

Therefore, the density of sea water is

$$\text{Density} = \frac{1.025 \times 10^{-3} \text{ kg}}{1 \times 10^{-6} \text{ m}^3} = 1.025 \times 10^3 \text{ kg/m}^3$$

Exercise 1.14—Using Dimensional Analysis

(a) The annual snowfall at Lake Otsego is 198 cm each year. What is this depth in meters? In feet (where 1 foot = 30.48 cm)?

(b) The area of Lake Otsego is 2.33×10^7 m^2. What is this area in square kilometers?

(c) The density of gold is 19,320 kg/m^3. What is this density in g/cm^3?

(d) See Figure 1.21. Show that 9.0×10^3 pc is 2.8×10^{17} km.

Graphing

In a number of instances in this text, graphs are used when analyzing experimental data with a goal of obtaining a mathematical equation. The procedure used will often result in a straight line, which has the equation

$$y = mx + b$$

In this equation, y is usually referred to as the dependent variable; its value is determined from (that is, is dependent on) the values of x, m, and b. In this equation x is called the independent variable and m is the slope of the line. The parameter b is the y-intercept—that is, the value of y when $x = 0$. Let us use an example to investigate two things: (a) how to construct a graph from a set of data points, and (b) how to derive an equation for the line generated by the data.

A set of data points to be graphed is presented in Figure 1.23. We first mark off each axis in increments of the values of x and y. Here our x-data range from -2 to 4, so the x-axis is marked off in increments of 1 unit. The y-data range from 0 to 2.5, so we mark off the y-axis in increments of 0.5. Each data set is marked as a circle on the graph.

After plotting the points on the graph (round circles), we draw a *straight line* that comes as close as possible to representing the trend in the data. (Do not connect the dots!) Because there is always some inaccuracy in experimental data, this line may not pass exactly through every point.

To identify the specific equation corresponding to our data, we must determine the y-intercept (b) and slope (m) for the equation $y = mx + b$. The y-intercept is the point at which $x = 0$. (In Figure 1.23, $y = 1.87$ when $x = 0$). The slope is determined by selecting two points on the line (marked with squares in Figure 1.23) and calculating the difference in values of y ($\Delta y = y_2 - y_1$) and x ($\Delta x = x_2 - x_1$). The slope of the line is then the ratio of these differences, $m = \Delta y / \Delta x$. Here the slope has the value -0.525. With the slope and intercept now known, we can write the equation for the line

$$y = -0.525x + 1.87$$

and we can use this equation to calculate y-values for points that are not part of our original set of x-y data. For example, when $x = 1.50$, $y = 1.08$.

■ **Determining the Slope with a Computer Program—Least-Squares Analysis**
Generally the easiest method of determining the slope and intercept of a straight line (and thus the line's equation) is to use a program such as Microsoft Excel. These programs perform a "least squares" or "linear regression" analysis and give the best straight line based on the data. (This line is referred to in Excel as a *trendline*.) The General ChemistryNow CD-ROM also has a useful plotting program that performs this analysis; see the "Plotting Tool" in the menu on any screen.

Figure 1.23 Plotting Data. Data for the variable x are plotted along the horizontal axis (abscissa), and data for y are plotted along the vertical axis (ordinate). The slope of the line, m in the equation $y = mx + b$, is given by $\Delta y/\Delta x$. The intercept of the line with the y-axis (when $x = 0$) is b in the equation.

Using Microsoft Excel with these data, and doing a linear regression (or least-squares) analysis, we find $y = -0.525x + 1.87$.

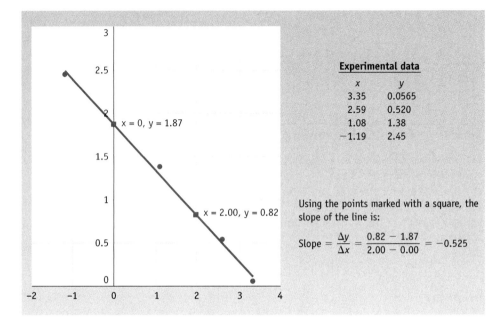

Experimental data

x	y
3.35	0.0565
2.59	0.520
1.08	1.38
−1.19	2.45

Using the points marked with a square, the slope of the line is:

$$\text{Slope} = \frac{\Delta y}{\Delta x} = \frac{0.82 - 1.87}{2.00 - 0.00} = -0.525$$

Exercise 1.15—Graphing

To find the mass of 50 jelly beans, we weighed several samples of beans.

Number of Beans	Mass (g)
5	12.82
11	27.14
16	39.30
24	59.04

Plot these data with the number of beans on the horizontal or x-axis, and the mass of beans on the vertical or y-axis. What is the slope of the line? Use your equation of a straight line to calculate the mass of exactly 50 jelly beans.

Problem Solving and Chemical Arithmetic

Problem-Solving Strategy

Some of the calculations in chemistry can be complex. Students frequently find it is helpful to follow a definite plan of attack as illustrated in examples throughout this book.

Step 1: Problem. State the problem. Read it carefully.

Step 2: Strategy. What key principles are involved? What information is known or not known? What information might be there just to place the question in the context of chemistry? Organize the information to see what is required and to discover the relationships among the data given. Try writing the information down in table form. If it is numerical information, be sure to include units.

One of the greatest difficulties for a student in introductory chemistry is picturing what is being asked for. Try sketching a picture of the situation involved. For example, we sketched a picture of the piece of metal whose density we wanted to calculate, and put the dimensions on the drawing (page 39).

Develop a plan. Have you done a problem of this type before? If not, perhaps the problem is really just a combination of several simpler ones you have seen before. Break it down into those simpler components. Try reasoning backward from the units of the answer. What data do you need to find an answer in those units?

Step 3: Solution. Execute the plan. Carefully write down each step of the problem, being sure to keep track of the units on numbers. (Do the units cancel to give you the answer in the desired units?) Don't skip steps. Don't do anything except the simplest steps in your head. Students often say they got a problem wrong because they "made a stupid mistake." Your instructor—and book authors—make them, too, and it is usually because they don't take the time to write down the steps of the problem clearly.

Step 4: Comment and Check Answer. As a final check, ask yourself whether the answer is reasonable.

Example 1.7—Problem Solving

Problem A mineral oil has a density of 0.875 g/cm^3. Suppose you spread 0.75 g of this oil over the surface of water in a large dish with an inner diameter of 21.6 cm. How thick is the oil layer? Express the thickness in centimeters.

Strategy It is often useful to begin solving such problems by sketching a picture of the situation.

This helps recognize that the solution to the problem is to find the volume of the oil on the water. If we know the volume, then we can find the thickness because

Volume of oil layer = (thickness of layer) × (area of oil layer)

So, we need two things: (a) the volume of the oil layer and (b) the area of the layer.

Solution First calculate the volume of oil. The mass of the oil layer is known, so combining the mass of oil with its density gives the volume of the oil used:

$$0.75 \text{ g} \times \frac{1 \text{ cm}^3}{0.875 \text{ g}} = 0.86 \text{ cm}^3$$

Next calculate the area of the oil layer. The oil is spread over a circular surface, whose area is given by

Area = π × (radius)^2

Study Questions

▲ denotes more challenging questions.

■ denotes questions available in the Homework and Goals section of the General ChemistryNow CD-ROM or website.

Blue numbered questions have answers in Appendix O and fully worked solutions in the *Student Solutions Manual.*

Structures of many of the compounds used in these questions are found on the General ChemistryNow CD-ROM or website in the Models folder.

GENERAL Chemistry•⦁•Now™ Assess your understanding of this chapter's topics with additional quizzing and conceptual questions at http://now.brookscole.com/kotz6e

Practicing Skills

Matter: Elements and Atoms, Compounds and Molecules
(See Exercise 1.2.)

1. Give the name of each of the following elements:
   (a) C   (c) Cl   (e) Mg
   (b) K   (d) P   (f) Ni

2. ■ Give the name of each of the following elements:
   (a) Mn   (c) Na   (e) Xe
   (b) Cu   (d) Br   (f) Fe

3. Give the symbol for each of the following elements:
   (a) barium   (d) lead
   (b) titanium   (e) arsenic
   (c) chromium   (f) zinc

4. Give the symbol for each of the following elements:
   (a) silver   (d) tin
   (b) aluminum   (e) technetium
   (c) plutonium   (f) krypton

5. In each of the following pairs, decide which is an element and which is a compound.
   (a) Na and NaCl
   (b) sugar and carbon
   (c) gold and gold chloride

6. In each of the following pairs, decide which is an element and which is a compound.
   (a) $Pt(NH_3)_2Cl_2$ and Pt
   (b) copper or copper(II) oxide
   (c) silicon or sand

Physical and Chemical Properties
(See Exercises 1.3 and 1.6.)

7. In each case, decide whether the underlined property is a physical or chemical property.

   (a) The normal color of elemental bromine is <u>orange</u>.
   (b) Iron <u>turns to rust</u> in the presence of air and water.
   (c) Hydrogen can <u>explode</u> when ignited in air.
   (d) The <u>density</u> of titanium metal is 4.5 g/cm^3.
   (e) Tin metal <u>melts</u> at 505 K.
   (f) Chlorophyll, a plant pigment, <u>is green</u>.

8. ■ In each case, decide whether the change is a chemical or physical change.
   (a) A cup of household bleach changes the color of your favorite T-shirt from purple to pink.
   (b) Water vapor in your exhaled breath condenses in the air on a cold day.
   (c) Plants use carbon dioxide from the air to make sugar.
   (d) Butter melts when placed in the sun.

9. Which part of the description of a compound or element refers to its physical properties and which to its chemical properties?
   (a) The colorless liquid ethanol burns in air.
   (b) The shiny metal aluminum reacts readily with orange, liquid bromine.

10. Which part of the description of a compound or element refers to its physical properties and which to its chemical properties?
    (a) Calcium carbonate is a white solid with a density of 2.71 g/cm^3. It reacts readily with an acid to produce gaseous carbon dioxide.
    (b) Gray, powdered zinc metal reacts with purple iodine to give a white compound.

Using Density
(See Example 1.1. and the General ChemistryNow Screen 1.10.)

11. ■ Ethylene glycol, $C_2H_6O_2$, is an ingredient of automobile antifreeze. Its density is 1.11 g/cm^3 at 20 °C. If you need exactly 500. mL of this liquid, what mass of the compound, in grams, is required?

12. A piece of silver metal has a mass of 2.365 g. If the density of silver is 10.5 g/cm^3, what is the volume of the silver?

13. ■ A chemist needs 2.00 g of a liquid compound with a density of 0.718 g/cm^3. What volume of the compound is required?

14. The *cup* is a volume measure widely used by cooks in the United States. One cup is equivalent to 237 mL. If 1 cup of olive oil has a mass of 205 g, what is the density of the oil (in grams per cubic centimeter)?

15. A sample of unknown metal is placed in a graduated cylinder containing water. The mass of the sample is 37.5 g, and the water levels before and after adding the sample to the cylinder are as shown in the figure. Which metal in the following list is most likely the sample? (*d* is the density of the metal.)

(a) Mg, $d = 1.74 \text{ g/cm}^3$ (d) Al, $d = 2.70 \text{ g/cm}^3$
(b) Fe, $d = 7.87 \text{ g/cm}^3$ (e) Cu, $d = 8.96 \text{ g/cm}^3$
(c) Ag, $d = 10.5 \text{ g/cm}^3$ (f) Pb, $d = 11.3 \text{ g/cm}^3$

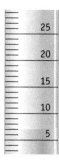

Graduated cylinders with unknown metal (*right*).

16. Iron pyrite is often called "fool's gold" because it looks like gold (see page 19). Suppose you have a solid that looks like gold, but you believe it to be fool's gold. The sample has a mass of 23.5 g. When the sample is lowered into the water in a graduated cylinder (see Study Question 15), the water level rises from 47.5 mL to 52.2 mL. Is the sample fool's gold ($d = 5.00 \text{ g/cm}^3$) or "real" gold ($d = 19.3 \text{ g/cm}^3$)?

Temperature Scales
(*See Exercise 1.7. and the General ChemistryNow Screen 1.15.*)

17. Many laboratories use 25 °C as a standard temperature. What is this temperature in kelvins?

18. The temperature on the surface of the sun is 5.5×10^3 °C. What is this temperature in kelvins?

19. ■ Make the following temperature conversions:

°C	K
(a) 16	——
(b) ——	370
(c) 40	——

20. Make the following temperature conversions:

°C	K
(a) ——	77
(b) 63	——
(c) ——	1450

Using Units
(*See Examples 1.2 and 1.3. and the General ChemistryNow Screen 1.14.*)

21. A marathon race covers a distance of 42.195 km. What is this distance in meters? In miles?

22. The average lead pencil, new and unused, is 19 cm long. What is its length in millimeters? In meters?

23. A standard U.S. postage stamp is 2.5 cm long and 2.1 cm wide. What is the area of the stamp in square centimeters? In square meters?

24. ■ A compact disk has a diameter of 11.8 cm. What is the surface area of the disk in square centimeters? In square meters? [Area of a circle $= (\pi)(\text{radius})^2$.]

25. A typical laboratory beaker has a volume of 250. mL. What is its volume in cubic centimeters? In liters? In cubic meters?

26. Some soft drinks are sold in bottles with a volume of 1.5 L. What is this volume in milliliters? In cubic centimeters? In cubic decimeters?

27. A book has a mass of 2.52 kg. What is this mass in grams?

28. A new U. S. dime has a mass of 2.265 g. What is this mass in kilograms? In milligrams?

Accuracy, Precision, and Error
(*See Example 1.4.*)

29. You and your lab partner are asked to determine the density of an aluminum bar. The mass is known accurately (to four significant figures). You use a simple metric ruler to determine its size and calculate the results in A. Your partner uses a precision micrometer and obtains the results in B.

Method A (g/cm³)	Method B (g/cm³)
2.2	2.703
2.3	2.701
2.7	2.705
2.4	5.811

The accepted density of aluminum is 2.702 g/cm^3.

(a) Calculate the average density for each method. Should all the experimental results be included in your calculations? If not, justify any omissions.

(b) Calculate the percent error for each method's average value.

(c) Which method's average value is more precise? Which method is more accurate?

30. ■ The accepted value of the melting point of pure aspirin is 135 °C. Trying to verify that value, you obtain the melting points of 134 °C, 136 °C, 133 °C, and 138 °C in four separate trials. Your partner finds melting points of 138 °C, 137 °C, 138 °C, and 138 °C.

(a) Calculate the average value and percent error for you and your partner.

(b) Which of you is more precise? More accurate?

General Questions

These questions are not designated as to type or location in the chapter. They may combine several concepts.

31. A piece of turquoise is a blue-green solid, and has a density of 2.65 g/cm^3 and a mass of 2.5 g.

(a) Which of these observations are qualitative and which are quantitative?

(b) Which of these observations are extensive and which are intensive?

(c) What is the volume of the piece of turquoise?

32. Give a physical property and a chemical property for the elements hydrogen, oxygen, iron, and sodium. (The elements listed are selected from examples given in Chapter 1.)

33. The gemstone called aquamarine is composed of aluminum, silicon, and oxygen.

Aquamarine is the bluish crystal. It is surrounded by aluminum foil and crystalline silicon.

(a) What are the symbols of the three elements that combine to make the gem aquamarine?

(b) Based on the photo, describe some of the physical properties of the elements and the mineral. Are any the same? Are any properties different?

34. Eight observations are listed below. Which of these observations identify chemical properties?

(a) Sugar is soluble in water.

(b) Water boils at 100 °C.

(c) Ultraviolet light converts O_3 (ozone) to O_2 (oxygen).

(d) Ice is less dense than water.

(e) Sodium metal reacts violently with water.

(f) CO_2 does not support combustion.

(g) Chlorine is a yellow gas.

(h) Heat is required to melt ice.

35. Neon, a gaseous element used in neon signs, has a melting point of −248.6 °C and a boiling point of −246.1 °C. Express these temperatures in kelvins.

36. You can identify a metal by carefully determining its density (d). An unknown piece of metal, with a mass of 2.361 g, is 2.35 cm long, 1.34 cm wide, and 1.05 mm thick. Which of the following is this element?

(a) Nickel, d = 8.90 g/cm^3

(b) Titanium, d = 4.50 g/cm^3

(c) Zinc, d = 7.13 g/cm^3

(d) Tin, d = 7.23 g/cm^3

37. Molecular distances are usually given in nanometers (1 nm = 1 × 10^{-9} m) or in picometers (1 pm = 1 × 10^{-12} m). However, the angstrom (Å) is sometimes used, where 1 Å = 1 × 10^{-10} m. (The angstrom is not an SI unit.) If the distance between the Pt atom and the N atom in the cancer chemotherapy drug cisplatin is 1.97 Å, what is this distance in nanometers? In picometers?

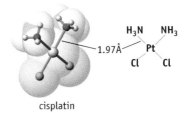

cisplatin

38. The separation between carbon atoms in diamond is 0.154 nm. What is their separation in meters? In picometers?

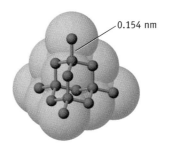

A portion of the diamond structure

39. ■ A red blood cell has a diameter of 7.5 μm (micrometers). What is this dimension in (a) meters, (b) nanometers, and (c) picometers?

40. ■ Which occupies a larger volume, 600. g of water (with a density of 0.995 g/cm^3) or 600. g of lead (with a density of 11.34 g/cm^3)?

41. The platinum-containing cancer drug cisplatin contains 65.0% platinum. If you have 1.53 g of the compound, what mass of platinum (in grams) is contained in this sample?

42. The solder once used by plumbers to fasten copper pipes together consists of 67% lead and 33% tin. What is the mass of lead in a 250-g block of solder?

43. ■ The anesthetic procaine hydrochloride is often used to deaden pain during dental surgery. The compound is packaged as a 10.% solution (by mass; d = 1.0 g/mL) in water. If your dentist injects 0.50 mL of the solution, what mass of procaine hydrochloride (in milligrams) is injected?

44. A cube of aluminum (density = 2.70 g/cm^3) has a mass of 7.6 g. What must be the length of the cube's edge (in centimeters)? (*See General ChemistryNow Screen 1.10, Tutorial 2, Density.*)

45. ■ You have a 100.0-mL graduated cylinder containing 50.0 mL of water. You drop a 154-g piece of brass (d = 8.56 g/cm^3) into the water. How high does the water rise in the graduated cylinder?

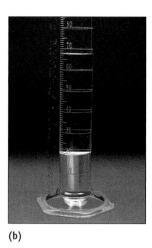

(a)   **(b)**

(a) A graduated cylinder with 50.0 ml of water. (b) A piece of brass is added to the cylinder.

46. You have a white crystalline solid, known to be one of the potassium compounds listed below. To determine which, you measure the solid's density. You measure out 18.82 g and transfer it to a graduated cylinder containing kerosene (in which salts will not dissolve). The level of liquid kerosene rises from 8.5 mL to 15.3 mL. Calculate the density of the solid, and identify the compound from the following list.

    (a) KF, $d = 2.48$ g/cm^3 (c) KBr, $d = 2.75$ g/cm^3

    (b) KCl, $d = 1.98$ g/cm^3 (d) KI, $d = 3.13$ g/cm^3

47. A distant acquaintance has offered to sell you a necklace, said to be pure (24-carat) gold, for $300. You have some doubts, however; perhaps it is gold plated. You decide to run a test. You have a graduated cylinder and a small balance. You partially fill the cylinder with water and immerse the necklace; the height of water rises from 22.5 mL to 26.0 mL. Then you determine the mass to be 67 g. You recall that the density of gold is 19.3 g/cm^3, and that no other element has a density near this value. (Silver has a density of 11.5 g/cm^3.) The price of gold on the open market is $380 per troy ounce (1 troy ounce = 31.1 g). Is the necklace gold? Explain your conclusion. Is $300 a good price?

Conceptual Questions

48. The mineral fluorite contains the elements calcium and fluorine. What are the symbols of these elements? How would you describe the shape of the fluorite crystals in the photo? What can this tell us about the arrangement of the atoms inside the crystal?

The mineral fluorite, calcium fluoride.

49. Small chips of iron are mixed with sand (see the photo). Is this a homogeneous or heterogeneous mixture? Suggest a way to separate the iron from the sand.

Chips of iron mixed with sand.

50. The following photo shows copper balls, immersed in water, floating on top of mercury. What are the liquids and the solids in this photo? Which substance is most dense? Which is least dense?

Water, copper, and mercury.

51. ■ Carbon tetrachloride, CCl$_4$, a liquid compound, has a density of 1.58 g/cm^3. If you place a piece of a plastic soda bottle ($d = 1.37$ g/cm^3) and a piece of aluminum ($d = 2.70$ g/cm^3) in liquid CCl$_4$, will the plastic and aluminum float or sink?

52. Figure 1.7 shows a piece of table salt and a representation of its internal structure. Which is the macroscopic view and which is the particulate view? How are the macroscopic and particulate views related?

53. ▲ You have a sample of a white crystalline substance from your kitchen. You know that it is either salt or sugar. Although you could decide by taste, suggest another property that you could use to determine the sample's identity. (*Hint:* You may use the World Wide Web or a handbook of chemistry in the library to find some pertinent information.)

54. Milk in a glass bottle was placed in the freezer compartment of a refrigerator overnight. By morning a column of frozen milk emerged from the bottle. Explain this observation.

Frozen milk in a glass bottle.

55. The element gallium has a melting point of 29.8 °C. If you held a sample of gallium in your hand, should it melt? Explain briefly.

Gallium metal.

56. ▲ The density of pure water is given at various temperatures.

T(°C)	d(g/cm³)
4	0.99997
15	0.99913
25	0.99707
35	0.99406

Suppose your laboratory partner tells you that the density of water at 20 °C is 0.99910 g/cm³. Is this a reasonable number? Why or why not?

57. You can figure out whether a substance floats or sinks if you know its density and the density of the liquid. In which of the liquids listed below will high-density polyethylene (HDPE, a common plastic whose density is 0.97 g/mL) float? (HDPE does not dissolve in these liquids.)

Substance	Density (g/cm³)	Properties, Uses
Ethylene glycol	1.1088	Toxic; the major component of automobile antifreeze
Water	0.9997	
Ethanol	0.7893	The alcohol in alcoholic beverages
Methanol	0.7914	Toxic; gasoline additive to prevent gas line freezing
Acetic acid	1.0492	Component of vinegar
Glycerol	1.2613	Solvent used in home care products.

58. Hexane (C_6H_{14}, $d = 0.766$ g/cm³), perfluorohexane (C_6F_{14}, $d = 1.669$ g/cm³), and water are immiscible liquids; that is, they do not dissolve in one another. You place 10 mL of each liquid in a graduated cylinder, along with pieces of high-density polyethylene (HDPE, $d = 0.97$ g/cm³), polyvinyl chloride (PVC, $d = 1.36$ g/cm³), and Teflon (density = 2.3 g/cm³). None of these common plastics dissolve in these liquids. Describe what you expect to see.

59. Make a drawing, based on the kinetic-molecular theory and the ideas about atoms and molecules presented in this chapter, of the arrangement of particles in each of the cases listed here. For each case draw ten particles of each substance. Your diagram can be two-dimensional. Represent each atom as a circle and distinguish each kind of atom by shading.

(a) a sample of solid iron (which consists of iron atoms)

(b) a sample of *liquid* water (which consists of H_2O molecules)

(c) a sample of water *vapor*

(d) a homogeneous mixture of water vapor and helium gas (which consists of helium atoms)

(e) a heterogeneous mixture consisting of liquid water and solid aluminum; show a region of the sample that includes both substances

(f) a sample of brass (which is a homogeneous mixture of copper and zinc)

60. You are given a sample of a silvery metal. What information would you seek to prove that the metal is silver?

61. Suggest a way to determine whether the colorless liquid in a beaker is water. If it is water, does it contain dissolved salt? How could you discover whether salt is dissolved in the water?

62. Describe an experimental method that can be used to determine the density of an irregularly shaped piece of metal.

63. Three liquids of different densities are mixed. Because they are not miscible (do not form a homogeneous solution with one another), they form discrete layers, one on top of the other. Sketch the result of mixing carbon tetrachloride (CCl_4, $d = 1.58$ g/cm^3), mercury ($d = 13.546$ g/cm^3), and water ($d = 1.00$ g/cm^3).

64. Diabetes can alter the density of urine, so urine density can be used as a diagnostic tool. People with diabetes may excrete too much sugar or too much water. What do you predict will happen to the density of urine under each of these conditions? (*Hint:* Water containing dissolved sugar has a higher density than pure water.)

65. The following photo shows the element potassium reacting with water to form the element hydrogen, a gas, and a solution of the compound potassium hydroxide.

Potassium reacting with water to produce hydrogen gas and potassium hydroxide.

(a) What states of matter are involved in the reaction?

(b) Is the observed change chemical or physical?

(c) What are the reactants in this reaction and what are the products?

(d) What qualitative observations can be made concerning this reaction?

66. A copper-colored metal is found to conduct an electric current. Can you say with certainty that it is copper? Why or why not? Suggest additional information that could provide unequivocal confirmation that the metal is copper.

67. What experiment can you use to:

(a) Separate salt from water?

(b) Separate iron filings from small pieces of lead?

(c) Separate elemental sulfur from sugar?

68. Four balloons (each with a volume of 10 L and a mass of 1.00 g) are filled with a different gas:

Helium, $d = 0.164$ g/L

Neon, $d = 0.825$ g/L

Argon, $d = 1.633$ g/L

Krypton, $d = 4.425$ g/L

If the density of dry air is 1.12 g/L, which balloon or balloons float in air?

69. Many foods are fortified with vitamins and minerals. For example, some breakfast cereals have elemental iron added. Iron chips are used instead of iron compounds because the compounds can be converted by the oxygen in air to a form of iron that is not biochemically useful. Iron chips, in contrast, are converted to useful iron compounds in the gut, and the iron can then be absorbed. Outline a method by which you could remove the iron (as iron chips) from a box of cereal and determine the mass of iron in a given mass of cereal. (*See General ChemistryNow Screens 1.1 and 1.18 Chemical Puzzler.*)

Some breakfast cereals contain iron in the form of elemental iron.

70. Describe what occurs when a hot object comes in contact with a cooler object. (*See General ChemistryNow Screen 1.15 Temperature.*)

71. Study the animation of the conversion of P_4 and Cl_2 molecules to PCl_3 molecules on General ChemistryNow CD-ROM or website Screen 1.13 Chemical Change on the Molecular Scale.

(a) What are the reactants in this chemical change? What are the products?

(b) Describe how the structures of the reactant molecules differ from the structures of the product molecules.

72. The photo below shows elemental iodine dissolving in ethanol to give a solution. Is this a physical or a chemical change?

Elemental iodine dissolving in ethanol.

(*See General ChemistryNow Screen 1.9 Exercise, Physical Properties of Matter.*)

Mathematics of Chemistry

These questions provide an additional review of the mathematical skill used in general chemistry as presented in Section 1.8.

Exponential Notation

73. Express the following numbers in exponential or scientific notation.
    (a) 0.054   (b) 5462   (c) 0.000792

74. Express the following numbers in fixed notation (e.g., $123 \times 10^2 = 123$).
    (a) 1.62×10^3
    (b) 2.57×10^{-4}
    (c) 6.32×10^{-2}

75. Carry out the following operations. Provide the answer with the correct number of significant figures.
    (a) $(1.52)(6.21 \times 10^{-3})$
    (b) $(6.21 \times 10^3) - (5.23 \times 10^2)$
    (c) $(6.21 \times 10^3) \div (5.23 \times 10^2)$

76. Carry out the following operations. Provide the answer with the correct number of significant figures.
    (a) $(6.25 \times 10^2)^3$
    (b) $\sqrt{2.35 \times 10^{-3}}$
    (c) $(2.35 \times 10^{-3})^{1/3}$

Significant Figures
(*See Exercise 1.13.*)

77. Give the number of significant figures in each of the following numbers:
    (a) 0.0123 g   (c) 1.6402 g
    (b) 3.40×10^3 mL   (d) 1.020 L

78. ■ Give the number of significant figures in each of the following numbers:
    (a) 0.00546 g   (c) 2.300×10^{-4} g
    (b) 1600 mL   (d) 2.34×10^9 atoms

79. Carry out the following calculation, and report the answer with the correct number of significant figures.

$$(0.0546)(16.0000)\left[\frac{7.779}{55.85}\right]$$

80. ■ Carry out the following calculation, and report the answer to the correct number of significant figures.

$$(1.68)\left[\frac{23.56 - 2.3}{1.248 \times 10^3}\right]$$

Graphing
(*See Exercise 1.15. Use the plotting program on the General ChemistryNow CD-ROM or website or Microsoft Excel.*)

81. You are asked to calibrate a spectrophotometer in the laboratory and collect the following data. Plot the data with concentration on the x-axis and absorbance on the y-axis. Draw the best straight line using the points on the graph (or do a least-squares or linear regression analysis using a computer program) and then write the equation for the resulting straight line. What is the slope of the line? What is the concentration when the absorbance is 0.635?

Concentration (M)	Absorbance
0.00	0.00
1.029×10^{-3}	0.257
2.058×10^{-3}	0.518
3.087×10^{-3}	0.771
4.116×10^{-3}	1.021

82. To determine the average mass of a popcorn kernel you collect the following data:

Number of kernels	Mass (g)
5	0.836
12	2.162
35	5.801

Plot the data with number of kernels on the x-axis and mass on the y-axis. Draw the best straight line using the points on the graph (or do a least-squares or linear regression analysis using a computer program) and then write the equation for the resulting straight line. What is the slope of the line? What does the slope of the line signify about the mass of a popcorn kernel? What is the mass of 50 popcorn kernels? How many kernels are there in a handful of popcorn (20.88 g)?

83. Using the graph below:
    (a) What is the value of x when y = 4.0?
    (b) What is the value of y when x = 0.30?
    (c) ■ What are the slope and the y-intercept of the line?
    (d) What is the value of y when x = 1.0?

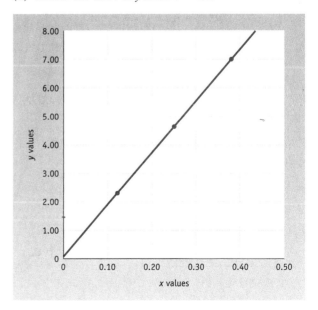

84. ■ Use the graph below to answer the following questions.

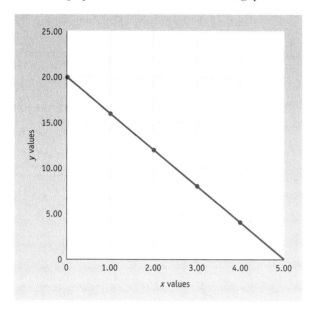

(a) Derive the equation for the straight line, $y = mx + b$.

(b) What is the value of y when $x = 6.0$?

Using Equations

85. Solve the following equation for the unknown value, C.
$$(0.502)(123) = (750.)C$$

86. Solve the following equation for the unknown value, n.
$$(2.34)(15.6) = n(0.0821)(273)$$

87. Solve the following equation for the unknown value, T.
$$(4.184)(244)(T - 292.0) + (0.449)(88.5)(T - 369.0) = 0$$

88. Solve the following equation for the unknown value, n.
$$-246.0 = 1312\left[\frac{1}{2^2} - \frac{1}{n^2}\right]$$

Problem Solving

89. ■ Diamond has a density of 3.513 g/cm³. The mass of diamonds is often measured in "carats," where 1 carat equals 0.200 g. What is the volume (in cubic centimeters) of a 1.50-carat diamond?

90. ▲ The smallest repeating unit of a crystal of common salt is a cube (called a unit cell) with an edge length of 0.563 nm.

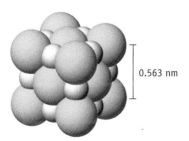

0.563 nm

sodium chloride, NaCl

(a) What is the volume of this cube in cubic nanometers? In cubic centimeters?

(b) The density of NaCl is 2.17 g/cm³. What is the mass of this smallest repeating unit ("unit cell")?

(c) Each repeating unit is composed of four NaCl "molecules." What is the mass of one NaCl molecule?

91. An ancient gold coin is 2.2 cm in diameter and 3.0 mm thick. It is a cylinder for which volume = $(\pi)(\text{radius})^2(\text{thickness})$. If the density of gold is 19.3 g/cm³, what is the mass of the coin in grams?

92. Copper has a density of 8.96 g/cm³. An ingot of copper with a mass of 57 kg (126 lb) is drawn into wire with a diameter of 9.50 mm. What length of wire (in meters) can be produced? [Volume of wire = $(\pi)(\text{radius})^2(\text{length})$].

93. ▲ In July 1983, an Air Canada Boeing 767 ran out of fuel over central Canada on a trip from Montreal to Edmonton. (The plane glided safely to a landing at an abandoned airstrip.) The pilots knew that 22,300 kg of fuel were required for the trip, and they knew that 7682 L of fuel were already in the tank. The ground crew added 4916 L of fuel, which was only about one fifth of what was required. The crew members used a factor of 1.77 for the fuel density—the problem is that 1.77 has units of *pounds* per liter and not *kilograms* per liter! What is the fuel density in units of kg/L? What mass of fuel should have been loaded? (1 lb = 453.6 g.)

94. When you heat popcorn, it pops because it loses water explosively. Assume a kernel of corn, with a mass of 0.125 g, has a mass of only 0.106 g after popping.

(a) What percentage of its mass did the kernel lose on popping?

(b) ■ Popcorn is sold by the pound in the United States. Using 0.125 g as the average mass of a popcorn kernel, how many kernels are there in a pound of popcorn? (1 lb = 453.6 g.)

95. ▲ The aluminum in a package containing 75 ft² of kitchen foil weighs approximately 12 ounces. Aluminum has a density of 2.70 g/cm³. What is the approximate thickness of the aluminum foil in millimeters? (1 oz = 28.4 g.)

96. ▲ The fluoridation of city water supplies has been practiced in the United States for several decades. It is done by continuously adding sodium fluoride to water as it comes from a reservoir. Assume you live in a medium-sized city of 150,000 people and that 660 L (170 gal) of water is consumed per person per day. What mass of sodium fluoride (in kilograms) must be added to the water supply each year (365 days) to have the required fluoride concentration of 1 ppm (part per million)—that is, 1 kilogram of fluoride per 1 million kilograms of water? (Sodium fluoride is 45.0% fluoride, and water has a density of 1.00 g/cm³.)

97. ■ ▲ About two centuries ago, Benjamin Franklin showed that 1 teaspoon of oil would cover about 0.5 acre of still water. If you know that 1.0×10^4 m² = 2.47 acres, and that there is approximately 5 cm³ in a teaspoon, what is the thickness of the layer of oil? How might this thickness be related to the sizes of molecules?

98. ▲ Automobile batteries are filled with an aqueous solution of sulfuric acid. What is the mass of the acid (in grams) in 500. mL of the battery acid solution if the density of the solution is 1.285 g/cm^3 and if the solution is 38.08% sulfuric acid by mass?

99. A piece of copper has a mass of 0.546 g. Show how to set up an expression to find the volume of this piece of copper in units of liters. (Copper density = 8.96 g/cm^3.) (*See General ChemistryNow Screen 1.17 Tutorial 1, Using Numerical Information.*)

100. Evaluate the value of x in the following expressions:
   (a) $x = [(9.345 \times 10^{-4})(6.23 \times 10^{6})]^{3}$
   (b) $x = \sqrt{(1.23 \times 10^{-2})(4.5 \times 10^{5})}$
   (c) $x = \sqrt[3]{(1.23 \times 10^{-2})(4.5 \times 10^{5})}$

   Show the answers to the correct number of significant figures. (*See General ChemistryNow CD-ROM or website Screen 1.17 Tutorial 4, Using Numerical Information.*)

101. A 26-meter tall statue of Buddha in Tibet is covered with 279 kg of gold. If the gold was applied to a thickness of 0.0015 mm, what surface area is covered (in square meters)? (Gold density = 19.3 g/cm^3.)

102. At 25 °C the density of water is 0.997 g/cm^3, whereas the density of ice at −10 °C is 0.917 g/cm^3.
   (a) If a soft-drink can (volume = 250. mL) is filled completely with pure water at 25 °C and then frozen at −10 °C, what volume does the solid occupy?
   (b) Can the ice be contained within the can?

103. Suppose your bedroom is 18 ft long, 15 ft wide, and the distance from floor to ceiling is 8 ft, 6 in. You need to know the volume of the room in metric units for some scientific calculations.
   (a) What is the room's volume in cubic meters? In liters?
   (b) What is the mass of air in the room in kilograms? In pounds? (Assume the density of air is 1.2 g/L and that the room is empty of furniture.)

104. ■ A spherical steel ball has a mass of 3.475 g and a diameter of 9.40 mm. What is the density of the steel? [The volume of a sphere = $(4/3)\pi r^3$ where r = radius.]

105. ▲ The substances listed below are clear liquids. You are asked to identify an unknown liquid that is known to be one of these liquids. You pipette a 3.50-mL sample into a beaker. The empty beaker had a mass of 12.20 g, and the beaker plus the liquid weighed 16.08 g.

Substance	Known Density at 25 °C (g/cm^3)
Ethylene glycol	1.1088 (the major component of antifreeze)
Water	0.9997
Ethanol	0.7893 (the alcohol in alcoholic beverages)
Acetic acid	1.0492 (the active component of vinegar)
Glycerol	1.2613 (a solvent, used in home care products)

   (a) Calculate the density and identify the unknown.

(b) If you were able to measure the volume to only two significant figures (that is, 3.5 mL, not 3.50 mL), will the results be sufficiently accurate to identify the unknown? Explain.

106. ▲ You have an irregularly shaped chunk of an unknown metal. To identify it, you determine its density and then compare this value with known values that you look up in the chemistry library. The mass of the metal is 74.122 g. Because of the irregular shape, you measure the volume by submerging the metal in water in a graduated cylinder. When you do this, the water level in the cylinder rises from 28.2 mL to 36.7 mL.
   (a) What is the density of the metal? (Use the correct number of significant figures in your answer.)
   (b) The unknown is one of the seven metals listed below. Is it possible to identify the metal based on the density you have calculated? Explain.

Metal	Density (g/cm^3)	Metal	Density (g/cm^3)
zinc	7.13	nickel	8.90
iron	7.87	copper	8.96
cadmium	8.65	silver	10.50
cobalt	8.90		

107. ▲ A 7.50×10^2-mL sample of an unknown gas has a mass of 0.9360 g.
   (a) What is the density of the gas? Express your answer in units of g/L.
   (b) Nine gases and their densities are listed below. Compare the experimentally determined density with these values. Can you determine the identity of the gas based on the experimentally determined density?
   (c) A more accurate measure of volume is made next, and the volume of this sample of gas is found to be 7.496×10^2 mL. Using a more accurate density calculated using this value, can you now determine the identity of the gas?

Gas	Density (g/L)	Gas	Density (g/L)
B_2H_6	1.2345	C_2H_4	1.2516
CH_2O	1.3396	CO	1.2497
Dry air	1.2920	C_2H_6	1.3416
N_2	1.2498	NO	1.2949
O_2	1.4276		

108. ▲ The density of a single, small crystal can be determined by the *flotation method.* This method is based on the idea that if a crystal and a liquid have precisely the same density, the crystal will hang suspended in the liquid. A crystal that is more dense will sink; one that is less dense will float. If the crystal neither sinks nor floats, then the density of the crystal equals the density of the liquid. Generally, mixtures of liquids are used to get the proper density. Chlorocarbons and bromocarbons (see

---

the list below) are often the liquids of choice. If the two liquids are similar, then volumes are usually additive and the density of the mixture relates directly to composition. (An example: 1.0 mL of $CHCl_3$, $d = 1.4832$ g/mL, and 1.0 mL of CCl_4, $d = 1.5940$ g/mL, when mixed, give 2.0 mL of a mixture with a density of 1.5386 g/mL. The density of the mixture is the average of the values of the two individual components.)

The problem: A small crystal of silicon, germanium, tin, or lead (Group 4A in the periodic table) will hang suspended in a mixture made of 61.18% (by volume) $CHBr_3$ and 38.82% (by volume) $CHCl_3$. Calculate the density and identify the element. (You will have to look up the values of the density of the elements in a manual such as the *The Handbook of Chemistry and Physics* in the library or in a World Wide Web site such as WebElements at, **www.webelements.com.**)

Liquid	Density (g/mL)	Liquid	Density (g/mL)
CH_2Cl_2	1.3266	CH_2Br_2	2.4970
$CHCl_3$	1.4832	$CHBr_3$	2.8899
CCl_4	1.5940	CBr_4	2.9609

109. ▲ Suppose you have a cylindrical glass tube with a thin capillary opening, and you wish to determine the diameter of the capillary. You can do this experimentally by weighing a piece of the tubing before and after filling a portion of the capillary with mercury. Using the following information, calculate the diameter of the capillary.

Mass of tube before adding mercury = 3.263 g

Mass of tube after adding mercury = 3.416 g

Length of capillary filled with mercury = 16.75 mm

Density of mercury = 13.546 g/cm³

Volume of cylindrical capillary filled with mercury
  = $(\pi)(radius)^2(length)$

Do you need a live tutor for homework problems?
Access vMentor at General ChemistryNow at
http://now.brookscole.com/kotz6e
for one-on-one tutoring from a chemistry expert

▲ More challenging   ■ In General ChemistryNow   Blue-numbered questions answered in Appendix O

2—Atoms and Elements

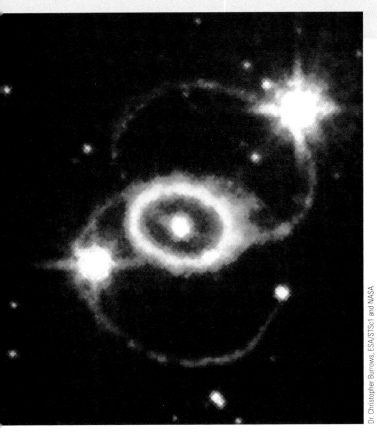

The supernova of 1987. When a star becomes more and more dense, and hotter and hotter, it can become a "red giant." The star is unstable and explodes as a "supernova." One such spectacular event occurred in a distant star in 1987. These explosions are the origin of the heavier elements, such as iron, nickel, and cobalt.

Dr. Christopher Burrows, ESA/STSc1 and NASA.

Stardust

A wide array of elements make up planet Earth and every living thing on it. What is science's view of the cosmic origin of these elements that we take for granted in our environment and in our lives?

The "big bang" theory is the generally accepted theory for the origin of the universe. This theory holds that an unimaginably dense, grapefruit-sized sphere of matter exploded about 15 billion years ago, spewing the products of that explosion as a rapidly expanding cloud with a temperature in the range of 10^{30} K. Within 1 second, the universe was populated with the particles we explore in this chapter: protons, electrons, and neutrons. Within a few more seconds, the universe had cooled by millions and millions of degrees, and protons and neutrons began to combine to form helium nuclei. After only about 8 minutes, scientists believe, the universe was about one-quarter helium and about three-quarters hydrogen. In fact, this is very close to the composition of the universe today, 15 billion years later. But humans, animals, and plants are built mainly from carbon, oxygen, nitrogen, sulfur, phosphorus, iron, and zinc—heavier elements that have only a trace abundance in the universe as a whole. Where do these heavier elements come from?

The cloud of hydrogen and helium cooled over a period of thousands of years and condensed into stars like our sun. There hydrogen atoms fuse into more helium atoms and energy streams outward. Every second on the sun, 700 million tons of hydrogen is converted to 695 million tons of helium, and 3.9×10^{26} joules of energy is evolved.

Gradually, over millions of years, a hydrogen-burning star becomes more and more dense and hotter and hotter. The helium atoms initially formed in the star begin to fuse into heavier atoms—first carbon, then oxygen, and then neon, magnesium, silicon, phosphorus, and argon. The star becomes even hotter and more dense. Hydrogen is forced to the outer reaches of the star, and the star becomes a red giant. Under certain circumstances, the star will explode, and earth-bound observers see it as a supernova. A supernova can be as much as 10^8 times brighter than the original star. A single supernova is comparable in brightness to the whole of the galaxy in which it is formed!

The supernova that appeared in 1987 gave astronomers an opportunity to study what happens in these element factories. It is here that the heavier atoms such as iron form. In fact, it is with iron that nature reaches its zenith of stability. To make heavier and heavier elements requires energy, rather than having energy as an outcome of element synthesis.

The elements spewing out of an exploding supernova move through space and gradually condense into planets, of which ours is just one.

The mechanism of element formation in stars is reasonably well understood, and much experimental evidence exists to support this theory. However, the way in which these elements are then assembled out of stardust into living organisms on our planet—and perhaps other planets—is not yet understood at all. (*See the General ChemistryNow Screen 2.2 Introduction to Atoms, to watch a video on the "big bang" theory.*)

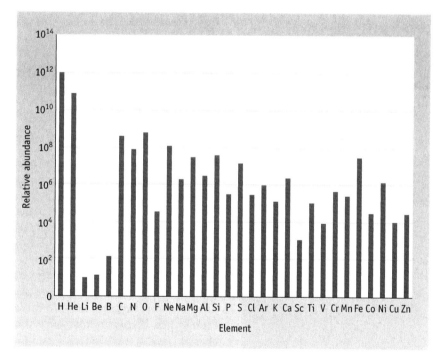

The abundance of the elements in the solar system from H to Zn. The chart shows a general decline in abundance with increasing mass among the first 30 elements. The decline continues above zinc. Notice that the scale on the vertical axis is logarithmic—that is, it progresses in powers of ten. The abundance of nitrogen, for example, is 1/10,000 ($1/10^4$) of the abundance of hydrogen. (All abundances are plotted as the number of atoms per 10^{12} atoms of H. The fact that the abundances of Li, Be, and B, as well as those of the elements near Fe, do not follow the general decline is a consequence of the way that elements are synthesized in stars.)

To Review Before You Begin
- Names and uses of SI units (Section 1.6)
- Solving numerical problems using units (Section 1.8)

GENERAL
Chemistry⚛Now™

Throughout the chapter this icon introduces a list of resources on the General ChemistryNow CD-ROM or website (http://now.brookscole.com/kotz6e) that will:
- help you evaluate your knowledge of the material
- provide homework problems
- allow you to take an exam-prep quiz
- provide a personalized Learning Plan targeting resources that address areas you should study

The chemical elements are forged in stars. What are the similarities among the elements? What are the differences? What are their physical and chemical properties? How can we tell them apart? This chapter begins our exploration of the chemistry of the elements, the building blocks of the science of chemistry.

2.1—Protons, Electrons, and Neutrons: Development of Atomic Structure

Around 1900 a series of experiments done by scientists such as Sir John Joseph Thomson (1856–1940) and Ernest Rutherford (1871–1937) in England established a model of the atom that is still the basis of modern atomic theory. Three **subatomic particles** make up all atoms: electrically positive protons, electrically neutral neutrons, and electrically negative electrons. The model places the more massive protons and neutrons in a very small nucleus, which contains all the positive charge and almost all the mass of an atom. Electrons, with a much smaller mass than protons or neutrons, surround the nucleus and occupy most of the volume (Figure 2.1). Atoms have no net charge; the positive and negative charges balance. *The number of electrons outside the nucleus equals the number of protons within the nucleus.*

What is the experimental basis of atomic structure? How did the work of Thomson, Rutherford, and others lead to this model?

Electricity

Electricity is involved in many of the experiments from which the theory of atomic structure was derived. The fact that objects can bear an electric charge was first observed by the ancient Egyptians, who noted that amber, when rubbed with wool or silk, attracted small objects. You can observe the same thing when you rub a balloon on your hair on a dry day—your hair is attracted to the balloon (Figure 2.2a). A bolt of lightning or the shock you get when touching a doorknob results when an electric charge moves from one place to another.

Two types of electric charge had been discovered by the time of Benjamin Franklin (1706–1790), the American statesman and inventor. He named them positive (+) and negative (−), because they appear as opposites and can neutralize each other. Experiments show that like charges repel each other and unlike charges attract each other. Franklin also concluded that charge is balanced: If a negative charge appears somewhere, a positive charge of the same size must appear somewhere else. The fact that a charge builds up when one substance is rubbed over another implies that the rubbing separates positive and negative charges. By the 19th century it was understood that positive and negative charges are somehow associated with matter—and perhaps with atoms.

Radioactivity

In 1896 the French physicist Henri Becquerel (1852–1908) discovered that a uranium ore emitted rays that could darken a photographic plate, even though the plate was covered by black paper to protect it from being exposed to light. In 1898 Marie

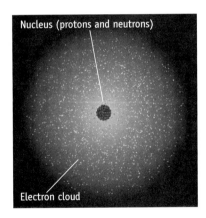

Figure 2.1   The structure of the atom. All atoms contain a nucleus with one or more protons (positive electric charge) and neutrons (no charge). Electrons (negative electric charge) are arranged in space as a "cloud" around the nucleus. In an electrically neutral atom, the number of electrons equals the number of protons. Note that this figure is not drawn to scale. If the nucleus were really the size depicted here, the electron cloud would extend about 800 feet. The atom is mostly empty space!

Nucleus (protons and neutrons)

Electron cloud

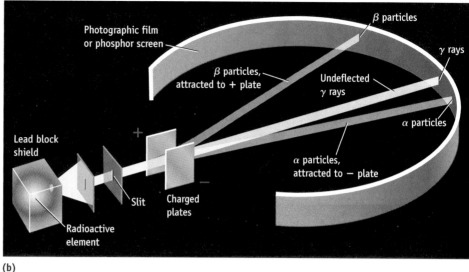

(a)   **(b)**

Figure 2.2   Electricity and radioactivity. (a) If you brush your hair with a balloon, a static electric charge builds up on the surface of the balloon. Experiments show that objects having opposite electric charges attract each other, whereas objects having the same electric charge repel each other. (*See the General ChemistryNow Screen 2.4 Electricity and Electric Charge, for an exercise on the effects of charge.*) (b) Alpha (α), beta (β), and gamma (γ) rays from a radioactive element are separated by passing them between electrically charged plates. Positively charged α particles are attracted to the negative plate, and negatively charged β particles are attracted to the positive plate. (Note that the heavier α particles are deflected less than the lighter β particles.) Gamma rays have no electric charge and pass undeflected between the charged plates. (*See the General ChemistryNow Screen 2.5 Evidence of Subatomic Particles, for an exercise on this experiment.*)

and Pierre Curie (1867–1934) isolated polonium and radium, which also emitted the same kind of rays, and in 1899 they suggested that atoms of certain substances emit these unusual rays when they disintegrate. They named this phenomenon **radioactivity**, and substances that display this property are said to be **radioactive**.

Early experiments identified three kinds of radiation: alpha (α), beta (β), and gamma (γ) rays. These rays behave differently when passed between electrically charged plates (Figure 2.2b). Alpha and β rays are deflected, but γ rays pass straight through. This implies that α and β rays are electrically charged particles, because they are attracted or repelled by the charged plates. Even though an α particle was found to have an electric charge (+2) twice as large as that of a β particle (−1), α particles are deflected less, which implies that α particles must be heavier than β particles. Gamma rays have no detectable charge or mass; they behave like light rays.

Marie Curie's suggestion that atoms disintegrate contradicted ideas put forward in 1803 by John Dalton (1766–1844) that atoms are indivisible. If atoms can break apart, there must be something smaller than an atom; that is, atoms must be composed of even smaller, subatomic particles.

Cathode-Ray Tubes and the Characterization of Electrons

Further evidence that atoms are composed of smaller particles came from experiments with cathode-ray tubes (Figure 2.3). These are glass tubes from which most of the air has been removed and that contain two metal electrodes. When a sufficiently high voltage is applied to the electrodes, a *cathode ray* flows from the negative electrode (cathode) to the positive electrode (anode). Experiments showed that cathode rays travel in straight lines, cause gases to glow, can heat metal objects red hot, can be deflected by a magnetic field, and are attracted toward positively charged

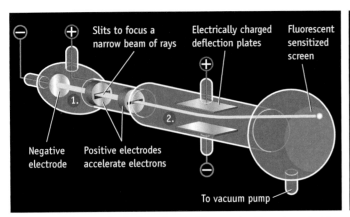

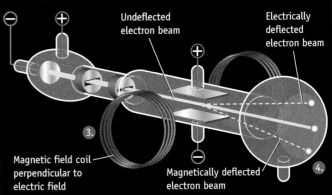

1. A beam of electrons (cathode rays) is accelerated through two focusing slits.

2. When passing through an electric field the beam of electrons is deflected.

3. The experiment is arranged so that the electric field causes the beam of electrons to be deflected in one direction. The magnetic field deflects the beam in the opposite direction.

4. By balancing the effects of the electrical and magnetic fields the charge-to-mass ratio of the electron can be determined.

Active Figure 2.3   **Measuring the electron's charge-to-mass ratio.** This experiment was done by J. J. Thomson in 1896–1897.

GENERAL
Chemistry·ᵢ·Now™ See the General ChemistryNow CD-ROM or website to explore an interactive version of this figure accompanied by an exercise.

plates. When cathode rays strike a fluorescent screen, light is given off in a series of tiny flashes. We can understand all of these observations if a cathode ray is assumed to be a beam of the negatively charged particles we now know as **electrons**.

You are already familiar with cathode rays. Television pictures and the images on some types of computer monitors are formed by using electrically charged plates to aim cathode rays onto the back of a phosphor screen on which we view the image. Sir Joseph John Thomson (1856–1940) used this principle to prove experimentally the existence of the electron and to study its properties. He applied electric and magnetic fields simultaneously to a beam of cathode rays (Figure 2.3). By balancing the effect of the electric field against that of the magnetic field and using basic laws of electricity and magnetism, Thompson calculated the ratio of the charge to the mass for the particles in the beam. He was not able to determine either charge or mass independently. However, he found the same charge-to-mass ratio in experiments using 20 different metals as cathodes and several different gases. These results suggested that electrons are present in atoms of all elements.

It remained for the American physicist Robert Andrews Millikan (1868–1953) to measure the charge on an electron and thereby enable scientists to calculate its mass (Figure 2.4). In his experiment tiny droplets of oil were sprayed into a chamber. As they settled slowly through the air, the droplets were exposed to x-rays, which caused them to acquire an electric charge. Millikan used a small telescope to observe individual droplets. If the electric charge on the plates above and below the droplets was adjusted, the electrostatic attractive force pulling a droplet upward could be balanced by the force of gravity pulling the droplet downward. From the equations describing these forces, Millikan calculated the charge on various droplets. Different droplets had different charges, but Millikan found that each was a whole-number multiple of the same smaller charge, 1.60×10^{-19} C (where C represents the coulomb, the SI unit of electric charge; Appendix C). Millikan assumed this to be the fundamental unit of charge, the charge on an electron. Because the

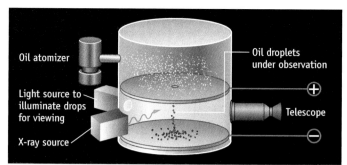 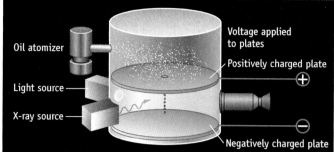

1. A fine mist of oil drops is introduced into one chamber.

2. The droplets fall one by one into the lower chamber under the force of gravity.

3. Gas molecules in the bottom chamber are ionized (split into electrons and a positive fragment) by a beam of x-rays. The electrons adhere to the oil drops, some droplets having one electron, some two, and so on.

These negatively charged droplets continue to fall due to gravity.

4. By carefully adjusting the voltage on the plates, the force of gravity on the droplet is exactly counterbalanced

by the attraction of the negative droplet to the upper, positively charged plate.

Analysis of these forces lead to a value for the charge on the electron.

Figure 2.4 Electron Charge. The experiment was done by R. A. Millikan in 1909. (*See the General ChemistryNow Screen 2.7 Charge and Mass of the Electron, for an exercise on this experiment.*)

charge-to-mass ratio of the electron was known, the mass of an electron could be calculated. The currently accepted value for the electron mass is 9.109383×10^{-28} g, and the electron charge is $-1.602176 \times 10^{-19}$ C. When describing the properties of fundamental particles, we always express charge relative to the charge on the electron, which is given the value of -1.

Additional experiments showed that cathode rays have the same properties as the β particles emitted by radioactive elements. This provided further evidence that the electron is a fundamental particle of matter.

Extensive studies with cathode ray tubes in the late nineteenth century provided another dividend. In addition to cathode rays, a second type of radiation was detected. A beam of positively charged particles called **canal rays** was observed using a specially designed cathode-ray tube with a perforated cathode (Figure 2.5).

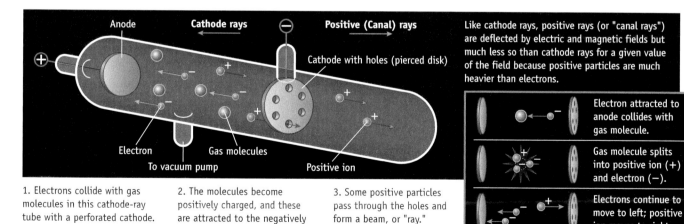

1. Electrons collide with gas molecules in this cathode-ray tube with a perforated cathode.

2. The molecules become positively charged, and these are attracted to the negatively charged, perforated cathode.

3. Some positive particles pass through the holes and form a beam, or "ray."

Like cathode rays, positive rays (or "canal rays") are deflected by electric and magnetic fields but much less so than cathode rays for a given value of the field because positive particles are much heavier than electrons.

Electron attracted to anode collides with gas molecule.

Gas molecule splits into positive ion (+) and electron (−).

Electrons continue to move to left; positive ion moves to right.

Figure 2.5 Canal rays. In 1886 E. Goldstein detected a stream of particles traveling in the direction opposite to that of the negatively charged cathode rays. We now know that these particles are positively charged ions, formed by collisions of electrons with gaseous molecules in the cathode-ray tube. (*See the General ChemistryNow Screen 2.8 Protons, to view an animation on this experiment.*)

These particles, which moved in the opposite direction to cathode rays, passed through the holes in the cathode and were detected on the opposite side. Charge-to-mass values for canal rays were much smaller than the corresponding values measured for cathode rays, indicating particles of higher mass. However, the values also varied depending on the nature of the gas in the tube. We now know that canal rays arise through collisions of cathode rays with gaseous atoms within the cathode-ray tube, which cause each atom to fragment into a positive ion and an electron. The positive particles are attracted to the negatively charged cathode.

GENERAL
Chemistry･⁑･Now™

See the General ChemistryNow CD-ROM or website:

- **Screen 2.6 Electrons,** for an exercise on cathode rays and an animation on cathode-ray deflection
- **Screen 2.8 Protons,** for an exercise on the properties of nuclei in a canal-ray tube

---

Protons

A century after these seminal studies on the structure of the atom, it is easy for us to recognize the proton as the fundamental positively charged particle in an atom. This understanding did not come so easily a hundred years ago, however. This basic fact was not established in one specific experiment or at one specific moment.

With the determination that negatively charged electrons were a component of the atom came recognition that positively charged atomic particles must also exist. One hypothesis suggested that there should be a complementary particle to the electron with a corresponding small mass and a +1 charge, but there was no experimental evidence for such a particle. The positive particles detected and studied in early experiments (α particles from radioactive elements and positive ions making up canal rays) were considerably more massive.

Ernest Rutherford (1871–1937) probably deserves most of the credit for the discovery of the proton. He carried out experiments in the early 1900s in which various elements were irradiated with α particles. One of his better-known experiments involved the irradiation of metals such as gold, which led to the conclusion that atoms contained a small positively charged nucleus with most of the mass of an atom [▶ The Nucleus of the Atom, page 65]. At the same time Rutherford was performing similar experiments using gaseous elements, and, in these experiments, he observed the deflection of α particles as a function of atomic mass. From these observations he concluded, in 1911, that "the hydrogen atom has the simplest possible structure with only one unit charge." However, the formal identification of the proton did not come until almost 10 years later. In experiments in which nitrogen was bombarded with α particles, Rutherford and his collaborators observed highly energetic particles. The values of their charge-to-mass ratio matched those for hydrogen, the positive particle known to have the lowest mass. Unexpectedly they had carried out the first artificial nuclear reaction. Expelling a proton from the nucleus was accepted as definitive evidence of the proton as a nuclear particle. The name "proton" for this particle appears to have been first used by Rutherford in a report in a scientific meeting in 1919.

Neutrons

Because atoms have no net electric charge, the number of positive protons must equal the number of negative electrons in an atom. Most atoms, however, have

Historical Perspectives

Uncovering Atomic Structure

The last few years of the 19th century and the first decades of the 20th century were among the most important in the history of science, in part because the structure of the atom was discovered, setting the stage for the explosion of developments in science in the 20th century.

The notion that matter was built of atoms and that this structure could be used

to explain chemical phenomena was first used by **John Dalton** (1766–1844). Dalton proposed not only that all matter is made of atoms, but also that all atoms of a given element are identical and that atoms are indivisible and indestructible. Dalton's ideas were generally accepted within a few years of his proposal, but we know now the last two postulates are not correct.

Marie Curie (1867–1934) understood the nature of radioactivity and its implications for the nature of the atom. She was born Marya Sklodovska in Poland. When she later lived in France she was known as Marie, but today she is often referred to as Madame Curie. With her husband Pierre she

isolated the previously unknown elements polonium and radium from a uranium-bearing ore. They shared the 1911 Nobel Prize in chemistry for this discovery. One of their daughters, Irène,

married Frédéric Joliot, and they shared the 1935 Nobel Prize in chemistry for their discovery of artificial radioactivity. (*See the General ChemistryNow Screen 2.5 Evidence of Subatomic Particles, to view an animation on separation of radiation by electric field.*)

Sir Joseph John Thomson (1856–1940) was Cavendish Professor of Experimental Physics at Cambridge University in England. In 1896 he gave a series of lectures at Princeton University in the United States titled the *Discharge of Electricity in Gases*. This work on cathode rays led to his discovery of the electron, which he announced at a lecture on the evening of Friday, April 30, 1897. Thomson later published a number of books on the electron and was awarded the Nobel Prize in physics in 1906. (*See the General ChemistryNow Screen 2.6 Electrons, to view an animation on cathode-ray deflection.*)

Ernest Rutherford (1871–1937) was born in New Zealand in 1871 but went to Cambridge University in England to pursue his Ph.D. in physics in 1895. There he worked with J. J. Thomson, and it was at Cambridge that he discovered α and β radiation. At McGill University in Canada in 1899 Rutherford did

further experiments to prove that α radiation is composed of helium nuclei and that β radiation consists of electrons. He received the Nobel Prize in chemistry for his work in 1908. His research on the structure of the atom was done after he moved to Manchester University in England. In 1919 he returned to Cambridge University, where he took up the position formerly held by Thomson. In his career, Rutherford guided the work of ten future recipients of the Nobel Prize. Element 104 has been named *rutherfordium* in his honor. (*See the General ChemistryNow Screen 2.10 The Nucleus of the Atom, to view an animation on Rutherford's α particle experiment.*)

Photos: (Left and Right) Oesper Collection in the History of Chemistry/University of Cincinnati; (Center Top) E. F. Smith Collection; (Center Bottom) Corbis.

masses greater than would be predicted on the basis of only protons and electrons, which suggested that atoms must also contain relatively massive particles with no electric charge. In 1932, the British physicist James Chadwick (1891–1974), a student of Rutherford, presented experimental evidence for the existence of such particles. Chadwick found very penetrating radiation was released when particles from radioactive polonium struck a beryllium target. This radiation was directed at a paraffin wax target, and Chadwick observed protons coming from that target. He reasoned that only a heavy, noncharged particle emanating from the beryllium could have caused this effect. This particle, now known as the **neutron**, has no electric charge and a mass of 1.674927×10^{-24} g, slightly greater than the mass of a proton.

The Nucleus of the Atom

J. J. Thomson had supposed that an atom was a uniform sphere of positively charged matter within which thousands of electrons were embedded. Thomson and his students thought the only question was the number of electrons

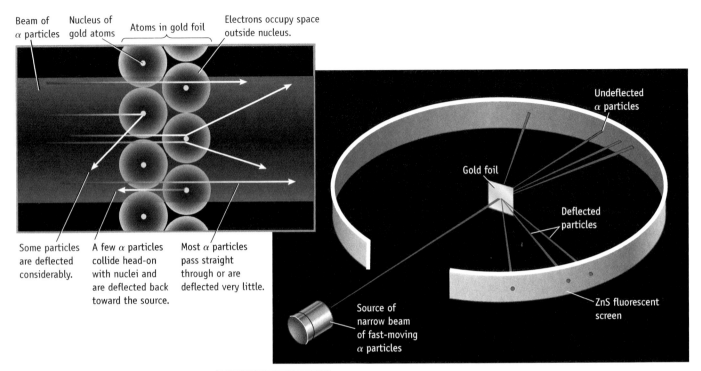

Beam of α particles | Nucleus of gold atoms | Atoms in gold foil | Electrons occupy space outside nucleus.

Some particles are deflected considerably. | A few α particles collide head-on with nuclei and are deflected back toward the source. | Most α particles pass straight through or are deflected very little.

Undeflected α particles | Gold foil | Deflected particles | Source of narrow beam of fast-moving α particles | ZnS fluorescent screen

Active Figure 2.6 **Rutherford's experiment to determine the structure of the atom.** A beam of positively charged α particles was directed at a thin gold foil. A fluorescent screen coated with zinc sulfide (ZnS) was used to detect particles passing through. Most of the particles passed through the foil, but some were deflected from their path. A few were even deflected backward.

GENERAL
Chemistry ·ᴏ· Now™ See the General ChemistryNow CD-ROM or website to explore an interactive version of this figure accompanied by an exercise.

circulating within this sphere. About 1910, Rutherford decided to test Thomson's model. Rutherford had discovered earlier that α rays (see Figure 2.2b) consisted of positively charged particles having the same mass as helium atoms. He reasoned that, if Thomson's atomic model were correct, a beam of such massive particles would be deflected very little as it passed through the atoms in a thin sheet of gold foil. Rutherford, with his associates Hans Geiger (1882–1945) and Ernst Marsden, set up the apparatus diagrammed in Figure 2.6 and observed what happened when α particles hit the foil. Most passed almost straight through, but a few were deflected at large angles, and some came almost straight back! Rutherford later described this unexpected result by saying, "It was about as credible as if you had fired a 15-inch [artillery] shell at a piece of paper and it came back and hit you."

The only way for Rutherford and his colleagues to account for their observations was to propose a new model of the atom, in which all of the positive charge and most of the mass of the atom is concentrated in a very small volume. Rutherford called this tiny core of the atom the **nucleus**. The electrons occupy the rest of the space in the atom. From their results Rutherford, Geiger, and Marsden calculated that the nucleus of a gold atom had a positive charge in the range of 100 ± 20 and a radius of about 10^{-12} cm. The currently accepted values are $+79$ for the charge and about 10^{-13} cm for the radius.

■ **How Small Is an Atom?**
The radius of the typical atom is between 30 and 300 pm (3×10^{-11} m to 3×10^{-10} m). To get a feeling for the incredible smallness of an atom, consider that one teaspoon of water (about 1 cm³) contains about three times as many atoms as the Atlantic Ocean contains teaspoons of water.

GENERAL
Chemistry ·⚛·Now™

See the General ChemistryNow CD-ROM or website:
- **Screen 2.10 The Nucleus of the Atom,** for an exercise on an experiment investigating the properties of the nuclei

---

Exercise 2.1—Describing Atoms

We know now that the radius of the nucleus is about 0.001 pm, and the radius of an atom is approximately 100 pm. If an atom were a macroscopic object with a radius of 100 m, it would approximately fill a small football stadium. What would be the radius of the nucleus of such an atom? Can you think of an object that is about that size?

2.2—Atomic Number and Atomic Mass

Atomic Number

All atoms of the same element have the same number of protons in the nucleus. Hydrogen is the simplest element, with one nuclear proton. All helium atoms have two protons, all lithium atoms have three protons, and all beryllium atoms have four protons. The number of protons in the nucleus of an element is its **atomic number**, generally given the symbol **Z**.

Currently known elements are listed in the periodic table inside the front cover of this book. The integer number at the top of the box for each element is its atomic number. A sodium atom, for example, has an atomic number of 11, so its nucleus contains 11 protons. A uranium atom has 92 nuclear protons and $Z = 92$.

■ **The Periodic Table Entry for Copper**

Copper
29 - - - - - Atomic number
Cu - - - - Symbol
63.546 - - - Atomic weight

Relative Atomic Mass and the Atomic Mass Unit

What is the mass of an atom? Chemists in the 18th and 19th centuries recognized that careful experiments could give *relative* atomic masses. For example, the mass of an oxygen atom was found to be 1.33 times the mass of a carbon atom, and a calcium atom has 2.5 times the mass of an oxygen atom.

Chemistry in the 21st century still uses a system of relative masses. After trying several standards, scientists settled on the current one: A carbon atom having six protons and six neutrons in the nucleus is assigned a mass value of exactly 12.000. An oxygen atom having eight protons and eight neutrons has 1.3333 times the mass of carbon, so it has a relative mass of 16.000. Masses of atoms of other elements have been assigned in a similar manner.

Masses of fundamental atomic particles are often expressed in **atomic mass units (u)**. *One atomic mass unit, 1 u, is one-twelfth of the mass of an atom of carbon with six protons and six neutrons.* Thus, such a carbon atom has a mass of 12.000 u. The atomic mass unit can be related to other units of mass using a conversion factor; that is, $1 \text{ u} = 1.661 \times 10^{-24}$ g.

Mass Number

Protons and neutrons have masses very close to 1 u (Table 2.1). The electron, in contrast, has a mass only about 1/2000 of this value. Because proton and neutron masses are so close to 1 u, the approximate mass of an atom can be estimated if the

Table 2.1   Properties of Subatomic Particles*

Particle	Mass		Charge	Symbol
	Grams	*Atomic Mass Units*		
Electron	9.109383×10^{-28}	0.0005485799	-1	^0_1e or e^-
Proton	1.672622×10^{-24}	1.007276	$+1$	^1_1p or p^+
Neutron	1.674927×10^{-24}	1.008665	0	^1_0n or n^0

* These values and others in the book are taken from the National Institute of Standards and Technology website at http://physics.nist.gov/cuu/Constants/index.html

number of neutrons and protons is known. The sum of the number of protons and neutrons for an atom is called its **mass number** and is given the symbol A.

$$A = \text{mass number} = \text{number of protons} + \text{number of neutrons}$$

For example, a sodium atom, which has 11 protons and 12 neutrons in its nucleus, has a mass number of $A = 23$. The most common atom of uranium has 92 protons and 146 neutrons, and a mass number of $A = 238$. Using this information, we often symbolize atoms with the notation

$$\begin{array}{l} \text{Mass number} \rightarrow \\ \text{Atomic number} \rightarrow \end{array} {}^A_Z X \leftarrow \text{Element symbol}$$

The subscript Z is optional because the element symbol tells us what the atomic number must be. For example, the atoms described previously have the symbols $^{23}_{11}\text{Na}$ or $^{238}_{92}\text{U}$, or just ^{23}Na or ^{238}U. In words, we say "sodium-23" or "uranium-238."

GENERAL
Chemistry Now™

See the General ChemistryNow CD-ROM or website:
- **Screen 2.11 Summary of Atomic Composition,** for a tutorial on the notation for symbolizing atoms

Example 2.1—Atomic Composition

Problem What is the composition of an atom of phosphorus with 16 neutrons? What is its mass number? What is the symbol for such an atom? If the atom has an actual mass of 30.9738 u, what is its mass in grams?

Strategy All P atoms have the same number of protons, 15, which is given by the atomic number (see the periodic table inside the front cover of this book). The mass number is the sum of the number of protons and neutrons. The mass of the atom in grams can be obtained from the mass in atomic mass units using the conversion factor $1\ u = 1.661 \times 10^{-24}$ g.

Solution A phosphorus atom has 15 protons and, because it is electrically neutral, also has 15 electrons.

Mass number = number of protons + number of neutrons = 15 + 16 = **31**

The atom's complete symbol is $^{31}_{15}P$.

Mass of one ^{31}P atom (g) = (30.9738 u) × (1.661 × 10^{-24} g/u) = 5.145 × 10^{-23} g

---

Exercise 2.2—Atomic Composition

(a) What is the mass number of an iron atom with 30 neutrons?

(b) A nickel atom with 32 neutrons has a mass of 59.930788 u. What is its mass in grams?

(c) How many protons, neutrons, and electrons are in a ^{64}Zn atom?

2.3—Isotopes

In only a few instances (for example, aluminum, fluorine, and phosphorus) do all atoms in a naturally occurring sample of a given element have the same mass. Most elements consist of atoms having several different mass numbers. For example, there are two kinds of boron atoms, one with a mass of about 10 u (^{10}B) and a second with a mass of about 11 u (^{11}B). Atoms of tin can have any of 10 different masses. Atoms with the same atomic number but different mass numbers are called **isotopes**.

All atoms of an element have the same number of protons—five in the case of boron. This means that, to have different masses, isotopes must have different numbers of neutrons. The nucleus of a ^{10}B atom ($Z = 5$) contains five protons and five neutrons, whereas the nucleus of a ^{11}B atom contains five protons and six neutrons.

Scientists often refer to a particular isotope by giving its mass number (for example, uranium-238, ^{238}U), but the isotopes of hydrogen are so important that they have special names and symbols. All hydrogen atoms have one proton. When that is the only nuclear particle, the isotope is called *protium*, or just "hydrogen." The isotope of hydrogen with one neutron, $^{2}_{1}H$, is called *deuterium*, or "heavy hydrogen" (symbol = D). The nucleus of radioactive hydrogen-3, $^{3}_{1}H$, or *tritium* (symbol = T), contains one proton and two neutrons.

The substitution of one isotope of an element for another isotope of the same element in a compound sometimes has an interesting effect (Figure 2.7). This is especially true when deuterium is substituted for hydrogen because the mass of deuterium is double that of hydrogen.

Isotope Abundance

A sample of water from a stream or lake will consist almost entirely of H_2O where the H atoms are the ^{1}H isotope. A few molecules, however, will have deuterium (^{2}H) substituted for ^{1}H. We can predict this outcome because we know that 99.985% of all hydrogen atoms on earth are ^{1}H atoms. That is, the **percent abundance** of ^{1}H atoms is 99.985%.

$$\text{Percent abundance} = \frac{\text{number of atoms of a given isotope}}{\text{total number of atoms of all isotopes of that element}} \times 100\%$$

(2.1)

The remainder of naturally occurring hydrogen is deuterium, whose abundance is only 0.015% of the total hydrogen atoms. Tritium, the radioactive ^{3}H isotope, does not occur naturally.

Solid H_2O

Liquid H_2O

Solid D_2O

Charles D. Winters

Figure 2.7 Ice made from "heavy water." Water containing ordinary hydrogen ($^{1}_{1}H$, protium) forms a solid that is less dense ($d = 0.917$ g/cm^3 at 0 °C) than liquid H_2O ($d = 0.997$ g/cm^3 at 25 °C) and so floats in the liquid. (Water is unique in this regard. The solid phase of virtually all other substances sinks in the liquid phase of that substance.) Similarly, "heavy ice" (D_2O, deuterium oxide) floats in "heavy water." D_2O-ice is denser than H_2O, however, so cubes made of D_2O sink in liquid H_2O.

Consider the two isotopes of boron. The boron-10 isotope has an abundance of 19.91%; the abundance of boron-11 is 80.09%. Thus, if you could count out 10,000 boron atoms from an "average" natural sample, 1991 of them would be boron-10 atoms and 8009 of them would be boron-11 atoms.

Example 2.2—Isotopes

Problem Silver has two isotopes, one with 60 neutrons (percent abundance = 51.839%) and the other with 62 neutrons. What are the mass numbers and symbols of these isotopes? What is the percent abundance of the isotope with 62 neutrons?

Strategy Recall that the mass number is the sum of the number of protons and neutrons. The symbol is written as $^A_Z X$, where X is the one or two-letter element symbol. The percent abundances of all isotopes must add up to 100%.

Solution Silver has an atomic number of 47, so each silver atom has 47 protons in its nucleus. The two isotopes, therefore, have mass numbers of 107 and 109.

Isotope 1, with 47 protons and 60 neutrons

$$A = 47 \text{ protons} + 60 \text{ neutrons} = 107$$

Isotope 2, with 47 protons and 62 neutrons

$$A = 47 \text{ protons} + 62 \text{ neutrons} = 109$$

The first isotope has a symbol $^{107}_{47}Ag$ and the second is $^{109}_{47}Ag$.

Silver-107 has a percent abundance of 51.839%. Therefore, the percent abundance of silver-109 is

$$\% \text{ Abundance of } ^{109}Ag = 100.000\% - 51.839\% = 48.161\%$$

■ Atomic Masses of Some Isotopes

Atom	Mass (u)
^{4}He	4.0092603
^{13}C	13.003355
^{16}O	15.994915
^{58}Ni	57.935348
^{60}Ni	59.930791
^{79}Br	78.918338
^{81}Br	80.916291
^{197}Au	196.966552
^{238}U	238.050783

Exercise 2.3—Isotopes

(a) Argon has three isotopes with 18, 20, and 22 neutrons, respectively. What are the mass numbers and symbols of these three isotopes?

(b) Gallium has two isotopes: ^{69}Ga and ^{71}Ga. How many protons and neutrons are in the nuclei of each of these isotopes? If the abundance of ^{69}Ga is 60.1%, what is the abundance of ^{71}Ga?

Determining Atomic Mass and Isotope Abundance

The masses of isotopes and their percent abundances are determined experimentally using a *mass spectrometer* (Figure 2.8). A gaseous sample of an element is introduced into the evacuated chamber of the spectrometer, and the molecules or atoms of the sample are converted to charged particles (ions). A beam of these ions is subjected to a magnetic field, which causes the paths of the ions to be deflected. The extent of deflection depends on particle mass: The less massive ions are deflected more, and the more massive ions are deflected less. The ions, now separated by mass, are detected at the end of the chamber. In early experiments, ions were detected using photographic film, but modern instruments measure the electric current in a detector. The darkness of a spot on photographic film, or the amount of

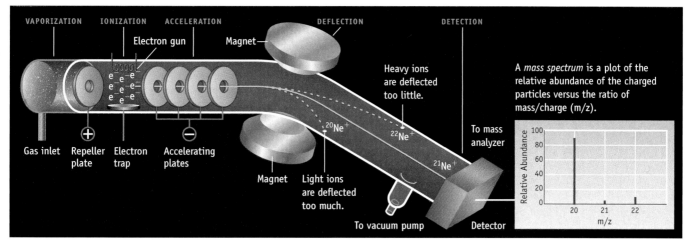

1. A sample is introduced as a vapor into the ionization chamber.

There it is bombarded with high-energy electrons that strip electrons from the atoms or molecules of the sample.

2. The resulting positive particles are accelerated by a series of negatively charged accelerator plates into an analyzing chamber.

3. This chamber is in a magnetic field, which is perpendicular to the direction of the beam of charged particles.

The magnetic field causes the beam to curve. The radius of curvature depends on the mass and charge of the particles (as well as the accelerating voltage and strength of the magnetic field).

4. Here particles of $^{21}Ne^+$ are focused on the detector, whereas beams of ions of $^{20}Ne^+$ and $^{22}Ne^+$ (of lighter or heavier mass) experience greater and lesser curvature, respectively, and so fail to be detected.

By changing the magnetic field, a beam of charged particles of different mass can be focused on the detector, and a spectrum of masses is observed.

Active Figure 2.8 Mass spectrometer.

GENERAL
Chemistry⚛Now™ See the General ChemistryNow CD-ROM or website to explore an interactive version of this figure accompanied by an exercise.

current measured, is related to the number of ions of a particular mass and hence to the abundance of the ion.

The mass-to-charge ratio for the ions can also be determined from the extent of curvature in the ion's path to the detector. Knowing that most of the ions within the spectrometer have a +1 charge allows us to derive a value for mass. Chemists using modern instruments can measure isotopic masses with as many as nine significant figures.

A Closer Look

Atomic Mass and the Mass Defect

You might expect that the mass of a deuterium nucleus, ^2H, would be the sum of the masses of its constituent particles, a proton and a neutron.

^1_1p (1.007276 u) + ^1_0n (1.008665 u)
$$\rightarrow \; ^2_1H \; (2.01355 \; u)$$

However, the mass of ^2H is *less* than the sum of its constituents!

Difference in mass, Δm

= mass of product − total mass of reactants
= 2.01355 u − 2.015941 u
= −0.00239 u

This "missing mass" is equated to energy, the binding energy for the nucleus. The binding energy can be calculated from Einstein's equation that relates the mass, m, to energy, E ($E = mc^2$, where c is the velocity of light).

Although the mass loss on forming an atomic nucleus from its constituent protons and neutrons seems small, the energy equivalent is enormous. In fact, it is the mass loss from fusing protons into helium nuclei on the sun that provides the energy for life on earth. (See the story, *Stardust*, on page 58 and see Chapter 23 for more details.)

Except for carbon-12, whose mass is defined to be exactly 12u, isotopic masses do not have integer values. However, the isotopic masses are always very close to the mass numbers for the isotope. For example, the mass of an atom of boron-11 (^{11}B, 5 protons and 6 neutrons) is 11.0093 u, and the mass of an atom of iron-58 (^{58}Fe, 26 protons and 32 neutrons) is 57.9333 u. Note also that the masses of individual isotopes are always slightly less than the sum of the masses of the protons, neutrons, and electrons making up the atom. This mass difference, called the "**mass defect**," is related to the energy binding the particles of the nucleus together. (See A Closer Look: Atomic Mass and the Mass Defect.)

2.4—Atomic Weight

Because every sample of boron has some atoms with a mass of 10.0129 u and others with a mass of 11.0093 u, the average atomic mass must be somewhere between these values. The **atomic weight** is the average weight of a representative sample of atoms. For boron, the atomic weight is 10.81. In general, the atomic weight of an element can be calculated using the equation

$$\text{Atomic weight} = \left(\frac{\%\ \text{abundance isotope 1}}{100}\right)(\text{mass of isotope 1})$$
$$+ \left(\frac{\%\ \text{abundance isotope 2}}{100}\right)(\text{mass of isotope 2}) + \cdots$$

(2.2)

■ **Atomic Weight and Units**
Values of atomic weight are relative to the mass of the carbon-12 isotope and so are unitless numbers.

For boron with two isotopes (^{10}B, 19.91% abundant; ^{11}B, 80.09% abundant), we find

$$\text{Atomic weight} = \left(\frac{19.91}{100}\right) \times 10.0129 + \left(\frac{80.09}{100}\right) \times 11.0093$$
$$= 10.81$$

Equation 2.2 gives an average, weighted in terms of the abundance of each isotope for the element. As illustrated by the data in Table 2.2, *the atomic weight of an element is always closer to the mass of the more abundant isotope or isotopes.*

■ **Fractional Abundance**
The percent abundance of an isotope divided by 100 is called its fractional abundance.

Table 2.2   Isotope Abundance and Atomic Weight

Element	Symbol	Atomic Weight	Mass Number	Isotopic Mass (u)	Natural Abundance (%)
Hydrogen	H	1.00794	1	1.0078	99.985
	D*		2	2.0141	0.015
	T†		3	3.0161	0
Boron	B	10.811	10	10.0129	19.91
			11	11.0093	80.09
Neon	Ne	20.1797	20	19.9924	90.48
			21	20.9938	0.27
			22	21.9914	9.25
Magnesium	Mg	24.305	24	23.9850	78.99
			25	24.9858	10.00
			26	25.9826	11.01

*D = deuterium; †T = tritium, radioactive.

The atomic weight of each stable element has been determined experimentally, and these numbers appear in the periodic table inside the front cover of this book. In the periodic table, each element's box contains the atomic number, the element symbol, and the atomic weight. For unstable (radioactive) elements, the atomic mass or mass number of the most stable isotope is given in parentheses.

■ **The Periodic Table Entry for Copper**

Copper
29 ----- Atomic number
Cu ---- Symbol
63.546 --- Atomic weight

Example 2.3—Calculating Atomic Weight from Isotope Abundance

Problem Bromine (used to make silver bromide, the important component of photographic film) has two naturally occurring isotopes. One has a mass of 78.918338 u and an abundance of 50.69%. The other isotope, of mass 80.916291 u, has an abundance of 49.31%. Calculate the atomic weight of bromine.

Strategy The atomic weight of any element is the weighted average of the masses of the isotopes in a representative sample. To calculate the atomic weight, multiply the mass of each isotope by its percent abundance divided by 100 (Equation 2.2). (*See the General ChemistryNow Screen 2.13 Atomic Mass, for a tutorial on calculating atomic weight from isotope abundance.*)

Solution

Average atomic mass of bromine = (50.69/100)(78.918338) + (49.31/100)(80.916291)

= 79.90

Exercise 2.4—Calculating Atomic Weight

Verify that the atomic weight of chlorine is 35.45, given the following information:

^{35}Cl mass = 34.96885; percent abundance = 75.77%

^{37}Cl mass = 36.96590; percent abundance = 24.23%

2.5—Atoms and the Mole

One of the most exciting aspects of chemical research is the discovery of some new substance, and part of this process of discovery involves quantitative experiments. When two chemicals react with each other, we want to know how many atoms of each are used so that formulas can be established for the reaction's products. To do so, we need some method of counting atoms. That is, we must discover a way of connecting the macroscopic world, the world we can see, with the particulate world of atoms, molecules, and ions. The solution to this problem is to define a convenient amount of matter that contains a known number of particles. That chemical unit is the **mole**.

The mole (abbreviated mol) is the SI base unit for measuring an *amount of a substance* (see Table 1.2) and is defined as follows:

A **mole** is the amount of a substance that contains as many elementary entities (atoms, molecules, or other particles) as there are atoms in exactly 12 g of the carbon-12 isotope.

The key to understanding the concept of the mole is recognizing that *one mole always contains the same number of particles, no matter what the substance.* One mole of sodium

■ **The "Mole"**

The term "mole" was introduced about 1896 by Wilhelm Ostwald (1853–1932), who derived the term from the Latin word *moles*, meaning a "heap" or "pile."

contains the same number of atoms as one mole of iron. How many particles? Many, many experiments over the years have established that number as

$$1 \text{ mole} = 6.0221415 \times 10^{23} \text{ particles}$$

This value is known as **Avogadro's number** in honor of Amedeo Avogadro, an Italian lawyer and physicist (1776–1856) who conceived the basic idea (but never determined the number).

■ **The Difference Between "Amount" and "Quantity"**
The terms "amount" and "quantity" are used in a specific sense by chemists. The *amount* of a substance is the number of moles of that substance. *Quantity* refers to the mass of the substance. See W. G. Davies and J. W. Moore: *Journal of Chemical Education*, Vol. 57, p. 303, 1980. See also http://physics.nist.gov on the Internet.

Molar Mass

The mass in grams of one mole of any element (6.0221415×10^{23} atoms of that element) is the **molar mass** of that element. Molar mass is conventionally abbreviated with a capital italicized M and has units of grams per mole (g/mol). An element's *molar mass is the amount in grams numerically equal to its atomic weight.* Using sodium and lead as examples,

$$
\begin{aligned}
\text{Molar mass of sodium (Na)} &= \text{mass of 1.000 mol of Na atoms} \\
&= 22.99 \text{ g/mol} \\
&= \text{mass of } 6.022 \times 10^{23} \text{ Na atoms} \\
\text{Molar mass of lead (Pb)} &= \text{mass of 1.000 mol of Pb atoms} \\
&= 207.2 \text{ g/mol} \\
&= \text{mass of } 6.022 \times 10^{23} \text{ Pb atoms}
\end{aligned}
$$

Figure 2.9 shows the relative sizes of a mole of some common elements. Although each of these "piles of atoms" has a different volume and different mass, each contains 6.022×10^{23} atoms.

The mole concept is the cornerstone of quantitative chemistry. It is essential to be able to convert from moles to mass and from mass to moles. Dimensional analysis, which is described in Section 1.8 (pages 41–43), shows that this can be done in the following way:

Figure 2.9 **One-mole of common elements.** (*left to right*) Sulfur powder, magnesium chips, tin, and silicon. (*above*) Copper beads.

Charles D. Winters

MASS $\longleftrightarrow$ MOLES CONVERSION

Moles to Mass	*Mass to Moles*

$$\text{Moles} \times \frac{\text{grams}}{1 \text{ mol}} = \text{grams} \qquad \text{Grams} \times \frac{1 \text{ mol}}{\text{grams}} = \text{moles}$$

$$\uparrow \qquad\qquad\qquad\qquad \uparrow$$
$$\text{molar mass} \qquad\qquad\quad 1/\text{molar mass}$$

For example, what mass, in grams, is represented by 0.35 mol of aluminum? Using the molar mass of aluminum (27.0 g/mol), you can determine that 0.35 mol of Al has a mass of 9.5 g.

$$0.35 \text{ mol Al} \times \frac{27.0 \text{ g Al}}{1 \text{ mol Al}} = 9.5 \text{ g Al}$$

Molar masses are generally known to at least four significant figures. The convention followed in calculations in this book is to use a value of the molar mass with one more significant figure than in any other number in the problem. For example, if you weigh out 16.5 g of carbon, you use 12.01 g/mol for the molar mass of C to find the amount of carbon present.

$$16.5 \text{ g C} \times \frac{1 \text{ mol C}}{12.01 \text{ g C}} = 1.37 \text{ mol C}$$

Note that four significant figures are used in the
molar mass, but there are three in the sample mass.

Using one more significant figure for the molar mass means the accuracy of this value will not affect the accuracy of the result.

Lead. A 150-mL beaker containing 2.50 mol or 518 g of lead.

Tin. A sample of tin having a mass of 36.5 g (1.85×10^{23} atoms).

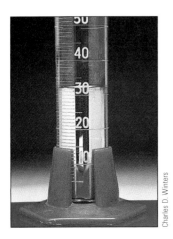

Mercury. A graduated cylinder containing 32.0 cm³ of mercury. This is equivalent to 433 g or 2.16 mol of mercury.

GENERAL
Chemistry••}•Now™

See the General ChemistryNow CD-ROM or website:

- **Screen 2.14 the Mole,** for a tutorial on moles and atoms conversion
- **Screen 2.15 Moles and Molar Mass of the Elements,** for two tutorials on molar mass conversion

Example 2.4—Mass, Moles, and Atoms

Problem Consider two elements in the same vertical column of the periodic table, lead and tin.

(a) What mass of lead, in grams, is equivalent to 2.50 mol of lead (Pb, atomic number = 82)?

(b) What amount of tin, in moles, is represented by 36.5 g of tin (Sn, atomic number = 50)? How many atoms of tin are in the sample?

Strategy The molar masses of lead (207.2 g/mol) and tin (118.7 g/mol) are required and can be found in the periodic table inside the front cover of this book. Avogadro's number is needed to convert the amount of each element to number of atoms.

Solution

(a) Convert the amount of lead in moles to mass in grams.

$$2.50 \text{ mol Pb} \times \frac{207.2 \text{ g}}{1 \text{ mol Pb}} = 518 \text{ g Pb}$$

(b) First convert the mass of tin to the amount in moles.

$$36.5 \text{ g Sn} \times \frac{1 \text{ mol Sn}}{118.7 \text{ g Sn}} = 0.308 \text{ mol Sn}$$

Finally, use Avogadro's number to find the number of atoms in the sample.

$$0.308 \text{ mol Sn} \times \frac{6.022 \times 10^{23} \text{ atoms Sn}}{1 \text{ mol Sn}}$$

$$= 1.85 \times 10^{23} \text{ atoms Sn}$$

Example 2.5—Mole Calculation

Problem The graduated cylinder in the photo contains 32.0 cm³ of mercury. If the density of mercury at 25 °C is 13.534 g/cm³, what amount of mercury, in moles, is in the cylinder?

Strategy Volume and moles of mercury are not directly connected. You must first use the density of mercury to find the mass of the metal and then use this value with the molar mass of mercury to calculate the amount in moles.

$$\text{Volume (cm}^3) \times \text{density (g/cm}^3) = \text{mass of mercury (g)}$$

$$\text{Amount of mercury (mol)} = \text{mass of mercury (g)} \times (1/\text{molar mass})(\text{mol/g})$$

Solution Combining the volume and density gives the mass of the mercury.

$$32.0 \text{ cm}^3 \times \frac{13.534 \text{ g Hg}}{1 \text{ cm}^3} = 433 \text{ g Hg}$$

Finally, the amount of mercury can be calculated from its mass and molar mass.

$$433 \text{ g Hg} \times \frac{1 \text{ mol Hg}}{200.6 \text{ g Hg}} = 2.16 \text{ mol Hg}$$

Example 2.6—Mass of an Atom

Problem What is the average mass of an atom of platinum (Pt)?

Strategy The mass of one mole of platinum is 195.08 g. Each mole contains Avogadro's number of atoms.

Solution Here we divide the mass of a mole by the number of objects in that unit.

$$\frac{195.08 \text{ g Pt}}{1 \text{ mol Pt}} \times \frac{1 \text{ mol Pt}}{6.02214 \times 10^{23} \text{ atoms Pt}} = \frac{3.2394 \times 10^{-22} \text{ g Pt}}{1 \text{ atom Pt}}$$

Comment Notice that the units "mol Pt" cancel and leave an answer with units of g/atom.

Exercise 2.5—Mass/Mole Conversions

(a) What is the mass, in grams, of 1.5 mol of silicon?
(b) What amount (moles) of sulfur is represented by 454 g? How many atoms?
(c) What is the average mass of one sulfur atom?

Exercise 2.6—Atoms

The density of gold, Au, is 19.32 g/cm^3. What is the volume (in cubic centimeters) of a piece of gold that contains 2.6×10^{24} atoms? If the piece of metal is a square with a thickness of 0.10 cm, what is the length (in centimeters) of one side of the piece?

2.6—The Periodic Table

The periodic table of elements is one of the most useful tools in chemistry. Not only does it contain a wealth of information, but it can also be used to organize many of the ideas of chemistry. It is important that you become familiar with its main features and terminology.

Features of the Periodic Table

The main organizational features of the periodic table are the following:

- Elements are arranged so that those with similar chemical and physical properties lie in vertical columns called **groups** or **families**. The periodic table commonly used in the United States has groups numbered 1 through 8, with each number followed by a letter: A or B. The A groups are often called the **main group elements** and the B groups are the **transition elements**.

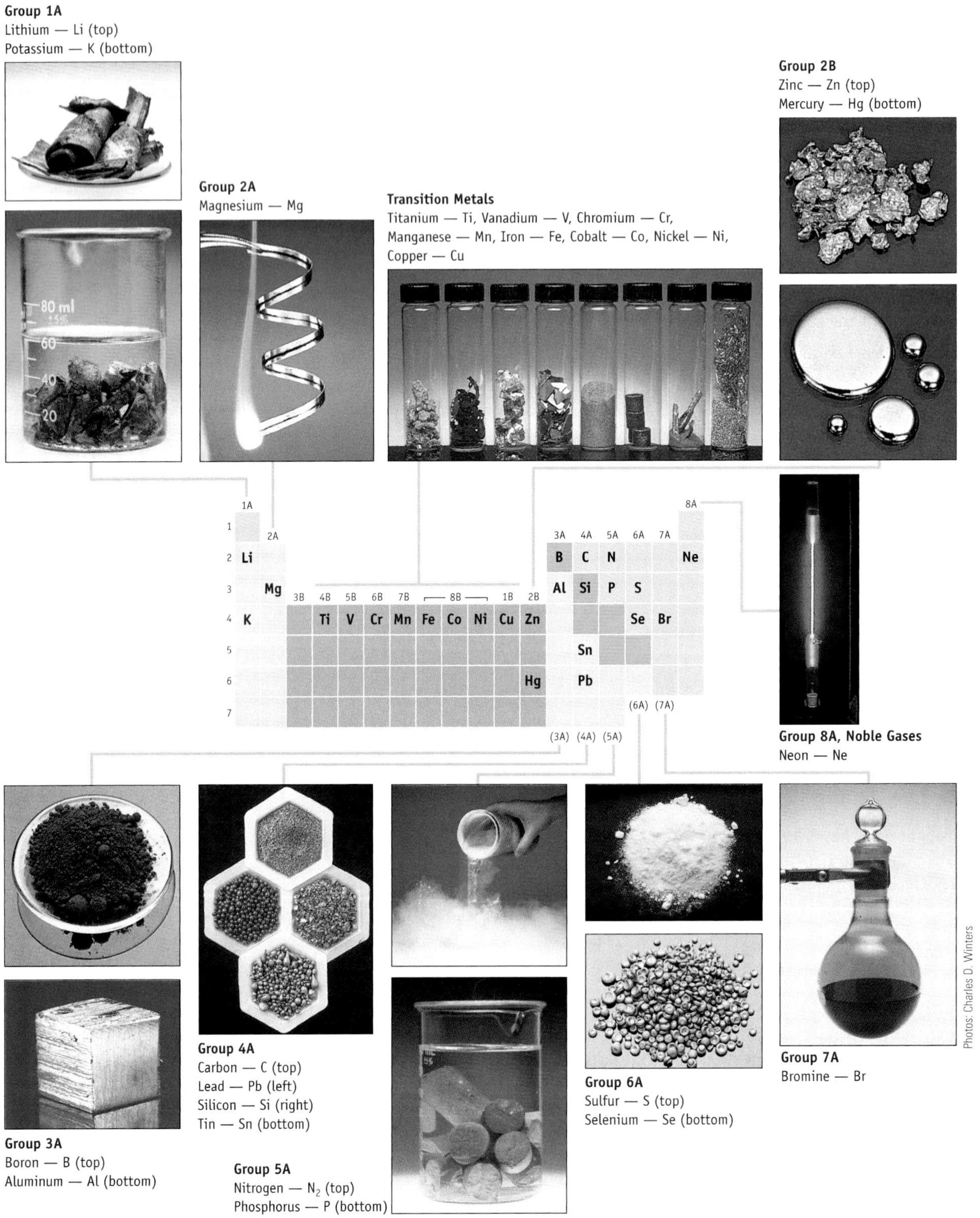
Group 1A
Lithium — Li (top)
Potassium — K (bottom)
Group 2A
Magnesium — Mg
Transition Metals
Titanium — Ti, Vanadium — V, Chromium — Cr,
Manganese — Mn, Iron — Fe, Cobalt — Co, Nickel — Ni,
Copper — Cu
Group 2B
Zinc — Zn (top)
Mercury — Hg (bottom)
1A
2A
3B 4B 5B 6B 7B 8B 1B 2B
3A 4A 5A 6A 7A 8A
1 2 3 4 5 6 7
Li Mg K
Ti V Cr Mn Fe Co Ni Cu Zn
Hg
B C N Ne
Al Si P S
Se Br
Sn
Pb
(6A) (7A)
(3A) (4A) (5A)
Group 8A, Noble Gases
Neon — Ne
Group 3A
Boron — B (top)
Aluminum — Al (bottom)
Group 4A
Carbon — C (top)
Lead — Pb (left)
Silicon — Si (right)
Tin — Sn (bottom)
Group 5A
Nitrogen — N_2 (top)
Phosphorus — P (bottom)
Group 6A
Sulfur — S (top)
Selenium — Se (bottom)
Group 7A
Bromine — Br
Photos: Charles D. Winters

Active Figure 2.10 **Some of the 116 known elements.**

GENERAL **ChemistryNow**™ See the General ChemistryNow CD-ROM or website to explore an interactive version of this figure accompanied by an exercise.

- The horizontal rows of the table are called **periods**, and they are numbered beginning with 1 for the period containing only H and He. For example, sodium, Na, in Group 1A, is the first element in the third period. Mercury, Hg, in Group 2B, is in the sixth period (or sixth row).

The periodic table can be divided into several regions according to the properties of the elements. On the table inside the front cover of this book, elements that behave as *metals* are indicated in purple, those that are *nonmetals* are indicated in yellow, and elements called *metalloids* appear in green. Elements gradually become less metallic as one moves from left to right across a period, and the metalloids lie along the metal–nonmetal boundary. Some elements are shown in Figure 2.10.

You are probably familiar with many properties of **metals** from everyday experience (Figure 2.11a). Metals are solids (except for mercury), can conduct electricity, are usually ductile (can be drawn into wires) and malleable (can be rolled into sheets), and can form alloys (solutions of one or more metals in another metal). Iron (Fe) and aluminum (Al) are used in automobile parts because of their ductility, malleability, and low cost relative to other metals. Copper (Cu) is used in electric wiring because it conducts electricity better than most other metals. Chromium (Cr) is plated onto automobile parts, not only because its metallic luster makes cars look better but also because chrome-plating protects the underlying metal from reacting with oxygen in the air.

The **nonmetals** lie to the right of a diagonal line that stretches from B to Te in the periodic table and have a wide variety of properties. Some are solids (carbon, sulfur, phosphorus, and iodine). Four elements are gases at room temperature (oxygen, nitrogen, fluorine, and chlorine). One element, bromine, is a liquid at room temperature (Figure 2.11b). With the exception of carbon in the form of graphite, nonmetals do not conduct electricity, which is one of the main features that distinguishes them from metals.

Some of the elements next to the diagonal line from boron (B) to tellurium (Te) have properties that make them difficult to classify as metals or nonmetals. Chemists call them metalloids or, sometimes, semimetals (Figure 2.11c). You should

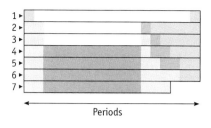

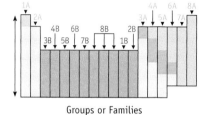

Periods

Groups or Families

■ **Two Ways to Designate Groups**
One way to designate periodic groups is to number them 1 through 18 from left to right. This method is generally used outside the United States. The system predominant in the United States labels main group elements as Groups 1A–8A and transition elements as Groups 1B–8B. This book uses the A/B system.

(a) Metals

(b) Nonmetals

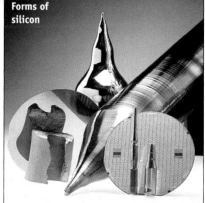

(c) Metalloids

Figure 2.11 Representative elements. (a) Magnesium, aluminum, and copper are metals. All can be drawn into wires and conduct electricity. (b) Only 15 or so elements can be classified as nonmetals. Here are orange liquid bromine and purple solid iodine. (c) Only 6 elements are generally classified as metalloids or semimetals. This photograph shows solid silicon in various forms, including a wafer that holds printed electronic circuits.

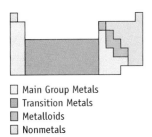

☐ Main Group Metals
■ Transition Metals
■ Metalloids
☐ Nonmetals

know, however, that chemists often disagree about what a metalloid is as well as which elements fit into this category. We will define a **metalloid** as an element that has some of the physical characteristics of a metal but some of the chemical characteristics of a nonmetal; we include only B, Si, Ge, As, Sb, and Te in this category. This distinction reflects the ambiguity in the behavior of these elements. Antimony (Sb), for example, conducts electricity as well as many elements that are true metals. Its chemistry, however, resembles that of a nonmetal such as phosphorus.

Developing the Periodic Table

Although the arrangement of elements in the periodic table can now be understood on the basis of atomic structure [▶ Chapter 8], the table was originally developed from many, many experimental observations of the chemical and physical properties of elements and is the result of the ideas of a number of chemists in the 18th and 19th centuries.

In 1869, at the University of St. Petersburg in Russia, Dmitri Ivanovitch Mendeleev (1834–1907) was pondering the properties of the elements as he wrote a textbook on chemistry. On studying the chemical and physical properties of the elements, he realized that, if the elements were arranged in order of increasing atomic mass, elements with similar properties appeared in a regular pattern. That is, he saw a **periodicity** or periodic repetition of the properties of elements. Mendeleev organized the known elements into a table by lining them up in a horizontal row in order of increasing atomic mass. Every time he came to an element with properties similar to one already in the row, he started a new row. For example, the elements Li, Be, B, C, N, O, and F were in a row. Sodium was the next element then known; because its properties closely resembled those of Li, Mendeleev started a new row. The columns, then, contained elements such as Li, Na, and K with similar properties.

An important feature of Mendeleev's table—and a mark of his genius—was that he left an empty space in a column when an element was not known but should exist and have properties similar to the element above it in his table. He deduced that these spaces would be filled by undiscovered elements. For example, a space was left between Si (silicon) and Sn (tin) in what is now Group 4A. Based on the progression of properties in this group, Mendeleev was able to predict the properties of this missing element. With the discovery of germanium in 1886, Mendeleev's prediction was confirmed.

In Mendeleev's table the elements were ordered by increasing mass. A glance at a modern table, however, shows that, on this basis, Ni and Co, Ar and K, and Te and I, should be reversed. Mendeleev assumed the atomic masses known at that time were inaccurate—not a bad assumption based on the analytical methods then in use. In fact, his order was correct and what was wrong was his assumption that element properties were a function of their mass.

In 1913 H. G. J. Moseley (1887–1915), a young English scientist working with Ernest Rutherford, corrected Mendeleev's assumption. Moseley was doing experiments in which he bombarded many different metals with electrons in a cathode-ray tube (Figure 2.3) and examined the x-rays emitted in the process. In seeking some order in his data, he realized that the wavelength of the x-rays emitted by a given element were related in a precise manner to the *atomic number* of the element. Indeed, chemists quickly recognized that organizing the elements in a table by increasing atomic number corrected the inconsistencies in the Mendeleev table. The **law of chemical periodicity** is now stated as "the properties of the elements are periodic functions of atomic number."

■ **About the Periodic Table**
For more information on the periodic table, the central icon of chemistry, we recommend the following:
• The American Chemical Society has a description of every element on its website at www.cen-online.org.
• J. Emsley: *Nature's Building Blocks—An A–Z Guide to the Elements,* New York, Oxford University Press, 2001.
• O. Sacks: *Uncle Tungsten—Memories of a Chemical Boyhood,* New York, Alfred A. Knopf, 2001.

■ **Placing H in the Periodic Table**
Where to place H? Tables often show it in Group 1A even though it is clearly not an alkali metal. However, in its reactions it forms a 1+ ion just like the alkali metals. For this reason, H is often placed in Group 1A.

Historical Perspectives

Periodic Table

In his book *Nature's Building Blocks* (p. 527, New York, Oxford University Press, 2001), John Emsley tells us that "As long as chemistry is studied, there will be a periodic table. Even if some day we communicate with another part of the Universe, we can be sure that one thing both cultures will have in common is an ordered system of the elements that will be instantly recognizable by both intelligent life forms."

The person credited with organizing the elements into a periodic table is Dmitri Mendeleev. However, other chemists had long recognized that groups of elements shared similar properties. In 1829 Johann Dobereiner (1780–1849) announced the Law of Triads. He showed that there were groups of three elements (triads), in which the middle element had an atomic weight that was the average of the other two. One such triad consisted of Li, Na, and K; another was made up of Cl, Br, and I.

Perhaps the first revelation of the periodicity of the elements was published by a French geologist, A. E. Béguyer de Chancourtois (1820–1886), in 1862. He listed the elements on a paper tape, and, according to Emsley, "then wound this, spiral like around a cylinder. The cylinder's surface was divided into 16 parts, based on the atomic weight of oxygen. De Chancourtois noted that certain triads came together down the cylinder, such as the alkali metals." He called his model the "telluric screw."

Another attempt at organizing the elements was proposed by John Newlands (1837–1898) in 1864. His "Law of Octaves" proposed that there was a periodic similarity every eight elements, just as the musical scale repeats every eighth note. Unfortunately, his proposal was ridiculed at the time.

Julius Lothar Meyer (1830–1895) came closer than any other to discovering the periodic table. He drew a graph of atomic volumes of elements plotted against their atomic weight. This clearly showed a periodic rise and fall in atomic volume on moving across what we now call the periods of the table. Before publishing the paper, Meyer passed it along to a colleague for comment. His colleague was slow to return the paper, and, unfortunately for Meyer, Mendeleev's paper was published in the interim. Because chemists quickly recognized the importance of Mendeleev's paper, Meyer was not given the recognition he perhaps deserves.

An essay on Mendeleev and his life appears at the beginning of Chapter 8 (page 320).

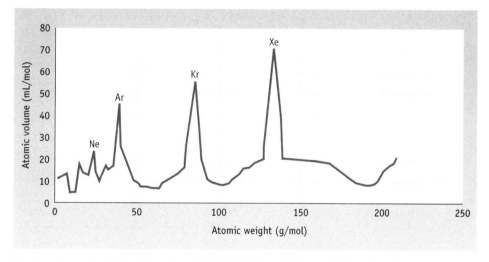

Atomic volume plot. Julius Lothar Meyer (1830–1895) illustrated the periodicity of the elements in 1868 by plotting atomic volume against atomic weight. (This plot uses current data.)

Source: C. N. Singman: *Journal of Chemical Education,* Vol. 61, p. 137, 1984.

GENERAL
Chemistry·⚛·Now™

See the General ChemistryNow CD-ROM or website:
- **Screen 2.16 The Periodic Table,** for an exercise on the periodic table organization

2.7—An Overview of the Elements, Their Chemistry, and the Periodic Table

The vertical columns, or groups, of the periodic table contain elements having similar chemical and physical properties, and several groups of elements have distinctive names that are useful to know.

Group 1A, Alkali Metals: Li, Na, K, Rb, Cs, Fr

Elements in the leftmost column, Group 1A, are known as the **alkali metals**. All are metals and are solids at room temperature. The metals of Group 1A are all reactive. For example, they react with water to produce hydrogen and alkaline solutions (Figure 2.12). Because of their reactivity, these metals are found in nature only combined in compounds, such as NaCl (Figure 1.7)—never as the free element.

Group 2A, Alkaline Earth Metals: Be, Mg, Ca, Sr, Ba, Ra

The second group in the periodic table, Group 2A, is composed entirely of metals that occur naturally only in compounds (Figure 2.13). Except for beryllium (Be), these elements also react with water to produce alkaline solutions, and most of their oxides (such as lime, CaO) form alkaline solutions; hence, they are known as the **alkaline earth metals**. Magnesium (Mg) and calcium (Ca) are the seventh and fifth most abundant elements in the earth's crust, respectively (Table 2.3). Calcium is one of the important elements in teeth and bones, and it occurs naturally in vast limestone deposits. Calcium carbonate ($CaCO_3$) is the chief constituent of limestone and of corals, sea shells, marble, and chalk (see Figure 2.13b). Radium (Ra), the heaviest alkaline earth element, is radioactive and is used to treat some cancers by radiation.

Group 3A: B, Al, Ga, In, Tl

Group 3A contains one element of great importance, aluminum (Figure 2.14). This element and three others (gallium, indium, and thallium) are metals, whereas boron (B) is a metalloid. Aluminum (Al) is the most abundant metal in the earth's crust at 8.2% by mass. It is exceeded in abundance only by the nonmetals oxygen and silicon. These three elements are found combined in clays and other common

---

■ **Alkali and Alkaline**
The word "alkali" comes from the Arabic language; ancient Arabian chemists discovered that ashes of certain plants, which they called al-qali, gave water solutions that felt slippery and burned the skin. These ashes contain compounds of Group 1A elements that produce alkaline (basic) solutions.

Table 2.3   The Ten Most Abundant Elements in the Earth's Crust

Rank	Element	Abundance (ppm)*
1	Oxygen	474,000
2	Silicon	277,000
3	Aluminum	82,000
4	Iron	41,000
5	Calcium	41,000
6	Sodium	23,000
7	Magnesium	23,000
8	Potassium	21,000
9	Titanium	5,600
10	Hydrogen	1,520

*ppm = g per 1000 kg.

Figure 2.12   Alkali metals. (a) Cutting a bar of sodium with a knife is about like cutting a stick of cold butter. (b) When an alkali metal such as potassium is treated with water, a vigorous reaction occurs, giving an alkaline solution and hydrogen gas, which burns in air. See also Figure 1.7, the reaction of sodium with chlorine.

(a) Cutting sodium.   **(b)** Potassium reacts with water.

Charles D. Winters

(a) Magnesium and strontium in fireworks.

(b) Calcium-containing compounds.

Figure 2.13 Alkaline earth metals. (a) When heated in air, magnesium burns to give magnesium oxide. The white sparks you see in burning fireworks are burning magnesium. (b) Some common calcium-containing substances: calcite (the clear crystal); a seashell; limestone; and an over-the-counter remedy for excess stomach acid.

minerals. Boron occurs in the mineral borax, a compound used as a cleaning agent, antiseptic, and flux for metal work.

As a metalloid, boron has a different chemistry than the other elements of Group 3A, all of which are metals. Nonetheless, all form compounds with analogous formulas such as BCl_3 and $AlCl_3$, and this similarity marks them as members of the same periodic group.

Group 4A: C, Si, Ge, Sn, Pb

All of the elements we have described so far, except boron, have been metals. Beginning with Group 4A, however, the groups contain more and more nonmetals. Group 4A includes one nonmetal, carbon (C); two metalloids, silicon (Si) and germanium (Ge), and two metals, tin (Sn) and lead (Pb). Because of the change from nonmetallic to metallic behavior, more variation occurs in the properties of the elements of this group than in most others. Nonetheless, these elements also form

(a) Wagons for hauling borax in Death Valley.

(b) Aluminum-containing minerals.

Figure 2.14 Group 3A elements. (a) Boron is mined as borax, a natural compound used in soap. Borax was mined in Death Valley, California, at the end of the 19th century and was hauled from the mines in wagons drawn by teams of 20 mules. Boron is also a component of borosilicate glass, which is used for laboratory glassware. (b) Aluminum is abundant in the earth's crust; it is found in all clays and in many minerals and gems. It has many commercial applications as the metal as well as in aluminum sulfate, which is used in water purification.

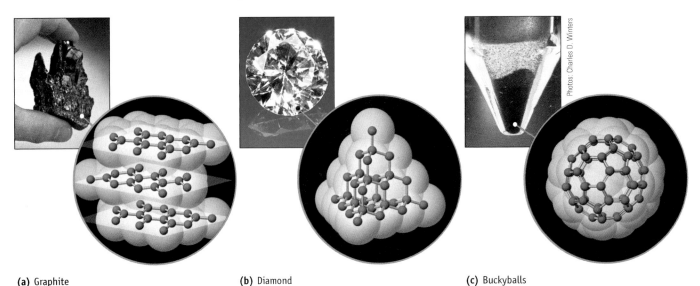

(a) Graphite   **(b)** Diamond   **(c)** Buckyballs

Figure 2.15   The allotropes of carbon. (a) Graphite consists of layers of carbon atoms. Each carbon atom is linked to three others to form a sheet of six-member, hexagonal rings. (b) In diamond the carbon atoms are also arranged in six-member rings, but the rings are not flat because each C atom is connected tetrahedrally to four other C atoms. (c) A member of the family called buckminsterfullerenes, C_{60} is an allotrope of carbon. Sixty carbon atoms are arranged in a spherical cage that resembles a hollow soccer ball. Notice that each six-member ring shares an edge with three other six-member rings and three five-member rings. Chemists call this molecule a "buckyball." C_{60} is a black powder; it is shown here in the tip of a pointed glass tube.

compounds with analogous formulas (such as CO_2, SiO_2, GeO_2, SnO_2, and PbO_2), so they are assigned to the same group.

Carbon is the basis for the great variety of chemical compounds that make up living things. It is found in the earth's atmosphere as CO_2, on the surface of the earth in carbonates like limestone and coral (see Figure 2.13b), and in coal, petroleum, and natural gas—the fossil fuels.

One of the most interesting aspects of the chemistry of the nonmetals is that a particular element can often exist in several different and distinct forms, called **allotropes**, each having its own properties. Carbon has at least three allotropes, the best known of which are graphite and diamond (Figure 2.15). The flat sheets of carbon atoms in graphite (Figure 2.15a) cling only weakly to one another. One layer can slip easily over another, which explains why graphite is soft, is a good lubricant, and is used in pencil lead. (Pencil "lead" is not the element lead, Pb, but rather a composite of clay and graphite that leaves a trail of graphite on the page as you write.)

In diamond each carbon atom is connected to four others at the corners of a tetrahedron, and this pattern extends throughout the solid (see Figure 2.15b). This structure causes diamonds to be extremely hard, denser than graphite (d = 3.51 g/cm^3 for diamond and d = 2.22 g/cm^3 for graphite), and chemically less reactive. Because diamonds are not only hard but are also excellent conductors of heat, they are used on the tips of metal- and rock-cutting tools.

In the late 1980s another form of carbon was identified as a component of black soot, the stuff that collects when carbon-containing materials are burned in a deficiency of oxygen. This substance is made up of molecules with 60 carbon atoms arranged as a spherical "cage" (Figure 2.15c). You may recognize that the surface is made up of five- and six-member rings and resembles a hollow soccer ball. The shape reminded its discoverers of an architectural dome invented several decades ago by the American philosopher and engineer, R. Buckminster Fuller. The official name of the allotrope is therefore buckminsterfullerene, and chemists often call these molecules "buckyballs."

Oxides of silicon are the basis of many minerals such as clay, quartz, and beautiful gemstones like amethyst (Figure 2.16). Tin and lead have been known for centuries because they are easily smelted from their ores. Tin alloyed with copper makes bronze, which was used in ancient times in utensils and weapons. Lead has been used in water pipes and paint, even though the element is toxic to humans.

Group 5A: N, P, As, Sb, Bi

Nitrogen, which occurs naturally in the form of N_2 (Figures 2.10 and 2.17), makes up about three-fourths of earth's atmosphere. It is also incorporated in biochemically important substances such as chlorophyll, proteins, and DNA. Scientists have long sought ways to make compounds from atmospheric nitrogen, a process referred to as "nitrogen fixation." Nature accomplishes this transformation easily in plants, but severe conditions (high temperatures, for example) must be used in the laboratory and in industry to cause N_2 to react with other elements (such as H_2 to make ammonia, NH_3, which is widely used as a fertilizer).

Phosphorus is also essential to life. For example, it is an important constituent in bones and teeth. The element glows in the dark if it is in the air, and its name, based on Greek words meaning "light-bearing," reflects this property. This element also has several allotropes, the most important being white (Figure 2.10) and red phosphorus. Both forms of phosphorus are used commercially. White phosphorus ignites spontaneously in air, so it is normally stored under water. When it does react with air, it forms P_4O_{10}, which can react with water to form phosphoric acid (H_3PO_4), a compound used in food products such as soft drinks. Red phosphorus also reacts with oxygen in the air and is used in the striking strips on match books.

As with Group 4A, we again see nonmetals (N and P), metalloids (As and Sb), and a metal (Bi) in Group 5A. In spite of these variations, all of the members of this group form analogous compounds such as the oxides N_2O_5, P_2O_5, and As_2O_5.

Group 6A: O, S, Se, Te, Po

Oxygen, which constitutes about 20% of earth's atmosphere and combines readily with most other elements, is found at the top of Group 6A. Most of the energy that

Figure 2.16 Compounds containing silicon. Ordinary clay, sand, and many gemstones are based on compounds of silicon and oxygen. Here clear, colorless quartz and dark purple amethyst lie in a bed of sand. All are made of silicon dioxide, SiO_2. The different colors are due to impurities.

Charles D. Winters

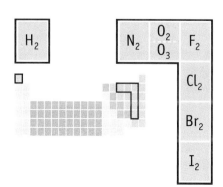

Figure 2.17 Elements that exist as diatomic molecules. Seven of the elements in the periodic table exist as diatomic, or two-atom, molecules. Oxygen has an additional allotrope, ozone, with three O atoms in each molecule.

powers life on earth is derived from reactions in which oxygen combines with other substances.

Sulfur has been known in elemental form since ancient times as brimstone or "burning stone" (Figure 2.18). Sulfur, selenium, and tellurium are referred to collectively as *chalcogens* (from the Greek word, *khalkos*, for copper) because most copper ores contain these elements. Their compounds can be foul-smelling and poisonous; nevertheless, sulfur and selenium are essential components of the human diet. By far the most important compound of sulfur is sulfuric acid (H_2SO_4), which is manufactured in larger amounts than any other compound.

As in Group 5A, the second- and third-period elements have different structures. Like nitrogen, oxygen is a diatomic molecule (see Figure 2.17). Unlike nitrogen, however, oxygen has an allotrope, the well-known ozone, O_3. Sulfur, which can be found in nature as a yellow solid, has many allotropes. The most common allotrope consists of eight-member, crown-shaped rings of sulfur atoms (see Figure 2.18).

Polonium, a radioactive element, was isolated in 1898 by Marie and Pierre Curie, who separated it from tons of a uranium-containing ore and named it for Madame Curie's native country, Poland.

With Group 6A, once again we observe variations in the properties in a group. Oxygen, sulfur, and selenium are nonmetals, tellurium is a metalloid, and polonium is a metal. Nonetheless, there is a family resemblance in their chemistries. All form oxygen-containing compounds such as SO_2, SeO_2, and TeO_2 and sodium-containing compounds such as Na_2O, Na_2S, Na_2Se, and Na_2Te.

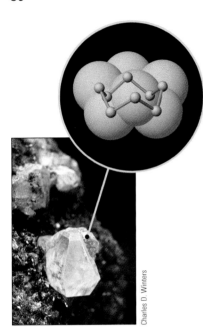

Figure 2.18 Sulfur. The most common allotrope of sulfur consists of eight-member, crown-shaped rings.

Group 7A, Halogens: F, Cl, Br, I, At

At the far right of the periodic table are two groups composed entirely of nonmetals. The Group 7A elements—fluorine, chlorine, bromine, and iodine—are nonmetals, all of which exist as diatomic molecules (see Figure 2.17). At room temperature, fluorine (F_2) and chlorine (Cl_2) are gases. Bromine (Br_2) is a liquid and iodine (I_2) is a solid, but bromine and iodine vapor are clearly visible over the liquid or solid.

The Group 7A elements are among the most reactive of all elements. All combine violently with alkali metals to form salts such as table salt, NaCl (Figure 1.7). The name for this group, the **halogens**, comes from the Greek words *hals*, meaning "salt," and *genes*, meaning "forming." The halogens also react with other metals and with most nonmetals to form compounds.

Group 8A, Noble Gases: He, Ne, Ar, Kr, Xe, Rn

The Group 8A elements—helium, neon, argon, krypton, xenon, and radon—are the least reactive elements (Figure 2.19). All are gases, and none is abundant on earth or in earth's atmosphere. Because of this, they were not discovered until the end of the 19th century. Helium, the second most abundant element in the universe after hydrogen, was detected in the sun in 1868 by analysis of the solar spectrum. (The name of the element comes from the Greek word for the sun, *helios*.) It was not found on earth until 1895, however. Until 1962, when a compound of xenon was first prepared, it was believed that none of these elements would combine chemically with any other element. The common name **noble gases** for this group, a term meant to denote their general lack of reactivity, derives from this fact.

The halogens bromine and iodine. Bromine is a liquid at room temperature and iodine is a solid. However, some of the element exists in the vapor state above the liquid or solid.

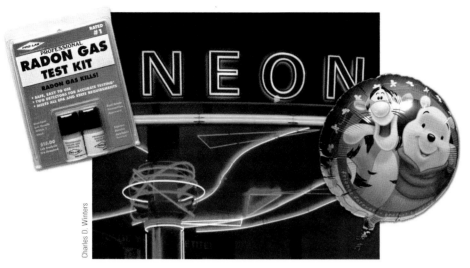

Figure 2.19 The noble gases. This kit is sold for detecting the presence of radon gas in the home. Neon gas is used in advertising signs, and helium-filled balloons are popular.

For the same reason they are sometimes called the *inert gases* or, because of their low abundance, the *rare gases*.

 The Transition Elements

Stretching between Groups 2A and 3A is a series of elements called the **transition elements**. These fill the B-groups (1B through 8B) in the fourth through the seventh periods in the center of the periodic table. All are metals (see Figure 2.10), and 13 of them are in the top 30 elements in terms of abundance in the earth's crust. Some, like iron (Fe), are abundant in nature (Table 2.4). Most occur naturally in combination with other elements, but a few—silver (Ag), gold (Au), and platinum (Pt)—are much less reactive and can be found in nature as pure elements.

Virtually all of the transition elements have commercial uses. They are used as structural materials (iron, titanium, chromium, copper); in paints (titanium, chromium); in the catalytic converters in automobile exhaust systems (platinum, rhodium); in coins (copper, nickel, zinc); and in batteries (manganese, nickel, cadmium, mercury).

A number of the transition elements play important biological roles. For example, iron, a relatively abundant element (see Table 2.3), is the central element in the chemistry of hemoglobin, the oxygen-carrying component of blood.

Two rows at the bottom of the table accommodate the **lanthanides** [the series of elements between the elements lanthanum ($Z = 57$) and hafnium ($Z = 72$)] and the **actinides** [the series of elements between actinium ($Z = 89$) and rutherfordium ($Z = 104$)]. Some lanthanide compounds are used in color television picture tubes, uranium ($Z = 92$) is the fuel for atomic power plants, and americium ($Z = 95$) is used in smoke detectors.

Table 2.4 Abundance of the Ten Most Abundant Transition Elements in the Earth's Crust

Rank	Element	Abundance (ppm)*
4	Iron	41,000
9	Titanium	5,600
12	Manganese	950
18	Zirconium	190
19	Vanadium	160
21	Chromium	100
23	Nickel	80
24	Zinc	75
25	Cerium	68
26	Copper	50

*ppm = g per 1000 kg.

Exercise 2.7—The Periodic Table

How many elements are in the third period of the periodic table? Give the name and symbol of each. Tell whether each element in the period is a metal, metalloid, or nonmetal.

Table 2.5   Relative Amounts of Elements in the Human Body

Element	Percent by Mass
Oxygen	65
Carbon	18
Hydrogen	10
Nitrogen	3
Calcium	1.5
Phosphorus	1.2
Potassium, sulfur, chlorine	0.2
Sodium	0.1
Magnesium	0.05
Iron, cobalt, copper, zinc, iodine	<0.05
Selenium, fluorine	<0.01

2.8—Essential Elements

As our knowledge of biochemistry—the chemistry of living systems—increases, we learn more and more about essential elements. These elements are so important to life that a deficiency in any one will result in either death, severe developmental abnormalities, or chronic ailments. No other element can take the place of an essential element.

Of the 116 known elements, 11 are predominant in many different biological systems and are present in approximately the same relative amounts (Table 2.5). In humans these 11 elements constitute 99.9% of the total number of atoms present, but 4 of these elements—C, H, N, and O—account for 99% of the total. These elements are found in the basic structure of all biochemical molecules. Additionally, H and O are present in water, a major component of all biological systems.

The other 7 elements of the group of 11 elements comprise only 0.9% of the total atoms in the body. These are sodium, potassium, calcium, magnesium, phosphorus, sulfur, and chlorine. These generally occur in the form of ions such as Na^+, K^+, Mg^{2+}, Ca^{2+}, Cl^-, and HPO_4^{2-}.

The 11 essential elements represent 6 of the groups of the periodic table, and all are "light" elements; they have atomic numbers less than 21. Another 17 elements are required by most but not all biological systems. Some may be required by plants, some by animals, and others by only certain plants or animals. With a few exceptions, these elements are generally "heavier" elements, elements having an atomic number greater than 18. They are about evenly divided between metals and nonmetals (or metalloids).

Elements in the Human Body

Major Elements	Trace Elements
99.9% of all atoms (99.5% by mass)	0.1% of all atoms (0.5% by mass)
C, H, N, O	V, Cr, Mo, Mn, Fe, Co, Ni, Cu, Zn
Na, Mg, P, S, Cl	B, Si, Se, F, Br, I, As, Sn
K, Ca	

Sources of Some Biologically Important Elements

Element	Source	mg/100 g
Iron	Brewer's yeast	17.3
	Eggs	2.3
Zinc	Brazil nuts	4.2
	Chicken	2.6
Copper	Oysters	13.7
	Brazil nuts	2.3
Calcium	Swiss cheese	925
	Whole milk	118
	Broccoli	103
Selenium	Butter	0.15
	Cider vinegar	0.09

Although many of the metals in this group are required only in trace amounts, they are often an integral part of specific biological molecules—such as hemoglobin (Fe), myoglobin (Fe), and vitamin B_{12} (Co)—and activate or regulate their functions.

Much of the 3 or 4 g of iron in the body is found in hemoglobin, the substance responsible for carrying oxygen to cells. Iron deficiency is marked by fatigue, infections, and mouth inflammation. The average person also contains about 2 g of zinc. A deficiency of this element will be evidenced as loss of appetite, failure to grow, and changes in the skin.

The human body has about 75 mg of copper, about one third of which is found in the muscles and the remainder in other tissues. Copper is involved in many biological functions, and a deficiency shows up in a variety of ways: anemia, degeneration of the nervous system, impaired immunity, and defects in hair color and structure.

Chapter Goals Revisited

Now that you have studied this chapter, you should ask whether you have met the chapter goals. In particular, you should be able to

Describe atomic structure and define atomic number and mass number
a. Explain the historical development of the atomic theory and identify some of the scientists who made important contributions (Section 2.1). General ChemistryNow homework: Study Question(s) 65

b. Describe electrons, protons, and neutrons, and the general structure of the atom (Section 2.1). General ChemistryNow homework: SQ(s) 6, 12

c. Understand the relative mass scale and the atomic mass unit (Section 2.2). General ChemistryNow homework: SQ(s) 64

Understand the nature of isotopes and calculate atomic weight from isotope abundances and isotopic masses
a. Define isotope and give the mass number and number of neutrons for a specific isotope (Sections 2.2 and 2.3). General ChemistryNow homework: SQ(s) 14

b. Do calculations that relate the atomic weight (atomic mass) of an element and isotopic abundances and masses (Section 2.4). General ChemistryNow homework: SQ(s) 20, 22, 25, 47

Explain the concept of the mole and use molar mass in calculations
a. Understand that the molar mass of an element is the mass in grams of Avogadro's number of atoms of that element (Section 2.5). General ChemistryNow homework: SQ(s) 27, 29, 31

b. Know how to use the molar mass of an element and Avogadro's number in calculations (Section 2.5). General ChemistryNow homework: SQ(s) 33, 57, 67, 77

Know the terminology of the periodic table
a. Identify the periodic table locations of groups, periods, metals, metalloids, nonmetals, alkali metals, alkaline earth metals, halogens, noble gases, and the transition elements (Sections 2.6 and 2.7). General ChemistryNow homework: SQ(s) 38, 39, 41, 49

b. Recognize similarities and differences in properties of some of the common elements of a group. General ChemistryNow homework: SQ(s) 56

GENERAL
Chemistry ⫶ Now™

See the General ChemistryNow CD-ROM or website to:
- Assess your understanding with homework questions keyed to each goal
- Check your readiness for an exam by taking the exam-prep quiz and exploring the resources in the personalized Learning Plan it provides

Foods rich in essential elements.

Key Equations

Equation 2.1 (page 69)
Calculate the percent abundance of an isotope.

$$\text{Percent abundance} = \frac{\text{number of atoms of a given isotope}}{\text{total number of atoms of all isotopes of that element}} \times 100\%$$

Equation 2.2 (page 72)
Calculate the atomic mass (atomic weight) from isotope abundances and the exact atomic mass of each isotope of an element.

$$\text{Atomic weight} = \left(\frac{\%\ \text{abundance isotope 1}}{100}\right)(\text{mass of isotope 1}) + \left(\frac{\%\ \text{abundance isotope 2}}{100}\right)(\text{mass of isotope 2}) + \cdots$$

Study Questions

▲ denotes more challenging questions.

■ denotes questions available in the Homework and Goals section of the General ChemistryNow CD-ROM or website.

Blue numbered questions have answers in Appendix O and fully worked solutions in the *Student Solutions Manual.*

Structures of many of the compounds used in these questions are found on the General ChemistryNow CD-ROM or website in the Models folder.

GENERAL
Chemistry⋅▲⋅Now™ Assess your understanding of this chapter's topics with additional quizzing and conceptual questions at http://now.brookscole.com/kotz6e

Practicing Skills

Atoms: Their Composition and Structure
(See Example 2.1, Exercise 2.2, and the General ChemistryNow Screen 2.11.)

1. What are the three fundamental particles from which atoms are built? What are their electric charges? Which of these particles constitute the nucleus of an atom? Which is the least massive particle of the three?

2. Around 1910 Rutherford carried out his now-famous alpha-particle scattering experiment. What surprising observation did he make in this experiment and what conclusion did he draw from it?

3. What did the discovery of radioactivity reveal about the structure of atoms?

4. What scientific instrument was used to discover that not all atoms of neon have the same mass?

5. If the nucleus of an atom were the size of a medium-sized orange (say, with a diameter of about 6 cm), what would be the diameter of the atom?

6. ■ If a gold atom has a radius of 145 pm, and you could string gold atoms like beads on a thread, how many atoms would you need to have a necklace 36 cm long?

7. The volcanic eruption of Mount St. Helens in the state of Washington in 1980 produced a considerable quantity of a radioactive element in the gaseous state. The element has atomic number 86. What are the symbol and name of this element?

8. Titanium and thallium have symbols that are easily confused with each other. Give the symbol, atomic number, atomic weight, and group and period number of each element. Are they metals, metalloids, or nonmetals?

9. Give the mass number of each of the following atoms: (a) magnesium with 15 neutrons, (b) titanium with 26 neutrons, and (c) zinc with 32 neutrons.

10. Give the mass number of (a) a nickel atom with 31 neutrons, (b) a plutonium atom with 150 neutrons, and (c) a tungsten atom with 110 neutrons.

11. Give the complete symbol (^A_ZX) for each of the following atoms: (a) potassium with 20 neutrons, (b) krypton with 48 neutrons, and (c) cobalt with 33 neutrons.

12. ■ Give the complete symbol (^A_ZX) for each of the following atoms: (a) fluorine with 10 neutrons, (b) chromium with 28 neutrons, and (c) xenon with 78 neutrons.

13. How many electrons, protons, and neutrons are there in an atom of (a) magnesium-24, ^{24}Mg; (b) tin-119, ^{119}Sn; and (c) thorium-232, ^{232}Th?

14. ■ How many electrons, protons, and neutrons are there in an atom of (a) carbon-13, ^{13}C; (b) copper-63, ^{63}Cu; and (c) bismuth-205, ^{205}Bi?

Isotopes
(See Example 2.2 and the General Chemistry Now Screen 2.12.)

15. The synthetic radioactive element technetium is used in many medical studies. Give the number of electrons, protons, and neutrons in an atom of technetium-99.

16. Radioactive americium-241 is used in household smoke detectors and in bone mineral analysis. Give the number of electrons, protons, and neutrons in an atom of americium-241.

17. Cobalt has three radioactive isotopes used in medical studies. Atoms of these isotopes have 30, 31, and 33 neutrons, respectively. Give the symbol for each of these isotopes.

18. Which of the following are isotopes of element X, the atomic number for which is 9: $^{19}_{9}$X, $^{20}_{9}$X, $^{9}_{18}$X, and $^{21}_{9}$X?

Isotope Abundance and Atomic Mass
(See Exercises 2.3 and 2.4 and the General ChemistryNow Screen 2.13.)

19. Thallium has two stable isotopes, ^{203}Tl and ^{205}Tl. Knowing that the atomic weight of thallium is 204.4, which isotope is the more abundant of the two?

20. ■ Strontium has four stable isotopes. Strontium-84 has a very low natural abundance, but ^{86}Sr, ^{87}Sr, and ^{88}Sr are all reasonably abundant. Knowing that the atomic weight of strontium is 87.62, which of the more abundant isotopes predominates?

21. Verify that the atomic mass of lithium is 6.94, given the following information:
^{6}Li, mass = 6.015121 u; percent abundance = 7.50%
^{7}Li, mass = 7.016003 u; percent abundance = 92.50%

22. ■ Verify that the atomic mass of magnesium is 24.31, given the following information:
^{24}Mg, mass = 23.985042 u; percent abundance = 78.99%
^{25}Mg, mass = 24.985837 u; percent abundance = 10.00%
^{26}Mg, mass = 25.982593 u; percent abundance = 11.01%

23. Silver (Ag) has two stable isotopes, ^{107}Ag and ^{109}Ag. The isotopic mass of ^{107}Ag is 106.9051 and the isotopic mass of ^{109}Ag is 108.9047. The atomic weight of Ag, from the periodic table, is 107.868. Estimate the percentage of ^{107}Ag in a sample of the element.
    (a) 0%   (b) 25%   (c) 50%   (d) 75%

24. Copper exists as two isotopes: ^{63}Cu (62.9298 u) and ^{65}Cu (64.9278 u). What is the approximate percentage of ^{63}Cu in samples of this element?
    (a) 10%   (c) 50%   (e) 90%
    (b) 30%   (d) 70%

25. ■ Gallium has two naturally occurring isotopes, ^{69}Ga and ^{71}Ga, with masses of 68.9257 u and 70.9249 u, respectively. Calculate the percent abundances of these isotopes of gallium.

26. Antimony has two stable isotopes, ^{121}Sb and ^{123}Sb, with masses of 120.9038 u and 122.9042 u, respectively. Calculate the percent abundances of these isotopes of antimony.

Atoms and the Mole

(See Examples 2.5–2.7 and the General ChemistryNow Screens 2.14 and 2.15.)

27. ■ Calculate the mass, in grams, of the following:
    (a) 2.5 mol of aluminum
    (b) 1.25×10^{-3} mol of iron
    (c) 0.015 mol of calcium
    (d) 653 mol of neon

28. Calculate the mass, in grams, of
    (a) 4.24 mol of gold
    (b) 15.6 mol of He
    (c) 0.063 mol of platinum
    (d) 3.63×10^{-4} mol of Pu

29. ■ Calculate the amount (moles) represented by each of the following:
    (a) 127.08 g of Cu
    (b) 0.012 g of lithium
    (c) 5.0 mg of americium
    (d) 6.75 g of Al

30. Calculate the amount (moles) represented by each of the following:
    (a) 16.0 g of Na
    (b) 0.876 g of tin
    (c) 0.0034 g of platinum
    (d) 0.983 g of Xe

31. ■ You are given 1.0-g samples of He, Fe, Li, Si, and C. Which sample contains the largest number of atoms? Which contains the smallest?

32. You are given 1.0-mol amounts of He, Fe, Li, Si, and C. Which sample has the largest mass?

33. ■ What is the average mass of one copper atom?

34. What is the average mass of one titanium atom?

The Periodic Table

(See Section 2.6 and Exercise 2.7. See also the Periodic Table Tool on the General ChemistryNow CD-ROM or website.)

35. Give the name and symbol of each of the Group 5A elements. Tell whether each is a metal, nonmetal, or metalloid.

36. Give the name and symbol of each of the fourth-period elements. Tell whether each is a metal, nonmetal, or metalloid.

37. How many periods of the periodic table have 8 elements, how many have 18 elements, and how many have 32 elements?

38. ■ How many elements occur in the seventh period? What is the name given to the majority of these elements and what well-known property characterizes them?

39. ■ Select answers to the questions listed below from the following list elements whose symbols start with the letter C: C, Ca, Cr, Co, Cd, Cl, Cs, Ce, Cm, Cu, and Cf. (You should expect to use some symbols more than once.)
    (a) Which are nonmetals?
    (b) Which are main group elements?
    (c) Which are lanthanides?
    (d) Which are transition elements?
    (e) Which are actinides?
    (f) Which are gases?

40. Give the name and chemical symbol for the following.
    (a) a nonmetal in the second period
    (b) an alkali metal
    (c) the third-period halogen
    (d) an element that is a gas at 20°C and 1 atmosphere pressure

41. ■ Classify the following elements as metals, metalloids, or nonmetals: N, Na, Ni, Ne, and Np.

42. Here are symbols for five of the seven elements whose names begin with the letter B: B, Ba, Bk, Bi, and Br. Match each symbol with one of the descriptions below.
    (a) a radioactive element
    (b) a liquid at room temperature
    (c) a metalloid
    (d) an alkaline earth element
    (e) a Group 5A element

43. Use the elements in the following list to answer the questions: sodium, silicon, sulfur, scandium, selenium, strontium, silver, and samarium. (Some elements will be entered in more than one category.)
    (a) Identify those that are metals.
    (b) Identify those that are main group elements
    (c) Identify those that are transition metals.

44. Compare the elements silicon (Si) and phosphorus (P) using the following criteria:
    (a) metal, metalloid, or nonmetal
    (b) possible conductor of electricity
    (c) physical state at 25 °C (solid, liquid, or gas)

▲ More challenging   ■ In General ChemistryNow   Blue-numbered questions answered in Appendix O

General Questions

These questions are not designed as to type or location in the chapter. They may combine several concepts. More challenging questions are marked with the icon ▲.

45. Fill in the blanks in the table (one column per element).

Symbol	^{58}Ni	^{33}S	_____	_____
Number of protons	_____	_____	10	_____
Number of neutrons	_____	_____	10	30
Number of electrons in the neutral atom	_____	_____	_____	25
Name of element	_____	_____	_____	_____

46. Fill in the blanks in the table (one column per element).

Symbol	^{65}Cu	^{86}Kr	_____	_____
Number of protons	_____	_____	78	_____
Number of neutrons	_____	_____	117	46
Number of electrons in the neutral atom	_____	_____	_____	35
Name of element	_____	_____	_____	_____

47. ■ Potassium has three naturally occurring isotopes (^{39}K, ^{40}K, and ^{41}K), but ^{40}K has a very low natural abundance. Which of the other two isotopes is the more abundant? Briefly explain your answer.

48. *Crossword Puzzle:* In the 2×2 box shown here, each answer must be correct four ways: horizontally, vertically, diagonally, and by itself. Instead of words, use symbols of elements. When the puzzle is complete, the four spaces will contain the overlapping symbols of ten elements. There is only one correct solution.

1	2
3	4

Horizontal

1–2: Two-letter symbol for a metal used in ancient times

3–4: Two-letter symbol for a metal that burns in air and is found in Group 5A

Vertical

1–3: Two-letter symbol for a metalloid

2–4: Two-letter symbol for a metal used in U.S. coins

Single squares: all one-letter symbols

1: A colorful nonmetal

2: Colorless gaseous nonmetal

3: An element that makes fireworks green

4: An element that has medicinal uses

Diagonal

1–4: Two-letter symbol for an element used in electronics

2–3: Two-letter symbol for a metal used with Zr to make wires for superconducting magnets

This puzzle first appeared in *Chemical & Engineering News*, p. 86, December 14, 1987 (submitted by S. J. Cyvin) and in *Chem Matters*, October 1988.

49. ■ The chart shown in the *Stardust* story (page 58) plots the logarithm of the abundance of elements 1 through 30 in the solar system on a logarithmic scale.
    (a) What is the most abundant main group metal?
    (b) What is the most abundant nonmetal?
    (c) What is the most abundant metalloid?
    (d) Which of the transition elements is most abundant?
    (e) Which halogens are included on this plot and which is the most abundant?

50. The molecule buckminsterfullerene, commonly called a "buckyball," is one of three common allotropes of a familiar element. Identify two other allotropes of this element.

51. Which of the following is impossible?
    (a) silver foil that is 1.2×10^{-4} m thick
    (b) a sample of potassium that contains 1.784×10^{24} atoms
    (c) a gold coin of mass 1.23×10^{-3} kg
    (d) 3.43×10^{-27} mol of S_8

52. Give the symbol for a metalloid in the third period and then identify a property of this element.

53. Reviewing the periodic table.
    (a) Name an element in Group 2A.
    (b) Name an element in the third period.
    (c) Which element is in the second period in Group 4A?
    (d) Which element is in the third period in Group 6A?
    (e) Which halogen is in the fifth period?
    (f) Which alkaline earth element is in the third period?
    (g) Which noble gas element is in the fourth period?
    (h) Name the nonmetal in Group 6A and the third period.
    (i) Name a metalloid in the fourth period.

54. Reviewing the periodic table:
    (a) Name an element in Group 2B.
    (b) Name an element in the fifth period.
    (c) Which element is in the sixth period in Group 4A?
    (d) Which element is in the third period in Group 6A?
    (e) Which alkali metal is in the third period?
    (f) Which noble gas element is in the fifth period?
    (g) Name the element in Group 6A and the fourth period. Is it a metal, nonmetal, or metalloid?
    (h) Name a metalloid in Group 5A.

55. The plot on the following page shows the variation in density with atomic number for the first 36 elements. Use this plot to answer the following questions:
    (a) Which three elements in this series have the highest density? What is their approximate density? Are these elements metals or nonmetals?

(b) Which element in the second period has the highest density? Which element in the third period has the highest density? What do these two elements have in common?

(c) Some elements have densities so low that they do not show up on the plot. What elements are these? What property do they have in common?

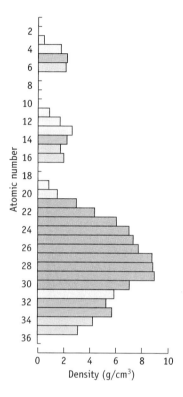

Density (g/cm^3)

56. ■ Give two examples of nonmetallic elements that have allotropes. Name those elements and describe the allotropes of each.

57. ■ In each case, decide which represents more mass:
(a) 0.5 mol of Na or 0.5 mol of Si
(b) 9.0 g of Na or 0.50 mol of Na
(c) 10 atoms of Fe or 10 atoms of K

58. A semiconducting material is composed of 52 g of Ga, 9.5 g of Al, and 112 g of As. Which element has the largest number of atoms in the final mixture?

59. You are given 15 g each of yttrium, boron, and copper. Which sample represents the largest number of atoms?

60. Lithium has two stable isotopes: ^{6}Li and ^{7}Li. One of them has an abundance of 92.5%, and the other has an abundance of 7.5%. Knowing that the atomic mass of lithium is 6.941, which is the more abundant isotope?

61. Superman comes from the planet Krypton. If you have 0.00789 g of the gaseous element krypton, how many moles does this represent? How many atoms?

62. The recommended daily allowance (RDA) of iron in your diet is 15 mg. How many moles is this? How many atoms?

63. Put the following elements in order from smallest to largest mass:
(a) 3.79×10^{24} atoms Fe   (e) 9.221 mol Na
(b) 19.921 mol H$_2$ (f) 4.07×10^{24} atoms Al
(c) 8.576 mol C   (g) 9.2 mol Cl$_2$
(d) 7.4 mol Si

64. ■ ▲ When a sample of phosphorus burns in air, the compound P$_4$O$_{10}$ forms. One experiment showed that 0.744 g of phosphorus formed 1.704 g of P$_4$O$_{10}$. Use this information to determine the ratio of the atomic masses of phosphorus and oxygen (mass P/mass O). If the atomic mass of oxygen is assumed to be 16.000 u, calculate the atomic mass of phosphorus.

65. ■ The data below were collected in a Millikan oil drop experiment.

Oil Drop	Measured Charge on Drop (C)
1	1.59×10^{-19}
2	11.1×10^{-19}
3	9.54×10^{-19}
4	15.9×10^{-19}
5	6.36×10^{-19}

(a) Use these data to calculate the charge on the electron (in coulombs).
(b) How many electrons have accumulated on each oil drop?
(c) The accepted value of the electron charge is 1.60×10^{-19} C. Calculate the percent and error for the value determined by the data in the table.

66. ▲ Although carbon-12 is now used as the standard for atomic masses, this has not always been the case. Early attempts at classification used hydrogen as the standard, with the mass of hydrogen being set equal to 1.0000 u. Later attempts defined atomic masses using oxygen (with a mass of 16.0000 u). In each instance, the atomic masses of the other elements were defined relative to these masses. (To answer this question, you need more precise data on current atomic masses: H, 1.00794 u; O, 15.9994 u.)
(a) If H = 1.0000 u was used as a standard for atomic masses, what would the atomic mass of oxygen be? What would be the value of Avogadro's number under these circumstances?
(b) Assuming the standard is O = 16.0000 u, determine the value for the atomic mass of hydrogen and the value of Avogadro's number.

67. ■ A reagent occasionally used in chemical synthesis is sodium-potassium alloy. (Alloys are mixtures of metals, and Na-K has the interesting property that it is a liquid.) One formulation of the alloy (the one that melts at the lowest temperature) contains 68 atom percent K; that is, out of every 100 atoms, 68 are K and 32 are Na. What is the weight percent of potassium in sodium-potassium alloy?

68. Mass spectrometric analysis showed that there are four isotopes of an unknown element having the following masses and abundances:

Isotope	Mass Number	Isotope Mass	Abundance (%)
1	136	135.9090	0.193
2	138	137.9057	0.250
3	140	139.9053	88.48
4	142	141.9090	11.07

Three elements in the periodic table that have atomic weights near these values are lanthanum (La), atomic number 57, atomic weight 139.9055; cerium (Ce), atomic number 58, atomic weight 140.115; and praeseodymium (Pr), atomic number 59, atomic weight 140.9076. Using the data above, calculate the atomic weight and identify the element if possible.

Summary and Conceptual Questions

The following questions use concepts from the preceding chapter (Chapter 1).

69. Draw a picture showing the approximate positions of all protons, electrons, and neutrons in an atom of helium-4. Make certain that your diagram indicates both the number and position of each type of particle.

70. Draw two boxes, each about 3 cm on a side. In one box, sketch a representation of iron metal. In the other box, sketch a representation of nitrogen gas. How do these drawings differ?

71. ▲ Identify, from the list below, the information needed to calculate the number of atoms in 1 cm³ of iron. Outline the procedure used in this calculation.
    (a) the structure of solid iron
    (b) the molar mass of iron
    (c) Avogadro's number
    (d) the density of iron
    (e) the temperature
    (f) iron's atomic number
    (g) the number of iron isotopes

72. Consider the plot of relative element abundances on page 58. Is there a relationship between abundance and atomic number? Is there any difference between the relative abundance of an element of even atomic number and the relative abundance of an element of odd atomic number?

73. The photo here depicts what happens when a coil of magnesium ribbon and a few calcium chips are placed in water.

Magnesium (*left*) and calcium (*right*) in water.

(a) Based on their relative reactivities, what might you expect to see when barium, another Group 2A element, is placed in water?

(b) Give the period in which each element (Mg, Ca, and Ba) is found. What correlation do you think you might find between the reactivity of these elements and their positions in the periodic table?

74. ▲ In an experiment, you need 0.125 mol of sodium metal. Sodium can be cut easily with a knife (Figure 2.12), so if you cut out a block of sodium, what should the volume of the block be in cubic centimeters? If you cut a perfect cube, what is the length of the edge of the cube? (The density of sodium is 0.97 g/cm³.)

75. ▲ Dilithium is the fuel for the *Starship Enterprise.* Because its density is quite low, however, you need a large space to store a large mass. To estimate the volume required, we shall use the element lithium. If you need 256 mol for an interplanetary trip, what must the volume of the piece of lithium be? If the piece of lithium is a cube, what is the dimension of an edge of the cube? (The density for the element lithium is 0.534 g/cm³ at 20 °C.)

76. An object is coated with a layer of chromium, 0.015 cm thick. The object has a surface area of 15.3 cm³. How many atoms of chromium are used in the coating? (The density of chromium = 7.19 g/cm³.)

77. ■ A cylindrical piece of sodium is 12.00 cm long and has a diameter of 4.5 cm. The density of sodium is 0.971 g/cm³. How many atoms does the piece of sodium contain? (The volume of a cylinder is $V = \pi \times r^2 \times$ length.)

78. ▲ Consider an atom of ^{64}Zn.
    (a) Calculate the density of the nucleus in grams per cubic centimeter, knowing that the nuclear radius is 4.8×10^{-6} nm and the mass of the ^{64}Zn atom is 1.06×10^{-22} g. Recall that the volume of a sphere is $(4/3)\pi r^3$.
    (b) Calculate the density of the space occupied by the electrons in the zinc atom, given that the atomic radius is 0.125 nm and the electron mass is 9.11×10^{-28} g.
    (c) Having calculated these densities, what statement can you make about the relative densities of the parts of the atom?

79. ▲ Most standard analytical balances can measure accurately to the nearest 0.0001 g. Assume you have weighed out a 2.0000-g sample of carbon. How many atoms are in this sample? Assuming the indicated accuracy of the measurement, what is the largest number of atoms that can be present in the sample?

80. ▲ To estimate the radius of a lead atom:
    (a) You are given a cube of lead that is 1.000 cm on each side. The density of lead is 11.35 g/cm³. How many atoms of lead are in the sample?
    (b) Atoms are spherical; therefore, the lead atoms in this sample cannot fill all the available space. As an approximation, assume that 60% of the space of the cube is filled with spherical lead atoms. Calculate the volume

of one lead atom from this information. From the calculated volume (V), and the formula $V = \frac{4}{3}(\pi r^3)$, estimate the radius (r) of a lead atom.

81. A jar contains some number of jelly beans. To find out precisely how many are in the jar you could dump them out and count them. How could you estimate their number without counting each one? (Chemists need to do just this kind of "bean counting" when we work with atoms and molecules. They are too small to count one by one, so we have worked out other methods to "count atoms.") (*See General ChemistryNow Screen 2.18, Chemical Puzzler.*)

Do you need a live tutor for homework problems?
Access vMentor at General ChemistryNow at
http://now.brookscole.com/kotz6e
for one-on-one tutoring from a chemistry expert

Charles D. Winters

How many jelly beans are in the jar?

3—Molecules, Ions, and Their Compounds

James D. Watson and Francis Crick. In a photo taken in 1953, Watson (*left*) and Crick (*right*) stand by their model of the DNA double helix. Together with Maurice Wilkins, Watson and Crick received the Nobel Prize in medicine and physiology in 1962.

A. Barrington Brown/Science Source/Photo Researchers, Inc.

DNA: The Most Important Molecule

DNA is the substance in every plant and animal that carries the exact blueprint of that plant or animal. The structure of this molecule, the cornerstone of life, was uncovered in 1953, and James D. Watson, Francis Crick, and Maurice Wilkins shared the 1962 Nobel Prize in medicine and physiology for the work. It was one of the most important scientific discoveries of the 20th century, and the story has recently been told by Watson in his book *The Double Helix*.

When Watson was a graduate student at Indiana University, he had an interest in the gene and said he hoped that its biological role might be solved "without my learning any chemistry." Later, however, he and Crick found out just how useful chemistry can be when they began to unravel the structure of DNA.

Solving important problems requires teamwork among scientists of many kinds so Watson went to Cambridge University in England in 1951. There he met Crick, who, Watson said, talked louder and faster than anyone else. Crick shared Watson's belief in the fundamental importance of DNA, and the pair soon learned that Maurice Wilkins and Rosalind Franklin at King's College in London were using a technique called x-ray crystallography to learn more about DNA's structure. Watson and Crick believed that understanding this structure was crucial to understanding genetics. To solve the structural problem, however, they needed experimental data of the type that could come from the experiments at King's College.

The King's College group was initially reluctant to share their data; and, what is more, they did not seem to share Watson and Crick's sense of urgency. There was also an ethical dilemma: Could Watson and Crick work on a problem that others had claimed as theirs? "The English sense of fair play would not allow Francis to move in on Maurice's problem," said Watson.

Watson and Crick approached the problem through a technique chemists now use frequently—model building. They built models of the pieces of the DNA chain, and they tried various chemically reasonable ways of fitting them together. Finally, they discovered that one arrangement was "too pretty not to be true." Ultimately, the experimental evidence of Wilkins and Franklin confirmed the "pretty structure" to be the real DNA structure. As you will see, chemists often use models to help guide them to experimental evidence that is definitive.

The story of how Watson, Crick, Wilkins, and Franklin ultimately came to share information and insight is an interesting human drama and illustrates how scientific progress is often made. For more on this interesting human and scientific drama, read *Rosalind Franklin: The Dark Lady of DNA* by Brenda Maddox and Watson's book *The Double Helix*.

Rosalind Franklin of King's College, London. She died in 1958 at the age of 37. Because Nobel Prizes are never awarded posthumously, she did not share in this honor with Watson, Crick, and Wilkins.

Watson and Crick recognized early on that the overall structure of DNA was a helix; that is, the atomic-level building blocks formed chains that twisted in space like the strands of a grapevine. They also knew which chemical elements it contained and roughly how they were grouped together. What they did not know was the detailed structure of the helix. By the spring of 1953, however, they had the answer. The atomic-level building blocks of DNA form two chains twisted together in a double helix.

Structure of DNA: Sugar, Phosphate, and Bases

DNA is a very large molecule that consists of two chains of atoms (P, C, and O) that twist together. The P, C, and O atoms are parts of phosphate ions (P) and sugar molecules. The chains are joined by four different molecules (adenine, thymine, guanine, and cytosine) belonging to a general class of molecules called bases.

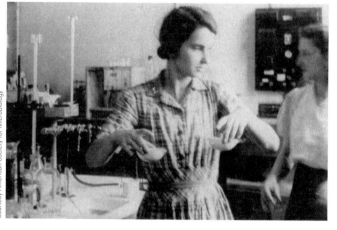

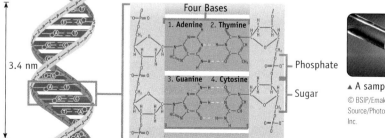

▲ A sample of DNA.
© BSIP/Emakoff/Science Source/Photo Researchers, Inc.

To Review Before You Begin

- Know how to calculate and use molar amounts (Section 2.5)

GENERAL
Chemistry⋅⚛⋅Now™

Throughout the chapter this icon introduces a list of resources on the General ChemistryNow CD-ROM or website (http://now .brookscole.com/kotz6e) that will:

- help you evaluate your knowledge of the material
- provide homework problems
- allow you to take an exam-prep quiz
- provide a personalized Learning Plan targeting resources that address areas you should study

In 1953 the structure of the giant molecule DNA, *deoxyribonucleic acid*, was finally understood (page 96). Chromosomes, which are present in the nuclei of almost all living cells, consist of DNA. Recently discovered knowledge of the human genome, which is the complete structure of the DNA in every one of our 23 chromosomes, is widely expected to revolutionize the practice of medicine. To comprehend modern molecular biology—indeed all of modern chemistry—we have to understand the structures and properties of molecules. This chapter marks the beginning of our attempt to acquaint you with this important subject.

3.1—Molecules, Compounds, and Formulas

A *molecule* is the smallest identifiable unit into which a pure substance like sugar and water can be divided and still retain the composition and chemical properties of the substance. Such substances are composed of identical molecules consisting of atoms of two or more elements bound firmly together. For example, atoms of the element aluminum, Al, combine with molecules of the element bromine, Br_2, to produce the compound aluminum bromide, Al_2Br_6 (Figure 3.1).

$$2\ Al(s) + 3\ Br_2(\ell) \longrightarrow Al_2Br_6(s)$$

aluminum + bromine ⟶ aluminum bromide

(a)

(b)

(c)

Photos: Charles D. Winters

Active Figure 3.1   **Reaction of the elements aluminum and bromine.** (a) Solid aluminum and (in the beaker) liquid bromine. (b) When the aluminum is added to the bromine, a vigorous chemical reaction produces white, solid aluminum bromide, Al_2Br_6 (c).

GENERAL
Chemistry⋅⚛⋅Now™ See the General ChemistryNow CD-ROM or website to explore an interactive version of this figure accompanied by an exercise.

NAME	MOLECULAR FORMULA	CONDENSED FORMULA	STRUCTURAL FORMULA	MOLECULAR MODEL
Ethanol	C_2H_6O	CH_3CH_2OH	H H | | H—C—C—O—H | | H H	
Dimethyl ether	C_2H_6O	CH_3OCH_3	H H | | H—C—O—C—H | | H H	

Figure 3.2 Four approaches to showing molecular formulas. Here the two molecules have the same molecular formula. However, once they are written as condensed or structural formulas, and illustrated with a molecular model, it is clear that these molecules are different. (*See the General ChemistryNow Screen 3.4 Representing Compounds, for a tutorial on identifying molecular representations.*)

To describe this chemical change (or chemical reaction) on paper, the composition of each element and compound is represented by a symbol or formula. Here one molecule of Al_2Br_6 is composed of two Al atoms and six Br atoms.

How do compounds differ from elements? When a compound is produced from its elements, the characteristics of the constituent elements are lost. Solid, metallic aluminum and red-orange liquid bromine, for example, react to form Al_2Br_6, a white solid (see Figure 3.1).

Formulas

For molecules more complicated than water, there is often more than one way to write the formula. For example, the formula of ethanol (also called ethyl alcohol) can be represented as C_2H_6O (Figure 3.2). This **molecular formula** describes the composition of ethanol molecules—two carbon atoms, six hydrogen atoms, and one atom of oxygen occur per molecule—but it gives us no structural information. Structural information—how the atoms are connected and how the molecule fills space—is important, however, because it helps us understand how a molecule can interact with other molecules, which is the essence of chemistry.

To provide some structural information, it is useful to write a **condensed formula**, which indicates how certain atoms are grouped together. For example, the condensed formula of ethanol, CH_3CH_2OH (see Figure 3.2), informs us that the molecule consists of three "groups": a CH_3 group, a CH_2 group, and an OH group. Writing the formula as CH_3CH_2OH also shows that the compound is not dimethyl ether, CH_3OCH_3, a compound with the same molecular formula but a different structure and distinctly different properties.

That ethanol and dimethyl ether are different molecules is further apparent from their **structural formulas** (see Figure 3.2). This type of formula gives us an even higher level of structural detail, showing how all of the atoms are attached within a molecule. The lines between atoms represent the chemical bonds that hold atoms together in this molecule [▶ Chapters 9 and 10].

■ **Standard Colors for Atoms in Molecular Models**
The colors listed here are used in this book and are generally used by chemists. The colors of some common atoms are:

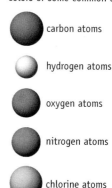

carbon atoms

hydrogen atoms

oxygen atoms

nitrogen atoms

chlorine atoms

■ **Isomers**
Compounds having the same molecular formula but different structures are called *isomers*. (*See Chapter 11 and General ChemistryNow Screen 3.4 Representing Compounds.*)

Example 3.1—Molecular Formulas

Problem The acrylonitrile molecule is the building block for acrylic plastics (such as Orlon and Acrilan). Its structural formula is shown here. What is the molecular formula for acrylonitrile?

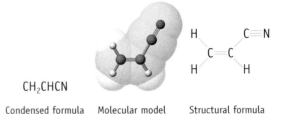

CH$_2$CHCN

Condensed formula   Molecular model   Structural formula

Strategy Count the number of atoms of each type.

Solution Acrylonitrile has three C atoms, three H atoms, and one N atom. Therefore, its molecular formula is C$_3$H$_3$N.

Comment When writing molecular formulas of organic compounds (compounds with C, H, and other elements) the convention is to write C first, then H, and finally other elements in alphabetical order.

Exercise 3.1—Molecular Formulas

The styrene molecule is the building block of polystyrene, a material used for drinking cups and building insulation. What is the molecular formula of styrene?

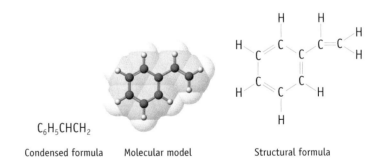

C$_6$H$_5$CHCH$_2$

Condensed formula   Molecular model   Structural formula

3.2—Molecular Models

Molecular structures are often beautiful in the same sense that art is beautiful. For example, there is something intrinsically beautiful about the pattern created by water molecules assembled in ice (Figure 3.3).

More important, however, is the fact that the physical and chemical properties of a molecular compound are often closely related to its structure. For example, two well-known features of ice are easily related to its structure. The first is the shape of ice crystals: The sixfold symmetry of macroscopic ice crystals also appears at the particulate level in the form of six-sided rings of hydrogen and oxygen atoms. The second is water's unique property of being less dense when solid than it is when liquid. The lower density of ice, which has enormous consequences for earth's climate, results from the fact that molecules of water are not packed together tightly.

Figure 3.3 **Ice.** Snowflakes are six-sided structures, reflecting the underlying structure of ice. Ice consists of six-sided rings formed by water molecules, in which each side of a ring consists of two O atoms and an H atom.

Because molecules are three-dimensional, it is often difficult to represent their shapes on paper. Certain conventions have been developed, however, that help represent three-dimensional structures on two-dimensional surfaces. Simple perspective drawings are often used (Figure 3.4).

Wood or plastic models are also a useful way of representing molecular structure. These models can be held in the hand and rotated to view all parts of the molecule.

Several kinds of molecular models exist. In the **ball-and-stick model**, spheres, usually in different colors, represent the atoms, and sticks represent the bonds holding them together. These models make it easy to see how atoms are attached to one another. Molecules can also be represented using **space-filling models**. These models are more realistic because they offer a better representation of relative sizes of atoms and their proximity to each other when in a molecule. A disadvantage of pictures of space-filling models is that atoms can often be hidden from view.

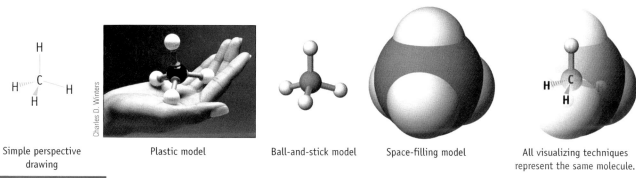

Simple perspective drawing

Plastic model

Ball-and-stick model

Space-filling model

All visualizing techniques represent the same molecule.

Active Figure 3.4 **Ways of depicting the methane (CH_4) molecule.**

Chemistry·ᵢ·Now™ See the General ChemistryNow CD-ROM or website to explore an interactive version of this figure accompanied by an exercise.

A Closer Look

Computer Resources for Molecular Modeling

With the availability of relatively low-cost, high-powered computers, the use of molecular modeling programs has become common. Although the computer screen is two-dimensional, the perspective drawings obtained from molecular-modeling programs are usually quite good. In addition, most programs offer an option to rotate the model on the computer screen to allow the viewer to see the structure from any desired angle. Both ball-and-stick and space-filling representations can be portrayed. Most of the drawings in this book were prepared with the commercial molecular modeling software from CAChe/Fujitsu.

The General ChemistryNow CD-ROM includes a program for visualizing molecules and for measuring atom–atom distances and angles.

The site on the World Wide Web for this textbook (**http://www.brookscole.com**) contains a link to RasMol and Chime, molecular visualization software. Models of many of the compounds mentioned in this book are available through the General ChemistryNow CD-ROM and website. You can visualize these molecules using the software on the CD-ROM, or, if you download RasMol or Chime and configure your browser properly, you can download files from the Internet will that allow you to visualize these models on your own computer.

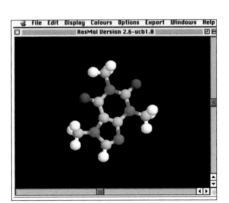
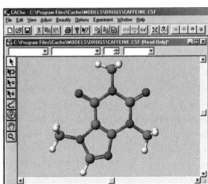

A model of caffeine as viewed with RasMol (*left*) and the CAChe/Fujitsu software (*right*).

Example 3.2—Using Molecular Models

Problem A model of uracil, an important biological molecule, is given here. Write its molecular formula.

Molecular model

Strategy The standard color codes used for the atoms are as follows: carbon atoms = gray; hydrogen atoms = white; nitrogen atoms = blue; and oxygen atoms = red.

Solution Uracil has four C atoms, four H atoms, two N atoms, and two O atoms, giving a formula of $C_4H_4N_2O_2$.

Exercise 3.2—Formulas of Molecules

Cysteine, whose molecular model and structural formula are illustrated here, is an important amino acid and a constituent of many living things. What is its molecular formula? See Example 3.2 and page 99 for the color coding of the model.

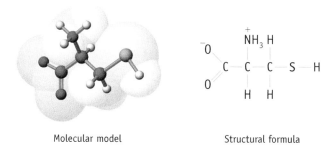

Molecular model   Structural formula

3.3—Ionic Compounds: Formulas, Names, and Properties

The compounds you have encountered so far in this chapter are molecular compounds—that is, compounds that consist of discrete molecules at the particulate level. **Ionic compounds** constitute another major class of compounds. They consist of *ions*, atoms or groups of atoms that bear a positive or negative electric charge. Many familiar compounds are composed of ions (Figure 3.5). Table salt, or sodium chloride ($NaCl$), and lime (CaO) are just two. To recognize ionic compounds, and to be able to write formulas for these compounds, it is important to know the formulas and charges of common ions. You also need to know the names of ions and be able to name the compounds they form.

GENERAL
Chemistry•⚛•Now™

See the General ChemistryNow CD-ROM or website:

- **Screen 3.5 Ions,** for tutorials on determining the number of protons and electrons in an ion and determining ionic charge

Common Name	Name	Formula	Ions Involved
Calcite	Calcium carbonate	$CaCO_3$	Ca^{2+}, CO_3^{2-}
Fluorite	Calcium fluoride	CaF_2	Ca^{2+}, F^-
Gypsum	Calcium sulfate dihydrate	$CaSO_4 \cdot 2\,H_2O$	Ca^{2+}, SO_4^{2-}
Hematite	Iron(III) oxide	Fe_2O_3	Fe^{3+}, O^{2-}
Orpiment	Arsenic sulfide	As_2S_3	As^{3+}, S^{2-}

Hematite, Fe_2O_3
Calcite, $CaCO_3$
Gypsum, $CaSO_4 \cdot 2\,H_2O$
Fluorite, CaF_2
Orpiment, As_2S_3

Charles D. Winters

Figure 3.5 Some common ionic compounds.

Ions

Atoms of many elements can lose or gain electrons in the course of a chemical reaction. To be able to predict the outcome of chemical reactions [▶ Section 5.6], you need to know whether an element will likely gain or lose electrons and, if so, how many.

Cations

If an atom loses an electron (which is transferred to an atom of another element in the course of a reaction), the atom now has one fewer negative electrons than it has positive protons in the nucleus. The result is a positively charged ion called a **cation** (see Figure 3.6). (The name is pronounced "cat'-ion.") Because it has an excess of one positive charge, we write the cation's symbol as, for example, Li^+:

$$Li\ atom \longrightarrow e^- + Li^+\ cation$$
$$(3\ protons\ and\ 3\ electrons) \quad (3\ protons\ and\ 2\ electrons)$$

■ **Writing Ion Formulas**
When writing the formula of an ion, the charge on the ion must be included.

Anions

Conversely, if an atom gains one or more electrons, there is now one or more negatively charged electrons than protons. The result is an **anion** (see Figure 3.6). (The name is pronounced "ann'-ion.")

$$O\ atom + 2\ e^- \longrightarrow O^{2-}\ anion$$
$$(8\ protons\ and\ 8\ electrons) \quad (8\ protons\ and\ 10\ electrons)$$

Here the O atom has gained two electrons so we write the anion's symbol as O^{2-}.

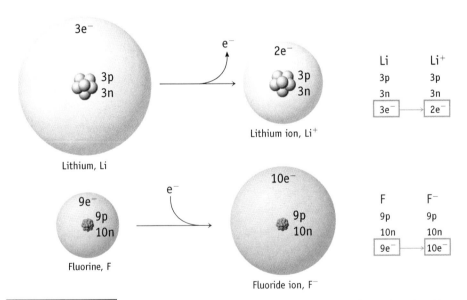

Active Figure 3.6   **Ions.** A lithium-6 atom is electrically neutral because the number of positive charges (three protons) and negative charges (three electrons) are the same. When it loses one electron, it has one more positive charge than negative charge, so it has a net charge of 1+. We symbolize the resulting lithium cation as Li^+. A fluorine atom is also electrically neutral, having nine protons and nine electrons. A fluorine atom can acquire an electron to produce a F^- anion. This anion has one more electron than it has protons, so it has a net charge of 1−.

GENERAL
Chemistry ⚛ Now™ See the General ChemistryNow CD-ROM or website to explore an interactive version of this figure accompanied by an exercise.

How do you know whether an atom is likely to form a cation or an anion? It depends on whether the element is a metal or a nonmetal.

- Metals generally lose electrons in the course of their reactions to form cations.
- Nonmetals frequently gain one or more electrons to form anions in the course of their reactions.

Monatomic Ions

Monatomic ions are single atoms that have lost or gained electrons. As indicated in Figure 3.7, metals typically lose electrons to form monatomic cations, and nonmetals typically gain electrons to form monatomic anions.

How can you predict the number of electrons gained or lost? Typical charges on such ions are indicated in Figure 3.7. *Metals of Groups 1A–3A form positive ions having a charge equal to the group number of the metal.*

Group	Metal Atom	Electron Change	Resulting Metal Cation
1A	Na (11 protons, 11 electrons)	$-1 \longrightarrow$	Na^+ (11 protons, 10 electrons)
2A	Ca (20 protons, 20 electrons)	$-2 \longrightarrow$	Ca^{2+} (20 protons, 18 electrons)
3A	Al (13 protons, 13 electrons)	$-3 \longrightarrow$	Al^{3+} (13 protons, 10 electrons)

Transition metals (B-group elements) also form cations. Unlike the A-group metals, however, no easily predictable pattern of behavior occurs for transition metal cations. In addition, many transition metals form several different ions. An iron-containing compound, for example, may contain either Fe^{2+} or Fe^{3+} ions. Indeed, 2+ and 3+ ions are typical of many transition metals (see Figure 3.7).

Group	Metal Atom	Electron Change	Resulting Metal Cation
7B	Mn (25 protons, 25 electrons)	$-2 \longrightarrow$	Mn^{2+} (25 protons, 23 electrons)
8B	Fe (26 protons, 26 electrons)	$-2 \longrightarrow$	Fe^{2+} (26 protons, 24 electrons)
8B	Fe (26 protons, 26 electrons)	$-3 \longrightarrow$	Fe^{3+} (26 protons, 23 electrons)

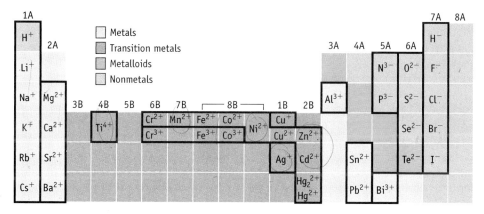

Figure 3.7 Charges on some common monatomic cations and anions. Metals usually form cations and nonmetals usually form anions. (The boxed areas show ions of identical charge.)

Nonmetals often form ions having a negative charge equal to 8 minus the group number of the element. For example, nitrogen is in Group 5A, so it forms an ion having a charge of 3− because a nitrogen atom can gain three electrons.

Group	Nonmetal Atom	Electron Change	Resulting Nonmetal Anion
5A	N (7 protons, 7 electrons)	+3 (= 8 − 5) ⟶	N^{3-} (7 protons, 10 electrons)
6A	S (16 protons, 16 electrons)	+2 (= 8 − 6) ⟶	S^{2-} (16 protons, 18 electrons)
7A	Br (35 protons, 35 electrons)	+1 (= 8 − 7) ⟶	Br^- (35 protons, 36 electrons)

Notice that hydrogen appears at two locations in Figure 3.7. The H atom can either lose or gain electrons, depending on the other atoms it encounters.

Electron lost: H (1 proton, 1 electron) ⟶ H^+ (1 proton, 0 electrons) + e^-
Electron gained: H (1 proton, 1 electron) + e^- ⟶ H^- (1 proton, 2 electrons)

Finally, the noble gases do not form monatomic cations or anions in chemical reactions.

Ion Charges and the Periodic Table

As illustrated in Figure 3.7, the metals of Groups 1A, 2A, and 3A form ions having 1+, 2+, and 3+ charges; that is, their atoms lose one, two, or three electrons, respectively. *For cations formed from A-group elements, the number of electrons remaining on the ion is the same as the number of electrons in an atom of the noble gas that precedes it in the periodic table.* For example, Mg^{2+} has 10 electrons, the same number as in an atom of the noble gas neon (atomic number 10).

An atom of a nonmetal near the right side of the periodic table would have to lose a great many electrons to achieve the same number as a noble gas atom of lower atomic number. (For instance, Cl, whose atomic number is 17, would have to lose 7 electrons to have the same number of electrons as Ne.) If a nonmetal atom were to gain just a few electrons, however, it would have the same number as a noble gas atom of higher atomic number. For example, an oxygen atom has eight electrons. By gaining two electrons per atom it forms O^{2-}, which has ten electrons, the same number as neon. *Anions having the same number of electrons as the noble gas atom succeeding it in the periodic table are commonly observed in chemical compounds.*

■ **Cation Charges and the Periodic Table**

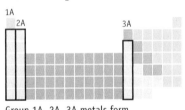

Group 1A, 2A, 3A metals form M^{n+} cations where n = group number.

Exercise 3.3—Predicting Ion Charges

Predict formulas for monatomic ions formed from (a) K, (b) Se, (c) Ba, and (d) Cs. In each case indicate the number of electrons gained or lost by an atom of the element in forming the anion or cation, respectively. For each ion, indicate the noble gas atom having the same total number of electrons.

Polyatomic Ions

Polyatomic ions are made up of two or more atoms, and the collection has an electric charge (Figure 3.8 and Table 3.1). For example, carbonate ion, CO_3^{2-}, a common polyatomic anion, consists of one C atom and three O atoms. The ion has two units of negative charge because there are two more electrons (a total of 32) in the ion than there are protons (a total of 30) in the nuclei of one C atom and three O atoms.

A common polyatomic cation is NH_4^+, the ammonium ion. In this case, four H atoms surround an N atom, and the ion has a 1+ electric charge. This ion has ten

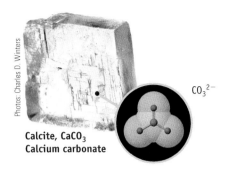

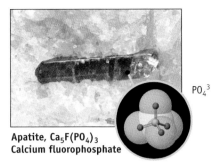

CO_3^{2-}

Calcite, $CaCO_3$
Calcium carbonate

PO_4^{3-}

Apatite, $Ca_5F(PO_4)_3$
Calcium fluorophosphate

SO_4^{2-}

Celestite, $SrSO_4$
Strontium sulfate

Active Figure 3.8 Common ionic compounds based on polyatomic ions.

Chemistry Now™ See the General ChemistryNow CD-ROM or website to explore an interactive version of this figure accompanied by an exercise.

electrons, but there are 11 positively charged protons in the nuclei of the N and H atoms (seven and one each, respectively).

Table 3.1 Formulas and Names of Some Common Polyatomic Ions

Formula	Name	Formula	Name
CATION: Positive Ion			
NH_4^+	ammonium ion		
ANIONS: Negative Ions			
Based on a Group 4A element		**Based on a Group 7A element**	
CN^-	cyanide ion	ClO^-	hypochlorite ion
$CH_3CO_2^-$	acetate ion	ClO_2^-	chlorite ion
CO_3^{2-}	carbonate ion	ClO_3^-	chlorate ion
HCO_3^-	hydrogen carbonate ion (or bicarbonate ion)	ClO_4^-	perchlorate ion
Based on a Group 5A element		**Based on a transition metal**	
NO_2^-	nitrite ion	CrO_4^{2-}	chromate ion
NO_3^-	nitrate ion	$Cr_2O_7^{2-}$	dichromate ion
PO_4^{3-}	phosphate ion	MnO_4^-	permanganate ion
HPO_4^{2-}	hydrogen phosphate ion		
$H_2PO_4^-$	dihydrogen phosphate ion		
Based on a Group 6A element			
OH^-	hydroxide ion		
SO_3^{2-}	sulfite ion		
SO_4^{2-}	sulfate ion		
HSO_4^-	hydrogen sulfate ion (or bisulfate ion)		

Formulas of Ionic Compounds

Ionic compounds are composed of ions. For an ionic compound to be electrically neutral—to have no net charge—the numbers of positive and negative ions must be such that the positive and negative charges balance. In sodium chloride, the sodium ion has a $1+$ charge (Na^+) and the chloride ion has a $1-$ charge (Cl^-). These ions must be present in a $1:1$ ratio, and the formula is NaCl.

Photos: Charles D. Winters

Balancing Ion Charges

Aluminum, a metal in Group 3A, loses three electrons to form the Al^{3+} cation. Oxygen, a nonmetal in Group 6A, gains two electrons to form an O^{2-} anion. Notice that the charge on the cation is the subscript on the anion, and vice versa.

$$2\ Al^{3+} + 3\ O^{2-} \longrightarrow Al_2O_3$$

This often works well, but be careful. The subscripts in $Ti^{4+} + O^{2-}$ are reduced to the simplest ratio (1 Ti to 2 O, rather than, 2 Ti to 4 O).

$$Ti^{4+} + 2\ O^{2-} \longrightarrow TiO_2$$

The gem ruby is largely the compound formed from aluminum ions (Al^{3+}) and oxide ions (O^{2-}). Here the ions have positive and negative charges that are of different absolute value. To have a compound with the same number of positive and negative charges, two Al^{3+} ions [total charge = $2 \times (3+) = 6+$] must combine with three O^{2-} ions [total charge = $3 \times (2-) = 6-$] to give a formula of Al_2O_3.

Calcium is a Group 2A metal, and it forms a cation having a 2+ charge. It can combine with a variety of anions to form ionic compounds such as those in the following table:

Compound	Ion Combination	Overall Charge on Compound
$CaCl_2$	$Ca^{2+} + 2\ Cl^-$	$(2+) + 2 \times (1-) = 0$
$CaCO_3$	$Ca^{2+} + CO_3^{2-}$	$(2+) + (2-) = 0$
$Ca_3(PO_4)_2$	$3\ Ca^{2+} + 2\ PO_4^{3-}$	$3 \times (2+) + 2 \times (3-) = 0$

In writing formulas, the convention is that the symbol of the cation is given first, followed by the anion symbol. Also notice the use of parentheses when more than one polyatomic ion is present.

Example 3.3—Ionic Compound Formulas

Problem For each of the following ionic compounds, write the symbols for the ions present and give the number of each: (a) $MgBr_2$, (b) Li_2CO_3, and (c) $Fe_2(SO_4)_3$.

Strategy Divide the formula of the compound into the cation and the anion. To accomplish this you will have to recognize, and remember, the composition and charges of common ions.

Solution

(a) $MgBr_2$ is composed of one Mg^{2+} ion and two Br^- ions. When a halogen such as bromine is combined only with a metal, you can assume the halogen is an anion with a charge of 1−. Magnesium is a metal in Group 2A and always has a charge of 2+ in its compounds.

(b) Li_2CO_3 is composed of two lithium ions, Li^+, and one carbonate ion, CO_3^{2-}. Li is a Group 1A element and always has a 1+ charge in its compounds. Because the two 1+ charges balance the negative charge of the carbonate ion, the latter must be 2−.

(c) $Fe_2(SO_4)_3$ contains two iron ions, Fe^{3+}, and three sulfate ions, SO_4^{2-}. The way to recognize this is to recall that sulfate has a 2− charge. Because three sulfate ions are present (with a total charge of 6−), the two iron cations must have a total charge of 6+. This is possible only if each iron cation has a charge of 3+.

Comment Remember that the formula for an ion must include its composition and its charge. Formulas for ionic compounds are *always* written with the cation first and then the anion.

Example 3.4—Ionic Compound Formulas

Problem Write formulas for ionic compounds composed of an aluminum cation and each of the following anions: (a) fluoride ion, (b) sulfide ion, and (c) nitrate ion.

Strategy First decide on the formula of the Al cation and the formula of each anion. Combine the Al cation with each type of anion to form an electrically neutral compound.

Solution An aluminum cation is predicted to have a charge of 3+ because Al is a metal in Group 3A.

(a) Fluorine is a Group 7A element. The charge of the fluoride ion is predicted to be 1− (from 8 − 7 = 1). Therefore, we need 3 F^- ions to combine with one Al^{3+}. The formula of the compound is AlF_3.

(b) Sulfur is a nonmetal in Group 6A, so it forms a 2− anion. Thus, we need to combine two Al^{3+} ions [total charge is 6+ = 2 × (3+)] with three S^{2-} ions [total charge is 6− = 3 × (2−)]. The compound has the formula Al_2S_3.

(c) The nitrate ion has the formula NO_3^- (see Table 3.1). The answer here is therefore similar to the AlF_3 case, and the compound has the formula $Al(NO_3)_3$. Here we place parentheses around NO_3 to show that three polyatomic NO_3^- ions are involved.

Comment The most common error students make is not knowing the correct charge on an ion.

Exercise 3.4—Formulas of Ionic Compounds

(a) Give the number and identity of the constituent ions in each of the following ionic compounds: NaF, $Cu(NO_3)_2$, and $NaCH_3CO_2$.

(b) Iron, a transition metal, forms ions having at least two different charges. Write the formulas of the compounds formed between two different iron cations and chloride ions.

(c) Write the formulas of all neutral ionic compounds that can be formed by combining the cations Na^+ and Ba^{2+} with the anions S^{2-} and PO_4^{3-}.

Names of Ions

Naming Positive Ions (Cations)

With a few exceptions (such as NH_4^+), the positive ions described in this text are metal ions. Positive ions are named by the following rules:

1. For a monatomic positive ion (that is, a metal cation) the name is that of the metal plus the word "cation." For example, we have already referred to Al^{3+} as the aluminum cation.

2. Some cases occur, especially in the transition series, in which a metal can form more than one type of positive ion. In these cases the charge of the ion is indicated by a Roman numeral in parentheses immediately following the ion's name. For example, Co^{2+} is the cobalt(II) cation, and Co^{3+} is the cobalt(III) cation.

Finally, you will encounter the ammonium cation, NH_4^+, many times in this book and in the laboratory. Do not confuse the ammonium cation with the ammonia molecule, NH_3, which has no electric charge and one less H atom.

■ **"-ous" and "-ic" Endings**
An older naming system for metal ions uses the ending *-ous* for the ion of lower charge and *-ic* for the ion of higher charge. For example, there are cobaltous (Co^{2+}) and cobaltic (Co^{3+}) ions, and ferrous (Fe^{2+}) and ferric (Fe^{3+}) ions. We do not use this system in this book, but some chemical manufacturers continue to use it.

Problem-Solving Tip 3.1

Formulas for Ions and Ionic Compounds

Writing formulas for ionic compounds takes practice, and it requires that you know the formulas and charges of the most common ions. The charges on monatomic ions are often evident from the position of the element in the periodic table, but you simply have to remember the formula and charges of polyatomic ions; especially the most common ones such as nitrate, sulfate, carbonate, phosphate, and acetate.

If you cannot remember the formula of a polyatomic ion, or if you encounter an ion you have not seen before, you may be able to figure out its formula and the name of one of its compounds. For example, suppose you are told that $NaCHO_2$ is sodium formate. You know that the sodium ion is Na^+, so the formate ion must be the remaining portion of the compound; it must have a charge of 1− to balance the 1+ charge on the sodium ion. Thus, the formate ion must be CHO_2^-.

Finally, when writing the formulas of ions, you must include the charge on the ion (except in an ionic compound formula). Writing Na when you mean sodium ion is incorrect. There is a vast difference in the properties of the element sodium (Na) and those of its ion (Na^+).

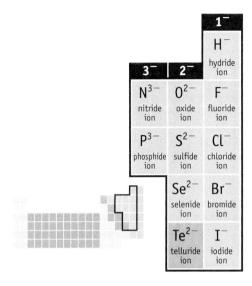

Figure 3.9   Names and charges of some common monatomic anions.

Naming Negative Ions (Anions)

There are two types of negative ions: those having only one atom (*monatomic*) and those having several atoms (*polyatomic*).

1. A monatomic negative ion is named by adding *-ide* to the stem of the name of the nonmetal element from which the ion is derived (Figure 3.9). The anions of the Group 7A elements, the halogens, are known as the fluoride, chloride, bromide, and iodide ions and as a group are called **halide ions**.

2. Polyatomic negative ions are common, especially those containing oxygen (called **oxoanions**). The names of some of the most common oxoanions are given in Table 3.1. Although most of these names must simply be learned, some guidelines can help. For example, consider the following pairs of ions:

NO_3^- is the nitrate ion; NO_2^- is the nitrite ion.

SO_4^{2-} is the sulfate ion; SO_3^{2-} is the sulfite ion.

■ **Naming Oxoanions**

The oxoanion having the *greater number of oxygen atoms* is given the suffix *-ate*, and the oxoanion having the *smaller number of oxygen atoms* has the suffix *-ite*. For a series of oxoanions having more than two members, the ion with the largest number of oxygen atoms has the prefix *per-* and the suffix *-ate*. The ion having the smallest number of oxygen atoms has the prefix *hypo-* and the suffix *-ite*. The oxoanions containing chlorine are good examples.

ClO_4^-	*per*chlor*ate* ion
ClO_3^-	chlor*ate* ion
ClO_2^-	chlor*ite* ion
ClO^-	*hypo*chlor*ite* ion

Oxoanions that contain hydrogen are named by adding the word "hydrogen" before the name of the oxoanion. If two hydrogens are in the compound, we say "dihydrogen." Many hydrogen-containing oxoanions have common names that are used as well. For example, the hydrogen carbonate ion, HCO_3^-, is called the bicarbonate ion.

Ion	Systematic Name	Common Name
HPO_4^{2-}	hydrogen phosphate ion	
$H_2PO_4^-$	dihydrogen phosphate ion	
HCO_3^-	hydrogen carbonate ion	bicarbonate ion
HSO_4^-	hydrogen sulfate ion	bisulfate ion
HSO_3^-	hydrogen sulfite ion	bisulfite ion

Names of Ionic Compounds

The name of an ionic compound is built from the names of the positive and negative ions in the compound. The name of the positive cation is given first, followed by the name of the negative anion. If an element such as titanium can form cations with more than one charge, the charge is indicated by a Roman numeral. Examples of ionic compound names are given below.

Ionic Compound	Ions Involved	Name
$CaBr_2$	Ca^{2+} and 2 Br^-	calcium bromide
$NaHSO_4$	Na^+ and HSO_4^-	sodium hydrogen sulfate
$(NH_4)_2CO_3$	2 NH_4^+ and CO_3^{2-}	ammonium carbonate
$Mg(OH)_2$	Mg^{2+} and 2 OH^-	magnesium hydroxide
$TiCl_2$	Ti^{2+} and 2 Cl^-	titanium(II) chloride
Co_2O_3	2 Co^{3+} and 3 O^{2-}	cobalt(III) oxide

GENERAL
Chemistry Now™

See the General ChemistryNow CD-ROM or website:
- **Screen 3.6 Polyatomic Ions,** for a tutorial on the names of polyatomic ions
- **Screen 3.9 Naming Ionic Compounds,** for a tutorial on naming ionic compounds

---

Exercise 3.5—Names of Ionic Compounds

1. Give the formula for each of the following ionic compounds. Use Table 3.1 and Figure 3.9.

(a) ammonium nitrate   **(d)** vanadium(III) oxide
(b) cobalt(II) sulfate   **(e)** barium acetate
(c) nickel(II) cyanide   **(f)** calcium hypochlorite

2. Name the following ionic compounds:

(a) $MgBr_2$ **(d)** $KMnO_4$
(b) Li_2CO_3 **(e)** $(NH_4)_2S$
(c) $KHSO_3$ **(f)** $CuCl$ and $CuCl_2$

Properties of Ionic Compounds

What is the "glue" that causes ions of opposite electric charge to be held together and to form an orderly arrangement of ions in an ionic compound? As described in

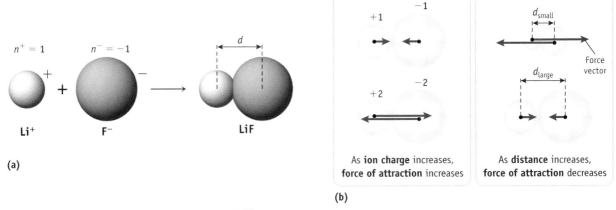

(a)

(b)

Active Figure 3.10   Coulomb's law and electrostatic forces. (a) Ions such as Li^+ and F^- are held together by an electrostatic force. Here a lithium ion is attracted to a fluoride ion, and the distance between the nuclei of the two ions is d. (b) Forces of attraction between ions of opposite charge increase with increasing ion charge and decrease with increasing distance (d). (*The force of attraction is proportional to the length of the arrow in this figure.*)

GENERAL
Chemistry Now™ See the General ChemistryNow CD-ROM or website to explore an interactive version of this figure accompanied by an exercise.

Section 2.1, when a substance having a negative electric charge is brought near a substance having a positive electric charge, a force of attraction occurs between them (Figure 3.10). In contrast, a force of repulsion occurs when two substances with the same charge—both positive or both negative—are brought together. These forces are called **electrostatic forces**, and the force of attraction or repulsion between ions is given by **Coulomb's law** (Equation 3.1).

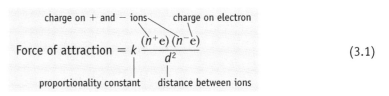

$$\text{Force of attraction} = k \frac{(n^+\text{e})\,(n^-\text{e})}{d^2} \tag{3.1}$$

charge on + and − ions · charge on electron · proportionality constant · distance between ions

where, for example, n^+ is 3 for Al^{3+} and n^- is 2 for O^{2-}. Based on Coulomb's law, the force of attraction between oppositely charged ions increases

- As the ion charges (n^+ and n^-) increase. Thus, the attraction between ions having charges of 2+ and 2− is greater than that between ions having 1+ and 1− charges (see Figure 3.10).
- As the distance between the ions becomes smaller [Figure 3.10; ▶ Chapter 9].

Ionic compounds do not consist of simple pairs or small groups of positive and negative ions. The simplest ratio of cations to anions in an ionic compound is represented by its formula, but an ionic solid consists of millions upon millions of ions arranged in an extended three-dimensional network called a **crystal lattice**. A portion of the lattice for NaCl, illustrated in Figure 3.11, represents a common way of arranging ions for compounds that have a 1 : 1 ratio of cations to anions.

Ionic compounds have characteristic properties that can be understood in terms of the charges of the ions and their arrangement in the lattice. Because each ion is surrounded by oppositely charged nearest neighbors, it is held tightly in its allotted location. At room temperature each ion can move just a bit around its aver-

Figure 3.11   Sodium chloride. A crystal of NaCl consists of an extended lattice of sodium ions and chloride ions in a 1:1 ratio. (*See General ChemistryNow Screen 3.8 Ionic Compounds, to view an animation on the formation of a sodium chloride crystal lattice.*)

Photo: Charles D. Winters; model, S. M. Young.

Problem-Solving Tip 3.2

Is a Compound Ionic?

Students often ask how to know whether a compound is ionic. No method works all of the time, but here are some useful guidelines.

1. Most metal-containing compounds are ionic. So, if a metal atom appears in the formula of a compound, a good first guess is that it is ionic. (There are interesting exceptions, but few come up in introductory chemistry.) It is helpful in this regard to recall trends in metallic behavior: All elements to the left of a diagonal line running from boron to tellurium in the periodic table are metallic.

2. If there is no metal in the formula, it is likely that the compound is not ionic. The exceptions here are compounds composed of polyatomic ions based on nonmetals (e.g., NH_4Cl or NH_4NO_3).

3. Learn to recognize the formulas of polyatomic ions (see Table 3.1). Chemists write the formula of ammonium nitrate as NH_4NO_3 (not as $N_2H_4O_3$) to alert others to the fact that it is an ionic compound composed of the common polyatomic ions NH_4^+ and NO_3^-.

As an example of these guidelines, you can be sure that $MgBr_2$ (Mg^{2+} with Br^-) and K_2S (K^+ with S^{2-}) are ionic compounds. On the other hand, the compound CCl_4, formed from two nonmetals, C and Cl, is not ionic.

---

age position. However, considerable energy must be added before an ion can move fast enough and far enough to escape the attraction of its neighboring ions. Only if enough energy is added will the lattice structure collapse and the substance melt. Greater attractive forces mean that ever more energy—higher and higher temperatures—is required to cause melting. Thus, Al_2O_3, a solid composed of Al^{3+} and O^{2-} ions, melts at a much higher temperature (2072 °C) than NaCl (801 °C), a solid composed of Na^+ and Cl^- ions.

Most ionic compounds are "hard" solids. That is, the solids are not pliable or soft. The reason for this characteristic is again related to the lattice of ions. The nearest neighbors of a cation in a lattice are anions, and the force of attraction makes the lattice rigid. However, a blow with a hammer can cause the lattice to cleave cleanly along a sharp boundary. The hammer blow displaces layers of ions just enough to cause ions of like charge to become nearest neighbors. The repulsion between like charges then forces the lattice apart (Figure 3.12).

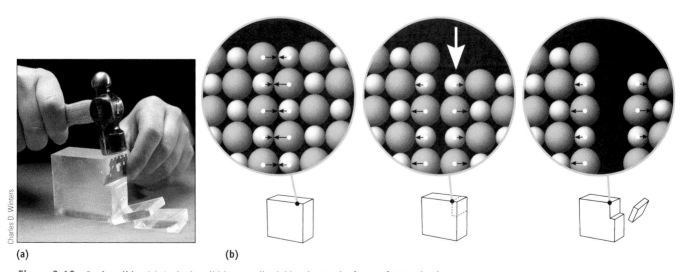

(a) (b)

Figure 3.12 Ionic solids. (a) An ionic solid is normally rigid owing to the forces of attraction between oppositely charged ions. When struck sharply, however, the crystal can cleave cleanly. (b) When a crystal is struck, layers of ions move slightly, and ions of like charge become nearest neighbors. Repulsions between ions of similar charge cause the crystal to cleave. (*See the General ChemistryNow Screen 3.10 Properties of Ionic Compounds, to watch a video of cleaving a crystal.*)

See the General ChemistryNow CD-ROM or website:

- **Screen 3.8 Ionic Compounds,** to watch a video of the sodium + chlorine reaction and for a simulation on the relationship between cations and anions in ionic compounds

---

Exercise 3.6—Coulomb's Law

Explain why the melting point of MgO (2830 °C), much higher than the melting point of NaCl (801 °C).

3.4—Molecular Compounds: Formulas, Names, and Properties

Many familiar compounds are not ionic, they are molecular: the water you drink, the sugar in your coffee or tea, or the aspirin you take for a headache.

Ionic compounds are generally solids, whereas molecular compounds can range from gases to liquids to solids at ordinary temperatures (see Figure 3.13). As size and molecular complexity increase, compounds generally exist as solids. We will explore some of the underlying causes of these general observations in Chapter 13.

Some molecular compounds have complicated formulas that you cannot, at this stage, predict or even decide if they are correct. However, there are many simple compounds you will encounter often, and you should understand how to name them and, in many cases, know their formulas.

Let us look first at molecules formed from combinations of two nonmetals. These "two-element" compounds of nonmetals, often called **binary compounds**, can be named in a systematic way.

Hydrogen forms binary compounds with all of the nonmetals except the noble gases. For compounds of oxygen, sulfur, and the halogens, the H atom is generally

Photo: Charles D. Winters

Figure 3.13 Molecular compounds. Ionic compounds are generally solids at room temperature. In contrast, molecular compounds can be gases, liquids, or solids. The models are of caffeine (in coffee), water, and citric acid (in lemons).

written first in the formula and is named first. The other nonmetal is named as if it were a negative ion.

Compound	Name
HF	hydrogen fluoride
HCl	hydrogen chloride
H_2S	hydrogen sulfide

Virtually all binary molecular compounds of nonmetals are a combination of elements from Groups 4A–7A with one another or with hydrogen. The formula is generally written by putting the elements in order of increasing group number. When naming the compound, the number of atoms of a given type in the compound is designated with a prefix, such as "di-," "tri-," "tetra-," "penta-," and so on.

Compound	Systematic Name
NF_3	nitrogen trifluoride
NO	nitrogen monoxide
NO_2	nitrogen dioxide
N_2O	dinitrogen monoxide
N_2O_4	dinitrogen tetraoxide
PCl_3	phosphorus trichloride
PCl_5	phosphorus pentachloride
SF_6	sulfur hexafluoride
S_2F_{10}	disulfur decafluoride

Finally, many of the binary compounds of nonmetals were discovered years ago and have common names.

Compound	Common Name
CH_4	methane
C_2H_6	ethane
C_3H_8	propane
C_4H_{10}	butane
NH_3	ammonia
N_2H_4	hydrazine
PH_3	phosphine
NO	nitric oxide
N_2O	nitrous oxide ("laughing gas")
H_2O	water

■ **Formulas of Binary Nonmetal Compounds Containing Hydrogen**
Simple hydrocarbons (compounds of C and H) such as methane and ethane have formulas written with H following C, and the formulas of ammonia and hydrazine have H following N. Water and the hydrogen halides, however, have the H atom preceding O or the halogen atom. Tradition is the only explanation for such irregularities in writing formulas.

■ **Hydrocarbons**
Compounds such as methane, ethane, propane, and butane belong to a class of hydrocarbons called alkanes. (*See Chapter 11 and General ChemistryNow Screen 3.13, Alkanes.*)

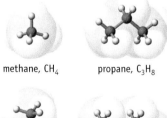

methane, CH_4 propane, C_3H_8

ethane, C_2H_6 butane, C_4H_{10}

Chemistry ⋅꙳⋅ Now™
GENERAL

See the General ChemistryNow CD-ROM or website:

- **Screen 3.12 Binary Compounds of the Nonmetals,** for a tutorial on naming compounds of the nonmetals

- **Screen 3.13 Alkanes,** for a simulation and exercise on naming alkanes

---

Exercise 3.7—Naming Compounds of the Nonmetals

1. Give the formula for each of the following binary, nonmetal compounds:

   (a) carbon dioxide   **(d)** boron trifluoride
   (b) phosphorus triiodide   **(e)** dioxygen difluoride
   (c) sulfur dichloride   **(f)** xenon trioxide

2. Name the following binary, nonmetal compounds:

   (a) N_2F_4 **(c)** SF_4 **(e)** P_4O_{10}
   (b) HBr   **(d)** BCl_3 **(f)** ClF_3

3.5—Formulas, Compounds, and the Mole

The formula of a compound tells you the type of atoms or ions in the compound and the relative number of each. For example, one molecule of methane, CH_4, is made up of one atom of C and four atoms of H. But suppose you have Avogadro's number of C atoms (6.022×10^{23}) combined with the proper number of H atoms. The compound's formula tells us that four times as many H atoms are required ($4 \times 6.022 \times 10^{23}$ H atoms) to give Avogadro's number of CH_4 molecules. What masses of atoms are combined, and what is the mass of this many CH_4 molecules?

C	**+**	**4 H**	$\longrightarrow$	**CH_4**
6.022×10^{23} C atoms		$4 \times 6.022 \times 10^{23}$ H atoms		6.022×10^{23} CH_4 molecules
= 1.000 mol of C		= 4.000 mol of H atoms		= 1.000 mol of CH_4 molecules
= 12.01 g of C atoms		= 4.032 g of H atoms		= 16.04 g of CH_4 molecules

Because we know the number of moles of C and H atoms, we know the masses of carbon and hydrogen that combine to form CH_4. It follows that the mass of CH_4 is the sum of these masses. That is, 1 mol of CH_4 has a mass equivalent to the mass of 1 mol of C atoms (12.01 g) plus 4 mol of H atoms (4.032 g). Thus, the *molar mass*, *M*, of CH_4 is 16.04 g/mol [◄ Section 2.5].

■ **Molar Mass or Molecular Weight**
Although chemists often use the term "molecular weight," we should more properly cite a compound's molar mass. The SI unit of molar mass is kg/mol, but chemists worldwide usually express it in units of g/mol. See "NIST Guide to SI Units" at www.NIST.gov

Molar and Molecular Masses

Element or Compound	Molar Mass, *M* (g/mol)	Average Mass of One Molecule* (g/molecule)
O_2	32.00	5.314×10^{-23}
P_4	123.9	2.057×10^{-22}
NH_3	17.03	2.828×10^{-23}
H_2O	18.02	2.992×10^{-23}
CH_2Cl_2	84.93	1.410×10^{-22}

*See text, page 117, for the calculation of the mass of one molecule.

Ionic compounds such as NaCl do not exist as individual molecules. Thus, we write the simplest formula that shows the relative number of each kind of atom in a "formula unit" of the compound, and the molar mass is calculated from this formula. To differentiate substances like NaCl that do not contain molecules, chemists sometimes refer to their *formula weight* instead of their molecular weight.

Figure 3.14 illustrates 1-mol quantities of several common compounds. To find the molar mass of any compound, you need only add up the atomic masses for each element in one formula unit. As an example, let us find the molar mass of aspirin, $C_9H_8O_4$. In one mole of aspirin there are 9 mol of carbon atoms, 8 mol of hydrogen atoms, and 4 mol of oxygen atoms, which add up to 180.2 g/mol of aspirin.

$$\text{Mass of C in 1 mol } C_9H_8O_4 = 9 \text{ mol C} \times \frac{12.01 \text{ g C}}{1 \text{ mol C}} = 108.1 \text{ g C}$$

$$\text{Mass of H in 1 mol } C_9H_8O_4 = 8 \text{ mol H} \times \frac{1.008 \text{ g H}}{1 \text{ mol H}} = 8.064 \text{ g H}$$

$$\text{Mass of O in 1 mol } C_9H_8O_4 = 4 \text{ mol O} \times \frac{16.00 \text{ g O}}{1 \text{ mol O}} = 64.00 \text{ g O}$$

$$\text{Total mass of 1 mol of } C_9H_8O_4 = \text{molar mass of } C_9H_8O_4 = 180.2 \text{ g}$$

As was the case with elements, it is important to be able to convert the mass of a compound to the equivalent number of moles (or moles to mass) [◀ Section 2.5]. For example, if you take 325 mg (0.325 g) of aspirin in one tablet, what amount of the compound have you ingested? Based on a molar mass of 180.2 g/mol, there are 0.00180 mol of aspirin per tablet.

$$0.325 \text{ g aspirin} \times \frac{1 \text{ mol aspirin}}{180.2 \text{ g aspirin}} = 0.00180 \text{ mol aspirin}$$

Using the molar mass of a compound it is possible to determine the number of molecules in any sample from the sample mass and to determine the mass of one molecule. For example, the number of aspirin molecules in one tablet is

$$0.00180 \text{ mol aspirin} \times \frac{6.022 \times 10^{23} \text{ molecules}}{1 \text{ mol aspirin}} = 1.08 \times 10^{21} \text{ molecules}$$

and the mass of one molecule is

$$\frac{180.2 \text{ g aspirin}}{1 \text{ mol aspirin}} \times \frac{1 \text{ mol aspirin}}{6.022 \times 10^{23} \text{ molecules}} = 2.99 \times 10^{-22} \text{ g/molecule}$$

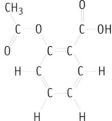

■ **Aspirin Formula**
Aspirin has the molecular formula $C_9H_8O_4$ and a molar mass of 180.2 g/mol. Aspirin is the common name of the compound acetyl-salicylic acid.

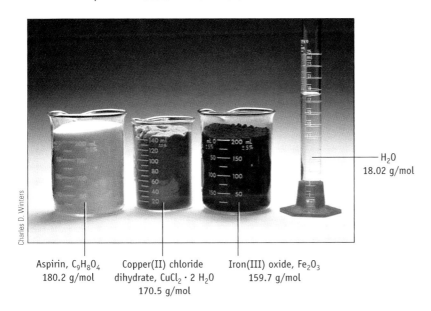

Figure 3.14 **One-mole quantities of some compounds.**

Aspirin, $C_9H_8O_4$
180.2 g/mol

Copper(II) chloride dihydrate, $CuCl_2 \cdot 2\ H_2O$
170.5 g/mol

Iron(III) oxide, Fe_2O_3
159.7 g/mol

H_2O
18.02 g/mol

Charles D. Winters

See the General ChemistryNow CD-ROM or website:

- **Screen 3.14 Compounds, Molecules, and the Mole,** for a simulation exploring the relationship between mass, moles, molecules, and atoms, and a tutorial on determining molar mass
- **Screen 3.15 Using Molar Mass,** for a tutorial on determining moles from mass and a second tutorial on determining mass from moles

Example 3.5—Molar Mass and Moles

Problem You have 16.5 g of oxalic acid, $H_2C_2O_4$.

(a) What amount is represented by 16.5 g of oxalic acid?

(b) How many molecules of oxalic acid are in 16.5 g?

(c) How many atoms of carbon are in 16.5 g of oxalic acid?

(d) What is the mass of one molecule of oxalic acid?

Strategy The first step in any problem involving the conversion of mass and moles is to find the molar mass of the compound in question. Then you can perform the other calculations as outlined by the scheme shown here to find the number of molecules from the amount of substance and the number of atoms of a particular kind:

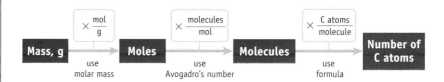

(*See the General ChemistryNow Screen 3.14 Compounds and Moles, and Screen 3.15 Molar Mass.*)

Solution

(a) *Moles represented by 16.5 g*

Let us first calculate the molar mass of oxalic acid:

$$2 \text{ mol C per mol } H_2C_2O_4 \times \frac{12.01 \text{ g C}}{1 \text{ mol C}} = 24.02 \text{ g C per mol } H_2C_2O_4$$

$$2 \text{ mol H per mol } H_2C_2O_4 \times \frac{1.008 \text{ g H}}{1 \text{ mol H}} = 2.016 \text{ g H per mol } H_2C_2O_4$$

$$4 \text{ mol O per mol } H_2C_2O_4 \times \frac{16.00 \text{ g O}}{1 \text{mol O}} = 64.00 \text{ g O per mol } H_2C_2O_4$$

$$\text{Molar mass of } H_2C_2O_4 = 90.04 \text{ g per mol } H_2C_2O_4$$

Now calculate the amount in moles. The molar mass (expressed in units of 1 mol/90.04 g) is the conversion factor in all mass-to-mole conversions.

$$16.5 \text{ g } H_2C_2O_4 \times \frac{1 \text{ mol}}{90.04 \text{ g } H_2C_2O_4} = 0.183 \text{ mol } H_2C_2O_4$$

(b) *Number of molecules*

Use Avogadro's number to find the number of oxalic acid molecules in 0.183 mol of $H_2C_2O_4$.

$$0.183 \text{ mol} \times \frac{6.022 \times 10^{23} \text{ molecules}}{1 \text{ mol}} = 1.10 \times 10^{23} \text{ molecules}$$

(c) *Number of C atoms*

Because each molecule contains two carbon atoms, the number of carbon atoms in 16.5 g of the acid is

$$1.10 \times 10^{23} \text{ molecules} \times \frac{2 \text{ C atoms}}{1 \text{ molecule}} = 2.20 \times 10^{23} \text{ C atoms}$$

(d) *Mass of one molecule*

The units of the desired answer are grams per molecule, which indicates that you should multiply the starting unit of molar mass (grams per mole) by (1/Avogadro's number) (units are mole/molecule), so that the unit "mol" cancels.

$$\frac{90.04 \text{ g}}{1 \text{ mol}} \times \frac{1 \text{ mol}}{6.0221 \times 10^{23} \text{ molecules}} = 1.495 \times 10^{-22} \text{ g/molecule}$$

Exercise 3.8—Molar Mass and Moles-to-Mass Conversions

(a) Calculate the molar mass of citric acid, $C_6H_8O_7$, and $MgCO_3$.

(b) If you have 454 g of citric acid, what amount (moles) does this represent?

(c) To have 0.125 mol of $MgCO_3$, what mass (g) must you have?

3.6—Describing Compound Formulas

Given a sample of an unknown compound, how can its formula be determined? The answer lies in chemical analysis, a major branch of chemistry that deals with the determination of formulas and structures.

Percent Composition

Any sample of a pure compound always consists of the same elements combined in the same proportion by mass. This means *molecular composition* can be expressed in at least three ways:

- In terms of the number of atoms of each type per molecule or per formula unit—that is, by giving the formula of the compound
- In terms of the mass of each element per mole of compound
- In terms of the mass of each element in the compound relative to the total mass of the compound—that is, as a **mass percent**

Suppose you have 1.000 mol of NH_3 or 17.03 g. This mass of NH_3 is composed of 14.01 g of N (1.000 mol) and 3.024 g of H (3.000 mol). If you compare the mass of N to the total mass of compound, 82.27% of the total mass is N (and 17.76% is H).

■ **Molecular Composition**
Molecular composition can be expressed as a percent (mass of an element in a 100-g sample). For example, NH_3 is 82.27% N. Therefore, it has 82.27 g of N in 100.0 g of compound.

82.27% of NH_3 mass
is **nitrogen**.

17.76% of NH_3 mass
is **hydrogen**.

Note that the %N and %H do not add up to exactly 100%. This is not unusual and does not mean there is an error. The last digit of the answer is limited by the accuracy of the data used.

$$\text{Mass of N per mole of NH}_3 = \frac{1 \text{ mol N}}{1 \text{ mol NH}_3} \times \frac{14.01 \text{ g N}}{1 \text{ mol N}} = 14.01 \text{ g N/1 mol NH}_3$$

$$\text{Mass percent N in NH}_3 = \frac{\text{mass of N in 1 mol NH}_3}{\text{mass of 1 mol NH}_3}$$

$$= \frac{14.01 \text{ g N}}{17.03 \text{ g NH}_3} \times 100\%$$

$$= 82.27\% \text{ (or 82.27 g N in 100.0 g NH}_3)$$

$$\text{Mass of H per mole of NH}_3 = \frac{3 \text{ mol H}}{1 \text{ mol NH}_3} \times \frac{1.008 \text{ g H}}{1 \text{ mol H}} = 3.024 \text{ g H/1 mol NH}_3$$

$$\text{Mass percent H in NH}_3 = \frac{\text{mass of H in 1 mol NH}_3}{\text{mass of 1 mol NH}_3} \times 100\%$$

$$= \frac{3.024 \text{ g H}}{17.03 \text{ g NH}_3} \times 100\%$$

$$= 17.76\% \text{ (or 17.76 g H in 100.0 g NH}_3)$$

These values represent the mass percent of each element, or percent composition by mass. They tell you that in a 100.0-g sample there are 82.27 g of N and 17.76 g of H.

GENERAL
Chemistry ⚛ Now™

See the General ChemistryNow CD-ROM or website:
- **Screen 3.16 Percent Composition,** for a tutorial on determining percent composition

---

Example 3.6—Using Percent Composition

Problem What is the mass percent of each element in propane, C_3H_8? What mass of carbon is contained in 454 g of propane?

Strategy First find the molar mass of C_3H_8 and then calculate the mass percent of C and H per mole of C_3H_8. Using the knowledge of the mass percent of C, calculate the mass of carbon in 454 g of C_3H_8.

Solution

(a) The molar mass of C_3H_8 is 44.10 g/mol.

(b) Mass percent of C and H in C_3H_8:

$$\frac{3 \text{ mol C}}{1 \text{ mol C}_3H_8} \times \frac{12.01 \text{ g C}}{1 \text{ mol C}} = 36.03 \text{ g C/1 mol C}_3H_8$$

$$\text{Mass percent of C in C}_3H_8 = \frac{36.03 \text{ g C}}{44.10 \text{ g C}_3H_8} \times 100\% = \textbf{81.70\% C}$$

$$\frac{8 \text{ mol H}}{1 \text{ mol C}_3H_8} \times \frac{1.008 \text{ g H}}{1 \text{ mol H}} = 8.064 \text{ g H/1 mol C}_3H_8$$

$$\text{Mass percent of H in C}_3H_8 = \frac{8.064 \text{ g H}}{44.10 \text{ g C}_3H_8} \times 100\% = \textbf{18.29\% H}$$

(c) Mass of C in 454 g of C_3H_8:

$$454 \text{ g C}_3H_8 \times \frac{81.70 \text{ g C}}{100.0 \text{ g C}_3H_8} = 371 \text{ g C}$$

Exercise 3.9—Percent Composition

(a) Express the composition of ammonium carbonate, $(NH_4)_2CO_3$, in terms of the mass of each element in 1.00 mol of compound and the mass percent of each element.

(b) What is the mass of carbon in 454 g of octane, C_8H_{18}?

Empirical and Molecular Formulas from Percent Composition

Now let us consider the *reverse* of the procedure just described: using relative mass or percent composition data to find a molecular formula. Suppose you know the identity of the elements in a sample and have determined the mass of each element in a given mass of compound by chemical analysis [▶ Section 4.6]. You can then calculate the relative amount (moles) of each element and from this the relative number of atoms of each element in the compound. For example, for a compound composed of atoms of A and B, the steps from percent composition to a formula are

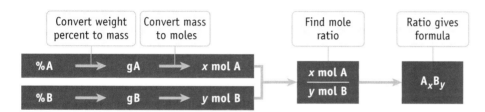

Let us derive the formula for hydrazine, a close relative of ammonia and a compound used to remove oxygen from water used for heating and cooling.

Step 1: *Convert mass percent to mass.* The mass percentages in a sample of hydrazine are 87.42% N and 12.58% H. Thus, in a 100.00-g sample of hydrazine, there are 87.42 g of N and 12.58 g of H.

Step 2: *Convert the mass of each element to moles.* The amount of each element in the 100.00-g sample is

$$87.42 \text{ g N} \times \frac{1 \text{ mol N}}{14.007 \text{ g N}} = 6.241 \text{ mol N}$$

$$12.58 \text{ g H} \times \frac{1 \text{ mol H}}{1.008 \text{ g H}} = 12.48 \text{ mol H}$$

■ **Deriving a Formula**
Percent composition gives the mass of an element in 100 g of sample. However, any amount of sample is appropriate if you know the mass of an element in that sample mass. See Example 3.8.

Step 3: *Find the mole ratio of elements.* Use the amount (moles) of each element in the 100.00-g of sample to find the amount of one element relative to the other. For hydrazine, this ratio is 2 mol of H to 1 mol of N,

$$\frac{12.48 \text{ mol H}}{6.241 \text{ mol N}} = \frac{2.00 \text{ mol H}}{1.00 \text{ mol N}} \longrightarrow NH_2$$

■ **Deriving a Formula—Mole Ratios**
When finding the ratio of moles of one element relative to another, *always* divide the larger number by the smaller one.

showing that there are 2 mol of H atoms for every 1 mol of N atoms in hydrazine. Thus, in one molecule, two atoms of H occur for every atom of N; that is, the formula is NH_2. This simplest whole-number atom ratio is called the **empirical formula**.

Percent composition data allow us to calculate the atom ratios in a compound. A *molecular formula*, however, must convey *two* pieces of information: (1) the relative numbers of atoms of each element in a molecule (the atom ratios) and (2) the total number of atoms in the molecule. For hydrazine there are twice as many H atoms as N atoms, so the molecular formula could be NH_2. Recognize, however, that percent composition data give only the *simplest possible ratio of atoms* in a molecule. The empirical formula of hydrazine is NH_2, but the true *molecular formula* could also be NH_2, N_2H_4, N_3H_6, N_4H_8, or any other formula having a 1:2 ratio of N to H.

To determine the molecular formula from the empirical formula, the molar mass must be obtained from experiment. For example, experiments show that the molar mass of hydrazine is 32.0 g/mol, twice the formula mass of NH_2, which is 16.0 g/mol. Thus, the molecular formula of hydrazine is two times the empirical formula of NH_2, that is, N_2H_4.

As another example of the usefulness of percent composition data, let us say that you collected the following information in the laboratory for isooctane, the compound used as the standard for determining the octane rating of a fuel: % carbon = 84.12; % hydrogen = 15.88; molar mass = 114.2 g/mol. These data can be used to calculate the empirical and molecular formulas for the compound. The data inform us that 84.12 g of C and 15.88 g of H occur in a 100.0-g sample. From this, we find the amount (moles) of each element in this sample.

$$84.12 \text{ g C} \times \frac{1 \text{ mol C}}{12.011 \text{ g C}} = 7.004 \text{ mol C}$$

$$15.88 \text{ g H} \times \frac{1 \text{ mol H}}{1.0079 \text{ g H}} = 15.76 \text{ mol H}$$

This means that, in any sample of isooctane, the ratio of moles of H to C is

$$\text{Mole ratio} = \frac{15.76 \text{ mol H}}{7.004 \text{ mol C}} = \frac{2.250 \text{ mol H}}{1.000 \text{ mol C}}$$

Now the task is to turn this decimal fraction into a whole-number ratio of H to C. To do this, recognize that 2.25 is the same as $2\frac{1}{4}$ or 9/4. Therefore, the ratio of C to H is

$$\text{Mole ratio} = \frac{2.25 \text{ mol H}}{1.00 \text{ mol C}} = \frac{2\frac{1}{4} \text{ mol H}}{1 \text{ mol C}} = \frac{9/4 \text{ mol H}}{1 \text{ mol C}} = \frac{9 \text{ mol H}}{4 \text{ mol C}}$$

■ **Isooctane and the Octane Rating**
Isooctane, C_8H_{18}, is the standard against which the octane rating of gasoline is determined. Octane numbers are assigned by comparing the burning performance of gasoline with the burning performance of mixtures of isooctane and heptane. Gasoline with an octane rating of 90 matches the burning characteristics of a mixture of 90% isooctane and 10% heptane.

You now know that nine H atoms occur for every four C atoms in isooctane. Thus, the simplest or *empirical formula* is C_4H_9. If C_4H_9 were the molecular formula, the molar mass would be 57.12 g/mol. However, we know from experiment that the actual molar mass is 114.2 g/mol, twice the value for the empirical formula.

$$\frac{114.2 \text{ g/mol of isooctane}}{57.12 \text{ g/mol of } C_4H_9} = 2.00 \text{ mol } C_4H_9 \text{ per mol of isooctane}$$

The molecular formula is therefore C_8H_{18}.

GENERAL
Chemistry•ᐧ•**Now**™

See the General ChemistryNow CD-ROM or website:

- **Screen 3.17 Determining Empirical Formulas,** for a tutorial on determining empirical formulas
- **Screen 3.18 Determining Molecular Formulas,** for a tutorial on determining molecular formulas

Example 3.7—Calculating a Formula from Percent Composition

Problem Eugenol is the major component in oil of cloves. It has a molar mass of 164.2 g/mol and is 73.14% C and 7.37% H; the remainder is oxygen. What are the empirical and molecular formulas of eugenol?

Strategy To derive a formula we need to know the mass percent of each element. Because the mass percents of all elements must add up to 100.0%, we find the mass percent of O from the difference between 100.0% and the mass percents of C and H. Next, we assume the mass percent of each element is equivalent to its mass in grams, and convert each mass to moles. Finally, the ratio of moles gives the empirical formula. The mass of a mole of compound having the calculated empirical formula is compared with the actual, experimental molar mass to find the true molecular formula.

Solution The mass of O in a 100.0-g sample is

$$100.0 \text{ g} = 73.14 \text{ g C} + 7.37 \text{ g H} + \text{mass of O}$$

$$\text{Mass of O} = 19.49 \text{ g}$$

The amount of each element is

$$73.14 \text{ g C} \times \frac{1 \text{ mol C}}{12.011 \text{ g C}} = 6.089 \text{ mol C}$$

$$7.37 \text{ g H} \times \frac{1 \text{ mol H}}{1.008 \text{ g H}} = 7.31 \text{ mol H}$$

$$19.49 \text{ g O} \times \frac{1 \text{ mol O}}{15.999 \text{ g O}} = 1.218 \text{ mol O}$$

To find the mole ratio, the best approach is to base the ratios on the smallest number of moles present—in this case, oxygen.

$$\frac{\text{mol C}}{\text{mol O}} = \frac{6.089 \text{ mol C}}{1.218 \text{ mol O}} = \frac{4.999 \text{ mol C}}{1.000 \text{ mol O}} = \frac{5 \text{ mol C}}{1 \text{ mol O}}$$

$$\frac{\text{mol H}}{\text{mol O}} = \frac{7.31 \text{ mol H}}{1.218 \text{ mol O}} = \frac{6.00 \text{ mol H}}{1.000 \text{ mol O}} = \frac{6 \text{ mol H}}{1 \text{ mol O}}$$

Now we know there are 5 mol of C and 6 mol of H per 1 mol of O. Thus, the empirical formula is C_5H_6O.

The experimentally determined molar mass of eugenol is 164.2 g/mol. This is twice the mass of C_5H_6O (82.1 g/mol).

$$\frac{164.2 \text{ g/mol of eugenol}}{82.10 \text{ g/mol of } C_5H_6O} = 2.00 \text{ mol } C_5H_6O \text{ per mol of eugenol}$$

The molecular formula is $C_{10}H_{12}O_2$.

Eugenol, $C_{10}H_{12}O_2$, is an important component in oil of cloves.

Figure 3.16   Gypsum wallboard. Gypsum is hydrated calcium sulfate, $CaSO_4 \cdot 2\,H_2O$.

the existence of isotopes and to measure their relative abundance [◀ Figure 2.8]. If a compound can be turned into a vapor, the vapor can be passed through an electron beam in a mass spectrometer where high-energy electrons collide with the gas-phase molecules. These high-energy collisions cause the molecule to lose electrons and turn the molecules into positive ions. These ions usually break apart or fragment into smaller pieces. As illustrated in Figure 3.15, the cation created from ethanol ($CH_3CH_2OH^+$) fragments (losing an H atom) to give another cation ($CH_3CH_2O^+$), which further fragments. A mass spectrometer detects and records the masses of the different particles. Analysis of the spectrum can help identify a compound and can give an accurate molar mass.

3.7—Hydrated Compounds

If ionic compounds are prepared in water solution and then isolated as solids, the crystals often have molecules of water trapped within the lattice. Compounds in which molecules of water are associated with the ions of the compound are called **hydrated compounds.** The beautiful blue copper(II) compound in Figure 3.14, for example, has a formula that is conventionally written as $CuCl_2 \cdot 2\,H_2O$. The dot between $CuCl_2$ and $2\,H_2O$ indicates that 2 mol of water is associated with every mole of $CuCl_2$; it is equivalent to writing the formula as $CuCl_2(H_2O)_2$. The name of the compound, copper(II) chloride dihydrate, reflects the presence of 2 mol of water per mole of $CuCl_2$. The molar mass of $CuCl_2 \cdot 2\,H_2O$ is 134.5 g/mol (for $CuCl_2$) plus 36.0 g/mol (for $2\,H_2O$) giving a total mass of 170.5 g/mol.

Hydrated compounds are common. The walls of your home may be covered with wallboard, or "plaster board" (Figure 3.16). These sheets contain hydrated calcium sulfate, or gypsum ($CaSO_4 \cdot 2\,H_2O$), as well as unhydrated $CaSO_4$, sandwiched between paper. Gypsum is a mineral that can be mined. Now, however, it is more commonly a byproduct in the manufacture of superphosphate fertilizer from $Ca_5F(PO_4)_3$ and sulfuric acid.

If gypsum is heated between 120 and 180 °C, the water is partly driven off to give $CaSO_4 \cdot \frac{1}{2}H_2O$, a compound commonly called "plaster of Paris." If you have ever broken an arm or leg and had to have a cast, the cast may have been made of this compound. It is an effective casting material because, when added to water, it forms a thick slurry that can be poured into a mold or spread out over a part of the body. As it takes on more water, the material increases in volume and forms a hard, inflexible solid. These properties also make plaster of Paris a useful material to artists, because the expanding compound fills a mold completely and makes a high-quality reproduction.

Hydrated cobalt(II) chloride is the red solid in Figure 3.17. When heated it turns first purple and then deep blue as it loses water to form anhydrous $CoCl_2$; "**anhydrous**" means a substance without water. On exposure to moist air, anhydrous $CoCl_2$ takes up water and is converted back into the red hydrated compound. It is this property that allows crystals of the blue compound to be used as a humidity indicator. You may have seen them in a small bag packed with a piece of electronic equipment. The compound also makes a good "invisible ink." A solution of cobalt(II) chloride in water is red, but if you write on paper with the solution it cannot be seen. When the paper is warmed, however, the cobalt compound dehydrates to give the deep blue anhydrous compound, and the writing becomes visible.

There is no simple way to predict how much water will be present in a hydrated compound, so it must be determined experimentally. Such an experiment may involve heating the hydrated material so that all the water is released from the solid

Active Figure 3.17 **Dehydrating hydrated cobalt(II) chloride, CoCl$_2$ · 6H$_2$O.** (*left*) Cobalt chloride hexahydrate, CoCl$_2$ · 6H$_2$O, is a deep red compound. (*left and center*) When it is heated, the compound loses the water of hydration and forms the deep blue compound CoCl$_2$.

GENERAL
Chemistry Now™ See the General ChemistryNow CD-ROM or website to explore an interactive version of this figure accompanied by and exercise.

and evaporated. Only the anhydrous compound is left. The formula of hydrated copper(II) sulfate, commonly known as "blue vitriol," is determined in this manner in Example 3.10.

GENERAL
Chemistry Now™

See the General ChemistryNow CD-ROM or website:
- **Screen 3.19 Hydrated Compounds,** for a tutorial on detemining mass and moles of compounds of a hydrated compound and an exercise on analyzing a mixture

White CuSO$_4$

Blue CuSO$_4$ · 5 H$_2$O

Heating a Hydrated Compound
The formula of a hydrated compound can be determined by heating a weighed sample enough to cause the compound to release its water of hydration. Knowing the mass of the hydrated compound before heating, and the mass of the anhydrous compound after heating, we can determine the mass of water in the original sample.

Example 3.10—Determining the Formula of a Hydrated Compound

Problem You want to know the value of x in blue, hydrated copper(II) sulfate, CuSO$_4$ · x H$_2$O—that is, the number of water molecules for each unit of CuSO$_4$. In the laboratory you weigh out 1.023 g of the solid. After heating the solid thoroughly in a porcelain crucible (see figure), 0.654 g of nearly white, anhydrous copper(II) sulfate, CuSO$_4$, remains.

$$1.023 \text{ g CuSO}_4 \cdot x \text{ H}_2\text{O} + \xrightarrow{\text{heat}} 0.654 \text{ g CuSO}_4 + ? \text{ g H}_2\text{O}$$

Strategy To find x we need to know the amount of H$_2$O per mole of CuSO$_4$. Therefore, first we find the mass of water lost by the sample from the difference between the mass of hydrated compound and the anhydrous form. Finally, we find the ratio of amount of water lost (moles) to the amount of anhydrous CuSO$_4$.

Solution Find the mass of water.

Mass of hydrated compound	1.023 g
Mass of anhydrous compound, $CuSO_4$	-0.654 g
Mass of water	0.369 g

Next convert the masses of $CuSO_4$ and H_2O to moles.

$$0.369 \text{ g } H_2O \cdot \frac{1 \text{ mol } H_2O}{18.02 \text{ g } H_2O} = 0.0205 \text{ mol } H_2O$$

$$0.654 \text{ g } CuSO_4 \cdot \frac{1 \text{ mol } CuSO_4}{159.6 \text{ g } CuSO_4} = 0.00410 \text{ mol } CuSO_4$$

The value of x is determined from the mole ratio.

$$\frac{0.0205 \text{ mol } H_2O}{0.00410 \text{ mol } CuSO_4} = \frac{5.00 \text{ mol } H_2O}{1.00 \text{ mol } CuSO_4}$$

The water-to-$CuSO_4$ ratio is 5 : 1, so the formula of the hydrated compound is $CuSO_4 \cdot 5\,H_2O$. Its name is **copper(II) sulfate pentahydrate**.

Exercise 3.14—Determining the Formula of a Hydrated Compound

Hydrated nickel(II) chloride is a beautiful green, crystalline compound. When heated strongly, the compound is dehydrated. If 0.235 g of $NiCl_2 \cdot x\,H_2O$ gives 0.128 g of $NiCl_2$ on heating, what is the value of x ?

Chapter Goals Revisited

Now that you have studied this chapter, you should ask whether you have met the chapter goals. In particular, you should be able to

Interpret, predict, and write formulas for ionic and molecular compounds

a. Recognize and interpret molecular formulas, condensed formulas, and structural formulas (Section 3.1).

b. Recognize that metal atoms commonly lose one or more electrons to form positive ions, called cations, and nonmetal atoms often gain electrons to form negative ions, called anions (see Figure 3.7).

c. Recognize that the charge on a metal cation in Groups 1A, 2A, and 3A is equal to the group number in which the element is found in the periodic table (M^{n+}, n = Group number) (Section 3.3). Charges on transition metal cations are often 2+ or 3+, but other charges are observed. General ChemistryNow homework: Study Question(s) 11

d. Recognize that the negative charge on a single-atom or monatomic anion, X^{n-}, is given by n = 8 − group number (Section 3.3).

   e. Write formulas for ionic compounds by combining ions in the proper ratio to give no overall charge (Section 3.4).

Name compounds

   a. Give the names or formulas of polyatomic ions, knowing their formulas or names, respectively (Table 3.1 and Section 3.3).

   b. Name ionic compounds and simple binary compounds of the nonmetals (Sections 3.3 and 3.4). General ChemistryNow homework: SQ(s) 7, 19, 21, 27, 29

Understand some properties of ionic compounds

   a. Understand the importance of Coulomb's law (Equation 3.1), which describes the electrostatic forces of attraction and repulsion of ions. Coulomb's law states that the force of attraction between oppositely charged species increases with electric charge and with decreasing distance between the species (Section 3.3). General ChemistryNow homework: SQ(s) 26

Calculate and use molar mass

   a. Understand that the molar mass of a compound (often called the molecular weight) is the mass in grams of Avogadro's number of molecules (or formula units) of a compound (Section 3.5). For ionic compounds, which do not consist of individual molecules, the sum of atomic masses is often called the formula mass (or formula weight).

   b. Calculate the molar mass of a compound from its formula and a table of atomic weights (Section 3.5). General ChemistryNow homework: SQ(s) 31, 33

   c. Calculate the number of moles of a compound that is represented by a given mass, and vice versa (Section 3.5). General ChemistryNow homework: SQ(s) 35

Calculate percent composition for a compound and derive formulas from experimental data

   a. Express the composition of a compound in terms of percent composition (Section 3.6). General ChemistryNow homework: SQ(s) 41, 45

   b. Use percent composition or other experimental data to determine the empirical formula of a compound (Section 3.6). General ChemistryNow homework: SQ(s) 47, 52, 53, 94

   c. Understand how mass spectrometry can be used to find a molar mass (Section 3.6).

   d. Use experimental data to find the number of water molecules in a hydrated compound (Section 3.7). General ChemistryNow homework: SQ(s) 55, 57. 59

Key Equations

Equation 3.1 (page 112)

Coulomb's law describes the dependence of the force of attraction between ions of opposite charge (or the force of repulsion between ions of like charge) on ion charge and the distance between ions.

$$\text{Force of attraction} = k \, \frac{(n^+e)(n^-e)}{d^2}$$

charge on + and − ions   charge on electron

proportionality constant   distance between ions

Study Questions

▲ denotes more challenging questions.

■ denotes questions available in the Homework and
   Goals section of the General ChemistryNow CD-ROM
   or website.

Blue numbered questions have answers in Appendix O and
   fully worked solutions in the *Student Solutions Manual.*

Structures of many of the compounds used in these
   questions are found on the General ChemistryNow
   CD-ROM or website in the Models folder.

Chemistry ⚛ Now™ Assess your understanding of this
   chapter's topics with additional quizzing and conceptual
   questions at http://now.brookscole.com/kotz6e

Practicing Skills

Molecular Formulas and Models
(See Examples 3.1 and 3.2 and Exercises 3.1 and 3.2.)

1. A ball-and-stick model of sulfuric acid is illustrated here.
   Write the molecular formula for sulfuric acid and draw the
   structural formula. Describe the structure of the molecule.
   Is it flat? That is, are all the atoms in the plane of the paper? (Color code: sulfur atoms are yellow; oxygen atoms
   are red; and hydrogen atoms are white.)

2. A ball-and-stick model of toluene is illustrated here. What
   is its molecular formula? Describe the structure of the
   molecule. Is it flat or is only a portion of it flat? (Color
   code: carbon atoms are gray and hydrogen atoms are
   white.)

3. A model of the cancer chemotherapy agent cisplatin is
   given here. Write the molecular formula for the
   compound and draw its structural formula.

4. The molecule illustrated here is methanol. Using Figure
   3.4 as a guide, decide which atoms are in the plane of the
   paper, which lie above the plane, and which lie below.
   Sketch a ball-and-stick model. If available to you, go to the
   General ChemistryNow CD-ROM or website and find the
   model of methanol.

Ions and Ion Charges
*(See Exercise 3.3, Figure 3.7, Table 3.1, and the General
ChemistryNow Screens 3.5 and 3.6.)*

5. What charges are most commonly observed for
   monatomic ions of the following elements?
   (a) magnesium   (c) nickel
   (b) zinc   (d) gallium

6. What charges are most commonly observed for
   monatomic ions of the following elements?
   (a) selenium   (c) iron
   (b) fluorine   (d) nitrogen

7. ■ Give the symbol, including the correct charge, for each
   of the following ions:
   (a) barium ion
   (b) titanium(IV) ion
   (c) phosphate ion
   (d) hydrogen carbonate ion
   (e) sulfide ion
   (f) perchlorate ion
   (g) cobalt(II) ion
   (h) sulfate ion

8. Give the symbol, including the correct charge, for each of
   the following ions:
   (a) permanganate ion
   (b) nitrite ion
   (c) dihydrogen phosphate ion
   (d) ammonium ion
   (e) phosphate ion
   (f) sulfite ion

9. When a potassium atom becomes a monatomic ion, how many electrons does it lose or gain? What noble gas atom has the same number of electrons as a potassium ion?

10. When oxygen and sulfur atoms become monatomic ions, how many electrons does each lose or gain? Which noble gas atom has the same number of electrons as an oxygen ion? Which noble gas atom has the same number of electrons as a sulfur ion?

Ionic Compounds

(See Examples 3.3 and 3.4 and the General ChemistryNow Screen 3.8.)

11. ■ Predict the charges of the ions in an ionic compound containing the elements barium and bromine. Write the formula for the compound.

12. What are the charges of the ions in an ionic compound containing cobalt(III) and fluoride ions? Write the formula for the compound.

13. For each of the following compounds, give the formula, charge, and the number of each ion that makes up the compound:
    (a) K_2S (d) $(NH_4)_3PO_4$
    (b) $CoSO_4$ (e) $Ca(ClO)_2$
    (c) $KMnO_4$

14. For each of the following compounds, give the formula, charge, and the number of each ion that makes up the compound:
    (a) $Mg(CH_3CO_2)_2$ (d) $Ti(SO_4)_2$
    (b) $Al(OH)_3$ (e) KH_2PO_4
    (c) $CuCO_3$

15. Cobalt forms Co^{2+} and Co^{3+} ions. Write the formulas for the two cobalt oxides formed by these transition metal ions.

16. Platinum is a transition element and forms Pt^{2+} and Pt^{4+} ions. Write the formulas for the compounds of each of these ions with (a) chloride ions and (b) sulfide ions.

17. Which of the following are correct formulas for ionic compounds? For those that are not, give the correct formula.
    (a) $AlCl_2$ (c) Ga_2O_3
    (b) KF_2 (d) MgS

18. Which of the following are correct formulas for ionic compounds? For those that are not, give the correct formula.
    (a) Ca_2O (c) Fe_2O_5
    (b) $SrBr_2$ (d) Li_2O

Naming Ionic Compounds

(See Exercise 3.5 and the General ChemistryNow Screen 3.9.)

19. ■ Name each of the following ionic compounds:
    (a) K_2S (c) $(NH_4)_3PO_4$
    (b) $CoSO_4$ (d) $Ca(ClO)_2$

20. Name each of the following ionic compounds:
    (a) $Ca(CH_3CO_2)_2$ (c) $Al(OH)_3$
    (b) $Ni_3(PO_4)_2$ (d) KH_2PO_4

21. ■ Give the formula for each of the following ionic compounds:
    (a) ammonium carbonate
    (b) calcium iodide
    (c) copper(II) bromide
    (d) aluminum phosphate
    (e) silver(I) acetate

22. Give the formula for each of the following ionic compounds:
    (a) calcium hydrogen carbonate
    (b) potassium permanganate
    (c) magnesium perchlorate
    (d) potassium hydrogen phosphate
    (e) sodium sulfite

23. Write the formulas for the four ionic compounds that can be made by combining each of the cations Na^+ and Ba^{2+} with the anions CO_3^{2-} and I^-. Name each of the compounds.

24. Write the formulas for the four ionic compounds that can be made by combining the cations Mg^{2+} and Fe^{3+} with the anions PO_4^{3-} and NO_3^-. Name each compound formed.

Coulomb's Law

(See Equation 3.1, Figure 3.10, and the General ChemistryNow Screen 3.7.)

25. Sodium ion, Na^+, forms ionic compounds with fluoride, F^-, and iodide, I^-. The radii of these ions are as follows: $Na^+ = 116$ pm; $F^- = 119$ pm; and $I^- = 206$ pm. In which ionic compound, NaF or NaI, are the forces of attraction between cation and anion stronger? Explain your answer.

26. ■ Consider the two ionic compounds NaCl and CaO. In which compound are the cation–anion attractive forces stronger? Explain your answer.

Naming Binary, Nonmetal Compounds

(See Exercise 3.6 and the General ChemistryNow Screen 3.12.)

27. ■ Name each of the following binary, nonionic compounds:
    (a) NF_3 (b) HI   (c) BI_3 (d) PF_5

28. Name each of the following binary, nonionic compounds:
    (a) N_2O_5 (b) P_4S_3 (c) OF_2 (d) XeF_4

29. ■ Give the formula for each of the following compounds:
    (a) sulfur dichloride
    (b) dinitrogen pentaoxide
    (c) silicon tetrachloride
    (d) diboron trioxide (commonly called boric oxide)

---

30. Give the formula for each of the following compounds:
    (a) bromine trifluoride
    (b) xenon difluoride
    (c) hydrazine
    (d) diphosphorus tetrafluoride
    (e) butane

Molecules, Compounds, and the Mole

(See Example 3.5 and the General ChemistryNow Screens 3.14 and 3.15.)

31. ■ Calculate the molar mass of each of the following compounds:
    (a) Fe_2O_3, iron(III) oxide
    (b) BCl_3, boron trichloride
    (c) $C_6H_8O_6$, ascorbic acid (vitamin C)

32. Calculate the molar mass of each of the following compounds:
    (a) $Fe(C_6H_{11}O_7)_2$, iron(II) gluconate, a dietary supplement
    (b) $CH_3CH_2CH_2CH_2SH$, butanethiol, has a skunk-like odor
    (c) $C_{20}H_{24}N_2O_2$, quinine, used as an antimalarial drug

33. ■ Calculate the molar mass of each hydrated compound. Note that the water of hydration is included in the molar mass. (See Section 3.7.)
    (a) $Ni(NO_3)_2 \cdot 6\ H_2O$
    (b) $CuSO_4 \cdot 5\ H_2O$

34. Calculate the molar mass of each hydrated compound. Note that the water of hydration is included in the molar mass. (See Section 3.7.)
    (a) $H_2C_2O_4 \cdot 2\ H_2O$
    (b) $MgSO_4 \cdot 7\ H_2O$, Epsom salts

35. ■ What mass is represented by 0.0255 mol of each of the following compounds?
    (a) C_3H_7OH, propanol, rubbing alcohol
    (b) $C_{11}H_{16}O_2$, an antioxidant in foods, also known as BHA (butylated hydroxyanisole)
    (c) $C_9H_8O_4$, aspirin

36. Assume you have 0.123 mol of each of the following compounds. What mass of each is present?
    (a) $C_{14}H_{10}O_4$, benzoyl peroxide, used in acne medications
    (c) $Pt(NH_3)_2Cl_2$, cisplatin, a cancer chemotherapy agent

37. Acetonitrile, CH_3CN, was found in the tail of Comet Hale-Bopp in 1997. What amount (moles) of acetonitrile is represented by 2.50 kg?

38. Acetone, $(CH_3)_2CO$, is an important industrial solvent. If 1260 million kg of this organic compound is produced annually, what amount (moles) is produced?

39. Sulfur trioxide, SO_3, is made industrially in enormous quantities by combining oxygen and sulfur dioxide, SO_2. What amount (moles) of SO_3 is represented by 1.00 kg of sulfur trioxide? How many molecules? How many sulfur atoms? How many oxygen atoms?

40. An Alka-Seltzer tablet contains 324 mg of aspirin ($C_9H_8O_4$), 1904 mg of $NaHCO_3$, and 1000. mg of citric acid ($H_3C_6H_5O_7$). (The last two compounds react with each other to provide the "fizz," bubbles of CO_2, when the tablet is put into water.)
    (a) Calculate the amount (moles) of each substance in the tablet.
    (b) If you take one tablet, how many molecules of aspirin are you consuming?

Percent Composition

(See Exercise 3.6 and the General ChemistryNow Screen 3.16.)

41. ■ Calculate the mass percent of each element in the following compounds.
    (a) PbS, lead(II) sulfide, galena
    (b) C_3H_8, propane
    (c) $C_{10}H_{14}O$, carvone, found in caraway seed oil

42. Calculate the mass percent of each element in the following compounds:
    (a) $C_8H_{10}N_2O_2$, caffeine
    (b) $C_{10}H_{20}O$, menthol
    (c) $CoCl_2 \cdot 6\ H_2O$

43. Calculate the weight percent of lead in PbS, lead(II) sulfide. What mass of lead (in grams) is present in 10.0 g of PbS?

44. Calculate the weight percent of iron in Fe_2O_3, iron(III) oxide. What mass of iron (in grams) is present in 25.0 g of Fe_2O_3?

45. ■ Calculate the weight percent of copper in CuS, copper(II) sulfide. If you wish to obtain 10.0 g of copper metal from copper(II) sulfide, what mass of the sulfide (in grams) must you use?

46. Calculate the weight percent of titanium in the mineral ilmenite, $FeTiO_3$. What mass of ilmenite (in grams) is required if you wish to obtain 750 g of titanium?

Empirical and Molecular Formulas

(See Example 3.7 and the General ChemistryNow Screens 3.16–3.18.)

47. ■ Succinic acid occurs in fungi and lichens. Its empirical formula is $C_2H_3O_2$ and its molar mass is 118.1 g/mol. What is its molecular formula?

48. An organic compound has the empirical formula C_2H_4NO. If its molar mass is 116.1 g/mol, what is the molecular formula of the compound?

49. Complete the following table:

	Empirical Formula	Molar Mass (g/mol)	Molecular Formula
(a)	CH	26.0	_____
(b)	CHO	116.1	_____
(c)	_____	_____	C_8H_{16}

---

50. Complete the following table:

Empirical Formula	Molar Mass (g/mol)	Molecular Formula
(a) $C_2H_3O_3$	150.0	_____
(b) C_3H_8	44.1	_____
(c) _____	_____	B_4H_{10}

51. Acetylene is a colorless gas used as a fuel in welding torches, among other things. It is 92.26% C and 7.74% H. Its molar mass is 26.02 g/mol. What are the empirical and molecular formulas of acetylene?

52. ■ A large family of boron-hydrogen compounds has the general formula B_xH_y. One member of this family contains 88.5% B; the remainder is hydrogen. Which of the following is its empirical formula: BH_2, BH_3, B_2H_5, B_5H_7, or B_5H_{11}?

53. ■ Cumene is a hydrocarbon, a compound composed only of C and H. It is 89.94% carbon, and its molar mass is 120.2 g/mol. What are the empirical and molecular formulas of cumene?

54. Nitrogen and oxygen form a series of oxides with the general formula N_xO_y. One of them, a blue solid, contains 36.84% N. What is the empirical formula of this oxide?

55. ■ Mandelic acid is an organic acid composed of carbon (63.15%), hydrogen (5.30%), and oxygen (31.55%). Its molar mass is 152.14 g/mol. Determine the empirical and molecular formulas of the acid.

56. Nicotine, a poisonous compound found in tobacco leaves, is 74.0% C, 8.65% H, and 17.35% N. Its molar mass is 162 g/mol. What are the empirical and molecular formulas of nicotine?

Determining Formulas from Mass Data

(See Examples 3.8–3.10 and the General ChemistryNow Screens 3.17–3.19.)

57. ■ If Epsom salt, $MgSO_4 \cdot x\,H_2O$, is heated to 250 °C, all the water of hydration is lost. On heating a 1.687-g sample of the hydrate, 0.824 g of $MgSO_4$ remains. How many molecules of water occur per formula unit of $MgSO_4$?

58. The "alum" used in cooking is potassium aluminum sulfate hydrate, $KAl(SO_4)_2 \cdot x\,H_2O$. To find the value of x, you can heat a sample of the compound to drive off all of the water and leave only $KAl(SO_4)_2$. Assume you heat 4.74 g of the hydrated compound and that the sample loses 2.16 g of water. What is the value of x?

59. ■ A new compound containing xenon and fluorine was isolated by shining sunlight on a mixture of Xe (0.526 g) and F_2 gas. If you isolate 0.678 g of the new compound, what is its empirical formula?

60. Elemental sulfur (1.256 g) is combined with fluorine, F_2, to give a compound with the formula SF_x, a very stable, colorless gas. If you have isolated 5.722 g of SF_x, what is the value of x?

61. Zinc metal (2.50 g) combines with 9.70 g of iodine to produce zinc iodide, Zn_xI_y. What is the formula of this ionic compound?

62. You combine 1.25 g of germanium, Ge, with excess chlorine, Cl_2. The mass of product, Ge_xCl_y, is 3.69 g. What is the formula of the product, Ge_xCl_y?

General Questions

These questions are not designated as to type or locations in the chapter. They may combine several concepts. More challenging questions are marked with the icon ▲.

63. Write formulas for all of the compounds that can be made by combining the cations NH_4^+ and Ni^{2+} with the anions CO_3^{2-} and SO_4^{2-}.

64. Using the General ChemistryNow CD-ROM or website, find a model for each of the following molecules. Write the molecular formula and draw the structural formula.
   (a) acetic acid
   (b) methylamine
   (c) formaldehyde

65. How many electrons are in a strontium atom (Sr)? Does an atom of Sr gain or lose electrons when forming an ion? How many electrons are gained or lost by the atom? When Sr forms an ion, the ion has the same number of electrons as which one of the noble gases?

66. The compound $(NH_4)_2SO_4$ consists of two polyatomic ions. What are the names and electric charges of these ions? What is the molar mass of this compound?

67. Which of the following compounds has the highest weight percent of chlorine?
   (a) BCl_3 (d) $AlCl_3$
   (b) $AsCl_3$ (e) PCl_3
   (c) $GaCl_3$

68. Which of the following samples has the largest number of ions?
   (a) 1.0 g of $BeCl_2$ (d) 1.0 g of $SrCO_3$
   (b) 1.0 g of $MgCl_2$ (e) 1.0 g of $BaSO_4$
   (c) 1.0 g of CaS

69. Which of the following compounds (NO, CO, MgO, or CaO) has the highest weight percent of oxygen?

70. The chemical compound alum has the formula $KAl(SO_4)_2 \cdot 12\,H_2O$. Give formulas for the ions that make up this ionic compound.

71. Knowing that the formula of sodium borate is Na_3BO_3, give the formula and charge of the borate ion. Is the borate ion a cation or an anion?

72. What is the difference between an empirical formula and a molecular formula? Use the compound ethane, C_2H_6, to illustrate your answer.

73. The structure of one of the bases in DNA, adenine, is shown here. Which represents the greater mass: 40.0 g of adenine or 3.0×10^{23} molecules of the compound?

74. Which has the larger mass, 0.5 mol of $BaCl_2$ or 0.5 mol of $SiCl_4$?

75. ■ A drop of water has a volume of about 0.05 mL. How many molecules of water are in a drop of water? (Assume water has a density of 1.00 g/cm³.)

76. Capsaicin, the compound that gives the hot taste to chili peppers, has the formula $C_{18}H_{27}NO_3$.
    (a) Calculate its molar mass.
    (b) If you eat 55 mg of capsaicin, what amount (moles) have you consumed?
    (c) Calculate the mass percent of each element in the compound.
    (d) What mass of carbon (in milligrams) is there in 55 mg of capsaicin?

77. Calculate the molar mass and the mass percent of each element in the blue solid $Cu(NH_3)_4SO_4 \cdot H_2O$. What are the mass (in grams) of copper and the mass of water in 10.5 g of the compound?

78. Write the molecular formula and calculate the molar mass for each of the molecules shown here. Which has the larger percentage of carbon? Of oxygen?
    (a) Ethylene glycol (used in antifreeze)

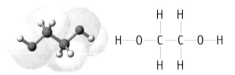

    (b) Dihydroxyacetone (used in artificial tanning lotions)

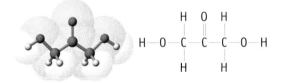

79. Malic acid, an organic acid found in apples, contains C, H, and O in the following ratios: $C_1H_{1.50}O_{1.25}$. What is the empirical formula of malic acid?

80. Your doctor has diagnosed you as being anemic—that is, as having too little iron in your blood. At the drugstore you find two iron-containing dietary supplements: one with iron(II) sulfate, $FeSO_4$, and the other with iron(II) gluconate, $Fe(C_6H_{11}O_7)_2$. If you take 100. mg of each compound, which will deliver more atoms of iron?

81. ▲ Spinach is high in iron (2 mg per 90-g serving). It is also a source of the oxalate ion, $C_2O_4{}^{2-}$; however, oxalate ion combines with iron ions to form iron oxalate, $Fe_x(C_2O_4)_y$, a substance that prevents your body from absorbing the iron. Analysis of a 0.109-g sample of iron oxalate shows that it contains 38.82% iron. What is the empirical formula of the compound?

82. A compound composed of iron and carbon monoxide, $Fe_x(CO)_y$, is 30.70% iron. What is the empirical formula for the compound?

83. *Ma huang*, an extract from the ephedra species of plants, contains ephedrine. The Chinese have used this herb more than 5000 years to treat asthma. More recently the substance has been used in diet pills that can be purchased over the counter in herbal medicine shops. However, very serious concerns have been raised regarding these pills following reports of serious heart problems with their use.
    (a) Write the molecular formula for ephedrine, draw its structural formula, and calculate its molar mass.
    (b) What is the weight percent of carbon in ephedrine?
    (c) Calculate the amount (moles) of ephedrine in a 0.125-g sample.
    (d) How many molecules of ephedrine are there in 0.125 g? How many C atoms?

84. Saccharin is more than 300 times sweeter than sugar. It was first made in 1897, a time when it was common practice for chemists to record the taste of any new substances they synthesized.
    (a) Write the molecular formula for the compound and draw its structural formula. (S atoms are yellow.)
    (b) If you ingest 125 mg of saccharin, what amount (moles) of saccharin have you ingested?
    (c) What mass of sulfur is contained in 125 mg of saccharin?

85. Which of the following pairs of elements are likely to form ionic compounds when allowed to react with each other? Write appropriate formulas for the ionic compounds you expect to form, and give the name of each.
    (a) chlorine and bromine
    (b) phosphorus and bromine
    (c) lithium and sulfur
    (d) indium and oxygen
    (e) sodium and argon
    (f) sulfur and bromine
    (g) calcium and fluorine

86. Name each of the following compounds, and tell which ones are best described as ionic:
    (a) ClF_3 (f) OF_2
    (b) NCl_3 (g) KI
    (c) $SrSO_4$ (h) Al_2S_3
    (d) $Ca(NO_3)_2$ (i) PCl_3
    (e) XeF_4 (j) K_3PO_4

87. Write the formula for each of the following compounds, and tell which ones are best described as ionic:
    (a) sodium hypochlorite
    (b) boron triiodide
    (c) aluminum perchlorate
    (d) calcium acetate
    (e) potassium permanganate
    (f) ammonium sulfite
    (g) potassium dihydrogen phosphate
    (h) disulfur dichloride
    (i) chlorine trifluoride
    (j) phosphorus trifluoride

88. Complete the table by placing symbols, formulas, and names in the blanks.

Cation	Anion	Name	Formula
		ammonium bromide	
Ba^{2+}			BaS
	Cl^-	iron(II) chloride	
	F^-		PbF_2
Al^{3+}	CO_3^{2-}		
		iron(III) oxide	

89. Complete the table by placing symbols, formulas, and names in the blanks.

Cation	Anion	Name	Formula
			$LiClO_4$
		aluminum phosphate	
	Br^-	lithium bromide	
			$Ba(NO_3)_2$
Al^{3+}		aluminum oxide	
		iron(III) carbonate	

90. Fluorocarbonyl hypofluorite is composed of 14.6% C, 39.0% O, and 46.3% F. If the molar mass of the compound is 82 g/mol, determine the empirical and molecular formulas of the compound.

91. Azulene, a beautiful blue hydrocarbon, is 93.71% C and has a molar mass of 128.16 g/mol. What are the empirical and molecular formulas of azulene?

92. Cacodyl, a compound containing arsenic, was reported in 1842 by the German chemist Robert Wilhelm Bunsen. It has an almost intolerable garlic-like odor. Its molar mass is 210 g/mol, and it is 22.88% C, 5.76% H, and 71.36% As. Determine its empirical and molecular formulas.

93. The action of bacteria on meat and fish produces a compound called cadaverine. As its name and origin imply, it stinks! (It is also present in bad breath and adds to the odor of urine.) It is 58.77% C, 13.81% H, and 27.40% N. Its molar mass is 102.2 g/mol. Determine the molecular formula of cadaverine.

94. ■ ▲ Transition metals can combine with carbon monoxide (CO) to form compounds such as $Fe(CO)_5$ (Study Question 3.82). Assume that you combine 0.125 g of nickel with CO and isolate 0.364 g of $Ni(CO)_x$. What is the value of x?

95. ▲ A major oil company has used a gasoline additive called MMT to boost the octane rating of its gasoline. What is the empirical formula of MMT if it is 49.5% C, 3.2% H, 22.0% O, and 25.2% Mn?

96. ▲ Elemental phosphorus is made by heating calcium phosphate with carbon and sand in an electric furnace. What is the weight percent of phosphorus in calcium phosphate? Use this value to calculate the mass of calcium phosphate (in kilograms) that must be used to produce 15.0 kg of phosphorus.

97. ▲ Chromium is obtained by heating chromium(III) oxide with carbon. Calculate the weight percent of chromium in the oxide and then use this value to calculate the quantity of Cr_2O_3 required to produce 850 kg of chromium metal.

98. ▲ Stibnite, Sb_2S_3, is a dark gray mineral from which antimony metal is obtained. What is the weight percent of antimony in the sulfide? If you have 1.00 kg of an ore that contains 10.6% antimony, what mass of Sb_2S_3 (in grams) is in the ore?

99. ▲ Direct reaction of iodine (I_2) and chlorine (Cl_2) produces an iodine chloride, I_xCl_y, a bright yellow solid. If you completely used up 0.678 g of iodine and produced 1.246 g of I_xCl_y, what is the empirical formula of the compound? A later experiment showed that the molar mass of I_xCl_y was 467 g/mol. What is the molecular formula of the compound?

100. ▲ In a reaction 2.04 g of vanadium combined with 1.93 g of sulfur to give a pure compound. What is the empirical formula of the product?

101. ▲ Iron pyrite, often called "fool's gold," has the formula FeS_2. If you could convert 15.8 kg of iron pyrite to iron metal, what mass of the metal would you obtain?

---

102. Which of the following statements about 57.1 g of octane, C_8H_{18}, is (are) not true?

(a) 57.1 g is 0.500 mol of octane.

(b) The compound is 84.1% C by weight.

(c) The empirical formula of the compound is C_4H_9.

(d) 57.1 g of octane contains 28.0 g of hydrogen atoms.

103. The formula of barium molybdate is $BaMoO_4$. Which of the following is the formula of sodium molybdate?

(a) Na_4MoO (c) Na_2MoO_3 (e) Na_4MoO_4

(b) $NaMoO$ (d) Na_2MoO_4

104. ▲ A metal M forms a compound with the formula MCl_4. If the compound is 74.75% chlorine, what is the identity of M?

105. Pepto-Bismol, which helps provide soothing relief for an upset stomach, contains 300. mg of bismuth subsalicylate, $C_{21}H_{15}Bi_3O_{12}$, per tablet. If you take two tablets for your stomach distress, what amount (in moles) of the "active ingredient" are you taking? What mass of Bi are you consuming in two tablets?

106. ▲ The weight percent of oxygen in an oxide that has the formula MO_2 is 15.2%. What is the molar mass of this compound? What element or elements are possible for M?

107. The mass of 2.50 mol of a compound with the formula ECl_4, in which E is a nonmetallic element, is 385 g. What is the molar mass of ECl_4? What is the identity of E?

108. ▲ The elements A and Z combine to produce two different compounds: A_2Z_3 and AZ_2. If 0.15 mol of A_2Z_3 has a mass of 15.9 g and 0.15 mole of AZ_2 has a mass of 9.3 g, what are the atomic masses of A and Z?

109. ▲ Polystyrene can be prepared by heating styrene with tribromobenzoyl peroxide in the absence of air. A sample prepared by this method has the empirical formula $Br_3C_6H_3(C_8H_8)_n$, where the value of n can vary from sample to sample. If one sample has 10.46% Br, what is the value of n?

110. A sample of hemoglobin is found to be 0.335% iron. If hemoglobin contains one iron atom per molecule, what is the molar mass of hemoglobin? What is the molar mass if there are four iron atoms per molecule?

Summary and Conceptual Questions

The following questions use concepts from the preceding chapters.

111. A piece of nickel foil, 0.550 mm thick and 1.25 cm square, is allowed to react with fluorine, F_2, to give a nickel fluoride.

(a) How many moles of nickel foil were used? (The density of nickel is 8.902 g/cm^3.)

(b) If you isolate 1.261 g of the nickel fluoride, what is its formula?

(c) What is its complete name?

112. An ionic compound can dissolve in water because the cations and anions are attracted to water molecules. The drawing here shows how a cation and a water molecule, which has a negatively charged O atom and positively charged H atoms, can interact. Which of the following cations should be most strongly attracted to water: Na^+, Mg^{2+}, or Al^{3+}? Explain briefly.

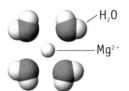

Water molecules interacting with a magnesium ion.

113. ▲ When analyzed, an unknown compound gave these experimental results: C, 54.0%; H, 6.00%; and O, 40.0%. Four different students used these values to calculate the empirical formulas shown here. Which answer is correct? Why did some students not get the correct answer?

(a) $C_4H_5O_2$ (c) $C_7H_{10}O_4$

(b) $C_5H_7O_3$ (d) $C_9H_{12}O_5$

114. ▲ Two general chemistry students working together in the lab weigh out 0.832 g of $CaCl_2 \cdot 2\,H_2O$ into a crucible. After heating the sample for a short time and allowing the crucible to cool, the students determine that the sample has a mass of 0.739 g. They then do a quick calculation. On the basis of this calculation, what should they do next?

(a) Congratulate themselves on a job well done.

(b) Assume the bottle of $CaCl_2 \cdot 2\,H_2O$ was mislabeled; it actually contained something different.

(c) Heat the crucible again, and then reweigh it.

115. ▲ Uranium is used as a fuel, primarily in the form of uranium(IV) oxide, in nuclear power plants. This question considers some uranium chemistry.

(a) A small sample of uranium metal (0.169 g) is heated to between 800 and 900 °C in air to give 0.199 g of a dark green oxide, U_xO_y. How many moles of uranium metal were used? What is the empirical formula of the oxide, U_xO_y? What is the name of the oxide? How many moles of U_xO_y must have been obtained?

(b) The naturally occurring isotopes of uranium are [234]U, [235]U, and [238]U. Knowing that uranium's atomic weight is 238.02 g/mol, which isotope must be the most abundant?

(c) If the hydrated compound $UO_2(NO_3)_2 \cdot z\,H_2O$ is heated gently, the water of hydration is lost. If you have 0.865 g of the hydrated compound and obtain 0.679 g of $UO_2(NO_3)_2$ on heating, how many molecules of water of hydration are in each formula unit of the original compound? (The oxide U_xO_y is obtained if the hydrate is heated to temperatures over 800 °C in the air.)

▲ More challenging   ■ In General ChemistryNow   Blue-numbered questions answered in Appendix O

116. The "simulation" section on General ChemistryNow Screen 3.7 Coulomb's Law, helps you explore Coulomb's law. You can change the charges on the ions and the distance between them. If the ions experience an attractive force, arrows point from one ion to the other. Repulsion is indicated by arrows pointing in opposite directions. Change the ion charges (from $1\pm$ to $2\pm$ to $3\pm$). How does this affect the attractive force? How close can the ions approach before significant repulsive forces set in? How does this distance vary with ion charge?

117. The common chemical compound alum has the formula $KAl(SO_4)_2 \cdot 12 H_2O$. An interesting characteristic of alum is that it is possible to grow very large crystals of this compound. Suppose you have a crystal of alum in the form of a cube that is 3.00 cm on each side. You want to know how many aluminum atoms are contained in this cube. Outline the steps to determine this value, and indicate the information that you need to carry out each step.

118. Cobalt(II) chloride hexahydrate dissolves readily in water to give a red solution. If we use this solution as an "ink," we can write secret messages on paper. The writing is not visible when the water evaporates from the paper. When the paper is heated, however, the message can be read. Explain the chemistry behind this observation. (*See General ChemistryNow Screen 3.20 Chemical Puzzler.*)

Charles D. Winters

A solution of CoCl₂ · 6 H₂O.   **Using the secret ink to write on paper.**   **Heating the paper reveals the writing.**

Do you need a live tutor for homework problems?
Access vMentor at General ChemistryNow at
http://now.brookscole.com/kotz6e
for one-on-one tutoring from a chemistry expert

4—Chemical Equations and Stoichiometry

A "black smoker" in the East Pacific Rise.

National Oceanic and Atmospheric Administration/Department of Commerce

Black Smokers and the Origin of Life

"The origin of life appears almost a miracle, so many are the conditions which would have had to be satisfied to get it going."

Francis Crick, quoted by John Horgan, "In the Beginning," *Scientific American,* pp. 116–125, February 1991.

The statement by Francis Crick on the origin of life does not mean that chemists and biologists have not tried to find the conditions under which life might have begun. Charles Darwin thought life might have begun when simple molecules combined to produce molecules of greater and greater complexity. Darwin's idea lives on in experiments such as those done by Stanley Miller in 1953. Attempting to recreate what was thought to be the atmosphere of the primeval earth, Miller filled a flask with the gases methane, ammonia, and hydrogen and added a bit of water. A discharge of electricity acted like lightning in the mixture. The inside of the flask was soon covered with a reddish slime, a mixture found to contain amino acids, the building blocks of proteins. Chemists thought they would soon know in more detail how living organisms began their development—but it was not to be. As Miller said recently, "The problem of the origin of life has turned out to be much more difficult than I, and most other people, envisioned."

Other theories have been advanced to account for the origin of life. The most recent conjecture relates to the discovery of geologically active sites on the ocean floor. Could life have originated in such exotic environments? The evidence is tenuous. As in Miller's experiments, this hypothesis relies on the creation of complex carbon-based molecules from simple ones.

In 1977 scientists were exploring the junction of two of the tectonic plates that form the floor of the Pacific Ocean. There they

See Chapter Goals Revisited (page 165). Test your knowledge of these goals by taking the exam-prep quiz on the General ChemistryNow CD-ROM or website.

- Balance equations for simple chemical reactions.
- Perform stoichiometry calculations using balanced chemical equations.
- Understand the meaning of a limiting reactant.
- Calculate the theoretical and percent yields of a chemical reaction.
- Use stoichiometry to analyze a mixture of compounds or to determine the formula of a compound.

National Oceanic and Atmospheric Administration/Department of Commerce

Black smoker chimney and shrimp on the Mid-Atlantic Ridge.

found thermal springs gushing a hot, black soup of minerals. Water seeping into cracks in the thin surface of the earth is superheated to between 300 and 400 °C by the magma of the earth's core. This superhot water dissolves minerals in the crust and provides conditions for the conversion of sulfate ions in sea water to hydrogen sulfide, H_2S. When this hot water, now laden with dissolved minerals and rich in sulfides, gushes through the surface, it cools. Metal sulfides, such as those of copper, manganese, iron, zinc, and nickel, then precipitate.

Many metal sulfides are black, and the plume of material coming from the sea bottom looks like black "smoke"; for this reason,

the vents have been called "black smokers." The solid sulfides settle around the edges of the vent on the sea floor, eventually forming a "chimney" of precipitated minerals.

Scientists were amazed to find that the black smoker vents were surrounded by primitive animals living in the hot, sulfide-rich environment. Because smokers lie under hundreds of meters of water and sunlight does not penetrate to these depths, the animals have developed a way to live without energy from sunlight. It is currently believed that they derive the energy needed to survive from the reaction of oxygen with hydrogen sulfide, H_2S:

$$H_2S(aq) + 2\ O_2(aq) \longrightarrow H_2SO_4(aq) + energy$$

The hypothesis that life might have originated in this inhospitable location developed out of laboratory experiments by a German lawyer and scientist, G. Wächtershäuser and a colleague, Claudia Huber. They found that metal sulfides such as iron sulfide promote reactions that convert simple carbon-containing molecules to more complex molecules. If this transformation could happen in the laboratory, perhaps similar chemistry might also occur in the exotic environment of black smokers.

To Review Before You Begin

- Review names and formulas of common compounds and ions (Chapter 3)
- Know how to convert mass to moles and moles to mass (Chapters 2 and 3)

Throughout the chapter this icon introduces a list of resources on the General ChemistryNow CD-ROM or website (http://now .brookscole.com/kotz6e) that will:

- help you evaluate your knowledge of the material
- provide homework problems
- allow you to take an exam-prep quiz
- provide a personalized Learning Plan targeting resources that address areas you should study

When you think about chemistry, you probably think of chemical reactions. The image of a medieval chemist mixing chemicals in hopes of turning lead into gold lingers in the imagination. Of course, there is much more to chemistry. Just reading this sentence involves an untold number of chemical reactions in your body. Indeed, every activity of living things depends on carefully regulated chemical reactions. Our objective in this chapter is to introduce the quantitative study of chemical reactions. Quantitative studies are needed to determine, for example, how much oxygen is required for the complete combustion of a given quantity of gasoline and what mass of carbon dioxide and water can be obtained. This part of chemistry is fundamental to much of what chemists, chemical engineers, biochemists, molecular biologists, geochemists, and many others do.

4.1—Chemical Equations

When a stream of chlorine gas, Cl_2, is directed onto solid phosphorus, P_4, the mixture bursts into flame, and a chemical reaction produces liquid phosphorus trichloride, PCl_3 (Figure 4.1). We can depict this reaction using a **balanced chemical equation**.

$$P_4(s) + 6\ Cl_2(g) \longrightarrow 4\ PCl_3(\ell)$$

Reactants   Products

■ **Information from Chemical Equations**
Chemical equations show the compounds involved in the chemical reaction and their physical state. Equations usually do *not* show the conditions of the experiment or indicate whether any energy (in the form of heat or light) is involved.

In a balanced equation, the formulas for the **reactants** (the substances combined in the reaction) are written to the left of the arrow and the formulas for the **products** (the substances produced) are written to the right of the arrow. The physical states of reactants and products can also be indicated. The symbol (s) indicates a solid, (g) a gas, and (ℓ) a liquid. A substance dissolved in water—that is, an *aqueous* solution of a substance—is indicated by (aq). The relative amounts of the reactants and products are shown by numbers, the *coefficients*, before the formulas.

Photos: Charles D. Winters

$$P_4(s) + 6\ Cl_2(g) \longrightarrow 4\ PCl_3(\ell)$$

Reactants   Products

Figure 4.1   Reaction of solid white phosphorus with chlorine gas. The product is liquid phosphorus trichloride.

Historical Perspectives

Antoine Laurent Lavoisier (1743–1794)

On Monday, August 7, 1774, the Englishman Joseph Priestley (1733–1804) became the first person to isolate oxygen. He heated solid mercury(II) oxide, HgO, causing the oxide to decompose to mercury and oxygen.

$$2 \text{ HgO(s)} \longrightarrow 2 \text{ Hg}(\ell) + \text{O}_2(g)$$

Priestley did not immediately understand the significance of his discovery, but he mentioned it to the French chemist Antoine Lavoisier in October 1774. One of Lavoisier's contributions to science was his recognition of the importance of exact scientific measurements and of carefully planned experiments, and he applied these methods to the study of oxygen. From this work he came to believe Priestley's gas was present in all acids, so he named it "oxygen," from the Greek words meaning "to form an acid." In addition, Lavoisier observed that the heat produced by a guinea pig when exhaling a given amount of carbon dioxide is similar to the quantity of heat produced by burning carbon to give the same amount of carbon dioxide. From this and other experiments he concluded that "Respiration is a combustion, slow it is true, but otherwise perfectly similar to that of charcoal." Although he did not understand the details of the process,

Lavoisier and his wife, as painted in 1788 by Jacques-Louis David. Lavoisier was then 45 and his wife, Marie Anne Pierrette Paulze, was 30.

Lavoisier's recognition marked an important step in the development of biochemistry.

Lavoisier was a prodigious scientist and the principles of naming chemical substances that he introduced are still in use today. Further, he wrote a textbook in which he applied for the first time the principles of the conservation of matter to chemistry and used the idea to write early versions of chemical equations.

Because Lavoisier was an aristocrat, he came under suspicion during the Reign of Terror of the French Revolution. He was an investor in the Ferme Générale, the infamous tax-collecting organization in 18th-century France. Tobacco was a monopoly product of the Ferme Générale, and it was a common occurrence to cheat the purchaser by adding water to the tobacco, a practice that Lavoisier opposed. Nonetheless, because of his involvement with the Ferme, his career was cut short by the guillotine on May 8, 1794, on the charge of "adding water to the people's tobacco."

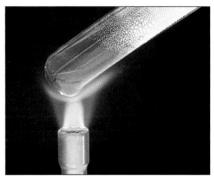

The decomposition of red mercury (II) oxide. The decomposition reaction gives mercury metal and oxygen gas. The mercury is seen as a film on the surface of the test tube.

In the 18th century, the great French scientist Antoine Lavoisier (1743–1794) introduced the **law of conservation of matter**, which states that *matter can be neither created nor destroyed.* This means that if the total mass of reactants is 10 g, and if the reaction completely converts reactants to products, you must end up with 10 g of products. It also means that if 1000 atoms of a particular element are contained in the reactants, then those 1000 atoms must appear in the products in some fashion.

When applied to the reaction of phosphorus and chlorine, the law of conservation of matter tells us that 1 molecule of phosphorus (with 4 phosphorus atoms) and 6 diatomic molecules of Cl_2 (with 12 atoms of Cl) are required to produce 4 molecules of PCl_3. Because each PCl_3 molecule contains 1 P atom and 3 Cl atoms, the 4 PCl_3 molecules are needed to account for 4 P atoms and 12 Cl atoms in the product.

■ **More Information from Chemical Equations**
The same number of atoms must exist after a reaction as before it takes place. However, these atoms are arranged differently. In the phosphorus/chlorine reaction, for example, the P atoms were in the form of P_4 molecules before reaction, but appear as PCl_3 molecules after reaction.

$$
\begin{array}{c}
\overbrace{}^{\substack{6 \times 2 = \\ 12 \text{ Cl atoms}}} \qquad \overbrace{}^{\substack{4 \times 3 = \\ 12 \text{ Cl atoms}}} \\
P_4(s) + 6 \text{ Cl}_2(g) \longrightarrow 4 \text{ PCl}_3(\ell) \\
\underbrace{}_{4 \text{ P atoms}} \qquad \underbrace{\phantom{4 \text{ PCl}_3}}_{4 \text{ P atoms}}
\end{array}
$$

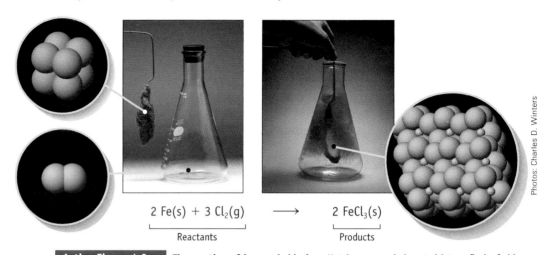

$2 \text{ Fe(s)} + 3 \text{ Cl}_2\text{(g)} \longrightarrow 2 \text{ FeCl}_3\text{(s)}$

Reactants   Products

Active Figure 4.2   **The reaction of iron and chlorine.** Hot iron gauze is inserted into a flask of chlorine gas. The heat from the reaction causes the iron gauze to glow, and brown iron(III) chloride forms.

GENERAL
Chemistry⚛Now™ See the General ChemistryNow CD-ROM or website to explore an interactive version of this figure accompanied by an exercise.

Photos: Charles D. Winters

The numbers in front of each formula in a *balanced* chemical equation are required by the law of conservation of matter. Review the equation for the reaction of phosphorus and chlorine, and then consider the balanced equation for the reaction of iron and chlorine (Figure 4.2).

$$\underset{\text{stoichiometric coefficients}}{2 \text{ Fe(s)} + 3 \text{ Cl}_2\text{(g)} \longrightarrow 2 \text{ FeCl}_3\text{(s)}}$$

The number in front of each chemical formula can be read as the number of atoms or molecules (2 atoms of Fe and 3 molecules of Cl_2 form 2 formula units of $FeCl_3$). It can refer equally well to amounts of reactants and products: 2 moles of solid iron combine with 3 moles of chlorine gas to produce 2 moles of solid $FeCl_3$. The relationship between the quantities of chemical reactants and products is called **stoichiometry** (pronounced "stoy-key-AHM-uh-tree"), and the coefficients in a balanced equation are the **stoichiometric coefficients**.

Balanced chemical equations are fundamentally important for understanding the quantitative basis of chemistry. *You must always begin with a balanced equation before carrying out a stoichiometry calculation.*

GENERAL
Chemistry⚛Now™

See the General ChemistryNow CD-ROM or website:

• **Screen 4.3 The Law of Conservation of Mass,** for two exercises on the conservation of mass in several reactions

Exercise 4.1—Chemical Reactions

The reaction of aluminum with bromine is shown on page 98. The equation for the reaction is

$$2 \ Al(s) + 3 \ Br_2(\ell) \longrightarrow Al_2Br_6(s)$$

(a) Name the reactants and products in this reaction and give their states.

(b) What are the stoichiometric coefficients in this equation?

(c) If you were to use 8000 atoms of Al, how many molecules of Br_2 are required to consume the Al completely?

4.2—Balancing Chemical Equations

Balancing an equation ensures that the same number of atoms of each element appear on both sides of the equation. Many chemical equations can be balanced by trial and error, although some will involve more trial than others.

One general class of chemical reactions is the reaction of metals or nonmetals with oxygen to give oxides of the general formula M_xO_y. For example, iron can react with oxygen to give iron(III) oxide (Figure 4.3a),

$$4 \ Fe(s) + 3 \ O_2(g) \longrightarrow 2 \ Fe_2O_3(s)$$

magnesium and oxygen react to form magnesium oxide (Figure 4.3b),

$$2 \ Mg(s) + O_2(g) \longrightarrow 2 \ MgO(s)$$

and phosphorus, P_4, reacts vigorously with oxygen to give tetraphosphorus decaoxide, P_4O_{10} (Figure 4.3c),

$$P_4(s) + 5 \ O_2(g) \longrightarrow P_4O_{10}(s)$$

The equations written above are balanced. The same number of metal or phosphorus atoms and oxygen atoms occurs on each side of these equations.

(a) Reaction of iron and oxygen to give iron(III) oxide, Fe_2O_3.

(b) Reaction of magnesium and oxygen to give magnesium oxide, MgO.

(c) Reaction of phosphorus and oxygen to give tetraphosphorus decaoxide, P_4O_{10}.

Figure 4.3 Reactions of metals and a nonmetal with oxygen. (*See General ChemistryNow Screen 4.4 Balancing Chemical Equations, for a video of the phosphorus and oxygen reaction.*)

A combustion reaction. Propane, C_3H_8, burns to give CO_2 and H_2O. These simple oxides are always the products of the complete combustion of a hydrocarbon. (*See General ChemistryNow Screen 4.4 Balancing Chemical Equations, for a animation of this reaction.*)

The **combustion**, or burning, of a fuel in oxygen is accompanied by the evolution of heat. You are familiar with combustion reactions such as the burning of octane, C_8H_{18}, a component of gasoline, in an automobile engine:

$$2\ C_8H_{18}(\ell) + 25\ O_2(g) \longrightarrow 16\ CO_2(g) + 18\ H_2O(g)$$

In all combustion reactions, some or all of the elements in the reactants end up as oxides, compounds containing oxygen. When the reactant is a hydrocarbon (a compound containing only C and H), the products of complete combustion are carbon dioxide and water.

When balancing chemical equations, there are two important things to remember:

- Formulas for reactants and products must be correct or the equation is meaningless.

- Subscripts in the formulas of reactants and products cannot be changed to balance equations. Changing the subscripts changes the identity of the substance. For example, you cannot change CO_2 to CO to balance an equation; carbon monoxide, CO, and carbon dioxide, CO_2, are different compounds.

As an example of equation balancing, let us write the balanced equation for the complete combustion of propane, C_3H_8.

Step 1. *Write correct formulas for the reactants and products.*

$$C_3H_8(g) + O_2(g) \xrightarrow{\text{unbalanced equation}} CO_2(g) + H_2O(g)$$

Here propane and oxygen are the reactants, and carbon dioxide and water are the products.

Step 2. *Balance the C atoms.* In combustion reactions such as this it is usually best to balance the carbon atoms first and leave the oxygen atoms until the end (because the oxygen atoms are often found in more than one product). In this case three carbon atoms are in the reactants, so three must occur in the products. Three CO_2 molecules are therefore required on the right side:

$$C_3H_8(g) + O_2(g) \xrightarrow{\text{unbalanced equation}} 3\ CO_2(g) + H_2O(g)$$

Step 3. *Balance the H atoms.* Propane, the reactant, contains 8 H atoms. Each molecule of water has two hydrogen atoms, so four molecules of water account for the required eight hydrogen atoms on the right side:

$$C_3H_8(g) + O_2(g) \xrightarrow{\text{unbalanced equation}} 3\ CO_2(g) + 4\ H_2O(g)$$

Step 4. *Balance the number of O atoms.* Ten oxygen atoms are on the right side ($3 \times 2 = 6$ in CO_2 plus $4 \times 1 = 4$ in water). Therefore, five O_2 molecules are needed to supply the required ten oxygen atoms:

$$C_3H_8(g) + 5\ O_2(g) \longrightarrow 3\ CO_2(g) + 4\ H_2O(g)$$

Step 5. *Verify that the number of atoms of each element is balanced.* The equation shows three carbon atoms, eight hydrogen atoms, and ten oxygen atoms on each side.

GENERAL
Chemistry·👤·Now™

See the General ChemistryNow CD-ROM or website:

- **Screen 4.4 Balancing Chemical Equations,** for a tutorial in which you balance a series of combustion reactions.

Example 4.1—Balancing an Equation for a Combustion Reaction

Problem Write the balanced equation for the combustion of ammonia ($NH_3 + O_2$) to give NO and H_2O.

Strategy First write the unbalanced equation. Next balance the N atoms, then balance the H atoms, and finally balance the O atoms.

Solution

Step 1. *Write correct formulas for reactants and products.* The unbalanced equation for the combustion is

$$NH_3(g) + O_2(g) \xrightarrow{\text{unbalanced equation}} NO(g) + H_2O(g)$$

Step 2. *Balance the N atoms.* There is one N atom on each side of the equation. The N atoms are in balance, at least for the moment.

$$NH_3(g) + O_2(g) \xrightarrow{\text{unbalanced equation}} NO(g) + H_2O(g)$$

Step 3. *Balance the H atoms.* There are three H atoms on the left and two on the right. To have the same number on each side, let us use two molecules of NH_3 on the left and three molecules of H_2O on the right (which gives us six H atoms on each side).

$$2\ NH_3(g) + O_2(g) \xrightarrow{\text{unbalanced equation}} NO(g) + 3\ H_2O(g)$$

Notice that when we balance the H atoms, the N atoms are no longer balanced. To bring them into balance, let us use two NO molecules on the right.

$$2\ NH_3(g) + O_2(g) \xrightarrow{\text{unbalanced equation}} 2\ NO(g) + 3\ H_2O(g)$$

Step 4. *Balance the O atoms.* After Step 3, there are two O atoms on the left side and five on the right. That is, there are an even number of O atoms on the left and an odd number on the right. Because there cannot be an odd number of O atoms on the left (O atoms are paired in O_2 molecules), multiply each coefficient on both sides of the equation by 2 so that an even number of oxygen atoms (ten) can now occur on the right side:

$$4\ NH_3(g) + O_2(g) \xrightarrow{\text{unbalanced equation}} 4\ NO(g) + 6\ H_2O(g)$$

Now the oxygen atoms can be balanced by having five O_2 molecules on the left side of the equation:

$$4\ NH_3(g) + 5\ O_2(g) \xrightarrow{\text{balanced equation}} 4\ NO(g) + 6\ H_2O(g)$$

Step 5. *Verify the result.* Four N atoms, 12 H atoms, and 10 O atoms occur on each side of the equation.

> **Comment** An alternative way to write this equation is
>
> $$2 NH_3(g) + \tfrac{5}{2} O_2(g) \longrightarrow 2 NO(g) + 3 H_2O(g)$$
>
> where a fractional coefficient has been used. This equation is correctly balanced and will be useful under some circumstances. In general, however, we balance equations with whole-number coefficients.

Exercise 4.2—Balancing the Equation for a Combustion Reaction

(a) Butane gas, C_4H_{10}, can burn completely in air [use $O_2(g)$ as the other reactant] to give carbon dioxide gas and water vapor. Write a balanced equation for this combustion reaction.

(b) Write a balanced chemical equation for the complete combustion of liquid tetraethyllead, $Pb(C_2H_5)_4$ (which was used until the 1970s as a gasoline additive). The products of combustion are $PbO(s)$, $H_2O(g)$, and $CO_2(g)$.

4.3—Mass Relationships in Chemical Reactions: Stoichiometry

A balanced chemical equation shows the quantitative relationship between reactants and products in a chemical reaction. Let us apply this concept to the reaction of phosphorus and chlorine (see Figure 4.1). Suppose you use 1.00 mol of phosphorus (P_4, 124 g/mol) in this reaction. The balanced equation shows that 6.00 mol (= 425 g) of Cl_2 must be used for complete reaction with 1.00 mol of P_4 and that 4.00 mol (= 549 g) of PCl_3 can be produced.

■ **Amounts Tables**
Amounts tables not only are useful here but will also be used extensively when you study chemical equilibria in Chapters 16–18.

Equation	$P_4(s)$	+	$6\ Cl_2$ (g)	$\longrightarrow$	$4\ PCl_3$ (ℓ)
Initial amount (mol)	1.00 mol (124 g)		6.00 mol (425 g)		0 mol (0 g)
Change in amount upon reaction (mol)	−1.00 mol		−6.00 mol		+4.00 mol
Amount after complete reaction (mol)	0 mol (0 g)		0 mol (0 g)		4.00 mol [549 g = 124 g + 425 g]

The mole and mass relationships of reactants and products in a reaction are summarized in an *amounts table*. Such tables identify the amounts of reactants and products and the changes that occur upon reaction.

The balanced equation for a reaction tells us the correct *mole ratios* of reactants and products. Therefore, the equation for the phosphorus and chlorine reaction, for example, applies no matter how much P_4 is used. Suppose 0.0100 mol of P_4 (1.24 g) is used. Now only 0.0600 mol of Cl_2 (4.25 g) is required, and 0.0400 mol of PCl_3 (5.49 g) can form.

Following this line of reasoning, let us decide (a) what mass of Cl_2 is required to react completely with 1.45 g of phosphorus and (b) what mass of PCl_3 is produced.

■ **Mass Balance**
Mass is always conserved in chemical reactions. The total mass before reaction is always the same as that after reaction. This does not mean, however, that the total amount (moles) of reactants is the same as that of the products. Atoms are rearranged into different "units" (molecules) in the course of a reaction. In the $P_4 + Cl_2$ reaction, 7 mol of reactants gives 4 mol of product.

Part (a): Mass of Cl_2 Required
Step 1. *Write the balanced equation* (using correct formulas for reactants and products). This is always the first step when dealing with chemical reactions.

$$P_4(s) + 6\ Cl_2(g) \longrightarrow 4\ PCl_3(\ell)$$

Problem-Solving Tip 4.1

Stoichiometry Calculations

You are asked to determine what mass of product can be formed from a given mass of reactant. Keep in mind that it is not possible to calculate the mass of product in a single step. Instead, you must follow a route such as that illustrated here for the reaction of a reactant A to give the product B according to an equation such as $x A \rightarrow y B$. Here the mass of reactant A is converted to moles of A. Then, using the stoichiometric factor, you find moles of B. Finally, the mass of B is obtained by multiplying moles of B by its molar mass.

When solving a chemical stoichiometry problem, remember that you will always use a stoichiometric factor at some point.

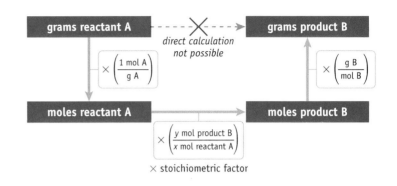

Step 2. *Calculate moles from masses.* From the mass of P_4, calculate the amount of P_4 available.

$$1.45 \text{ g } P_4 \times \frac{1 \text{ mol } P_4}{123.9 \text{ g } P_4} = 0.0117 \text{ mol } P_4$$

Step 3. *Use a stoichiometric factor.* The amount of P_4 available is related to the amount of the other reactant (Cl_2) required by the balanced equation.

$$0.0117 \text{ mol } P_4 \times \frac{6 \text{ mol } Cl_2 \text{ required}}{1 \text{ mol } P_4 \text{ available}} = 0.0702 \text{ mol } Cl_2 \text{ required}$$

⇑ stoichiometric factor (from balanced equation)

To perform this calculation the amount of phosphorus available has been multiplied by a **stoichiometric factor,** a *mole ratio based on the coefficients for the two chemicals in the balanced equation.* This is the reason you must balance chemical equations before proceeding with calculations. Here the balanced equation specifies that 6 mol of Cl_2 is required for each mole of P_4, so the stoichiometric factor is (6 mol Cl_2/ 1 mol P_4). Calculation shows that 0.0702 mol of Cl_2 is required to react with all the available phosphorus (1.45 g, 0.0117 mol).

■ **Stoichiometric Factor**
The stoichiometric factor is a conversion factor (see page 42). Thus, a stoichiometric factor can also relate moles of a reactant to moles of a product, and vice versa.

Step 4. *Calculate mass from moles.* Convert amount (moles) of Cl_2 calculated in Step 3 to quantity (mass in grams) of Cl_2 required.

$$0.0702 \text{ mol } Cl_2 \times \frac{70.91 \text{ g } Cl_2}{1 \text{ mol } Cl_2} = 4.98 \text{ g } Cl_2$$

Part (b) Mass of PCl₃ Produced from P₄ and Cl₂

What mass of PCl_3 can be produced from the reaction of 1.45 g of phosphorus with 4.98 g of Cl_2? Because matter is conserved, the answer can be obtained in this case

■ **Amount and Quantity**
When doing stoichiometry problems, recall from Chapter 2 that the terms "amount" and "quantity" are used in a specific sense by chemists. The *amount* of a substance is the number of moles of that substance. *Quantity* refers to the mass of the substance.

by adding the masses of P_4 and Cl_2 used (giving 1.45 g + 4.98 g = 6.43 g of PCl_3 produced). Alternatively, Steps 3 and 4 can be repeated, but with the appropriate stoichiometric factor and molar mass.

Step 3b. *Use a stoichiometric factor.* Convert the amount of available P_4 to the amount of PCl_3 produced. Here the balanced equation specifies that 4 mol PCl_3 is produced for each mole of P_4 used, so the stoichiometric factor is (4 mol PCl_3/1 mol P_4).

$$0.0117 \text{ mol } P_4 \times \frac{4 \text{ mol } PCl_3 \text{ produced}}{1 \text{ mol } P_4 \text{ available}} = 0.0468 \text{ mol } PCl_3 \text{ produced}$$

⇑ stoichiometric factor (from balanced equation)

Step 4b. *Calculate mass from moles.* Convert the amount of PCl_3 produced to a mass in grams.

$$0.0468 \text{ mol } PCl_3 \times \frac{137.3 \text{ g } PCl_3}{1 \text{ mol } PCl_3} = 6.43 \text{ g } PCl_3$$

GENERAL
Chemistry ⚛ Now™

See the General ChemistryNow CD-ROM or website:
- **Screen 4.5 Weight Relations in Chemical Reactions**
   (a) for a video and animation of the phosphorus and chlorine reaction discussed in this section
   (b) for an exercise that examines the reaction between chlorine and elemental phosphorus
- **Screen 4.6 Calculations in Stoichiometry,** for a tutorial on yield

---

Example 4.2—Mass Relations in Chemical Reactions

Problem Glucose reacts with oxygen to give CO_2 and H_2O.

$$C_6H_{12}O_6(s) + 6 O_2(g) \longrightarrow 6 CO_2(g) + 6 H_2O(\ell)$$

What mass of oxygen (in grams) is required for complete reaction of 25.0 g of glucose? What masses of carbon dioxide and water (in grams) are formed?

Strategy After referring to the balanced equation, you can perform the stoichiometric calculations using the scheme in Problem-Solving Tip 4.1.

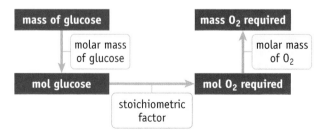

First find the amount of glucose available, then relate it to the amount of O_2 required using the stoichiometric factor based on the coefficients in the balanced equation. Finally, find the mass of O_2 required from the amount of O_2. Follow the same procedure to find the masses of carbon dioxide and water.

Solution

Step 1. *Write a balanced equation.*

$$C_6H_{12}O_6(s) + 6\ O_2(g) \longrightarrow 6\ CO_2(g) + 6\ H_2O(\ell)$$

Step 2. *Convert the mass of glucose to moles.*

$$25.0\ \text{g glucose} \times \frac{1\ \text{mol}}{180.2\ \text{g}} = 0.139\ \text{mol glucose}$$

Step 3. *Use the stoichiometric factor.* Here we calculate the amount of O_2 required.

$$0.139\ \text{mol glucose} \times \frac{6\ \text{mol } O_2}{1\ \text{mol glucose}} = 0.832\ \text{mol } O_2$$

Step 4. *Calculate mass from moles.* Convert the required amount of O_2 to a mass in grams.

$$0.832\ \text{mol } O_2 \times \frac{32.00\ \text{g}}{1\ \text{mol } O_2} = 26.6\ \text{g } O_2$$

Repeat Steps 3 and 4 to find the mass of CO_2 produced in the combustion. First, relate the amount (moles) of glucose available to the amount of CO_2 produced using a stoichiometric factor. Then convert the amount of CO_2 to the mass in grams.

$$0.139\ \text{mol glucose} \times \frac{6\ \text{mol } CO_2}{1\ \text{mol glucose}} \times \frac{44.01\ \text{g } CO_2}{1\ \text{mol } CO_2} = 36.6\ \text{g } CO_2$$

Now, how can you find the mass of H_2O produced? You could go through Steps 3 and 4 again. However, recognize that the total mass of reactants

$$25.0\ \text{g } C_6H_{12}O_6 + 26.6\ \text{g } O_2 = 51.6\ \text{g of reactants}$$

must be the same as the total mass of products. The mass of water that can be produced is therefore

$$\text{Total mass of products} = 51.6\ \text{g} = 36.6\ \text{g } CO_2\ \text{produced} + ?\ \text{g } H_2O$$

$$\text{Mass of } H_2O\ \text{produced} = 15.0\ \text{g}$$

The amounts table for this problem is

Equation	$C_6H_{12}O_6(s)$	+	$6\ O_2(g)$	$\longrightarrow$	$6\ CO_2(g)$	+	$6\ H_2O(\ell)$
Initial amount (mol)	0.139 mol		6(0.139 mol)		0		0
			= 0.832 mol				
Change (mol)	−0.139 mol		−0.832 mol		+0.832 mol		+0.832 mol
Amount after reaction (mol)	0		0		0.832 mol		0.832 mol

> **Comment** When you know the mass of all but one of the chemicals in a reaction, you can find the unknown mass using the principle of mass conservation (the total mass of reactants must equal the total mass of products).

Exercise 4.3—Mass Relations in Chemical Reactions

What mass of oxygen, O_2, is required to completely combust 454 g of propane, C_3H_8? What masses of CO_2 and H_2O are produced?

$$C_3H_8(g) + 5\ O_2(g) \longrightarrow 3\ CO_2(g) + 4\ H_2O(\ell)$$

4.4—Reactions in Which One Reactant Is Present in Limited Supply

You may have observed in your laboratory experiments that reactions are often carried out with an excess of one reactant over that required by stoichiometry. This is usually done to ensure that one of the reactants in the reaction is consumed completely, even though some of another reactant remains unused.

Suppose you burn a toy "sparkler," a wire coated with magnesium (Figure 4.3b). The magnesium burns in air, consuming oxygen and producing magnesium oxide, MgO.

$$Mg(s) + O_2(g) \longrightarrow 2\ MgO(s)$$

The sparkler burns until the magnesium is consumed completely. What about the oxygen? Two moles of magnesium require one mole of oxygen, but there is much, much more O_2 available in the air than is needed to consume the magnesium in a sparkler. How much MgO is produced? That depends on the quantity of magnesium in the sparkler, not on the quantity of O_2 in the atmosphere. A substance such as the magnesium in this example is called the **limiting reactant** because its amount determines, or limits, the amount of product formed.

Let us look at an example of a limiting reactant situation using the reaction of oxygen and carbon monoxide to give carbon dioxide. The balanced equation for the reaction is

$$2\ CO(g) + O_2(g) \longrightarrow 2\ CO_2(g)$$

Suppose you have a mixture of four CO molecules and three O_2 molecules.

Reactants: 4 CO and 3 O_2 Products: 4 CO_2 and 1 O_2

■ **Comparing Reactant Ratios**
For the CO/O_2 reaction, the stoichiometric ratio of reactants should be (2 mol CO/ 1 mol O_2). However, the ratio of amounts of reactants available is (4 mol CO/3 mol O_2) or (1.33 mol CO/1 mol O_2). Clearly, there is not sufficient CO to react with all of the available O_2. Carbon monoxide is the limiting reactant, and some O_2 will be left over when all of the CO is consumed.

The four CO molecules require only two O_2 molecules (and produce four CO_2 molecules). This means that one O_2 molecule remains after reaction is complete. Because more O_2 molecules are available than are required, the number of CO_2 molecules produced is determined by the number of CO molecules available. Carbon monoxide, CO, is therefore the limiting reactant in this case.

(a)   (b)

Active Figure 4.4   **Oxidation of ammonia.** (a) Burning ammonia on the surface of a platinum wire produces so much heat that the wire glows bright red. (b) Billions of kilograms of HNO_3 are made annually starting with the oxidation of ammonia over a wire gauze containing platinum.

GENERAL
Chemistry⚛Now™ See the General ChemistryNow CD-ROM or website to explore an interactive version of this figure accompanied by an exercise.

A Stoichiometry Calculation with a Limiting Reactant

The first step in the manufacture of nitric acid is the oxidation of ammonia to NO over a platinum-wire gauze (Figure 4.4).

$$4\ NH_3(g) + 5\ O_2(g) \longrightarrow 4\ NO(g) + 6\ H_2O(\ell)$$

Suppose that equal masses of NH_3 and of O_2 are mixed (750. g of each). Are these reactants mixed in the correct stoichiometric ratio or is one of them in short supply? That is, will one of them limit the quantity of NO that can be produced? How much NO can be formed if the reaction using this reactant mixture goes to completion? And how much of the excess reactant is left over when the maximum amount of NO has been formed?

Step 1. Find the amount of each reactant.

$$750.\ g\ NH_3 \times \frac{1\ mol\ NH_3}{17.03\ g\ NH_3} = 44.0\ mol\ NH_3\ available$$

$$750.\ g\ O_2 \times \frac{1\ mol\ O_2}{32.00\ g\ O_2} = 23.4\ mol\ O_2\ available$$

Step 2. What is the limiting reactant? Examine the ratio of amounts of reactants.
Are the reactants present in the correct stoichiometric ratio as given by the balanced equation?

Stoichiometric ratio of reactants **required** by balanced equation

$$= \frac{5\ mol\ O_2}{4\ mol\ NH_3} = \frac{1.25\ mol\ O_2}{1\ mol\ NH_3}$$

Ratio of reactants **actually available** $= \dfrac{23.4\ mol\ O_2}{44.0\ mol\ NH_3} = \dfrac{0.532\ mol\ O_2}{1\ mol\ NH_3}$

Problem-Solving Tip 4.2

More on Reactions with a Limiting Reactant

There is another method of solving limiting reactant problems: Calculate the mass of product expected based on each reactant. The limiting reactant is the reactant that gives the smallest quantity of product. For example, refer to the $NH_3 + O_2$ reaction on page 153. To confirm that O_2 is the limiting reactant, calculate the quantity of NO that can be formed starting with (a) 44.1 mol of NH_3 and unlimited O_2 and (b) with 23.4 mol of O_2 and unlimited NH_3.

1. Quantity of NO produced from 44.1 mol of NH_3 and unlimited O_2

$$44.0 \text{ mol } NH_3 \times \frac{4 \text{ mol NO}}{4 \text{ mol } NH_3} \times \frac{30.01 \text{ g NO}}{1 \text{ mol NO}} = 1320 \text{ g NO}$$

2. Quantity of NO produced from 23.4 mol O_2 and unlimited NH_3

$$23.4 \text{ mol } O_2 \times \frac{4 \text{ mol NO}}{5 \text{ mol } O_2} \times \frac{30.01 \text{ g NO}}{1 \text{ mol NO}} = 562 \text{ g NO}$$

3. Compare the quantities of NO produced. The available O_2 is capable of producing less NO (562 g) than the available NH_3 (1320 g), which confirms that O_2 is the limiting reactant.

As a final note, you may find this approach easier to use when there are more than two reactants, each present initially in some designated quantity.

Dividing moles of O_2 available by moles of NH_3 available shows that the ratio of available reactants is much smaller than the 5 mol O_2/4 mol NH_3 ratio required by the balanced equation. Thus there is not sufficient O_2 available to react with all of the NH_3. In this case, *oxygen, O_2, is the limiting reactant.* That is, 1 mol of NH_3 requires 1.25 mol of O_2, but we have only 0.532 mol of O_2 available for each mole of NH_3.

Step 3. Calculate the mass of product.

We can now calculate the mass of product, NO, expected based on the amount of the limiting reactant, O_2.

$$23.4 \text{ mol } O_2 \times \frac{4 \text{ mol NO}}{5 \text{ mol } O_2} \times \frac{30.01 \text{ g NO}}{1 \text{ mol NO}} = 562 \text{ g NO}$$

Step 4. Calculate the mass of excess reactant.

Ammonia is the "excess reactant" in this NH_3/O_2 reaction because more than enough NH_3 is available to react with 23.4 mol of O_2. Let us calculate the quantity of NH_3 remaining after all the O_2 has been used. To do so, we first need to know the amount of NH_3 required to consume all the limiting reactant, O_2.

$$23.4 \text{ mol } O_2 \text{ available} \times \frac{4 \text{ mol } NH_3 \text{ required}}{5 \text{ mol } O_2} = 18.8 \text{ mol } NH_3 \text{ required}$$

Because 44.0 mol of NH_3 is available, the amount of excess NH_3 can be calculated,

$$\text{Excess } NH_3 = 44.0 \text{ mol } NH_3 \text{ available} - 18.8 \text{ mol } NH_3 \text{ required}$$

$$= 25.2 \text{ mol } NH_3 \text{ remaining}$$

and then converted to a mass,

$$25.2 \text{ mol } NH_3 \times \frac{17.03 \text{ g } NH_3}{1 \text{ mol } NH_3} = 429 \text{ g } NH_3 \text{ in excess of that required}$$

Finally, because 429 g of NH_3 is left over, this means that 321 g of NH_3 has been consumed (= 750. g − 429 g).

It is helpful in limiting reactant problems to summarize your results in an amounts table.

Equation	4 NH_3(g)	+	5 O_2(g)	→	4 NO(g)	+	6 H_2O(g)
Initial amount (mol)	44.0		23.4		0		0
Change in amount (mol)	−(4/5)(23.4)		−23.4		+(4/5)(23.4)		+(6/5)(23.4)
	= −18.8		−23.4		= +18.8		= +28.1
After complete reaction (mol)	25.2		0		18.8		28.1

All of the limiting reactant, O_2, has been consumed. Of the original 44.0 mol of NH_3, 18.8 mol has been consumed and 25.2 mol remains. The balanced equation indicates that the amount of NO produced is equal to the amount of NH_3 consumed, so 18.8 mol of NO is produced from 18.8 mol of NH_3. In addition, 28.1 mol of H_2O has been produced.

GENERAL
Chemistry•Now™

See the General ChemistryNow CD-ROM or website:

- **Screen 4.7 Reactions Controlled by the Supply of One Reactant** for a video and animation of the limiting reactant in the methanol and oxygen reaction

- **Screen 4.8 Limiting Reactants**

  (a) for an exercise on zinc and hydrochloric acid in aqueous solution

  (b) for a simulation using limiting reactants

Example 4.3—A Reaction with a Limiting Reactant

Problem Methanol, CH_3OH, which is used as a fuel, can be made by the reaction of carbon monoxide and hydrogen.

$$CO(g) + 2 H_2(g) \longrightarrow CH_3OH(\ell)$$
$$\text{methanol}$$

Suppose 356 g of CO and 65.0 g of H_2 are mixed and allowed to react.

(a) Which is the limiting reactant?

(b) What mass of methanol can be produced?

(c) What mass of the excess reactant remains after the limiting reactant has been consumed?

Strategy There are usually two steps to a limiting reactant problem:

(a) After calculating the amount of each reactant, compare the ratio of reactant amounts to the required stoichiometric ratio, 2 mol H_2/1 mol CO.

- If [mol H_2 available/mol CO available] > 2/1, then CO is the limiting reactant.

- If [mol H_2 available/mol CO available] < 2/1, then H_2 is the limiting reactant.

(b) Use the amount of limiting reactant to find the amount of product.

■ **Conservation of Mass**
Mass is conserved in the NH_3 + O_2 reaction. The total mass present before reaction (1500. g) is the same as the total mass produced in the reaction plus the mass of NH_3 remaining. That is, 562 g of NO (18.8 mol) and 506 g of H_2O (28.1 mol) are produced. Because 429 g of NH_3 (25.2 mol) remains, the total mass after reaction (562 g + 506 g + 429 g) is the same as the total mass before reaction.

A car that uses methanol as a fuel. In this car methanol is converted to hydrogen, which is then combined with oxygen in a fuel cell. The fuel cell generates electric energy to run the car (see Chapter 20). See Example 4.3.

Solution

(a) *What is the limiting reactant?* The amount of each reactant is

$$\text{Amount of CO} = 356 \text{ g CO} \times \frac{1 \text{ mol CO}}{28.01 \text{ g CO}} = 12.7 \text{ mol CO}$$

$$\text{Amount of H}_2 = 65.0 \text{ g H}_2 \times \frac{1 \text{ mol H}_2}{2.016 \text{ g H}_2} = 32.2 \text{ mol H}_2$$

Are these reactants present in a perfect stoichiometric ratio?

$$\frac{\text{Mol H}_2 \text{ available}}{\text{Mol CO available}} = \frac{32.2 \text{ mol H}_2}{12.7 \text{ mol CO}} = \frac{2.54 \text{ mol H}_2}{1.00 \text{ mol CO}}$$

The required mole ratio is 2 mol of H_2 to 1 mol of CO. Here we see that more hydrogen is available than is required to consume all the O_2. It follows that not enough CO is present to use up all of the hydrogen. *CO is the limiting reactant.*

(b) *What is the maximum mass of CH_3OH that can be formed?* This calculation must be based on the amount of limiting reactant.

$$12.7 \text{ mol CO} \times \frac{1 \text{ mol CH}_3\text{OH formed}}{1 \text{ mol CO available}} \times \frac{32.04 \text{ g CH}_3\text{OH}}{1 \text{ mol CH}_3\text{OH}} = 407 \text{ g CH}_3\text{OH}$$

(c) *What amount of H_2 remains when all the CO has been converted to product?* First, we must find the amount of H_2 required to react with all the CO.

$$12.7 \text{ mol CO} \times \frac{2 \text{ mol H}_2}{1 \text{ mol CO}} = 25.4 \text{ mol H}_2 \text{ required}$$

Because 32.2 mol of H_2 is available, but only 25.4 mol is required by the limiting reactant, 32.2 mol − 25.4 mol = 6.8 mol of H_2 is in excess. This is equivalent to 14 g of H_2.

$$6.8 \text{ mol H}_2 \times \frac{2.02 \text{ g H}_2}{1 \text{ mol H}_2} = 14 \text{ g H}_2 \text{ remaining}$$

Comment The amounts table for this reaction is

Equation	CO(g)	+	2 H₂(g)	⟶	CH₃OH(ℓ)
Initial amount (mol)	12.7		32.2		0
Change (mol)	−12.7		−2(12.7)		+12.7
After complete reaction (mol)	0		6.8		12.7

The mass of product formed plus the mass of H_2 remaining after reaction (407 g CH_3OH produced + 14 g H_2 remaining = 421 g) is equal to the mass of reactants present before reaction (356 g CO + 65.0 g H_2 = 421 g).

Exercise 4.4—A Reaction With a Limiting Reactant

Titanium is an important structural metal, and a compound of titanium, TiO_2, is the white pigment in paint. In the refining process, titanium ore (impure TiO_2) is first converted to liquid $TiCl_4$ by the following reaction.

$$TiO_2(s) + 2 Cl_2(g) + C(s) \longrightarrow TiCl_4(\ell) + CO_2(g)$$

Using 125 g each of Cl_2 and C, but plenty of TiO_2-containing ore, which is the limiting reactant in this reaction? What mass of $TiCl_4$, in grams, can be produced?

Exercise 4.5—A Reaction with a Limiting Reactant

The thermite reaction produces iron metal and aluminum oxide from a mixture of powdered aluminum metal and iron(III) oxide.

$$Fe_2O_3(s) + 2\ Al(s) \longrightarrow 2\ Fe(s) + Al_2O_3(s)$$

A mixture of 50.0 g each of Fe_2O_3 and Al is used.

(a) Which is the limiting reactant?
(b) What mass of iron metal can be produced?

Thermite reaction Iron(III) oxide reacts with aluminum metal to produce aluminum oxide and iron metal. The reaction produces so much heat that the iron melts and spews out of the reaction vessel. See Exercise 4.5.

4.5—Percent Yield

The maximum quantity of product we calculate can be obtained from a chemical reaction is the **theoretical yield**. Frequently, however, the **actual yield** of a compound—the quantity of material that is actually obtained in the laboratory or a chemical plant—is less than the theoretical yield. Some loss of product often occurs during the isolation and purification steps. In addition, some reactions do not go completely to products, and reactions are sometimes complicated by giving more than one set of products. For all these reasons, the actual yield is likely to be less than the theoretical yield (Figure 4.5).

To provide information to other chemists who might want to carry out a reaction, it is customary to report a **percent yield**. Percent yield, which specifies how much of the theoretical yield was obtained, is defined as

$$\text{Percent yield} = \frac{\text{actual yield}}{\text{theoretical yield}} \times 100\% \qquad (4.1)$$

(a)

Suppose you made aspirin in the laboratory by the following reaction:

$$C_6H_4(OH)CO_2H(s) + (CH_3CO)_2O(\ell) \longrightarrow C_6H_4(OCOCH_3)CO_2H(s) + CH_3CO_2H(\ell)$$

salicylic acid   acetic anhydride   aspirin   acetic acid

and that you began with 14.4 g of salicylic acid and an excess of acetic anhydride. That is, salicylic acid is the limiting reactant. If you obtain 6.26 g of aspirin, what is the percent yield of this product? The first step is to find the amount of the limiting reactant, salicylic acid ($C_6H_4(OH)CO_2H$).

$$14.4\ \text{g}\ C_6H_4(OH)CO_2H \times \frac{1\ \text{mol}\ C_6H_4(OH)CO_2H}{138.1\ \text{g}\ C_6H_4(OH)CO_2H} = 0.104\ \text{mol}\ C_6H_4(OH)CO_2H$$

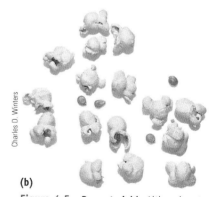

(b)

Figure 4.5 **Percent yield.** Although not a chemical reaction, popping corn is a good analogy to the difference between a theoretical yield and an actual yield. Here we began with 20 popcorn kernels and found that only 16 of them popped. The percent yield from our "reaction" was (16/20) × 100%, or 80%.

Next, use the stoichiometric factor from the balanced equation to find the amount of aspirin expected based on the limiting reactant, $C_6H_4(OH)CO_2H$.

$$0.104 \text{ mol } C_6H_4(OH)CO_2H \times \frac{1 \text{ mol aspirin}}{1 \text{ mol } C_6H_4(OH)CO_2H} = 0.104 \text{ mol aspirin}$$

The maximum amount of aspirin that can be produced—the theoretical yield—is 0.104 mol. Because the quantity you measure in the laboratory is the mass of the product, it is customary to express the theoretical yield as a mass in grams.

$$0.104 \text{ mol aspirin} \times \frac{180.2 \text{ g aspirin}}{1 \text{ mol aspirin}} = 18.7 \text{ g aspirin}$$

Finally, with the actual yield known to be only 6.26 g, the percent yield of aspirin can be calculated.

$$\text{Percent yield} = \frac{6.26 \text{ g aspirin obtained (actual yield)}}{18.7 \text{ g aspirin expected (theoretical yield)}} \times 100\% = 33.5\% \text{ yield}$$

GENERAL
Chemistry Now™

See the General ChemistryNow CD-ROM or website:
- **Screen 4.9 Percent Yield**
  - **(a)** for a tutorial on determining the theoretical yield of a reaction
  - **(b)** for a tutorial on determining the percent yield of a reaction

---

Exercise 4.6—Percent Yield

Methanol, CH_3OH, can be burned in oxygen to provide energy, or it can be decomposed to form hydrogen gas, which can then be used as a fuel (see Example 4.3).

$$CH_3OH(\ell) \longrightarrow 2 H_2(g) + CO(g)$$

If 125 g of methanol is decomposed, what is the theoretical yield of hydrogen? If only 13.6 g of hydrogen is obtained, what is the percent yield of this gas?

4.6—Chemical Equations and Chemical Analysis

Analytical chemists use a variety of approaches to identify substances as well as to measure the quantities of components of mixtures. Analytical chemistry is often done now using instrumental methods (Figure 4.6), but classical chemical reactions and stoichiometry play a central role.

Quantitative Analysis of a Mixture

Quantitative chemical analyses generally depend on one or the other of two basic ideas:

- A substance, present in unknown amount, can be allowed to react with a known quantity of another substance. If the stoichiometric ratio for their reaction is known, the unknown amount can be determined.

Figure 4.6   A modern analytical instrument. This nuclear magnetic resonance (NMR) spectrometer is closely related to a magnetic resonance imaging (MRI) instrument found in a hospital. The NMR is used to analyze compounds and to decipher their structure.

Charles D. Winters

- A material of unknown composition can be converted to one or more substances of known composition. Those substances can be identified, their amounts determined, and these amounts related to the amount of the original, unknown substance.

An example of the first type of analysis is the analysis of a sample of vinegar containing an unknown amount of acetic acid, the ingredient that makes vinegar acidic. The acid reacts readily and completely with sodium hydroxide.

$$CH_3CO_2H(aq) + NaOH(aq) \longrightarrow CH_3CO_2Na(aq) + H_2O(\ell)$$
  acetic acid

If the exact amount of sodium hydroxide used in the reaction can be measured, the amount of acetic acid present is also known. This type of analysis is the subject of a major portion of Chapter 5 [▶ Section 5.10].

The second type of analysis is exemplified by the analysis of a sample of a mineral, thenardite, which is largely sodium sulfate, Na_2SO_4, (Figure 4.7). Sodium sulfate is soluble in water. Therefore, to find the quantity of Na_2SO_4 in an impure mineral sample, we would crush the rock and then wash it thoroughly with water to dissolve the sodium sulfate. Next, we would treat this solution with barium chloride to form the water-insoluble compound barium sulfate. The barium sulfate is collected on a filter and weighed (Figure 4.8).

$$Na_2SO_4(aq) + BaCl_2(aq) \longrightarrow BaSO_4(s) + 2\ NaCl(aq)$$

Figure 4.7   Thenardite. The mineral thenardite is sodium sulfate, Na_2SO_4. It is named after the French chemist Louis Thenard (1777–1857), a co-discoverer (with Gay-Lussac and Davy) of boron. Sodium sulfate is used in making detergents, glass, and paper.

■ **Analysis and 100% Yield**
Quantitative analysis requires reactions in which the yield is 100%.

(a)

Na₂SO₄(aq), clear solution   BaCl₂(aq), clear solution

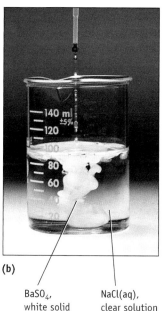

(b)

BaSO₄, white solid   NaCl(aq), clear solution

(c)

NaCl(aq), clear solution   BaSO₄, white solid caught in filter

(d)

Filter paper weighed

Active Figure 4.8   **Analysis for the sulfate content of a sample.** The sulfate ion in a solution of Na₂SO₄ reacts with barium ion (Ba²⁺) to form BaSO₄. The solid precipitate, barium sulfate (BaSO₄), is collected on a filter and weighed. The amount of BaSO₄ obtained can be related to the amount of Na₂SO₄ in the sample.

GENERAL
Chemistry꞉Now™ See the General ChemistryNow CD-ROM or website to explore an interactive version of this figure accompanied by an exercise.

We can then find the amount of sulfate in the mineral sample because it is directly related to the amount of $BaSO_4$.

$$1 \text{ mol } Na_2SO_4 \longrightarrow 1 \text{ mol } BaSO_4$$

This approach to the analysis of a mineral is one of many examples of the use of stoichiometry in chemical analysis. Examples 4.4 and 4.5 further illustrate this method.

Example 4.4—Analysis of a Lead-Containing Mineral

Problem The mineral cerussite is mostly lead carbonate, $PbCO_3$, but other substances are present. To analyze for the $PbCO_3$ content, a sample of the mineral is first treated with nitric acid to dissolve the lead carbonate.

$$PbCO_3(s) + 2 HNO_3(aq) \longrightarrow Pb(NO_3)_2(aq) + H_2O(\ell) + CO_2(g)$$

On adding sulfuric acid to the resulting solution, lead sulfate precipitates.

$$Pb(NO_3)_2(aq) + H_2SO_4(aq) \longrightarrow PbSO_4(s) + 2 HNO_3(aq)$$

Solid lead sulfate is isolated and weighed (as in Figure 4.8). Suppose a 0.583-g sample of mineral produced 0.628 g of $PbSO_4$. What is the mass percent of $PbCO_3$ in the mineral sample?

Strategy The key is to recognize that 1 mol of $PbCO_3$ will ultimately yield 1 mol of $PbSO_4$. Based on the amount of $PbSO_4$ isolated, we can calculate the amount of $PbCO_3$ (in moles), and its mass, in the original sample. When the mass of $PbCO_3$ is known, this is compared with the mass of the mineral sample to give the percent composition.

Solution Let us first calculate the amount of $PbSO_4$.

$$0.628 \text{ g } PbSO_4 \times \frac{1 \text{ mol } PbSO_4}{303.3 \text{ g } PbSO_4} = 0.00207 \text{ mol } PbSO_4$$

From stoichiometry, we can relate the amount of $PbSO_4$ to the amount of $PbCO_3$. (Here the two stoichiometric factors are based on the two balanced equations describing the chemical reactions.)

$$0.00207 \text{ mol } PbSO_4 \times \frac{1 \text{ mol } Pb(NO_3)_2}{1 \text{ mol } PbSO_4} \times \frac{1 \text{ mol } PbCO_3}{1 \text{ mol } Pb(NO_3)_2} = 0.00207 \text{ mol } PbCO_3$$

The mass of $PbCO_3$ is

$$0.00207 \text{ mol } PbCO_3 \times \frac{267.2 \text{ g } PbCO_3}{1 \text{ mol } PbCO_3} = 0.553 \text{ g } PbCO_3$$

Finally, the mass percent of $PbCO_3$ in the mineral sample is

$$\text{Mass percent of } PbCO_3 = \frac{0.553 \text{ g } PbCO_3}{0.583 \text{ g sample}} \times 100\% = 94.9\%$$

Example 4.5—Mineral Analysis

Problem Nickel(II) sulfide, NiS, occurs naturally as the relatively rare mineral millerite. One of its occurrences is in meteorites. To analyze a mineral sample for the quantity of NiS, the sample is digested in nitric acid to form a solution of $Ni(NO_3)_2$.

$$NiS(s) + 4\ HNO_3(aq) \longrightarrow Ni(NO_3)_2(aq) + S(s) + 2\ NO_2(g) + 2\ H_2O(\ell)$$

The aqueous solution of $Ni(NO_3)_2$ is then treated with the organic compound dimethylglyoxime ($C_4H_8N_2O_2$, DMG) to give the red solid $Ni(C_4H_7N_2O_2)_2$.

$$Ni(NO_3)_2(aq) + 2\ C_4H_8N_2O_2(aq) \longrightarrow Ni(C_4H_7N_2O_2)_2(s) + 2\ HNO_3(aq)$$

Suppose a 0.468-g sample containing millerite produces 0.206 g of red, solid $Ni(C_4H_7N_2O_2)_2$. What is the mass percent of NiS in the sample?

Strategy The balanced equations show the following "road map":

$$1\ mol\ NiS \longrightarrow 1\ mol\ Ni(NO_3)_2 \longrightarrow 1\ mol\ Ni(C_4H_7N_2O_2)_2$$

Thus, if we know the mass of $Ni(C_4H_7N_2O_2)_2$, we can calculate its amount and thus the amount of NiS. The amount of NiS allows us to calculate the mass and mass percent of NiS.

Solution The molar mass of $Ni(C_4H_7N_2O_2)_2$ is 288.9 g/mol. Thus, the amount of the red solid is

$$0.206\ g\ Ni(C_4H_7N_2O_2)_2 \times \frac{1\ mol\ Ni(C_4H_7N_2O_2)_2}{288.9\ g\ Ni(C_4H_7N_2O_2)_2} = 7.13 \times 10^{-4}\ mol\ Ni(C_4H_7N_2O_2)_2$$

Because 1 mol of $Ni(C_4H_7N_2O_2)_2$ is ultimately produced from 1 mol of NiS, the amount of NiS in the sample must have been 7.13×10^{-4} mol.

With the amount of NiS known, we calculate the mass of NiS.

$$7.13 \times 10^{-4}\ mol\ NiS \times \frac{90.76\ g\ NiS}{1\ mol\ NiS} = 0.0647\ g\ NiS$$

Finally, the mass percent of NiS in the 0.468-g sample is

$$\text{Mass percent NiS} = \frac{0.0647\ g\ NiS}{0.468\ g\ sample} \times 100\% = 13.8\%\ NiS$$

A precipitate of nickel. Red, insoluble $Ni(C_4H_7N_2O_2)_2$ precipitates when dimethylglyoxime ($C_4H_8N_2O_2$) is added to an aqueous solution of nickel(II) ions. (See Example 4.5.)

Exercise 4.7—Analysis of a Mixture

One method for determining the purity of a sample of titanium(IV) oxide, TiO_2, an important industrial chemical, is to combine the sample with bromine trifluoride.

$$3\ TiO_2(s) + 4\ BrF_3(\ell) \longrightarrow 3\ TiF_4(s) + 2\ Br_2(\ell) + 3\ O_2(g)$$

This reaction is known to occur completely and quantitatively. That is, all of the oxygen in TiO_2 is evolved as O_2. Suppose 2.367 g of a TiO_2-containing sample evolves 0.143 g of O_2. What is the mass percent of TiO_2 in the sample?

Determining the Formula of a Compound by Combustion

The empirical formula of a compound can be determined if the percent composition of the compound is known [◀ Section 3.6]. But where do the percent composition data come from? One chemical method that works well for compounds that burn in oxygen is *analysis by combustion*. In this technique, each element in the compound combines with oxygen to produce the appropriate oxide.

Consider an analysis of the hydrocarbon methane, CH_4, as an example of combustion analysis. A balanced equation for the combustion of methane shows that every mole of carbon in the original compound is converted to a mole of CO_2. Every mole of hydrogen in the original compound gives *half* a mole of H_2O. (Here the four moles of H atoms in one mole of CH_4 give two moles of H_2O.)

■ **Finding an Empirical Formula by Chemical Analysis**
Finding the empirical formula of a compound by chemical analysis always uses the following procedure:
1. The unknown but pure compound is decomposed into known products.
2. The reaction products are isolated in pure form and the amount of each is determined.
3. The amount of each product is related to the amount of each element in the original compound to give the empirical formula.

$$CH_4(g) \ + \ 2\,O_2(g) \ \longrightarrow \ CO_2(g) \ + \ 2\,H_2O(\ell)$$

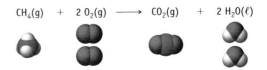

The gaseous carbon dioxide and water are separated (as illustrated in Figure 4.9) and their masses determined. From these masses it is possible to calculate the amounts of C and H in CO_2 and H_2O, respectively. The ratio of amounts of C and H in a sample of the original compound can then be found. This ratio gives the empirical formula:

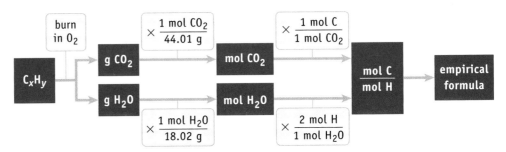

When using this procedure, a key observation is that every atom of C in the original compound appears as CO_2 and every atom of H appears in the form of water. In other words, for every mole of CO_2 observed, there must have been one mole of carbon in the unknown compound. Similarly, for every mole of H_2O observed from combustion, there must have been *two* moles of H atoms in the unknown carbon-hydrogen compound.

Example 4.6—Using Combustion Analysis to Determine the Formula of a Hydrocarbon

Problem When 1.125 g of a liquid hydrocarbon, C_xH_y, was burned in an apparatus like that shown in Figure 4.9, 3.447 g of CO_2 and 1.647 g of H_2O were produced. The molar mass of the compound was found to be 86.2 g/mol in a separate experiment. Determine the empirical and molecular formulas for the unknown hydrocarbon, C_xH_y.

Strategy As outlined in the preceding diagram, we first calculate the amounts of CO_2 and H_2O. These are then converted to amounts of C and H. The ratio (mol H/mol C) gives the empirical formula of the compound.

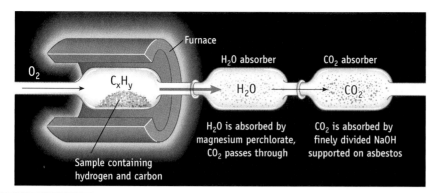

Active Figure 4.9 **Combustion analysis of a hydrocarbon.** If a compound containing C and H is burned in oxygen, CO_2 and H_2O are formed, and the mass of each can be determined. The H_2O is absorbed by magnesium perchlorate, and the CO_2 is absorbed by finely divided NaOH supported on asbestos. The mass of each absorbent before and after combustion gives the masses of CO_2 and H_2O. Only a few milligrams of a combustible compound are needed for analysis.

GENERAL
Chemistry·[•]·Now™ See the General ChemistryNow CD-ROM or website to explore an interactive version of this figure accompanied by an exercise.

Solution The amounts of CO_2 and H_2O isolated from the combustion are

$$3.447 \text{ g CO}_2 \times \frac{1 \text{ mol CO}_2}{44.010 \text{ g CO}_2} = 0.07832 \text{ mol CO}_2$$

$$1.647 \text{ g H}_2\text{O} \times \frac{1 \text{ mol H}_2\text{O}}{18.015 \text{ g H}_2\text{O}} = 0.09142 \text{ mol H}_2\text{O}$$

For every mole of CO_2 isolated, 1 mol of C must have been present in the compound C_xH_y.

$$0.07832 \text{ mol CO}_2 \times \frac{1 \text{ mol C in C}_x\text{H}_y}{1 \text{ mol CO}_2} = 0.07832 \text{ mol C}$$

For every mole of H_2O isolated, 2 mol of H must have been present in C_xH_y.

$$0.09142 \text{ mol H}_2\text{O} \times \frac{2 \text{ mol H in C}_x\text{H}_y}{1 \text{ mol H}_2\text{O}} = 0.1828 \text{ mol H in C}_x\text{H}_y$$

The original 1.125-g sample of compound therefore contained 0.07832 mol of C and 0.1828 mol of H. To determine the empirical formula of C_xH_y, we find the ratio of moles of H to moles of C [◄ Section 3.6].

$$\frac{0.1828 \text{ mol H}}{0.07832 \text{ mol C}} = \frac{2.335 \text{ mol H}}{1.000 \text{ mol C}}$$

Atoms combine to form molecules in whole-number ratios. The translation of this ratio (2.335/1) to a whole-number ratio can usually be done quickly by trial and error. Multiplying the numerator and denominator by 3 gives 7/3. So, we know the ratio is 7 mol H to 3 mol C, which means the *empirical formula* of the hydrocarbon is C_3H_7.

Comparing the experimental molar mass with the molar mass calculated for the empirical formula,

$$\frac{\text{Experimental molar mass}}{\text{Molar mass of } C_3H_7} = \frac{86.2 \text{ g/mol}}{43.1 \text{ g/mol}} = \frac{2}{1}$$

we find that the molecular formula is twice the empirical formula. That is, the *molecular formula* is $(C_3H_7)_2$, or C_6H_{14}.

Comment As noted in Problem-Solving Tip 3.3 (page 124), for problems of this type be sure to use data with enough significant figures to give accurate atom ratios. Finally, note that the determination of the molecular formula does not end the problem for a chemist. In this case, the formula C_6H_{14} is appropriate for several distinctly different compounds. Two of the five compounds having this formula are shown here:

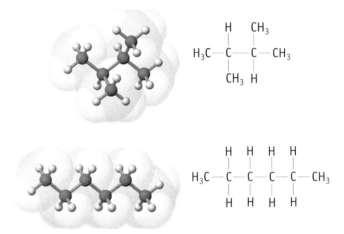

To decide finally the identity of the unknown compound, more laboratory experiments will have to be done.

Exercise 4.8—Determining the Empirical and Molecular Formulas for a Hydrocarbon

A 0.523-g sample of the unknown compound C_xH_y was burned in air to give 1.612 g of CO_2 and 0.7425 g of H_2O. A separate experiment gave a molar mass for C_xH_y of 114 g/mol. Determine the empirical and molecular formulas for the hydrocarbon.

Exercise 4.9—Determining the Empirical and Molecular Formulas for a Compound Containing C, H, and O

A 0.1342-g sample of a compound with C, H, and O ($C_xH_yO_z$) was burned in oxygen, and 0.240 g of CO_2 and 0.0982 g of H_2O were isolated. What is the empirical formula of the compound? If the experimentally determined molar mass was 74.1 g/mol, what is the molecular formula of the compound? (*Hint:* The carbon atoms in the compound are converted to CO_2 and the hydrogen atoms are converted to H_2O. The O atoms are found in both CO_2 and H_2O. To find the mass of O in the original sample, use the masses of CO_2 and H_2O to find the masses of C and H in the 0.1342 g-sample. Whatever of the 0.1342-g sample is not C and H is the mass of O.)

Chapter Goals Revisited

When you have finished studying this chapter, you should ask whether you have met the chapter goals. In particular, you should be able to

Balance equations for simple chemical reactions
a. Understand the information conveyed by a balanced chemical equation (Section 4.1).

b. Balance simple chemical equations (Section 4.2). General ChemistryNow homework: Study Question(s) 2, 12b

Perform stoichiometry calculations using balanced chemical equations
a. Understand the principle of the conservation of matter, which forms the basis of chemical stoichiometry (Section 4.3).

b. Calculate the mass of one reactant or product from the mass of another reactant or product by using the balanced chemical equation (Section 4.3). General ChemistryNow homework: SQ(s) 8, 16, 47, 53, 70, 72

c. Use amounts tables to organize stoichiometric information. General ChemistryNow homework: SQ(s) 16

Understand the impact of a limiting reactant on a chemical reaction
a. Determine which of two reactants is the limiting reactant (Section 4.4). General ChemistryNow homework: SQ(s) 22

b. Determine the yield of a product based on the limiting reactant. General ChemistryNow homework: SQ(s) 20, 24, 26

Calculate the theoretical and percent yields of a chemical reaction
a. Explain the differences among actual yield, theoretical yield, and percent yield, and calculate percent yield (Section 4.5). General ChemistryNow homework: SQ(s) 27

Use stoichiometry to analyze a mixture of compounds or to determine the formula of a compound
a. Use stoichiometry principles to analyze a mixture (Section 4.6). General ChemistryNow homework: SQ(s) 31, 69, 77

b. Find the empirical formula of an unknown compound using chemical stoichiometry (Section 4.6). General ChemistryNow homework: SQ(s) 37, 42, 66

General Chemistry Now™
See the General ChemistryNow CD-ROM or website to:
- Assess your understanding with homework questions keyed to each goal
- Check your readiness for an exam by taking the exam-prep quiz and exploring the resources in the personalized Learning Plan it provides

Key Equation

Equation 4.1 (page 157) Calculating percent yield.

$$\text{Percent yield} = \frac{\text{actual yield (g)}}{\text{theoretical yield (g)}} \times 100\%$$

One beaker contains a solution of KI; the other contains a solution of $Pb(NO_3)_2$. When the solution in one beaker is poured completely into the other, the following reaction occurs:

$$2\ KI(aq) + Pb(NO_3)_2(aq) \longrightarrow 2\ KNO_3(aq) + PbI_2(s)$$

Solutions after reaction.

What is the total mass of the beakers and solutions after reaction? Explain completely. (*See the General ChemistryNow Screen 4.3, Exercise 1.*)

53. ■ Some metal halides react with water to produce the metal oxide and the appropriate hydrogen halide (see photo). For example,

$$TiCl_4(\ell) + 2\ H_2O(\ell) \longrightarrow TiO_2(s) + 4\ HCl(g)$$

(a) Name the four compounds involved in this reaction.
(b) If you begin with 14.0 mL of $TiCl_4$ ($d = 1.73$ g/mL), what mass of water, in grams, is required for complete reaction?
(c) What mass of each product is expected?

54. The reaction of 750. g each of NH_3 and O_2 was found to produce 562 g of NO (see pages 153–155).

$$4\ NH_3(g) + 5\ O_2(g) \longrightarrow 4\ NO(g) + 6\ H_2O(g)$$

(a) What mass of water is produced by this reaction?
(b) What quantity of O_2 is required to consume 750. g of NH_3?

55. Sodium azide, the explosive chemical used in automobile airbags, is made by the following reaction:

$$NaNO_3 + 3\ NaNH_2 \longrightarrow NaN_3 + 3\ NaOH + NH_3$$

If you combine 15.0 g of $NaNO_3$ (85.0 g/mol) with 15.0 g of $NaNH_2$, what mass of NaN_3 is produced?

56. Iodine is made by the reaction

$$2\ NaIO_3(aq) + 5\ NaHSO_3(aq) \longrightarrow$$
$$3\ NaHSO_4(aq) + 2\ Na_2SO_4(aq) + H_2O(\ell) + I_2(aq)$$

(a) Name the two reactants.
(b) If you wish to prepare 1.00 kg of I_2, what mass of $NaIO_3$ is required? What mass of $NaHSO_3$?

57. Copper(I) sulfide reacts with O_2 upon heating to give copper metal and sulfur dioxide.
(a) Write a balanced equation for the reaction.
(b) What mass of copper metal can be obtained from 500. g of copper(I) sulfide?

58. Saccharin, an artificial sweetener, has the formula $C_7H_5NO_3S$. Suppose you have a sample of a saccharin-containing sweetener with a mass of 0.2140 g. After decomposition to free the sulfur and convert it to the SO_4^{2-} ion, the sulfate ion is trapped as water-insoluble $BaSO_4$ (see Figure 4.8). The quantity of $BaSO_4$ obtained is 0.2070 g. What is the mass percent of saccharin in the sample of sweetener?

59. ▲ Boron forms an extensive series of compounds with hydrogen, all with the general formula B_xH_y.

$$B_xH_y(s) + \text{excess } O_2(g) \longrightarrow \tfrac{x}{2} B_2O_3(s) + \tfrac{y}{2} H_2O(g)$$

If 0.148 g of B_xH_y gives 0.422 g of B_2O_3 when burned in excess O_2, what is the empirical formula of B_xH_y?

60. ▲ Silicon and hydrogen form a series of compounds with the general formula Si_xH_y. To find the formula of one of them, a 6.22-g sample of the compound is burned in oxygen. All of the Si is converted to 11.64 g of SiO_2, and all of the H is converted to 6.980 g of H_2O. What is the empirical formula of the silicon compound?

61. ▲ Menthol, from oil of mint, has a characteristic odor. The compound contains only C, H, and O. If 95.6 mg of menthol burns completely in O_2, and gives 269 mg of CO_2 and 110 mg of H_2O, what is the empirical formula of menthol?

62. ▲ Quinone, a chemical used in the dye industry and in photography, is an organic compound containing only C, H, and O. What is the empirical formula of the compound if 0.105 g of the compound gives 0.257 g of CO_2 and 0.0350 g of H_2O when burned completely in oxygen?

63. ▲ In the Simulation portion of Screen 4.8 of the General ChemistryNow CD-ROM or website, choose the reaction of $FeCl_2$ and Na_2S.
(a) Write the balanced equation for the reaction.
(b) Choose 40 g of Na_2S as one reactant and add 40 g of $FeCl_2$. What is the limiting reactant?

▲ More challenging   ■ In General ChemistryNow   Blue-numbered questions answered in Appendix O

(c) What mass of FeS is produced?

(d) What mass of Na_2S or $FeCl_2$ remains after the reaction?

(e) What mass of $FeCl_2$ is required to react completely with 40 g of Na_2S?

64. Sulfuric acid can be prepared starting with the sulfide ore, cuprite (Cu_2S). If each S atom in Cu_2S leads to one molecule of H_2SO_4, what mass of H_2SO_4 can be produced from 3.00 kg of Cu_2S?

65. ▲ In an experiment 1.056 g of a metal carbonate, containing an unknown metal M, is heated to give the metal oxide and 0.376 g CO_2.

$$MCO_3(s) + heat \longrightarrow MO(s) + CO_2(g)$$

What is the identity of the metal M?

(a) M = Ni   (c) M = Zn

(b) M = Cu   (d) M = Ba

66. ■ ▲ An unknown metal reacts with oxygen to give the metal oxide, MO_2. Identify the metal based on the following information:

Mass of metal = 0.356 g

Mass of sample after converting metal completely to oxide = 0.452 g

67. ▲ Titanium(IV) oxide, TiO_2, is heated in hydrogen gas to give water and a new titanium oxide, Ti_xO_y. If 1.598 g of TiO_2 produces 1.438 g of Ti_xO_y, what is the formula of the new oxide?

68. ▲ Thioridazine, $C_{21}H_{26}N_2S_2$, is a pharmaceutical used to regulate dopamine. (Dopamine, a neurotransmitter, affects brain processes that control movement, emotional response, and ability to experience pleasure and pain.) A chemist can analyze a sample of the pharmaceutical for the thioridazine content by decomposing it to convert the sulfur in the compound to sulfate ion. This is then "trapped" as water-insoluble barium sulfate (see Figure 4.8).

$$SO_4^{2-}(aq, from\ thioridazine) + BaCl_2(aq) \longrightarrow$$
$$BaSO_4(s) + 2\ Cl^-(aq)$$

Suppose a 12-tablet sample of the drug yielded 0.301 g of $BaSO_4$. What is the thioridazine content, in milligrams, of each tablet?

69. ■ ▲ A herbicide contains 2,4-D (2,4-dichlorophenoxyacetic acid), $C_8H_6Cl_2O_3$. A 1.236-g sample of the herbicide was decomposed to liberate the chlorine as Cl^- ion. This was precipitated as AgCl, with a mass of 0.1840 g. What is the mass percent of 2,4-D in the sample?

70. ■ ▲ Potassium perchlorate is prepared by the following sequence of reactions:

$$Cl_2(g) + 2\ KOH(aq) \longrightarrow KCl(aq) + KClO(aq) + H_2O(\ell)$$
$$3\ KClO(aq) \longrightarrow 2\ KCl(aq) + KClO_3(aq)$$
$$4\ KClO_3(aq) \longrightarrow 3\ KClO_4(aq) + KCl(aq)$$

What mass of $Cl_2(g)$ is required to produce 234 kg of $KClO_4$?

71. ▲ Commercial sodium "hydrosulfite" is 90.1% pure $Na_2S_2O_4$. The sequence of reactions used to prepare the compound is

$$Zn(s) + 2\ SO_2(g) \longrightarrow ZnS_2O_4(s)$$
$$ZnS_2O_4(s) + Na_2CO_3(aq) \longrightarrow ZnCO_3(s) + Na_2S_2O_4(aq)$$

(a) What mass of pure $Na_2S_2O_4$ can be prepared from 125 kg of Zn, 500 g of SO_2, and an excess of Na_2CO_3?

(b) What mass of the commercial product would contain the $Na_2S_2O_4$ produced using the amounts of reactants in part (a)?

72. ■ What mass of lime, CaO, can be obtained by heating 125 kg of limestone that is 95.0% by mass $CaCO_3$?

$$CaCO_3(s) \longrightarrow CaO(s) + CO_2(g)$$

73. Sulfuric acid can be produced from a sulfide ore such as iron pyrite by the following sequence of reactions:

$$4\ FeS_2(s) + 11\ O_2(g) \longrightarrow 2\ Fe_2O_3(s) + 8\ SO_2(g)$$
$$2\ SO_2(g) + O_2(g) \longrightarrow 2\ SO_3(g)$$
$$SO_3(g) + H_2O(\ell) \longrightarrow H_2SO_4(\ell)$$

Starting with 525 kg of FeS_2 (and an excess of other reactants), what mass of pure H_2SO_4 can be prepared?

74. ▲ The elements silver, molybdenum, and sulfur combine to form Ag_2MoS_4. What is the maximum mass of Ag_2MoS_4 that can be obtained if 8.63 g of silver, 3.36 g of molybdenum, and 4.81 g of sulfur are combined?

75. ▲ A mixture of butene, C_4H_8, and butane, C_4H_{10}, is burned in air to give CO_2 and water. Suppose you burn 2.86 g of the mixture and obtain 8.80 g of CO_2 and 4.14 g of H_2O. What is the weight percents of butene and butane in the mixture?

76. ▲ Cloth can be waterproofed by coating it with a silicone layer. This is done by exposing the cloth to $(CH_3)_2SiCl_2$ vapor. The silicon compound reacts with OH groups on the cloth to form a waterproofing film (density = $1.0\ g/cm^3$) of $[(CH_3)_2SiO]_n$, where n is a large integer number.

$$n(CH_3)_2SiCl_2 + 2n\ OH^- \longrightarrow$$
$$2n\ Cl^- + n\ H_2O + [(CH_3)_2SiO]_n$$

The coating is added layer by layer, each layer of $[(CH_3)_2SiO]_n$ being 0.60 nm thick. Suppose you want to waterproof a piece of cloth that is 3.00 m square, and you want 250 layers of waterproofing compound on the cloth. What mass of $(CH_3)_2SiCl_2$ do you need?

▲ More challenging   ■ In General ChemistryNow   Blue-numbered questions answered in Appendix O

77. ▲ Sodium hydrogen carbonate, NaHCO₃, can be decomposed quantitatively by heating.

$$2\,NaHCO_3(s) \longrightarrow Na_2CO_3(s) + CO_2(g) + H_2O(g)$$

A 0.682-g sample of impure NaHCO₃ yielded a solid residue (consisting of Na₂CO₃ and other solids) with a mass of 0.467 g. What was the mass percent of NaHCO₃ in the sample?

78. ▲ Copper metal can be prepared by roasting copper ore, which can contain cuprite (Cu₂S) and copper(II) sulfide.

$$Cu_2S(s) + O_2(g) \longrightarrow 2\,Cu(s) + SO_2(g)$$
$$CuS(s) + O_2(g) \longrightarrow Cu(s) + SO_2(g)$$

Suppose an ore sample contains 11.0% impurity in addition to a mixture of CuS and Cu₂S. Heating 100.0 g of the mixture produces 75.4 g of copper metal with a purity of 89.5%. What is the weight percent of CuS in the ore? The weight percent of Cu₂S?

Summary and Conceptual Questions

The following questions use concepts from the preceding chapters.

79. ▲ A weighed sample of iron (Fe) is added to liquid bromine (Br₂) and allowed to react completely. The reaction produces a single product, which can be isolated and weighed. The experiment was repeated a number of times with different masses of iron but with the same mass of bromine. (See the graph below.)

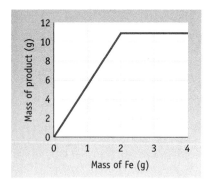

(a) What mass of Br₂ is used when the reaction consumes 2.0 g of Fe?
(b) What is the mole ratio of Br₂ to Fe in the reaction?
(c) What is the empirical formula of the product?
(d) Write the balanced chemical equation for the reaction of iron and bromine.
(e) What is the name of the reaction product?
(f) Which statement or statements best describe the experiments summarized by the graph?
   (i) When 1.00 g of Fe is added to the Br₂, Fe is the limiting reagent.
   (ii) When 3.50 g of Fe is added to the Br₂, there is an excess of Br₂.

   (iii) When 2.50 g of Fe is added to the Br₂, both reactants are used up completely.
   (iv) When 2.00 g of Fe is added to the Br₂, 10.0 g of product is formed. The percent yield must therefore be 20.0%.

80. Chlorine and iodine react according to the balanced equation

$$I_2(g) + 3\,Cl_2(g) \longrightarrow 2\,ICl_3(g)$$

Suppose that you mix I₂ and Cl₂ in a flask and that the mixture is represented by the diagram below.

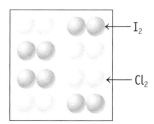

When the reaction between the Cl₂ and I₂ (according to the balanced equation above) is complete, which panel below represents the outcome? Which compound is the limiting reactant?

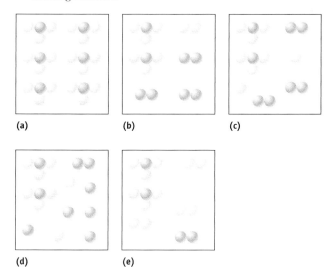

81. Cisplatin [Pt(NH₃)₂Cl₂] is a cancer chemotherapy agent. Notice that it contains NH₃ groups attached to platinum.
(a) What is the weight percent of Pt, N, and Cl in the cisplatin?
(b) Cisplatin is made by reacting K₂PtCl₄ with ammonia.

$$K_2PtCl_4(aq) + 2\,NH_3(aq) \longrightarrow Pt(NH_3)_2Cl_2(aq) + 2\,KCl(aq)$$

If you begin with 16.0 g of K₂PtCl₄, what mass of ammonia should be used to completely consume the K₂PtCl₄? What mass of cisplatin will be produced?

82. Iron(III) chloride is produced by the reaction of iron and chlorine (Figure 4.2).
    (a) If you place 1.54 g of iron gauze in chlorine gas, what mass of chlorine is required for complete reaction? What mass of iron(III) chloride is produced?
    (b) Iron(III) chloride reacts readily with NaOH to produce iron(III) hydroxide and sodium chloride. If you mix 2.0 g of iron(III) chloride with 4.0 g of NaOH, what mass of iron(III) hydroxide is produced? (*See the General ChemistryNow Screen 4.8 Simulation.*)

83. Let us explore a reaction with a limiting reactant. (*See the General ChemistryNow Screen 4.8.*) Here zinc metal is added to a flask containing aqueous HCl, and H_2 gas is a product.

    $$Zn(s) + 2\,HCl(aq) \longrightarrow ZnCl_2(aq) + H_2(g)$$

    The three flasks each contain 0.100 mol of HCl. Zinc is added to each flask in the following quantities.

    Flask 1: 7.00 g Zn

    Flask 2: 3.27 g Zn

    Flask 3: 1.31 g Zn

When the reactants are combined, the H_2 inflates the balloon attached to the flask. The results are as follows:

Flask 1: Balloon inflates completely but some Zn remains when inflation ceases.

Flask 2: Balloon inflates completely. No Zn remains.

Flask 3: Balloon does not inflate completely. No Zn remains.

Explain these results completely. Perform calculations that support your explanation.

84. The reaction of aluminum and bromine is pictured in Figure 3.1 and below. The white solid on the lip of the beaker at the end of the reaction is Al_2Br_6. In the reaction pictured below, which was the limiting reactant, Al or Br_2? (*See General ChemistryNow Screen 4.2.*)

Before reaction   After reaction

Do you need a live tutor for homework problems?
Access vMentor at General ChemistryNow at
http://now.brookscole.com/kotz6e
for one-on-one tutoring from a chemistry expert

5—Reactions in Aqueous Solution

API/Explorer/Photo Researchers, Inc.

Volcanoes are the chief source of chloride ion in the earth's oceans.

Salt

There is a French legend about a princess who told her father, the king, that she loved him as much as she loved salt. Thinking that this was not a great measure of love, he banished her from the kingdom. Only later did he realize how much he needed, and valued, salt.

Salt has played a key role in history. The earliest written record of salt production dates from around 800 B.C., but the sea has always been a source of salt. Indeed, there is evidence of the Chinese harvesting salt from sea water by 6000 B.C.

Saltiness is one of the basic taste sensations, and a taste of sea water quickly reveals its nature. How did the oceans become salty? What, in addition to salt, is dissolved in sea water?

Sea water contains enormous amounts of dissolved salts. Ions of virtually every element are present as well as dozens of polyatomic ions. What is their origin? And why is chloride ion the most abundant ion?

The carbonate ion and its close relative HCO_3^-, the bicarbonate ion, can come from the interaction of atmospheric CO_2 with water.

$$CO_2(g) + H_2O(\ell) \longrightarrow H_2CO_3(aq)$$

$$H_2CO_3(aq) \longrightarrow H^+(aq) + HCO_3^-(aq)$$

The reaction of CO_2 and H_2O is the reason rain is normally acidic. The slightly acidic rainwater then causes substances such as limestone or corals to dissolve, producing calcium ions and more bicarbonate ions.

$$CaCO_3(s) + CO_2(g) + H_2O(\ell) \longrightarrow Ca^{2+}(aq) + 2\ HCO_3^-(aq)$$

Magnesium ions come from a similar reaction with the mineral dolomite (a mixture of $CaCO_3$ and $MgCO_3$), which is found in terrestrial rocks such as those in Arizona's Grand Canyon and Italy's Dolomite Mountains. ($MgCl_2$ is often found with sea salt and gives the salt a bitter taste.)

Chapter Goals

See Chapter Goals Revisited (page 221). Test your knowledge of these goals by taking the exam-prep quiz on the General ChemistryNow CD-ROM or website.

- Understand the nature of ionic substances dissolved in water.
- Recognize common acids and bases and understand their behavior in aqueous solution.
- Recognize and write equations for the common types of reactions in aqueous solution.
- Recognize common oxidizing and reducing agents and identify oxidation–reduction reactions.
- Define and use molarity in solution stoichiometry.

Sodium ions arrive in the oceans by a similar reaction with sodium-bearing minerals such as albite, $NaAlSi_3O_8$. Acidic rain falling on the land extracts sodium ions that rivers then carry to the ocean.

The average chloride content of rocks in the earth's crust is only 0.01%, so only a minute proportion of the chloride ions in the oceans can come from the weathering of rocks and minerals. What, then, is the origin of the chloride ions in sea water? Volcanoes. Hydrogen chloride gas, HCl, is an important constituent of volcanic gases. Early in earth's history, the planet was much hotter, and volcanoes were much more widespread. The HCl gas emitted from these volcanoes is very soluble in water and quickly dissolves to give a dilute solution of hydrochloric acid.

$$HCl(g) \longrightarrow H^+(aq) + Cl^-(aq)$$

The chloride ions from dissolved HCl gas and the sodium ions from weathered rocks are the source of the salt in the sea.

The average human body contains about 230 g of salt. Because we continually lose salt in urine, sweat, and other excretions, salt must be a part of our diet. Early humans recognized that salt deficiency causes headaches, cramps, loss of appetite, and, in extreme cases, death. Consuming meat provides salt, but consuming vegetables does not. This is the reason why herbivorous animals seek out "salt licks."

Early humans also learned that salt preserves other materials. Egyptians used salt to make mummies, and fish and meat are often preserved by salting. This ability to protect against decay led to the Jewish tradition of bringing salt to a new home. In medieval France, salt was placed on the tongue of a newborn child and a young child was salted.

The importance of salt in society is reflected in a 16th-century book of table etiquette. It was written that salt could be handled safely only with the middle two fingers. If a person were to use a thumb, his children will die, and using the index finger would cause one to become a murderer.

Salt is so indispensible that it has been, not surprisingly, a source of revenue for governments. One example, which led to an extremely abusive tax, occurred in India in the 20th century. In colonial times the British established a salt tax and outlawed the production of salt from sea water. Salt could only be purchased from British government agents at a price established by the British. What is more, even though the salt tax was eliminated in Great Britain in the 18th century, the tax on salt was doubled in India in 1923. To protest this tax, in March 1930 Mahatma Gandhi led a pilgrimage to the sea, joined by thousands, to collect salt. Thousands were jailed, but strikes and demonstrations continued. A year later the salt tax was relaxed, and Britain's monopoly on salt was broken. This event marked the beginning of the end of British rule in India, and the country became independent in 1947.

For an account of salt in history and society, read *Salt, A World History*, by Mark Kurlansky (New York, Penguin Books, 2003).

Paul Stephan-Vierow/Photo Researchers, Inc.

The Dead Sea. This sea in the Middle East has the highest salt content of any body of water.

To Review Before You Begin
- Review the names of common ions (Section 3.3 and Table 3.1)
- Know how to do mass-to-moles and moles-to-mass calculations

The human body is two-thirds water. Water is essential because it is involved in every function of the body. It assists in transporting nutrients and waste products in and out of cells and is necessary for all digestive, absorption, circulatory, and excretory functions. We turn now to the study of **aqueous** solutions, chemical systems in which water plays a major role.

5.1—Properties of Compounds in Aqueous Solution

A **solution** is a homogeneous mixture of two or more substances. One substance is generally considered the **solvent**, the medium in which another substance, the **solute**, is dissolved. To understand reactions occurring in aqueous solution, it is important first to understand something about the behavior of compounds in water. The focus here is on compounds that produce ions when dissolved in water.

Ions in Aqueous Solution: Electrolytes

The water you drink every day and the oceans of the world contain many ions, most of which result from dissolving solid materials present in the environment (Table 5.1).

Dissolving an ionic solid requires separating each ion from the oppositely charged ions that surround it in the solid state. Water is especially good at dissolving ionic compounds, because each water molecule has a positively charged end and a negatively charged end. When an ionic compound dissolves in water, each negative ion becomes surrounded by water molecules with their positive ends pointing toward it, and each positive ion becomes surrounded by the negative ends of several water molecules (Figure 5.1).

Table 5.1   **Concentrations of Some Cations and Anions in the Environment and in Living Cells**

Element	Dissolved Species	Sea Water	*Valonia**	Red-Blood Cells	Blood Plasma
Chlorine	Cl^-	550	50	50	100
Sodium	Na^+	460	80	11	160
Magnesium	Mg^{2+}	52	50	2.5	2
Calcium	Ca^{2+}	10	1.5	10^{-4}	2
Potassium	K^+	10	400	92	10
Carbon	HCO_3^-, CO_3^{2-}	30	<10	<10	30
Phosphorus	HPO_4^{2-}	<1	5	3	<3

Data are taken from J. J. R. Fraústo da Silva and R. J. P. Williams: *The Biological Chemistry of the Elements*, Oxford, UK, Clarendon Press, 1991. Concentrations are given in millimoles per liter. (A millimole is 1/1000 of a mole.)

Valonia are single-celled algae that live in sea water.

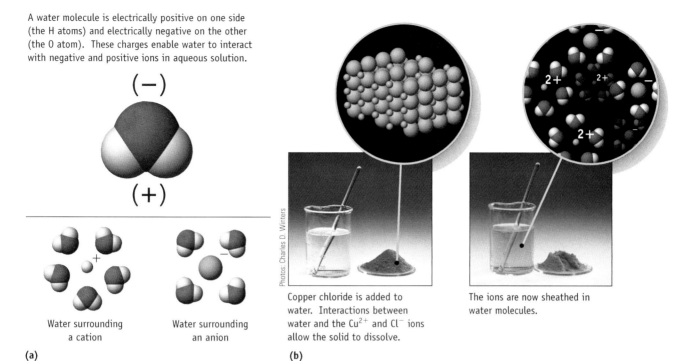

A water molecule is electrically positive on one side (the H atoms) and electrically negative on the other (the O atom). These charges enable water to interact with negative and positive ions in aqueous solution.

(−)

(+)

Water surrounding a cation

Water surrounding an anion

(a)

Photos: Charles D. Winters

Copper chloride is added to water. Interactions between water and the Cu^{2+} and Cl^- ions allow the solid to dissolve.

The ions are now sheathed in water molecules.

(b)

Figure 5.1 **Water as a solvent for ionic substances.** (a) Water can bind to both positive cations and negative anions in aqueous solution. (b) When an ionic substance dissolves in water, each ion is surrounded by a sheath of water molecules. (The number of H_2O molecules around an ion is often 6.)

The water-encased ions produced by dissolving an ionic compound are free to move about in solution. Under normal conditions, the movement of ions is random, and the cations and anions from a dissolved ionic compound are dispersed uniformly throughout the solution. However, if two **electrodes** (conductors of electricity such as copper wire) are placed in the solution and connected to a battery, ion movement is no longer random. Positive cations move through the solution to the negative electrode, and negative anions move to the positive electrode (Figure 5.2). If a light bulb is inserted into the circuit, the bulb lights, showing that ions are available to conduct charge in the solution, just as electrons conduct charge in the wire part of the circuit. Compounds whose aqueous solutions conduct electricity are called **electrolytes**. *All ionic compounds that are soluble in water are electrolytes.*

Types of Electrolytes

For every mole of NaCl that dissolves, 1 mol of Na^+ ions and 1 mol of Cl^- ions enter the solution.

$$NaCl(s) \longrightarrow Na^+(aq) + Cl^-(aq)$$
$$100\% \text{ Dissociation} \equiv \text{strong electrolyte}$$

Because the solute has dissociated completely into ions, the solution will be a good conductor of electricity. Substances whose solutions are good electrical conductors owing to the presence of ions are called **strong electrolytes** (see Figure 5.2).

Other substances dissociate only partially in solution and so are poor conductors of electricity; they are known as **weak electrolytes** (see Figure 5.2). For example,

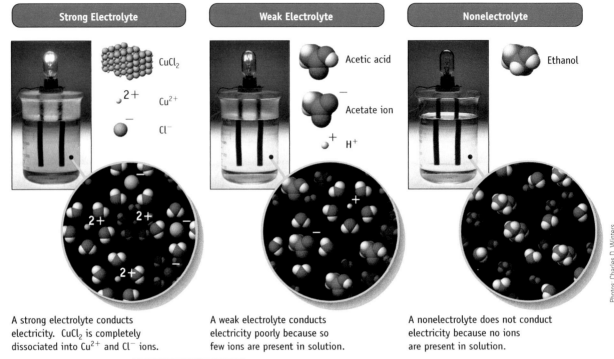

| Strong Electrolyte | Weak Electrolyte | Nonelectrolyte |

A strong electrolyte conducts electricity. $CuCl_2$ is completely dissociated into Cu^{2+} and Cl^- ions.

A weak electrolyte conducts electricity poorly because so few ions are present in solution.

A nonelectrolyte does not conduct electricity because no ions are present in solution.

Active Figure 5.2   Classifying solutions by their ability to conduct electricity.

GENERAL
Chemistry ⚛ Now™ See the General ChemistryNow CD-ROM or website to explore an interactive version of this figure accompanied by an exercise.

when acetic acid—an important ingredient in vinegar—dissolves in water, only a few molecules in every 100 molecules of acetic acid are ionized to form acetate and hydrogen ions.

■ **Double arrows,** ⇌
The double arrows in the equation for the ionization of acetic acid, and in many other chemical equations, indicate the reactant produces the product, but also that the product ions recombine to produce the original reactant. This is the subject of chemical equilibrium [▶Chapters 16–18].

$$CH_3CO_2H(aq) \rightleftharpoons CH_3CO_2^-(aq) + H^+(aq)$$

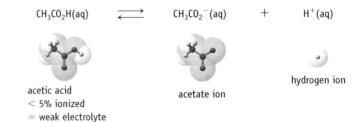

acetic acid
< 5% ionized
= weak electrolyte

acetate ion

hydrogen ion

Many other substances dissolve in water but do not ionize. They are called **nonelectrolytes** because their solutions do not conduct electricity (see Figure 5.2). Examples of nonelectrolytes include sucrose ($C_{12}H_{22}O_{11}$), ethanol (CH_3CH_2OH), and antifreeze (ethylene glycol, $HOCH_2CH_2OH$).

GENERAL
Chemistry ⚛ Now™

See the General ChemistryNow CD-ROM or website:
• **Screen 5.2 Solutions,** for a video and an animation on the dissolving of an ionic compound
• **Screen 5.3 Compounds in Aqueous Solution,** for an animation on the types of electrolytes

Photos: Charles D. Winters

SILVER COMPOUNDS

AgNO$_3$ AgCl   AgOH

(a) Nitrates are generally soluble, as are chlorides (except AgCl). Hydroxides are generally not soluble.

SULFIDES

(NH$_4$)$_2$S   CdS   Sb$_2$S$_3$ PbS

(b) Sulfides are generally not soluble (exceptions include salts with NH$_4^+$ and Na$^+$).

HYDROXIDES

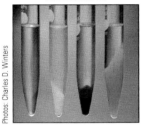

Photos: Charles D. Winters

NaOH   Ca(OH)$_2$ Fe(OH)$_3$ Ni(OH)$_2$

(c) Hydroxides are generally not soluble except when the cation is a Group 1A metal.

SOLUBLE COMPOUNDS	EXCEPTIONS
Almost all salts of Na$^+$, K$^+$, NH$_4^+$	

Salts of nitrate, NO$_3^-$
     chlorate, ClO$_3^-$
     perchlorate, ClO$_4^-$
     acetate, CH$_3$CO$_2^-$	

Almost all salts of Cl$^-$, Br$^-$, I$^-$	Halides of Ag$^+$, Hg$_2^{2+}$, Pb^{2+}
Compounds containing F$^-$	Fluorides of Mg^{2+}, Ca^{2+}, Sr^{2+}, Ba^{2+}, Pb^{2+}
Salts of sulfate, SO$_4^{2-}$	Sulfates of Ca^{2+}, Sr^{2+}, Ba^{2+}, Pb^{2+}

INSOLUBLE COMPOUNDS	EXCEPTIONS

Most salts of carbonate, CO$_3^{2-}$
     phosphate, PO$_4^{3-}$
     oxalate, C$_2$O$_4^{2-}$
     chromate, CrO$_4^{2-}$
	Salts of NH$_4^+$ and the alkali metal cations

Most metal sulfides, S^{2-}	
Most metal hydroxides and oxides	Ba(OH)$_2$ is soluble

Active Figure 5.3   **Guidelines to predict the solubility of ionic compounds.** If a compound contains one of the ions in the column to the left in the top chart, it is predicted to be at least moderately soluble in water. There are a few exceptions, which are noted at the right. Most ionic compounds formed by the anions listed at the bottom of the chart are poorly soluble (with exceptions such as compounds with NH$_4^+$ and the alkali metal cations).

GENERAL
Chemistry·ᵢ·Now™ See the General ChemistryNow CD-ROM or website to explore an interactive version of this figure accompanied by an exercise.

Exercise 5.1—Electrolytes

Epsom salt, MgSO$_4$ · 7 H$_2$O, is sold in drugstores and, as a solution in water, is used for various medical purposes. Methanol, CH$_3$OH, is dissolved in gasoline in the winter in colder climates to prevent the formation of ice in automobile fuel lines. Which of these compounds is an electrolyte and which is a nonelectrolyte?

Solubility of Ionic Compounds in Water

Not all ionic compounds dissolve completely in water. Many dissolve only to a small extent, and still others are essentially insoluble. Fortunately, we can make some general statements about which ionic compounds are water soluble.

Figure 5.3 lists broad guidelines that help predict whether a particular ionic compound will be soluble in water. For example, sodium nitrate, NaNO$_3$, contains

■ **Solubility Guidelines**
Observations such as those shown in
Figure 5.3 were used to create the
solubility guidelines. Note, however, that
these are general guidelines and not rules
followed under any circumstance. Some
exceptions do exist, but the guidelines are
a good place to begin. See B. Blake,
Journal of Chemical Education, Vol. 80,
pp. 1348–1350, 2003.

both an alkali metal cation, Na^+, and the nitrate anion, NO_3^-. The presence of either of these ions ensures that the compound is soluble in water. By contrast, calcium hydroxide is poorly soluble in water (Figure 5.3c). If a spoonful of solid $Ca(OH)_2$ is added to 100 mL of water, only 0.17 g, or 0.0023 mol, will dissolve at 10 °C. Very few Ca^{2+} and OH^- ions are present in solution. Nearly all of the $Ca(OH)_2$ remains as a solid.

0.0023 mol $Ca(OH)_2$ dissolves in 100 mL water at 10 °C $\longrightarrow$
$$0.0023 \text{ mol } Ca^{2+}(aq) + (2 \times 0.0023) \text{ mol } OH^-(aq)$$

GENERAL
Chemistry ·⚛·Now™

See the General ChemistryNow CD-ROM or website:
● **Screen 5.4 Solubility of Ionic Compounds**

(a) for a simulation exploring the rules for predicting whether a compound is soluble or insoluble

(b) for a tutorial on determining whether a compound is soluble in water

■ **Soluble Ionic Compounds =
Electrolytes**
Ionic compounds that dissolve in water are
electrolytes. For example, an aqueous solution of $AgNO_3$ (Figure 5.3a) consists of the
separated ions $Ag^+(aq)$ and $NO_3^-(aq)$ and
is a good conductor of electricity.

Example 5.1—Solubility Guidelines

Problem Predict whether the following ionic compounds are likely to be water-soluble. List the ions present in solution for soluble compounds.

(a) KCl   (c) Fe_2O_3

(b) $MgCO_3$ (d) $Cu(NO_3)_2$

Strategy You must first recognize the cation and anion involved and then decide the probable water solubility based on the guidelines outlined in Figure 5.3.

Solution

(a) KCl is composed of K^+ and Cl^- ions. The presence of *either* of these ions means that the compound is likely to be soluble in water. The solution consists of K^+ and Cl^- ions.

$$KCl(s) \longrightarrow K^+(aq) + Cl^-(aq)$$

(The solubility of KCl is about 35 g in 100 mL of water at 20 °C.)

(b) Magnesium carbonate is composed of Mg^{2+} and CO_3^{2-} ions. Salts containing the carbonate ion are usually insoluble, unless combined with an ion like Na^+ or NH_4^+. Therefore, $MgCO_3$ is predicted to be insoluble in water. (The experimental solubility of $MgCO_3$ is less than 0.2 g/100 mL of water.)

(c) Iron(III) oxide is composed of Fe^{3+} and O^{2-} ions. Oxides are soluble only when O^{2-} is combined with an alkali metal ion; Fe^{3+} is a transition metal ion, so Fe_2O_3 is insoluble.

(d) Copper(II) nitrate is composed of Cu^{2+} and NO_3^- ions. Almost all nitrates are soluble in water, so $Cu(NO_3)_2$ is water-soluble and produces copper(II) cations and nitrate anions in water.

$$Cu(NO_3)_2(s) \longrightarrow Cu^{2+}(aq) + 2 \text{ } NO_3^-(aq)$$

Comment Notice that $Cu(NO_3)_2$ gives one Cu^{2+} ion and *two* NO_3^- ions on dissolving in water.

Exercise 5.2—Solubility of Ionic Compounds

Predict whether each of the following ionic compounds is likely to be soluble in water. If it is soluble, write the formulas of the ions present in aqueous solution.

(a) $LiNO_3$ **(b)** $CaCl_2$ **(c)** CuO **(d)** $NaCH_3CO_2$

5.2—Precipitation Reactions

A **precipitation reaction** produces a water-insoluble product, known as a **precipitate**. The reactants in such reactions are generally water-soluble ionic compounds. When these substances dissolve in water, they dissociate to give the appropriate cations and anions. If the cation from one compound can form an insoluble compound with the anion from the other compound in the solution, precipitation occurs. For example, silver nitrate and potassium chloride, both of which are water-soluble ionic compounds, form insoluble silver chloride and soluble potassium nitrate (Figure 5.4).

$$AgNO_3(aq) + KCl(aq) \longrightarrow AgCl(s) + KNO_3(aq)$$

Reactants	Products
$Ag^+(aq) + NO_3^-(aq)$	Insoluble AgCl
$K^+(aq) + Cl^-(aq)$	$K^+(aq) + NO_3^-(aq)$

Many combinations of positive and negative ions give insoluble substances (see Figures 5.4 and 5.5). For example, lead(II) chromate precipitates when a water soluble lead(II) compound is combined with a water-soluble chromate compound (Figure 5.5a).

$$Pb(NO_3)_2(aq) + K_2CrO_4(aq) \longrightarrow PbCrO_4(s) + 2\ KNO_3(aq)$$

Reactants	Products
$Pb^{2+}(aq) + 2\ NO_3^-(aq)$	Insoluble $PbCrO_4$
$2\ K^+(aq) + CrO_4^{2-}(aq)$	$2\ K^+(aq) + 2\ NO_3^-(aq)$

■ **Exchange Reactions**
When two ionic compounds in aqueous solution react to form a solid precipitate, they do so by exchanging ions. For example, silver(I) ions exchange nitrate ions for chloride ions, and potassium ions exchange chloride ions for nitrate ions.

$$Ag^+ + NO_3^-$$
$$K^+ + Cl^-$$

Photo, *a*, Charles D. Winters; *b–d*, model from an animation by Roy Tasker, University of Western Sydney, Australia

(a)

(b) Initially the Ag^+ ions (silver color) and Cl^- ions (green) are widely separated.

(c) Ag^+ and Cl^- ions approach and form ion pairs.

(d) As more and more Ag^+ and Cl^- ions come together, a precipitate of solid AgCl forms.

Figure 5.4 **Precipitation of silver chloride.** (a) Mixing aqueous solutions of silver nitrate and potassium chloride produces white, insoluble silver chloride, AgCl. In (b) through (d) you see a model of the process.

Figure 5.5   Precipitation reactions.
Many ionic compounds are insoluble in water. Guidelines for predicting the solubilities of ionic compounds are given in Figure 5.3.

(a) $Pb(NO_3)_2$ and K_2CrO_4 produce yellow, insoluble $PbCrO_4$ and soluble KNO_3.

(b) $Pb(NO_3)_2$ and $(NH_4)_2S$ produce black, insoluble PbS and soluble NH_4NO_3.

(c) $FeCl_3$ and $NaOH$ produce red, insoluble $Fe(OH)_3$ and soluble $NaCl$.

(d) $AgNO_3$ and K_2CrO_4 produce red, insoluble Ag_2CrO_4 and soluble KNO_3. See Example 5.2.

Charles D. Winters

Almost all metal sulfides are insoluble in water (Figure 5.5b). In nature, if a soluble metal compound comes in contact with a source of sulfide ions, the metal sulfide precipitates.

$$Pb(NO_3)_2(aq) + (NH_4)_2S(aq) \longrightarrow PbS(s) + 2\,NH_4NO_3(aq)$$

Reactants	Products
$Pb^{2+}(aq) + 2\,NO_3^{-}(aq)$	Insoluble PbS
$2\,NH_4^{+}(aq) + S^{2-}(aq)$	$2\,NH_4^{+}(aq) + 2\,NO_3^{-}(aq)$

In fact, this process is how many sulfur-containing minerals such as iron pyrite (see page 19) are believed to have been formed. (The black "smoke" from undersea volcanoes consists of precipitated metal sulfides arising from sulfide anions and metal cations in the volcanic emissions; see page 140.)

Finally, with the exception of the alkali metal cations (and Ba^{2+}), all metal cations form insoluble hydroxides. Thus, water-soluble iron(III) chloride and sodium hydroxide react to give insoluble iron(III) hydroxide (Figures 5.3c and 5.5c).

$$FeCl_3(aq) + 3\,NaOH(aq) \longrightarrow Fe(OH)_3(s) + 3\,NaCl(aq)$$

Reactants	Products
$Fe^{3+}(aq) + 3\,Cl^{-}(aq)$	Insoluble $Fe(OH)_3$
$3\,Na^{+}(aq) + 3\,OH^{-}(aq)$	$3\,Na^{+}(aq) + 3\,Cl^{-}(aq)$

Example 5.2—Writing the Equation for a Precipitation Reaction

Problem Is an insoluble product formed when aqueous solutions of potassium chromate and silver nitrate are mixed? If so, write the balanced equation.

Strategy First decide which ions are formed in solution when the reactants dissolve. Then use information in Figure 5.3 to determine whether a cation from one reactant will combine with an anion from the other reactant to form an insoluble compound.

Solution Both reactants—$AgNO_3$ and K_2CrO_4—are water-soluble. The ions Ag^+, NO_3^-, K^+, and CrO_4^{2-} are released into solution when these compounds are dissolved.

$$AgNO_3(s) \longrightarrow Ag^+(aq) + NO_3^-(aq)$$

$$K_2CrO_4(s) \longrightarrow 2\ K^+(aq) + CrO_4^{2-}(aq)$$

Here Ag^+ could combine with CrO_4^{2-}, and K^+ could combine with NO_3^-. The former combination, Ag_2CrO_4, is an insoluble compound, whereas KNO_3 is soluble in water. Thus, the balanced equation for the reaction of silver nitrate and potassium chromate is

$$2\ AgNO_3(aq) + K_2CrO_4(aq) \longrightarrow Ag_2CrO_4(s) + 2\ KNO_3(aq)$$

Comment This reaction is illustrated in Figure 5.5d.

Exercise 5.3—Precipitation Reactions

In each of the following cases, does a precipitation reaction occur when solutions of two water-soluble reactants are mixed? Give the formula of any precipitate that forms, and write a balanced chemical equation for the precipitation reactions that occur.

(a) Sodium carbonate is mixed with copper(II) chloride.
(b) Potassium carbonate is mixed with sodium nitrate.
(c) Nickel(II) chloride is mixed with potassium hydroxide.

Net Ionic Equations

An aqueous solution of silver nitrate contains Ag^+ and NO_3^- ions, and an aqueous solution of potassium chloride contains K^+ and Cl^- ions. When these solutions are mixed (Figure 5.4), insoluble AgCl precipitates, and the ions K^+ and NO_3^- remain in solution.

$$\underset{\text{before reaction}}{Ag^+(aq) + NO_3^-(aq) + K^+(aq) + Cl^-(aq)} \longrightarrow \underset{\text{after reaction}}{AgCl(s) + K^+(aq) + NO_3^-(aq)}$$

■ **Net ionic equations**
1. All chemical equations must be balanced. The same number of atoms of each kind must appear on both the product and reactant sides. In addition, the sum of positive and negative charges must be the same on both sides of the equation.
2. See Problem-Solving Tip 5.1, page 185.

The K^+ and NO_3^- ions are present in solution before and after reaction, so they appear on both the reactant and product sides of the balanced chemical equation. Such ions are often called **spectator ions** because they do not participate in the net reaction; they merely "look on" from the sidelines. Little chemical information is lost if the equation is written without them, and so we can simplify the equation to

$$Ag^+(aq) + Cl^-(aq) \longrightarrow AgCl(s)$$

The balanced equation that results from leaving out the spectator ions is the **net ionic equation** for the reaction. *Only the aqueous ions and nonelectrolytes* (which can be insoluble compounds, soluble molecular compounds such as sugar, weak acids or bases (page 177), or gases) *that participate in a chemical reaction need to be included in the net ionic equation.*

Leaving out the spectator ions does not imply that K^+ and NO_3^- ions are unimportant in the $AgNO_3$ + KCl reaction. Indeed, Ag^+ and Cl^- ions cannot exist alone in solution; a negative ion must be present to balance the positive ion charge of Ag^+, for example. Any anion will do, however, as long as it forms water-soluble compounds with Ag^+. Thus, we could have used $AgClO_4$ instead of $AgNO_3$ and NaCl instead of KCl. The net ionic equation would have been the same.

Finally, notice that there must be a *charge balance* as well as a mass balance in a balanced chemical equation. In the $Ag^+ + Cl^-$ net ionic equation, the cation and anion charges on the left add together to give a net charge of 0, the same as the 0 charge on $AgCl(s)$ on the right.

GENERAL
Chemistry•♦•Now™

See the General ChemistryNow CD-ROM or website:
- **Screen 5.7 Net Ionic Equations,** for a tutorial on writing net ionic equations

---

◼ Dissolving Halides
When an ionic compound with halide ions dissolves in water, the halide ions are released into aqueous solution. Thus, $BaCl_2$ produces one Ba^{2+} ion and two Cl^- ions for each Ba^{2+} ion (and not Cl_2 or Cl_2^{2-} ions).

Precipitation reaction. The reaction of barium chloride and sodium sulfate produces insoluble barium sulfate and water-soluble sodium chloride. See Example 5.3.

Example 5.3—Writing and Balancing Net Ionic Equations

Problem Write a balanced, net ionic equation for the reaction of aqueous solutions of $BaCl_2$ and Na_2SO_4 to give $BaSO_4$ and $NaCl$.

Strategy First, write a balanced equation for the overall reaction. Next, decide which compounds are soluble in water (Figure 5.3) and determine the ions that these compounds produce in solution. Finally, eliminate ions that appear on both the reactant and product sides of the equation.

Solution

Step 1. Write the balanced equation.

$$BaCl_2 + Na_2SO_4 \longrightarrow BaSO_4 + 2\,NaCl$$

Step 2. Decide on the solubility of each compound. Compounds containing sodium ions are always water-soluble, and those containing chloride ions are almost always soluble. Sulfate salts are also usually soluble, with one important exception being $BaSO_4$. We can therefore write

$$BaCl_2(aq) + Na_2SO_4(aq) \longrightarrow BaSO_4(s) + 2\,NaCl(aq)$$

Step 3. Identify the ions in solution. All soluble ionic compounds dissociate to form ions in aqueous solution. (All are electrolytes.)

$$BaCl_2(s) \longrightarrow Ba^{2+}(aq) + 2\,Cl^-(aq)$$
$$Na_2SO_4(s) \longrightarrow 2\,Na^+(aq) + SO_4^{2-}(aq)$$
$$NaCl(s) \longrightarrow Na^+(aq) + Cl^-(aq)$$

This results in the following ionic equation:

$$Ba^{2+}(aq) + 2\,Cl^-(aq) + 2\,Na^+(aq) + SO_4^{2-}(aq) \longrightarrow BaSO_4(s) + 2\,Na^+(aq) + 2\,Cl^-(aq)$$

Step 4. Identify and eliminate the spectator ions (Na^+ and Cl^-) to give the net ionic equation.

$$Ba^{2+}(aq) + SO_4^{2-}(aq) \longrightarrow BaSO_4(s)$$

Notice that the sum of ion charges is the same on both sides of the equation. On the left, 2+ and 2− give zero; on the right, the charge on $BaSO_4$ is also zero.

Comment The steps followed in this example represent a general approach to writing net ionic equations.

Problem-Solving Tip 5.1

Writing Net Ionic Equations

Net ionic equations are commonly written for chemical reactions in aqueous solution because they describe the actual chemical species involved in a reaction. To write net ionic equations we must know which compounds exist as ions in solution.

1. Strong acids, soluble strong bases, and soluble salts exist as ions in solution. Examples include the acids HCl and HNO_3, a base such as NaOH, and salts such as NaCl and $CuCl_2$ (see Figures 5.1–5.3).

2. All other species should be represented by their complete formulas. Weak acids such as acetic acid (CH_3CO_2H) exist in solutions primarily as molecules. (See Section 5.3.) Insoluble salts such as $CaCO_3$(s) or insoluble bases such as $Mg(OH)_2$(s) should not be written in ionic form, even though they are ionic compounds.

The best way to approach writing net ionic equations is to follow precisely a set of steps.

1. Write a complete, balanced equation. Indicate the state of each substance (aq, s, ℓ, g).

2. Rewrite the equation, writing all strong acids, strong soluble bases, and soluble salts as ions. Look carefully at species labeled with an "(aq)" suffix.

3. Some ions may remain unchanged in the reaction (the ions that appear in the equation as both reactants and products). These spectator ions are not part of the chemistry that is going on. You can cancel them from each side of the equation.

Here are three general net ionic equations it is helpful to remember:

- The net ionic equation for the reaction between any strong acid and any soluble strong base is $H^+(aq) + OH^-(aq) \longrightarrow H_2O(\ell)$.

- The equation for the reaction of any weak acid HX (such as HCN, HF, HOCl, CH_3CO_2H) and a soluble strong base is $HX + OH^-(aq) \longrightarrow H_2O(\ell) + X^-(aq)$. (See Section 5.3.)

- The net ionic equation for the reaction of ammonia with any weak acid HX is $NH_3(aq) + HX(aq) \longrightarrow NH_4^+(aq) + X^-(aq)$ and with a strong acid it is $NH_3(aq) + H^+(aq) \longrightarrow NH_4^+(aq)$. (See Section 5.3.)

Finally, like molecular equations, net ionic equations must be balanced. The same number of atoms appears on each side of the arrow. But, an additional requirement applies. The sum of the ion charges on the two sides must be equal.

Exercise 5.4—Net Ionic Equations

Write balanced net ionic equations for each of the following reactions:

(a) $AlCl_3 + Na_3PO_4 \longrightarrow AlPO_4 + NaCl$ (not balanced)
(b) Solutions of iron(III) chloride and potassium hydroxide give iron(III) hydroxide and potassium chloride when combined. See Figure 5.5c.
(c) Solutions of lead(II) nitrate and potassium chloride give lead(II) chloride and potassium nitrate when combined.

5.3—Acids and Bases

Acids and bases, two important classes of compounds, have some related properties. Solutions of acids or bases, for example, can change the colors of vegetable pigments (Figure 5.6). You may have seen acids change the color of litmus, a dye derived from certain lichens, from blue to red. If an acid has made blue litmus paper turn red, then adding a base reverses the effect, making the litmus blue again. Thus, acids and bases seem to be opposites. A base can neutralize the effect of an acid, and an acid can neutralize the effect of a base.

Acids

Acids have characteristic properties. They produce bubbles of CO_2 gas when added to a metal carbonate such as $CaCO_3$, and they react with many metals to produce hydrogen gas, H_2, (Figure 5.6). Although tasting substances is *never* done in a chemistry laboratory, you have probably experienced the sour taste of acids such as acetic

(a) The juice of a red cabbage is normally blue-purple. On adding acid, the juice becomes more red. Adding base produces a yellow color.

(b) A piece of coral (mostly CaCO₃) dissolves in acid to give CO₂ gas.

(c) Zinc reacts with hydrochloric acid to produce zinc chloride and hydrogen gas.

Figure 5.6   Some properties of acids and bases. (a) The colors of natural dyes, such as the juice from a red cabbage, are affected by acids and bases. (b) Acids react readily with coral ($CaCO_3$) and other metal carbonates to produce gaseous CO_2 (and a salt). (c) Acids react with many metals to produce hydrogen gas (and a metal salt).

acid (in vinegar) or citric acid (commonly found in fruits and added to candies and soft drinks).

The properties of acids can be interpreted in terms of a feature common to all acid molecules:

> An acid is a substance that, when dissolved in water, increases the concentration of hydrogen ions, H^+, in the solution.

Hydrochloric acid, an aqueous solution of gaseous HCl, is a common acid. In water, hydrogen chloride ionizes to form a hydrogen ion, $H^+(aq)$, and a chloride ion, $Cl^-(aq)$.

$$HCl(aq) \longrightarrow H^+(aq) + Cl^-(aq)$$

hydrochloric acid
strong electrolyte
= 100% ionized

Because it is completely converted to ions in aqueous solution, HCl is a **strong acid** (and a strong electrolyte). See Table 5.2 for a list of other common acids.

Many acids, such as sulfuric acid, can provide more than 1 mol of H^+ per mole of acid. This occurs in two steps.

Strong Acid: $$H_2SO_4(aq) \longrightarrow H^+(aq) + HSO_4^-(aq)$$
sulfuric acid   hydrogen ion   hydrogen
100% ionized   sulfate ion

Weak Acid: $$HSO_4^-(aq) \rightleftharpoons H^+(aq) + SO_4^{2-}(aq)$$
hydrogen sulfate ion   hydrogen ion   sulfate ion
<100% ionized

■ **Weak Acids**
Common acids and bases are listed in Table 5.2. There are numerous other weak acids and bases, many of which are natural substances. Oxalic acid and acetic acid are among them. All of these natural acids contain CO_2H groups. (The H of this group is lost as H^+.) This structural feature is characteristic of hundreds of organic acids. (See Chapter 11.)

Oxalic acid
$H_2C_2O_4$

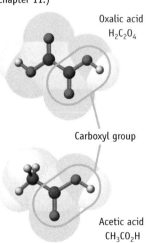

Carboxyl group

Acetic acid
CH_3CO_2H

Chemical Perspectives

Sulfuric Acid

For some years sulfuric acid has been the chemical produced in the largest quantity in the United States (and in many other industrialized countries). Approximately 40–50 billion kilograms (40–50 million metric tons) are made annually in the United States. The acid is so important to the economy of industrialized nations that some economists have said sulfuric acid production is a measure of a nation's industrial strength.

Sulfuric acid is a colorless, syrupy liquid with a density of 1.84 g/mL and a boiling point of 337 °C. It has several desirable properties that have led to its widespread use: It is generally less expensive to produce than other acids, is a strong acid, can be handled in steel containers, reacts readily with many organic compounds to produce useful products, and reacts readily with lime (CaO), the least expensive and most readily available base, to give calcium sulfate.

The first step in the industrial preparation of sulfuric acid is combustion of sulfur in air to give sulfur dioxide.

$$S_8(s) + 8\ O_2(g) \longrightarrow 8\ SO_2(g)$$

This gas is then combined with more oxygen, in the presence of a catalyst, to give sulfur trioxide,

$$2\ SO_2(g) + O_2(g) \longrightarrow 2\ SO_3(g)$$

which can give sulfuric acid when absorbed in water.

$$SO_3(g) + H_2O(\ell) \longrightarrow H_2SO_4(aq)$$

Currently more than two thirds of the production is used in the fertilizer industry, which makes "superphosphate" fertilizer by treating phosphate rock with sulfuric acid.

Sulfur is found in pure form in underground deposits along the coast of the United States in the Gulf of Mexico. It is recovered by pumping superheated steam into the sulfur beds to melt the sulfur. The molten sulfur is brought to the surface by means of compressed air.

$$2\ Ca_5F(PO_4)_3(s) + 7\ H_2SO_4(aq) + 3\ H_2O(\ell)$$
$$\longrightarrow 3\ Ca(H_2PO_4)_2 \cdot H_2O(s) + 7\ CaSO_4(s)$$
$$+ 2\ HF(g)$$

The remainder is used to make pigments, explosives, alcohol, pulp and paper, and detergents, and is employed as a component in storage batteries.

A sulfuric acid plant.

Some products that depend on sulfuric acid for their manufacture or use.

Table 5.2 Common Acids and Bases

Strong Acids (Strong Electrolytes)		Strong Bases (Strong Electrolytes)	
HCl	Hydrochloric acid	LiOH	Lithium hydroxide
HBr	Hydrobromic acid	NaOH	Sodium hydroxide
HI	Hydroiodic acid	KOH	Potassium hydroxide
HNO_3	Nitric acid		
$HClO_4$	Perchloric acid		
H_2SO_4	Sulfuric acid		
Weak Acids (Weak Electrolytes)*		**Weak Base (Weak Electrolyte)**	
H_3PO_4	Phosphoric acid	NH_3	Ammonia
H_2CO_3	Carbonic acid		
CH_3CO_2H	Acetic acid		
$H_2C_2O_4$	Oxalic acid		
$H_2C_4H_4O_6$	Tartaric acid		
$H_3C_6H_5O_7$	Citric acid		
$HC_9H_8O_4$	Aspirin		

* These are representative of hundreds of weak acids.

A Closer Look

The H⁺ Ion in Water

The H^+ ion is a hydrogen atom that has lost its electron. Only the nucleus, a proton, remains. Because a proton is only about 1/100,000 as large as the average atom or ion, water molecules can approach closely, and the proton and the water molecules are strongly attracted. In fact, the H^+ ion in water is better represented as H_3O^+, called the *hydronium ion*. This ion is formed by combining H^+ and H_2O. Experiments also show that other forms of the ion exist in water, one example being $[H_3O(H_2O)_3]^+$.

For simplicity we will use $H^+(aq)$ in this text for the hydronium and similar ions. When discussing the functions of acids in detail, however, we will use H_3O^+ [▶Chapters 17–18].

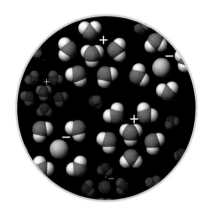

When an acid ionizes in water, it produces a hydronium ion, H_3O^+, which is surrounded by water molecules.

$$HCl(aq) \quad + \quad H_2O(\ell) \quad \longrightarrow \quad H_3O^+(aq) \quad + \quad Cl^-(aq)$$

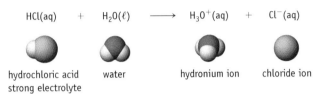

hydrochloric acid   water   hydronium ion   chloride ion
strong electrolyte
= 100% ionized

The first ionization reaction is essentially complete, so sulfuric acid is a strong acid (and, therefore, a strong electrolyte). However, the hydrogen sulfate ion (HSO_4^-), like acetic acid (Figure 5.2), is only partially ionized in aqueous solution. Both the hydrogen sulfate ion and acetic acid are therefore classified as **weak acids**.

Bases

The hydroxide ion is characteristic of bases so we can immediately recognize metal hydroxides as bases from their formulas. Although most metal hydroxides are insoluble (see Figure 5.3c), a few dissolve in water, which leads to an increase in the concentration of OH^- ions in solution.

> A base is a substance that, when dissolved in water, increases the concentration of hydroxide ions, OH^-, in the solution.

Compounds that contain hydroxide ions, such as sodium hydroxide or potassium hydroxide, are obvious bases. These water-soluble ionic compounds are **strong bases** (and strong electrolytes).

$$NaOH(s) \longrightarrow Na^+(aq) + OH^-(aq)$$

sodium hydroxide, soluble   hydroxide ion
base, strong electrolyte
= 100% dissociated

Ammonia, NH_3, another common base, does not have an OH^- ion as part of its formula. Instead, the OH^- ion is a result of the reaction with water.

$$NH_3(aq) + H_2O(\ell) \quad \rightleftharpoons \quad NH_4^+(aq) + OH^-(aq)$$

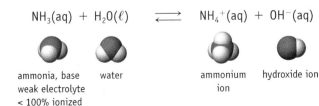

ammonia, base   water   ammonium   hydroxide ion
weak electrolyte   ion
< 100% ionized

Only a small concentration of ammonium and hydroxide ions is present in a solution of NH_3. Therefore, ammonia is a **weak base** (and a weak electrolyte). (See Figure 5.7.)

GENERAL
Chemistry·✦·Now™

See the General ChemistryNow CD-ROM or website:

- **Screen 5.8 Acids**

  (a) for a simulation exploring the degree to which different acids ionize to give H^+ ions in aqueous solution

  (b) for animations on weak and strong acids

- **Screen 5.9 Bases**, for animations of weak and strong bases

---

Exercise 5.5—Acids and Bases

(a) What ions are produced when nitric acid dissolves in water?

(b) Barium hydroxide is moderately soluble in water. What ions are produced when it dissolves in water?

Figure 5.7 Ammonia, a weak electrolyte. Ammonia, NH_3, interacts with water to produce a very small number of NH_4^+ and OH^- ions per mole of ammonia molecules.

Oxides of Nonmetals and Metals

Each acid shown in Table 5.2 has one or more H atoms in the molecular formula that dissociate in water to form H^+ ions. There are, however, less obvious compounds that form acidic solutions. Oxides of nonmetals, such as carbon dioxide and sulfur trioxide, have no H atoms but react with water to produce H^+ ions. Carbon dioxide, for example, dissolves in water to a small extent, and some of the dissolved molecules react with water to form the weak acid, carbonic acid. This acid then ionizes to a small extent to form the hydrogen ion, H^+, and the hydrogen carbonate (bicarbonate) ion, HCO_3^-.

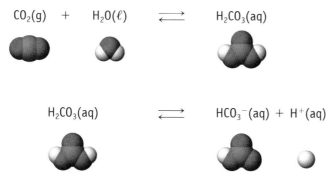

$$CO_2(g) \ + \ H_2O(\ell) \ \rightleftharpoons \ H_2CO_3(aq)$$

$$H_2CO_3(aq) \ \rightleftharpoons \ HCO_3^-(aq) \ + \ H^+(aq)$$

Like the HSO_4^- ion, the HCO_3^- ion can also function as an acid, and it can ionize to produce H^+ and the carbonate ion, CO_3^{2-}.

$$HCO_3^-(aq) \ \rightleftharpoons \ CO_3^{2-}(aq) \ + \ H^+(aq)$$

These reactions are important in our environment and in the human body. Carbon dioxide is normally found in small amounts in the atmosphere, so rainwater is always slightly acidic. In the human body, carbon dioxide is dissolved in body fluids where the HCO_3^- and CO_3^{2-} ions perform an important "buffering" action [▶ Chapter 18].

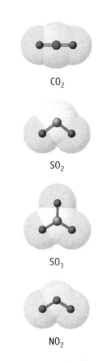

CO_2

SO_2

SO_3

NO_2

Some common nonmetal oxides that form acids in water.

Chemical Perspectives

Limelight and Metal Oxides

In the 1820s, Lt. Thomas Drummond (1797–1840) of the Royal Engineers was involved in a survey of Great Britain. During the winters he attended the famous public chemistry lectures and demonstrations by the great chemist Michael Faraday at the Royal Institution in London. There he apparently heard about the bright light that is emitted when a piece of lime, CaO, is heated to a high temperature. It occurred to him that this phenomenon could be used to make distant surveying stations visible, especially at night. Soon he developed an apparatus in which a ball of lime was heated by an alcohol flame in a stream of oxygen gas. It was reported at the time that the light from a "ball of lime not larger than a boy's marble" could be seen at a distance of 70 miles! Such lights were adapted to lighthouses and became known as Drummond lights.

Many inventions are soon adapted to warfare, and such was the case with limelights. They were used to illuminate targets in the battle of Charleston, South Carolina, during the U.S. Civil War in the 1860s. The public came to know about limelights when they moved into theaters. Gaslights were used in the early 1800s to illuminate the stage, but they were clearly not adequate. Soon after Drummond's invention, though, actors trod the boards "in the limelight."

Limelight. Metal oxides such as CaO and ThO$_2$ [thorium(IV) oxide] emit a brilliant white light when heated to incandescence.

Oxides like CO$_2$ that can react with water to produce H$^+$ ions are known as **acidic oxides**. Other acidic oxides include those of sulfur and nitrogen, which can be present in significant amounts in polluted air and can ultimately lead to acids and other pollutants. For example, sulfur dioxide, SO$_2$, from human and natural sources can react with oxygen to give sulfur trioxide, SO$_3$, which then forms sulfuric acid with water.

$$2\ SO_2(g) + O_2(g) \longrightarrow 2\ SO_3(g)$$

$$SO_3(g) + H_2O(\ell) \longrightarrow H_2SO_4(aq)$$

Nitrogen dioxide, NO$_2$, reacts with water to give nitric and nitrous acids.

$$2\ NO_2(g) + H_2O(\ell) \longrightarrow \underset{\text{nitric acid}}{HNO_3(aq)} + \underset{\text{nitrous acid}}{HNO_2(aq)}$$

These reactions are the origin of the acid in so-called acid rain. The acidic oxides arise from the burning of fossil fuels such as coal and gasoline in the United States, Canada, and other industrialized countries. The gaseous oxides mix with water and other chemicals in the troposphere, and the rain that falls is more acidic than if it contained only dissolved CO$_2$. When the rain falls on areas that cannot easily tolerate this greater than normal acidity, such as the northeastern parts of the United States and the eastern provinces of Canada, serious environmental problems can occur.

Oxides of metals are **basic oxides**, so called because they give basic solutions if they dissolve appreciably in water. Perhaps the best example is calcium oxide, CaO, often called *lime*, or *quicklime*. Almost 20 billion kg of lime is produced annually in the United States for use in the metals and construction industries, in sewage and pollution control, in water treatment, and in agriculture. This metal oxide reacts with water to give calcium hydroxide, commonly called *slaked lime*. This compound, although only slightly soluble in water (0.17 g/100 g H$_2$O at 10 °C), is widely used in industry as a base because it is inexpensive.

$$\underset{\text{lime}}{CaO(s)} + H_2O(\ell) \longrightarrow \underset{\text{slaked lime}}{Ca(OH)_2(s)}$$

See the General ChemistryNow CD-ROM or website:
• **Screen 5.9 Bases,** for a description of strong and weak bases

---

Exercise 5.6—Acidic and Basic Oxides

For each of the following, indicate whether you expect an acidic or basic solution when the compound dissolves in water. Remember that compounds based on elements in the same group usually behave similarly.

(a) SeO_2 **(b)** MgO **(c)** P_4O_{10}

5.4—Reactions of Acids and Bases

Acids and bases in aqueous solution react to produce a salt and water. For example (Figure 5.8),

$$HCl(aq) \quad + \quad NaOH(aq) \quad \longrightarrow \quad H_2O(\ell) + \quad NaCl(aq)$$

hydrochloric acid   sodium hydroxide   water   sodium chloride

The word "salt" has come into the language of chemistry as a description for any ionic compound whose cation comes from a base (here Na^+ from NaOH) and

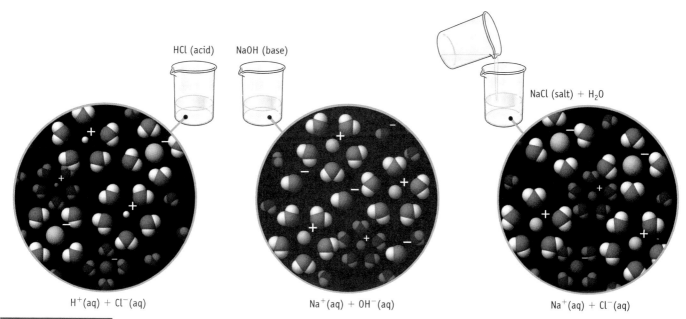

HCl (acid)   NaOH (base) $NaCl$ (salt) $+ H_2O$

$H^+(aq) + Cl^-(aq)$ $Na^+(aq) + OH^-(aq)$ $Na^+(aq) + Cl^-(aq)$

Active Figure 5.8   **An acid–base reaction, HCl and NaOH.** The acid and base consist of ions in solution. On mixing, the H^+ and OH^- ions combine to produce H_2O, whereas the ions Na^+ and Cl^- remain in solution.

whose anion comes from an acid (here Cl^- from HCl). Reaction of any of the acids listed in Table 5.2 with any of the hydroxide-containing bases listed there produces a salt and water. (The reaction of an acid with the weak base NH_3 produces only a salt [▶ Example 5.4].)

Hydrochloric acid and sodium hydroxide are strong electrolytes in water (see Figure 5.8 and Table 5.2), so the complete ionic equation for the reaction of HCl(aq) and NaOH(aq) should be written as

$$\underbrace{H^+(aq) + Cl^-(aq)}_{\text{from HCl(aq)}} + \underbrace{Na^+(aq) + OH^-(aq)}_{\text{from NaOH(aq)}} \longrightarrow \underset{\text{water}}{H_2O(\ell)} + \underbrace{Na^+(aq) + Cl^-(aq)}_{\text{from salt}}$$

Because Na^+ and Cl^- ions appear on both sides of the equation, the *net ionic equation* is just the combination of the ions H^+ and OH^- to give water.

$$H^+(aq) + OH^-(aq) \longrightarrow H_2O(\ell)$$

This is always the net ionic equation when a strong acid reacts with a strong base.

Reactions between *strong acids* and *strong bases* are called **neutralization reactions** because, on completion of the reaction, the solution is neutral; that is, it is neither acidic nor basic. The other ions (the cation of the base and the anion of the acid) remain unchanged. If the water is evaporated, however, the cation and anion form a solid salt. In the preceding example, NaCl can be obtained, whereas nitric acid, HNO_3, and NaOH give the salt sodium nitrate, $NaNO_3$ (and water).

$$HNO_3(aq) + NaOH(aq) \longrightarrow H_2O(\ell) + NaNO_3(aq)$$

One of the major uses of the basic oxide calcium oxide (lime) is in "scrubbing" sulfur oxides from the exhaust gases of power plants fueled by coal and oil. The oxides of sulfur dissolve in water to produce acids (page 190), and these acids can react with a base. Lime produces the base calcium hydroxide when added to water. A water suspension of lime is sprayed into the exhaust stack of the power plant, where it reacts with acids such as H_2SO_4 to produce $CaSO_4 \cdot 2H_2O$.

$$Ca(OH)_2(s) + H_2SO_4(aq) \longrightarrow CaSO_4 \cdot 2\ H_2O(s)$$

Hydrated calcium sulfate, $CaSO_4 \cdot 2\ H_2O$, is also found in the earth as the mineral gypsum. Assuming the gypsum from a coal-burning power plant is not contaminated with compounds that are pollutants, it is environmentally acceptable to put this substance into the earth.

Acetic acid, CH_3CO_2H, is the substance that gives the taste and odor to vinegar. Fermentation of carbohydrates such as sugar produces ethanol, CH_3CH_2OH, and the action of bacteria on the alcohol results in acetic acid. Even a trace of acetic acid will ruin the taste of wine. This characteristic is the source of the name "vinegar," which comes from the French *vin egar* meaning "sour wine." In addition to its use in food products such as salad dressings, mayonnaise, and pickles, acetic acid is used in hair-coloring products and in the manufacture of cellulose acetate, a commonly used synthetic fiber.

Acetic acid is a weak acid. Only a few acetic acid molecules are ionized to form H^+ and $CH_3CO_2^-$ ions in water (Figure 5.2).

$$CH_3CO_2H(aq) \rightleftharpoons H^+(aq) + CH_3CO_2^-(aq)$$

Nonetheless, like all acids, acetic acid will react with metal carbonates such as calcium carbonate. This carbonate is a common residue from hard water in home heating systems and cooking utensils, so washing with vinegar is a good way to clean the system or utensils because the insoluble calcium carbonate is turned into water-soluble calcium acetate (Figure 5.9).

$$2\ CH_3CO_2H(aq) + CaCO_3(s) \longrightarrow Ca(CH_3CO_2)_2(aq) + H_2O(\ell) + CO_2(g)$$

What is the net ionic equation for this reaction? Acetic acid is a weak acid, so it produces only a trace of ions in solution. Calcium carbonate is insoluble in water. Therefore, the two reactants are simply $CH_3CO_2H(aq)$ and $CaCO_3(s)$. The product, calcium acetate, is water-soluble and forms calcium and acetate ions.

$$2\ CH_3CO_2H(aq) + CaCO_3(s) \longrightarrow Ca^{2+}(aq) + 2\ CH_3CO_2{}^-(aq) + H_2O(\ell) + CO_2(g)$$

There are no spectator ions in this reaction. (See Problem Solving Tip 5.1, Writing Net Ionic Equations, page 185.)

Figure 5.9 Dissolving limestone (calcium carbonate, $CaCO_3$) in vinegar. This reaction shows why vinegar can be used as a household cleaning agent. It can be used, for example, to clean the calcium carbonate deposited from hard water in the filter in an electric coffee maker.

Charles D. Winters

Example 5.4—Net Ionic Equation for an Acid–Base Reaction

Problem Ammonia, NH_3, is one of the most important chemicals in industrial economies. Not only is it used directly as a fertilizer but it is also the raw material for the manufacture of nitric acid. As a base, it reacts with acids such as hydrochloric acid. Write a balanced, net ionic equation for this reaction.

Strategy First, write a complete balanced equation for the reaction. Next, indicate whether each reactant and product is a solid, liquid, gas, or soluble in water (aq). Then, write each water-soluble salt or any strong acids and bases as the ions they produce in water. *Insoluble solids and weak acids and bases are not written as ions.* Finally, eliminate any spectator ions to give the net ionic equation.

Solution The complete balanced equation is

$$\underset{\text{ammonia}}{NH_3(aq)} + \underset{\text{hydrochloric acid}}{HCl(aq)} \longrightarrow \underset{\text{ammonium chloride}}{NH_4Cl(aq)}$$

Notice that the reaction produces a salt, NH_4Cl. An H^+ ion from the acid transfers directly to ammonia, a weak base, to give the ammonium ion. To write the net ionic equation, start with the facts that hydrochloric acid is a strong acid and produces H^+ and Cl^- ions and that NH_4Cl is a soluble, ionic compound.

$$NH_3(aq) + H^+(aq) + Cl^-(aq) \longrightarrow NH_4{}^+(aq) + Cl^-(aq)$$

Eliminating the spectator ion, Cl^-, we have

$$NH_3(aq) + H^+(aq) \longrightarrow NH_4{}^+(aq)$$

Comment The net ionic equation shows that the important aspect of the reaction between the weak base ammonia and the strong acid HCl is the transfer of an H^+ ion from the acid to the NH_3. Any strong acid could be used here (HBr, HNO_3, $HClO_4$, H_2SO_4) and the net ionic equation would be the same.

Figure 5.10 Muffins rise because of a gas-forming reaction. The acid and sodium bicarbonate in baking powder produce carbon dioxide gas. The acid used in many baking powders is $CaHPO_4$, but $NaAl(SO_4)_2$ is also common. (The aluminum-containing compound forms an acidic solution when placed in water; See Chapter 17.)

Exercise 5.7—Acid–base Reactions

Write the balanced, overall equation and the net ionic equation for the reaction of magnesium hydroxide with hydrochloric acid.

5.5—Gas-Forming Reactions

Have you ever made biscuits or muffins? As you bake the dough, it rises in the oven (Figure 5.10). But what makes it rise? A gas-forming reaction occurs between an acid and baking soda, sodium hydrogen carbonate (bicarbonate of soda, $NaHCO_3$). One acid used for this purpose is tartaric acid, a weak acid found in many foods. The net ionic equation for a typical reaction would be

$$H_2C_4H_4O_6(aq) + HCO_3^-(aq) \longrightarrow HC_4H_4O_6^-(aq) + H_2O(\ell) + CO_2(g)$$

tartaric acid   hydrogen carbonate ion   tartrate ion

In dry baking powder, the acid and $NaHCO_3$ are kept apart by using starch as a filler. When mixed into the moist batter, however, the acid and sodium hydrogen carbonate dissolve and come into contact. Now they can react to produce CO_2, causing the dough to rise.

Several different chemical reactions lead to gas formation (Table 5.3), but the most common are those leading to CO_2 formation. All metal carbonates (and bicarbonates) react with acids to produce a salt and carbonic acid, H_2CO_3, which in turn decomposes rapidly to carbon dioxide and water (Figure 5.6b).

$$CaCO_3(s) + 2\ HCl(aq) \longrightarrow CaCl_2(aq) + H_2CO_3(aq)$$
$$H_2CO_3(aq) \longrightarrow H_2O(\ell) + CO_2(g)$$

Overall reaction: $CaCO_3(s) + 2\ HCl(aq) \longrightarrow CaCl_2(aq) + H_2O(\ell) + CO_2(g)$

If the reaction is done in an open beaker, most of the CO_2 gas bubbles out of the solution.

GENERAL
Chemistry·⚛·Now™

See the General ChemistryNow CD-ROM or website:
- **Screen 5.11 Gas Forming Reactions,** for a tutorial on identifying the type of reaction that will result from the mixing of solutions and to watch videos about four of the most important gases produced in reactions

---

■ **Gas-Forming Reactions**
Metal carbonates such as $CaCO_3$ react with acids to produce a salt and CO_2 gas. See Figure 5.6.

Table 5.3   Gas-Forming Reactions

Metal carbonate or bicarbonate + acid $\longrightarrow$ metal salt + CO_2(g) + $H_2O(\ell)$
$Na_2CO_3(aq) + 2\ HCl(aq) \longrightarrow 2\ NaCl(aq) + CO_2(g) + H_2O(\ell)$
Metal sulfide + acid $\longrightarrow$ metal salt + H_2S(g)
$Na_2S(aq) + 2\ HCl(aq) \longrightarrow 2\ NaCl(aq) + H_2S(g)$
Metal sulfite + acid $\longrightarrow$ metal salt + SO_2(g) + $H_2O(\ell)$
$Na_2SO_3(aq) + 2\ HCl(aq) \longrightarrow 2\ NaCl(aq) + SO_2(g) + H_2O(\ell)$
Ammonium salt + strong base $\longrightarrow$ metal salt + NH_3(g) + $H_2O(\ell)$
$NH_4Cl(aq) + NaOH(aq) \longrightarrow NaCl(aq) + NH_3(g) + H_2O(\ell)$

Example 5.5—Gas-Forming Reactions

Problem Write a balanced equation for the reaction that occurs when nickel(II) carbonate is treated with sulfuric acid.

Strategy First, identify the reactants and write their formulas (here $NiCO_3$ and H_2SO_4). Next, recognize this case as a typical gas-forming reaction (Table 5.3) between a metal carbonate (or metal hydrogen carbonate) and an acid. According to Table 5.3, the products are water, CO_2, and a metal salt. The anion of the metal salt is the anion from the acid (SO_4^{2-}), and the cation is from the metal carbonate (Ni^{2+}).

Solution The complete, balanced equation is

$$NiCO_3(s) + H_2SO_4(aq) \longrightarrow NiSO_4(aq) + H_2O(\ell) + CO_2(g)$$

Exercise 5.8—Gas-Forming Reactions

(a) Barium carbonate, $BaCO_3$, is used in the brick, ceramic, glass, and chemical manufacturing industries. Write a balanced equation that shows what happens when barium carbonate is treated with nitric acid. Give the name of each of the reaction products.

(b) Write a balanced equation for the reaction of ammonium sulfate with sodium hydroxide.

5.6—Classifying Reactions in Aqueous Solution

One goal of this chapter is to explore the most common types of reactions that can occur in aqueous solution. This helps you decide, for example, that a gas forming reaction occurs when an Alka-Seltzer tablet (containing citric acid and $NaHCO_3$) is dropped into water (Figure 5.11).

$$H_3C_6H_5O_7(aq) + HCO_3^-(aq) \longrightarrow H_2C_6H_5O_7^-(aq) + H_2O(\ell) + CO_2(g)$$

citric acid · · hydrogen carbonate ion · · dihydrogen citrate ion

Reactions in aqueous solution are important not only because they provide a way to make useful products, but also because these kinds of reactions occur on the earth and in plants and animals. Therefore, it is useful to look for common reaction patterns to see what their "driving forces" might be and how to predict the products. Most of the reactions described thus far in this chapter are **exchange reactions**, in which *the ions of the reactants changed partners.*

$$A^+B^- + C^+D^- \longrightarrow A^+D^- + C^+B^-$$

Recognizing that cations exchange anions gives us a good way to predict the products of precipitation, acid–base, and gas-forming reactions.

Precipitation Reactions (see Figure 5.5): Ions combine in solution to form an insoluble reaction product.

Overall Equation

$$Pb(NO_3)_2(aq) + 2\ KI(aq) \longrightarrow PbI_2(s) + 2\ KNO_3(aq)$$

Net Ionic Equation

$$Pb^{2+}(aq) + 2\ I^-(aq) \longrightarrow PbI_2(s)$$

Charles D. Winters

Figure 5.11 A Gas-Forming Reaction. An Alka-Seltzer tablet contains an acid (citric acid) and sodium hydrogen carbonate ($NaHCO_3$), the reactants in a gas-forming reaction.

Acid–Base Reactions(see Figure 5.6): Water is a product of an acid–base reaction, and the cation of the base and the anion of the acid form a salt.

Overall Equation for the Reaction of a Strong Acid and a Strong Base

$$HNO_3(aq) + KOH(aq) \longrightarrow HOH(\ell) + KNO_3(aq)$$

Net Ionic Equation for the Reaction of a Strong Acid and a Strong Base

$$H^+(aq) + OH^-(aq) \longrightarrow H_2O(\ell)$$

Overall Equation for the Reaction of a Weak Acid and a Strong Base

$$CH_3CO_2H(aq) + NaOH(aq) \longrightarrow NaCH_3CO_2(aq) + HOH(\ell)$$

Net Ionic Equation for the Reaction of a Weak Acid and a Strong Base

$$CH_3CO_2H(aq) + OH^-(aq) \longrightarrow CH_3CO_2^-(aq) + H_2O(\ell)$$

Gas-Forming Reactions (see Figures 5.9 and 5.11): The most common examples involve metal carbonates and acids but other gas-forming reactions exist (see Table 5.3). One product with a metal carbonate is always carbonic acid, H_2CO_3, most of which decomposes to H_2O and CO_2. Carbon dioxide is the gas in the bubbles you see during these reactions.

$$CuCO_3(s) + 2\ HNO_3(aq) \longrightarrow Cu(NO_3)_2(aq) + H_2CO_3(aq)$$
$$H_2CO_3(aq) \longrightarrow CO_2(g) + H_2O(\ell)$$

Overall Equation

$$CuCO_3(s) + 2\ HNO_3(aq) \longrightarrow Cu(NO_3)_2(aq) + CO_2(g) + H_2O(\ell)$$

Net Ionic Equation

$$CuCO_3(s) + 2\ H^+(aq) \longrightarrow Cu^{2+}(aq) + CO_2(g) + H_2O(\ell)$$

A Summary of Common Reaction Types in Aqueous Solution

Three common "driving forces" responsible for reactions in aqueous solution were outlined above. A fourth, to be discussed in the next section (Section 5.7), is the transfer of electrons from one substance to another. Such reactions are called oxidation–reduction processes.

Reaction Type	Driving Force
Precipitation	Formation of an insoluble compound (Section 5.2)
Acid–base; neutralization	Formation of a salt and water; proton transfer (Section 5.4)
Gas-forming	Evolution of a water-insoluble gas such as CO_2 (Section 5.5)
Oxidation–reduction	Electron transfer (Section 5.7)

These four types of reactions are usually easy to recognize, but keep in mind that a reaction may have more than one driving force. For example, barium hydroxide reacts readily with sulfuric acid to give barium sulfate and water, a reaction that is both a precipitation reaction and an acid–base reaction.

$$Ba(OH)_2(aq) + H_2SO_4(aq) \longrightarrow BaSO_4(s) + 2\ H_2O(\ell)$$

A Closer Look

Product-Favored and Reactant-Favored Reactions

The driving force for a precipitation, acid–base, or gas-forming reaction is, in each case, the formation of a product that removes ions from solution: a solid precipitate, a water molecule, or a gas molecule. These, and all other reactions in which reactants are completely or largely converted to products, are said to be **product-favored**.

The opposite of a product-favored reaction is one that is **reactant-favored.** Such reactions lead to the conversion of little, if any, of the reactants to products. An example would be the formation of hydrochloric acid and sodium hydroxide in a solution of sodium chloride in water.

Reactant-favored:

$$NaCl(aq) + H_2O(\ell)$$

$$— \textbf{X} \longrightarrow NaOH(aq) + HCl(aq)$$

This reaction, which does not occur to any measurable extent, is the opposite of an acid–base reaction.

The title of this book is *Chemistry and Chemical Reactivity*. One aspect of chemical reactivity, and a goal of this book, is to be able to predict whether a chemical reaction is product- or reactant-favored. Thus far you have learned that certain common reactions are generally product-favored. We will use this idea to organize chemistry many more times in this book, particularly in Chapters 6 and 16–19.

GENERAL
Chemistry · · Now™

See the General ChemistryNow CD-ROM or website:

- **Screen 5.5 Types of Aqueous Solutions,** to watch videos on the four reaction types

---

Exercise 5.9—Classifying Reactions

Classify each of the following reactions as a precipitation, acid–base, or gas-forming reaction. Predict the products of the reaction, and then balance the completed equation. Write the net ionic equation for each.

(a) $CuCO_3(s) + H_2SO_4(aq) \longrightarrow$

(b) $Ba(OH)_2(s) + HNO_3(aq) \longrightarrow$

(c) $CuCl_2(aq) + (NH_4)_2S(aq) \longrightarrow$

5.7—Oxidation–Reduction Reactions

The terms "oxidation" and "reduction" come from reactions that have been known for centuries. Ancient civilizations learned how to change metal oxides and sulfides to the metal—that is, how to "reduce" ore to the metal. A modern example is the reduction of iron(III) oxide with carbon monoxide to give iron metal (Figure 5.12a).

Fe_2O_3 loses oxygen and is reduced.

$$Fe_2O_3(s) + 3 CO(g) \longrightarrow 2 Fe(s) + 3 CO_2(g)$$

CO is the reducing agent. It gains oxygen and is oxidized.

In this reaction carbon monoxide is the agent that brings about the reduction of iron ore to iron metal, so it is called the **reducing agent**.

Figure 5.12   Oxidation–reduction.
(a) Iron ore, which is largely Fe_2O_3, is reduced to metallic iron with carbon or carbon monoxide in a blast furnace, a process done on a massive scale. (b) Burning magnesium metal in air produces magnesium oxide.

(a)   (b)

When Fe_2O_3 is reduced by carbon monoxide, oxygen is removed from the iron ore and added to the carbon monoxide. The carbon monoxide, therefore, is "oxidized" by the addition of oxygen to give carbon dioxide. *Any process in which oxygen is added to another substance is an oxidation.* In the reaction of oxygen with magnesium, for example (see Figure 5.12b), oxygen is the **oxidizing agent** because it is responsible for the oxidation of magnesium.

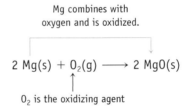

Mg combines with
oxygen and is oxidized.

$$2 \ Mg(s) + O_2(g) \longrightarrow 2 \ MgO(s)$$

O_2 is the oxidizing agent

The observations outlined here lead to several important conclusions:

- If one substance is oxidized, another substance in the same reaction must be reduced. For this reason, such reactions are often called oxidation–reduction reactions, or **redox reactions** for short.
- The reducing agent is itself oxidized, and the oxidizing agent is reduced.
- Oxidation is the opposite of reduction. For example, the removal of oxygen is reduction and the addition of oxygen is oxidation.

Redox Reactions and Electron Transfer

Not all redox reactions involve oxygen, but *all oxidation and reduction reactions involve transfer of electrons between substances.* When a substance accepts electrons, it is said to be **reduced** because there is a reduction in the positive charge on an atom of the substance. In the net ionic equation for the reaction of a silver salt with copper metal, for example, positively charged Ag^+ ions are reduced to uncharged silver atoms when they accept electrons from copper metal (Figure 5.13).

Ag$^+$ ions accept electrons from Cu and are
reduced to Ag. Ag$^+$ is the oxidizing agent.
$$Ag^+(aq) + e^- \longrightarrow Ag(s)$$

$$2\ Ag^+(aq) + Cu(s) \longrightarrow 2\ Ag(s) + Cu^{2+}(aq)$$

Cu donates electrons to Ag$^+$ and is oxidized to Cu^{2+}.
Cu is the reducing agent.
$$Cu(s) \longrightarrow Cu^{2+}(aq) + 2\ e^-$$

Because copper metal supplies the electrons and causes Ag$^+$ ions to be reduced, Cu is the *reducing agent*.

When a substance *loses electrons*, the positive charge on an atom of the substance increases. The substance is said to have been **oxidized**. In our example, copper metal releases electrons on going to Cu^{2+}, so the metal is oxidized. For this to happen, something must be available to accept the electrons from copper. In this case, Ag$^+$ is the electron acceptor, and its charge is reduced to zero in silver metal. Therefore, Ag$^+$ is the "agent" that causes Cu metal to be oxidized; that is, Ag$^+$ is the *oxidizing agent*.

In every oxidation–reduction reaction, one reactant is reduced (and is therefore the oxidizing agent) and one reactant is oxidized (and is therefore the reducing agent). We can show this by dividing the general redox reaction $X + Y \rightarrow X^{n+} + Y^{n-}$ into two parts or half-reactions:

Half Reaction	Electron Transfer	Result
$X \longrightarrow X^{n+} + ne^-$	X transfers electrons to Y.	X is **oxidized** to X^{n+}. X is the **reducing agent**.
$Y + ne^- \longrightarrow Y^{n-}$	Y accepts electron from X.	Y is **reduced** to Y^{n-}. Y is the **oxidizing agent**.

■ **Balancing Equations for Redox Reactions**
The notion that a redox reaction can be divided into an oxidizing portion and a reducing portion will lead us to a method of balancing more complex equations for redox reactions, described in Chapter 20.

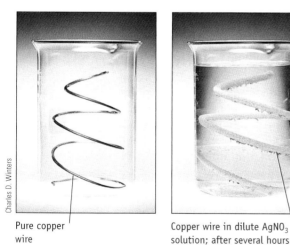

Pure copper wire

Copper wire in dilute AgNO$_3$ solution; after several hours

Blue color due to Cu^{2+} ions formed in redox reaction

Silver crystals formed after several weeks

Charles D. Winters

Figure 5.13 The oxidation of copper metal by silver ions. A clean piece of copper wire is placed in a solution of silver nitrate, AgNO$_3$. Over time, the copper reduces Ag$^+$ ions, forming silver crystals, and the copper metal is oxidized to copper ions, Cu^{2+}. The blue color of the solution is due to the presence of aqueous copper(II) ions. (*See General ChemistryNow Screen 5.12 Redox Reactions and Electron Transfer, to watch a video of the reaction.*)

In the reaction of magnesium and oxygen (see Figure 5.12b), O_2 is reduced because it gains electrons (four electrons per molecule) on going to two oxide ions. Thus, O_2 is the oxidizing agent.

Mg releases 2 e^- per atom. Mg is oxidized to Mg^{2+}
and is the reducing agent.

$$2\ Mg(s)\ +\ O_2(g) \longrightarrow 2\ MgO(s)$$

O_2 gains 4 e^- per molecule to form 2 O^{2-}. O_2 is
reduced and is the oxidizing agent.

In the same reaction, magnesium is the reducing agent because it releases two electrons per atom on being oxidized to the Mg^{2+} ion (and so two Mg atoms are required to supply the four electrons required by one O_2 molecule). All redox reactions can be analyzed in a similar manner.

Oxidation Numbers

How can you tell an oxidation–reduction reaction when you see one? How can you tell which substance has gained (or lost) electrons and so decide which substance is the oxidizing (or reducing) agent? Sometimes it is obvious. For example, if an uncombined element becomes part of a compound (Mg becomes part of MgO, for example), the reaction is definitely a redox process. If it's not obvious, then the answer is to *look for a change in the oxidation number of an element in the course of the reaction*. The **oxidation number** of an atom in a molecule or ion is defined as the charge an atom has, *or appears to have*, as determined by the following guidelines for assigning oxidation numbers.

■ **Why Use Oxidation Numbers?**
The reason for learning about oxidation numbers at this point is to be able to identify which reactions are oxidation-reduction processes and to know which is the oxidizing agent and which is the reducing agent in a reaction. We return to a more detailed discussion of redox reactions in Chapter 20.

1. **Each atom in a pure element has an oxidation number of zero.** The oxidation number of Cu in metallic copper is 0, and it is 0 for each atom in I_2 or S_8.

2. **For monatomic ions, the oxidation number is equal to the charge on the ion.** Elements of Groups 1A–3A form monatomic ions with a positive charge and an oxidation number equal to the group number. Magnesium forms Mg^{2+}, and its oxidation number is therefore +2. (See Section 3.3.)

3. **Fluorine always has an oxidation number of −1 in compounds with all other elements**.

4. **Cl, Br, and I always have oxidation numbers of −1 in compounds, except when combined with oxygen or fluorine.** This means that Cl has an oxidation number of −1 in NaCl (in which Na is +1, as predicted by the fact that it is a member of Group 1A). In the ion ClO^-, however, the Cl atom has an oxidation number of +1 (and O has an oxidation number of −2; see Guideline 5).

5. **The oxidation number of H is +1 and of O is −2 in most compounds.** Although this statement applies to most compounds, a few important exceptions occur.

   • When H forms a binary compound with a metal, the metal forms a positive ion and H becomes a hydride ion, H^-. Thus, in CaH_2 the oxidation number of Ca is +2 (equal to the group number) and that of H is −1.

   • Oxygen can have an oxidation number of −1 in a class of compounds called peroxides. For example, in H_2O_2, hydrogen peroxide, H is assigned its usual oxidation number of +1, so O is −1.

6. **The algebraic sum of the oxidation numbers for the atoms in a neutral compound must be zero; in a polyatomic ion, the sum must be equal to the ion**

A Closer Look

Are Oxidation Numbers "Real"?

Do oxidation numbers reflect the actual electric charge on an atom in a molecule or ion? With the exception of monatomic ions such as Cl^- or Na^+, the answer is no.

Oxidation numbers assume that the atoms in a molecule are positive or negative ions, which is not true. For example, in H_2O, the H atoms are not H^+ ions and the O atoms are not O^{2-} ions. This is not to say, however, that atoms in molecules do not bear an electric charge of any kind. In

Charge on O atom $= -0.4$

Charge on each H atom $= +0.2$

water, for example, calculations indicate the O atom has a charge of about -0.4 (or 40% of the electron charge) and the

H atoms are each about $+0.2$. (The partial charges on H and O in water are responsible for water molecules' ability to solvate ions in solution. See Figure 5.1.)

So why use oxidation numbers? These numbers provide a way of dividing up the electrons among the atoms in a molecule or polyatomic ion. Because the division of electrons changes in a redox reaction, we use this method as a way to decide whether a redox reaction has occurred, to distinguish the oxidizing and reducing agents, and, as you will see in Chapter 20, to balance equations for redox reactions.

charge. For example, in $HClO_4$ the H atom is assigned $+1$ and the O atom is assigned -2. This means the Cl atom must be $+7$. Additional examples are found in Example 5.6.

Example 5.6—Determining Oxidation Numbers

Problem Determine the oxidation number of the indicated element in each of the following compounds or ions:

(a) aluminum in aluminum oxide, Al_2O_3

(b) phosphorus in phosphoric acid, H_3PO_4

(c) sulfur in the sulfate ion, SO_4^{2-}

(d) each Cr atom in the dichromate ion, $Cr_2O_7^{2-}$

Strategy Follow the guidelines in the text, paying particular attention to Guidelines 5 and 6.

Solution

(a) Al_2O_3 is a neutral compound. Assuming that O has its usual oxidation number of -2, the oxidation number of Al must be $+3$, in agreement with its position in the periodic table.

$$\begin{aligned} \text{Net charge on } Al_2O_3 &= 0 \\ &= \text{sum of oxidation numbers of Al atoms} \\ &\quad + \text{sum of oxidation numbers of O atoms} \\ &= 2(+3) + 3(-2) \end{aligned}$$

(b) H_3PO_4 has an overall charge of 0. If each of the oxygen atoms has an oxidation number of -2 and each of the H atoms is $+1$, the oxidation number of P must be $+5$.

$$\begin{aligned} \text{Net charge on } H_3PO_4 &= 0 \\ &= \text{sum of oxidation numbers for H atoms} + \text{oxidation number of P} \\ &\quad + \text{sum of oxidation numbers for O atoms} \\ &= 3(+1) + (+5) + 4(-2) \end{aligned}$$

(c) The sulfate ion, SO_4^{2-}, has an overall charge of $2-$. Because this compound is not a peroxide, O is assigned an oxidation number of -2, which means that S has an oxidation number of $+6$.

$$\begin{aligned} \text{Net charge on } SO_4^{2-} &= -2 \\ &= \text{oxidation number of S} + \text{sum of oxidation numbers for O atoms} \\ &= (+6) + 4(-2) \end{aligned}$$

■ **Writing Charges on Ions**
By convention, charges on ions are written as (number, sign), whereas oxidation numbers are written as (sign, number). For example, the oxidation number of the Cu^{2+} ion is $+2$ and it charge is $2+$.

(d) The net charge on $Cr_2O_7^{2-}$ ion is $2-$. Assigning each O atom an oxidation number of -2 means that each Cr atom must have an oxidation number of $+6$.

Net charge on $Cr_2O_7^{2-} = -2$

$\qquad$ = sum of oxidation numbers for Cr atoms + sum of oxidation numbers for O atoms

$\qquad$ = $2(+6) + 7(-2)$

Exercise 5.10—Determining Oxidation Numbers

Assign an oxidation number to the underlined atom in each ion or molecule.

(a) $\underline{Fe}_2O_3$ $\qquad$ **(b)** $H_2\underline{S}O_4$ $\qquad$ **(c)** $\underline{C}O_3^{2-}$ $\qquad$ **(d)** $\underline{N}O_2^+$

Recognizing Oxidation–Reduction Reactions

You can tell whether a reaction involves oxidation and reduction by assessing the oxidation number of each element and noting whether any of these numbers change in the course of the reaction. In many cases, however, this analysis will not be necessary. It will be obvious that a redox reaction has occurred if an uncombined element is converted to a compound or involves a well-known oxidizing or reducing agent (Table 5.4).

Like oxygen, O_2, the halogens (F_2, Cl_2, Br_2, and I_2) are always oxidizing agents in their reactions with metals and nonmetals. An example is the reaction of chlorine with sodium metal (see Figure 1.7).

■ **Sodium/Chlorine Reaction**
Sodium metal reduces chlorine gas. See Figure 1.7, page 19.

Charles D. Winters

Na releases 1 e^- per atom.
Oxidation number increases.
Na is oxidized to Na^+ and is the reducing agent.

$$2\ Na(s)\ +\ Cl_2(g)\ \longrightarrow\ 2\ NaCl(s)$$

Cl_2 gains 2 e^- per molecule.
Oxidation number decreases by 1 per Cl.
Cl_2 is reduced to Cl^- and is the oxidizing agent.

A chlorine molecule ends up as two Cl^- ions, having acquired two electrons (from two Na atoms). Thus, the oxidation number of each Cl atom has decreased from 0 to -1. This means Cl_2 has been reduced and so is the oxidizing agent.

Figure 5.14 illustrates the chemistry of another excellent oxidizing agent, nitric acid, HNO_3. Here copper metal is oxidized to give copper(II) nitrate, and the nitrate ion is reduced to the brown gas NO_2. The net ionic equation for the reaction is

Oxidation number of Cu changes from 0 to $+2$. Cu is oxidized to Cu^{2+} and is the reducing agent.

$$Cu(s) + 2\ NO_3^-(aq) + 4\ H^+(aq)\ \longrightarrow\ Cu^{2+}(aq) + 2\ NO_2(g) + 2\ H_2O(\ell)$$

N in NO_3^- changes from $+5$ to $+4$ in NO_2. NO_3^- is reduced to NO_2 and is the oxidizing agent.

Table 5.4 Common Oxidizing and Reducing Agents

Oxidizing Agent	Reaction Product	Reducing Agent	Reaction Product
O_2, oxygen	O^{2-}, oxide ion or O combined in H_2O	H_2, hydrogen	$H^+(aq)$, hydrogen ion or H combined in H_2O or other molecule
Halogen, F_2, Cl_2, Br_2, or I_2	Halide ion, F^-, Cl^-, Br^-, or I^-	M, metals such as Na, K, Fe, and Al	M^{n+}, metal ions such as Na^+, K^+, Fe^{2+} or Fe^{3+}, and Al^{3+}
HNO_3, nitric acid	Nitrogen oxides* such as NO and NO_2	C, carbon (used to reduce metal oxides)	CO and CO_2
$Cr_2O_7^{2-}$, dichromate ion	Cr^{3+}, chromium(III) ion (in acid solution)		
MnO_4^-, permanganate ion	Mn^{2+}, manganese(II) ion (in acid solution)		

* NO is produced with dilute HNO_3, whereas NO_2 is a product of concentrated acid.

Nitrogen has been reduced from +5 (in the NO_3^- ion) to +4 (in NO_2); therefore, the nitrate ion in acid solution is an oxidizing agent. Copper metal is the reducing agent; here each metal atom has given up two electrons to produce the Cu^{2+} ion.

In the reactions of sodium with chlorine and copper with nitric acid, the metals are oxidized. This is typical of metals, which are generally good reducing agents. Indeed, the alkali and alkaline earth metals are especially good reducing agents. An example is the reaction of potassium with water. Here potassium reduces the hydrogen in water to H_2 gas (page 82).

$$2\ K(s)\ +\ 2\ H_2O(\ell)\ \longrightarrow\ 2\ KOH(aq)\ +\ H_2(g)$$

reducing   oxidizing
agent   agent

Aluminum metal, a good reducing agent, is capable of reducing iron(III) oxide to iron metal in a reaction called the *thermite reaction* (Figure 5.15).

$$Fe_2O_3(s)\ +\ 2\ Al(s)\ \longrightarrow\ 2\ Fe(\ell)\ +\ Al_2O_3(s)$$

oxidizing   reducing
agent   agent

Such a large quantity of heat is evolved in this reaction that the iron is produced in the molten state.

Tables 5.4 and 5.5 may help you organize your thinking as you look for oxidation–reduction reactions and use their terminology.

Table 5.5 Recognizing Oxidation–Reduction Reactions

	Oxidation	Reduction
In terms of oxidation number	Increase in oxidation number of an atom	Decrease in oxidation number of an atom
In terms of electrons	Loss of electrons by an atom	Gain of electrons by an atom
In terms of oxygen	Gain of one or more O atoms	Loss of one or more O atoms

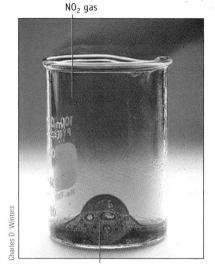

NO_2 gas

Copper metal oxidized to green $Cu(NO_3)_2$

Active Figure 5.14 The reaction of **copper with nitric acid.** Copper (a reducing agent) reacts vigorously with concentrated nitric acid, an oxidizing agent, to give the brown gas NO_2 and a deep green solution of copper(II) nitrate.

GENERAL
Chemistry·⚛·Now™ See the General ChemistryNow CD-ROM or website to explore an interactive version of this figure accompanied by an exercise.

Figure 5.15 **Thermite reaction.** Here Fe_2O_3 is reduced by aluminum metal to produce iron metal and aluminum oxide.

GENERAL
Chemistry •Now™

See the General ChemistryNow CD-ROM or website:

- **Screen 5.13 Oxidation Numbers,** for an exercise that examines the reaction between bromine and elemental aluminum, for an exercise that explores the electron transfer aspects of reactions with hydrogen peroxide, and for a tutorial on assigning oxidation numbers

- **Screen 5.14 Recognizing Oxidation–Reduction Reactions,** for an exercise examining the process of a redox reaction

---

Example 5.7—Oxidation–Reduction Reaction

Problem For the reaction of iron(II) ion with permanganate ion in aqueous acid,

$$5\ \text{Fe}^{2+}(aq) + \text{MnO}_4^-(aq) + 8\ \text{H}^+(aq) \longrightarrow 5\ \text{Fe}^{3+}(aq) + \text{Mn}^{2+}(aq) + 4\ \text{H}_2\text{O}(\ell)$$

decide which atoms are undergoing a change in oxidation number and identify the oxidizing and reducing agents.

Strategy Determine the oxidation numbers of the atoms in each ion or molecule involved in the reaction. Decide which atoms have increased in oxidation number (oxidation) and which have decreased in oxidation number (reduction).

Solution The Mn oxidation number in MnO_4^- is $+7$, and it decreases to $+2$ in the product, the Mn^{2+} ion. Thus, the MnO_4^- ion has been reduced and is the oxidizing agent (see Table 5.4).

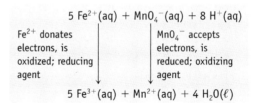

The oxidation number of iron has increased from $+2$ to $+3$, so the Fe^{2+} ion has lost electrons upon being oxidized to Fe^{3+} (see Table 5.5). This means the Fe^{2+} ion is the reducing agent.

Comment If one of the reactants in a redox reaction is a simple substance (here Fe^{2+}), it is usually obvious whether its oxidation number has increased or decreased. Once a species has been established as having been reduced (or oxidized), you know another species has been oxidized (or reduced). It is also helpful to recognize common oxidizing and reducing agents (Table 5.4).

The reaction of iron(II) ion and permanganate ion. The reaction of purple permanganate ion (MnO_4^-, the oxidizing agent) with the iron(II) ion (Fe^{2+}, the reducing agent) in acidified aqueous solution gives the nearly colorless manganese(II) ion (Mn^{2+}) and the iron(III) ion (Fe^{3+}).

Charles D. Winters

Example 5.8—Types of Reactions

Problem Classify each of the following reactions as precipitation, acid–base, gas-forming, or oxidation–reduction.

(a) $2\ \text{HNO}_3(aq) + \text{Ca(OH)}_2(s) \longrightarrow \text{Ca(NO}_3)_2(aq) + 2\ \text{H}_2\text{O}(\ell)$

(b) $\text{SO}_4^{2-}(aq) + 2\ \text{CH}_2\text{O}(aq) + 2\ \text{H}^+(aq) \longrightarrow \text{H}_2\text{S}(aq) + 2\ \text{CO}_2(g) + 2\ \text{H}_2\text{O}(\ell)$

Strategy A good strategy is first to check whether a reaction is one of the three types of exchange reactions. An acid–base reaction is usually easy to distinguish. Next, check the oxidation numbers of each element. If they change, then it is a redox reaction. If there is no change, then it is a simple precipitation or gas-forming process.

Solution Reaction (a) involves a common acid (nitric acid, HNO_3) and a common base [calcium hydroxide, $Ca(OH)_2$]; it produces a salt, calcium nitrate, and water. It is an acid–base reaction. Reaction (b) is a redox reaction because the oxidation numbers of S and C change.

$$SO_4^{2-}(aq) + 2\ CH_2O(aq) + 2\ H^+(aq) \longrightarrow H_2S(aq) + 2\ CO_2(g) + 2\ H_2O(\ell)$$

$$\ \ \ +6, -2 \qquad\quad 0, +1, -2 \qquad +1 \qquad\qquad +1, -2 \quad +4, -2 \quad +1, -2$$

The oxidation number of S changes from $+6$ to -2, and that of C changes from 0 to $+4$. Therefore, sulfate, SO_4^{2-}, has been reduced (and is the oxidizing agent), and CH_2O has been oxidized (and is the reducing agent).

No changes occur in the oxidation numbers of the elements in reaction (a).

$$HNO_3(aq) + Ca(OH)_2(s) \longrightarrow Ca(NO_3)_2(aq) + 2\ H_2O(\ell)$$

$$\ \ +1, +5, -2 \quad +2, -2, +1 \qquad +2, +5, -2 \qquad +1, -2$$

Comment If an uncombined element is a reactant or product, the reaction is a redox reaction.

Exercise 5.11—Oxidation–Reduction Reactions

The following reaction occurs in a device for testing the breath for the presence of ethanol. Identify the oxidizing and reducing agents and the substances oxidized and reduced (Figure 5.16).

$$3\ C_2H_5OH(aq) + 2\ Cr_2O_7^{2-}(aq) + 16\ H^+(aq) \longrightarrow 3\ CH_3CO_2H(aq) + 4\ Cr^{3+}(aq) + 11\ H_2O(\ell)$$

ethanol   dichromate ion;   acetic acid   chromium(III)
               orange-red   ion; green

Exercise 5.12—Oxidation–Reduction and Other Reactions

Decide which of the following reactions are oxidation–reduction reactions. In each case explain your choice and identify the oxidizing and reducing agents.

(a) $NaOH(aq) + HNO_3(aq) \longrightarrow NaNO_3(aq) + H_2O(\ell)$
(b) $Cu(s) + Cl_2(g) \longrightarrow CuCl_2(s)$
(c) $Na_2CO_3(aq) + 2\ HClO_4(aq) \longrightarrow CO_2(g) + H_2O(\ell) + 2\ NaClO_4(aq)$
(d) $2\ S_2O_3^{2-}(aq) + I_2(aq) \longrightarrow S_4O_6^{2-}(aq) + 2\ I^-(aq)$

5.8—Measuring Concentrations of Compounds in Solution

Most chemical studies require quantitative measurements, including experiments involving aqueous solutions. When doing such experiments, we continue to use balanced equations and moles, but we measure volumes of solution rather than masses of solids, liquids, or gases. Solution concentration expressed as molarity relates the volume of solution in liters to the amount of substance in moles.

Solution Concentration: Molarity

The concept of concentration is useful in many contexts. For example, about 5,500,000 people live in Wisconsin, and the state has a land area of roughly 56,000 square miles; therefore, the average concentration of people is about $(5.5 \times 10^6$ people$/5.6 \times 10^4$ square miles) or 96 people per square mile. In chemistry the

■ **Chemical Safety and Redox Reactions** It is not a good idea to mix a strong oxidizing agent with a strong reducing agent; a violent reaction—even an explosion—may take place. This reason explains why chemicals are not necessarily stored on shelves in alphabetical order. This practice can be unsafe, because such an ordering may place a strong oxidizing agent next to a strong reducing agent.

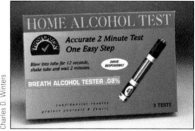

Figure 5.16 **The redox reaction of ethanol and dichromate ion is the basis of the test used in a Breathalyzer.** When ethanol, an alcohol, is poured into a solution of orange-red dichromate ion, it reduces the dichromate ion to green chromium(III) ion. The bottom photo is a breath-tester that can be purchased in grocery or drug stores. See Exercise 5.11.

Charles D. Winters

■ **Molar and Molarity**
Chemists use "molar" as an adjective to describe a solution. We use "molarity" as a noun. For example, we refer to a 0.1 molar solution or say the solution has a molarity of 0.1 mole per liter.

amount of solute dissolved in a given volume of solution, the concentration of the solution, can be found in the same way. A useful unit of solute concentration, c, is **molarity**, which is defined as *amount of solute per liter of solution.*

$$\text{Concentration } (c_{\text{molarity}}) = \frac{\text{amount of solute (mol)}}{\text{volume of solution (L)}} \qquad (5.1)$$

■ **Volumetric Flask**
A volumetric flask is a special flask with a line marked on its neck (see Figures 5.17 and 5.18). If the flask is filled with a solution to this line (at a given temperature), it contains precisely the volume of solution specified.

For example, if 58.4 g, or 1.00 mol, of NaCl is dissolved in enough water to give a total solution volume of 1.00 L, the concentration, c, is 1.00 mol/L, or 1.00 molar. This is often abbreviated as 1.00 M, where the capital "M" stands for "moles per liter." Another common notation is to place the formula of the compound in square brackets; this implies that the concentration of the solute in moles of compound per liter of solution is being specified.

$$c_{\text{molarity}} = 1.00 \text{ M} = [\text{NaCl}]$$

It is important to notice that molarity refers to the amount of solute per liter of *solution* and not per liter of *solvent*. If one liter of water is added to one mole of a solid compound, the final volume probably will not be exactly one liter, and the final concentration will not be exactly one molar (Figure 5.17). When making solutions of a given molarity, it is almost always the case that we dissolve the solute in a volume of solvent smaller than the desired volume of solution, then add solvent until the final solution volume is reached.

Potassium permanganate, $KMnO_4$, which was used at one time as a germicide in the treatment of burns, is a shiny, purple-black solid that dissolves readily in water to give a deep purple solution. Suppose 0.435 g of $KMnO_4$ has been dissolved in enough water to give 250. mL of solution (Figure 5.18). What is the molar concen-

Figure 5.17   Volume of solution versus volume of solvent.
To make a 0.100 M solution of $CuSO_4$, 25.0 g or 0.100 mol of $CuSO_4 \cdot 5\ H_2O$ (the blue crystalline solid) was placed in a 1.00-L volumetric flask.

For this photo we measured out exactly 1.00 L of water, which was slowly added to the volumetric flask containing $CuSO_4 \cdot 5\ H_2O$. When enough water had been added so that the solution volume was exactly 1.00 L, approximately 8 mL (the quantity in the small graduated cylinder) was left over from the original 1.00 L of water.

This emphasizes that molar concentrations are defined as moles of solute per liter of solution and not per liter of water or other solvent.

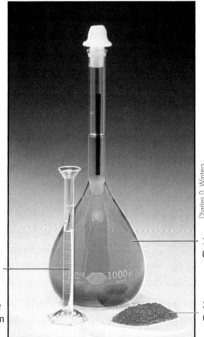

Volume of water remaining when 1.0 L of water was used to make 1.0 L of a solution

1.0 L of 0.100 M $CuSO_4$

25.0 g or 0.100 mol of $CuSO_4 \cdot 5\ H_2O$

250 mL
volumetric flask   0.435g KMnO₄

The KMnO₄ is first dissolved in a small amount of water.

Distilled water is added to fill the flask with solution just to the mark on the flask.

A mark on the neck of a volumetric flask indicates a volume of exactly 250 mL at 25 °C.

Active Figure 5.18 **Making a solution.** A 0.0110 M solution of KMnO₄ is made by adding enough water to 0.435 g of KMnO₄ to make 0.250 L of solution.

Chemistry Now™ See the General ChemistryNow CD-ROM or website to explore an interactive version of this figure accompanied by an exercise.

tration of KMnO₄? The first step is to convert the mass of KMnO₄ to an amount (moles) of solute.

$$0.435 \text{ g KMnO}_4 \times \frac{1 \text{ mol KMnO}_4}{158.0 \text{ g KMnO}_4} = 0.00275 \text{ mol KMnO}_4$$

Now that the amount of KMnO₄ is known, this information can be combined with the volume of solution—which must be in liters—to give the molarity. Because 250. mL is equivalent to 0.250 L,

$$\text{Concentration of KMnO}_4 = \frac{0.00275 \text{ mol KMnO}_4}{0.250 \text{ L}} = 0.0110 \text{ M}$$

$$[\text{KMnO}_4] = 0.0110 \text{ M}$$

The KMnO₄ concentration is 0.0110 mol/L, or 0.0110 M. This is useful information, but it is often equally useful to know the concentration of each type of ion in a solution. Like all soluble ionic compounds, KMnO₄ dissociates completely into its ions, K⁺ and MnO₄⁻, when dissolved in water.

$$\text{KMnO}_4(aq) \longrightarrow \text{K}^+(aq) + \text{MnO}_4^-(aq)$$
$$\text{100\% dissociation}$$

One mole of KMnO₄ provides 1 mol of K⁺ ions and 1 mol of MnO₄⁻ ions. Accordingly, 0.0110 M KMnO₄ gives a concentration of K⁺ in the solution of 0.0110 M; similarly, the concentration of MnO₄⁻ is also 0.0110 M.

Another example of ion concentrations is provided by the dissociation of an ionic compound such as $CuCl_2$.

$$CuCl_2(aq) \longrightarrow Cu^{2+}(aq) + 2\ Cl^-(aq)$$
100% dissociation

If 0.10 mol of $CuCl_2$ is dissolved in enough water to make 1.0 L of solution, the concentration of the copper(II) ion is $[Cu^{2+}] = 0.10$ M. The concentration of chloride ions, $[Cl^-]$, is 0.20 M because the compound dissociates in water to provide 2 mol of Cl^- ions for each mole of $CuCl_2$.

Ion concentrations for a soluble ionic compound. Here 1 mol of $CuCl_2$ dissociates to 1 mol of Cu^{2+} ions and 2 mol of Cl^- ions. Therefore, the Cl^- concentration is twice the stated concentration of $CuCl_2$.

Photo: Charles D. Winters

GENERAL
Chemistry·⚛·Now™

See the General ChemistryNow CD-ROM or website:
- **Screen 5.15 Solution Concentrations,** for a tutorial on determining solution concentration and for a tutorial on determining ion concentration

---

Example 5.9—Concentration

Problem If 25.3 g of sodium carbonate, Na_2CO_3, is dissolved in enough water to make 250. mL of solution, what is the molar concentration of Na_2CO_3? What are the concentrations of the Na^+ and CO_3^{2-} ions?

Strategy The molar concentration of Na_2CO_3 is defined as the amount of Na_2CO_3 per liter of solution. We know the volume of solution (0.250 L). We need the amount of Na_2CO_3. To find the concentrations of the individual ions, recognize that the dissolved salt dissociates completely.

$$Na_2CO_3(s) \longrightarrow 2\ Na^+(aq) + CO_3^{2-}(aq)$$

Thus, a 1 M solution of Na_2CO_3 is really a 2 M solution of Na^+ ions and 1 M solution of CO_3^{2-} ions.

Solution Let us first find the amount of Na_2CO_3,

$$25.3 \text{ g Na}_2\text{CO}_3 \times \frac{1 \text{ mol Na}_2\text{CO}_3}{106.0 \text{ g Na}_2\text{CO}_3} = 0.239 \text{ mol Na}_2\text{CO}_3$$

and then the molar concentration of Na_2CO_3,

$$\text{Concentration} = \frac{0.239 \text{ mol Na}_2\text{CO}_3}{0.250 \text{ L}} = 0.955 \text{ M}$$

$$[Na_2CO_3] = 0.955 \text{ M}$$

The ion concentrations follow from the concentration of Na_2CO_3 and the knowledge that each mole of Na_2CO_3 produces 2 mol of Na^+ ions and 1 mol of CO_3^{2-} ions.

$$0.955 \text{ M Na}_2\text{CO}_3(aq) \equiv 2 \times 0.955 \text{ M Na}^+(aq) + 0.955 \text{ M CO}_3^{2-}(aq)$$

That is, $[Na^+] = 1.91$ M and $[CO_3^{2-}] = 0.955$ M.

Exercise 5.13—Concentration

Sodium bicarbonate, $NaHCO_3$, is used in baking powder formulations and in the manufacture of plastics and ceramics, among other things. If 26.3 g of the compound is dissolved in enough water to make 200. mL of solution, what is the molar concentration of $NaHCO_3$? What are the concentrations of the ions in solution?

Preparing Solutions of Known Concentration

A task chemists often must perform is preparing a given volume of solution of known concentration. There are two commonly used ways to do this.

Combining a Weighed Solute with the Solvent

Suppose you wish to prepare 2.00 L of a 1.50 M solution of Na_2CO_3. You have some solid Na_2CO_3 and distilled water. You also have a 2.00-L volumetric flask (see Figures 5.17 and 5.18). To make the solution, you must weigh the necessary quantity of Na_2CO_3 as accurately as possible, carefully place all the solid in the volumetric flask, and then add some water to dissolve the solid. After the solid has dissolved completely, more water is added to bring the solution volume to 2.00 L. The solution then has the desired concentration and the volume specified.

But what mass of Na_2CO_3 is required to make 2.00 L of 1.50 M Na_2CO_3? First, calculate the amount of substance required,

$$2.00 \text{ L} \times \frac{1.50 \text{ mol } Na_2CO_3}{1.00 \text{ L solution}} = 3.00 \text{ mol } Na_2CO_3 \text{ required}$$

and then the mass in grams,

$$3.00 \text{ mol } Na_2CO_3 \times \frac{106.0 \text{ g } Na_2CO_3}{1 \text{ mol } Na_2CO_3} = 318 \text{ g } Na_2CO_3$$

Thus, to prepare the desired solution, you should dissolve 318 g of Na_2CO_3 in enough water to make 2.00 L of solution.

Exercise 5.14—Preparing Solutions of Known Concentration

An experiment in your laboratory requires 250. mL of a 0.0200 M solution of $AgNO_3$. You are given solid $AgNO_3$, distilled water, and a 250.-mL volumetric flask. Describe how to make up the required solution.

Diluting a More Concentrated Solution

Another method of making a solution of a given concentration is to *begin with a concentrated solution and add water until the desired, lower concentration is reached* (Figure 5.19). Many of the solutions prepared for your laboratory course are probably made by this dilution method. It is more efficient to store a small volume of a concentrated solution and then, when needed, add water to make a much larger volume of a dilute solution.

Suppose you need 500. mL of 0.0010 M potassium dichromate, $K_2Cr_2O_7$, for use in chemical analysis. You have some 0.100 M $K_2Cr_2O_7$ solution available. To make the required 0.0010 M solution, place a measured volume of the more concentrated

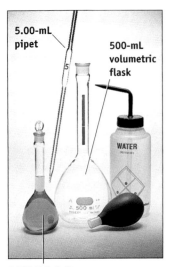

5.00-mL pipet

500-mL volumetric flask

WATER

0.100 M K₂Cr₂O₇

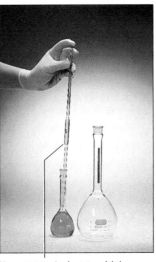

Use a 5.00-mL pipet to withdraw 5.00 mL of 0.100 M K₂Cr₂O₇ solution.

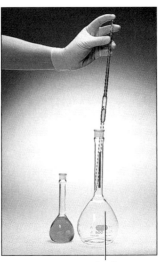

Add the 5.00-mL sample of 0.100 M K₂Cr₂O₇ solution to a 500-mL volumetric flask.

Fill the flask to the mark with distilled water to give 0.00100 M K₂Cr₂O₇ solution.

Charles D. Winters

Figure 5.19   Making a solution by dilution. Here 5.00 mL of a K₂Cr₂O₇ solution is diluted to 500. mL. This means the solution is diluted by a factor of 100, from 0.100 M to 0.00100 M *(See General ChemistryNow Screen 5.16 for a video of this procedure.)*

K₂Cr₂O₇ solution in a flask and then add water until the K₂Cr₂O₇ is contained in a larger volume of water—that is, until it is less concentrated (or more dilute) (Figure 5.19).

What volume of a 0.100 M K₂Cr₂O₇ solution must be diluted to make the 0.0010 M solution? In general, if the volume and concentration of a solution are known, the amount of solute is also known. Therefore, the amount of K₂Cr₂O₇ that must be in the final dilute solution is

$$\text{Amount of } K_2Cr_2O_7 \text{ in dilute solution } = (0.500 \text{ L})\left(\frac{0.0010 \text{ mol}}{L}\right)$$
$$= 0.00050 \text{ mol } K_2Cr_2O_7$$

Problem Solving Tips 5.2

Preparing a Solution by Dilution

The preparation of the K₂Cr₂O₇ solution and Example 5.10 suggest a way to do the calculations for dilutions. The central idea is that the amount of solute in the final, dilute solution has to be equal to the amount of solute taken from the more concentrated solution. If c is the concentration (molarity) and V is the volume (and the subscripts d and c identify the dilute and concentrated solutions, respectively), then the amount of solute in either solution (in the case of the K₂Cr₂O₇ example in the text) can be calculated as follows:

Amount of K₂Cr₂O₇ in the final dilute solution $= c_d V_d$
$$= 0.00050 \text{ mol}$$

Amount of K₂Cr₂O₇ taken from the more concentrated solution =
$$c_c V_c = 0.00050 \text{ mol}$$

Because both cV products are equal to the same amount of solute, we can use the following equation:
$$c_c V_c = c_d V_d$$
Amount of reagent in concentrated solution =
            Amount of reagent in dilute solution

This equation is valid for all cases in which a more concentrated solution is used to make a more dilute one. It can be used to find, for example, the molarity of the dilute solution, c_d, when the values of c_c, V_c, and V_d are known.

A more concentrated solution containing this amount of $K_2Cr_2O_7$ must be placed in a 500.-mL flask and then diluted to the final volume. The volume of 0.100 M $K_2Cr_2O_7$ that must be transferred and diluted is 5.0 mL.

$$0.00050 \text{ mol } K_2Cr_2O_7 \times \frac{1.00 \text{ L}}{0.100 \text{ mol } K_2Cr_2O_7} = 0.0050 \text{ L or } 5.0 \text{ mL}$$

Thus, to prepare 500. mL of 0.0010 M $K_2Cr_2O_7$, place 5.0 mL of 0.100 M $K_2Cr_2O_7$ in a 500.-mL flask and add water until a volume of 500. mL is reached (Figure 5.19).

■ **Diluting Concentrated Sulfuric Acid** The direction that one can prepare a solution by adding water to a more concentrated solution is correct except for sulfuric acid solutions. When mixing water and sulfuric acid, the resulting solution becomes quite warm. If water is added to concentrated sulfuric acid, so much heat is evolved that the solution may boil over or splash and burn someone nearby. To avoid this problem, chemists always add concentrated sulfuric acid to water to make a dilute solution.

GENERAL
Chemistry·⚛·Now™

See the General ChemistryNow CD-ROM or website:
- **Screen 5.16 Preparing Solutions of Known Concentrations,** for an exercise and a tutorial on the direct addition method of preparing a solution and for an exercise and tutorial on the dilution method of preparing a solution

EXAMPLE 5.10—Preparing a Solution by Dilution

Problem You need a 2.36×10^{-3} M solution of iron(III) ion. A lab procedure suggests this can be done by placing 1.00 mL of 0.236 M iron(III) nitrate in a volumetric flask and diluting to exactly 100.0 mL. Show that this method will work.

Strategy First calculate the amount of iron(III) ion in the 1.00-mL sample. The concentration of the ion in the final, dilute solution is equal to this amount of iron(III) divided by the new volume.

Solution The amount of iron(III) ion in the 1.00 mL sample is

$$c \times V = \frac{0.236 \text{ mol Fe}^{3+}}{L} \times 1.00 \times 10^{-3} \text{ L}$$

$$= 2.36 \times 10^{-4} \text{ mol Fe}^{3+}$$

This amount of iron(III) ion is distributed in the new volume of 100.0 mL, so the final concentration of the diluted solution is

$$[Fe^{3+}] = \frac{2.36 \times 10^{-4} \text{ mol Fe}^{3+}}{0.100 \text{ L}} = 2.36 \times 10^{-3} \text{ M}$$

Comment The experimental procedure is illustrated in Figure 5.19.

Exercise 5.15—Preparing a Solution by Dilution

In one of your laboratory experiments you are given a solution of $CuSO_4$ that has a concentration of 0.15 M. If you mix 6.0 mL of this solution with enough water to have a total volume of 10.0 mL, what is the concentration of $CuSO_4$ in the new solution?

Exercise 5.16—Preparing a Solution by Dilution

An experiment calls for you to use 250. mL of 1.00 M NaOH, but you are given a large bottle of 2.00 M NaOH. Describe how to make the 1.00 M NaOH in the desired volume.

5.9—pH, a Concentration Scale for Acids and Bases

Vinegar, which contains the weak acid acetic acid, has a hydrogen ion concentration of only 1.6×10^{-3} M and "pure" rainwater has $[H^+] = 2.5 \times 10^{-6}$ M. These extremely small values can be expressed using scientific notation, but this is awkward. A more convenient way to express such numbers is the logarithmic pH scale.

The **pH** of a solution is the negative of the base-10 logarithm of the hydrogen ion concentration.

$$pH = -\log [H^+] \tag{5.2}$$

■ **Logarithms**
Numbers less than 1 have negative logs. Defining pH as $-\log [H^+]$ produces a positive number. See Appendix A for a discussion of logs.

Taking vinegar, pure water, blood, and ammonia as examples,

pH of vinegar	$= -\log (1.6 \times 10^{-3} \text{ M}) = -(-2.80) = 2.80$
pH of pure water (at 25 °C)	$= -\log (1.0 \times 10^{-7} \text{ M}) = -(7.00) = 7.00$
pH of blood	$= -\log (4.0 \times 10^{-8} \text{ M}) = -(-7.40) = 7.40$
pH of ammonia	$= -\log (1.0 \times 10^{-11} \text{ M}) = -(-11.00) = 11.00$

■ **pH of Pure Water**
Highly purified water, which is said to be "neutral," has a pH of exactly 7 at 25 °C. This is the "dividing line" between acidic substances (pH < 7) and basic substances (pH > 7).

you see that something you recognize as acidic has a relatively low pH, whereas ammonia, a common base, has a *very* low hydrogen ion concentration and a high pH. Blood, which your common sense tells you is likely to be neither acidic nor basic, has a pH near 7. Indeed, for aqueous solutions at 25 °C, we can say that acids will have pH values less than 7, bases will have values greater than 7, and a pH of 7 represents a neutral solution (Figure 5.20).

Suppose you know the pH of a solution. To find the hydrogen ion concentration you take the antilog of the pH. That is,

$$[H^+] = 10^{-pH} \tag{5.3}$$

■ **Logs and Your Calculator**
All scientific calculators have a key marked "log." To find an antilog, use the key marked "10^x" or the inverse log. When you enter the value of x for 10^x, make sure it has a negative sign.

0   7   14

pH = 3.8
Orange

pH = 2.8   pH = 2.9
Vinegar   Soda

pH = 7.4
Blood

pH = 11.0
Ammonia

pH = 11.7
Oven cleaner

Photos: Charles D. Winters

Active Figure 5.20   pH values of some common substances. Here the "bar" is colored red at one end and blue at the other. These are the colors of litmus paper, commonly used in the laboratory to decide whether a solution is acidic (litmus is red) or basic (litmus is blue).

GENERAL
Chemistry⋅᛭⋅Now™ See the General ChemistryNow CD-ROM or website to explore an interactive version of this figure accompanied by an exercise.

(a)

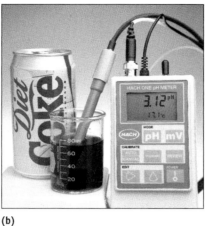

(b)

Figure 5.21 Determining pH. (a) Some household products. Each solution contains a few drops of a dye called a pH indicator (in this case a "universal indicator"). A color of yellow or red indicates a pH less than 7. A green to purple color indicates a pH greater than 7. (b) The pH of a soda is measured with a modern pH meter. Soft drinks are often quite acidic owing to the dissolved CO_2 and other ingredients.

For example, the pH of a diet soda is 3.12, and the hydrogen ion concentration of the solution is

$$[H^+] = 10^{-3.12} = 7.6 \times 10^{-4} \text{ M}$$

The approximate pH of a solution may be determined using any of a variety of dyes. The litmus paper you use in the laboratory contains a dye extracted from a variety of lichen, but many other dyes are also available (Figure 5.21a). A more accurate measurement of pH is done with a pH meter such as that shown in Figure 5.21b. Here a pH electrode is immersed in the solution to be tested, and the pH is read from the instrument.

■ **pH Indicating Dyes**
Many natural substances change color in solution as pH changes. See the extract of red cabbage in Figure 5.6a and of red rose petals on page 849. Tea changes color when acidic lemon juice is added.

GENERAL Chemistry·⚛·Now™

See the General ChemistryNow CD-ROM or website:

• **Screen 5.17 The pH Scale,** for a tutorial on determining the pH of a solution

Example 5.11—pH of Solutions

Problem

(a) Lemon juice has $[H^+]$ = 0.0032 M. What is its pH?

(b) Sea water has a pH of 8.30. What is the hydrogen ion concentration of this solution?

(c) A solution of nitric acid has $[HNO_3]$ = 0.0056 M. What is the pH of this solution?

Strategy Use Equation 5.2 to calculate pH from the H^+ concentration. Use Equation 5.3 to find $[H^+]$ from the pH.

Solution

(a) *Lemon juice:* Because the hydrogen ion concentration is known, the pH is found using Equation 5.2.

$$pH = -\log[H^+] = -\log(3.2 \times 10^{-3}) = -(-2.49) = 2.49$$

(b) *Sea water:* Here pH = 8.30. Therefore,

$$[H^+] = 10^{-pH} = 10^{-8.30} = 5.0 \times 10^{-9} \text{ M}$$

(c) *Nitric acid:* Nitric acid is a strong acid (Table 5.2, page 187) and is completely ionized in aqueous solution. Because $[HNO_3] = 0.0056$ M, the ion concentrations are

$$[H^+] = [NO_3^-] = 0.0056 \text{ M}$$

$$pH = -\log[H^+] = -\log(0.0056 \text{ M}) = 2.25$$

Comment A comment on logarithms and significant figures (Appendix A) is useful. The number to the left of the decimal point in a logarithm is called the *characteristic*, and the number to the right is the *mantissa*. The mantissa has as many significant figures as the number whose log was found. For example, the logarithm of 3.2×10^{-3} (two significant figures) is 2.49 (two numbers to the right of the decimal point).

Exercise 5.17—pH of Solutions

(a) What is the pH of a solution of HCl in which $[HCl] = 2.6 \times 10^{-2}$ M?

(b) What is the hydrogen ion concentration in orange juice with a pH of 3.80?

5.10—Stoichiometry of Reactions in Aqueous Solution

General Solution Stoichiometry

Suppose we want to know what mass of $CaCO_3$ is required to react completely with 25 mL of 0.750 M HCl. The first step in finding the answer is to write a balanced equation. In this case, we have an exchange reaction involving a metal carbonate and an aqueous acid (Figure 5.22).

$$CaCO_3(s) + 2 \text{ HCl(aq)} \longrightarrow CaCl_2(aq) + H_2O(\ell) + CO_2(g)$$

metal carbonate +   acid $\longrightarrow$ salt   + water + carbon dioxide

This problem can be solved in the same way as all the stoichiometry problems you have seen so far, except that the quantity of one reactant is given in volume and concentration units instead of as a mass in grams. The first step is to find the amount of HCl.

$$0.025 \text{ L HCl} \times \frac{0.750 \text{ mol HCl}}{1 \text{ L HCl}} = 0.019 \text{ mol HCl}$$

same step as w/ grams

This is then related to the amount of $CaCO_3$ required.

$$0.019 \text{ mol HCl} \times \frac{1 \text{ mol CaCO}_3}{2 \text{ mol HCl}} = 0.0094 \text{ mol CaCO}_3$$

Finally, the amount of $CaCO_3$ is converted to a mass in grams.

$$0.0094 \text{ mol CaCO}_3 \times \frac{100. \text{ g CaCO}_3}{1 \text{ mol CaCO}_3} = 0.94 \text{ g CaCO}_3$$

Chemists are likely to do such calculations many times in the course of their work. If you follow the general scheme outlined in Problem-Solving Tip 5.3, and pay attention to the units on the numbers, you can successfully carry out any kind of stoichiometry calculations involving concentrations.

Figure 5.22 A commercial remedy for excess stomach acid. The tablet contains calcium carbonate, which reacts with hydrochloric acid, the acid present in the digestive system. The most obvious product is CO_2 gas.

Charles D. Winters

Problem-Solving Tip 5.3

Stoichiometry Calculations Involving Solutions

In Problem-Solving Tip 4.1, you learned about a general approach to stoichiometry problems. We can now modify that scheme for a reaction involving solutions such as x A(aq) + y B(aq) ⟶ products.

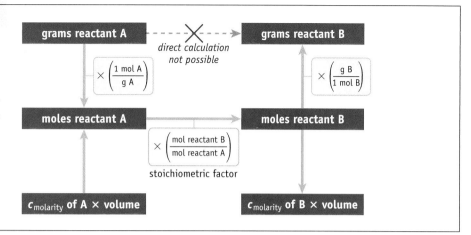

GENERAL
Chemistry Now™

See the General ChemistryNow CD-ROM or website:

- **Screen 5.18 Stoichiometry of Reactions in Solution,** for an exercise on solution stoichiometry, for a tutorial on determining the mass of a product, and for a tutorial on determining the volume of a reactant

Example 5.12—Stoichiometry of a Reaction in Solution

Problem Metallic zinc reacts with aqueous HCl (see Figure 5.6c).

$$Zn(s) + 2\ HCl(aq) \longrightarrow ZnCl_2(aq) + H_2(g)$$

What volume of 2.50 M HCl, in milliliters, is required to convert 11.8 g of Zn completely to products?

Strategy Here the mass of zinc is known, so you first calculate the amount of zinc. Next, use a stoichiometric factor (= 2 mol HCl/1 mol Zn) to relate amount of HCl required to amount of Zn available. Finally, calculate the volume of HCl from the amount of HCl and its concentration.

Solution Begin by calculating the amount of Zn.

$$11.8\ g\ Zn \times \frac{1\ mol\ Zn}{65.39\ g\ Zn} = 0.180\ mol\ Zn$$

Use the stoichiometric factor to calculate the amount of HCl required.

$$0.180\ mol\ Zn \times \frac{2\ mol\ HCl}{1\ mol\ Zn} = 0.360\ mol\ HCl$$

Use the amount of HCl and the solution concentration to calculate the volume.

$$0.360\ mol\ HCl \times \frac{1.00\ L\ solution}{2.50\ mol\ HCl} = 0.144\ L\ HCl$$

The answer is requested in units of milliliters, so we convert the volume to milliliters and find that 144 mL of 2.50 M HCl is required to convert 11.8 g of Zn completely to products.

Exercise 5.18—Solution Stoichiometry

If you combine 75.0 mL of 0.350 M HCl and an excess of Na_2CO_3, what mass of CO_2, in grams, is produced?

$$Na_2CO_3(s) + 2\ HCl(aq) \longrightarrow 2\ NaCl(aq) + H_2O(\ell) + CO_2(g)$$

Titration: A Method of Chemical Analysis

Oxalic acid, $H_2C_2O_4$, is a naturally occurring acid. Suppose you are asked to determine the mass of this acid in an impure sample. Because the compound is an acid, it reacts with a base such as sodium hydroxide (see Section 5.4).

$$H_2C_2O_4(aq) + 2\ NaOH(aq) \longrightarrow Na_2C_2O_4(aq) + 2\ H_2O(\ell)$$

You can use this reaction to determine the quantity of oxalic acid present in a given mass of sample if the following conditions are met:

- You can determine when the amount of sodium hydroxide added is just enough to react with all the oxalic acid present in solution.

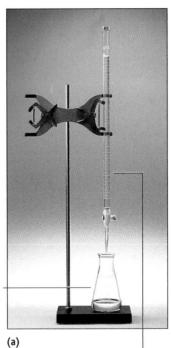

(a)
Buret containing aqueous NaOH of accurately known concentration.

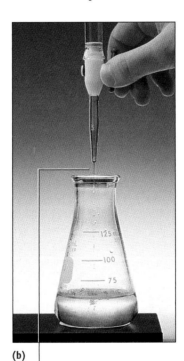

(b)
A solution of NaOH is added slowly to the sample being analyzed.

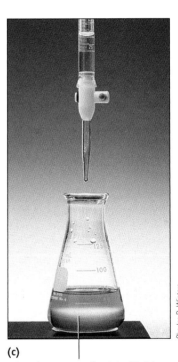

(c)
When the amount of NaOH added from the buret exactly equals the amount of H^+ supplied by the acid being analyzed, the dye (indicator) changes color.

Charles D. Winters

Active Figure 5.23 **Titration of an acid in aqueous solution with a base.** (a) A buret, a volumetric measuring device calibrated in divisions of 0.1 mL, is filled with an aqueous solution of a base of known concentration. (b) Base is added slowly from the buret to the solution containing the acid being analyzed and an indicator. (c) A change in the color of an indicator signals the equivalence point. (The indicator used here is phenolphthalein.)

GENERAL
Chemistry ⋅⚛⋅Now™ See the General ChemistryNow CD-ROM or website to explore an interactive version of this figure accompanied by an exercise.

Flask containing aqueous solution of sample being analyzed

- You know the concentration of the sodium hydroxide solution and volume that has been added at exactly the point of complete reaction.

These conditions are fulfilled in a **titration**, a procedure illustrated in Figure 5.23. The solution containing oxalic acid is placed in a flask along with an **acid–base indicator**, a dye that changes color when the pH of the reaction solution changes. It is common practice to use a dye that has one color in acid solution and another color in basic solution. Aqueous sodium hydroxide of accurately known concentration is placed in a buret. The sodium hydroxide in the buret is added slowly to the acid solution in the flask. As long as some acid is present in solution, all the base supplied from the buret is consumed, the solution remains acidic, and the indicator color is unchanged. At some point, however, the amount of OH^- added exactly equals the amount of H^+ that can be supplied by the acid. This is called the **equivalence point**. As soon as the slightest excess of base has been added beyond the equivalence point, the solution becomes basic, and the indicator changes color (see Figure 5.23).

When the equivalence point has been reached in a titration, the volume of base added is determined by reading the calibrated buret. From this volume and the concentration of the base, the amount of base used can be found:

Amount of base added (mol) = concentration of base (mol/L) × volume of base (L)

Then, using the stoichiometric factor from the balanced equation, the amount of base added is related to the amount of acid present in the original sample. For the specific problem of finding the mass of oxalic acid in an impure sample, we would convert the amount of acid to a mass. If the mass of oxalic acid is divided by the mass of the impure sample (and the quotient multiplied by 100%), we can express the purity of the sample in terms of a mass percent.

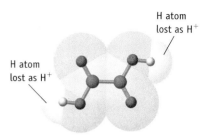

Oxalic acid $H_2C_2O_4$

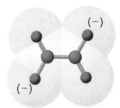

Oxalate anion $C_2O_4{}^{2-}$

Oxalic acid. Oxalic acid has two groups that can supply an H^+ ion to solution. Hence, 1 mol of the acid requires 2 mol of NaOH for complete reaction.

GENERAL
Chemistry ⚛ Now™

See the General ChemistryNow CD-ROM or website:

- **Screen 5.19 Titration,** for a tutorial on the volume of titrant used, for a tutorial on determining the concentration of acid solution, and for a tutorial on determining the concentration of an unknown acid

Example 5.13—Acid–Base Titration

Problem A 1.034-g sample of impure oxalic acid is dissolved in water and an acid–base indicator added. The sample requires 34.47 mL of 0.485 M NaOH to reach the equivalence point. What is the mass of oxalic acid and what is its mass percent in the sample?

Strategy The balanced equation for the reaction of NaOH and $H_2C_2O_4$ is

$$H_2C_2O_4(aq) + 2\ NaOH(aq) \longrightarrow Na_2C_2O_4(aq) + 2\ H_2O(\ell)$$

The concentration of NaOH and the volume used in the titration are used to determine the amount of NaOH. Use a stoichiometric factor to relate the amount of NaOH to the amount of $H_2C_2O_4$. Finally, the amount of $H_2C_2O_4$ is converted to a mass. The mass percent of acid in the sample is then calculated. See Problem Solving Tip 5.3.

Solution The amount of NaOH is given by

$$c_{NaOH} \times V_{NaOH} = \frac{0.485\ \text{mol NaOH}}{L} \times 0.03447\ L = 0.0167\ \text{mol NaOH}$$

The balanced equation for the reaction shows that 1 mol of oxalic acid requires 2 mol of sodium hydroxide. This is the required stoichiometric factor to obtain the amount of oxalic acid present.

$$0.0167 \text{ mol NaOH} \times \frac{1 \text{ mol } H_2C_2O_4}{2 \text{ mol NaOH}} = 0.00836 \text{ mol } H_2C_2O_4$$

The mass of oxalic acid is found from the amount of the acid.

$$0.00836 \text{ mol } H_2C_2O_4 \times \frac{90.04 \text{ g } H_2C_2O_4}{1 \text{ mol } H_2C_2O_4} = 0.753 \text{ g } H_2C_2O_4$$

This mass of oxalic acid represents 72.8% of the total sample mass.

$$\frac{0.753 \text{ g } H_2C_2O_4}{1.034 \text{ g sample}} \times 100\% = 72.8\% \ H_2C_2O_4$$

Exercise 5.19—Acid–Base Titration

A 25.0-mL sample of vinegar (which contains the weak acid acetic acid, CH_3CO_2H) requires 28.33 mL of a 0.953 M solution of NaOH for titration to the equivalence point. What mass of acetic acid, in grams, is in the vinegar sample, and what is the concentration of acetic acid in the vinegar?

$$CH_3CO_2H(aq) + NaOH(aq) \longrightarrow NaCH_3CO_2(aq) + H_2O(\ell)$$

Standardizing an Acid or Base

In Example 5.13 the concentration of the base used in the titration was given. In actual practice this usually has to be found by a prior measurement. The procedure by which the concentration of an analytical reagent is determined accurately is called **standardization**, and there are two general approaches.

One approach is to weigh accurately a sample of a pure, solid acid or base (known as a **primary standard**) and then titrate this sample with a solution of the base or acid to be standardized (Example 5.14). An alternative approach to standardizing a solution is to titrate it with another solution that is already standardized (Exercise 5.20). This is often done with standard solutions purchased from chemical supply companies.

Example 5.14—Standardizing an Acid by Titration

Problem A sample of sodium carbonate, a base (Na_2CO_3, 0.263 g), requires 28.35 mL of aqueous HCl for titration to the equivalence point. What is the molarity of the HCl?

Strategy The balanced equation for the reaction is written first.

$$Na_2CO_3(aq) + 2 \ HCl(aq) \longrightarrow 2 \ NaCl(aq) + H_2O(\ell) + CO_2(g)$$

The amount of Na_2CO_3 can be calculated from its mass and then, using the stoichiometric factor, the amount of HCl in 28.35 mL can be calculated. The amount of HCl divided by the volume of solution (in liters) gives its molar concentration.

Solution Convert the mass of Na_2CO_3 used as the standard to amount of the base.

$$0.263 \text{ g } Na_2CO_3 \times \frac{1 \text{ mol } Na_2CO_3}{106.0 \text{ g } Na_2CO_3} = 0.00248 \text{ mol } Na_2CO_3$$

Use the stoichiometric factor to calculate amount of HCl in 28.35 mL.

$$0.00248 \text{ mol Na}_2\text{CO}_3 \times \frac{2 \text{ mol HCl required}}{1 \text{ mol Na}_2\text{CO}_3 \text{ available}} = 0.00496 \text{ mol HCl}$$

The 28.35-mL (0.02835-L) sample of aqueous HCl contains 0.00496 mol of HCl, so the concentration of the HCl solution is 0.175 M.

$$[\text{HCl}] = \frac{0.00496 \text{ mol HCl}}{0.02835 \text{ L}} = 0.175 \text{ M}$$

Comment In this example Na_2CO_3 is a primary standard. Sodium carbonate can be obtained in pure form, which can be weighed accurately, and which reacts completely with a strong acid.

Exercise 5.20—Standardization of a Base

Hydrochloric acid, HCl, can be purchased from chemical supply houses with a concentration of 0.100 M, and this solution can be used to standardize the solution of a base. If titrating 25.00 mL of a sodium hydroxide solution to the equivalence point requires 29.67 mL of 0.100 M HCl, what is the concentration of the base?

Determining Molar Mass by Titration

In Chapters 3 and 4 we used analytical data to determine the empirical formula of a compound. The molecular formula could then be derived if the molar mass were known. If the unknown substance is an acid or a base, it is possible to determine the molar mass by titration.

Example 5.15—Determining the Molar Mass of an Acid by Titration

Problem To determine the molar mass of an organic acid, HA, we titrate 1.056 g of HA with standardized NaOH. Calculate the molar mass of HA assuming the acid reacts with 33.78 mL of 0.256 M NaOH according to the equation

$$\text{HA(aq)} + \text{OH}^-\text{(aq)} \longrightarrow \text{A}^-\text{(aq)} + \text{H}_2\text{O}(\ell)$$

Strategy The key to this problem is to recognize that the molar mass of a substance is the ratio of the mass of a sample (g) to the amount of substance (mol) in the sample. Here molar mass of HA = 1.056 g HA/x mol HA. Because 1 mol of HA reacts with 1 mol of NaOH in this case, the amount of acid (x mol) is equal to the amount of NaOH used in the titration, which is given by its concentration and volume.

Solution Let us first calculate the amount of NaOH used in the titration.

$$c_{\text{NaOH}}V_{\text{NaOH}} = (0.256 \text{ mol/L})(0.03378 \text{ L}) = 8.65 \times 10^{-3} \text{ mol NaOH}$$

Next, recognize that the amount of NaOH used in the titration is the same as the amount of acid titrated. That is,

$$8.65 \times 10^{-3} \text{ mol NaOH} \left(\frac{1 \text{ mol HA}}{1 \text{ mol NaOH}} \right) = 8.65 \times 10^{-3} \text{ mol HA}$$

Finally, calculate the molar mass of HA.

$$\text{Molar mass of acid} = \frac{1.056 \text{ g HA}}{8.65 \times 10^{-3} \text{ mol HA}} = 122 \text{ g/mol}$$

Using an oxidation–reduction reaction for analysis by titration. Purple, aqueous $KMnO_4$ is added to a solution containing Fe^{2+}. As $KMnO_4$ drops into the solution, colorless Mn^{2+} and pale yellow Fe^{3+} form. Here an area of the solution containing unreacted $KMnO_4$ is seen. As the solution is mixed, this disappears until the equivalence point is reached.

Charles D. Winters

Exercise 5.21—Determining the Molar Mass of an Acid by Titration

An acid reacts with NaOH according to the net ionic equation

$$HA(aq) + OH^-(aq) \longrightarrow A^-(aq) + H_2O(\ell)$$

Calculate the molar mass of HA if 0.856 g of the acid requires 30.08 mL of 0.323 M NaOH.

Titrations Using Oxidation–Reduction Reactions

Oxidation–reduction reactions (see Section 5.7) also lend themselves to chemical analysis by titration. Many of these reactions go rapidly to completion in aqueous solution, and methods exist to determine their equivalence point.

Example 5.16—Using an Oxidation–Reduction Reaction in a Titration

Problem We wish to analyze an iron ore for its iron content. The iron in the sample can be converted quantitatively to the iron(II) ion, Fe^{2+}, in aqueous solution, and this solution can then be titrated with aqueous potassium permanganate, $KMnO_4$. The balanced, net ionic equation for the reaction occurring in the course of this titration is

$$MnO_4^-(aq) + 5\ Fe^{2+}(aq) + 8\ H^+(aq) \longrightarrow Mn^{2+}(aq) + 5\ Fe^{3+}(aq) + 4\ H_2O(\ell)$$

purple   colorless   colorless   pale yellow

A 1.026-g sample of iron-containing ore requires 24.35 mL of 0.0195 M $KMnO_4$ to reach the equivalence point. What is the mass percent of iron in the ore?

Strategy Because the volume and molar concentration of the $KMnO_4$ solution are known, the amount of $KMnO_4$ used in the titration can be calculated. Using the stoichiometric factor, the amount of $KMnO_4$ is related to the amount of iron(II) ion. The amount of iron(II) is converted to its mass, and the mass percent of iron in the sample is determined.

Solution First, calculate the amount of $KMnO_4$.

$$c_{KMnO_4} \times V_{KMnO_4} = \frac{0.0195\ mol\ KMnO_4}{L} \times 0.02435\ L$$
$$= 0.000475\ mol\ KMnO_4$$

Use the stoichiometric factor to calculate the amount of iron(II) ion.

$$0.000475\ mol\ KMnO_4^- \times \frac{5\ mol\ Fe^{2+}}{1\ mol\ KMnO_4} = 0.00237\ mol\ Fe^{2+}$$

The mass of iron can now be calculated,

$$0.00237\ mol\ Fe^{2+} \times \frac{55.85\ g\ Fe^{2+}}{1\ mol\ Fe^{2+}} = 0.133\ g\ Fe^{2+}$$

Finally, the mass percent can be determined.

$$\frac{0.133\ g\ Fe^{2+}}{1.026\ g\ sample} \times 100\% = 12.9\%\ iron$$

Comment This is a useful analytical reaction because it is easy to detect when all the iron(II) ion has reacted. The MnO_4^- ion is a deep purple color, but when it reacts with Fe^{2+} the color disappears because the reaction product, Mn^{2+}, is colorless. Thus, as $KMnO_4$ is added from a buret, the purple color disappears as the solutions mix. When all the Fe^{2+} has been converted to Fe^{3+}, any additional $KMnO_4$ will give the solution a permanent purple color. Therefore, $KMnO_4$ solution is added from the buret until the initially colorless, Fe^{2+}-containing solution just turns a faint purple color, the signal that the equivalence point has been reached.

Exercise 5.22—Using an Oxidation–Reduction Reaction in a Titration

Vitamin C, ascorbic acid ($C_6H_8O_6$), is a reducing agent. One way to determine the ascorbic acid content of a sample is to mix the acid with an excess of iodine,

$$C_6H_8O_6(aq) + I_2(aq) \longrightarrow C_6H_6O_6(aq) + 2\ H^+(aq) + 2\ I^-(aq)$$

and then titrate the iodine that did *not* react with the ascorbic acid with sodium thiosulfate. The balanced, net ionic equation for the reaction occurring in this titration is

$$I_2(aq) + 2\ S_2O_3^{2-}(aq) \longrightarrow 2\ I^-(aq) + S_4O_6^{2-}(aq)$$

Suppose 50.00 mL of 0.0520 M I_2 was added to the sample containing ascorbic acid. After the ascorbic acid/I_2 reaction was complete, the I_2 not used in the reaction required 20.30 mL of 0.196 M $Na_2S_2O_3$ for titration to the equivalence point. Calculate the mass of ascorbic acid in the unknown sample.

Chapter Goals Revisited

Now that you have studied this chapter, you should ask if you have met the chapter goals. In particular, you should be able to

Understand the nature of ionic substances dissolved in water

a. Explain the difference between electrolytes and nonelectrolytes and recognize examples of each (Section 5.1 and Figure 5.2).

b. Predict the solubility of ionic compounds in water (Section 5.1 and Figure 5.3). General ChemistryNow homework: Study Question(s) 7

c. Recognize which ions are formed when an ionic compound or acid or base dissolves in water (Sections 5.1–5.3). General ChemistryNow homework: SQ(s) 13

Recognize common acids and bases and understand their behavior in aqueous solution (Section 5.3 and Table 5.2)

a. Know the names and formulas of common acids and bases. General ChemistryNow homework: SQ(s) 13, 18

b. Categorize acids and bases as strong or weak.

Recognize and write equations for the common types of reactions in aqueous solution

a. Predict the products of precipitation reactions (Section 5.2), which involve the formation of an insoluble reaction product by the exchange of anions between the cations of the reactants. General ChemistryNow homework: SQ(s) 11

b. Write net ionic equations and show how to arrive at such an equation for a given reaction (Sections 5.2 and 5.6). General ChemistryNow homework: SQ(s) 11

GENERAL Chemistry•Now™

See the General ChemistryNow CD-ROM or website to:

- Assess your understanding with homework questions keyed to each goal
- Check your readiness for an exam by taking the exam-prep quiz and exploring the resources in the personalized Learning Plan it provides

c. Predict the products of acid–base reactions involving common acids and strong bases (Section 5.4). General ChemistryNow homework: SQ(s) 19

d. Understand that the net ionic equation for the reaction of a strong acid with a strong base is $H^+(aq) + OH^-(aq) \longrightarrow H_2O(\ell)$ (Section 5.4).

e. Predict the products of gas-forming reactions (Section 5.5), the most common of which are those between a metal carbonate and an acid.

$$NiCO_3(s) + 2\ HNO_3(aq) \longrightarrow Ni(NO_3)_2(aq) + CO_2(g) + H_2O(\ell)$$

f. Use the ideas developed in Sections 5.2–5.7 as an aid in recognizing four of the common types of reactions that occur in aqueous solution, and write balanced equations for such reactions (Section 5.6).

Reaction Type	Driving Force
Precipitation	Formation of an insoluble compound
Acid–strong base	Formation of a salt and water
Gas-forming	Evolution of a water-insoluble gas such as CO_2
Oxidation–reduction	Transfer of electrons

The first three of these reaction types involve the exchange of anions between the cations involved, and so are called exchange reactions. The fourth type (redox reactions) involves the transfer of electrons. General ChemistryNow homework: SQ(s) 29

g. Identify reactant- and product-favored reactions. General ChemistryNow homework: SQ(s) 33

Recognize common oxidizing and reducing agents and identify oxidation–reduction reactions

a. Determine oxidation numbers of elements in a compound and understand that these numbers represent the charge an atom has, or appears to have, when the electrons of the compound are counted according to a set of guidelines (Section 5.7). General ChemistryNow homework: SQ(s) 35

b. Identify oxidation–reduction reactions (redox reactions) and identify the oxidizing and reducing agents and substances oxidized and reduced in the reaction (Section 5.7 and Tables 5.4 and 5.5). General ChemistryNow homework: SQ(s) 39

Define and use molarity in solution stoichiometry

a. Calculate the concentration of a solute in a solution in units of moles per liter (molarity), and use concentrations in calculations (Section 5.8). General ChemistryNow homework: SQ(s) 41, 43, 45

b. Describe how to prepare a solution of a given molarity from the solute and a solvent or by dilution from a more concentrated solution (Section 5.8). General ChemistryNow homework: SQ(s) 50, 51

c. Calculate the pH of a solution containing an acid or a base and know what this means in terms of the relative amount of hydrogen ion in the solution. Calculate the hydrogen ion concentration of a solution from the pH (Section 5.9). General ChemistryNow homework: SQ(s) 56, 57

d. Solve stoichiometry problems using solution concentrations (Section 5.10). General ChemistryNow homework: SQ(s) 61, 64

e. Explain how a titration is carried out, explain the procedure of standardization, and calculate concentrations or amounts of reactants from titration data (Section 5.10). General ChemistryNow homework: SQ(s) 69, 73

Key Equations

Equation 5.1 (page 206)

Definition of molarity, a measure of the concentration of a solute in a solution.

$$\text{Concentration } (c_{\text{molarity}}) = \frac{\text{amount of solute (mol)}}{\text{volume of solution (L)}}$$

A useful form of this equation is

$$\text{Amount of solute (moles)} = c_{\text{molarity}} \times \text{volume of solution (L)}$$

Related to this equation is the "shortcut" used when diluting a concentrated solution to obtain a more dilute solution. The product of the concentration and volume of a more concentrated solution (c) must be the same as that for the diluted solution (d).

$$c_c \times V_c = c_d \times V_d$$

If any three of these parameters is known (say c_c, V_c, and c_d), the fourth may be calculated (say V_d).

Equation 5.2 (page 212)

The pH of a solution is the negative logarithm of the hydrogen ion concentration.

$$\text{pH} = -\log [\text{H}^+]$$

Equation 5.3 (page 212)

The equation for calculating the hydrogen ion concentration of a solution from the pH of the solution.

$$[\text{H}^+] = 10^{-\text{pH}}$$

Study Questions

▲ denotes more challenging questions.

■ denotes questions available in the Homework and Goals section of the General ChemistryNow CD-ROM or website.

Blue numbered questions have answers in Appendix O and fully worked solutions in the *Student Solutions Manual*.

Structures of many of the compounds used in these questions are found on the General ChemistryNow CD-ROM or website in the Models folder.

GENERAL
Chemistry•⚛•Now™ Assess your understanding of this chapter's topics with additional quizzing and conceptual questions at **http://now.brookscole.com/kotz6e**

Practicing Skills

Electrolytes and Solubility of Compounds
(See Exercise 5.1, Example 5.1, and General ChemistryNow Screens 5.3 and 5.4.)

1. What is an electrolyte? How can you differentiate experimentally between a weak electrolyte and a strong electrolyte? Give an example of each.

2. Name two acids that are strong electrolytes and one acid that is a weak electrolyte. Name two bases that are strong electrolytes and one base that is a weak electrolyte.

3. Which compound or compounds in each of the following groups is (are) expected to be soluble in water?

   (a) CuO, CuCl$_2$, FeCO$_3$

   (b) AgI, Ag$_3$PO$_4$, AgNO$_3$

   (c) K$_2$CO$_3$, KI, KMnO$_4$

4. Which compound or compounds in each of the following groups is (are) expected to be soluble in water?
   (a) $BaSO_4$, $Ba(NO_3)_2$, $BaCO_3$
   (b) Na_2SO_4, $NaClO_4$, $NaCH_3CO_2$
   (c) $AgBr$, KBr, Al_2Br_6

5. The following compounds are water-soluble. What ions are produced by each compound in aqueous solution?
   (a) KOH
   (b) K_2SO_4
   (c) $LiNO_3$
   (d) $(NH_4)_2SO_4$

6. The following compounds are water-soluble. What ions are produced by each compound in aqueous solution?
   (a) KI
   (b) $Mg(CH_3CO_2)_2$
   (c) K_2HPO_4
   (d) $NaCN$

7. ■ Decide whether each of the following is water-soluble. If soluble, tell what ions are produced.
   (a) Na_2CO_3
   (b) $CuSO_4$
   (c) NiS
   (d) $BaBr_2$

8. Decide whether each of the following is water-soluble. If soluble, tell what ions are produced.
   (a) $NiCl_2$
   (b) $Cr(NO_3)_3$
   (c) $Pb(NO_3)_2$
   (d) $BaSO_4$

Precipitation Reactions and Net Ionic Equations
(See Examples 5.2 and 5.3 and General ChemistryNow Screens 5.5–5.7.)

9. Balance the equation for the following precipitation reaction, and then write the net ionic equation. Indicate the state of each species (s, ℓ, aq, or g).

$$CdCl_2 + NaOH \longrightarrow Cd(OH)_2 + NaCl$$

10. Balance the equation for the following precipitation reaction, and then write the net ionic equation. Indicate the state of each species (s, ℓ, aq, or g).

$$Ni(NO_3)_2 + Na_2CO_3 \longrightarrow NiCO_3 + NaNO_3$$

11. ■ Predict the products of each precipitation reaction. Balance the completed equation, and then write the net ionic equation.
    (a) $NiCl_2(aq) + (NH_4)_2S(aq) \longrightarrow$?
    (b) $Mn(NO_3)_2(aq) + Na_3PO_4(aq) \longrightarrow$?

12. Predict the products of each precipitation reaction. Balance the completed equation, and then write the net ionic equation.
    (a) $Pb(NO_3)_2(aq) + KBr(aq) \longrightarrow$?
    (b) $Ca(NO_3)_2(aq) + KF(aq) \longrightarrow$?
    (c) $Ca(NO_3)_2(aq) + Na_2C_2O_4(aq) \longrightarrow$?

Acids and Bases
(See Exercises 5.5 and 5.6 and General ChemistryNow Screens 5.8 and 5.9.)

13. ■ Write a balanced equation for the ionization of nitric acid in water.

14. Write a balanced equation for the ionization of perchloric acid in water.

15. Oxalic acid, $H_2C_2O_4$, which is found in certain plants, can provide two hydrogen ions in water. Write balanced equations (like those for sulfuric acid on page 186) to show how oxalic acid can supply one and then a second H^+ ion.

16. Phosphoric acid can supply one, two, or three H^+ ions in aqueous solution. Write balanced equations (like those for sulfuric acid on page 186) to show this successive loss of hydrogen ions.

17. Write a balanced equation for reaction of the basic oxide, magnesium oxide, with water.

18. ■ Write a balanced equation for the reaction of sulfur trioxide with water.

Reactions of Acids and Bases
(See Example 5.4, Exercise 5.7, and General ChemistryNow Screens 5.5 and 5.10.)

19. ■ Complete and balance the following acid–base reactions. Name the reactants and products.
    (a) $CH_3CO_2H(aq) + Mg(OH)_2(s) \longrightarrow$
    (b) $HClO_4(aq) + NH_3(aq) \longrightarrow$

20. Complete and balance the following acid–base reactions. Name the reactants and products.
    (a) $H_3PO_4(aq) + KOH(aq) \longrightarrow$
    (b) $H_2C_2O_4(aq) + Ca(OH)_2(s) \longrightarrow$

    ($H_2C_2O_4$ is oxalic acid, an acid capable of donating two H^+ ions.)

21. Write a balanced equation for the reaction of barium hydroxide with nitric acid.

22. Write a balanced equation for the reaction of aluminum hydroxide with sulfuric acid.

Writing Net Ionic Equations
(See Example 5.3 and General ChemistryNow Screen 5.7.)

23. Balance the following equations, and then write the net ionic equation.
    (a) $(NH_4)_2CO_3(aq) + Cu(NO_3)_2(aq) \longrightarrow$
        $CuCO_3(s) + NH_4NO_3(aq)$
    (b) $Pb(OH)_2(s) + HCl(aq) \longrightarrow PbCl_2(s) + H_2O(\ell)$
    (c) $BaCO_3(s) + HCl(aq) \longrightarrow$
        $BaCl_2(aq) + H_2O(\ell) + CO_2(g)$

24. Balance the following equations, and then write the net ionic equation:
    (a) $Zn(s) + HCl(aq) \longrightarrow H_2(g) + ZnCl_2(aq)$
    (b) $Mg(OH)_2(s) + HCl(aq) \longrightarrow MgCl_2(aq) + H_2O(\ell)$
    (c) $HNO_3(aq) + CaCO_3(s) \longrightarrow$
        $Ca(NO_3)_2(aq) + H_2O(\ell) + CO_2(g)$

25. Balance the following equations, and then write the net ionic equation. Show states for all reactants and products (s, ℓ, g, aq).
    (a) the reaction of silver nitrate and potassium iodide to give silver iodide and potassium nitrate
    (b) the reaction of barium hydroxide and nitric acid to give barium nitrate and water

---

▲ More challenging   ■ In General ChemistryNow   *Blue-numbered questions* answered in Appendix O

(c) the reaction of sodium phosphate and nickel(II) nitrate to give nickel(II) phosphate and sodium nitrate

26. Balance each of the following equations, and then write the net ionic equation. Show states for all reactants and products (s, ℓ, g, aq).

    (a) the reaction of sodium hydroxide and iron(II) chloride to give iron(II) hydroxide and sodium chloride
    (b) the reaction of barium chloride with sodium carbonate to give barium carbonate and sodium chloride

Gas-Forming Reactions

(See Example 5.5 and General ChemistryNow Screens 5.5 and 5.11.)

27. Siderite is a mineral consisting largely of iron(II) carbonate. Write an overall, balanced equation for its reaction with nitric acid, and name each reactant and product.

28. The beautiful red mineral rhodochrosite is manganese(II) carbonate. Write an overall, balanced equation for the reaction of the mineral with hydrochloric acid. Name each reactant and product.

Charles D. Winters

Rhodochrosite, a mineral consisting largely of $MnCO_3$

Types of Reactions in Aqueous Solution

(See Exercise 5.9, Example 5.8, and General ChemistryNow Screen 5.5.)

29. ■ Balance the following reactions and then classify each as a precipitation, acid–base, or gas-forming reaction.

    (a) $Ba(OH)_2(aq) + HCl(aq) \longrightarrow BaCl_2(aq) + H_2O(\ell)$
    (b) $HNO_3(aq) + CoCO_3(s) \longrightarrow$
    $Co(NO_3)_2(aq) + H_2O(\ell) + CO_2(g)$
    (c) $Na_3PO_4(aq) + Cu(NO_3)_2(aq) \longrightarrow$
    $Cu_3(PO_4)_2(s) + NaNO_3(aq)$

30. Balance the following reactions and then classify each as a precipitation, acid–base reaction, or a gas-forming reaction.

    (a) $K_2CO_3(aq) + Cu(NO_3)_2(aq) \longrightarrow$
    $CuCO_3(s) + KNO_3(aq)$
    (b) $Pb(NO_3)_2(aq) + HCl(aq) \longrightarrow PbCl_2(s) + HNO_3(aq)$
    (c) $MgCO_3(s) + HCl(aq) \longrightarrow$
    $MgCl_2(aq) + H_2O(\ell) + CO_2(g)$

31. Balance the following reactions and then classify each as a precipitation, acid–base reaction, or gas-forming reaction. Show states for the products (s, ℓ, g, aq) and then balance the completed equation. Write the net ionic equation.

    (a) $MnCl_2(aq) + Na_2S(aq) \longrightarrow MnS + NaCl$
    (b) $K_2CO_3(aq) + ZnCl_2(aq) \longrightarrow ZnCO_3 + KCl$

32. Balance the following reactions and then classify each as a precipitation, acid–base, or gas-forming reaction. Write the net ionic equation.

    (a) $Fe(OH)_3(s) + HNO_3(aq) \longrightarrow Fe(NO_3)_3 + H_2O$
    (b) $FeCO_3(s) + HNO_3(aq) \longrightarrow Fe(NO_3)_2 + CO_2 + H_2O$

Product- or Reactant-Favored Reactions

33. ■ What feature causes the following reactions to be product-favored?

    (a) $CuCl_2(aq) + H_2S(aq) \longrightarrow CuS(s) + 2 HCl(aq)$
    (b) $H_3PO_4(aq) + 3 KOH(aq) \longrightarrow 3 H_2O(\ell) + K_3PO_4(aq)$

34. Which of the following reactions is predicted to be product-favored?

    (a) $Zn(s) + 2 HCl(aq) \longrightarrow H_2(g) + ZnCl_2(aq)$
    (b) $MgCl_2(aq) + 2 H_2O(\ell) \longrightarrow Mg(OH)_2(s) + 2 HCl(aq)$

Oxidation Numbers

(See Example 5.6 and General ChemistryNow Screen 5.13.)

35. ■ Determine the oxidation number of each element in the following ions or compounds.

    (a) BrO_3^- (d) CaH_2
    (b) $C_2O_4^{2-}$ (e) H_4SiO_4
    (c) F^- (f) HSO_4^-

36. Determine the oxidation number of each element in the following ions or compounds.

    (a) PF_6^- (d) N_2O_5
    (b) $H_2AsO_4^-$ (e) $POCl_3$
    (c) UO^{2+} (f) XeO_4^{2-}

Oxidation–Reduction Reactions

(See Example 5.7 and General ChemistryNow Screens 5.12–5.14.)

37. Which two of the following reactions are oxidation–reduction reactions? Explain your answer in each case. Classify the remaining reaction.

    (a) $Zn(s) + 2 NO_3^-(aq) + 4 H^+(aq) \longrightarrow$
    $Zn^{2+}(aq) + 2 NO_2(g) + 2 H_2O(\ell)$
    (b) $Zn(OH)_2(s) + H_2SO_4(aq) \longrightarrow ZnSO_4(aq) + 2 H_2O(\ell)$
    (c) $Ca(s) + 2 H_2O(\ell) \longrightarrow Ca(OH)_2(s) + H_2(g)$

38. Which two of the following reactions are oxidation–reduction reactions? Explain your answer briefly. Classify the remaining reaction.

    (a) $CdCl_2(aq) + Na_2S(aq) \longrightarrow CdS(s) + 2 NaCl(aq)$
    (b) $2 Ca(s) + O_2(g) \longrightarrow 2 CaO(s)$
    (c) $4 Fe(OH)_2(s) + 2 H_2O(\ell) + O_2(g) \longrightarrow 4 Fe(OH)_3(aq)$

39. ■ In the following reactions, decide which reactant is oxidized and which is reduced. Designate the oxidizing agent and the reducing agent.

    (a) $C_2H_4(g) + 3 O_2(g) \longrightarrow 2 CO_2(g) + 2 H_2O(g)$
    (b) $Si(s) + 2 Cl_2(g) \longrightarrow SiCl_4(\ell)$

40. In the following reactions, decide which reactant is oxidized and which is reduced. Designate the oxidizing agent and the reducing agent.
    (a) $Cr_2O_7^{2-}(aq) + 3\,Sn^{2+}(aq) + 14\,H^+(aq) \longrightarrow$
    $2\,Cr^{3+}(aq) + 3\,Sn^{4+}(aq) + 7\,H_2O(\ell)$
    (b) $FeS(s) + 3\,NO_3^-(aq) + 4\,H^+(aq) \longrightarrow$
    $3\,NO(g) + SO_4^{2-}(aq) + Fe^{3+}(aq) + 2\,H_2O(\ell)$

Solution Concentration

(See Example 5.9 and General ChemistryNow Screen 5.15.)

41. ■ If 6.73 g of Na_2CO_3 is dissolved in enough water to make 250. mL of solution, what is the molar concentration of the sodium carbonate? What are the molar concentrations of the Na^+ and CO_3^{2-} ions?

42. Some potassium dichromate ($K_2Cr_2O_7$), 2.335 g, is dissolved in enough water to make exactly 500. mL of solution. What is the molar concentration of the potassium dichromate? What are the molar concentrations of the K^+ and $Cr_2O_7^{2-}$ ions?

43. ■ What is the mass of solute, in grams, in 250. mL of a 0.0125 M solution of $KMnO_4$?

44. What is the mass of solute, in grams, in 125 mL of a 1.023×10^{-3} M solution of Na_3PO_4? What are the molar concentrations of the Na^+ and PO_4^{3-} ions?

45. ■ What volume of 0.123 M NaOH, in milliliters, contains 25.0 g of NaOH?

46. What volume of 2.06 M $KMnO_4$, in liters, contains 322 g of solute?

47. For each solution, identify the ions that exist in aqueous solution, and specify the concentration of each ion.
    (a) 0.25 M $(NH_4)_2SO_4$
    (b) 0.123 M Na_2CO_3
    (c) 0.056 M HNO_3

48. For each solution, identify the ions that exist in aqueous solution, and specify the concentration of each ion.
    (a) 0.12 M $BaCl_2$
    (b) 0.0125 M $CuSO_4$
    (c) 0.500 M $K_2Cr_2O_7$

Preparing Solutions

(See Exercise 5.14, Example 5.10, and General ChemistryNow Screen 5.16.)

49. An experiment in your laboratory requires exactly 500. mL of a 0.0200 M solution of Na_2CO_3. You are given solid Na_2CO_3, distilled water, and a 500.-mL volumetric flask. Describe how to prepare the required solution.

50. ■ What mass of oxalic acid, $H_2C_2O_4$, is required to prepare 250. mL of a solution that has a concentration of 0.15 M $H_2C_2O_4$?

51. ■ If you dilute 25.0 mL of 1.50 M hydrochloric acid to 500. mL, what is the molar concentration of the dilute acid?

52. If 4.00 mL of 0.0250 M $CuSO_4$ is diluted to 10.0 mL with pure water, what is the molar concentration of copper(II) sulfate in the diluted solution?

53. Which of the following methods would you use to prepare 1.00 L of 0.125 M H_2SO_4?
    (a) Dilute 20.8 mL of 6.00 M H_2SO_4 to a volume of 1.00 L.
    (b) Add 950. mL of water to 50.0 mL of 3.00 M H_2SO_4.

54. Which of the following methods would you use to prepare 300. mL of 0.500 M $K_2Cr_2O_7$?
    (a) Add 30.0 mL of 1.50 M $K_2Cr_2O_7$ to 270. mL of water.
    (b) Dilute 250. mL of 0.600 M $K_2Cr_2O_7$ to a volume of 300. mL.

pH

(See Example 5.11 and General ChemistryNow Screen 5.17.)

55. A table wine has a pH of 3.40. What is the hydrogen ion concentration of the wine? Is it acidic or basic?

56. ■ A saturated solution of milk of magnesia, $Mg(OH)_2$, has a pH of 10.5. What is the hydrogen ion concentration of the solution? Is the solution acidic or basic?

57. ■ What is the hydrogen ion concentration of a 0.0013 M solution of HNO_3? What is its pH?

58. What is the hydrogen ion concentration of a 1.2×10^{-4} M solution of $HClO_4$? What is its pH?

59. Make the following conversions. In each case, tell whether the solution is acidic or basic.

	pH	$[H^+]$
(a)	1.00	_____
(b)	10.50	_____
(c)	_____	1.3×10^{-5} M
(d)	_____	2.3×10^{-8} M

60. Make the following conversions. In each case, tell whether the solution is acidic or basic.

	pH	$[H^+]$
(a)	_____	6.7×10^{-10} M
(b)	_____	2.2×10^{-6} M
(c)	5.25	_____
(d)	_____	2.5×10^{-2} M

Stoichiometry of Reactions in Solution

(See Example 5.12 and General ChemistryNow Screen 5.18.)

61. ■ What volume of 0.109 M HNO_3, in milliliters, is required to react completely with 2.50 g of $Ba(OH)_2$?

    $2\,HNO_3(aq) + Ba(OH)_2(s) \longrightarrow 2\,H_2O(\ell) + Ba(NO_3)_2(aq)$

62. What mass of Na_2CO_3, in grams, is required for complete reaction with 50.0 mL of 0.125 M HNO_3?

    $Na_2CO_3(aq) + 2\,HNO_3(aq) \longrightarrow$
    $2\,NaNO_3(aq) + CO_2(g) + H_2O(\ell)$

63. When an electric current is passed through an aqueous solution of NaCl, the valuable industrial chemicals $H_2(g)$, $Cl_2(g)$, and NaOH are produced.

    $2\,NaCl(aq) + 2\,H_2O(\ell) \longrightarrow H_2(g) + Cl_2(g) + 2\,NaOH(aq)$

▲ More challenging   ■ In General ChemistryNow   Blue-numbered questions answered in Appendix O

What mass of NaOH can be formed from 15.0 L of 0.35 M NaCl? What mass of chlorine is obtained?

64. ■ Hydrazine, N_2H_4, a base like ammonia, can react with an acid such as sulfuric acid.

$$2 N_2H_4(aq) + H_2SO_4(aq) \longrightarrow 2 N_2H_5^+(aq) + SO_4^{2-}(aq)$$

What mass of hydrazine reacts with 250. mL of 0.146 M H_2SO_4?

65. In the photographic developing process, silver bromide is dissolved by adding sodium thiosulfate:

$$AgBr(s) + 2 Na_2S_2O_3(aq) \longrightarrow Na_3Ag(S_2O_3)_2(aq) + NaBr(aq)$$

If you want to dissolve 0.225 g of AgBr, what volume of 0.0138 M $Na_2S_2O_3$, in milliliters, should be used?

(a)   **(b)**

Silver Chemistry. (a) A precipitate of AgBr formed by adding $AgNO_3(aq)$ to KBr(aq). (b) On adding $Na_2S_2O_3(aq)$, sodium thiosulfate, the solid AgBr dissolves.

66. You can dissolve an aluminum soft-drink can in an aqueous base such as potassium hydroxide.

$$2 Al(s) + 2 KOH(aq) + 6 H_2O(\ell) \longrightarrow 2 KAl(OH)_4(aq) + 3 H_2(g)$$

If you place 2.05 g of aluminum in a beaker with 185 mL of 1.35 M KOH, will any aluminum remain? What mass of $KAl(OH)_4$ is produced?

67. What volume of 0.750 M $Pb(NO_3)_2$, in milliliters, is required to react completely with 1.00 L of 2.25 M NaCl solution? The balanced equation is

$$Pb(NO_3)_2(aq) + 2 NaCl(aq) \longrightarrow PbCl_2(s) + 2 NaNO_3(aq)$$

68. What volume of 0.125 M oxalic acid, $H_2C_2O_4$ is required to react with 35.2 mL of 0.546 M NaOH?

$$H_2C_2O_4(aq) + 2 NaOH(aq) \longrightarrow Na_2C_2O_4(aq) + 2 H_2O(aq)$$

Titrations
(See Examples 5.13–5.16 and General ChemistryNow Screen 5.19.)

69. ■ What volume of 0.812 M HCl, in milliliters, is required to titrate 1.45 g of NaOH to the equivalence point?

$$NaOH(aq) + HCl(aq) \longrightarrow H_2O(\ell) + NaCl(aq)$$

70. What volume of 0.955 M HCl, in milliliters, is required to titrate 2.152 g of Na_2CO_3 to the equivalence point?

$$Na_2CO_3(aq) + 2 HCl(aq) \longrightarrow H_2O(\ell) + CO_2(g) + 2 NaCl(aq)$$

71. If 38.55 mL of HCl is required to titrate 2.150 g of Na_2CO_3 according to the following equation, what is the molarity of the HCl solution?

$$Na_2CO_3(aq) + 2 HCl(aq) \longrightarrow 2 NaCl(aq) + CO_2(g) + H_2O(\ell)$$

72. Potassium hydrogen phthalate, $KHC_8H_4O_4$, is used to standardize solutions of bases. The acidic anion reacts with strong bases according to the following net ionic equation:

$$HC_8H_4O_4^-(aq) + OH^-(aq) \longrightarrow C_8H_4O_4^{2-}(aq) + H_2O(\ell)$$

If a 0.902-g sample of potassium hydrogen phthalate is dissolved in water and titrated to the equivalence point with 26.45 mL of NaOH, what is the molar concentration of the NaOH?

73. ■ You have 0.954 g of an unknown acid, H_2A, which reacts with NaOH according to the balanced equation

$$H_2A(aq) + 2 NaOH(aq) \longrightarrow Na_2A(aq) + 2 H_2O(\ell)$$

If 36.04 mL of 0.509 M NaOH is required to titrate the acid to the equivalence point, what is the molar mass of the acid?

74. ▲ An unknown solid acid is either citric acid or tartaric acid. To determine which acid you have, you titrate a sample of the solid with NaOH. The appropriate reactions are as follows:

Citric acid:
$$H_3C_6H_5O_7(aq) + 3 NaOH(aq) \longrightarrow 3 H_2O(\ell) + Na_3C_6H_5O_7(aq)$$

Tartaric acid:
$$H_2C_4H_4O_6(aq) + 2 NaOH(aq) \longrightarrow 2 H_2O(\ell) + Na_2C_4H_4O_6(aq)$$

A 0.956-g sample requires 29.1 mL of 0.513 M NaOH for titration to the equivalence point. What is the unknown acid?

75. To analyze an iron-containing compound, you convert all the iron to Fe^{2+} in aqueous solution and then titrate the solution with standardized $KMnO_4$. The balanced, net ionic equation is

$$MnO_4^-(aq) + 5 Fe^{2+}(aq) + 8 H^+(aq) \longrightarrow Mn^{2+}(aq) + 5 Fe^{3+}(aq) + 4 H_2O(\ell)$$

A 0.598-g sample of the iron-containing compound requires 22.25 mL of 0.0123 M $KMnO_4$ for titration to the equivalence point. What is the mass percent of iron in the sample?

76. Vitamin C is the simple compound $C_6H_8O_6$. Besides being an acid, it is a reducing agent. One method for determining the amount of vitamin C in a sample is therefore to titrate it with a solution of bromine, Br_2, an oxidizing agent.

$$C_6H_8O_6(aq) + Br_2(aq) \longrightarrow 2 HBr(aq) + C_6H_6O_6(aq)$$

Charles D. Winters

A 1.00-g "chewable" vitamin C tablet requires 27.85 mL of 0.102 M Br_2 for titration to the equivalence point. What is the mass of vitamin C in the tablet?

General Questions

These questions are not designated as to type or location in the chapter. They may combine several concepts.

77. Give the formula for the following:
    (a) a soluble compound containing the bromide ion
    (b) an insoluble hydroxide
    (c) an insoluble carbonate
    (d) a soluble nitrate-containing compound

78. Give the formula for the following:
    (a) a soluble compound containing the acetate ion
    (b) an insoluble sulfide
    (c) a soluble hydroxide
    (d) an insoluble chloride

79. Which of the following copper(II) salts are soluble in water and which are insoluble: $Cu(NO_3)_2$, $CuCO_3$, $Cu_3(PO_4)_2$, $CuCl_2$?

80. Name two anions that combine with Al^{3+} ion to produce water-soluble compounds.

81. Identify the spectator ion or ions in the reaction of nitric acid and magnesium hydroxide, and write the net ionic equation. What type of exchange reaction is this?

$$2\,H^+(aq) + 2\,NO_3^-(aq) + Mg(OH)_2(s) \longrightarrow$$
$$2\,H_2O(\ell) + Mg^{2+}(aq) + 2\,NO_3^-(aq)$$

82. Identify the water-insoluble product in each reaction and write the net ionic equation:
    (a) $CuCl_2(aq) + H_2S(aq) \longrightarrow CuS + 2\,HCl$
    (b) $CaCl_2(aq) + K_2CO_3(aq) \longrightarrow 2\,KCl + CaCO_3$
    (c) $AgNO_3(aq) + NaI(aq) \longrightarrow AgI + NaNO_3$

83. Bromine is obtained from sea water by the following reaction:

$$Cl_2(g) + 2\,NaBr(aq) \longrightarrow 2\,NaCl(aq) + Br_2(\ell)$$

    (a) What has been oxidized? What has been reduced?
    (b) Identify the oxidizing and reducing agents.
    (c) What mass of Cl_2 is required to react completely with 125 mL of 0.153 M NaBr?

84. Identify each of the following substances as an oxidizing or reducing agent: HNO_3, Na, Cl_2, O_2, $KMnO_4$.

85. Which contains the greater mass of solute: 1 L of 0.1 M NaCl or 1 L of 0.06 M Na_2CO_3?

86. Describe each of the following as product- or reactant-favored.
    (a) $BaBr_2(aq) + 2\,H_2O(\ell) \longrightarrow$
    $$Ba(OH)_2(aq) + 2\,HBr(aq)$$
    (b) $NaOH(aq) + FeCl_3(aq) \longrightarrow NaCl(aq) + Fe(OH)_3(s)$

87. You have a bottle of solid Na_2CO_3 and a 500.0-mL volumetric flask. Explain how you would make a 0.20 M solution of sodium carbonate.

88. You have 0.500 mol of KCl, some distilled water, and a 250.-mL volumetric flask. Describe how you would make a 0.500 M solution of KCl.

89. Which has the larger concentration of hydrogen ions, 0.015 M HCl or a hydrochloric acid solution with a pH of 1.2?

90. What volume of 0.054 M H_2SO_4 is required to react completely with 1.56 g of KOH?

91. The mineral dolomite contains magnesium carbonate.

$$MgCO_3(s) + 2\,HCl(aq) \longrightarrow CO_2(g) + MgCl_2(aq) + H_2O(\ell)$$

    (a) Write the net ionic equation for the reaction of magnesium carbonate and hydrochloric acid, and name the spectator ions.
    (b) What type of reaction is this?
    (c) What mass of $MgCO_3$ will react with 125 mL of HCl(aq) with a pH of 1.56?

92. Ammonium sulfide, $(NH_4)_2S$, reacts with $Hg(NO_3)_2$ to produce HgS and NH_4NO_3.
    (a) Write the overall balanced equation for the reaction. Indicate the state (s, aq) for each compound.
    (b) Name each compound.
    (c) What type of reaction is this?

93. What species (atoms, molecules, or ions) are present in an aqueous solution of each of the following compounds?
    (a) NH_3 (c) NaOH
    (b) CH_3CO_2H (d) HBr

94. Suppose an Alka-Seltzer tablet contains exactly 100 mg of citric acid, $H_3C_6H_5O_7$, plus some sodium bicarbonate. If the following reaction occurs, what mass of sodium bicarbonate must the tablet also contain?

$$H_3C_6H_5O_7(aq) + 3\,NaHCO_3(aq) \longrightarrow$$
$$3\,H_2O(\ell) + 3\,CO_2(g) + Na_3C_6H_5O_7(aq)$$

95. ▲ Sodium bicarbonate and acetic acid react according to the equation

$$NaHCO_3(aq) + CH_3CO_2H(aq) \longrightarrow$$
$$NaCH_3CO_2(aq) + CO_2(g) + H_2O(\ell)$$

What mass of sodium acetate can be obtained from mixing 15.0 g of $NaHCO_3$ with 125 mL of 0.15 M acetic acid?

96. A noncarbonated soft drink contains an unknown amount of citric acid, $H_3C_6H_5O_7$. If 100. mL of the soft drink requires 33.51 mL of 0.0102 M NaOH to neutralize the citric acid completely, what mass of citric acid does the soft drink contain per 100. mL? The reaction of citric acid and NaOH is

$$H_3C_6H_5O_7(aq) + 3\,NaOH(aq) \longrightarrow Na_3C_6H_5O_7(aq) + 3\,H_2O(\ell)$$

97. Sodium thiosulfate, $Na_2S_2O_3$, is used as a "fixer" in black-and-white photography. Suppose you have a bottle of sodium thiosulfate and want to determine its purity. The

thiosulfate ion can be oxidized with I_2 according to the balanced, net ionic equation

$$I_2(aq) + 2 S_2O_3^{2-}(aq) \longrightarrow 2 I^-(aq) + S_4O_6^{2-}(aq)$$

If you use 40.21 mL of 0.246 M I_2 in a titration, what is the weight percent of $Na_2S_2O_3$ in a 3.232-g sample of impure material?

98. You have a 4.554-g sample that is a mixture of oxalic acid, $H_2C_2O_4$, and another solid that does not react with sodium hydroxide. If 29.58 mL of 0.550 M NaOH is required to titrate the oxalic acid in the 4.554-g sample to the equivalence point, what is the weight percent of oxalic acid in the mixture? Oxalic acid and NaOH react according to the equation

$$H_2C_2O_4(aq) + 2 NaOH(aq) \longrightarrow Na_2C_2O_4(aq) + 2 H_2O(\ell)$$

99. (a) Name two water-soluble compounds containing the Cu^{2+} ion. Name two water-insoluble compounds based on the Cu^{2+} ion.

(b) Name two water-soluble compounds containing the Ba^{2+} ion. Name two water-insoluble compounds based on the Ba^{2+} ion.

100. Balance these reactions and then classify each one as a precipitation, acid–base, or gas-forming reaction. Show states for the products (s, ℓ, g, aq), and write the net ionic equation.

(a) $K_2CO_3(aq) + HClO_4(aq) \longrightarrow KClO_4 + CO_2 + H_2O$

(b) $FeCl_2(aq) + (NH_4)_2S(aq) \longrightarrow FeS + NH_4Cl$

(c) $Fe(NO_3)_2(aq) + Na_2CO_3(aq) \longrightarrow FeCO_3 + NaNO_3$

101. For each reaction, write an overall, balanced equation and the net ionic equation.

(a) the reaction of aqueous lead(II) nitrate and aqueous potassium hydroxide

(b) the reaction of aqueous copper(II) nitrate and aqueous sodium carbonate

102. (a) What is the pH of a 0.105 M HCl solution?

(b) What is the hydrogen ion concentration in a solution with a pH of 2.56? Is the solution acidic or basic?

(c) A solution has a pH of 9.67. What is the hydrogen ion concentration in the solution? Is the solution acidic or basic?

(d) A 10.0-mL sample of 2.56 M HCl is diluted with water to 250. mL. What is the pH of the dilute solution?

103. A solution of hydrochloric acid has a volume of 125 mL and a pH of 2.56. What mass of $NaHCO_3$ must be added to completely consume the HCl?

104. ▲ One-half liter (500. mL) of 2.50 M HCl is mixed with 250. mL of 3.75 M HCl. Assuming the total solution volume after mixing is 750. mL, what is the concentration of hydrochloric acid in the resulting solution? What is its pH?

105. A solution of hydrochloric acid has a volume of 250. mL and a pH of 1.92. Exactly 250. mL of 0.0105 M NaOH is added. What is the pH of the resulting solution?

106. Suppose you dilute 25.0 mL of a 0.110 M solution of Na_2CO_3 to exactly 100.0 mL. You then take exactly 10.0 mL of this diluted solution and add it to a 250-mL volumetric flask. After filling the volumetric flask to the mark with distilled water (indicating the volume of the new solution is exactly 250 mL), what is the concentration of the diluted Na_2CO_3 solution?

107. On General ChemistryNow CD-ROM or website Screen 4.12, Chemical Puzzler, you can explore the reaction of baking soda ($NaHCO_3$) with the acetic acid in vinegar. Suppose you place exactly 200 mL of vinegar in the beaker and add baking soda. The reaction occurring is

$$CH_3CO_2H(aq) + NaHCO_3(aq) \longrightarrow$$
$$NaCH_3CO_2(aq) + CO_2(g) + H_2O(\ell)$$

How many spoonfuls of baking soda is required to consume the acetic acid in the 200-mL sample? (Assume there is 50.0 g of acetic acid per liter of vinegar and a spoonful of baking soda has a mass of 3.8 g.) Are three spoonfuls sufficient? Are four spoonfuls enough?

108. The following reaction can be used to prepare iodine in the laboratory. (See photos.)

$$2 NaI(s) + 2 H_2SO_4(aq) + MnO_2(s) \longrightarrow$$
$$Na_2SO_4(aq) + MnSO_4(aq) + I_2(g) + 2 H_2O(\ell)$$

(a) Determine the oxidation number of each atom in the equation.

(b) What is the oxidizing agent and what has been oxidized? What is the reducing agent and what has been reduced?

(c) What quantity of iodine can be obtained if 20.0 g of NaI is mixed with 10.0 g of MnO_2 (and a stoichiometric excess of sulfuric acid)?

Charles D. Winters

Preparation of iodine. A mixture of sodium iodide and manganese(IV) oxide was placed in a flask in a hood (*left*). On adding concentrated sulfuric acid (*right*), brown gaseous I_2 was involved.

109. ▲ You place 2.56 g of $CaCO_3$ in a beaker containing 250. mL of 0.125 M HCl (Figure 5.5). When the reaction has ceased, does any calcium carbonate remain? What mass of $CaCl_2$ can be produced?

$$CaCO_3(s) + 2 HCl(aq) \longrightarrow CaCl_2(aq) + CO_2(g) + H_2O(\ell)$$

110. ▲ A compound has been isolated that can have either of two possible formulas: (a) $K[Fe(C_2O_4)_2(H_2O)_2]$ or (b) $K_3[Fe(C_2O_4)_3]$. To find which is correct, you dissolve a weighed sample of the compound in acid and then titrate the oxalate ion ($C_2O_4^{2-}$) that comes from the compound with potassium permanganate, $KMnO_4$ (the source of the MnO_4^- ion). The balanced, net ionic equation for the titration is

$$5\ C_2O_4^{2-}(aq) + 2\ MnO_4^-(aq) + 16\ H^+(aq) \longrightarrow$$
$$2\ Mn^{2+}(aq) + 10\ CO_2(g) + 8\ H_2O(\ell)$$

Titration of 1.356 g of the compound requires 34.50 mL of 0.108 M $KMnO_4$. Which is the correct formula of the iron-containing compound: (a) or (b)?

111. ▲ Chromium(III) ion forms many compounds with ammonia. To find the formula of one of these compounds, you titrate the NH_3 in the compound with standardized acid.

$$Cr(NH_3)_xCl_3(aq) + x\ HCl(aq) \longrightarrow$$
$$x\ NH_4^+(aq) + Cr^{3+}(aq) + (x+3)\ Cl^-(aq)$$

Assume that 24.26 mL of 1.500 M HCl is used to titrate 1.580 g of $Co(NH_3)_xCl_3$. What is the value of x?

112. ▲ The cancer chemotherapy drug cisplatin, $Pt(NH_3)_2Cl_2$, can be made by reacting $(NH_4)_2PtCl_4$ with ammonia in aqueous solution. Besides cisplatin, the other product is NH_4Cl.

(a) Write a balanced equation for this reaction.

(b) To obtain 12.50 g of cisplatin, what mass of $(NH_4)_2PtCl_4$ is required? What volume of 0.125 M NH_3 is required?

(c) Cisplatin can react with the organic compound pyridine, C_5H_5N, to form a new compound.

$$Pt(NH_3)_2Cl_2(aq) + x\ C_5H_5N(aq) \longrightarrow Pt(NH_3)_2Cl_2(C_5H_5N)_x(s)$$

Suppose you treat 0.150 g of cisplatin with what you believe is an excess of liquid pyridine (1.50 mL; $d = 0.979$ g/mL). When the reaction is complete, you can find out how much pyridine was not used by titrating the solution with standardized HCl. If 37.0 mL of 0.475 M HCl is required to titrate the excess pyridine,

$$C_5H_5N(aq) + HCl(aq) \longrightarrow C_5H_5NH^+(aq) + Cl^-(aq)$$

what is the formula of the unknown compound $Pt(NH_3)_2Cl_2(C_5H_5N)_x$?

113. You need to know the volume of water in a small swimming pool, but, owing to the pool's irregular shape, it is not a simple matter to determine its dimensions and calculate the volume. To solve the problem you stir in a solution of a dye (1.0 g of methylene blue, $C_{16}H_{18}ClN_3S$, in 50.0 mL of water). After the dye has mixed with the water in the pool, you take a sample of the water. Using an instrument such as a spectrophotometer, you determine that the concentration of the dye in the pool is 4.1×10^{-8} M. What is the volume of water in the pool?

114. ▲ In some laboratory analyses the preferred technique is to dissolve a sample in an excess of acid or base and then "back-titrate" the excess with a standard base or acid. This technique is used to assess the purity of a sample of $(NH_4)_2SO_4$. Suppose you dissolve a 0.475-g sample of impure $(NH_4)_2SO_4$ in aqueous KOH.

$$(NH_4)_2SO_4(aq) + KOH(aq) \longrightarrow$$
$$NH_3(aq) + K_2SO_4(aq) + 2\ H_2O(\ell)$$

The NH_3 liberated in the reaction is distilled from the solution into a flask containing 50.0 mL of 0.100 M HCl. The ammonia reacts with the acid to produce NH_4Cl, but not all of the HCl is used in this reaction. The amount of excess acid is determined by titrating the solution with standardized NaOH. This titration consumes 11.1 mL of 0.121 M NaOH. What is the weight percent of $(NH_4)_2SO_4$ in the 0.475-g sample?

115. You wish to determine the weight percent of copper in a copper-containing alloy. After dissolving a 0.251-g sample of the alloy in acid, an excess of KI is added, and the Cu^{2+} and I^- ions undergo the reaction

$$2\ Cu^{2+}(aq) + 5\ I^-(aq) \longrightarrow 2\ CuI(s) + I_3^-(aq)$$

The liberated I_3^- is titrated with sodium thiosulfate according to the equation

$$I_3^-(aq) + 2\ S_2O_3^{2-}(aq) \longrightarrow S_4O_6^{2-}(aq) + 3\ I^-(aq)$$

(a) Designate the oxidizing and reducing agents in the two reactions above.

(b) If 26.32 mL of 0.101 M $Na_2S_2O_3$ is required for titration to the equivalence point, what is the weight percent of Cu in 0.251-g sample of the alloy?

116. ▲ Calcium and magnesium carbonates occur together in the mineral dolomite. Suppose you heat a sample of the mineral to obtain the oxides, CaO and MgO, and then treat the oxide sample with hydrochloric acid. If 7.695 g of the oxide sample requires 125 mL of 2.55 M HCl,

$$CaO(s) + 2\ HCl(aq) \longrightarrow CaCl_2(aq) + H_2O(\ell)$$
$$MgO(s) + 2\ HCl(aq) \longrightarrow MgCl_2(aq) + H_2O(\ell)$$

What is the weight percent of each oxide (CaO and MgO) in the sample?

117. Gold can be dissolved from gold-bearing rock by treating the rock with sodium cyanide in the presence of oxygen.

$$4\ Au(s) + 8\ NaCN(aq) + O_2(g) + 2\ H_2O(\ell) \longrightarrow$$
$$4\ NaAu(CN)_2(aq) + 4\ NaOH(aq)$$

(a) Name the oxidizing and reducing agents in this reaction. What has been oxidized and what has been reduced?

(b) If you have exactly one metric ton (1 metric ton = 1000 kg) of gold-bearing rock, what volume of 0.075 M NaCN, in liters, do you need to extract the gold if the rock is 0.019% gold?

118. ▲ You mix 25.0 mL of 0.234 M FeCl₃ with 42.5 mL of 0.453 M NaOH.

    (a) What mass of Fe(OH)₃ (in grams) will precipitate from this reaction mixture?

    (b) On of the reactants (FeCl₃ or NaOH) is present in a stoichiometric excess. What is the molar concentra tion of the excess reactant remaining in solution after Fe(OH)₃ has been precipitated?

Summary and Conceptual Questions

The following questions use concepts from the preceding chapters.

119. ▲ Two students titrate different samples of the same solution of HCl using 0.100 M NaOH solution and phe-nolphthalein indicator (see Figure 5.23). The first student pipets 20.0 mL of the HCl solution into a flask, adds 20 mL of distilled water and a few drops of phenolph-thalein solution, and titrates until a lasting pink color appears. The second student pipets 20.0 mL of the HCl solution into a flask, adds 60 mL of distilled water and a few drops of phenolphthalein solution, and titrates to the first lasting pink color. Each student correctly calculates the molarity of a HCl solution. What will the second student's result be?

    (a) four times less than the first student's result

    (b) four times greater than the first student's result

    (c) two times less than the first student's result

    (d) two times greater than the first student's result

    (e) the same as the first student's result

120. On General ChemistryNow CD-ROM or website Screen 5.18, Exercise, Stoichiometry of Reactions in Solution, the video shows the reaction of Fe²⁺ with MnO₄⁻ in aque-ous solution.

    (a) What is the balanced equation for the reaction that occurred?

    (b) What is the oxidizing agent and what is the reducing agent?

    (c) Equal volumes of Fe²⁺-containing solution and MnO₄⁻-containing solution were mixed. The amount of Fe²⁺ was just sufficient to consume all of the MnO₄⁻. Which ion (Fe²⁺ or MnO₄⁻) was initially present in larger concentration?

121. ▲ General ChemistryNow CD-ROM or website Screen 4.8 Limiting Reactants, explores the reaction of zinc and hydrochloric acid.

$$Zn(s) + 2 HCl(aq) \longrightarrow ZnCl_2(aq) + H_2(g)$$

Different quantities of zinc are added to three flasks, each containing exactly 100 mL of 0.10 M HCl.

Flask 1: 7.00 g Zn

Flask 2: 3.27 g Zn

Flask 3: 1.31 g Zn

The same amount of H₂ gas was generated in Flasks 1 and 2, but a smaller amount was generated in Flask 3. The zinc was completely consumed in Flasks 2 and 3, but some remained in Flask 1. Explain these observations.

122. ▲ You want to prepare barium chloride, BaCl₂, using an exchange reaction of some type. To do so, you have the following reagents from which to select the reactants: BaSO₄, BaBr₂, BaCO₃, Ba(OH)₂, HCl, HgSO₄, AgNO₃, and HNO₃. Write a complete, balanced equation for the reaction chosen. *(Note: There are several possibilities.)*

123. Describe how to prepare BaSO₄, barium sulfate, by (a) a precipitation reaction and (b) a gas-forming reaction. To do so, you have the following reagents from which to select the reactants: BaCl₂, BaCO₃, Ba(OH)₂, H₂SO₄, and Na₂SO₄. Write complete, balanced equations for the reac-tions chosen. (See Figure 4.8 for an illustration of the preparation of a compound.)

124. Describe how to prepare zinc chloride by (a) an acid–base reaction, (b) a gas-forming reaction, and (c) an oxidation–reduction reaction. The available start-ing materials are ZnCO₃, HCl, Cl₂, HNO₃, Zn(OH)₂, NaCl, Zn(NO₃)₂, and Zn. Write complete, balanced equations for the reactions chosen.

125. In some states a person will receive a "driving while intoxicated" (DWI) ticket if the blood alcohol level (BAL) is 100 mg per deciliter (dL) of blood or higher. Suppose a person is found to have a BAL of 0.033 mol of ethanol (C₂H₅OH) per liter of blood. Will the person receive a DWI ticket?

Do you need a live tutor for homework problems?
Access vMentor at General ChemistryNow at
http://now.brookscole.com/kotz6e
for one-on-one tutoring from a chemistry expert

6—Principles of Reactivity: Energy and Chemical Reactions

The Rolex Awards for Enterprise/Tomas Bertelsen/Scientific American, Nov. 2000, p. 26

Mohammed Bah Abba. Abba's family were potmakers. As a boy Abba was fascinated by earthenware objects and their ability to absorb water yet remain structurally intact. Abba earned a college degree in business and, while still in his twenties, became an instructor in a college of business in Jigwa, Nigeria, and a consultant to the United Nations Development Program. That brought him back into close contact with rural communities in northern Nigeria and made him aware of the hardships of the families there. Making and distributing the pot-in-pot device for safe food storage closed this interesting circle.

Abba's Refrigerator

If you put a pot of water on a kitchen stove or a campfire, or if you put the pot in the sun, the water will evaporate. You must supply energy in some form because evaporation requires the input of energy. This well-known principle was applied in a novel way by a young African teacher, Mohammed Bah Abba, to improve the life of his people in Nigeria.

Life is hard in northern Nigerian communities. In this rural semi-desert area, most people eke out a living through subsistence

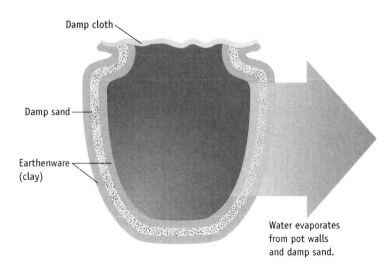

Damp cloth

Damp sand

Earthenware (clay)

Water evaporates from pot walls and damp sand.

The pot-in-pot refrigerator. Water seeps through the outer pot from the damp sand layer separating the pots, or from food stored in the inner pot. As the water evaporates from the surface of the outer pot, the food is cooled.

Chapter Goals

See Chapter Goals Revisited (page 270). Test your knowledge of these goals by taking the exam-prep quiz on the General ChemistryNow CD-ROM or website.

- Assess heat transfer associated with changes in temperature and changes of state.
- Apply the first law of thermodynamics.
- Define and understand the state functions enthalpy and internal energy.
- Calculate the energy changes occurring in chemical reactions and learn how these changes are measured.

James Cowlin/Image Enterprises, Pheonix, AZ.

Swamp coolers. These inexpensive air-conditioners work on the same principle as Abba's pot. A trickle of water washes over a bed of straw or other porous material. As air is drawn over the moist material, the air is cooled as the water takes energy from the air to evaporate.

farming. Because of the dearth of modern refrigeration, food spoilage is a major problem. Using a simple thermodynamic principle, Abba developed a refrigerator that cost about 30 cents to make and does not use electricity.

Abba's refrigerator consists of two earthen pots, one inside the other, separated by a layer of sand. The pots are covered with a damp cloth and placed in a well ventilated area. Water seeps

through the pot's outer wall and rapidly evaporates in the dry desert air. The water remaining in the pot and its contents drop in temperature. Food in the inner pot can stay cool for days and not spoil.

In the 1990s, at his own expense, Abba made and distributed almost 10,000 pots in the villages of northern Nigeria. He estimates that about 75% of the families in this area are now using his refrigerator. The impact of this simple device has implications not only for the health of his people but also for their economy and their social structure. Prior to the development of the pot-in-pot device for food storage, it was necessary to sell produce immediately upon harvesting it. The young girls in the family who sold food on the street daily could now be released from this chore to attend school and improve their lives.

Every two years, the Rolex Company, the Swiss maker of timepieces, gives a series of awards for enterprise. For his pot-in-pot refrigerator, Abba was one of the five recipients of a Rolex Award in 2000.

Charles D. Winters

Evaporative cooling. The same principle that cools Abba's refrigerator cools you down if you wear a strip of damp cloth, a "neck cooler," around your neck on a hot day.

To Review Before You Begin

- Know how to write balanced chemical equations (Chapter 4)
- Review product-favored and reactant-favored reactions (page 197)
- Know how to use Kelvin and Celsius temperature scales (Section 1.6)
- Review states of matter and changes of state (Section 1.5)

GENERAL
Chemistry⬩Now™

Throughout the chapter this icon introduces a list of resources on the General ChemistryNow CD-ROM or website (http://now.brookscole.com/kotz6e) that will:

- help you evaluate your knowledge of the material
- provide homework problems
- allow you to take an exam-prep quiz
- provide a personalized Learning Plan targeting resources that address areas you should study

Energy transfer accompanies both chemical and physical changes. Our bodies are cooled when we perspire—the evaporation of water in sweat, a physical change, draws energy from our body and causes us to feel cooler. When water vapor condenses, heat is given off, a process that has a significant impact on the weather (Figure 6.1). The sun's energy can be stored as chemical energy by the formation of carbohydrates and oxygen from carbon dioxide and water in the process of photosynthesis, a chemical change.

$$6 \ CO_2(g) + 6 \ H_2O(g) + energy \longrightarrow C_6H_{12}O_6(s) + 6 \ O_2(g)$$

This chemical energy can be released in a chemical reaction of carbohydrate and oxygen, whether in the laboratory (Figure 6.2), in living tissue, or in a forest fire.

$$C_6H_{12}O_6(s) + 6 \ O_2(g) \longrightarrow 6 \ CO_2(g) + 6 \ H_2O(g) + energy$$

When studying chemistry, it is important to know something about energy. The most common form of energy we see in chemical processes is heat. Changes of heat

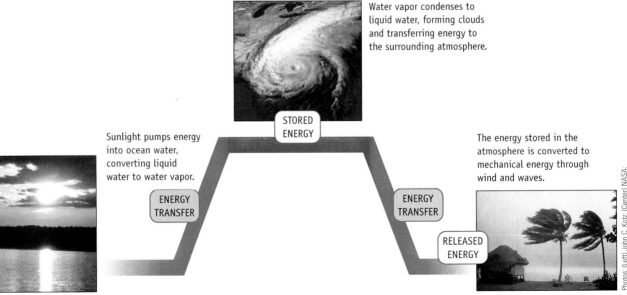

Water vapor condenses to liquid water, forming clouds and transferring energy to the surrounding atmosphere.

STORED ENERGY

Sunlight pumps energy into ocean water, converting liquid water to water vapor.

ENERGY TRANSFER

The energy stored in the atmosphere is converted to mechanical energy through wind and waves.

ENERGY TRANSFER

RELEASED ENERGY

Photos: (Left) John C. Kotz; (Center) NASA; (Right) Frederick Ayer/Photo Researchers, Inc.

Figure 6.1   Energy transfer in nature. Hurricanes and other forms of violent weather involve the storage and release of energy. The average hurricane releases energy equivalent to the annual U.S. production of electricity.

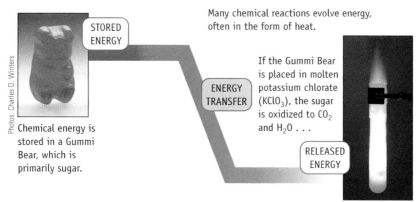

Many chemical reactions evolve energy, often in the form of heat.

STORED ENERGY

Photos: Charles D. Winters

Chemical energy is stored in a Gummi Bear, which is primarily sugar.

ENERGY TRANSFER

If the Gummi Bear is placed in molten potassium chlorate ($KClO_3$), the sugar is oxidized to CO_2 and H_2O . . .

RELEASED ENERGY

. . . and the energy evolved in the chemical reaction is observed as heat and light.

Figure 6.2 Energy transfer in a chemical reaction. (*See General ChemistryNow Screen 6.17 Product-Favored Systems, to watch a video of this reaction.*)

content and the transfer of heat between objects are major themes of **thermodynamics**, the science of heat and work—and the subject of this chapter and a later one (Chapter 19). As described in the "The Chemistry of Fuels and Energy Sources" (pages 282–293), the principles of thermodynamics apply to energy use in your home, to ways of conserving energy, to recycling of materials, and to problems of current and future energy availability and use in our economy.

a. Bruce Roberts/Photo Researchers, Inc.; *b.* Royalty-Free/Corbis; *c.* William James Warren/Corbis

(a) Gravitational energy   **(b)** Chemical potential energy   **(c)** Electrostatic energy

Active Figure 6.3   Energy and its conversion (a) Water at the top of a water wheel represents stored, or potential, energy. As water flows over the wheel, its potential energy is converted to mechanical energy. (b) Chemical potential energy is converted to heat and then to work. (c) Lightning converts electrostatic energy into radiant and thermal energy.

GENERAL
Chemistry ⚛ Now™ See the General ChemistryNow CD-ROM or website to explore an interactive version of this figure accompanied by an exercise.

6.1—Energy: Some Basic Principles

Energy is defined as the capacity to do work. You do work against the force of gravity when carrying yourself and hiking equipment up a mountain. You can do this work because you have the energy to do so, the energy having been provided by the food you have eaten. Food energy is chemical energy—energy stored in chemical compounds and released when the compounds undergo the chemical reactions of metabolism in your body.

Energy can be classified as kinetic or potential. **Kinetic energy,** as noted in the discussion of kinetic-molecular theory (Section 1.5), **is energy associated with motion,** such as

- *Thermal energy* of atoms, molecules, or ions in motion at the submicroscopic level. All matter has thermal energy.
- *Mechanical energy* of a macroscopic object like a moving tennis ball or automobile.
- *Electrical energy* of electrons moving through a conductor.
- *Sound,* which corresponds to compression and expansion of the spaces between molecules.

Potential energy, energy that results from an object's position (Figure 6.3), includes:

- *Gravitational energy,* such as that possessed by a ball held above the floor and by water at the top of a waterfall (Figure 6.3a).
- *Chemical potential energy.* The energy stored in coal, for example, is converted to heat when burned, and the heat is converted to work (Figure 6.3b). All chemical reactions involve a change in chemical potential energy.
- *Electrostatic energy,* potential energy associated with the separation of two dissimilar electrical charges. The energy is released (as light, heat, and sound) when the opposite charges are neutralized, as happens when a bolt of lightning darts between clouds and the ground (Figure 6.3c).

Potential energy is stored energy and can be converted into kinetic energy. For example, as water falls over a waterfall, its potential energy is converted into kinetic energy. Similarly, kinetic energy can be converted into potential energy: The kinetic energy of falling water can turn a turbine to produce electricity, which can then be used to convert water into H_2 and O_2 (Figure 1.6, page 18). The H_2 gas represents stored chemical potential energy because it can be burned to produce heat and light (Figure 1.13, page 25) or used in a fuel cell (as in the Space Shuttle) to produce electrical energy.

Conservation of Energy

Standing on a diving board, you have considerable potential energy because of your position above the water. Once you jump off the board, some of that potential energy is converted into kinetic energy (Figure 6.4). During the dive, the force of gravity accelerates your body so that it moves faster and faster. Your kinetic energy increases and your potential energy decreases. At the moment you hit the water, your velocity is abruptly reduced, and much of your kinetic energy is converted to mechanical energy; the water splashes as your body moves it aside by doing work on it. Eventually you float on the surface, and the water becomes still again. If you could see them, however, you would find that the water molecules are moving a lit-

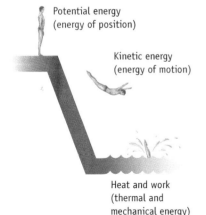

Potential energy
(energy of position)

Kinetic energy
(energy of motion)

Heat and work
(thermal and
mechanical energy)

Figure 6.4   The law of energy conservation. The diver's potential energy is converted to kinetic energy and then to thermal energy, illustrating the law of conservation of energy (*See General ChemistryNow Screen 6.2 Energy to view an animation based on this figure.*)

tle faster in the vicinity of your dive; that is, the kinetic energy of the water is slightly higher.

This series of energy conversions illustrates the **law of conservation of energy,** otherwise known as the **first law of thermodynamics.** These terms are synonymous; both state that energy can neither be created nor destroyed. Or, to state this law differently, *the total energy of the universe is constant.* These statements summarize the results of a great many experiments in which heat, work, and other forms of energy transfer have been measured and the total energy content found to be the same before and after an event.

Another example of the law of energy conservation is burning oil or coal to heat your house or to drive a locomotive (Figure 6.3b). These fuels are an energy resource. When burned, the chemical energy present in the oil or gas is converted to an equal quantity of energy, now in the form of heat for your home and the thermal energy of the gases going up the chimney.

GENERAL
Chemistry∿Now™

See the General ChemistryNow CD-ROM or website:

- **Screen 6.3 Forms of Energy,** for an exercise that examines the energy conversions in several situations

---

Exercise 6.1—Energy

A battery stores chemical potential energy. Into what types of energy can this potential energy be converted?

Temperature and Heat

The temperature of an object is a measure of its heat energy content and of its ability to transfer heat. One way to measure temperature is with a thermometer containing mercury or some other liquid (Figure 6.5). When the thermometer is placed in hot water, heat is transferred from the water to the thermometer. The increased energy causes the mercury atoms, for example, to move about more rapidly and the space between atoms to increase slightly. You observe this effect as an expansion in the volume of the mercury, such that the column of mercury rises higher in the thermometer tube.

Three important aspects of thermal energy and temperature should be understood:

- Heat is not the same as temperature.
- The more thermal energy a substance has, the greater the motion of its atoms and molecules.
- The total thermal energy in an object is the sum of the individual energies of all the atoms, molecules, or ions in that object.

The thermal energy of a given substance depends not only on temperature but also on the amount of substance. Thus, a cup of hot coffee may contain less thermal energy than a bathtub full of warm water, even though the coffee is at a higher temperature.

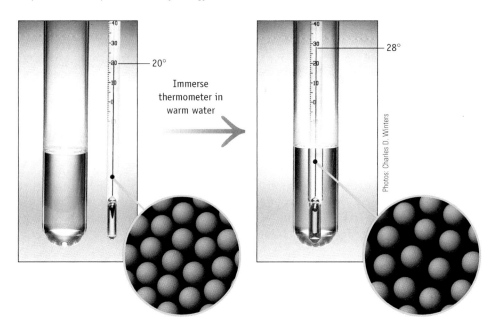

Immerse thermometer in warm water

20°

28°

Photos: Charles D. Winters

Figure 6.5   Measuring temperature. The volume of liquid mercury in a thermometer increases slightly when immersed in warm water. The volume increase causes the mercury to rise in the thermometer, which is calibrated to indicate the temperature.

Systems and Surroundings

In thermodynamics, the terms "system" and "surroundings" have precise and important scientific meanings. A **system** is defined as the object, or collection of objects, being studied (Figure 6.6). The **surroundings** include everything outside the system that can exchange energy with the system. In the discussions that follow, we will need to identify systems precisely. If we are studying the heat evolved in a chemical reaction, for example, the system might be defined as a reaction vessel and its contents. The surroundings would be the air in the room and anything else in contact with the vessel. At the atomic level, the system could be a single atom or molecule and the surroundings would be the atoms or molecules in its vicinity. In general, we choose how we define the system and its surroundings for each situation, depending on the information we are trying to obtain.

This concept of a system and its surroundings applies to nonchemical situations as well. If we want to study the energy balance on this planet, we might choose to define the earth as the system and outer space as the surroundings. On a cosmic level, the solar system might be defined as the system being studied, and the rest of the galaxy would be the surroundings.

Directionality of Heat Transfer: Thermal Equilibrium

Heat transfer occurs when two objects at different temperatures are brought into contact. In Figure 6.7, for example, the beaker of water and the piece of metal being heated in a Bunsen burner flame have different temperatures. When the hot metal is plunged into the cold water, heat is transferred from the metal to the water. Eventually, the two objects reach the same temperature. At that point, the system has reached **thermal equilibrium**. The distinguishing feature of thermal equilibrium is that, on the macroscopic scale, no further temperature change occurs and the temperature throughout the entire system (metal plus water) is the same.

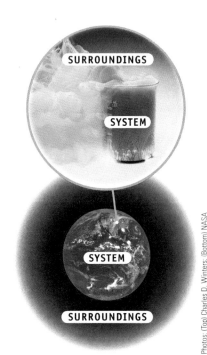

Photos: (Top) Charles D. Winters; (Bottom) NASA

SURROUNDINGS

SYSTEM

SYSTEM

SURROUNDINGS

Figure 6.6   Systems and their surroundings. Earth can be considered a thermodynamic system, with the rest of the universe as its surroundings. A chemical reaction occurring in a laboratory is also a system, with the laboratory as its surroundings.

Figure 6.7 Energy transfer. Heat is transferred from the hotter metal bar to the cooler water. Eventually the water and metal reach the same temperature and are said to be in thermal equilibrium. *(See General ChemistryNow Screen 6.9 Heat Transfer Between Substances, for a simulation and tutorial.)*

The manipulation of the hot metal bar and the beaker of water may seem like a rather simple experiment. Embedded in the experiment, however, are some principles that will be very important in our further discussion:

- Heat transfer always occurs from an object at a higher temperature to an object at a lower temperature. The directionality of heat transfer is an important principle of thermodynamics. (See "A Closer Look: Why Doesn't the Heat in a Room Cause Your Cup of Coffee to Boil?")

- Transfer of heat continues until both objects are at the same temperature (thermal equilibrium).

For the specific case where heat transfer occurs within a system, we can also say that the quantity of heat lost by a hotter object and the quantity of heat gained by a cooler object when they are in contact are numerically equal. (This is required by the law of conservation of energy.)

When heat transfer occurs across the boundary between system and surroundings, we can describe the directionality of heat transfer as exothermic or endothermic (Figure 6.8).

- In an **exothermic** process, heat is transferred from a system to the surroundings.

- An **endothermic** process is the opposite of an exothermic process: Heat is transferred from surroundings to the system.

■ **Thermal Equilibrium**
Although no change is evident at the macroscopic level when thermal equilibrium is reached, on the molecular level transfer of energy between individual molecules will continue to occur. This feature—no change visible on a macroscopic level, but processes still occurring at the particulate level—is a general feature of equilibria that we will encounter again (Chapters 16–18).

A Closer Look

Why Doesn't the Heat in a Room Cause Your Cup of Coffee to Boil?

If a cup of coffee or tea is hotter than its surroundings, heat is transferred to the surroundings until the hot coffee cools off and the surroundings warm up a bit. It is interesting and useful to think about why the opposite process doesn't occur. Why doesn't the heat in a room cause a cup of cold coffee to boil? The law of energy conservation would not be violated

if the coffee got hotter and hotter and the surroundings in the room got cooler and cooler. However, we know from experience that this will never happen. The directionality in heat transfer—heat energy always transfers from a hotter object to a cooler one, never the reverse—corresponds to a spreading out of energy over the greatest possible number of atoms, ions, or molecules. Energy transfers from a relatively small number of molecules in a hot cup of coffee to a large number of atoms and molecules surrounding the cup.

Similarly, the large number of particles in the surrounding environment will heat a glass of ice water by transferring some of their energy to the glass, ice, and water molecules. As in the previous example, the end result is to spread thermal energy more evenly over the maximum number of particles. The opposite process, concentrating energy in only a few particles at the expense of many, is never observed.

The directionality of energy transfer, which plays an important role in thermodynamics, will be discussed further in Chapter 19.

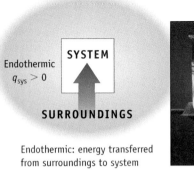

Endothermic
$q_{sys} > 0$

SYSTEM

SURROUNDINGS

Endothermic: energy transferred
from surroundings to system

Exothermic
$q_{sys} < 0$

SYSTEM

SURROUNDINGS

Exothermic: energy transferred
from system to surroundings

Photos: Charles D. Winters

Active Figure 6.8   **Exothermic and endothermic processes.** The symbol q represents heat transferred, and the subscript "sys" refers to the system.

GENERAL
Chemistry ⚛ Now™ See the General ChemistryNow CD-ROM or website to explore an interactive version of this figure accompanied by an exercise.

GENERAL
Chemistry ⚛ Now™

See the General ChemistryNow CD-ROM or website:
- **Screen 6.4 Directionality of Heat Transfer,** to view an animation on endothermic and exothermic systems

■ Kinetic Energy

Kinetic energy is calculated by the equation $KE = \frac{1}{2} mv^2$. One joule is the kinetic energy of a 2.0 kg mass (m) moving at 1.0 m/s (v).

$$KE = \frac{1}{2}(2.0 \text{ kg})(1.0 \text{ m/s})^2$$
$$= 1.0 \text{ kg} \cdot \text{m}^2/\text{s}^2 = 1.0 \text{ J}$$

Energy Units

When expressing energy quantities, most chemists (and much of the world outside the United States) use the **joule** (J), the SI unit of thermal energy. The joule is preferred in scientific study because it is related directly to the units used for mechanical energy: 1 J equals 1 kg · m²/s². However, the joule can be inconveniently small as a unit for use in chemistry, so the kilojoule (kJ), equivalent to 1000 joules, is often used.

To give you some feeling for joules, suppose you drop a six-pack of soft-drink cans, each full of liquid, on your foot. Although you probably will not take time to calculate the kinetic energy at the moment of impact, it is between 4 J and 10 J.

An older unit for measuring heat is the calorie (cal). It is defined as the heat required to raise the temperature of 1.00 g of pure liquid water from 14.5 °C to 15.5 °C. A kilocalorie (kcal) is equivalent to 1000 calories. The conversion factor relating joules and calories is

$$1 \text{ calorie (cal)} = 4.184 \text{ joules (J)}$$

The dietary Calorie (with a capital C) is often used in the United States to represent the energy content of foods. This unit is encountered when reading the nutritional information on a food label. The dietary Calorie (Cal) is equivalent to the kilocalorie or 1000 calories. Thus, a breakfast cereal that gives you 100.0 Calories of nutritional energy per serving provides 100.0 kcal or 418.4 kJ.

Chemical Perspectives

Food and Calories

The U.S. Food and Drug Administration (FDA) mandates that nutritional data, including energy content, be included on almost all packaged food. The Nutrition Labeling and Education Act of 1990 requires that the total energy from protein, carbohydrates, fat, and alcohol be specified. How is this determined? Initially the method used was calorimetry. In this method, which is described in Section 6.6, a food product is burned and the heat evolved in the combustion is measured. Now, however, all energy content is estimated using the Atwater system. This method specifies the following average values for energy sources in foods:

1 g protein = 4 kcal (17 kJ)

1 g carbohydrate = 4 kcal (17 kJ)

1 g fat = 9 kcal (38 kJ)

1 g alcohol = 7 kcal (29 kJ)

Because carbohydrates contain some indigestible fiber, the mass of fiber is subtracted from the mass of carbohydrate when calculating the energy from carbohydrates.

As an example, one serving of cashew nuts (about 28 g) has

14 g fat = 126 kcal

6 g protein = 24 kcal

7 g carbohydrates − 1 g fiber = 24 kcal

Total = 174 kcal (728 kJ)

A value of 170 kcal is reported on the package.

You can find data on more than 6000 foods at the Nutrient Data Laboratory Website (**www.nal.usda.gov/fnic/foodcomp/**). See also **nat.crgq.com** for an online tool that allows you to find the energy content of foods.

Nutrition Facts			
Serving Size 1 cup (30g)			
Children Under 4 - ¾ cup (20g)			
Servings Per Container About 19			
Children Under 4 - About 28			

Amount Per Serving	Cheerios	with ½ cup skim milk	Cereal for Children Under 4
Calories	110	150	70
Calories from Fat	15	20	10
	% Daily Value**		
Total Fat 2g*	3%	3%	1g
Saturated Fat 0g	0%	3%	0g
Polyunsaturated Fat 0.5g			0g
Monounsaturated Fat 0.5g			0g
Cholesterol 0mg	0%	1%	0mg
Sodium 210mg	9%	12%	140mg
Potassium 200mg	6%	12%	130mg

Charles D. Winters

Energy and food labels. All packaged foods must have labels specifying nutritional values, with energy given in Calories (where 1 Cal = 1 kilocalorie).

GENERAL
Chemistry Now™

See the General ChemistryNow CD-ROM or website:

• **Screen 6.5 Energy Units,** for a tutorial on converting energy units

Exercise 6.2—Energy Units

(a) In an old textbook you read that the oxidation of 1.00 g of hydrogen to form liquid water produces 3800 calories. What is this energy in units of joules?

(b) The label on a cereal box indicates that one serving (with skim milk) provides 250 Cal. What is this energy in kilojoules (kJ)?

6.2—Specific Heat Capacity and Heat Transfer

The quantity of heat transferred to or from an object depends on three things:

• The quantity of material
• The size of the temperature change
• The identity of the material gaining or losing heat

The **specific heat capacity** *(C)* is related to these three parameters. The specific heat capacity is the quantity of heat required to raise the temperature of 1 gram of a substance by one kelvin. It has units of joules per gram per kelvin (J/g · K).

Energy content of foods. In many countries that use standardized SI units, food energy is also measured in joules. The diet soda in this can (from Australia) is said to have an energy content of only 1 joule.

The quantity of heat gained or lost when a given mass of a substance is warmed or cooled is calculated using Equation 6.1.

Specific heat capacity (J/g · K)   Change in temperature (K)

$$q = C \times m \times \Delta T \tag{6.1}$$

Heat transferred (J)   Mass of substance (g)

Here, q is the quantity of heat transferred to or from a given mass of substance (m), C is the specific heat capacity, and ΔT is the change in temperature. The capital Greek letter delta, Δ, means "change in." The change in temperature, ΔT, is calculated as the final temperature minus the initial temperature.

$$\Delta T = T_{final} - T_{initial} \tag{6.2}$$

Calculating a change in temperature as in Equation 6.2 will give a result with an algebraic sign that indicates the direction of heat transfer ("A Closer Look, Sign Conventions"). For example, we can use the specific heat capacity of copper, $0.385 \, J/g \cdot K$, to calculate the change in heat content of a 10.0-g sample of copper if its temperature is raised from 298 K (25 °C) to 598 K (325 °C).

$$q = \left(0.385 \, \frac{J}{g \cdot K}\right)(10.0 \, g)(598 \, K - 298 \, K) = +1160 \, J$$

T_{final}
Final temp. $T_{initial}$
Initial temp.

■ **Change In Temperature, ΔT**

Sign of ΔT	Meaning
Positive	$T_{final} > T_{initial}$, so T has increased, and q will be positive. Heat has been transferred to the object under study.
Negative	$T_{final} < T_{initial}$, so T has decreased, and q will be negative. Heat has been transferred out of the object under study.

■ **Molar Heat Capacity**
Heat capacities can be expressed on a per mole basis. The molar heat capacity is the amount of heat required to raise the temperature of one mole of a substance by one degree kelvin. The molar heat capacity of metals at room temperature is always near 25 kJ/mol · K.

Notice that the answer has a positive sign. This indicates that the heat content of the sample of copper has *increased* by 1160 J because heat has transferred *to* the copper (the system) *from* the surroundings.

Specific heat capacities of some metals, compounds, and common substances are listed in Table 6.1. Notice that water has one of the highest values, $4.184 \, J/g \cdot K$. In contrast, the specific heat capacities of most common materials are considerably smaller. For example, the specific heat capacity of iron is $0.449 \, J/g \cdot K$; to raise the temperature of a gram of water by 1 K requires about nine times as much heat as is required to cause a 1 K change in temperature for a gram of iron.

The high specific heat capacity of water has major significance. A great deal of energy must be absorbed by a large body of water to raise its temperature by just a degree or so. Thus, large bodies of water have a profound influence on our weather. In spring, lakes tend to warm up more slowly than the air. In autumn, the heat given off by a large lake moderates the drop in air temperature.

The greater the specific heat and the larger the mass, the more thermal energy a substance can store. This relationship has numerous implications. For example, you might wrap some bread in aluminum foil and heat it in an oven. You can remove the foil with your fingers after taking the bread from the oven, even though the bread is very hot. A small quantity of aluminum foil is used and the metal has a low specific heat capacity; thus, when you touch the hot foil, only a small quantity of heat will be transferred to your fingers (which have a larger mass and a higher specific heat capacity). This is also the reason why a chain of fast-food restaurants warns you that the filling of an apple pie can be much warmer than the paper wrapper or the pie crust (Figure 6.9). Although the wrapper, pie crust, and filling are at the same temperature, the heat content of the filling is greater than that of the wrapper and crust.

Charles D. Winters

Figure 6.9   A practical example of specific heat capacity. The filling of the apple pie has a higher specific heat capacity (and larger mass) than the pie crust and wrapper. Notice the warning on the wrapper.

A Closer Look

Sign Conventions

Whenever you take the difference between two quantities in chemistry, you should always subtract the initial quantity from the final quantity. A consequence of this convention is that the algebraic sign of the result indicates an increase (+) or a decrease (−) in the quantity being studied. This is an important point, as you will see not only in this chapter but also in other chapters of this book.

Thus far, we have described temperature changes and the direction of heat transfer. The table below summarizes the conventions used.

When discussing the quantity of heat, we use an *unsigned number*. If we want to indicate the *direction of transfer* in a process, however, we attach a sign, either negative (heat transferred from the substance) or positive (heat transferred to the substance), to q. The sign of q "signals" the direction of heat transfer. Heat, a quantity of energy, cannot be negative but the heat content of an object can increase

or decrease, depending on the direction of heat transfer.

An analogy might make this point clearer. Consider your bank account. Assume you have $260 in your account ($A_{initial}$) and after a withdrawal you have $200 ($A_{final}$). The cash flow is thus

$$\text{Cash flow} = A_{final} - A_{initial}$$
$$= \$200 - \$260$$
$$= -\$60$$

The negative sign on the $60 indicates that a withdrawal has been made; the cash itself is not a negative quantity.

ΔT of System	Sign of ΔT	Sign of q	Direction of Heat Transfer
Increase	+	+	Heat transferred from surroundings to system (an endothermic process)
Decrease	−	−	Heat transferred from system to surroundings (an exothermic process)

GENERAL
Chemistry Now™

See the General ChemistryNow CD-ROM or website:

- **Screen 6.7 Heat Capacity of Pure Substances,** for a simulation and exercise on the change in thermal energy when various material are heated

Table 6.1   Specific Heat Capacity Values for Some Elements, Compounds, and Common Solids

Substance	Specific Heat Capacity (J/g · K)	Molar Heat Capacity (J/mol · K)
Elements		
Al, aluminum	0.897	24.2
C, graphite	0.685	8.23
Fe, iron	0.449	25.1
Cu, copper	0.385	24.5
Au, gold	0.129	25.4
Compounds		
$NH_3(\ell)$, ammonia	4.70	80.0
$H_2O(\ell)$, water (liquid)	4.184	75.4
$C_2H_5OH(\ell)$, ethanol	2.44	11.2
$HOCH_2CH_2OH(\ell)$, ethylene glycol (antifreeze)	2.39	14.8
$H_2O(s)$, water (ice)	2.06	37.1
Common Solids		
wood	1.8	
cement	0.9	
glass	0.8	
granite	0.8	

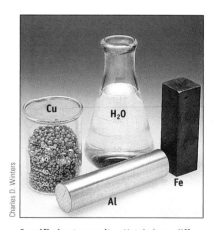

Specific heat capacity. Metals have different values of specific heat capacity on a per-gram basis. However, their molar heats capacities are all in the range of 25 J/mol · K. Among common substances, liquid water has the highest specific heat capacity on a per gram basis, a fact that plays a significant role in the earth's weather and climate.

Example 6.1—Specific Heat Capacity

Problem Determine the quantity of heat that must be added to raise the temperature of a cup of coffee (250 mL) from 20.5 °C (293.7 K) to 95.6 °C (368.8 K). Assume that water and coffee have the same density (1.00 g/mL) and the same specific heat capacity.

Strategy Use Equation 6.1. For this calculation, you will need the specific heat capacity for H_2O from Table 6.1 (4.184 J/g · K), the mass of the coffee (calculated from its density and volume), and the change in temperature ($T_{final} - T_{initial}$).

Solution

$$\text{Mass of coffee} = (250 \text{ mL})(1.00 \text{ g/mL}) = 250 \text{ g}$$

$$\Delta T = T_{final} - T_{initial} = 368.8 \text{ K} - 293.7 \text{ K} = 75.1 \text{ K}$$

$$q = C \times m \times \Delta T$$

$$q = (4.184 \text{ J/g} \cdot \text{K})(250 \text{ g})(75.1 \text{ K})$$

$$\Delta T = 79,000 \text{ J (or 79 kJ)}$$

Comment Notice that heat has been transferred to the coffee from the surroundings. The heat content of the coffee has increased.

Hot metal (55.0 g iron)

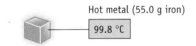

99.8 °C

Cool water (225 g)

21.0 °C

Immerse hot metal in water

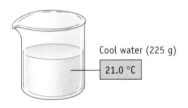

Metal cools in exothermic process.

ΔT of metal is negative.

q_{metal} is negative.

23.1 °C

Water is warmed in endothermic process.

ΔT of water is positive.

q_{water} is positive.

Active Figure 6.10   Heat transfer.
When heat transfers from a hot metal to cool water, the heat transferred from the metal, q_{metal}, has a negative value. The heat transferred to the water, q_{water}, is positive. (See also Figure 6.7.)

 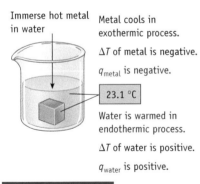

Exercise 6.3—Specific Heat Capacity

In an experiment it was determined that 59.8 J was required to change the temperature of 25.0 g of ethylene glycol (a compound used as antifreeze in automobile engines) by 1.00 K. Calculate the specific heat capacity of ethylene glycol from these data.

Quantitative Aspects of Heat Transfer

Like melting point, boiling point, and density, the specific heat capacity is a characteristic property of a pure substance. The specific heat capacity of a substance can be determined experimentally by accurately measuring temperature changes that occur when heat is transferred from the substance to a known quantity of water (whose specific heat capacity is known).

Suppose a 55.0-g piece of metal is heated in boiling water to 99.8 °C and then dropped into cool water in an insulated beaker (Figure 6.10). Assume the beaker contains 225 g of water and its initial temperature (before the metal was dropped in) was 21.0 °C. The final temperature of the metal and water is 23.1 °C. What is the specific heat capacity of the metal? Here are the most important aspects of this experiment:

- The metal and the water are the system, and the beaker and environment are the surroundings. We assume that heat is transferred only within the system and not between the system and the surroundings. (For an accurate calculation, we would want to include heat transfer to the surroundings.)

- The water and the metal bar end up at the same temperature. (T_{final} is the same for both.)

- The heat transferred from the metal to the water, q_{metal}, has a negative value because the temperature of the metal dropped as heat was transferred out of it to the water. Conversely, q_{water} has a positive value because its temperature increased as heat was transferred into the water from the metal.

- The values of q_{water} and q_{metal} are numerically equal but of opposite sign; that is, $q_{water} = -q_{metal}$. Expressed another way, $q_{water} + q_{metal} = 0$. To paraphrase this equation: *The sum of thermal energy changes in this system is zero.*

Problems involving heat transfer can be approached by assuming that the sum of the heat content changes within a given system is zero (Equation 6.3).

$$q_1 + q_2 + q_3 \cdots = 0 \qquad (6.3)$$

The quantities q_1, q_2, and so on represent the changes in thermal energy for the individual parts of the system. For this specific problem, there are two heat content changes, q_{water} and q_{metal}, and

$$q_{water} + q_{metal} = 0$$

Each of these quantities is related to the specific heat capacities, mass, and change of temperature of the water and metal, as defined by Equation 6.1. Thus

$$\left[C_{water} \times m_{water} \times (T_{final} - T_{initial,\ water}) \right] + \left[C_{metal} \times m_{metal} \times (T_{final} - T_{initial,\ metal}) \right] = 0$$

The specific heat capacity of the metal is the unknown in this problem. Using the specific heat capacity of water from Table 6.1 and converting Celsius temperatures to kelvin gives

$$[(4.184\ \text{J/g} \cdot \text{K})(225\ \text{g})(296.3\ \text{K} - 294.2\ \text{K})]$$
$$+ \left[(C_{metal})(55.0\ \text{g})(296.3\ \text{K} - 373.0\ \text{K}) \right] = 0$$
$$C_{metal} = 0.469\ \text{J/g} \cdot \text{K}$$

■ **Heat Transfer**
Remember that $T_{initial}$ for the metal and $T_{initial}$ for the water in this problem have different values.

GENERAL
Chemistry Now™

See the General ChemistryNow CD-ROM or website:
- **Screen 6.8 Calculating Heat Transfer**

    (a) for a tutorial on calculations using specific heat capacity

    (b) for a simulation and exercise on determining the temperature of thermal equilibrium when two objects are in contact

    (c) for a tutorial on calculating the final temperature of thermal equilibrium

Example 6.2—Using Specific Heat Capacity

Problem A 88.5-g piece of iron whose temperature is 78.8 °C (352.0 K) is placed in a beaker containing 244 g of water at 18.8 °C (292.0 K). When thermal equilibrium is reached, what is the final temperature? (Assume no heat is lost to warm the beaker and surroundings.)

Strategy First, define the system as the iron and water. Two changes occur within the system: Iron gives up heat and water gains heat. The sum of the heat quantities of these changes must equal zero. Each quantity of heat is related to the specific heat capacity, mass, and temperature change of the substance using Equation 6.1 [$q = C \times m \times (T_{final} - T_{initial})$]. The final temperature is unknown. The specific heat capacities of iron and water are given in Table 6.1. The change in temperature, ΔT, may be in °C or K. See Problem-Solving Tip 6.1.

Problem Solving Tip 6.1

Units for *T* and Specific Heat Capacity

(a) *Calculating ΔT.* Notice that specific heat values are given in units of joules per-gram per kelvin (J/g · K). Virtually all calculations that involve temperature in chemistry are expressed in kelvins. In calculating ΔT, however, we could use Celsius temperatures because a kelvin and a Celsius degree are the

same size, so that the difference between two temperatures is the same on both scales. For example, the difference between the boiling and freezing points of water is

$$\Delta T, \text{Celsius} = 100\ °C − 0\ °C = 100\ °C$$

$$\Delta T, \text{kelvin} = 373\ K − 273\ K = 100\ K$$

(b) *Units of Specific Heat Capacity.* Specific heat capacities are given in this book in units of joules per gram per kelvin (J/g · K). Often, however, specific heat

capacity values found in handbooks (such as the *CRC Handbook of Chemistry and Physics*) or the NIST Webbook (**webbook.nist.gov**) will have units of J/mol · K; that is, they are molar heat capacities. For example, liquid water has a specific heat capacity of 4.184 J/g · K or 75.40 J/mol · K. The values are related as follows:

$$(4.184\ \text{J/g} · K)(18.02\ \text{g/mol}) =$$
$$75.40\ \text{J/mol} · K$$

Solution

$$q_{\text{water}} + q_{\text{metal}} = 0$$

$$\left[C_{\text{water}} \times m_{\text{water}} \times (T_{\text{final}} − T_{\text{initial, water}}) \right] + \left[C_{\text{Fe}} \times m_{\text{Fe}} \times (T_{\text{final}} − T_{\text{initial, Fe}}) \right] = 0$$

$$\left[(4.184\ \text{J/g} · K)(244\ g)(T_{\text{final}} − 292.0\ K) \right] + \left[(0.449\ \text{J/g} · K)(88.5\ g)(T_{\text{final}} − 352.0\ K) \right] = 0$$

$$T_{\text{final}} = 295\ K\ (22\ °C)$$

Comment The low specific heat capacity of iron and the small quantity of iron result in the temperature of iron being reduced by about 60 degrees whereas the temperature of water has been raised by only a few degrees.

Exercise 6.4—Using Specific Heat Capacity

A 15.5-g piece of chromium, heated to 100.0 °C, is dropped into 55.5 g of water at 16.5 °C. The final temperature of the metal and the water is 18.9 °C. What is the specific heat capacity of chromium? (Assume no heat is lost to the container or to the surrounding air.)

Exercise 6.5—Heat Transfer Between Substances

A piece of iron (400. g) is heated in a flame and then dropped into a beaker containing 1000. g of water. The original temperature of the water was 20.0 °C, and the final temperature of the water and iron is 32.8 °C after thermal equilibrium has been attained. What was the original temperature of the hot iron bar? (Assume no heat is lost to the beaker or to the surrounding air.)

6.3—Energy and Changes of State

When a solid melts, its atoms, molecules, or ions move about vigorously enough to break free of the constraints imposed by their neighbors in the solid. When a liquid boils, particles move much farther apart from one another. A change between solid and liquid or between liquid and gas is called a **change of state**. In both cases, energy must be furnished to overcome attractive forces among the particles. The heat required to convert a substance from a solid at its melting point to a liquid is called the **heat of fusion**. The heat required to convert liquid at its boiling point to gas is called the **heat of vaporization**. Heats of fusion and vaporization for many pure substances are provided along with other physical properties in reference books.

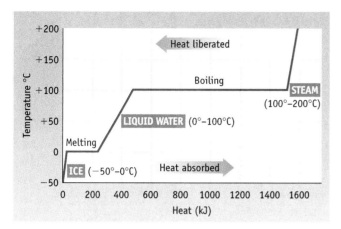

Active Figure 6.11 **Heat transfer and the temperature change for water.** This graph shows the quantity of heat absorbed and the consequent temperature change as 500. g of water warms from −50 °C to 200 °C.

GENERAL
Chemistry⚛Now™ See the General ChemistryNow CD-ROM or website to explore an interactive version of this figure accompanied by an exercise.

For water, the heat of fusion at 0 °C is 333 J/g and the heat of vaporization at 100 °C is 2256 J/g. These values are used when calculating the quantity of heat required or evolved when water boils or freezes. For example, the heat required to convert 500. g of water from the liquid to gaseous state at 100 °C is

$$(2256 \text{ J/g})(500. \text{ g}) = 1.13 \times 10^6 \text{ J (or 1130 kJ)}$$

If the same quantity of liquid water at 0 °C freezes to ice, the quantity of heat evolved is

$$(333 \text{ J/g})(500. \text{ g}) = 1.67 \times 10^5 \text{ J (or 167 kJ)}$$

Figure 6.11 illustrates the quantity of heat absorbed and the consequent temperature change as 500. g of water is warmed from −50 °C to 200 °C. First, the temperature of the ice increases as heat is added. On reaching 0 °C, however, the temperature remains constant as sufficient heat (167 kJ) is absorbed to melt the ice to liquid water. When all the ice has melted, the liquid absorbs heat and is warmed to 100 °C, the boiling point of water. The temperature again remains constant as enough heat is absorbed (1130 kJ) to convert the liquid completely to vapor. Any further heat added raises the temperature of the water vapor. The heat absorbed at other steps in Figure 6.11 and the total heat absorbed are calculated in Example 6.3.

It is important to notice that *temperature is constant throughout a change of state* (see Figure 6.11). During a change of state, the added energy is used to overcome the forces holding one molecule to another, not to increase the temperature of the substance (see Figures 6.11 and 6.12).

■ **Heats of Fusion and Vaporization for H₂O**

Heat of fusion = 333 J/g
= 6.00 kJ/mol

Heat of vaporization = 2256 J/g
= 40.65 kJ/mol

Example 6.3—Energy and Changes of State

Problem Calculate the quantity of heat involved in each step shown in Figure 6.11 and the total quantity of heat required to convert 500. g of ice at −50.0 °C to steam at 200.0 °C. The heat of fusion of water is 333 J/g and the heat of vaporization is 2256 J/g. The specific heat capacity of steam at 200 °C is 1.92 J/g · K. See also Table 6.1.

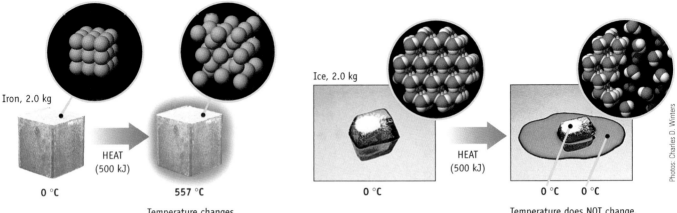

Iron, 2.0 kg

HEAT
(500 kJ)

0 °C   557 °C

Temperature changes.
State does NOT change.

Ice, 2.0 kg

HEAT
(500 kJ)

0 °C   0 °C   0 °C

Temperature does NOT change.
State changes.

Active Figure 6.12   **Changes of state.** Adding 500 kJ of heat to 2.0 kg of iron at 0 °C will cause the iron's temperature to increase to 557 °C (and the metal expands slightly). In contrast, adding 500 kJ of heat to 2.0 kg of ice will cause 1.5 kg of ice to melt to water at 0 °C (and 0.5 kg of ice will remain). No temperature change occurs.

GENERAL
Chemistry ⋅⁘⋅ Now™ See the General ChemistryNow CD-ROM or website to explore an interactive version of this figure accompanied by an exercise.

Strategy The problem is broken down into a series of steps: (1) warm the ice from −50 °C to 0 °C; (2) melt the ice at 0 °C; (3) raise the temperature of the liquid water from 0 °C to 100 °C; (4) evaporate the water at 100 °C; (5) raise the temperature of the steam from 100 °C to 200 °C. Use Equation 6.1 to calculate the heats associated with temperature changes. Use the heat of fusion and the heat of vaporization for heats associated with changes of state. The total heat required is the sum of the heats of the individual steps.

Solution

Step 1.

$$q \text{ (to warm ice from } -50 \text{ °C to 0 °C)} = (2.06 \text{ J/g} \cdot \text{K})(500. \text{ g})(273.2 \text{ K} - 223.2 \text{ K})$$
$$= 5.15 \times 10^4 \text{ J}$$

Step 2.

$$q \text{ (to melt ice at 0 °C)} = (500. \text{ g})(333 \text{ J/g}) = 1.67 \times 10^5 \text{ J}$$

Step 3.

q (to raise temperature of water from 0 °C to 100 °C)

$$= (4.184 \text{ J/g} \cdot \text{K})(500. \text{ g})(373.2 \text{ K} - 273.2 \text{ K})$$
$$= 2.09 \times 10^5 \text{ J}$$

Step 4.

$$q \text{ (to evaporate water at 100 °C)} = (2256 \text{ J/g})(500. \text{ g}) = 1.13 \times 10^6 \text{ J}$$

Step 5.

q (to raise temperature of steam from 100 °C to 200 °C)

$$= (1.92 \text{ J/g} \cdot \text{K})(500. \text{ g})(473.2 \text{ K} - 373.2 \text{ K})$$
$$= 9.60 \times 10^4 \text{ J}$$

The total thermal energy required is the sum of the thermal energy required in each step.

$$q_{total} = q_1 + q_2 + q_3 + q_4 + q_5$$
$$q_{total} = 1.60 \times 10^6 \text{ J (or 1600 kJ)}$$

Comment The conversion of liquid water to steam is the largest increment of energy added by a considerable margin. (You may have noticed that it does not take much time to heat water to boiling on a stove, but to boil off the water takes a much greater time.)

Example 6.4—Change of State

Problem What is the minimum amount of ice at 0 °C that must be added to the contents of a can of diet cola (340. mL) to cool it from 20.5 °C to 0 °C? Assume that the specific heat capacity and density of diet cola are the same as for water, and that no heat is gained or lost to the surroundings.

Strategy The system here is defined as the ice and the cola, and heat is transferred between these substances. Two energy quantities, the heat change in cooling the soda and the heat change in melting the ice, are needed. The first is calculated using the specific heat capacity and Equation 6.1 ($q_{cola} = C_{cola} \times m_{cola} \times \Delta T$); the second uses the heat of fusion of water [$q_{ice} = $ (heat of fusion)(mass of ice)]. The law of conservation of energy requires that the sum of these two quantities of energy is zero (Equation 6.3).

Solution The mass of cola is

$$(340.\ mL)(1.00\ g/mL) = 340.\ g$$

and its temperature changes from 293.7 K to 273.2 K. The heat of fusion of water is 333 J/g, and the mass of ice is the unknown.

$$q_{cola} + q_{ice} = 0$$
$$C_{cola} \times m_{cola} \times (T_{final} - T_{initial}) + q_{ice} = 0$$
$$[(4.184\ J/g \cdot K)(340.\ g)(273.2\ K - 293.7\ K)] + [(333\ J/g)\ (m_{ice})] = 0$$
$$m_{ice} = 87.6\ g$$

Comment This quantity of ice is just sufficient to cool the cola to 0 °C. If more than 87.6 g of ice is added then, when thermal equilibrium is reached, the temperature will be 0 °C and some ice will remain (see Exercise 6.7). If less than 87.6 g of ice is added, the final temperature will be greater than 0 °C. In this case, all the ice will melt and the liquid water formed by melting the ice will absorb additional heat to warm up to the final temperature (an example is given in Study Question 77, page 277).

Exercise 6.6—Changes of State

How much heat must be absorbed to warm 25.0 g of liquid methanol, CH_3OH, from 25.0 °C to its boiling point (64.6 °C) and then to evaporate the methanol completely at that temperature? The specific heat capacity of liquid methanol is 2.53 J/g · K. The heat of vaporization of methanol is 2.00×10^3 J/g.

Exercise 6.7—Changes of State

To make a glass of ice tea, you pour 250 mL of tea, whose temperature is 18.2 °C, into a glass containing 5 ice cubes. Each cube has a mass of about 15 g. What quantity of ice will melt, and how much ice will remain to float at the surface in this beverage? Ice tea has a density of 1.0 g/cm^3 and a specific heat capacity of 4.2 J/g · K. Assume that no heat is lost in cooling the glass or the surroundings.

6.4—The First Law of Thermodynamics

To this point, we have considered only energy in the form of heat. Now we need to broaden the discussion. Recall the definition given on page 235: *Thermodynamics is the science of heat and work.* Let us first note that *work* is done whenever something is moved against an opposing force. If a system does work on its surroundings, energy must be expended and the energy content of the system will decrease. Conversely, if work is done by the surroundings on a system, the energy content of the system increases. As with heat gained or lost, work done by a system or on a system will change its energy content (see "Historical Perspectives: Heat, Cannons, Soup, and Beer"). Therefore, we next want to introduce work into the equations for energy transfer.

An example of a system doing work on its surroundings is illustrated by the experiment shown in Figure 6.13. A small quantity of dry ice [$CO_2(s)$] is sealed inside a plastic bag, and a weight (a book in Figure 6.13) is placed on top of the bag. Dry ice has the interesting property that when it absorbs heat from its surroundings it changes directly from solid to gas at $-78\ °C$, in a process called **sublimation**:

$$CO_2(s,\ -78\ °C) \xrightarrow[+\ heat]{} CO_2(gas,\ -78\ °C)$$

As the experiment proceeds, the gaseous CO_2 expands within the plastic bag, lifting the book. To lift the book against the force of gravity requires that work be done. The system (the CO_2 inside the bag) is expending energy to do this work.

Even if the book had not been on top of the plastic bag, work would have been done by the expanding gas. This is because a gas must push back the atmosphere when it expands. Instead of raising a book, the expanding gas moves a part of the atmosphere.

Now let us recast this example as an experiment in thermodynamics. First, we must precisely identify the system and the surroundings. The system is the CO_2, a solid initially, and later a mixture of solid and gas. The surroundings consist of the objects that exchange energy with the system—that is, those objects in contact with the CO_2. They include the plastic bag, the book, the table-top, and the surrounding air. Thermodynamics focuses on the energy transfer that is occurring in the experi-

(a) Pieces of dry ice [$CO_2(s), -78°C$] are placed in a plastic bag. The dry ice will sublime (change directly from a solid to a gas) upon the input of heat.

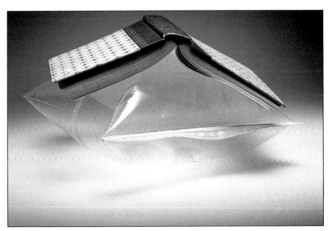

(b) Heat is absorbed by $CO_2(s)$ when it sublimes and the system (the contents of the bag) does work on its surroundings by lifting the book against the force of gravity.

Charles D. Winters

Active Figure 6.13 **Energy changes in a physical process.**

GENERAL
Chemistry Now™ See the General ChemistryNow CD-ROM or website to explore an interactive version of this figure accompanied by an exercise.

Historical Perspectives

Work, Heat, Cannons, Soup, and Beer

Benjamin Thompson (1753–1814) is one of the more colorful characters in the history of science. He was born in the state of Massachusetts, but fled to London, England, before the American Revolution because of his sympathy with the royalists. Thompson later moved to Munich, Germany, where he contributed so greatly to society that he was given the title of Count Rumford by the King of Bavaria in 1792. Among his contributions in Munich were the famous English Gardens and a unique system to care for the poor. He also created a candle so consistent in its light level that it became the international standard for measuring "candle power." Thompson became a nutritional expert, stressing the potato, and concocted a soup still known as *Rumfordsuppe*. He invented the modern kitchen range and convection oven, a double boiler, and a pressure cooker. And the efficient fireplace he designed is still known as a "Rumford fireplace."

Count Rumford is best known today for the experiments he did on heat. When visiting a cannon-boring factory, he

Work and Heat. A classic experiment that showed the relationship between work and heat was performed by Benjamin Thompson, Count Rumford, using the apparatus shown here. Thompson measured the rise in temperature of water (in the vessel mostly hidden at the back of the apparatus) that resulted from the energy expended to turn the crank.

noticed that the cannon barrels were hot, and the bore-hole shavings were even hotter. This had been observed for centuries, but Thompson was interested in what caused the heat and how it was passed along. Convinced that heat could not be a substance, as some then believed, he set up experiments to answer these questions.

Thompson eventually returned to London, and settled finally in Paris where he was acclaimed by Napoleon and elected to the French Academy. There he also met and married Madame Lavoisier, the widow of Antoine Lavoisier (page 143). He first described her as an "incarnation of goodness," but they divorced in 1809. Thompson died in France in 1814.

The unit of heat, the joule, is named for **James P. Joule** (1818–1889), the son of a wealthy brewer in Manchester, England, and a student of John Dalton. The family wealth, and a workshop in the brewery, gave Joule the opportunity to pursue scientific studies. Among the topics Joule studied was the issue of whether heat was a massless fluid, which some scientists called the caloric hypothesis. This had been a source of controversy for several decades, and it had not been resolved by the early experiments and advocacy of Rumford. Joule's careful experiments convincingly showed that heat and mechanical work can be interconverted and that heat is not a fluid. The caloric hypothesis was finally abandoned.

See G. I. Brown, *Count Rumford, The Extraordinary Life of a Scientific Genius,* Trowbridge, England, Sutton Publishing, 1999.

ment. Sublimation of CO_2 requires heat, which is transferred to the CO_2 from the surroundings. At the same time, the system does work on the surroundings by lifting the book. An energy balance for the system will include both quantities; that is, the change in energy content for the system (ΔE) will equal the sum of heat transferred (q) to or from the system and the work done by or to the system (w). We can express this explicitly as an equation:

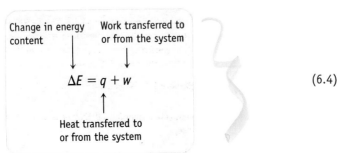

Change in energy content

Work transferred to or from the system

$$\Delta E = q + w \qquad (6.4)$$

Heat transferred to or from the system

A Closer Look

P-V Work

Work is done when an object of some mass is moved against an external resisting force. We know this well from common experience, such as when we use a pump to blow up a bicycle tire.

To evaluate the work done when a gas is compressed we can use, for example, a cylinder with a movable piston, as would occur in a bicycle pump (see figure). The drawing on the left shows the initial position of the piston, and the one on the right shows its final position. To depress the piston, we would have to expend some energy (the energy of this process coming from the energy obtained by food metabolism in our body.) The work required to depress the piston is calculated from a law of physics, $w = F \times d$, or work equals the magnitude of the force applied times the distance (d) over which the force is applied.

Pressure is defined as a force divided by the area over which the force is applied: $P = F/A$. In this example, the force is being applied to a piston with an area A. Substituting $P \times A$ for F in the equation gives $w = (P \times A) \times d$. However, since the product of $A \times d$ is the change of volume, ΔV, we can rewrite our equation for work as $w = -P\Delta V$.

Pushing down on the piston means we have done work on the system, the gas contained within the cylinder. The gas is now compressed to a smaller volume and has attained a higher energy as a consequence. The additional energy is equal to $-P\Delta V$.

Notice how we have allowed energy to be converted from one form to another—from chemical energy in food to mechanical energy used to depress the piston, to potential energy stored in a system of a gas at a higher pressure. In each step, energy was conserved, not lost, and the total energy of the universe remained constant.

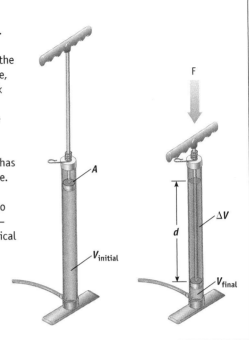

Equation 6.4 is a mathematical statement of the *first law of thermodynamics:* The energy change for a system is the sum of heat transferred between the system and its surroundings and the work done on the system by the surroundings or on the surroundings by the system. You will notice that this equation is a version of the general principle of conservation of energy applied specifically to the system.

The quantity E in Equation 6.4 has a formal name and a precise meaning in thermodynamics: **internal energy.** The internal energy in a chemical system is the sum of the potential and kinetic energies of the atoms, molecules, or ions in the system. Potential energy is the energy associated with the attractive and repulsive forces between all the nuclei and electrons in the system. It includes the energy associated with bonds in molecules, forces between ions, and forces between molecules in the liquid and solid state. Kinetic energy is the energy of motion of the atoms, ions, and molecules in the system. A value of internal energy is extremely difficult to determine but fortunately this step is not necessary. As the equation indicates, we are evaluating the *change* of internal energy, ΔE, which is a measurable quantity. In fact, the equation tells us how *to determine ΔE: Measure the heat transferred and the work done to or by the system.*

The work in the example involving the sublimation of CO_2 (Figure 6.13) is of a specific type, called **P-V (pressure-volume) work**. It is the work associated with a change in volume (ΔV) that occurs against a resisting external pressure (P). For a system in which the external pressure is constant, the value of P-V work can be calculated by Equation 6.5:

■ **Work**
Electrical work is another type of work commonly encountered in chemistry.

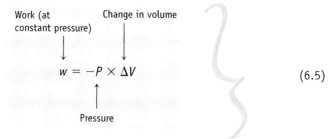

$$w = -P \times \Delta V \qquad (6.5)$$

The origin of this relationship is explained in "A Closer Look: P-V Work".

The sign convention for Equation 6.4 is important. The following table summarizes how the internal energy of a system is affected by heat and work.

Sign Conventions for q and w of the System

Change	Sign Convention	Effect on E_{system}
Heat transferred to system from surroundings	$q > 0$ (+)	E increases
Heat transferred from system to surroundings	$q < 0$ (−)	E decreases
Work done on system by surroundings	$w > 0$ (+)	E increases
Work done by system on surroundings	$w < 0$ (−)	E decreases

See the General ChemistryNow CD-ROM or website:

- **Screen 6.11 The First Law of Thermodynamics,** for a video of energy change in a physical change

Enthalpy

Most experiments in a chemical laboratory are carried out in beakers or flasks open to the atmosphere. Similarly, chemical processes that occur in living systems are open to the atmosphere. The fact that pressure is constant under these conditions is an important consideration when applying the first law of thermodynamics to heat measurements.

Because heat at constant pressure is so frequently the focus of attention in chemistry and biology, it is useful to have a specific measure of heat transfer under these conditions. The heat content of a substance at constant pressure is called **enthalpy** and is given the symbol H. In experiments at constant pressure, the **enthalpy change, ΔH,** is the difference between the final and initial enthalpy content. With enthalpy, as with internal energy, attention is focused on changes (that is, ΔH) rather than on the value of H itself. It is the value of the enthalpy change, ΔH, that is measured in chemical and physical processes.

Similar sign and symbol conventions apply to both ΔE and ΔH.

- Negative values of ΔE and ΔH specify that energy is transferred from the system to the surroundings.

- Positive values of ΔE and ΔH refer to energy transferred from the surroundings to the system.

- The heat transferred at constant pressure is often symbolized by q_p and is equivalent to ΔH. The heat transferred at constant volume, symbolized by q_v, is equivalent to ΔE. The two heat values, q_p and q_v, differ by the amount of work, w, done on or by the system.

Changes in internal energy and enthalpy are mathematically related by the general equation $\Delta E = \Delta H + w$, showing that ΔE and ΔH differ by the quantity of energy transferred to or from a system as work. Taking work to be $-P\Delta V$, we observe that in many processes—such as the melting of ice—ΔV is small and hence the amount of work is small. Under these circumstances, ΔE and ΔH are of similar magnitude. The amount of work can be significant, however, in processes in which the volume change is large. This usually occurs when gases are formed or consumed. In the evaporation or condensation of water, the sublimation of CO_2 (see Figure 6.13), and chemical reactions in which gas volumes change, for example, ΔE and ΔH are significantly different.

State Functions

Both internal energy and enthalpy share a significant characteristic—namely, changes in these quantities that accompany chemical or physical changes do not depend on the path chosen to go from the initial state to the final state. No matter how you go from reactants to products in a reaction, for example, the value of ΔH or ΔE for the reaction is always the same. A quantity that has this characteristic property is called a **state function**.

Many commonly measured quantities, such as the pressure of a gas, the volume of a gas or liquid, the temperature of a substance, and the size of your bank account, are state functions. For example, you could have arrived at a current bank balance of $25 by having deposited $25, or you could have deposited $100 and then withdrawn $75.

The volume of a balloon is also a state function. You can blow up a balloon to a large volume and then let some air out to arrive at the desired volume. Alternatively, you can blow up the balloon in stages, adding tiny amounts of air at each stage. The final volume does not depend on how you got there. For bank balances and balloons, there are an infinite number of ways to arrive at the final state, but the final value depends only on the size of the bank balance or the balloon, not on the path taken from the initial state to the final state.

Not all quantities are state functions. For instance, distance traveled is not a state function (Figure 6.14). The travel distance from Oneonta, New York, to Madison, Wisconsin, depends on the route taken. Nor is the elapsed time of travel between these two locations a state function. In contrast, the altitude above sea level is a state function; in going from Oneonta (538 m above sea level) to Madison (280 m above sea level), there is an altitude change of 258 m, regardless of the route followed. Interestingly, neither heat nor work individually is a state functions but their sum, the change in internal energy, ΔE, is. The value of ΔE is fixed by $E_{initial}$ and E_{final}, but a transition between the initial and final states can be accomplished by different routes having different values of q and w. Enthalpy is also a state function. The enthalpy change occurring when 1.0 g of water is heated from 20 °C to 50 °C, or when 1.0 g of water is evaporated at 100 °C, is independent of the way in which the process is carried out.

Taxi/Getty Images

Figure 6.14 State functions. There are many ways to climb a mountain, but the change in altitude from the base of the mountain to its summit is the same. The change in altitude is a state function. The distance traveled to reach the summit is not.

6.5—Enthalpy Changes for Chemical Reactions

Enthalpy changes accompany chemical reactions. For example, for the decomposition of 1 mol of water vapor to its elements, 1 mol of H_2 and $\frac{1}{2}$ mol of O_2, the enthalpy change $\Delta H = +241.8$ kJ at 25 °C.

$$H_2O(g) \longrightarrow H_2(g) + \tfrac{1}{2} O_2(g) \qquad \Delta H = +241.8 \text{ kJ}$$

(a) A lighted candle is brought up to a balloon filled with hydrogen gas.

(b) When the balloon breaks, the candle flame ignites the hydrogen.

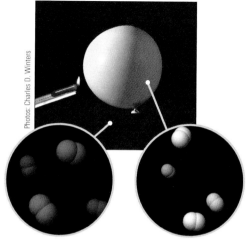

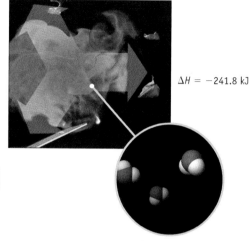

$\Delta H = -241.8$ kJ

O_2 (surroundings) H_2 (system)

$H_2(g) + \frac{1}{2}O_2(g) \longrightarrow H_2O(g)$

Active Figure 6.15   **The exothermic combustion of hydrogen in air.** The reaction transfers energy to the surroundings in the form of heat, work, and light.

GENERAL
Chemistry · Now™ See the General ChemistryNow CD-ROM or website to explore an interactive version of this figure accompanied by an exercise.

The positive sign of ΔH indicates that the decomposition is an endothermic process. That is, the reaction requires that 241.8 kJ be transferred to the system, $H_2O(g)$, from the surroundings.

Now consider the opposite reaction, the combination of hydrogen and oxygen to form water. The quantity of heat energy evolved in this reaction is the same as is required for the decomposition reaction, except that the sign of ΔH is reversed. The exothermic formation of 1 mol of water vapor from H_2 and $\frac{1}{2}$ mol of O_2 transfers 241.8 kJ to the surroundings (Figure 6.15).

$$H_2(g) + \tfrac{1}{2} O_2(g) \longrightarrow H_2O(g) \qquad \Delta H = -241.8 \text{ kJ}$$

The quantity of heat transferred during a chemical change depends on the amounts of reactants used or products formed. Thus, the formation of 2 mol of water vapor from the elements produces twice as much heat as the formation of 1 mol of water.

$$2\,H_2(g) + O_2(g) \longrightarrow 2\,H_2O(g) \qquad \Delta H = -483.6 \text{ kJ} \,(= 2 \times -241.8 \text{ kJ})$$

It is important to identify the states of reactants and products in a reaction because the magnitude of ΔH also depends on whether they are solids, liquids, or gases. Formation of 1 mol of *liquid* water from the elements is accompanied by the evolution of 285.8 kJ of energy.

$$H_2(g) + \tfrac{1}{2} O_2(g) \longrightarrow H_2O(\ell) \qquad \Delta H = -285.8 \text{ kJ}$$

The additional energy evolved relative to the formation of water vapor arises from the energy released when 1 mol of water vapor condenses to 1 mol of liquid water.

■ **Fractional Stoichiometric Coefficients** When writing balanced equations to define thermodynamic quantities, chemists often use fractional stoichiometric coefficients. For example, when we wish to define ΔH for the decomposition or formation of 1 mol of H_2O, the coefficient for O_2 must be $\frac{1}{2}$.

Photos: Charles D. Winters

These examples illustrate several features of the enthalpy changes for chemical reactions.

- Enthalpy changes are specific to the reactants and products and their amounts. Both the identities of reactants and products and their states (s, ℓ, g) are important.
- ΔH has a negative value if heat is evolved (an exothermic reaction). It has a positive value if heat is required (an endothermic reaction.)
- Values of ΔH are numerically the same, but opposite in sign, for chemical reactions that are the reverse of each other.
- The enthalpy change depends on the molar amounts of reactants and products. The formation of 2 mol of $H_2O(g)$ from the elements, for example, results in an enthalpy change that is twice as large as the enthalpy change in forming 1 mol of $H_2O(g)$.

Enthalpies of reactions are usually provided in one of two ways. They may be expressed as energy per mole of a reactant or per mole of a product. Alternatively, the enthalpy change may be given along with a balanced chemical equation, as was done earlier. In this case the value of ΔH is given for the equation as it is written. Whichever way the enthalpy change is presented, the value can be used to calculate the quantity of heat transferred by any given mass of a reactant or product. Suppose, for example, you want to know the enthalpy change if 454 g of propane, C_3H_8, is burned, given the equation for the exothermic combustion and the enthalpy change for the reaction.

$$C_3H_8(g) + 5\ O_2(g) \longrightarrow 3\ CO_2(g) + 4\ H_2O(\ell) \qquad \Delta H = -2220\ \text{kJ}$$

Two steps are needed. First, find the amount of propane present in the sample:

$$454\ \text{g } C_3H_8 \left(\frac{1\ \text{mol } C_3H_8}{44.10\ \text{g } C_3H_8} \right) = 10.3\ \text{mol } C_3H_8$$

Second, multiply the quantity of heat transferred per mole of propane by the amount of propane:

$$\Delta H = 10.3\ \text{mol } C_3H_8 \left(\frac{-2220\ \text{kJ}}{1\ \text{mol } C_3H_8} \right) = -22,900\ \text{kJ}$$

GENERAL
Chemistry ⋅⋅ Now™

See the General ChemistryNow CD-ROM or website:
- **Screen 6.13 Enthalpy Changes for Chemical Reactions,** for a tutorial on calculating the enthalpy change for a reaction

■ **Chemical Potential Energy** Gummi Bears are mostly sugar, and you can see in Figure 6.2 that their oxidation is highly exothermic. The enthalpy change for the oxidation of 1 teaspoonful of sugar, such as you might have in a large Gummi Bear, is about 100 kJ. See Example 6.5.

Charles D. Winters

Example 6.5—Enthalpy Calculation

Problem Sucrose (sugar, $C_{12}H_{22}O_{11}$) is oxidized to CO_2 and H_2O. The enthalpy change for the reaction can be measured in the laboratory

$$C_{12}H_{22}O_{11}(s) + 12\ O_2(g) \longrightarrow 12\ CO_2(g) + 11\ H_2O(\ell) \qquad \Delta H = -5645\ \text{kJ}$$

What is the enthalpy change for the oxidation of 5.00 g (1 teaspoonful) of sugar?

Strategy We will first determine the amount of sucrose in 5.00 g, then use this with the value given for the enthalpy change for the oxidation of 1 mol of sucrose.

Solution

$$5.00 \text{ g sucrose} \times \frac{1 \text{ mol sucrose}}{342.3 \text{ g sucrose}} = 1.46 \times 10^{-2} \text{ mol sucrose}$$

$$q = 1.46 \times 10^{-2} \text{ mol sucrose}\left(\frac{-5645 \text{ kJ}}{1 \text{ mol sucrose}}\right)$$

$$q = -82.5 \text{ kJ}$$

Comment Persons concerned about their diets might be interested to note that a (level) teaspoonful of sugar supplies about 25 Calories (dietary Calories; the conversion is 4.184 kJ = 1 Cal). As diets go, a single spoonful of sugar doesn't have a large caloric content. But will you use a level teaspoonful? And will you stop with just one?

Exercise 6.8—Enthalpy Calculation

(a) What quantity of heat energy is required to decompose 12.6 g of liquid water to the elements?

(b) The combustion of ethane, C_2H_6, has an enthalpy change of -2857.3 kJ for the reaction as written below. Calculate the value of ΔH when 15.0 g of C_2H_6 is burned.

$$2 \text{ } C_2H_6(g) + 7 \text{ } O_2(g) \longrightarrow 4 \text{ } CO_2(g) + 6 \text{ } H_2O(g) \qquad \Delta H = -2857.3 \text{ kJ}$$

6.6—Calorimetry

The heat transferred in a chemical or physical process is measured by an experimental technique called **calorimetry.** The apparatus used in this kind of experiment is a calorimeter, of which there are two basic types. A **constant pressure calorimeter** allows measurement of heats evolved or required under constant pressure conditions. In a **constant volume calorimeter**, the volume cannot change. The two types of calorimetry highlight the differences between enthalpy and internal energy. Heat transferred at constant pressure, q_p, is, by definition, ΔH, whereas the heat transferred at constant volume, q_v, is ΔE.

Constant Pressure Calorimetry: Measuring ΔH

Heat changes at constant pressure are often measured in the general chemistry laboratory by using a "coffee-cup calorimeter." This inexpensive device consists of two nested Styrofoam coffee cups with a loose-fitting lid and a temperature-measuring device such as a thermometer (Figure 6.16). The cup contains a solution of the reactants. The mass and specific heat capacity of the solution, and the amount of reactants, must be known. If heat is evolved in the process under study, the temperature of the solution rises. If heat is required, it is furnished by the solution and a decrease in temperature will be seen. In each case the change in temperature is measured. From mass, specific heat capacity, and temperature change, the heat change for the contents of the calorimeter can be calculated.

In the terminology of thermodynamics, the contents of the coffee-cup calorimeter are the system, and the cup and the immediate environment around the apparatus are the surroundings. Two heat changes occur within the system. One is the change that takes place as the chemical (potential) energy stored in the reactants is released as heat during the reaction. We label this heat quantity q_{rxn} (where

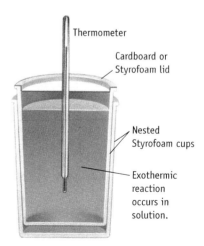

Figure 6.16 **A coffee-cup calorimeter.** A chemical reaction produces a change in the temperature of the solution in the calorimeter. The Styrofoam container is fairly effective in preventing heat transfer between the solution and its surroundings. Because the cup is open to the atmosphere, this is a constant pressure measurement.

Thermometer

Cardboard or Styrofoam lid

Nested Styrofoam cups

Exothermic reaction occurs in solution.

"rxn" is an abbreviation for "reaction"). The other is the heat gained or lost by the solution ($q_{solution}$). Assuming no heat transfer between the system and the surroundings, the sum of the heat changes within the system is zero.

$$q_{rxn} + q_{solution} = 0$$

The change in heat content of the solution ($q_{solution}$) can be calculated from its heat capacity, mass, and change in temperature. The quantity of heat evolved or required for the reaction (q_{rxn}) is the unknown in the equation. Because the reaction is carried out at constant pressure, the heat being measured is an enthalpy change, ΔH.

The accuracy of a calorimeter experiment depends on the accuracy of the measured quantities (temperature, mass, specific heat capacity). In addition, it depends on how closely the assumption of no heat transfer between system and surroundings is followed. A coffee-cup calorimeter is an unsophisticated apparatus and the results obtained with it are not highly accurate, largely because the latter assumption is poorly met. In research laboratories, scientists utilize calorimeters that more effectively limit the heat transfer between system and surroundings, and they may also estimate and correct for any minimal heat transfer that does occur between the system and the surroundings.

Example 6.6—Using a Coffee-Cup Calorimeter

Problem Suppose you place 0.500 g of magnesium chips in a coffee-cup calorimeter and then add 100.0 mL of 1.00 M HCl. The reaction that occurs is

$$Mg(s) + 2\ HCl(aq) \longrightarrow H_2(g) + MgCl_2(aq)$$

The temperature of the solution increases from 22.2 °C (295.4 K) to 44.8 °C (318.0 K). What is the enthalpy change for the reaction per mole of Mg? (Assume that the specific heat capacity of the solution is 4.20 J/g · K and the density of the HCl solution is 1.00 g/mL.)

Strategy Two changes in heat content take place within the system: the heat evolved in the reaction (q_{rxn}) and the heat gained by the solution to increase its temperature ($q_{solution}$). The problem solution has three steps. First, calculate $q_{solution}$ from the values of the mass, specific heat capacity, and ΔT using Equation 6.1. Second, calculate q_{rxn}, assuming no energy transfer occurs between the system and the surroundings (so the sum of heat changes in the system $q_{rxn} + q_{solution} = 0$). Third, use the value of q_{rxn} and the amount of Mg to calculate the enthalpy change per mole.

Solution

Step 1. *Calculate $q_{solution}$.* The mass of the solution is approximately the mass of the 100.0 mL of HCl plus the mass of magnesium, or 100.5 g.

$$q_{solution} = (100.5\ g)(4.20\ J/g \cdot K)(318.0\ K - 295.4\ K)$$
$$= 9.54 \times 10^3\ J$$

Step 2. *Calculate q_{rxn}.*

$$q_{rxn} + q_{solution} = 0$$
$$q_{rxn} + 9.54 \times 10^3\ J = 0$$
$$q_{rxn} = -9.54 \times 10^3\ J$$

Step 3. *Calculate the value of ΔH per mole.* The quantity of heat found in Step 2 is produced by the reaction of 0.500 g of Mg. The heat produced by the reaction of 1.00 mol of Mg is therefore

$$\Delta H = \left(\frac{-9.54 \times 10^3 \text{ J}}{0.500 \text{ g Mg}}\right)\left(\frac{24.31 \text{ g Mg}}{1 \text{ mol Mg}}\right)$$

$$\Delta H = -4.64 \times 10^5 \text{ J/mol Mg}$$

Comment The calculation will give the correct sign of q_{rxn} and ΔH. The negative sign indicates that this is an exothermic reaction.

Exercise 6.9—Using a Coffee-Cup Calorimeter

Assume you mix 200. mL of 0.400 M HCl with 200. mL of 0.400 M NaOH in a coffee-cup calorimeter. The temperature of the solutions before mixing was 25.10 °C; after mixing and allowing the reaction to occur, the temperature is 27.78 °C. What is the molar enthalpy of neutralization of the acid? (Assume that the densities of all solutions are 1.00 g/mL and their specific heat capacities are 4.20 J/g · K.)

Constant Volume Calorimetry: Measuring ΔE

Constant volume calorimetry is often used to evaluate heats of combustion of fuels and the caloric value of foods. A weighed sample of a combustible solid or liquid is placed inside a "bomb," often a cylinder about the size of a large fruit-juice can with thick steel walls and ends (Figure 6.17). The bomb is placed in a water-filled container

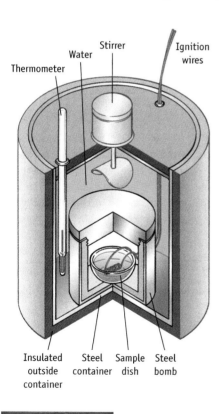
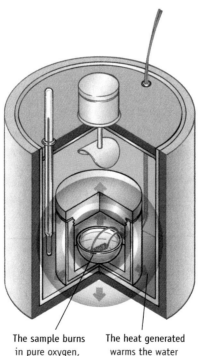

Thermometer Water Stirrer Ignition wires

Insulated outside container — Steel container — Sample dish — Steel bomb

The sample burns in pure oxygen, warming the bomb

The heat generated warms the water and ΔT is measured by the thermometer

Active Figure 6.17 **Constant volume calorimeter.** A combustible sample is burned in pure oxygen in a sealed metal container or "bomb." The heat generated warms the bomb and the water surrounding it. By measuring the increase in temperature, the heat evolved in the reaction can be determined.

GENERAL
Chemistry•Now™ See the General ChemistryNow CD-ROM or website to explore an interactive version of this figure accompanied by an exercise.

with well-insulated walls. After filling the bomb with pure oxygen, the sample is ignited, usually by an electric spark. The heat generated by the combustion reaction warms the bomb and the water around it. The bomb, its contents, and the water are defined as the system. Assessment of heat transfer within the system shows that

$$q_{rxn} + q_{bomb} + q_{water} = 0$$

Because the volume does not change in a constant volume calorimeter, energy transfer as work cannot occur. Therefore, the heat measured at constant volume (q_v) is the change in internal energy, ΔE.

GENERAL
Chemistry‑❄‑Now™

See the General ChemistryNow CD-ROM or website:
- **Screen 6.14 Measuring Heats of Reactions**
  - **(a)** for a simulation and exercise exploring reactions in a bomb calorimeter
  - **(b)** for a tutorial on calculating the heat of a reaction using a calorimeter

Example 6.7—Constant Volume Calorimetry

Problem Octane, C_8H_{18}, a primary constituent of gasoline, burns in air:

$$C_8H_{18}(\ell) + 25/2\ O_2(g) \longrightarrow 8\ CO_2(g) + 9\ H_2O(\ell)$$

A 1.00-g sample of octane is burned in a constant volume calorimeter similar to that shown in Figure 6.17. The calorimeter is in an insulated container with 1.20 kg of water. The temperature of the water and the bomb rises from 25.00 °C (298.15 K) to 33.20 °C (306.35 K). The heat capacity of the bomb, C_{bomb}, is 837 J/K.

(a) What is the heat of combustion per gram of octane?

(b) What is the heat of combustion per mole of octane?

Strategy (a) The sum of all heat changes in the system will be zero; that is, $q_{rxn} + q_{bomb} + q_{water} = 0$. The first term, q_{rxn}, is the unknown. The second and third terms in the equation can be calculated from the data given: q_{bomb} is calculated from the bomb's heat capacity and ΔT, and q_{water} is determined from the specific heat capacity, mass, and ΔT for water. (b) The value of q_{rxn} calculated in part (a) is the heat evolved in the combustion of 1.00 g of octane. Use this value and the molar mass of octane (114.2 g/mol) to calculate the heat evolved per mole of octane.

Solution

(a) $q_{water} = C_{water} \times m_{water} \times \Delta T$

$$ - (4.184\ \text{J/g} \cdot \text{K})(1.20 \times 10^3\ \text{g})(306.35\ \text{K} - 298.15\ \text{K}) = +41.2 \times 10^3\ \text{J}$$

$q_{bomb} = C_{bomb} \times \Delta T = 837\ \text{J/K}\ (306.35\ \text{K} - 298.15\ \text{K})$

$$= 6.86 \times 10^3\ \text{J}$$

$$q_{rxn} + q_{water} + q_{bomb} = 0$$

$$q_{rxn} + 41.2 \times 10^3\ \text{J} + 6.86 \times 10^3\ \text{J} = 0$$

$$q_{rxn} = -48.1 \times 10^3\ \text{J}$$

$$(\text{or} -48.1\ \text{kJ})$$

Heat of combustion per gram = −48.1 kJ

(b) Heat of combustion per mol = (−48.1 kJ/g)(114.2 g/mol) = −5.49 × 10³ kJ/mol

Comment Because the volume does not change, no energy transfer in the form of work occurs. The change of internal energy, ΔE, for the combustion of $C_8H_{18}(\ell)$ is -5.49×10^3 kJ/mol. Also note that C_{bomb} has no mass units. It is the heat required to warm the whole object by 1 kelvin.

Exercise 6.10—Constant Volume Calorimetry

A 1.00-g sample of ordinary table sugar (sucrose, $C_{12}H_{22}O_{11}$) is burned in a bomb calorimeter. The temperature of 1.50×10^3 g of water in the calorimeter rises from 25.00 °C to 27.32 °C. The heat capacity of the bomb is 837 J/K, and the specific heat capacity of the water is 4.20 J/g · K.

(a) Calculate the heat evolved per gram of sucrose.
(b) Calculate the heat evolved per mole of sucrose.

6.7—Hess's Law

Measuring a heat of reaction using a calorimeter is not possible for many chemical reactions. Consider, for example, the oxidation of carbon to carbon monoxide.

$$C(s) + \tfrac{1}{2} O_2(g) \longrightarrow CO(g)$$

Some CO_2 will always form in reactions of carbon and oxygen, even if there is a deficiency of oxygen. The reaction of CO and O_2 is very favorable; thus, as soon as CO is formed, it will react with O_2 to form CO_2. Therefore, using calorimetry to measure the heat evolved in the formation of CO is not possible.

Fortunately, the heat evolved in the reaction forming CO(g) from C(s) and O_2(g) can be calculated from heats measured for other reactions. The calculation is based on **Hess's law**, which states that *if a reaction is the sum of two or more other reactions, ΔH for the overall process is the sum of the ΔH values of those reactions.*

The oxidation of C(s) to CO_2(g) can be viewed as occurring in two steps: first the oxidation of C(s) to CO(g) (Equation 1), and then the oxidation of CO(g) to CO_2(g) (Equation 2). Adding these two equations gives the equation for the oxidation of C(s) to CO_2(g) (Equation 3).

Equation 1: $C(s) + \tfrac{1}{2} O_2(g) \longrightarrow CO(g)$ $\Delta H_1 = ?$

Equation 2: $\underline{CO(g) + \tfrac{1}{2} O_2(g) \longrightarrow CO_2(g)}$ $\Delta H_2 = -283.0$ kJ

Equation 3: $C(s) + O_2(g) \longrightarrow CO_2(g)$ $\Delta H_3 = -393.5$ kJ

Hess's law tells us that the enthalpy change for overall reaction (ΔH_3) will equal the sum of the enthalpy changes for reactions 1 and 2 ($\Delta H_1 + \Delta H_2$). Both ΔH_2 and ΔH_3 can be measured, and these values are then used to determine the enthalpy change for reaction 1.

$$\Delta H_3 = \Delta H_1 + \Delta H_2$$
$$-393.5 \text{ kJ} = \Delta H_1 + (-283.0 \text{ kJ})$$
$$\Delta H_1 = -110.5 \text{ kJ}$$

Hess's law applies to physical processes, too. The enthalpy change for the reaction of H_2(g) and O_2(g) to form 1 mol of liquid H_2O is different from the enthalpy change to form 1 mol of H_2O vapor (page 255). The difference is the heat of vaporization of water, ΔH_2.

Equation 1:	$H_2(g) + \frac{1}{2} O_2(g) \longrightarrow H_2O(\ell)$	$\Delta H_1 = -285.8$ kJ
Equation 2:	$H_2O(\ell) \longrightarrow H_2O(g)$	$\Delta H_2 = ?$
Equation 3:	$H_2(g) + \frac{1}{2} O_2(g) \longrightarrow H_2O(g)$	$\Delta H_3 = -241.8$ kJ

The relationship $\Delta H_3 = \Delta H_1 + \Delta H_2$ makes it possible to calculate the value of ΔH_2, the heat of vaporization of water (44.0 kJ, with all substances at 25 °C).

Energy Level Diagrams

When using Hess's law, it is often helpful to represent enthalpy data schematically in an **energy level diagram**. In such a drawing, various substances—for example, the reactants and products in a chemical reaction—are placed on an arbitrary (potential) energy scale. The relative energy of each substance is given by its position on the vertical axis, and numerical differences in energy between them are shown by the vertical arrows. Such diagrams provide an easy-to-read perspective on the magnitude and direction of energy changes and show how energy of the substances are related.

Energy level diagrams that summarize the two examples of Hess's law discussed earlier appear in Figure 6.18. In Figure 6.18a, the elements, C(s) and $O_2(g)$ are at

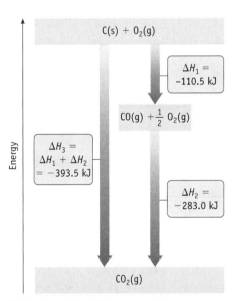

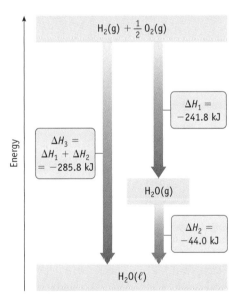

(a) The formation of CO_2 can occur in a single step or in a succession of steps. ΔH for the overall process is -393.5 kJ, no matter which path is followed.

(b) The formation of $H_2O(\ell)$ can occur in a single step or in a succession of steps. ΔH for the overall process is -285.8 kJ, no matter which path is followed.

Active Figure 6.18   **Energy level diagrams.** (a) Relating enthalpy changes in the formation of $CO_2(g)$. (b) Relating enthalpy changes in the formation of $H_2O(\ell)$. Enthalpy changes associated with changes between energy levels are given alongside the vertical arrows.

GENERAL
Chemistry ⋅ Now™ See the General ChemistryNow CD-ROM or website to explore an interactive version of this figure accompanied by an exercise.

the highest potential energy. Converting carbon and oxygen to CO_2 lowers the potential energy by 393.5 kJ. This can occur either in a single step, shown on the left in Figure 6.18a, or in two steps, shown on the right. Similarly, in Figure 6.18b, the potential energy of the elements is at the highest potential energy. The product, liquid or gaseous water, has a lower potential energy, with the difference between the two being the heat of vaporization.

GENERAL
Chemistry•ᢏ•Now™

See the General ChemistryNow CD-ROM or website:
- **Screen 6.15 Hess's Law,** for a simulation and exercise on "adding" reactions

---

Example 6.8—Using Hess's Law

Problem Suppose you want to know the enthalpy change for the formation of methane, CH_4, from solid carbon (as graphite) and hydrogen gas:

$$C(s) + 2\ H_2(g) \longrightarrow CH_4(g) \qquad \Delta H = ?$$

The enthalpy change for this reaction cannot be measured in the laboratory because the reaction is very slow. We can, however, measure enthalpy changes for the combustion of carbon, hydrogen, and methane.

Equation 1:	$C(s) + O_2(g) \longrightarrow CO_2(g)$	$\Delta H_1 = -393.5$ kJ
Equation 2:	$H_2(g) + \frac{1}{2} O_2(g) \longrightarrow H_2O(\ell)$	$\Delta H_2 = -285.8$ kJ
Equation 3:	$CH_4(g) + 2\ O_2(g) \longrightarrow CO_2(g) + 2\ H_2O(\ell)$	$\Delta H_3 = -890.3$ kJ

Use these energies to obtain ΔH for the formation of methane from its elements.

Strategy The three reactions (1, 2, and 3), as they are written, cannot be added together to obtain the equation for the formation of CH_4 from its elements. Methane, CH_4, is a product in a reaction whose enthalpy is sought, but it is a reactant in Equation 3. Water appears in two of these equations although it is not a component of the reaction forming CH_4 from carbon and hydrogen. To use Hess's law to solve this problem, we will have to manipulate the equations and adjust the heats accordingly. Recall, from Section 6.5, that writing an equation in the reverse direction changes the sign of ΔH, and that doubling the amount of reactants and products doubles the value of ΔH. Adjustments to Equations 2 and 3 will produce new equations that, along with Equation 1, can be combined to give the desired net reaction.

Solution To make CH_4 a product in the overall reaction, we reverse Equation 3 while changing the sign of ΔH. (If a reaction is exothermic in one direction, its reverse must be endothermic):

Equation 3': $\quad CO_2(g) + 2\ H_2O(\ell) \longrightarrow CH_4(g) + 2\ O_2(g) \qquad \Delta H_3' = -\Delta H_3 = +890.3$ kJ

Next, we see that 2 mol of $H_2(g)$ is on the reactant side in our desired equation. Equation 2 is written for only 1 mol of $H_2(g)$ as a reactant, however. We therefore multiply the stoichiometric coefficients in Equation 2 by 2 and multiply the value of ΔH by 2.

Equation 2: $\quad 2\ H_2(g) + O_2(g) \longrightarrow 2\ H_2O(\ell) \qquad 2\ \Delta H_2 = 2(-285.8$ kJ$) = -571.6$ kJ

With these modifications, we rewrite the three equations. When added together, $O_2(g)$, $H_2O(\ell)$, and $CO_2(g)$ all cancel to give the equation for the formation of methane from its elements.

Equation 1: $C(s) + O_2(g) \longrightarrow CO_2(g)$ $\Delta H_1 = -393.5$ kJ

Equation 2: $2 H_2(g) + O_2(g) \longrightarrow 2 H_2O(\ell)$ $2 \Delta H_2 = 2(-285.8 \text{ kJ}) = -571.6$ kJ

Equation 3: $CO_2(g) + 2 H_2O(\ell) \longrightarrow CH_4(g) + O_2(g)$ $\Delta H_3' = -\Delta H_3 = +890.3$ kJ

Net Equation: $C(s) + 2 H_2(g) \longrightarrow CH_4(g)$ $\Delta H_{net} = -74.8$ kJ

$\Delta H_{net} = \Delta H_1 + 2 \Delta H_2 + (-\Delta H_3)$

Comment You can construct an energy level diagram summarizing the energies of this process.

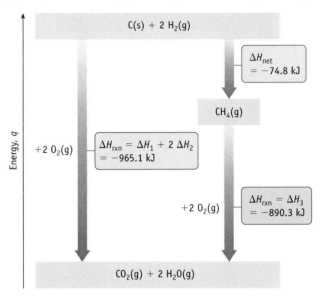

This diagram shows there are two ways to go from $C(g) + 2 H_2(g)$ to $CO_2(g) + 2 H_2O(g)$. The enthalpy changes along these two paths were $\Delta H_1 + 2 \Delta H_2$ and $\Delta H_{net} + \Delta H_3$. According to Hess's law,

$$\Delta H_1 + 2 \Delta H_2 = \Delta H_{net} + \Delta H_3$$

so

$$\Delta H_{net} = \Delta H_1 + 2 \Delta H_2 + (-\Delta H_3).$$

Exercise 6.11—Using Hess's Law

Graphite and diamond are two allotropes of carbon. The enthalpy change for the process

$$C(\text{graphite}) \longrightarrow C(\text{diamond})$$

cannot be measured directly, but it can be evaluated using Hess's law.

(a) Determine this enthalpy change, using experimentally measured heats of combustion of graphite (-393.5 kJ/mol) and diamond (-395.4 kJ/mol).

(b) Draw an energy level diagram for this system.

Exercise 6.12—Using Hess's Law

Use Hess's law to calculate the enthalpy change for the formation of $CS_2(\ell)$ from $C(s)$ and $S(s)$ from the following enthalpy values.

$C(s) + O_2(g) \longrightarrow CO_2(g)$ $\Delta H = -393.5$ kJ

$S(s) + O_2(g) \longrightarrow SO_2(g)$ $\Delta H = -296.8$ kJ

$CS_2(\ell) + 3 O_2(g) \longrightarrow CO_2(g) + 2 SO_2(g)$ $\Delta H = -1103.9$ kJ

$C(s) + 2 S(s) \longrightarrow CS_2(\ell)$ $\Delta H = ?$

Problem-Solving Tip 6.2

Using Hess's Law

How did we know how the three equations should be adjusted in Example 6.8? Here is a general strategy for solving this type of problem.

Step 1. Inspect the equation whose ΔH you wish to calculate, identifying the reactants and products, and locate those substances in the equations available to be added. In Example 6.8 the reactants, C(s)

and $H_2(g)$, are reactants in Equations 1 and 2, and the product, $CH_4(g)$, is a reactant in Equation 3. Equation 3 was reversed to get CH_4 on the product side where it is located in the target equation.

Step 2. Get the correct amount of the reagents on each side. In Example 6.8 only one adjustment was needed. There was 1 mol of H_2 on the left (reactant side) in Equation 2. We needed 2 mol of H_2 in the overall equation; this required doubling the quantities in Equation 2.

Step 3. Make sure other reagents in the equations will cancel when the equations are added. In Example 6.8, equal amounts of O_2 and H_2O appeared on the left and right sides in the three equations, so they cancelled when the equations were added together.

Each manipulation requires adjustment of the energy quantities. Summing the equations and the adjusted enthalpies gives the overall equation and its enthalpy change.

6.8—Standard Enthalpies of Formation

Calorimetry and the application of Hess's law have made available a great many ΔH values for chemical reactions. Often, these values are assembled into tables to make it easy to retrieve and use the data (see Table 6.2 or Appendix L). A very useful table contains **standard molar enthalpies of formation, ΔH_f°.** *The standard molar enthalpy of formation is the enthalpy change for the formation of 1 mol of a compound directly from its component elements in their standard states.* The **standard state** of an element or a compound is defined as the most stable form of the substance in the physical state that exists at a pressure of 1 bar and at a specified temperature. Most tables report standard molar enthalpies of formation at 25 °C (298 K).

Several examples of standard molar enthalpies of formation will be helpful to illustrate the meaning of these definitions.

ΔH_f° **for $CO_2(g)$:** At 25 °C and 1 bar, the standard states of carbon and oxygen are solid graphite and $O_2(g)$, respectively. The standard enthalpy of formation of $CO_2(g)$ is defined as the enthalpy change that occurs in the formation of 1 mol of $CO_2(g)$ from 1 mol of C(s, graphite) and 1 mol of $O_2(g)$; that is, it is the enthalpy change for the process

$$C(s) + O_2(g) \longrightarrow CO_2(g) \qquad \Delta H_f^\circ = -393.5 \text{ kJ}$$

ΔH_f° **for NaCl(s):** At 25 °C and 1 bar, Na is a solid and Cl_2 is a gas. The standard enthalpy of formation of NaCl(s) is defined as the enthalpy change that occurs if 1 mol of NaCl(s) is formed from 1 mol of Na(s) and $\frac{1}{2}$ mol of $Cl_2(g)$.

$$Na(s) + \tfrac{1}{2} Cl_2(g) \longrightarrow NaCl(s) \qquad \Delta H_f^\circ = -411.12 \text{ kJ}$$

ΔH_f° **for $C_2H_5OH(\ell)$:** At 25 °C and 1 bar, the standard states of the elements are C(s, graphite), $H_2(g)$, and $O_2(g)$. The standard enthalpy of formation of $C_2H_5OH(\ell)$ is defined as the enthalpy change that occurs if 1 mol of $C_2H_5OH(\ell)$ is formed from 2 mol of C(s), 3 mol of $H_2(g)$, and $\frac{1}{2}$ mol of $O_2(g)$.

$$2 C(s) + 3 H_2(g) + \tfrac{1}{2} O_2(g) \longrightarrow C_2H_5OH(\ell) \qquad \Delta H_f^\circ = -277.0 \text{ kJ}$$

Notice that the reaction defining the heat of formation need not be (and most often is not) a reaction that a chemist is likely to carry out in the laboratory. Ethanol, for example, is not made by a reaction of the elements.

■ **ΔH Under Standard Conditions** The superscript ° indicating standard conditions is applied to other types of thermodynamic data, such as the heat of fusion and vaporization (ΔH_{fus}° and ΔH_{vap}°) and the heat of a reaction (ΔH_{rxn}°).

■ **ΔH_f° Pressure and Standard Conditions** The bar is the unit of pressure for thermodynamic quantities. One bar is approximately one atmosphere. (1 atm = 1.013 bar; see Appendix B).

Table 6.2 (and Appendix L) list values of ΔH_f° for some common substances. These values are for the formation of one mole of the compound in its standard state from its elements in their standard states. A review of these values leads to some important observations.

- The standard enthalpy of formation for an element in its standard state is zero.

- Values for compounds in solution refer to the enthalpy change for the formation of a 1 M solution of the compound from the elements making up the compound plus the enthalpy change occurring when the substance dissolves in water.

- Most ΔH_f° values are negative, indicating that formation of most compounds from the elements is exothermic. Heat evolution generally indicates that forming compounds from their elements (under standard conditions) is product-favored (see Section 6.9, page 269).

- Values of ΔH_f° can be used to compare thermal stabilities of related compounds. Consider the values of ΔH_f° for the hydrogen halides in Table 6.3. Hydrogen fluoride is the most stable of these compounds, whereas HI is the least stable.

■ **ΔH_f° Values**
Enthalpy of formation values are found in this book in Table 6.2 or Appendix L. Consult the National Institute for Standards and Technology website (**webbook.nist.gov**) for an extensive compilation of data.

Table 6.2   Selected Standard Molar Enthalpies of Formation at 298 K

Substance	Name	Standard Molar Enthalpy of Formation (kJ/mol)
C(graphite)	graphite	0
C(diamond)	diamond	+1.8
$CH_4(g)$	methane	−74.87
$C_2H_6(g)$	ethane	−83.85
$C_3H_8(g)$	propane	−104.7
$C_2H_4(g)$	ethene (ethylene)	+52.47
$CH_3OH(\ell)$	methanol	−238.4
$C_2H_5OH(\ell)$	ethanol	−277.0
$C_{12}H_{22}O_{11}(s)$	sucrose	−2221.2
$CO(g)$	carbon monoxide	−110.53
$CO_2(g)$	carbon dioxide	−393.51
$CaCO_3(s)*$	calcium carbonate	−1207.6
$CaO(s)$	calcium oxide	−635.1
$H_2(g)$	hydrogen	0
$H_2O(\ell)$	liquid water	−285.83
$H_2O(g)$	water vapor	−241.83
$N_2(g)$	nitrogen	0
$NH_3(g)$	ammonia	−45.90
$NH_4Cl(s)$	ammonium chloride	−314.55
$NO(g)$	nitrogen monoxide	+90.29
$NO_2(g)$	nitrogen dioxide	+33.10
$NaCl(s)$	sodium chloride	−411.12
$S_8(s)$	sulfur	0
$SO_2(g)$	sulfur dioxide	−296.81
$SO_3(g)$	sulfur trioxide	−395.77

Data from the NIST Webbook (**http://webbook.nist.gov**).
* Data not in NIST database. Value is from J. Dean (editor): *Lange's Handbook of Chemistry*, 14th edition, New York, McGraw-Hill, 1992.

Exercise 6.13—Standard States

What are the standard states of the following elements or compounds (at 25 °C): bromine, mercury, sodium sulfate, ethanol?

Exercise 6.14—Standard Heats of Formation

Write equations for the reactions that define the standard enthalpy of formation of $FeCl_3(s)$ and sucrose (sugar, $C_{12}H_{22}O_{11}$). What are the standard states of the reactants in each equation?

Table 6.3   Standard Molar Enthalpies of Formation of the Hydrogen Halides (at 298 K)

Compound	ΔH_f° (kJ/mol)
HF(g)	−273.3
HCl(g)	−92.3
HBr(g)	−36.3
HI(g)	+26.5

Enthalpy Change for a Reaction

The enthalpy change for a reaction under standard conditions can be calculated using Equation 6.6 *if* the standard molar enthalpies of formation are known for *all* reactants and products.

$$\Delta H_{rxn}^\circ = \sum [\Delta H_f^\circ (\text{products})] - \sum [\Delta H_f^\circ (\text{reactants})] \qquad (6.6)$$

In this equation, the symbol Σ (the Greek capital letter sigma) means "take the sum." To find ΔH_{rxn}°, add up the molar enthalpies of formation of the products and subtract from this the sum of the molar enthalpies of formation of the reactants. This equation is a logical consequence of the definition of ΔH_f° and Hess's law (see "A Closer Look: Hess's Law and Equation 6.6").

Suppose you want to know how much heat is required to decompose one mole of calcium carbonate (limestone) to calcium oxide (lime) and carbon dioxide under standard conditions:

$$CaCO_3(s) \longrightarrow CaO(s) + CO_2(g) \qquad \Delta H_{rxn}^\circ = ?$$

To do so, you would use the following enthalpies of formation from Table 6.2 (or Appendix L):

Compound	ΔH_f° (kJ/mol)
$CaCO_3(s)$	−1207.6
$CaO(s)$	− 635.1
$CO_2(g)$	− 393.5

■ **Δ = Final − Initial** Equation 6.6 is another example of the principle that a change (Δ) is always calculated by subtracting the initial state (the reactants) from the final state (the products).

and then use Equation 6.6 to find the standard enthalpy change for the reaction, ΔH_{rxn}°.

$$\begin{aligned}
\Delta H_{rxn}^\circ &= \Delta H_f^\circ [CaO(s)] + \Delta H_f^\circ [CO_2(g)] - \Delta H_f^\circ [CaCO_3(s)] \\
&= [1 \text{ mol } (-635.1 \text{ kJ/mol}) + 1 \text{ mol } (-393.5 \text{ kJ/mol})] \\
&\qquad\qquad\qquad\qquad\qquad - [1 \text{ mol } (-1207.6 \text{ kJ/mol})] \\
&= +179.0 \text{ kJ}
\end{aligned}$$

The decomposition of limestone to lime and CO_2 is endothermic. That is, energy (179.0 kJ/mol of $CaCO_3$) must be supplied to decompose $CaCO_3(s)$ to $CaO(s)$ and $CO_2(g)$.

A Closer Look

Hess's Law and Equation 6.6

Equation 6.6 is an application of Hess's law. To illustrate this, let us look again at the decomposition of calcium carbonate.

$$CaCO_3(s) \longrightarrow CaO(s) + CO_2(g) \qquad \Delta H^\circ_{rxn} = ?$$

We know the enthalpy changes for the decomposition of both the reactant and the products to the elements. These correspond to ΔH°_1, ΔH°_2, and ΔH°_3 on the diagram, and each of these is the negative of the enthalpy of formation of the respective compound. We do not know ΔH°_{rxn} for $CaCO_3$'s decomposition to CaO and CO_2. However, we do know from Hess's law that

$$\Delta H^\circ_1 = \Delta H^\circ_{rxn} + \Delta H^\circ_2 + \Delta H^\circ_3$$

$$-\Delta H^\circ_f[CaCO_3(s)] = \Delta H^\circ_{rxn} - \Delta H^\circ_f[CaO(s)] - \Delta H^\circ_f[CO_2(g)]$$

$$\Delta H^\circ_{rxn} = \Delta H^\circ_f[CaO(s)] + \Delta H^\circ_f[CO_2(g)]$$
$$- \Delta H^\circ_f[CaCO_3(s)]$$

$$= (-635.1 \text{ kJ}) + (-393.5 \text{ kJ}) - (-1207.6 \text{ kJ})$$

$$\Delta H^\circ_{rxn} = +179.0 \text{ kJ}$$

This is exactly the result we obtain by applying Equation 6.6. The enthalpy change for the reaction is indeed the sum of the enthalpies of formation of the products minus that of the reactant.

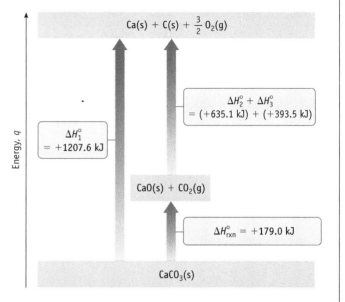

Energy level diagram for the decomposition of $CaCO_3(s)$

Energy, q

$Ca(s) + C(s) + \frac{3}{2} O_2(g)$

$\Delta H^\circ_2 + \Delta H^\circ_3$
$= (+635.1 \text{ kJ}) + (+393.5 \text{ kJ})$

ΔH°_1
$= +1207.6 \text{ kJ}$

$CaO(s) + CO_2(g)$

$\Delta H^\circ_{rxn} = +179.0 \text{ kJ}$

$CaCO_3(s)$

GENERAL
Chemistry ·⁜· Now ™

See the General ChemistryNow CD-ROM or website:

- **Screen 6.16 Standard Enthalpy of Formation,** for a tutorial on calculating the standard enthalpy change for a reaction

---

Example 6.9—Using Enthalpies of Formation

Problem Nitroglycerin is a powerful explosive that forms four different gases when detonated:

$$2\, C_3H_5(NO_3)_3(\ell) \longrightarrow 3\, N_2(g) + \tfrac{1}{2} O_2(g) + 6\, CO_2(g) + 5\, H_2O(g)$$

Calculate the enthalpy change when 10.0 g of nitroglycerin is detonated. The enthalpy of formation of nitroglycerin, ΔH°_f, is -364 kJ/mol. Use Table 6.2 or Appendix L to find other ΔH°_f values that are needed.

Strategy Use values of ΔH°_f for the reactants and products in Equation 6.6 to calculate the enthalpy change produced by the detonation of 2 moles of nitroglycerin (ΔH°_{rxn}). From Table 6.2, $\Delta H^\circ_f[CO_2(g)] = -393.5$ kJ/mol, $\Delta H^\circ_f[H_2O(g)] = -241.8$ kJ/mol, and $\Delta H^\circ_f = 0$ for $N_2(g)$ and $O_2(g)$. Determine the amount represented by 10.0 g of nitroglycerin, then use this value with ΔH°_{rxn} to obtain the answer.

Solution Using Equation 6.6, we find the enthalpy change for the explosion of 2 mol of nitroglycerin is

$$\Delta H^\circ_{rxn} = 6 \text{ mol} \times \Delta H^\circ_f[CO_2(g)] + 5 \text{ mol} \times \Delta H^\circ_f[H_2O(g)] - 2 \text{ mol} \times \Delta H^\circ_f[C_3H_5(NO_3)_3(\ell)]$$

$$= 6 \text{ mol}(-393.5 \text{ kJ/mol}) + 5 \text{ mol}(-241.8 \text{ kJ/mol}) - 2 \text{ mol}(-364 \text{ kJ/mol})$$

$$= -2842 \text{ kJ for 2 mol nitroglycerin}$$

The problem asks for the enthalpy change using 10.0 g of nitroglycerin. We next need to determine the amount of nitroglycerin in 10.0 g.

$$10.0 \text{ g nitroglycerin} \times \frac{1 \text{ mol nitroglycerin}}{227.1 \text{ g nitroglycerin}} = 0.0440 \text{ mol nitroglycerin}$$

The enthalpy change for the detonation of 0.0440 mol is

$$\Delta H^{\circ}_{\text{rxn}} = 0.0440 \text{ mol nitroglycerin} \left(\frac{-2842 \text{ kJ}}{2 \text{ mol nitroglycerin}} \right)$$

$$= -62.6 \text{ kJ}$$

Comment The large exothermic value of $\Delta H^{\circ}_{\text{rxn}}$ is in accord with the fact that this reaction is highly energetic.

Exercise 6.15—Using Enthalpies of Formation

Calculate the standard enthalpy of combustion for benzene, C_6H_6.

$$C_6H_6(\ell) + 7.5 \, O_2(g) \longrightarrow 6 \, CO_2(g) + 3 \, H_2O(\ell) \qquad \Delta H^{\circ}_{\text{rxn}} = ?$$

$\Delta H^{\circ}_f [C_6H_6(\ell)] = +49.0$ kJ/mol. Other values needed can be found in Table 6.2 and Appendix L.

6.9—Product- or Reactant-Favored Reactions and Thermochemistry

Reactions in which reactants are largely converted to products are said to be product-favored [◄ page 197]. One aspect of chemical reactivity, and a goal of this book, is to be able to predict whether a chemical reaction will be product- or reactant-favored. In Chapter 5 you learned that certain common reactions occurring in aqueous solution—precipitation, acid–base, and gas-forming reactions—and reactions such as combustions are generally product-favored. Our discussion of the energy changes in chemical reactions allows us to begin to understand more about predicting which reactions may be product-favored.

The oxidation reactions of hydrogen and carbon (see Figures 6.15 and 6.18), Gummi Bears (Figure 6.2), and iron (Figure 6.19)

$$4 \, Fe(s) + 3 \, O_2(g) \longrightarrow 2 \, Fe_2O_3(s)$$
$$\Delta H^{\circ}_{\text{rxn}} = 2 \, \Delta H^{\circ}_f [Fe_2O_3(s)] = 2(-825.5 \text{ kJ}) = -1651.0 \text{ kJ}$$

are exothermic. All have negative values for $\Delta H^{\circ}_{\text{rxn}}$, and transfer energy to their surroundings. They are also all product-favored reactions.

Conversely, the reactant-favored decomposition of calcium carbonate is endothermic. Heat is required for the reaction to occur, and $\Delta H^{\circ}_{\text{rxn}}$ is positive.

$$CaCO_3(s) \longrightarrow CaO(s) + CO_2(g) \qquad \Delta H^{\circ}_{\text{rxn}} = +179.0 \text{ kJ}$$

Are all exothermic reactions product-favored and all endothermic reactions reactant-favored? From these examples, we might formulate this idea as a hypothesis that can be tested by experiment and by examination of many other examples. We would find that *in most cases product-favored reactions have negative values of $\Delta H^{\circ}_{\text{rxn}}$ and reactant-favored reactions have positive values of $\Delta H^{\circ}_{\text{rxn}}$.* But this is not *always* true; there are exceptions, and we shall return to the issue in Chapter 19.

■ **Reactant- or Product-Favored?**
In most cases exothermic reactions are product-favored and endothermic reactions are reactant-favored.

Figure 6.19 **The product-favored oxidation of iron.** Iron powder, sprayed into a bunsen burner flame, is rapidly oxidized. The reaction is exothermic and is product-favored.

Charles D. Winters

Study Questions

▲ denotes more challenging questions.

■ denotes questions available in the Homework and Goals section of the General ChemistryNow CD-ROM or website.

Blue numbered questions have answers in Appendix O and fully worked solutions in the *Student Solutions Manual.*

Structures of many of the compounds used in these questions are found on the General ChemistryNow CD-ROM or website in the Models folder.

GENERAL
Chemistry•ᴧ•Now™ Assess your understanding of this chapter's topics with additional quizzing and conceptual questions at http://now.brookscole.com/kotz6e

Practicing Skills

Energy
(*See Exercise 6.1 and General ChemistryNow Screen 6.3.*)

1. The flashlight in the photo does not use batteries. Instead you move a lever, which turns a geared mechanism and results finally in light from the bulb. What type of energy is used to move the lever? What type or types of energy are produced?

A hand-operated flashlight.

2. A solar panel is pictured in the photo. When light shines on the panel, a small electric motor propels the car. What types of energy are involved in this setup?

A solar panel operates a toy car.

Energy Units
(*See Exercise 6.2 and General ChemistryNow Screen 6.5.*)

3. You are on a diet that calls for eating no more than 1200 Cal/day. How many joules would this be?

4. A 2-in. piece of chocolate cake with frosting provides 1670 kJ of energy. What is this in dietary Calories (Cal)?

5. ■ One food product has an energy content of 170 kcal per serving and another has 280 kJ per serving. Which food has a greater energy content per serving?

6. Which has a greater energy content, a raw apple or a raw apricot? Go to the USDA Nutrient Database on the World Wide Web for the information (http://www.nal.usda.gov/fnic/foodcomp/). Report the energy content of the fruit in kcal and kJ.

Specific Heat Capacity
(*See Examples 6.1 and 6.2 and General ChemistryNow Screens 6.7–6.9.*)

7. The molar heat capacity of mercury is 28.1 J/mol · K. What is the specific heat capacity of this metal in J/g · K?

8. ■ The specific heat capacity of benzene (C_6H_6) is 1.74 J/g · K. What is its molar heat capacity (in J/mol · K)?

9. The specific heat capacity of copper is 0.385 J/g · K. What quantity of heat is required to heat 168 g of copper from $-12.2\ ^\circ C$ to $+25.6\ ^\circ C$?

10. ■ What quantity of heat is required to raise the temperature of 50.00 mL of water from 25.52 °C to 28.75 °C? The density of water at this temperature is 0.997 g/mL.

11. The initial temperature of a 344-g sample of iron is 18.2 °C. If the sample absorbs 2.25 kJ of heat, what is its final temperature?

12. ■ After absorbing 1.850 kJ of heat, the temperature of a 0.500-kg block of copper is 37 °C. What was its initial temperature?

13. A 45.5-g sample of copper at 99.8 °C is dropped into a beaker containing 152 g of water at 18.5 °C. What is the final temperature when thermal equilibrium is reached?

14. A 182-g sample of gold at some temperature is added to 22.1 g of water. The initial water temperature is 25.0 °C, and the final temperature is 27.5 °C. If the specific heat capacity of gold is 0.128 J/g · K, what was the initial temperature of the gold?

15. One beaker contains 156 g of water at 22 °C and a second beaker contains 85.2 g of water at 95 °C. The water in the two beakers is mixed. What is the final water temperature?

16. ■ When 108 g of water at a temperature of 22.5 °C is mixed with 65.1 g of water at an unknown temperature, the final temperature of the resulting mixture is 47.9 °C. What was the initial temperature of the second sample of water?

17. A 13.8-g piece of zinc was heated to 98.8 °C in boiling water and then dropped into a beaker containing 45.0 g of water at 25.0 °C. When the water and metal come to ther-

mal equilibrium, the temperature is 27.1 °C. What is the specific heat capacity of zinc?

18. ■ A 237-g piece of molybdenum, initially at 100.0 °C, is dropped into 244 g of water at 10.0 °C. When the system comes to thermal equilibrium, the temperature is 15.3 °C. What is the specific heat capacity of molybdenum?

Changes of State

(*See Examples 6.3 and 6.4 and General ChemistryNow Screen 6.10.*)

19. What quantity of heat is evolved when 1.0 L of water at 0 °C solidifies to ice? The heat of fusion of water is 333 J/g.

20. The heat energy required to melt 1.00 g of ice at 0 °C is 333 J. If one ice cube has a mass of 62.0 g, and a tray contains 16 ice cubes, what quantity of energy is required to melt a tray of ice cubes to form liquid water at 0 °C?

21. What quantity of heat is required to vaporize 125 g of benzene, C_6H_6, at its boiling point, 80.1 °C? The heat of vaporization of benzene is 30.8 kJ/mol.

22. ■ Chloromethane, CH_3Cl, arises from the oceans and from microbial fermentation and is found throughout the environment. It is used in the manufacture of various chemicals and has been used as a topical anesthetic. What quantity of heat must be absorbed to convert 92.5 g of liquid to a vapor at its boiling point, −24.09 °C? The heat of vaporization of CH_3Cl is 21.40 kJ/mol.

23. The freezing point of mercury is −38.8 °C. What quantity of heat energy, in joules, is released to the surroundings if 1.00 mL of mercury is cooled from 23.0 °C to −38.8 °C and then frozen to a solid? (The density of liquid mercury is 13.6 g/cm^3. Its specific heat capacity is 0.140 J/g · K and its heat of fusion is 11.4 J/g.)

24. What quantity of heat energy, in joules, is required to raise the temperature of 454 g of tin from room temperature, 25.0 °C, to its melting point, 231.9 °C, and then melt the tin at that temperature? The specific heat capacity of tin is 0.227 J/g · K, and the heat of vaporization of this metal is 59.2 J/g.

25. Ethanol, C_2H_5OH, boils at 78.29 °C. What quantity of heat energy, in joules, is required to raise the temperature of 1.00 kg of ethanol from 20.0 °C to the boiling point and then to change the liquid to vapor at that temperature? (The specific heat capacity of liquid ethanol is 2.44 J/g · K and its enthalpy of vaporization is 855 J/g.)

26. ■ A 25.0-mL sample of benzene at 19.9 °C was cooled to its melting point, 5.5 °C, and then frozen. How much heat was given off in this process? The density of benzene is 0.80 g/mL, its specific heat capacity is 1.74 J/g · K, and its heat of fusion is 127 J/g.

Enthalpy

(*See Example 6.5 and General ChemistryNow Screens 6.12 and 6.13.*)

27. Nitrogen monoxide, a gas recently found to be involved in a wide range of biological processes, reacts with oxygen to give brown NO_2 gas.

$$2 NO(g) + O_2(g) \longrightarrow 2 NO_2(g) \qquad \Delta H^\circ_{rxn} = -114.1 \text{ kJ}$$

Is this reaction endothermic or exothermic? If 1.25 g of NO is converted completely to NO_2, what quantity of heat is absorbed or evolved?

28. Calcium carbide, CaC_2, is manufactured by the reaction of CaO with carbon at a high temperature. (Calcium carbide is then used to make acetylene.)

$$CaO(s) + 3 C(s) \longrightarrow CaC_2(s) + CO(g)$$
$$\Delta H^\circ_{rxn} = +464.8 \text{ kJ}$$

Is this reaction endothermic or exothermic? If 10.0 g of CaO is allowed to react with an excess of carbon, what quantity of heat is absorbed or evolved by the reaction?

29. Isooctane (2,2,4-trimethylpentane), one of the many hydrocarbons that make up gasoline, burns in air to give water and carbon dioxide.

$$2 C_8H_{18}(\ell) + 25 O_2(g) \longrightarrow 16 CO_2(g) + 18 H_2O(\ell)$$
$$\Delta H^\circ_{rxn} = -10,922 \text{ kJ}$$

If you burn 1.00 L of isooctane (density = 0.69 g/mL), what quantity of heat is evolved?

30. ■ Acetic acid, CH_3CO_2H, is made industrially by the reaction of methanol and carbon monoxide.

$$CH_3OH(\ell) + CO(g) \longrightarrow CH_3CO_2H(\ell)$$
$$\Delta H^\circ_{rxn} = -355.9 \text{ kJ}$$

If you produce 1.00 L of acetic acid (density = 1.044 g/mL) by this reaction, what quantity of heat is evolved?

Calorimetry

(*See Examples 6.6 and 6.7 and General ChemistryNow Screens 6.8, 6.9, and 6.14.*)

31. Assume you mix 100.0 mL of 0.200 M CsOH with 50.0 mL of 0.400 M HCl in a coffee-cup calorimeter. The following reaction occurs:

$$CsOH(aq) + HCl(aq) \longrightarrow CsCl(aq) + H_2O(\ell)$$

The temperature of both solutions before mixing was 22.50 °C, and it rises to 24.28 °C after the acid–base reaction. What is the enthalpy change for the reaction per mole of CsOH? Assume the densities of the solutions are all 1.00 g/mL and the specific heat capacities of the solutions are 4.2 J/g · K.

32. ■ You mix 125 mL of 0.250 M CsOH with 50.0 mL of 0.625 M HF in a coffee-cup calorimeter, and the temperature of both solutions rises from 21.50 °C before mixing to 24.40 °C after the reaction.

$$CsOH(aq) + HF(aq) \longrightarrow CsF(aq) + H_2O(\ell)$$

What is the enthalpy of reaction per mole of CsOH? Assume the densities of the solutions are all 1.00 g/mL and the specific heats of the solutions are 4.2 J/g · K.

33. A piece of titanium metal with a mass of 20.8 g is heated in boiling water to 99.5 °C and then dropped into a coffee-cup calorimeter containing 75.0 g of water at 21.7 °C. When thermal equilibrium is reached, the final temperature is 24.3 °C. Calculate the specific heat capacity of titanium.

34. ■ A piece of chromium metal with a mass of 24.26 g is heated in boiling water to 98.3 °C and then dropped into a coffee-cup calorimeter containing 82.3 g of water at 23.3 °C. When thermal equilibrium is reached, the final temperature is 25.6 °C. Calculate the specific heat capacity of chromium.

35. Adding 5.44 g of $NH_4NO_3(s)$ to 150.0 g of water in a coffee-cup calorimeter (with stirring to dissolve the salt) resulted in a decrease in temperature from 18.6 °C to 16.2 °C. Calculate the enthalpy change for dissolving $NH_4NO_3(s)$ in water, in kJ/mol. Assume that the solution (whose mass is 155.4 g) has a specific heat capacity of 4.2 J/g · K. (Cold packs take advantage of the fact that dissolving ammonium nitrate in water is an endothermic process.)

A cold pack uses the endothermic heat of solution of ammonium nitrate.

36. You should use care when dissolving H_2SO_4 in water because the process is highly exothermic. To measure the enthalpy change, 5.2 g $H_2SO_4(\ell)$ was added (with stirring) to 135 g of water in a coffee-cup calorimeter. This resulted in an increase in temperature from 20.2 °C to 28.8 °C. Calculate the enthalpy change for the process $H_2SO_4(\ell) \longrightarrow H_2SO_4(aq)$, in kJ/mol.

37. Sulfur (2.56 g) is burned in a constant volume calorimeter with excess $O_2(g)$. The temperature increases from 21.25 °C to 26.72 °C. The bomb has a heat capacity of 923 J/K, and the calorimeter contains 815 g of water. Calculate the heat evolved, per mole of SO_2 formed, for the reaction

$$S_8(s) + 8\,O_2(g) \longrightarrow 8\,SO_2(g)$$

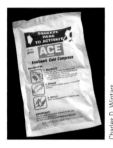

Sulfur burns in oxygen with a bright blue flame to give $SO_2(g)$.

38. ■ Suppose you burn 0.300 g of C(graphite) in an excess of $O_2(g)$ in a constant volume calorimeter to give $CO_2(g)$.

$$C(graphite) + O_2(g) \longrightarrow CO_2(g)$$

The temperature of the calorimeter, which contains 775 g of water, increases from 25.00 °C to 27.38 °C. The heat capacity of the bomb is 893 J/K. What quantity of heat is evolved per mole of carbon?

39. Suppose you burn 1.500 g of benzoic acid, $C_6H_5CO_2H$, in a constant volume calorimeter and find that the temperature increases from 22.50 °C to 31.69 °C. The calorimeter contains 775 g of water, and the bomb has a heat capacity of 893 J/K. What quantity of heat is evolved in this combustion reaction, per mole of benzoic acid?

Benzoic acid, $C_6H_5CO_2H$, occurs naturally in many berries. Its heat of combustion is well known so it is used as a standard to calibrate calorimeters.

40. A 0.692-g sample of glucose, $C_6H_{12}O_6$, is burned in a constant volume calorimeter. The temperature rises from 21.70 °C to 25.22 °C. The calorimeter contains 575 g of water and the bomb has a heat capacity of 650 J/K. What quantity of heat is evolved per mole of glucose?

41. An "ice calorimeter" can be used to determine the specific heat capacity of a metal. A piece of hot metal is dropped onto a weighed quantity of ice. The quantity of heat transferred from the metal to the ice can be determined from the amount of ice melted. Suppose you heat a 50.0-g piece of silver to 99.8 °C and then drop it onto ice. When the metal's temperature has dropped to 0.0 °C, it is found that 3.54 g of ice has melted. What is the specific heat capacity of silver?

42. ■ A 9.36-g piece of platinum is heated to 98.6 °C in a boiling water bath and then dropped onto ice. (See Study Question 41.) When the metal's temperature has dropped to 0.0 °C, it is found that 0.37 g of ice has melted. What is the specific heat capacity of platinum?

Hess's Law

(See Example 6.8 and General ChemistryNow Screen 6.15.)

43. The enthalpy changes for the following reactions can be measured:

$$CH_4(g) + 2\,O_2(g) \longrightarrow CO_2(g) + 2\,H_2O(g)$$
$$\Delta H° = -802.4 \text{ kJ}$$

$$CH_3OH(g) + \tfrac{3}{2}\,O_2(g) \longrightarrow CO_2(g) + H_2O(g)$$
$$\Delta H° = -676 \text{ kJ}$$

(a) Use these values and Hess's law to determine the enthalpy change for the reaction

$$CH_4(g) + \tfrac{1}{2}O_2(g) \longrightarrow CH_3OH(g)$$

(b) Draw an energy level diagram that shows the relationship between the energy quantities involved in this problem.

44. The enthalpy changes of the following reactions can be measured:

$$C_2H_4(g) + 3\,O_2(g) \longrightarrow 2\,CO_2(g) + 2\,H_2O(\ell)$$
$$\Delta H° = -1411.1 \text{ kJ}$$

$$C_2H_5OH(\ell) + 3\,O_2(g) \longrightarrow 2\,CO_2(g) + 3\,H_2O(\ell)$$
$$\Delta H° = -1367.5 \text{ kJ}$$

(a) ■ Use these values and Hess's law to determine the enthalpy change for the reaction

$$C_2H_4(g) + H_2O(\ell) \longrightarrow C_2H_5OH(\ell)$$

(b) Draw an energy level diagram that shows the relationship between the energy quantities involved in this problem.

45. Enthalpy changes for the following reactions can be determined experimentally:

$$N_2(g) + 3\,H_2(g) \longrightarrow 2\,NH_3(g) \qquad \Delta H° = -91.8 \text{ kJ}$$

$$4\,NH_3(g) + 5\,O_2(g) \longrightarrow 4\,NO(g) + 6\,H_2O(g)$$
$$\Delta H° = -906.2 \text{ kJ}$$

$$H_2(g) + \tfrac{1}{2}O_2(g) \longrightarrow H_2O(g) \qquad \Delta H° = -241.8 \text{ kJ}$$

Use these values to determine the enthalpy change for the formation of NO(g) from the elements (an enthalpy change that cannot be measured directly because the reaction is reactant-favored).

$$\tfrac{1}{2}N_2(g) + \tfrac{1}{2}O_2(g) \longrightarrow NO(g) \qquad \Delta H° = ?$$

46. You wish to know the enthalpy change for the formation of liquid PCl_3 from the elements.

$$P_4(s) + 6\,Cl_2(g) \longrightarrow 4\,PCl_3(\ell) \qquad \Delta H° = ?$$

The enthalpy change for the formation of PCl_5 from the elements can be determined experimentally, as can the enthalpy change for the reaction of $PCl_3(\ell)$ with more chlorine to give $PCl_5(s)$:

$$P_4(s) + 10\,Cl_2(g) \longrightarrow 4\,PCl_5(s) \qquad \Delta H° = -1774.0 \text{ kJ}$$

$$PCl_3(\ell) + Cl_2(g) \longrightarrow PCl_5(s) \qquad \Delta H° = -123.8 \text{ kJ}$$

Use these data to calculate the enthalpy change for the formation of 1.00 mol of $PCl_3(\ell)$ from phosphorus and chlorine.

Standard Enthalpies of Formation

(See Example 6.9 and General ChemistryNow Screen 6.16.)

47. Write a balanced chemical equation for the formation of $CH_3OH(\ell)$ from the elements in their standard states. Find the value for $\Delta H_f°$ for $CH_3OH(\ell)$ in Appendix L.

48. Write a balanced chemical equation for the formation of $CaCO_3(s)$ from the elements in their standard states. Find the value for $\Delta H_f°$ for $CaCO_3(s)$ in Appendix L.

49. (a) Write a balanced chemical equation for the formation of 1 mol of $Cr_2O_3(s)$ from Cr and O_2 in their standard states. Find the value for $\Delta H_f°$ for $Cr_2O_3(s)$ in Appendix L.

(b) ■ What is the standard enthalpy change if 2.4 g of chromium is oxidized to $Cr_2O_3(s)$?

50. (a) Write a balanced chemical equation for the formation of 1 mol of MgO(s) from the elements in their standard states. Find the value for $\Delta H_f°$ for MgO(s) in Appendix L.

(b) What is the standard enthalpy change for the reaction of 2.5 mol of Mg with oxygen?

51. Use standard heats of formation in Appendix L to calculate standard enthalpy changes for the following:

(a) 1.0 g of white phosphorus burns, forming $P_4O_{10}(s)$

(b) 0.20 mol of NO(g) decomposes to $N_2(g)$ and $O_2(g)$

(c) 2.40 g of NaCl is formed from Na(s) and excess $Cl_2(g)$

(d) 250 g of iron is oxidized with oxygen to $Fe_2O_3(s)$

52. Use standard heats of formation in Appendix L to calculate standard enthalpy changes for the following:

(a) 0.054 g of sulfur burns, forming $SO_2(g)$

(b) 0.20 mol of HgO(s) decomposes to $Hg(\ell)$ and $O_2(g)$

(c) 2.40 g of $NH_3(g)$ is formed from $N_2(g)$ and excess $H_2(g)$

(d) 1.05×10^{-2} mol of carbon is oxidized to $CO_2(g)$

53. The first step in the production of nitric acid from ammonia involves the oxidation of NH_3.

$$4\,NH_3(g) + 5\,O_2(g) \longrightarrow 4\,NO(g) + 6\,H_2O(g)$$

(a) Use standard enthalpies of formation to calculate the standard enthalpy change for this reaction.

(b) ■ What quantity of heat is evolved or absorbed in the formation of 10.0 g of NH_3?

54. The Romans used calcium oxide, CaO, to produce a strong mortar to build stone structures. The CaO was mixed with water to give $Ca(OH)_2$, which reacted slowly with CO_2 in the air to give $CaCO_3$.

$$Ca(OH)_2(s) + CO_2(g) \longrightarrow CaCO_3(s) + H_2O(g)$$

(a) Calculate the standard enthalpy change for this reaction.

(b) What quantity of heat is evolved or absorbed if 1.00 kg of $Ca(OH)_2$ reacts with a stoichiometric amount of CO_2?

55. The standard enthalpy of formation of solid barium oxide, BaO, is -553.5 kJ/mol, and the enthalpy of formation of barium peroxide, BaO_2, is -634.3 kJ/mol.

(a) Calculate the standard enthalpy change for the following reaction. Is the reaction exothermic or endothermic?

$$BaO_2(s) \longrightarrow BaO(s) + \tfrac{1}{2}O_2(g)$$

(b) Draw an energy level diagram that shows the relationship between the enthalpy change of the decomposition of BaO_2 to BaO and O_2 and the enthalpies of formation of BaO(s) and $BaO_2(s)$.

56. An important step in the production of sulfuric acid is the oxidation of SO_2 to SO_3.

$$SO_2(g) + \tfrac{1}{2} O_2(g) \longrightarrow SO_3(g)$$

Formation of SO_3 from the air pollutant SO_2 is also a key step in the formation of acid rain.

(a) Use standard enthalpies of formation to calculate the enthalpy change for the reaction. Is the reaction exothermic or endothermic?

(b) Draw an energy level diagram that shows the relationship between the enthalpy change for the oxidation of SO_2 to SO_3 and the enthalpies of formation of $SO_2(g)$ and $SO_3(g)$.

57. The enthalpy change for the oxidation of naphthalene, $C_{10}H_8$, is measured by calorimetry.

$$C_{10}H_8(s) + 12 O_2(g) \longrightarrow 10 CO_2(g) + 4 H_2O(\ell)$$
$$\Delta H^\circ_{rxn} = -5156.1 \text{ kJ}$$

Use this value, along with the standard heats of formation of $CO_2(g)$ and $H_2O(\ell)$, to calculate the enthalpy of formation of naphthalene, in kJ/mol.

58. ■ The enthalpy change for the oxidation of styrene, C_8H_8, is measured by calorimetry.

$$C_8H_8(\ell) + 10 O_2(g) \longrightarrow 8 CO_2(g) + 4 H_2O(\ell)$$
$$\Delta H^\circ_{rxn} = -4395.0 \text{ kJ}$$

Use this value, along with the standard heats of formation of $CO_2(g)$ and $H_2O(\ell)$, to calculate the enthalpy of formation of styrene, in kJ/mol.

Product- and Reactant-Favored Reactions

59. Use your "chemical sense" to decide whether each of the following reactions is product- or reactant-favored. Calculate ΔH°_{rxn} in each case, and draw an energy level diagram like those in Figure 6.18.

(a) the reaction of aluminum and chlorine to produce $AlCl_3(s)$

(b) the decomposition of mercury(II) oxide to produce liquid mercury and oxygen gas

60. Use your "chemical sense" to decide whether each of the following reactions is product- or reactant-favored. Calculate ΔH°_{rxn} in each case, and draw an energy level diagram like those in Figure 6.18.

(a) the decomposition of ozone, O_3, to oxygen molecules

(b) the decomposition of $MgCO_3(s)$ to give $MgO(s)$ and $CO_2(g)$

General Questions on Thermochemistry

These questions are not designated as to type or location in the chapter. They may combine several concepts.

61. The following terms are used extensively in thermodynamics. Define each and give an example.

(a) exothermic and endothermic

(b) system and surroundings

(c) specific heat capacity

(d) state function

(e) standard state

(f) enthalpy change, ΔH

(g) standard enthalpy of formation

62. For each of the following, tell whether the process is exothermic or endothermic. (No calculations are required.)

(a) $H_2O(\ell) \longrightarrow H_2O(s)$

(b) $2 H_2(g) + O_2(g) \longrightarrow 2 H_2O(g)$

(c) $H_2O(\ell, 25\,^\circ C) \longrightarrow H_2O(\ell, 15\,^\circ C)$

(d) $H_2O(\ell) \longrightarrow H_2O(g)$

63. For each of the following, define a system and its surroundings and give the direction of heat transfer between system and surroundings.

(a) Methane is burning in a gas furnace in your home.

(b) Water drops, sitting on your skin after a dip in a swimming pool, evaporate.

(c) Water, at 25 °C, is placed in the freezing compartment of a refrigerator, where it cools and eventually solidifies.

(d) Aluminum and $Fe_2O_3(s)$ are mixed in a flask sitting on a laboratory bench. A reaction occurs, and a large quantity of heat is evolved.

64. Which of the following are state functions?

(a) the volume of a balloon

(b) the time it takes to drive from your home to your college or university

(c) the temperature of the water in a coffee cup

(d) the potential energy of a ball held in your hand

65. Define the first law of thermodynamics using a mathematical equation and explain the meaning of each term in the equation.

66. What does the term "standard state" mean? What are the standard states of the following substances at 298 K: H_2O, NaCl, Hg, CH_4?

67. Use Appendix L to find the standard enthalpies of formation of oxygen atoms, oxygen molecules (O_2), and ozone (O_3). What is the standard state of oxygen? Is the formation of oxygen atoms from O_2 exothermic? What is the enthalpy change for the formation of 1 mol of O_3 from O_2?

68. See General ChemistryNow CD-ROM or website Screen 6.9 Heat Transfer Between Substances. Use the Simulation section of this screen to do the following experiment: Add 10.0 g of Al at 80 °C to 10.0 g of water at 20 °C. What is the final temperature when equilibrium is achieved? Use this value to estimate the specific heat capacity of aluminum.

69. See General ChemistryNow CD-ROM or website Screen 6.15 Hess's Law. Use the Simulation section of this screen to find the value of ΔH°_{rxn} for

$$SnBr_2(s) + TiCl_4(\ell) \longrightarrow SnCl_4(\ell) + TiBr_2(s)$$

▲ More challenging   ■ In General ChemistryNow   Blue-numbered questions answered in Appendix O

70. A piece of lead with a mass of 27.3 g was heated to 98.90 °C and then dropped into 15.0 g of water at 22.50 °C. The final temperature was 26.32 °C. Calculate the specific heat capacity of lead from these data.

71. Which gives up more heat on cooling from 50 °C to 10 °C, 50.0 g of water or 100. g of ethanol (specific heat capacity of ethanol = 2.46 J/g · K)?

72. A 192-g piece of copper is heated to 100.0 °C in a boiling water bath and then dropped into a beaker containing 751 g of water (density = 1.00 g/cm^3) at 4.0 °C. What is the final temperature of the copper and water after thermal equilibrium is reached? (The specific heat capacity of copper is 0.385 J/g · K).

73. You determine that 187 J of heat is required to raise the temperature of 93.45 g of silver from 18.5 °C to 27.0 °C. What is the specific heat capacity of silver?

74. Calculate the quantity of heat required to convert 60.1 g of $H_2O(s)$ at 0.0 °C to $H_2O(g)$ at 100.0 °C. The heat of fusion of ice at 0 °C is 333 J/g; the heat of vaporization of liquid water at 100 °C is 2260 J/g.

75. ■ You add 100.0 g of water at 60.0 °C to 100.0 g of ice at 0.00 °C. Some of the ice melts and cools the water to 0.00 °C. When the ice and water mixture has come to a uniform temperature of 0 °C, how much ice has melted?

76. ▲ Three 45-g ice cubes at 0 °C are dropped into 5.00 × 10^2 mL of tea to make ice tea. The tea was initially at 20.0 °C; when thermal equilibrium was reached, the final temperature was 0 °C. How much of the ice melted and how much remained floating in the beverage? Assume the specific heat capacity of tea is the same as that of pure water.

77. ▲ ■ Suppose that only two 45-g ice cubes had been added to your glass containing 5.00 × 10^2 mL of tea (See Study Question 76). When thermal equilibrium is reached, all of the ice will have melted and the temperature of the mixture will be somewhere between 20.0 °C and 0 °C. Calculate the final temperature of the beverage. (Note: The 90 g of water formed when the ice melts must be warmed from 0 °C to the final temperature.)

78. You take a diet cola from the refrigerator, and pour 240 mL of it into a glass. The temperature of the beverage is 10.5 °C. You then add one ice cube (45 g). Which of the following describes the system when thermal equilibrium is reached?

    (a) The temperature is 0 °C and some ice remains.

    (b) The temperature is 0 °C and no ice remains.

    (c) The temperature is higher than 0 °C and no ice remains.

    Determine the final temperature and the amount of ice remaining, if any.

79. Insoluble AgCl(s) precipitates when solutions of $AgNO_3(aq)$ and NaCl(aq) are mixed.

    $$AgNO_3(aq) + NaCl(aq) \longrightarrow AgCl(s) + NaNO_3(aq)$$
    $$\Delta H°_{rxn} = ?$$

To measure the heat evolved in this reaction, 250. mL of 0.16 M $AgNO_3(aq)$ and 125 mL of 0.32 M NaCl(aq) are mixed in a coffee-cup calorimeter. The temperature of the mixture rises from 21.15 °C to 22.90 °C. Calculate the enthalpy change for the precipitation of AgCl(s), in kJ/mol. (Assume the density of the solution is 1.0 g/mL and its specific heat capacity is 4.2 J/g · K.)

80. Insoluble $PbBr_2(s)$ precipitates when solutions of $Pb(NO_3)_2(aq)$ and NaBr(aq) are mixed.

    $$Pb(NO_3)_2(aq) + 2 NaBr(aq) \longrightarrow PbBr_2(s) + 2 NaNO_3(aq)$$
    $$\Delta H°_{rxn} = ?$$

To measure the heat evolved, 200. mL of 0.75 M $Pb(NO_3)_2(aq)$ and 200 mL of 1.5 M NaBr(aq) are mixed in a coffee-cup calorimeter. The temperature of the mixture rises by 2.44 °C. Calculate the enthalpy change for the precipitation of $PbBr_2(s)$, in kJ/mol. (Assume the density of the solution is 1.0 g/mL and its specific heat capacity is 4.2 J/g · K.)

81. The heat evolved in the decomposition of 7.647 g of ammonium nitrate can be measured in a bomb calorimeter. The reaction that occurs is

    $$NH_4NO_3(s) \longrightarrow N_2O(g) + 2 H_2O(g)$$

The temperature of the calorimeter, which contains 415 g of water, increases from 18.90 °C to 20.72 °C. The heat capacity of the bomb is 155 J/K. What quantity of heat is evolved in this reaction, in kJ/mol?

82. A bomb calorimetric experiment was run to determine the heat of combustion of ethanol (a common fuel additive). The reaction is

    $$C_2H_5OH(\ell) + 3 O_2(g) \longrightarrow 2CO_2(g) + 3 H_2O(\ell)$$

The bomb had a heat capacity of 550 J/K, and the calorimeter contained 650 g of water. Burning 4.20 g of ethanol, $C_2H_5OH(\ell)$ resulted in a rise in temperature from 18.5 °C to 22.3 °C. Calculate the heat of combustion of ethanol, in kJ/mol.

83. ▲ The standard molar enthalpy of formation of diborane, $B_2H_6(g)$, cannot be determined directly because the compound cannot be prepared by the reaction of boron and hydrogen. It can be calculated from other enthalpy changes, however. The following enthalpy changes can be measured.

    $$4 B(s) + 3 O_2(g) \longrightarrow 2 B_2O_3(s) \qquad \Delta H°_{rxn} = -2543.8 \text{ kJ}$$
    $$H_2(g) + \tfrac{1}{2} O_2(g) \longrightarrow H_2O(g) \qquad \Delta H°_{rxn} = -241.8 \text{ kJ}$$
    $$B_2H_6(g) + 3 O_2(g) \longrightarrow B_2O_3(s) + 3 H_2O(g)$$
    $$\Delta H°_{rxn} = -2032.9 \text{ kJ}$$

    (a) Show how these equations can be added together to give the equation for the formation of $B_2H_6(g)$ from B(s) and $H_2(g)$ in their standard states. Assign enthalpy changes to each reaction.

    (b) Calculate $\Delta H°_f$ for $B_2H_6(g)$.

    (c) Draw an energy level diagram that shows how the various enthalpies in this problem are related.

(d) Is the formation of $B_2H_6(g)$ from its elements product- or reactant-favored?

84. Chloromethane, CH_3Cl, a compound found ubiquitously in the environment, is formed in the reaction of chlorine atoms with methane.

$$CH_4(g) + 2\ Cl(g) \longrightarrow CH_3Cl(g) + HCl(g)$$

(a) ■ Calculate the enthalpy change for the reaction of $CH_4(g)$ and Cl atoms to give $CH_3Cl(g)$ and $HCl(g)$. Is the reaction product- or reactant-favored?

(b) Draw an energy level diagram that shows how the various enthalpies in this problem are related.

85. The meals-ready-to-eat (MREs) in the military can be heated on a flameless heater. The source of energy in the heater is

$$Mg(s) + 2\ H_2O(\ell) \longrightarrow Mg(OH)_2(s) + H_2(g)$$

Calculate the enthalpy change under standard conditions, in joules, for this reaction. What quantity of magnesium is needed to supply the heat required to warm 25 mL of water ($d = 1.00$ g/mL) from 25 °C to 85 °C? (See W. Jensen: *Journal of Chemical Education*, Vol. 77, pp. 713–717, 2000.)

86. Hydrazine, $N_2H_4(\ell)$, is an efficient oxygen scavenger. It is sometimes added to steam boilers to remove traces of oxygen that can cause corrosion in these systems. Combustion of hydrazine gives the following information:

$$N_2H_4(\ell) + O_2(g) \longrightarrow N_2(g) + 2\ H_2O(g)$$
$$\Delta H^\circ_{rxn} = -534.3 \text{ kJ}$$

(a) Is the reaction product- or reactant-favored?

(b) Use the value for ΔH°_{rxn} with the enthalpy of formation of $H_2O(g)$ to calculate the molar enthalpy of formation of $N_2H_4(\ell)$.

87. When heated to a high temperature, coke (mainly carbon, obtained by heating coal in the absence of air) and steam produce a mixture called water gas, which can be used as a fuel or as a chemical feedstock for other reactions. The equation for the production of water gas is

$$C(s) + H_2O(g) \longrightarrow CO(g) + H_2(g)$$

(a) Use standard heats of formation to determine the enthalpy change for this reaction.

(b) Is the reaction product- or reactant-favored?

(c) What quantity of heat is involved if 1.0 metric ton (1000.0 kg) of carbon is converted to water gas?

88. Camping stoves are fueled by propane (C_3H_8), butane [$C_4H_{10}(g)$, $\Delta H^\circ_f = -127.1$ kJ/mol], gasoline, or ethanol (C_2H_5OH). Calculate the heat of combustion per gram of each of these fuels. [Assume that gasoline is represented by isooctane, $C_8H_{18}(\ell)$, with $\Delta H^\circ_f = -259.2$ kJ/mol.] Do you notice any great differences among these fuels? Are these differences related to their composition?

89. Methanol, CH_3OH, a compound that can be made relatively inexpensively from coal, is a promising substitute for gasoline. The alcohol has a smaller energy content than gasoline, but, with its higher octane rating, it burns more

efficiently than gasoline in combustion engines. (It has the added advantage of contributing to a lesser degree to some air pollutants.) Compare the heat of combustion per gram of CH_3OH and C_8H_{18} (isooctane), the latter being representative of the compounds in gasoline. ($\Delta H^\circ_f = -259.2$ kJ/mol for isooctane.)

90. Hydrazine and 1,1-dimethylhydrazine both react spontaneously with O_2 and can be used as rocket fuels.

$$\underset{\text{hydrazine}}{N_2H_4(\ell)} + O_2(g) \longrightarrow N_2(g) + 2\ H_2O(g)$$

$$\underset{\text{1,1-dimethylhydrazine}}{N_2H_2(CH_3)_2(\ell)} + 4\ O_2(g) \longrightarrow$$
$$2\ CO_2(g) + 4\ H_2O(g) + N_2(g)$$

The molar enthalpy of formation of $N_2H_4(\ell)$ is $+50.6$ kJ/mol, and that of $N_2H_2(CH_3)_2(\ell)$ is $+48.9$ kJ/mol. Use these values, with other ΔH°_f values, to decide whether the reaction of hydrazine or, -dimethylhydrazine with oxygen gives more heat per gram.

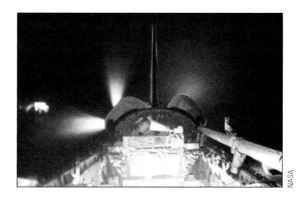

A control rocket in the Space Shuttle uses hydrazine as the fuel.

91. (a) Calculate the enthalpy change, ΔH°, for the formation of 1.00 mol of strontium carbonate (the material that gives the red color in fireworks) from its elements.

$$Sr(s) + C(graphite) + \tfrac{3}{2}O_2(g) \longrightarrow SrCO_3(s)$$

The experimental information available is

$$Sr(s) + \tfrac{1}{2}O_2(g) \longrightarrow SrO(s) \qquad \Delta H^\circ_f = -592 \text{ kJ}$$
$$SrO(s) + CO_2(g) \longrightarrow SrCO_3(s) \qquad \Delta H^\circ_{rxn} = -234 \text{ kJ}$$
$$C(graphite) + O_2(g) \longrightarrow CO_2(g) \qquad \Delta H^\circ_f = -394 \text{ kJ}$$

(b) Draw an energy level diagram relating the energy quantities in this problem.

92. You drink 350 mL of diet soda that is at a temperature of 5 °C.

(a) How much energy will your body expend to raise the temperature of this liquid to body temperature (37 °C)? Assume that the density and specific heat capacity of diet soda are the same as for water.

(b) Compare the value in part (a) with the caloric content of the beverage. (The label says that it has a caloric content of 1 Calorie.) What is the net energy change in your body resulting from drinking this beverage?

(c) Carry out a comparison similar to that in part (b) for a nondiet beverage whose label indicates a caloric content of 240 Calories.

93. Chloroform, $CHCl_3$, is formed from methane and chlorine in the following reaction.

$$CH_4(g) + 3\ Cl_2(g) \longrightarrow 3\ HCl(g) + CHCl_3(g)$$

Calculate ΔH°_{rxn}, the enthalpy change for this reaction, using the enthalpy of formation of $CHCl_3(g)$, $\Delta H^\circ_f = -103.1$ kJ/mol), and the enthalpy changes for the following reactions:

$$CH_4(g) + 2\ O_2(g) \longrightarrow 2\ H_2O(\ell) + CO_2(g)$$
$$\Delta H^\circ_{rxn} = -890.4\ \text{kJ}$$

$$2\ HCl(g) \longrightarrow H_2(g) + Cl_2(g) \qquad \Delta H^\circ_{rxn} = +184.6\ \text{kJ}$$

$$C(graphite) + O_2(g) \longrightarrow CO_2(g) \qquad \Delta H^\circ_f = -393.5\ \text{kJ}$$

$$H_2(g) + \tfrac{1}{2}\ O_2(g) \longrightarrow H_2O(\ell) \qquad \Delta H^\circ_f = -285.8\ \text{kJ}$$

94. Water gas, a mixture of carbon monoxide and hydrogen, is produced by treating carbon (in the form of coke or coal) with steam at high temperatures. (See Question 87.)

$$C(s) + H_2O(g) \longrightarrow CO(g) + H_2(g)$$

Not all of the carbon available is converted to water gas as some is burned to provide the heat for the endothermic reaction of carbon and water. What mass of carbon must be burned (to CO_2 gas) to provide the heat to convert 1.00 kg of carbon to water gas?

95. Compare the heat evolved by burning 1.00 kg of carbon (to CO_2 gas) with the heat evolved by the water gas [$CO(g) + H_2(g)$] obtained from 1.00 kg of carbon (assuming a 100% yield). (See Question 94.) Which provides more energy?

96. ▲ Isomers are molecules with the same elemental composition but a different atomic arrangement. Three isomers with the formula C_4H_8 are shown in the models below. The enthalpy of combustion of each isomer, determined using a calorimeter, is:

Compound	$\Delta H_{combustion}$ (kJ/mol)
cis-2 butene	−2687.5
trans-2-butene	−2684.2
1-butene	−2696.7

(a) Draw an energy level diagram relating the energy content of the three isomers to the energy content of the combustion products, $CO_2(g)$ and $H_2O(g)$.

(b) Use the $\Delta H_{combustion}$ data in part (a), along with the enthalpies of formation of $CO_2(g)$ and $H_2O(g)$ from Appendix L, to calculate the enthalpy of formation for each of the isomers.

(c) Draw an energy level diagram that relates the heats of formation of the three isomers to the energy of the elements in their standard states.

(d) What is the enthalpy change for the conversion of *cis*-2-butene to *trans*-2-butene?

Summary and Conceptual Questions

The following questions may use concepts from preceding chapters.

97. The first law of thermodynamics is often described as another way of stating the law of conservation of energy. Discuss whether this is an accurate portrayal.

98. Many people have tried to make a perpetual motion machine, but none have been successful although some have claimed success. Use the law of conservation of energy to explain why such a device is impossible.

99. Without doing calculations, decide whether each of the following is product- or reactant-favored.

(a) the combustion of natural gas

(b) the decomposition of glucose, $C_6H_{12}O_6$, to carbon and water

100. See General ChemistryNow CD-ROM or website Screen 6.18 Control of Chemical Reactions. What is the difference between thermodynamics and kinetics?

101. See General ChemistryNow CD-ROM or website Screen 6.9 Heat Transfer Between Substances.

(a) Explain what happens in terms of molecular motions when a hotter object comes in contact with a cooler one.

(b) What does it mean when two objects have come to thermal equilibrium?

$\Delta H_{combustion} = -2687.5$ kJ/mol

cis-2-butene

$\Delta H_{combustion} = -2684.2$ kJ/mol

trans-2-butene

$\Delta H_{combustion} = -2696.7$ kJ/mol

1-butene

102. The photograph here shows a toy car. A solar panel collects light, which generates electricity. This energy is used to electrolyze water to H_2 and O_2 gas, and these gases are recombined in a fuel cell (a special type of battery) to drive the car.

Charles D. Winters

A toy car that uses a solar panel to collect light.
The electricity generated by the panel generates hydrogen and oxygen gases, which are used in a fuel cell.

Describe the form of energy involved in the various processes in the toy car.

103. ▲ You want to determine the value for the enthalpy of formation of $CaSO_4(s)$.

$$Ca(s) + \tfrac{1}{8} S_8(s) + 2 O_2(g) \longrightarrow CaSO_4(s)$$

This reaction cannot be done directly. You know, however, that both calcium and sulfur react with oxygen to produce oxides in reactions that can be studied calorimetrically. You also know that the basic oxide CaO reacts with the acidic oxide SO_3 (g) to produce $CaSO_4(s)$ with $\Delta H^\circ_{rxn} = -402.7$ kJ. Outline a method for determining ΔH°_f for $CaSO_4(s)$ and identify the information that must be collected by experiment. Using information in Table 6.2, confirm that ΔH°_f for $CaSO_4(s) = -1433.5$ kJ/mol.

104. Prepare a graph of heat capacities for metals versus their atomic weights. Combine the data in Table 6.1 and the values in the following table. What is the relationship between specific heat capacity and atomic weight? Use this relationship to predict the specific heat capacity of platinum. The specific heat capacity for platinum is given in the literature as 0.133 J/g · K. How good is the agreement between the predicted and actual values?

Metal	Specific Heat Capacity (J/g · K)
Chromium	0.450
Lead	0.127
Silver	0.236
Tin	0.227
Titanium	0.522

105. Observe the molar heat capacity values for the metals in Table 6.1. What observation can you make about these values—specifically, are they widely different or very similar? Using this information, estimate the specific heat capacity for silver. Compare this estimate with the correct value for silver, 0.236 J/g · K.

106. ▲ Suppose you are attending summer school and are living in a very old dormitory. The day is oppressively hot. There is no air-conditioner, and you can't open the windows of your room because they are stuck shut from layers of paint. There is a refrigerator in the room, however. In a stroke of genius you open the door of the refrigerator, and cool air cascades out. The relief does not last long, though. Soon the refrigerator motor and condenser begin to run, and not long thereafter the room is hotter than it was before. Why did the room warm up?

107. You want to heat the air in your house with natural gas (CH_4). Assume your house has 275 m^2 (about 2800 ft^2) of floor area and that the ceilings are 2.50 m from the floors. The air in the house has a molar heat capacity of 29.1 J/mol · K. (The number of moles of air in the house can be found by assuming that the average molar mass of air is 28.9 g/mol and that the density of air at these temperatures is 1.22 g/L.) What mass of methane do you have to burn to heat the air from 15.0 °C to 22.0 °C?

108. Water can be decomposed to its elements, H_2 and O_2, using electrical energy or in a series of chemical reactions. The following sequence of reactions is one possibility:

$$CaBr_2(s) + H_2O(g) \longrightarrow CaO(s) + 2 HBr(g)$$
$$Hg(\ell) + 2 HBr(g) \longrightarrow HgBr_2(s) + H_2(g)$$
$$HgBr_2(s) + CaO(s) \longrightarrow HgO(s) + CaBr_2(s)$$
$$HgO(s) \longrightarrow Hg(\ell) + \tfrac{1}{2} O_2(g)$$

(a) Show that the net result of this series of reactions is the decomposition of water to its elements.

(b) If you use 1000. kg of water, what mass of H_2 can be produced?

(c) Calculate the value of ΔH°_{rxn} for each step in the series. Are the reactions predicted to be product- or reactant-favored?

$$\Delta H^\circ_f [CaBr_2(s)] = -683.2 \text{ kJ/mol}$$
$$\Delta H^\circ_f [HgBr_2(s)] = -169.5 \text{ kJ/mol}$$

(e) Comment on the commercial feasibility of using this series of reactions to produce $H_2(g)$ from water.

109. Suppose that an inch of rain falls over a square mile of ground. (A density of 1.0 g/cm^3 is assumed.) The heat of vaporization of water at 25 °C is 44.0 kJ/mol. Calculate the quantity of heat transferred to the surroundings from the condensation of water vapor in forming this quantity of liquid water. (The huge number tells you how much energy is "stored" in water vapor and why we think of storms as such great forces of energy in nature. It is interesting to compare this result with the energy given off, 4.2×10^6 kJ, when a ton of dynamite explodes.)

110. ▲ Peanuts and peanut oil are organic materials and burn in air. How many burning peanuts does it take to provide the energy to boil a cup of water (250 mL of water)? To solve this problem we assume each peanut, with an average mass of 0.73 g, is 49% peanut oil and 21% starch; the remainder is noncombustible. We further assume peanut oil is palmitic acid, $C_{16}H_{32}O_2$, with an enthalpy of formation of -848.4 kJ/mol. Starch is a long chain of $C_6H_{10}O_5$ units, each unit having an enthalpy of formation of -960 kJ. (*See General ChemistryNow Screens 6.1 and 6.19: Chemical Puzzler.*)

Charles D Winters

How many burning peanuts are required to provide the heat to boil 250 mL of water?

Mentor

Do you need a live tutor for homework problems?
Access vMentor at General ChemistryNow at
http://now.brookscole.com/kotz6e
for one-on-one tutoring from a chemistry expert

The Chemistry of Fuels and Energy Sources

Gabriela C. Weaver

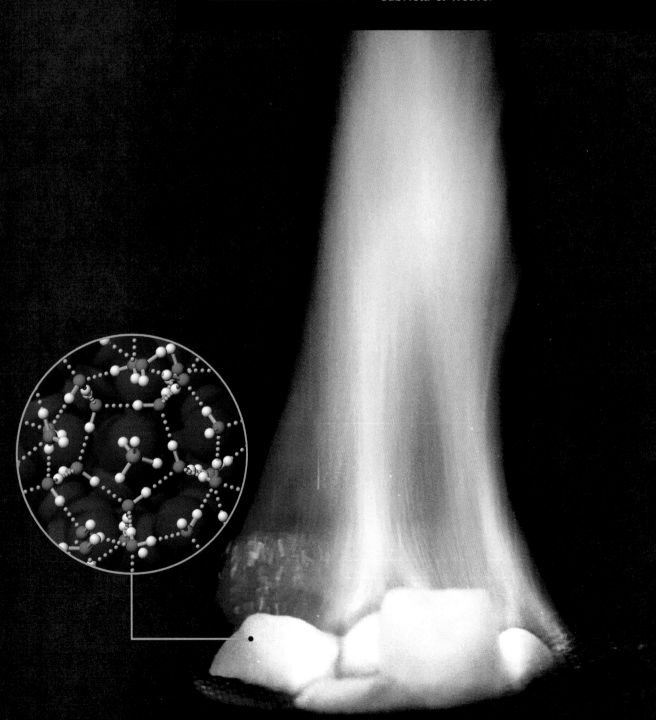

Energy is necessary for everything we do. Look around you—energy is involved in anything that is moving or is emitting light, sound, or heat (Figure 1). Heating and lighting your home, propelling your automobile, powering your portable CD player—all are commonplace examples in which energy is consumed and all are, at their origin, based on chemical processes. In this part of the text, we will examine how chemistry is fundamental to understanding and addressing current energy issues.

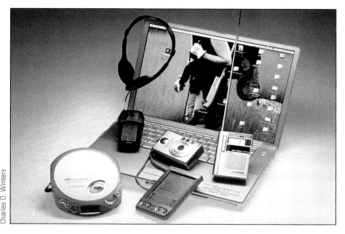

Figure 1   Energy-consuming devices. Our lives would not be the same without the heat and light in our homes and without our automobiles, computers, cell phones, music players, stoves, and refrigerators.

Supply and Demand: The Balance Sheet on Energy

We take for granted that energy is available and that it will always be there to use. But will it? Recently, chemist and Nobel Prize winner Richard Smalley stated that among the top 10 problems humanity will face over the next 50 years, the *energy supply* ranks as number one. What is the source of this dire prediction?

Information such as the following is often quoted in the popular press:

- Global demand for energy has tripled in the past 50 years and may triple again in the next 50 years. Most of the demand comes from industrialized nations.

- Fossil fuels account for 85% of the total energy used on our planet. Nuclear and hydroelectric power each contribute about 6% of the total energy budget. The remaining 3% derives from biomass, solar, wind, and geothermal energy-generating facilities.

◀ **Methane hydrate, a potential fuel source.** Methane, CH_4, can be trapped in a lattice of water molecules, but the methane is released when the pressure is reduced. See Figure 6 on page 287.

- With only 4.6% of the world's population, the United States consumes 25% of all the energy used in the world. This usage is equivalent to the consumption of 7 gallons of oil or 70 pounds of coal per person per day.

Two basic issues, *energy consumption* and *energy resources*, instantly leap out from these statistics. They form the basis for this discussion of energy.

Energy Consumption

Data indicate that *energy consumption* is related to the degree to which a country has industrialized. The more industrialized a country, the more energy is consumed on a per capita basis. Although some people express worries about the disproportionate use of energy by developed nations, an equally serious concern is the rate of growth of consumption worldwide. As a higher degree of industrialization occurs in developing nations, energy consumption worldwide will increase proportionally. The rapid growth in energy usage over the last half-century is strong evidence in support of predictions of similar growth in the next half-century.

One way to alter consumption is through energy conservation. Energy conservation is a small part of today's energy equation, although it has drawn greater attention recently (Figure 2). Some examples where energy conservation is already important are described here:

- Aluminum is recycled because recycling requires only one third of the energy needed to produce aluminum from its ore.

- Light-emitting diodes (LEDs) are being used in streetlights and compact fluorescent lights are finding wider use in the home. Both use a fraction of the energy required for incandescent bulbs (in which only 5% of the energy used is returned in the form of light; the remaining 95% is wasted as heat).

- Hybrid cars offer twice the gas mileage available with conventional cars.

We can be sure that energy conservation will continue to contribute to the world's energy balance sheet. Science and technology can be expected to introduce a variety of new energy-saving devices in coming years.

One of the exciting areas of current research in chemistry relating to energy conservation focuses on **superconductivity.** Superconductors are materials that, at temperatures of 90–150 K, offer virtually no resistance to electrical conductivity (see "The Chemistry of Modern Materials," page 642). When an electric current passes through a typical conductor such as a copper wire, some of the energy is inevitably lost as heat. As a result, there is substantial energy loss in power transmission lines. Substituting a superconducting wire for copper has the potential to greatly decrease this loss, so the search is on for materials that act as superconductors at moderate temperatures.

Figure 2   Energy-conserving devices. Energy efficient home appliances, hybrid automobiles, and compact fluorescent bulbs all provide alternatives that consume less energy than their conventional counterparts.

Energy Resources

On the other side of the energy balance sheet are *energy resources*, of which many exist. The data cited earlier make it obvious that we are hugely dependent on fossil fuels as a source of energy. The percentage of energy obtained and used from all other sources is small relative to that obtained from fossil fuels. We rely almost entirely on gasoline and diesel fuel in transportation. Fuel oil and natural gas are the standards for heating, and approximately 70% of the electricity in the United States is generated using fossil fuels, mostly coal (Table 1).

Why is there such a dominance of fossil fuels on the resource side of the equation? An obvious reason is that fossil fuels are cheap raw materials compared to other energy sources. In addition, humans have made an immense investment in the infrastructure needed to distribute and use this energy. Power plants using coal or natural gas cannot be converted readily to accommodate another fuel. The infrastructure for distribution of energy—gas pipelines, gasoline dispensing for cars, and the grid distributing electricity to users—is already set in place. Much of this infrastructure may have to change if the source of energy changes. Some countries already have energy distribution systems that do not depend nearly as much as the U.S. system on fossil fuels. For example, countries in Europe (such as France) make much greater use of nuclear power, and certain regions on the planet (such as Iceland and New Zealand) are able to exploit geothermal power as an energy source.

Table 1   Producing Electricity in the United States

Coal	52%
Nuclear	21%
Natural gas	12%
Renewable sources	7%
Petroleum	3%
Combining heat and power*	5%

*Cogeneration facilities using fossil fuels that yield both electricity and heat. See *Chemical and Engineering News*, p. 21, February 23, 2004.

In addition, we have become accustomed to an energy system based on fossil fuels. The internal combustion engine is the result of years of engineering. It is now well understood and can be produced in large quantities quickly and for a relatively low cost. The electric grid is well established to supply our buildings and roads. Natural gas supply to our homes is nearly invisible. The system works well.

But here is the root of the problem alluded to by Richard Smalley: Fossil fuels are nonrenewable energy sources. **Nonrenewable resources** are those in which the energy source is used and not concurrently replenished. Fossil fuels are the obvious example. Nuclear energy is also in this category (although the supply of nuclear fuels appears, for the moment, not likely to be used up in the conceivable future and breeder reactors can use other, even more abundant sources to create nuclear fuel). Conversely, energy sources that involve the sun's energy are **renewable resources.** These include solar energy and energy derived from winds, biomass, and moving water. Likewise, geothermal energy is a renewable resource.

There is a limited supply of fossil fuels. No more sources are being created. As a consequence, we must ask how long our fossil fuels will last. Regrettably, there is not an exact answer to this question. One current estimate suggests that the world's oil reserves will be depleted in 30–80 years. Natural gas and coal supplies are projected to last longer. The estimated life of natural gas reserves is 80–200 years, whereas coal reserves are projected to last from 150 to several hundred years. These numbers are highly uncertain, however. In part, this is because the estimates are based on guesses regarding fuel reserves not yet discovered; in part, it is because assumptions must be made about the rate of consumption in future years.

Despite our current state of comfort with our energy system, we cannot ignore the fact that a change away from fossil fuels must occur someday. As supply diminishes and demand increases, it will become necessary to expand the use of other fuel types. The technologies for doing so, and the answers regarding which alternative fuel types will be the most efficient and cost-effective, can be provided by chemistry research.

Fossil Fuels

Fossil fuels originate from organic matter that was trapped under the earth's surface for many millennia. Due to the particular combination of temperature, pressure, and available oxygen, the decomposition process from the basic compounds that constitute organic matter resulted in the hydrocarbons that we extract and use today: coal, crude oil, and natural gas—the solid, liquid, and gaseous forms of fossil fuels, respectively. These hydrocarbons have varying ratios of carbon to hydrogen.

Fossil fuels are simple to use and relatively inexpensive to extract, compared with the current cost requirements of other sources for the equivalent amount of energy. To use the energy stored in fossil fuels, these materials are burned. The combustion process, when it goes to completion, converts hydrocarbons to CO_2 and H_2O (Section 4.2). The heat evolved is then converted to mechanical and electrical energy (Chapter 6).

Energy output from burning fossil fuels varies among these fuels (Table 2). The heat evolved on burning is related to the carbon-to-hydrogen ratio. We can analyze this relationship by considering data on heats of formation and by looking at an example that is 100% carbon and another that is 100% hydrogen. The oxidation of 1.0 mol (12.01 g) of pure carbon produces 393.5 kJ of heat or 32.8 kJ per gram.

$$C(s) + O_2(g) \longrightarrow CO_2(g)$$
$$\Delta H° = -393.5 \text{ kJ/mol C} \quad \text{or} \quad -32.8 \text{ kJ/g C}$$

Burning hydrogen to form water is much more exothermic, with about 120 kJ evolved per gram of hydrogen consumed.

$$H_2(g) + \tfrac{1}{2} O_2(g) \longrightarrow H_2O(g)$$
$$\Delta H° = -241.8 \text{ kJ/mol H}_2 \quad \text{or} \quad -119.9 \text{ kJ/g H}_2$$

Coal is mostly carbon, so its heat output is similar to that of pure carbon. In contrast, methane is 25% hydrogen (by weight) and the higher-molecular-weight hydrocarbons in petroleum and products refined from petroleum average 16–17% hydrogen content. Therefore, their heat output on a per-gram basis is greater than that of pure carbon, but less than that of hydrogen itself.

While the basic chemical principles for extracting energy from fossil fuels are simple, complications arise in practice. Let us look at each of these fuels in turn.

Coal

The solid rock-like substance that we call coal began to form almost 290 million years ago, when swamp plants died. Decomposition occurred to a sufficient extent that the primary component of coal is carbon. Describing coal simply as carbon is a simplification, however. Samples of coal vary considerably in their composition and characteristics. Carbon content may range from 60%

to 95%, with variable amounts of hydrogen, oxygen, sulfur, and nitrogen being bound up in the coal in various forms.

Sulfur is a common constituent in some coals. The element was incorporated into the mixture partly from decaying plants and partly from hydrogen sulfide, H_2S, which is the waste product from certain bacteria. In addition, coal is likely to contain traces of many other elements, including some that are hazardous (such as arsenic, mercury, cadmium, and lead) and some that are not (such as iron).

When coal is burned, some of the impurities are dispersed into the air and some end up in the ash that is formed. In the United States, coal-fired power plants are responsible for 60% of the emissions of SO_2 and 25% of mercury emissions into the environment. SO_2 reacts with water and O_2 in the atmosphere to form sulfuric acid, which contributes (along with nitric acid) to the phenomenon known as acid rain.

$$2\,SO_2(g) + O_2(g) \longrightarrow 2\,SO_3(g)$$
$$SO_3(g) + H_2O(\ell) \longrightarrow H_2SO_4(aq)$$

Because these acids are harmful to the environment, legislation limits the extent of sulfur oxide emissions from coal-fired plants. Chemical scrubbers have been developed that can be attached to the smokestacks of power plants to reduce sulfur-based emissions. However, these devices are expensive and can increase the cost of the energy produced from these facilities.

Coal is classified into three categories (Table 3). Anthracite, or hard coal, is the highest-quality coal. Among the forms of coal, anthracite has the highest heat content per gram and a low sulfur content. Unfortunately, anthracite coal is fairly uncommon, with only 2% of the U.S. coal reserves occurring in this form (Figure 3). Bituminous coal, also referred to as soft coal, accounts for about 45% of the U.S. coal reserves and is the coal most widely used in electric power generation. Soft coal typically has the highest sulfur content. Lignite, also called brown coal because of its paler color, is geologically the "youngest" form of coal. It has a lower heat content than the other forms of coal, often contains a significant amount of water, and is the least popular as a fuel.

Table 2   Energy Released by Combustion of Fossil Fuels

Substance	Energy Released (kJ/g)
Coal	29–37
Crude petroleum	43
Gasoline (refined petroleum)	47
Natural gas (methane)	50

Table 3   Types of Coal

Type	Consistency	Sulfur Content	Heat Content (kJ/g)
Lignite	Very soft	Very low	28–30
Bituminous coal	Soft	High	29–37
Anthracite	Hard	Low	36–37

Figure 3 Anthracite coal. This form of coal has the highest energy content of the various forms of coal.

Coal can be converted to coke by heating in the absence of air. Coke is almost pure carbon and an excellent fuel. In the process of coke formation, a variety of organic compounds are driven off. These compounds are used as raw materials in the chemical industry for the production of polymers, pharmaceuticals, synthetic fabrics, waxes, tar, and numerous other products.

Technology to convert coal into gaseous fuels (*coal gasification*) (Figure 4) or liquid fuels (*liquefaction*) has also been developed. These processes provide fuels that will burn more cleanly than coal, albeit with a loss of 30–40% of the net energy content per gram of coal along the way. As petroleum and natural gas reserves dwindle, and the costs of these fuels increase, liquid and gaseous fuels derived from coal are likely to become more important.

Natural Gas

Natural gas is found deep under the earth's surface, where it was formed by bacteria working on organic matter in an anaerobic environment (in which no oxygen is present). The major component of natural gas (70–95%) is methane (CH_4). Lesser quantities of other gases such as ethane (C_2H_6), propane (C_3H_8), and butane (C_4H_{10}) are also present, along with other gases including N_2, He, CO_2, and H_2S. The impurities and higher-molecular-weight components of natural gas are separated out during the refining process, so that the gas piped through gas mains into our homes is primarily methane.

Natural gas is an increasingly popular choice as a fuel. It burns more cleanly than the other fossil fuels, emits fewer pollutants, and produces relatively more energy than the other fossil fuels. Natural gas can be transported by pipelines over land and piped into buildings such as your home to be used directly to heat ambient air, to heat water for washing and bathing, or for cooking.

Petroleum

Petroleum is a complicated mixture of hydrocarbons, whose molar masses range from low to very high (page 495). The hydrocarbons may have anywhere from one carbon atom to 20 or more such atoms in their structures, and compounds containing sulfur, nitrogen, and oxygen may also be present in small amounts.

Petroleum goes through extensive processing at refineries to separate the various components and convert less valuable compounds into more valuable components. Nearly 85% of the crude petroleum pumped from the ground ends up being used as a fuel, either for transportation (gasoline and diesel fuel) or for heating (fuel oils).

The high temperature and pressure used in the combustion process in automobile engines have the unfortunate consequence of also causing a reaction between atmospheric nitrogen and oxygen that results in some NO formation. In a series of exothermic reactions, the NO can then react further with oxygen to produce nitrogen dioxide. This poisonous, brown gas is further oxidized to form nitric acid, HNO_3, in the presence of water.

$$N_2(g) + O_2(g) \longrightarrow 2\ NO(g) \qquad \Delta H^\circ_{rxn} = 180.58\ \text{kJ}$$
$$2\ NO(g) + O_2(g) \longrightarrow 2\ NO_2(g) \qquad \Delta H^\circ_{rxn} = -114.4\ \text{kJ}$$
$$3\ NO_2(g) + H_2O(\ell) \longrightarrow 2\ HNO_3(\ell) + NO(g) \qquad \Delta H^\circ_{rxn} = -71.4\ \text{kJ}$$

To some extent, the amounts of pollutants released can be limited by use of catalytic converters. Catalytic converters are high-surface-area metal grids that are coated with platinum or palladium. These very expensive metals can catalyze a complete combustion reaction, helping to combine oxygen in the air with unburned hydrocarbons or other byproducts in the vehicle exhaust. As a result, the combustion products can be converted to

Figure 4 Coal gasification plant. Advanced coal-fired power plants, such as this 2544-ton-per-day coal gasification demonstration pilot plant, will have energy conversion efficiencies 20% to 35% higher than those of conventional pulverized-coal steam power plants.

water and carbon dioxide (or other oxides), provided they land on the grid of the catalytic converter before exiting the vehicle's tailpipe. Some nitric acid and NO_2 inevitably remain in automobile exhaust, however, and they are major contributors to environmental pollution in the form of acid rain and smog. The brown, acidic atmospheres in highly congested cities such as Los Angeles, Mexico City, and Houston largely result from the emissions from automobiles (Figure 5). The pollution problems have led to stricter emission standards for automobiles, and a high priority in the automobile industry is the development of low-emission or emission-free vehicles. Another approach is provided by the increasing popularity of hybrid vehicles, which use a combination of gasoline and electricity to run, thereby reducing the gasoline consumption per mile.

Other Fossil Fuel Sources

When natural gas pipelines were laid across the United States and Canada, pipeline operators soon found that, unless water was carefully kept out of the line, chunks of methane hydrate would form and clog the pipes. Methane hydrate was a completely unexpected substance because it is made up of methane and water, two chemicals that would appear to have little affinity for each other. In methane hydrate, methane becomes trapped in cavities in the molecular structure of ice (Figure 6). Methane hydrate is stable only at temperatures below the freezing point of water. If a sample of methane hydrate is warmed above 0°C, it melts and methane is released. The volume of gas released (at

Figure 5 Smog. The brown cloud that hangs over Santiago, Chile contains nitrogen oxides emitted by millions of automobiles in that city. Other compounds are also present, such as ozone (O_3), nitric oxide (NO_2), carbon monoxide (CO), and water.

normal pressure and temperature) is about 165 times larger than the volume of the hydrate.

If methane hydrate forms in a pipeline, is it found in nature as well? In May 1970, oceanographers drilling into the seabed off the coast of South Carolina pulled up samples of a whitish solid that fizzed and oozed when it was removed from the drill casing. They quickly realized it was methane hydrate. Since this original

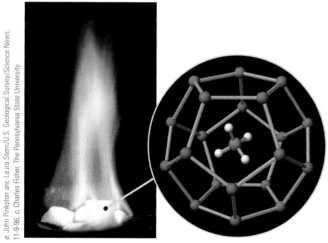

(a) Methane hydrate burns as methane gas escapes from the solid hydrate.

(b) Methane hydrate consists of a lattice of water molecules with methane molecules trapped in the cavity.

(c) A colony of worms on an outcropping of methane hydrate in the Gulf of Mexico.

Figure 6 Methane hydrate. (a) This interesting substance is found in huge deposits hundreds of feet down on the floor of the ocean. When a sample is brought to the surface, the methane oozes out of the solid, and the gas readily burns. (b) The structure of the solid hydrate consists of methane molecules trapped within a lattice of water molecules. Each point of the lattice shown here is an oxygen atom of a water molecule. The edges are O—H—O bonds. Such structures are often called "clathrates" and are mined for substances other than methane. (c) An outcropping of methane hydrate on the floor of the Gulf of Mexico. See E. Suess, G. Bohrmann, J. Greinert, and E. Lausch: *Scientific American*, pp. 76–83, November 1999.

discovery, methane hydrate has been found in many parts of the oceans as well as under permafrost in the Arctic. It is estimated that 1.5×10^{13} tons of methane hydrate is buried under the sea floor around the world. In fact, the energy content of this gas may surpass that of all the other known fossil fuel reserves by as much as a factor of 2! Clearly, this is a potential source of an important fuel in the future. Today, however, the technology to extract methane from these hydrate deposits is very expensive, especially in comparison to the well-developed technologies used to extract crude oil, coal, and gaseous methane.

There are other sources of methane in our environment. For example, methane is generated in swamps, where it is called *swamp gas* or *marsh gas*. Here, methane is formed by bacteria working on organic matter in an anaerobic environment—namely, sedimentary layers of coastal waters and in marshes. The process of formation is similar to the processes occurring eons ago that generated the natural gas deposits that we currently use for fuel. In a marsh, the gas can escape if the sediment layer is thin. You see it as bubbles rising to the surface. Unfortunately, because of the relatively small amounts generated, it is impractical to collect and use this gas as a fuel.

In a striking analogy to what occurs in nature, the formation of methane also occurs in human-made landfill sites. A great deal of organic matter is buried in landfills. Because it remains out of contact with oxygen in the air, this material is degraded by bacteria. In the past, landfill gases have been deemed a nuisance. Today, it is possible to collect this methane and use it as a fuel. In a pilot plant at the Rodefeld Landfill site near Madison, Wisconsin, a collection system for the methane produced in the landfill has been set up. The gas is used to generate electricity that is sold back to the local electric utility. In 2002, the methane gas collected at this facility was used to produce approximately 12 million kilowatt-hours of electricity, enough to power about 1700 homes for a year.

Energy in the Future: Choices and Alternatives

Fuel Cells

To generate electricity from the combustion of fossil fuels, the energy is used to create high-pressure steam, which spins a turbine in a generator. Unfortunately, not all of the energy from combustion can be converted to usable work. Some of the energy stored in the chemical bonds of a fuel is lost as heat to the surroundings, making this an inherently inefficient process. A much more efficient process would be possible if mobile electrons, the carriers of electricity, could be generated directly from the chemical bonds themselves, rather than going through an energy conversion process from heat to mechanical work to electricity.

Fuel cell technology makes direct conversion of chemical potential energy to electricity possible. Fuel cells are similar to batteries, except that fuel is supplied from an external source (Figure 7 and Section 20.3). They are more efficient than com-

bustion-based energy production, with up to 60% energy conversion efficiency compared to 20–25% for electricity generation from combustion.

Fuel cells are not a new discovery. In fact, the first fuel cell was demonstrated in 1839, and fuel cells have been used in the Space Shuttle. Fuel cells are currently under investigation for use in homes and in automobiles.

The basic design of fuel cells is quite simple. Oxidation and reduction take place in two separate compartments. [Recall the definitions of oxidation and reduction (page 197): Oxidation is the loss of electrons from a species, whereas reduction occurs when a species gains electrons.] These compartments are connected in a way that allows electrons to flow from the oxidation compartment to the reduction compartment through a conductor such as a wire. In one compartment, a fuel is oxidized, producing positive ions and electrons. The electrons move to the other compartment, where they react with an oxidizing agent, typically O_2. The spontaneous flow of electrons in the electrical circuit constitutes the electric current. While electrons flow through the external circuit, ions move between the two compartments so that the charges in each compartment remain in balance.

The net reaction is the oxidation of the fuel and the consumption of the oxidizing agent. Because the fuel and the oxidant never come directly in contact with each other, there is no combustion and no loss of energy as heat. The energy of the reaction is converted directly into electricity.

Hydrogen is the fuel employed in the fuel cells on board the Space Shuttle. The overall reaction in these fuel cells involves the combination of hydrogen and oxygen to form water (Figure 7). Hydrocarbon-based fuels such as methane (CH_4) and methanol

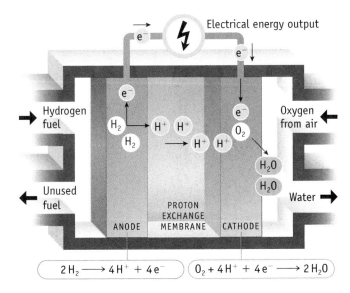

$$2\,H_2 \longrightarrow 4\,H^+ + 4\,e^- \qquad O_2 + 4\,H^+ + 4\,e^- \longrightarrow 2\,H_2O$$

Figure 7   Hydrogen-oxygen fuel cell. The cell uses hydrogen gas, which is converted to hydrogen ions and produces electrons. The electrons flow through the external circuit and are consumed by the oxygen, which, along with H^+ ions, produces water.

(CH₃OH) are also candidates for use as the fuel in fuel cells; for these compounds the reaction products are CO_2 and H_2O. When methanol is used in fuel cells, for example, the net reaction in the cell is

$$2\ CH_3OH(\ell) + 3\ O_2(g) \longrightarrow 2\ CO_2(g) + 4\ H_2O(\ell)$$
$$\Delta H^\circ_{rxn} = -727 \text{ kJ/mol } CH_3OH \quad \text{or} \quad -23 \text{ kJ/g } CH_3OH$$

Using heat of formation data (Section 6.8), we can calculate that the energy generated is 727 kJ/mol (or 23 kJ/g) of liquid methanol. That is equivalent to 200 watt-hours (W-h) of energy per mol of methanol (1 W = 1 J/s), or 5.0 kW-h per liter of methanol. This means that oxidation of one liter of methanol in a fuel cell could theoretically provide more than 5000 W of power over a 24-hour period, enough to keep about 70 standard desk lamps lit.

Prototypes of phones and laptop computers powered by fuel cells have been developed recently. Small methanol cartridges are used to fuel them. These devices are no bigger than a standard AA battery, yet they last up to 10 times longer than standard rechargeable batteries.

Note, however, that fuel cells do not provide a new source of energy. They require fuel to produce energy and are constructed to use currently available fuels. The merits of fuel cells derive from their greater efficiency of use and from their environmentally friendly nature.

A Hydrogen Economy

Predictions about the diminished supply of fossil fuels have led some people to speculate about other alternative fuels. In particular, hydrogen, H_2, has been suggested as a possible choice. The term *hydrogen economy* has been coined to describe the overall strategy using this fuel. As was the case with fuel cells, the hydrogen economy does not rely on a new energy resource; it merely provides a different scheme for use of existing resources.

There are reasons to consider hydrogen an attractive option, however. Oxidation of hydrogen yields almost three times as much energy per gram as the oxidation of fossil fuels. Comparing hydrogen combustion with combustion of propane, a fuel used in some cars, we find that H_2 produces about 2.6 times more heat per gram than propane.

$$H_2(g) + \tfrac{1}{2}\ O_2(g) \longrightarrow H_2O(g)$$
$$\Delta H^\circ_{rxn} = -241.83 \text{ kJ/mol } H_2 \quad \text{or} \quad -119.95 \text{ kJ/g } H_2$$

$$C_3H_8(g) + 5\ O_2(g) \longrightarrow 3\ CO_2(g) + 4\ H_2O(g)$$
$$\Delta H^\circ_{rxn} = -2043.15 \text{ kJ/mol } C_3H_8 \quad \text{or} \quad -46.37 \text{ kJ/g } C_3H_8$$

Another advantage of using hydrogen instead of a hydrocarbon fuel is that the only product of H_2 oxidation is H_2O, which is environmentally benign.

Thus, for several reasons it is relatively easy to imagine hydrogen replacing gasoline in automobiles and replacing natural gas in heating homes. It is similarly easy to imagine using hydrogen to generate electricity or as a fuel for industry.

Of course, there are many practical problems, including the following as-yet-unmet needs:

- An inexpensive method of producing hydrogen
- A practical means of storing hydrogen
- A distribution system (hydrogen refueling stations)

Perhaps the most serious problem in the hydrogen economy is the task of producing hydrogen. Hydrogen is abundant on earth, but not as the free element. Thus, elemental hydrogen has to be obtained from its compounds. Currently, most hydrogen is produced industrially from the reaction of natural gas and water by *steam-reforming* at high temperature (Figure 8).

Steam re-forming $CH_4(g) + H_2O(g) \longrightarrow 3H_2(g) + CO(g)$
$$\Delta H^\circ_{rxn} = +206.2 \text{ kJ/mol } CH_4$$

Hydrogen can also be obtained from the reaction of coal and water at high temperature (*water gas reaction*).

Water gas reaction $C(s) + H_2O(g) \longrightarrow H_2(g) + CO(g)$
$$\Delta H^\circ_{rxn} = +131.3 \text{ kJ/mol } C$$

Both reactions are highly endothermic, however, and both rely on use of a fossil fuel as a raw material. This, of course, makes no sense if the overriding goal is to replace fossil fuels.

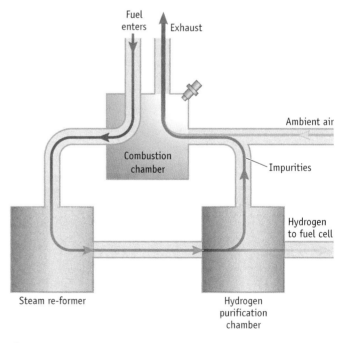

Figure 8 Steam re-forming. A fuel such as methanol (CH₃OH) or a hydrocarbon and water are heated and then passed into a steam re-former chamber. There a catalyst promotes their decomposition to hydrogen and other compounds such as CO. The hydrogen gas passes out to a fuel cell, and the CO and unused carbon-based compounds are burned in a combustion chamber. A small unit may be suitable for a car or light truck.

If the hydrogen economy is ever to take hold, the logical source of hydrogen is water.

$$H_2O(\ell) \longrightarrow H_2(g) + \tfrac{1}{2} O_2(g)$$
$$\Delta H^\circ_{rxn} = +285.83 \text{ kJ/mol } H_2O(\ell)$$

The electrolysis of water provides hydrogen but also requires considerable energy. The first law of thermodynamics tells us that we can get no more energy from the oxidation of hydrogen than we expended to obtain H_2 from H_2O. Hence, the only way to obtain hydrogen in the amounts that would be needed is to use a cheap and abundant source of energy to drive this process. A logical candidate is solar energy. Unfortunately, the technology to use solar energy in this way has yet to become practical. Here is a problem for chemists and engineers of the future to solve.

Hydrogen storage represents another problem to be solved. A number of ways to accomplish this storage in a vehicle, in your home, or at a distribution point have been proposed. An obvious way to store hydrogen is as the gas under moderate conditions, but this approach would be impractical because the volumes occupied would be too large (Figure 9). In addition, storing hydrogen at high pressure or as a liquid (bp = -252.87 °C) would require special equipment, and safety is a key issue.

One possibility known to chemists relies on the fact that certain metals will absorb hydrogen reversibly (Figure 10). When a metal absorbs hydrogen, H atoms fill the holes, called *interstices*, between metal atoms in a metallic crystal lattice. Palladium, for example, will absorb up to 935 times its volume of hydrogen. This hydrogen can be released upon heating, and the process of absorption and release can be repeated.

Another reversible system under study involves hydrogen storage in carbon nanotubes (page 31). Researchers have found that the carbon tubes absorb 4.2 weight percent of H_2; that is, they achieve an H : C atom ratio of 0.52 under a moderately high

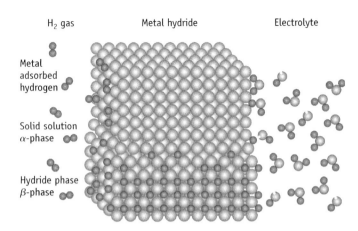

Figure 10   Hydrogen adsorbed onto a metal or metal alloy. Many metals and metal alloys reversibly absorb large quantities of hydrogen. On the left side of the diagram, H_2 molecules are adsorbed onto the surface of a metal. The H_2 molecules can dissociate into H atoms, which form a solid solution with the metal (α-phase). Under higher hydrogen pressures, a true hydride forms in which H atoms become H^- ions (β-phase). On the right side, H atoms can also be adsorbed from solution if the metal is used as an electrode in an electrochemical device.

pressure. Just as importantly, 78.3% of the hydrogen can be released under ambient pressure at room temperature, and the remainder can be released with heating.

There are several chemical methods of reversible hydrogen storage as well. For example, heating $NaAlH_4$, doped with titanium dioxide, releases hydrogen and the $NaAlH_4$ can be rejuvenated by adding hydrogen under pressure.

$$2 \text{ NaAlH}_4(s) \longrightarrow 2 \text{ NaH} + 2 \text{ Al}(s) + 3 \text{ H}_2(g)$$

No matter how hydrogen is used, it has to be delivered to vehicles and homes in a safe and practical manner. Work has also been done in this area (Figure 11), but many problems remain to be solved. European researchers have found that a tanker truck that can deliver 2400 kg of compressed natural gas (mostly methane) can deliver only 288 kg of H_2 at the same pressure. Although hydrogen oxidation delivers about 2.4 times more energy per gram (119.95 kJ/g) than methane,

$$CH_4(g) + 2 O_2(g) \longrightarrow CO_2(g) + 2 H_2O(g)$$
$$\Delta H^\circ_{rxn} = -802.30 \text{ kJ/mol} \quad \text{or} \quad -50.14 \text{ kJ/g}$$

the tanker can carry about 8 times more methane than H_2. That is, it will take more tanker trucks to deliver the hydrogen needed to power the same number of cars or homes running on hydrogen than those running on methane.

How close are we to the realization of a hydrogen economy? Not very near, and it is not clear whether it will ever come to pass.

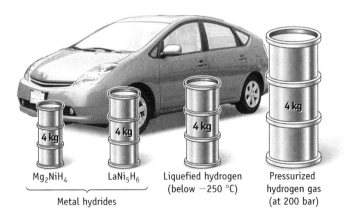

Figure 9   Comparison of the volumes required to store 4 kg of hydrogen relative to the size of a typical car. (L. Schlapbach and A. Züttel: *Nature*, Vol. 414, pp. 353–358, 2001.)

Figure 11   A prototype hydrogen-powered BMW. The car is being refueled with hydrogen at a distribution center in Germany. Note the solar panels in the background.

There is one interesting example in which the hydrogen economy has gained a real toehold. In 2001, Iceland announced that the country would become a "carbon-free economy." Icelanders plan to rely on hydrogen-powered electric fuel cells to run vehicles and fishing boats. Iceland is fortunate in that two thirds of its energy already comes from renewable sources—hydroelectric and geothermal energy (Figure 12). The country has decided to use the electricity produced by geothermal heat or hydroelectric power to separate water into hydrogen and oxygen. The hydrogen will then be used in fuel cells or combined with CO_2 to make methanol, CH_3OH, a liquid fuel that can either be burned or be used in different types of fuel cells.

Biosources of Energy

Gasoline sold today often contains ethanol, C_2H_5OH. In addition to being a fuel, ethanol serves to improve the burning characteristics of gasoline.

Ethanol is readily made by fermentation of glucose from renewable resources such as corn or sugar cane. While it may not emerge as the sole fuel of the future, this material is likely to contribute to the phasing-out process of fossil fuels and may be one of multiple fuel sources in the future.

There are several interesting points to make about ethanol as a fuel. Green plants use the sun's energy to create biomass from CO_2 and H_2O by photosynthesis. The sun is a renewable resource, as, in principle, is the ethanol derived from biomass. In addition, the process recycles CO_2. Plants use CO_2 to create biomass, which is in turn used to make ethanol. In the final step in this cycle, oxidation of ethanol returns CO_2 to the atmosphere.

Recent research on ethanol has taken this topic in a new direction. Namely, ethanol can be used as a source of hydrogen. It is possible to create hydrogen gas from ethanol by using a steam re-forming process like the methane-related process. The recently developed method involves the partial oxidation of ethanol mixed with water in a small fuel injector, like those used in cars to deliver gasoline, along with rhodium and cerium catalysts to create hydrogen gas exothermically (Figure 13). The net reaction is

$$C_2H_5OH(g) + 2\ H_2O(g) + \tfrac{1}{2}\ O_2(g) \longrightarrow 2\ CO_2(g) + 5\ H_2(g)$$

The heat of this reaction is approximately -70 kJ per mole of ethanol (or about 1.5 kJ/g).

Figure 12   Iceland, a "carbon-free," hydrogen-based economy. A geothermal field in Iceland. The country plans to use such renewable resources to produce hydrogen from water and then to use the hydrogen to produce electricity in fuel cells.

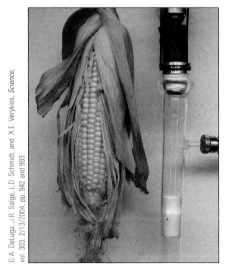

Figure 13   Hydrogen from ethanol. Ethanol can be obtained by fermentation from corn. In a prototype reactor (right), ethanol, water, and oxygen are converted by a catalyst (glowing white solid) to hydrogen (and CO_2).

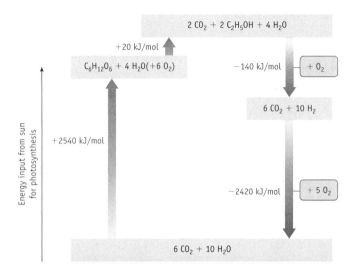

Figure 14   An energy-level diagram for the reactions leading from the production of biomass (glucose) to CO_2 and H_2. (Based on a Figure in G. A. DeLuga, J. R. Salge, L. D. Schmidt, and X. E. Verykios: *Science*, Vol. 303, pp. 942 and 993, 2004).

To examine the efficiency of this process, we must analyze the overall energy cycle, starting with the photosynthesis of CO_2 and water to generate glucose (Figure 14). The sun provides the initial 2540 kJ input of energy for this cycle to produce 1 mol of glucose ($C_6H_{12}O_6$). The sugar is then converted 2 mol of ethanol per 1 mol of sugar. This conversion process requires a small energy input, 20 kJ. At this point, hydrogen can be generated exothermically using the catalytic fuel-injector method described earlier. Once the hydrogen is generated, it can be used in a hydrogen fuel cell to produce energy and water.

Solar Energy

Every year the earth's surface receives about 10 times as much energy from sunlight as is contained in all the known reserves of coal, oil, natural gas, and uranium combined! The amount of solar energy incident on the earth's surface is equivalent to about 15,000 times the world's annual consumption of energy. Although solar energy is a renewable resource, today we are making very inefficient use of the sun's energy. Less than 2% of the electricity produced in the United States is generated using solar energy.

How might the sun's energy be exploited more efficiently? One strategy is to produce electricity using solar radiation. We already know how to do this. The direct conversion of solar energy to electricity can be carried out using photovoltaic cells (see "The Chemistry of Modern Materials," page 648). These devices are made from specific metal and metalloid combinations (often gallium and arsenic) that absorb light from the sun and produce an electric current. They are now used in applications as diverse as spacecraft and pocket calculators, and they have also been tested for large-scale commercial use.

Before solar energy can be a viable alternative, a number of issues need to be addressed, including the collection, storage, and transmission of energy. Furthermore, electricity generated from solar power stations is intermittent. (The output fluctuation results from the normal cycles of daylight and changing weather conditions.) Our current power grid cannot handle intermittent energy, so solar energy would need to be stored in some way and then doled out at a steady rate.

Likewise, we need to find ways to make solar cells cost-effective. Research has produced photovoltaic cells that can convert 20–30% of the energy that falls on them. However, even higher efficiency is necessary to offset the high cost of making the devices. Currently, 1 kW-h of energy generated from solar cells costs about 35 cents, compared to about 2 cents per kW-h generated from fossil fuels.

What Does the Future Hold for Energy?

Our society is at an energy crossroads. The modern world is increasingly reliant on energy, but we have built an energy infrastructure that depends primarily on a type of fuel that is limited. While fossil fuels provide an inexpensive and simple approach for providing power, they have several drawbacks, among them atmospheric contamination and diminishing supplies.

Alternative fuels, especially from renewable sources, and new ways of generating energy do exist. A great deal more research and resources must be put into them to make them affordable and reliable, however. This is where the study of chemistry fits squarely into the picture. Chemists will have a great deal of work to do in coming years to develop new means of generating and delivering energy. Meanwhile, numerous ways exist to conserve the resources we have. Ultimately, it will be necessary to bear in mind the various benefits and drawbacks of each technology so that they can be combined in the most rational ways, rather than remaining in a system that is dependent on a single form of energy.

Suggested Readings

1. R. A. Hinrichs and M. Kleinbach: *Energy—Its Use and the Environment*, 3rd ed. Orlando, Harcourt, 2002.

2. M. L. Wald: "Questions About a Hydrogen Economy," *Scientific American*, pp. 67–73, May 2004.

3. U.S. Department of Energy: Energy Efficiency and Renewable Energy, **www.eere.energy.gov/hydrogenandfuelcells**. Accessed May 2004.

4. G. T. Miller: *Living in the Environment*, 12th ed. Philadelphia, Brooks/Cole, 2001.

5. L. D. Burns, J. B. McCormick, and C. E. Borroni-Bird: "Vehicle of Change," *Scientific American*, pp. 64–73, October 2002.

6. M. S. Dresselhaus and I. L. Thomas: "Alternative Energy Technologies," *Nature*, Vol. 414, pp. 332–337, November 15, 2001.

Study Questions

Blue numbered questions have answers in Appendix P and fully worked solutions in the *Student Solutions Manual.*

1. Hydrogen can be produced using the reaction of steam (H_2O) with various hydrocarbons. Compare the mass of H_2 expected from the reaction of steam with 100. g each of methane, petroleum, and coal. (Assume complete reaction in each case. Use CH_2 and C as the representative formulas for petroleum and coal, respectively.)

2. Use the value for "energy released" in kilojoules per gram from gasoline in Table 2. Estimate the percentage of carbon, by weight, by comparing this value to the $\Delta H°$ values for burning pure C and H_2.

3. Per capita energy consumption in the United States was equated to the energy obtained by burning 70. lb of coal per day. Use enthalpy of formation data to calculate the energy evolved, in kilojoules, when 70 lb of coal is burned. (Assume the heat of combustion of coal is 33 kJ/g.)

4. Some gasoline contains 10% (by volume) ethanol. Using enthalpy of formation data in Appendix L, calculate the heat evolved from the combustion of 1.00 g of ethanol to $CO_2(g)$ and $H_2O(g)$. Compare this value to the heat evolved from the combustion of ethane to the same products. Why should you expect that the energy evolved in the combustion of ethanol is less than the energy evolved in the combustion of ethane?

5. Energy consumption in the United States amounts to the equivalent of the energy obtained by burning 7.0 gal of oil or 70. lb of coal per day per person. Carry out calculations to show that these energy quantities are approximately equivalent using data in Table 2. The density of fuel oil is approximately 0.8 g/mL.

6. The energy required to recycle aluminum is one third of the energy required to prepare aluminum from Al_2O_3 (bauxite). Calculate the energy required to recycle 1.0 lb (= 454 g) of aluminum.

7. The heat of combustion of isooctane (C_8H_{18}) is 5.45×10^3 kJ/mol. Calculate the heat evolved per gram of isooctane and per liter of isooctane ($d = 0.688$ g/mL). (Isooctane is one of the many hydrocarbons in gasoline, and its heat of combustion will approximate the energy obtained when gasoline burns.)

Isooctane
C_8H_{18}

8. Calculate the energy used, in kilojoules, to power a 100-W lightbulb continuously over a 24-h period. How much coal would have to be burned to provide this quantity of energy, assuming that the heat of combustion of coal is 33. kJ/g? [Electrical energy for home use is measured in kilowatt-hours (kW-h). One watt is defined as 1 J/s, so 1 kW-h is the quantity of energy involved when 1000 W is dispensed over a 1.0-h period.]

9. Major home appliances purchased in the United States are now labeled (with bright yellow "Energy Guide" tags) showing anticipated energy consumption. The tag on a recently purchased washing machine indicated the anticipated energy use would be 940 kW-h per year. Calculate the anticipated annual energy use in kilojoules. (See Question 8 for a definition of kilowatt-hour.)

10. Define the terms *renewable* and *nonrenewable* as applied to energy resources. Which of the following energy resources are renewable: solar energy, coal, natural gas, geothermal energy, wind power?

11. Confirm the statement in the text that oxidation of 1.0 L of methanol to form $CO_2(g)$ and $H_2O(\ell)$ in a fuel cell will provide at least 5.0 kW-h of energy. (The density of methanol is 0.787 g/mL.)

12. List the following substances in order of energy content per gram: C_8H_{18}, H_2, $C(s)$, CH_4. (See Question 7 for the heat of combustion of C_8H_{18}.)

13. A parking lot in Los Angeles, California, receives an average of 2.6×10^7 J/m^2 of solar energy per day in the summer. If the parking lot is 325 m long and 50.0 m wide, what is the total quantity of energy striking the area per day?

14. Your home loses heat in the winter through doors, windows, and any poorly insulated walls. A sliding glass door (6 ft × 6.5 ft with 0.5 in. of insulating glass) allows 1.0×10^6 J/h to pass through the glass if the inside temperature is 22 °C (72 °F) and the outside temperature is 0 °C (32 °F). What quantity of heat, expressed in kilojoules, is lost per day? Assume that your house is heated by electricity. How many kilowatt-hours of energy are lost per day through the door? (See Question 8.)

15. Palladium metal can absorb up to 935 times its volume in hydrogen, H_2. Assuming that 1.0 cm^3 of Pd metal can absorb 0.084 g of the gas, what is the approximate formula of the substance? (The α-form of hydrogen-saturated palladium has about the same density as palladium metal, 12.0 g/cm^3.)

16. Microwave ovens are highly efficient, compared to other means of cooking. A 1100 watt microwave oven, running at full power for 90 sec will raise the temperature of 1 cup of water (225 mL) from 20 °C to 67 °C. As a rough measure of the efficiency of the microwave oven, compare its energy consumption with the heat required to raise the water temperature.

17. New fuel-efficient hybrid cars are rated at 55.0 miles per gallon of gasoline. Calculate the energy consumed to drive 1.00 mile if gasoline produces 48.0 kJ/g and the density of gasoline is 0.737 g/cm^3.

7—Atomic Structure

A fireworks display by the famous Grucci Company of New York.

Colors in the Sky

The discovery of black powder, the predecessor of gunpowder, occurred well before 1000 A.D., most likely in China. It was not until the Middle Ages, however, that black powder was known in the Western world. In 1252, Roger Bacon in England described the preparation of black powder from "saltpetre [potassium nitrate], young willow, and sulfur," and its use by the military and for fireworks spread to the European continent. By the time of the American Revolution, fireworks formulations and manufacturing methods had been worked out that are still in use today.

Typical fireworks have several important chemical components. For example, there must be an oxidizer. Today this is usually potassium perchlorate ($KClO_4$), potassium chlorate ($KClO_3$), or potassium nitrate (KNO_3). Potassium salts are used instead of sodium salts because the latter have two important drawbacks. They are hygroscopic—they absorb water from the air—and so do not remain dry on storage. Also, when heated, sodium salts give off an intense, yellow light that is so bright it can mask other colors.

The parts of any fireworks display we remember best are the vivid colors and brilliant flashes. White light can be produced by oxidizing magnesium or aluminum metal at high temperatures. The flashes you see at rock concerts or similar events, for example, are typically $Mg/KClO_4$ mixtures.

Yellow light is easiest to produce because sodium salts give an intense light with a wavelength of 589 nm. Fireworks mixtures usually contain sodium in the form of nonhygroscopic compounds such as cryolite, Na_3AlF_6. Strontium salts are most often used to produce a red light, and green is produced by barium salts such as $Ba(NO_3)_2$.

Chapter Goals

See Chapter Goals Revisited (page 324). Test your knowledge of these goals by taking the exam-prep quiz on the General ChemistryNow CD-ROM or website.

- Describe the properties of electromagnetic radiation.
- Understand the origin of light from excited atoms and its relationship to atomic structure.
- Describe the experimental evidence for wave-particle duality.
- Describe the basic ideas of quantum mechanics.
- Define the three quantum numbers (n, ℓ, and m_ℓ) and their relationship to atomic structure.

Chapter Outline

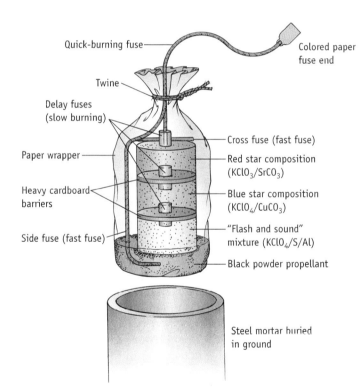

The design of an aerial rocket for a fireworks display. When the fuse is ignited, it burns quickly to the delay fuse at the top of the red star mixture as well as to the black powder propellant at the bottom. The propellant ignites, sending the shell into the air. Meanwhile, the delay fuses burn. If the timing is correct, the shell bursts high in the sky into a red star. This is followed by a blue burst and then a flash and sound.

The next time you see a fireworks display, watch for the ones that are blue. Blue has always been the most difficult color to produce. Recently, however, fireworks designers have learned that the best way to get a really good "blue" is to decompose copper(I) chloride at low temperatures. To achieve this effect, CuCl is mixed with $KClO_4$, copper powder, and the organic chlorine-containing compound hexachloroethane, C_2Cl_6.

Why are chemists—and many others—interested in fireworks? Because their colors arise from energetically excited atoms and molecules. The way that atoms can produce colored light provides insight into the structure of the atom, the subject of this chapter.

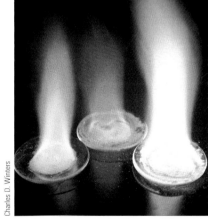

Charles D. Winters

Colors in fireworks. Fireworks displays are usually colored. Here the white, solid salts sodium chloride (*left*), strontium chloride (*center*), and boric acid (*right*) were soaked in methanol and then the methanol was ignited. The compounds were entrained in the burning fuel, and the energy from the combustion excited the atoms. The colors you observe are characteristic of sodium, strontium, and boron. (See *General ChemistryNow Screen 7.1 Chemical Puzzler*, for a description of colors in fireworks.)

To Review Before You Begin

- Review metric units of measurements (Chapter 1)
- Review the structure of the atom (Section 2.1)

Chemical elements that exhibit similar properties are found in the same column of the periodic table. But why should there be similarities among elements? The discovery of the electron, proton, and neutron [◄ Section 2.1] prompted scientists to look for relationships between atomic structure and chemical behavior. As early as 1902 Gilbert N. Lewis (1875–1946) suggested the idea that electrons in atoms might be arranged in shells, starting close to the nucleus and building outward. Lewis explained the similarity of chemical properties for elements in a given group by assuming that all the elements of that group have the same number of electrons in the outer shell.

Lewis's model of the atom raises a number of questions. Where are the electrons located? Do they have different energies? What experimental evidence supports this model? These questions were the reason for many of the experimental and theoretical studies that began around 1900 and continue to this day. This chapter and the next one outline the current theories of electronic structure.

7.1—Electromagnetic Radiation

You are familiar with water waves, and you may also know that some properties of radiation such as visible light and radio waves can be explained by wave motion. Our understanding of light as waves came from the experiments of physicists in the 19th century, among them a Scot, James Clerk Maxwell (1831–1879). In 1864, he developed an elegant mathematical theory to describe all forms of radiation in terms of oscillating, or wave-like, electric and magnetic fields (Figure 7.1). Hence, radiation, such as light, microwaves, television and radio signals, and x-rays, is collectively called **electromagnetic radiation**.

Wave Properties

The distance between successive crests or high points of a wave (or between successive troughs or low points) is the **wavelength** of a wave. This distance can be given in meters, nanometers, or whatever unit of length is convenient. The symbol for wavelength is the Greek letter λ (lambda).

Waves are also characterized by their **frequency**, symbolized by the Greek letter ν (nu). For any wave motion—whether water waves or electromagnetic radiation—the frequency is the number of waves that pass a given point in some unit of time, usually per second (Figure 7.1). The unit for frequency is often written s^{-1}, which stands for 1 per second, $1/s$, and is now called the **hertz**.

If you enjoy water sports, you are familiar with the height of waves. In more scientific terms, the maximum height of a wave is its **amplitude**. In Figure 7.1, notice that the wave has zero amplitude at certain intervals along the wave. Points of zero amplitude, called **nodes**, occur at intervals of $\lambda/2$.

Finally, the speed of a moving wave is an important factor. As an analogy, consider cars in a traffic jam traveling bumper to bumper. If each car is 5 m long, and if a car passes you every 4 s (that is, the frequency is 1 per 4 seconds, or $\frac{1}{4} s^{-1}$), then

■ **Hertz**
Heinrich Hertz (1957–1894) was the first to send and receive radio waves. He showed that they could be reflected and refracted the same as light, confirming Maxwell's prediction that light waves were electromagnetic radiation. In his honor scientists use "hertz" as the unit of frequency (number of cycles per second, s^{-1}) of radiation.

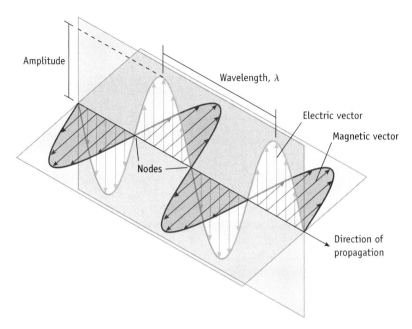

Figure 7.1 Electromagnetic radiation.   In the 1860s James Clerk Maxwell developed the currently accepted theory that all forms of radiation are propagated through space as vibrating electric and magnetic fields at right angles to one another. Each of the fields is described by a sine wave (the mathematical function describing the wave). Such oscillating fields emanate from vibrating charges in a source such as a light bulb or radio antenna.

the traffic is "moving" at the speed of $(5 \text{ m}) \times (\frac{1}{4} \text{ s}^{-1})$, or $1.25 \text{ m} \cdot \text{s}^{-1}$. The speed for any periodic motion, including a wave, is the product of the wavelength and the frequency of the wave:

$$\text{Speed (m} \cdot \text{s}^{-1}) = \text{wavelength (m)} \times \text{frequency (s}^{-1})$$

This equation also applies to electromagnetic radiation, where the speed of light, c, is the product of the wavelength and frequency of a light wave.

Speed of light (m · s⁻¹)

$$c = \lambda \times \nu \qquad\qquad (7.1)$$

Wavelength (m)   Frequency (s⁻¹)

The speed of visible light and all other forms of electromagnetic radiation in a vacuum is a constant, c ($= 2.99792458 \times 10^8 \text{ m} \cdot \text{s}^{-1}$; approximately $186{,}000$ miles $\cdot$ s^{-1}). Given this value, and knowing the wavelength of a light wave, you can calculate the frequency, and vice versa. For example, what is the frequency of orange light, which has a wavelength of 625 nm? Because the speed of light is expressed in meters

■ **Speed of Light**
The speed of light passing through a substance (air, glass, water, and so on) depends on the chemical constitution of the substance and the wavelength of the light. This is the basis for using a glass prism to disperse light and is the explanation for rainbows. The speed of sound also depends on the material through which it passes.

per second, the wavelength in nanometers must be changed to meters before substituting into Equation 7.1:

$$625 \text{ nm} \times \frac{1 \times 10^{-9} \text{ m}}{1 \text{ nm}} = 6.25 \times 10^{-7} \text{ m}$$

$$\nu = \frac{c}{\lambda} = \frac{2.998 \times 10^{8} \text{ m} \cdot \text{s}^{-1}}{6.25 \times 10^{-7} \text{ m}} = 4.80 \times 10^{14} \text{ s}^{-1}$$

Standing Waves

The wave motion described so far is that of traveling waves such as sound or water waves. Another type of wave motion, called **standing** or stationary waves, is relevant to modern atomic theory. If you tie down a string at both ends, as you would the string of a guitar, and then pluck it, the string vibrates as a standing wave (Figure 7.2). Several important points about standing waves are relevant to our discussion of electrons in atoms:

- A standing wave is characterized by having two or more points of no amplitude; that is, the wave amplitude is zero at the nodes.

- As with traveling waves, the distance between consecutive nodes is always $\lambda/2$.

- Only certain wavelengths are possible for standing waves. The only allowed vibrations have wavelengths of $n(\lambda/2)$, where n is an integer.

Figure 7.2 illustrates the third point. In the first of the vibrations illustrated, the distance between the ends of the string is half a wavelength, or $\lambda/2$. In the second vibration, the string length equals one complete wavelength, or $2(\lambda/2)$. In the third vibration, the string length is $3(\lambda/2)$, or $(3/2)\lambda$. Could the distance between the ends of a standing wave vibration ever be $(3/4)\lambda$? The answer is no. For standing waves, only certain wavelengths are possible. Because the ends of a standing wave must be nodes, the only allowed vibrations are those in which the distance from one end, or "boundary," to the other is $n(\lambda/2)$, where n is an integer (1, 2, 3, . . .).

■ **Standing Waves**
Only certain wavelengths are allowed for standing waves. This is an example of *quantization*, a concept we describe in the sections that follow.

Figure 7.2   Standing waves. In the first wave, the end-to-end distance is $(1/2)\lambda$, in the second wave it is λ, and in the third wave it is $(3/2)\lambda$.

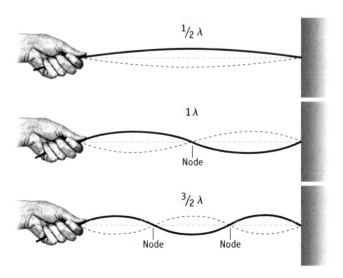

Exercise 7.1—Standing Waves

The line shown here is 10 cm long.

Using this line,

(a) Draw a standing wave with one node between the ends. What is the wavelength of this wave?

(b) Draw a standing wave with three evenly spaced nodes between the ends. What is its wavelength?

(c) If the wavelength of the standing wave is 2.5 cm, how many waves fit within the boundaries? How many nodes are there between the ends?

The Visible Spectrum of Light

Visible light consists of a spectrum of colors, ranging from red light at the long-wavelength end of the spectrum to violet light at the short-wavelength end (Figure 7.3). Visible light is, however, merely a small portion of the total electromagnetic spectrum. Ultraviolet (UV) radiation, the radiation that can lead to sunburn, has wavelengths shorter than those of visible light; x-rays and γ rays, the latter emitted in the process of radioactive disintegration of some atoms, have even shorter wavelengths. At longer wavelengths than those of visible light, we first encounter infrared radiation (IR), the type that is sensed as heat. Longer still is the wavelength of the radiation used in microwave ovens and in television and radio transmissions.

■ **ROY G BIV**
You can remember the colors of visible light, in order of decreasing wavelength, by the well-known mnemonic phrase ROY G BIV: red, orange, yellow, green, blue, indigo, and violet.

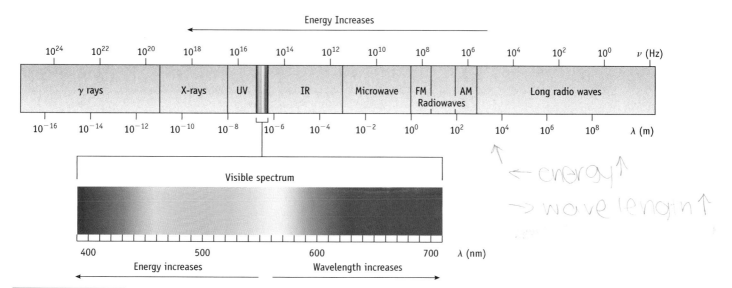

Active Figure 7.3 **The electromagnetic spectrum.** Visible light (enlarged portion) is a very small part of the entire spectrum. The radiation's energy increases from the radiowave end of the spectrum (low frequency, ν, and long wavelength, λ) to the γ-ray end (high frequency and short wavelength).

GENERAL
Chemistry·❀·Now™ See the General ChemistryNow CD-ROM or website to explore an interactive version of this figure accompanied by an exercise.

See the General ChemistryNow CD-ROM or website:

- **Screen 7.3 Electromagnetic Radiation**

   (a) for a tutorial on calculating the frequency of ultraviolet light

   (b) for a tutorial on calculating the wavelength of visible light

- **Screen 7.4 Electromagnetic Spectrum,** for a simulation exploring the wavelength and frequency of the visible portion of the electromagnetic spectrum

---

Example 7.1—Wavelength–Frequency Conversions

Problem The frequency of the radiation used in all microwave ovens sold in the United States is 2.45 GHz. (The unit GHz stands for "gigahertz"; 1 GHz is 1 billion cycles per second, or 10^9 s^{-1}.) What is the wavelength, in meters, of this radiation? Compare the wavelength of microwave radiation with the wavelength of light in the visible region—say, orange light with $\lambda = 625$ nm.

Strategy The wavelength of microwave radiation in meters can be calculated directly from Equation 7.1. Convert 625 nm to a wavelength in meters so that units are compatible.

Solution

$$\lambda = \frac{c}{\nu} = \frac{2.998 \times 10^8 \text{ m} \cdot \text{s}^{-1}}{2.45 \times 10^9 \text{ s}^{-1}} = 0.122 \text{ m}$$

Orange light has a wavelength, in meters, of

$$625 \text{ nm} \times \frac{1 \times 10^{-9} \text{ m}}{1 \text{ nm}} = 6.25 \times 10^{-7} \text{ m}$$

The wavelength of microwave radiation is about 200,000 times *longer* than that of orange light.

Exercise 7.2—Radiation, Wavelength, and Frequency

(a) Which color in the visible spectrum has the highest frequency? Which has the lowest frequency?

(b) Is the frequency of the radiation used in a microwave oven higher or lower than that from your favorite FM radio station (91.7 MHz), where MHz (megahertz) $= 10^6$ s^{-1}?

(c) Is the wavelength of x-rays longer or shorter than that of ultraviolet light?

7.2—Planck, Einstein, Energy, and Photons

Planck's Equation

If you heat a piece of metal, it emits electromagnetic radiation with wavelengths that depend on temperature. At first its color is a dull red. At higher temperatures the red color brightens (Figure 7.4a), and at still higher temperatures the redness turns to a brilliant white light. For example, the heating element of a toaster becomes "red hot," and the filament of an incandescent light bulb glows "white hot."

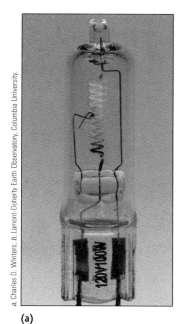 

(a) **(b)**

Figure 7.4 Infrared radiation. IR radiation has longer wavelengths than visible light. (a) The filament of an incandescent light bulb emitting radiation at the long-wavelength or red end of the visible spectrum. (b) A photo of the New York City area taken from a satellite using film sensitive to infrared light. Water is dark blue, pavement is light blue, and vegetation is red.

Your eyes detect the radiation from a piece of heated metal that occurs in the visible region of the electromagnetic spectrum. Of course, these are not the only wavelengths of the light emitted by the metal. **Radiation is also emitted with wavelengths both shorter (in the ultraviolet region) and longer (in the infrared region; Figure 7.4b) than those of visible light.** That is, a spectrum of electromagnetic radiation is emitted (Figure 7.5), with some wavelengths being more intense than others. As the metal is heated, the maximum in the curve of light intensity versus wavelength is shifted more and more to the ultraviolet region. The color of the glowing object shifts from red to yellow, and, if it does not melt, it will finally glow white hot.

At the end of the 19th century, scientists were trying to explain the relationship between the intensity and the wavelength for radiation given off by heated objects.

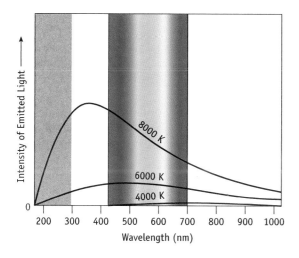

Figure 7.5 The spectrum of the radiation given off by a heated body. When an object is heated, it emits radiation covering a spectrum of wavelengths. For a given temperature, some of the radiation is emitted at long wavelengths and some at short wavelengths. Most, however, is emitted at some intermediate wavelength, the maximum in the curve. As the temperature of the object increases, the maximum moves from the red end of the spectrum to the violet end. At still higher temperatures intense light is emitted at all wavelengths in the visible region, and the maximum in the curve is in the ultraviolet region. Such an object is described as "white hot." (Stars are often referred to as "red giants" or "white dwarfs," a reference to their temperatures and relative sizes.) (*See the General ChemistryNow Screen 7.5 Planck's Equation, to watch a video on the light emitted by a heated metal bar.*)

All of their attempts were unsuccessful, however. Theories available at the time predicted that the intensity of radiation should increase continuously with decreasing wavelength (instead of declining with decreasing wavelength as is observed experimentally; Figure 7.5). This perplexing situation became known as the "ultraviolet catastrophe" because predictions failed in the ultraviolet region. Classical physics did not provide a satisfactory explanation, so a new way to look at matter and energy was needed.

In 1900, a German physicist, Max Planck (1858–1947), offered an explanation. Following classical theory, he assumed that vibrating atoms in a heated object give rise to the emitted electromagnetic radiation. He also introduced an important new assumption: These vibrations are **quantized**. In Planck's model, quantization means that only certain vibrations, with specific frequencies, are allowed.

Planck introduced an important equation, now called Planck's equation, that states that the energy of a vibrating system is proportional to the frequency of vibration. The proportionality constant h is called **Planck's constant** in his honor. It has the value $6.6260693 \times 10^{-34}$ J $\cdot$ s.

$$\text{Energy (J)} \quad \text{Planck's constant (J} \cdot \text{s)}$$

$$E = h\nu \tag{7.2}$$

$$\text{Frequency (s}^{-1})$$

Now, assume as Planck did that there must be a *distribution* of vibrations of atoms in an object—some atoms are vibrating at a high frequency, some are vibrating at a low frequency, but most have some intermediate frequency. The few atoms with high-frequency vibrations are responsible for some of the light, as are those few with low-frequency vibrations. Nevertheless, most of the light must come from the majority of the atoms that have intermediate vibrational frequencies. That is, a spectrum of light is emitted with a maximum intensity at some wavelength, in accord with experiment. The intensity should not become greater and greater on approaching the ultraviolet region. With this realization, the ultraviolet catastrophe was solved.

Einstein and the Photoelectric Effect

As almost always occurs, the explanation of a fundamental phenomenon—such as the spectrum of light from a hot object—leads to another fundamental discovery. A few years after Planck's work, Albert Einstein (1879–1955) incorporated Planck's ideas into an explanation of the photoelectric effect.

Photoelectric cells are commonly used in automatic door openers in stores and elevators. They depend on the **photoelectric effect**, the ejection of electrons when light strikes the surface of a metal. In the cell in Figure 7.6, an electric potential is applied to the cell. When light strikes the cathode of the cell, electrons are ejected from the cathode surface and move to a positively charged anode. A stream of electrons—a current—flows through the cell. Thus, the cell can act as a light-activated switch in an electric circuit.

Experiments with photoelectric cells show that electrons are ejected from the surface *only* if the frequency of the light is high enough. If lower-frequency light is

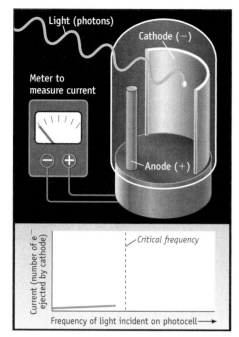

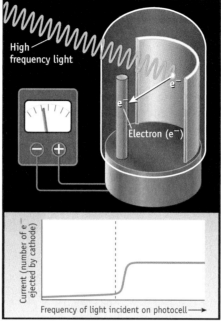

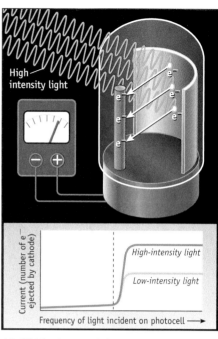

(a) A photocell operates by the photoelectric effect. The main part of the cell is a light-sensitive cathode. This is a material, usually a metal, that ejects electrons if struck by photons of light of sufficient energy. No current is observed until the critical frequency is reached.

(b) When light of higher frequency than the minimum is used, the excess energy of the photon allows the electron to escape the atom with greater velocity. The ejected electrons move to the anode and a current flows in the cell. Such a device can be used as a switch in electric circuits.

(c) If higher intensity light is used, the only effect is to cause more electrons to be released from the surface. The onset of current is observed at the same frequency as with lower intensity light, but more current flows.

Figure 7.6 **A photoelectric cell.**

used, no effect is observed, regardless of the light's intensity (brightness). In contrast, if the frequency is above the minimum, increasing the light intensity causes a higher current to flow because more and more electrons are ejected.

Einstein decided the experimental observations could be explained by combining Planck's equation ($E = h\nu$) with a new idea: Light has particle-like properties. Einstein assumed these massless "particles," now called **photons**, are packets of energy. The energy of each photon is proportional to the frequency of the radiation, as given by Planck's relation.

Einstein's proposal helps us understand the photoelectric effect. It is reasonable to suppose that a high-energy particle would have to bump into an atom to cause the atom to lose an electron. It is also reasonable to suppose that an electron can be torn away from the atom only if the photon has enough energy. If electromagnetic radiation is described as a stream of photons, as Einstein said, then the greater the intensity of light, the more photons are available to strike a surface per unit of time. However, the atoms of a metal surface will not lose electrons when the metal is bombarded by millions of photons if no individual photon has enough energy to remove an electron from an atom. Only if the critical minimum energy (that is, minimum light frequency) is exceeded will the energy content be sufficient to displace an electron from a metal atom. The greater the number of photons with this energy that strike the surface, the greater the number of electrons dislodged. Thus, the connection is made between light intensity and the number of electrons ejected.

Energy and Chemistry: Using Planck's Equation

Compact disc players use lasers that emit red light with a wavelength of 685 nm. What is the energy of one photon of this light? What is the energy of one mole of photons of red light? To answer these questions, first convert the wavelength to the frequency of the radiation and then use the frequency to calculate the energy per photon. Finally, calculate the energy of a mole of photons by multiplying the energy per photon by Avogadro's number:

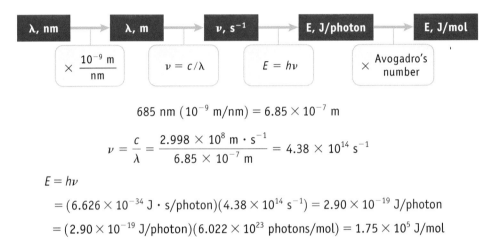

$$685 \text{ nm} \left(10^{-9} \text{ m/nm}\right) = 6.85 \times 10^{-7} \text{ m}$$

$$\nu = \frac{c}{\lambda} = \frac{2.998 \times 10^8 \text{ m} \cdot \text{s}^{-1}}{6.85 \times 10^{-7} \text{ m}} = 4.38 \times 10^{14} \text{ s}^{-1}$$

$$E = h\nu$$

$$= \left(6.626 \times 10^{-34} \text{ J} \cdot \text{s/photon}\right)\left(4.38 \times 10^{14} \text{ s}^{-1}\right) = 2.90 \times 10^{-19} \text{ J/photon}$$

$$= \left(2.90 \times 10^{-19} \text{ J/photon}\right)\left(6.022 \times 10^{23} \text{ photons/mol}\right) = 1.75 \times 10^5 \text{ J/mol}$$

The energy of a mole of photons of red light is equivalent to 175 kJ. A mole of photons of blue light ($\lambda = 400$ nm) has an energy of about 300 kJ. These energies are in a range that can affect the bonds between atoms in molecules. It should not be surprising, therefore, that light can cause chemical reactions to occur. For example, you may have seen paint or dye that has faded or even decomposed from exposure to light.

The previous calculation shows that, as the frequency of radiation increases, the energy of the radiation also increases (see Figure 7.3). Similarly, the energy increases as the wavelength of radiation decreases:

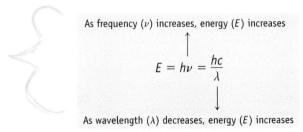

Photons of ultraviolet radiation—with wavelengths shorter than those of visible light—have higher energy than visible light. **Because visible light has enough energy to affect the bonds between atoms, obviously ultraviolet light does as well.** That is the reason ultraviolet radiation can cause a sunburn. In contrast, photons of infrared radiation—with wavelengths longer than those of visible light—have lower

Chemical Perspectives

UV Radiation, Skin Damage, and Sunscreens

Most of us are well aware of the effects of exposure to the sun. A sunburn results, and over the long term permanent skin damage can occur. Most of this problem results from the damage to organic molecules caused by ultraviolet (UV) radiation.

UV radiation is often divided into three categories: UVA (315–400 nm), UVB (290–315 nm), and UVC (100–290 nm). UVC radiation has a high energy, but it is absorbed by the earth's ozone layer. UVB light is responsible for your sunburn. Tanning occurs when the light strikes your skin and activates the melanocytes in the skin so that they produce melanin. UVA light also produces damage such as the alteration of connective tissue in the dermis.

We can calculate that the energy of a mole of photons in the ultraviolet region (at 300 nm) is about 400 kJ.

$$\text{For } \lambda = 300.\text{ nm}, \nu = 9.99 \times 10^{14}\text{ s}^{-1}$$
$$E = h\nu$$
$$= (6.626 \times 10^{-34}\text{ J}\cdot\text{s/photon}) \times (9.99 \times 10^{14}\text{ s}^{-1})$$
$$= 6.62 \times 10^{-19}\text{ J/photon}$$
$$E = 3.99 \times 10^{5}\text{ J/mol}$$

This energy is significantly greater than the energy of light in the visible region. Indeed, the energy of UV light is in the range of the energies necessary to break the chemical bonds in proteins.

Various manufacturers have developed mixtures of compounds that protect skin from UVA and UVB radiation. These sunscreens are given "sun protection factor" (SPF) labels that indicate how long the user can stay in the sun without burning.

Lowell Georgia/Corbis

Sunscreens produced by Coppertone, for example, contain the organic compounds 2-ethylhexyl-p-methoxycinnamate and oxybenzene. These molecules absorb UV radiation, preventing it from reaching your skin.

energy than visible light. They are generally not energetic enough to cause chemical reactions, but they can affect the vibrations of molecules. We sense infrared radiation as heat, such as the heat given off by a glowing burner on an electric stove.

GENERAL Chemistry Now™

See the General ChemistryNow CD-ROM or website:

- **Screen 7.5 Planck's Equation**
  - **(a)** for a simulation exploring the relationship between wavelength, frequency, and photon energy
  - **(b)** for an exercise on using Planck's equation to calculate wavelength

Exercise 7.3—Photon Energies

Compare the energy of a mole of photons of orange light (625 nm) with the energy of a mole of photons of microwave radiation having a frequency of 2.45 GHz (1 GHz = 10^9 s^{-1}). Which has the greater energy? By what factor is one greater than the other? (See Example 7.1.)

7.3—Atomic Line Spectra and Niels Bohr

Atomic Line Spectra

If a high voltage is applied to atoms of an element in the gas phase at low pressure, the atoms absorb energy and are said to be "excited." The excited atoms emit light. This light is different, however, from the *continuous spectrum* of wavelengths from

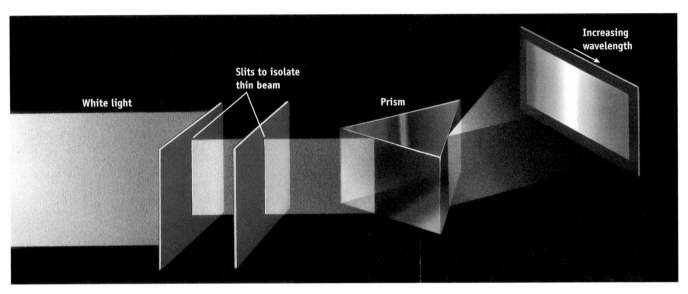

Figure 7.7   A spectrum of white light, produced by refraction in a prism. The light is first passed through a narrow slit to isolate a thin beam, or line, of light. The beam is then passed through a prism (or, in modern instruments, a diffraction grating is used). See the spectrum of visible light in Figure 7.3.

white light (Figure 7.7). **Excited atoms in the gas phase emit only certain wavelengths of light.** We know this because when this light is passed through a prism, only a few colored lines are seen. This phenomenon is called a **line emission spectrum** (Figure 7.8). A familiar example is the light from a neon advertising sign, in which excited neon atoms emit orange-red light.

The line emission spectra of hydrogen, mercury, and neon are shown in Figure 7.9. Every element has a unique line spectrum. Indeed, the characteristic lines in

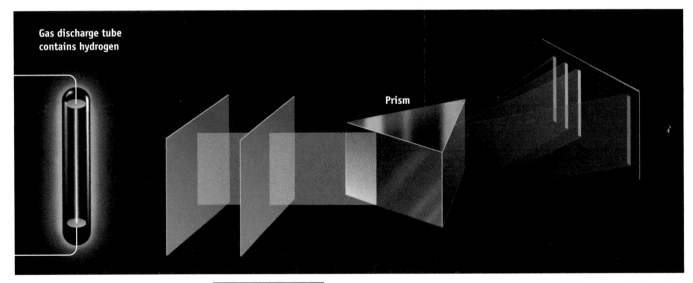

Active Figure 7.8   **The line emission spectrum of hydrogen.** The emitted light is passed through a series of slits to create a narrow beam of light, which is then separated into its component wavelengths by a prism. A photographic plate or photocell can be used to detect the separate wavelengths as individual lines. Hence, the name "line spectrum" is given to the light emitted by a glowing gas.

GENERAL
Chemistry Now™ See the General ChemistryNow CD-ROM or website to explore an interactive version of this figure accompanied by an exercise.

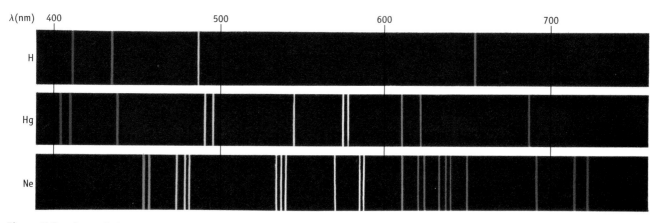

Figure 7.9 Line emission spectra of hydrogen, mercury, and neon. Excited gaseous elements produce characteristic spectra that can be used to identify the elements as well as to determine how much of each element is present in a sample.

the emission spectrum of an element can be used in chemical analysis, both to identify the element and to determine how much of it is present.

A goal of scientists in the late 19th century was to explain why gaseous atoms emitted light of only certain frequencies and to find a mathematical relationship among the observed frequencies. (It is always useful if experimental data can be related by a mathematical equation, because a regular pattern of information implies a logical explanation.) The first steps in this direction were made by Johann Balmer (1825–1898) and later Johannes Rydberg (1854–1919). They developed an equation—now called the **Rydberg equation**—from which it was possible to calculate the wavelength of the red, green, and blue lines in the visible emission spectrum of hydrogen atoms (Figure 7.9).

$$\frac{1}{\lambda} = R\left(\frac{1}{2^2} - \frac{1}{n^2}\right) \quad \text{when } n > 2 \tag{7.3}$$

In this equation n is an integer, and R, now called the **Rydberg constant**, has the value 1.0974×10^7 m^{-1}. If $n = 3$, the wavelength of the red line in the hydrogen spectrum is obtained (6.563×10^{-7} m, or 656.3 nm). If $n = 4$, the wavelength for the green line is obtained. The value $n = 5$ gives the wavelength of the blue line. This group of visible lines in the spectrum of hydrogen atoms (and others for which $n = 6, 7, 8, \ldots$) is now called the **Balmer series**.

The Bohr Model of the Hydrogen Atom

Niels Bohr (1885–1962), a Danish physicist, provided the first connection between the spectra of excited atoms and the quantum ideas of Planck and Einstein. From Rutherford's work [◀ Section 2.1], it was known that electrons are arranged in space outside the atom's nucleus. For Bohr the simplest model of a hydrogen atom was one in which the electron moved in a circular orbit around the nucleus, just as the planets revolve about the sun. In proposing this hypothesis, however, he had to contradict the laws of classical physics. According to the theories at the time, a charged electron moving in the positive electric field of the nucleus should lose energy. Eventually the electron should crash into the nucleus. This is clearly not the case; if it were so, all matter would eventually self-destruct.

To solve the contradiction with the laws of classical physics, Bohr postulated that an electron could occupy only *certain* orbits or energy levels in which it is stable. That is, the energy of the electron in the atom is quantized. By combining this

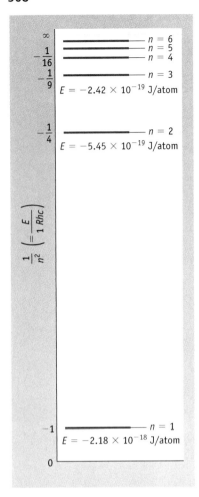

Active Figure 7.10   **Energy levels for the H atom in the Bohr model.** The energy of the electron in the hydrogen atom depends on the value of the principal quantum number n ($E_n = -Rhc/n^2$). The larger the value of n the larger the Bohr radius and the less negative the value of the energy. Energies are given in joules per atom (J/atom). Notice that the difference between successive energy levels becomes smaller as n becomes larger.

GENERAL

Chemistry•↓•Now™ See the General ChemistryNow CD-ROM or website to explore an interactive version of this figure accompanied by an exercise.

quantization postulate with the laws of motion from classical physics, Bohr showed that the energy possessed by the single electron in the nth orbit or energy level of the H atom is given by the equation

Potential energy of electron in the nth level = $E_n = -\dfrac{Rhc}{n^2}$ (7.4)

(Rydberg constant, Planck's constant, Speed of light, Principal quantum number)

which gives the energy in units of J/atom. Each allowed orbit was assigned a value of n, a unitless integer having values of 1, 2, 3, and so on. This integer is now known as the **principal quantum number**.

Equation 7.4 has several important features. First, the energy of the electron has a negative value. This follows from Coulomb's law [◀ Section 3.3]. The energy of attraction between oppositely charged bodies (a negative electron and the positive nuclear proton) has a negative value, and that value becomes more negative (the attraction increases) as the bodies move closer together. Bohr's equation shows that as the value of n increases, the value of the energy becomes *less* negative (Figure 7.10). Bohr also showed that as n increases (and the energy becomes less negative), the distance of the electron from the nucleus increases. An electron in the $n = 1$ orbit is closest to the nucleus and has the lowest or most negative energy. The electron of the hydrogen atom is normally in this energy level.

An atom with its electrons in the lowest possible energy levels is said to be in its **ground state**. When the electron of a hydrogen atom occupies an orbit with n greater than 1, the electron is farther from the nucleus, the value of its energy is less negative, and it is said to be in an **excited state**. The energies of the ground state and an excited state are calculated in Example 7.2.

Example 7.2—Energies of the Ground and Excited States of the H Atom

Problem Calculate the energies of the $n = 1$ and $n = 2$ states of the hydrogen atom in joules per atom and in kilojoules per mole. What is the difference in energy of these two states?

Strategy Here we use Equation 7.4 with the following constants: $R = 1.097 \times 10^7$ m^{-1}, $h = 6.626 \times 10^{-34}$ J · s, and $c = 2.998 \times 10^8$ m · s^{-1}.

Solution When $n = 1$, the energy of an electron in a single H atom is

$$E_1 = -\frac{Rhc}{n^2} = -\frac{Rhc}{1^2} = -Rhc$$

$$= -(1.097 \times 10^7 \text{ m}^{-1})(6.626 \times 10^{-34} \text{ J} \cdot \text{s})(2.998 \times 10^8 \text{ m} \cdot \text{s}^{-1})$$

$$= -2.179 \times 10^{-18} \text{ J/atom}$$

In units of kJ/mol, we have

$$E_1 = \frac{-2.179 \times 10^{-18} \text{ J}}{\text{atom}} \times \frac{6.022 \times 10^{23} \text{ atoms}}{\text{mol}} \times \frac{1 \text{ kJ}}{1000 \text{ J}}$$

$$= -1312 \text{ kJ/mol}$$

When n = 2, the energy is

$$E_2 = -\frac{Rhc}{2^2} = -\frac{E_1}{4} = -\frac{2.179 \times 10^{-18} \text{ J/atom}}{4}$$

$$= -5.448 \times 10^{-19} \text{ J/atom}$$

Finally, because $E_2 = E_1/4$, we calculate E_2 to be -328.1 kJ/mol.

The difference in energy, ΔE, between the first two energy states of the H atom is

$$\Delta E = E_2 - E_1 = (-328.1 \text{ kJ/mol}) - (-1312 \text{ kJ/mol}) = 984 \text{ kJ/mol}$$

Comment Notice that the calculated energies are negative for an electron in the $n = 1$ or $n = 2$ state, with E_1 more negative than E_2. Notice also that the $n = 2$ state is higher in energy than the $n = 1$ state by 984 kJ/mol. For more on this point, see Figure 7.11 and the discussion below.

Exercise 7.4—Electron Energies

Calculate the energy of the $n = 3$ state of the H atom in (a) joules per atom and (b) kilojoules per mole.

You can think of the energy levels in the Bohr model as the rungs of a ladder climbing out of the basement of an "atomic building," where the energy of the H atom is -2.18×10^{-18} J/atom (Example 7.2), to the ground level, where the energy is 0 (Figure 7.10). Each step represents a quantized energy level; as you climb the ladder, you can stop on any rung but not between them. Unlike the rungs of a real ladder, however, Bohr's energy levels get closer and closer together as n increases.

The Bohr Theory and the Spectra of Excited Atoms

A major assumption of Bohr's theory was that an electron in an atom would remain in its lowest energy level unless disturbed. Energy must be absorbed or evolved if the electron changes from one energy level to another, in agreement with the first law of thermodynamics [◄ Section 6.4]. This idea allowed Bohr to explain the spectra of excited gases.

When the H atom electron has $n = 1$ and so is in its ground state, the energy is a large negative value. As we climb the ladder (see Figure 7.10) to the $n = 2$ level, the electron is less strongly attracted to the nucleus, and the energy of an $n = 2$ electron is less negative. Therefore, to move an electron in the $n = 1$ state to the $n = 2$ state, the atom must absorb energy, just as energy must be expended in climbing a ladder. The electron must be excited (Figure 7.11).

Using Bohr's equation we can calculate the energy required to carry the H atom from the ground state ($n = 1$) to its first excited state ($n = 2$). As you learned in Chapter 6, the difference in energy between two states is always

$$\Delta E = E_{\text{final state}} - E_{\text{initial state}}$$

When E_{final} has $n = 2$, and E_{initial} has $n = 1$, we can calculate ΔE from the equation

$$\Delta E = E_{\text{final}} - E_{\text{initial}} = \left(-\frac{Rhc}{2^2}\right) - \left(-\frac{Rhc}{1^2}\right)$$

where Rhc has the value 1312 kJ/mol (as calculated in Example 7.2).

$$\Delta E = E_{\text{final}} - E_{\text{initial}} = (-Rhc/2^2) - (-Rhc/1^2) = \left(\tfrac{3}{4}\right)Rhc = 984 \text{ kJ/mol}$$

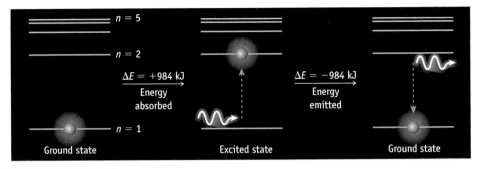

Active Figure 7.11   Absorption of energy by the atom as the electron moves to an excited state. Energy is absorbed when an electron moves from the $n = 1$ state to the $n = 2$ state ($\Delta E > 0$). When the electron returns to the $n = 1$ state from $n = 2$, energy is evolved ($\Delta E < 0$). The change in energy is 984 kJ/mol, as calculated in Example 7.2.

GENERAL
Chemistry ❄ Now ™ See the General ChemistryNow CD-ROM or website to explore an interactive version of this figure accompanied by an exercise.

The amount of energy that must be absorbed by the atom so that an electron can move from the first to the second energy state is $0.75Rhc$ or 984 kJ/mol of atoms— no more and no less. If $0.7Rhc$ or $0.8Rhc$ is provided, no transition between states is possible. *Energy levels in the H atom are quantized*, with the consequence that only certain amounts of energy may be absorbed or emitted.

Moving an electron from a state of low n to one of higher n requires that energy is absorbed, and the sign of the value of ΔE is positive. The opposite process, in which an electron "falls" from a level of higher n to one of lower n, emits energy (Figure 7.11). For example, for a transition from $n = 2$ to $n = 1$,

$$\Delta E = E_{\text{final state}} - E_{\text{initial state}}$$

$$\Delta E = E_{\text{final}} - E_{\text{initial}} = \left(-\frac{Rhc}{1^2}\right) - \left(-\frac{Rhc}{2^2}\right) = -\left(\frac{3}{4}\right)Rhc = -984 \text{ kJ/mol}$$

The negative sign indicates energy is evolved; that is, 984 kJ must be *emitted* per mole of H atoms.

Depending on how much energy is added to a collection of H atoms, some atoms have their electrons excited from the $n = 1$ to the $n = 2$ or 3 or higher states. After absorbing energy, these electrons naturally move back down to lower levels (either directly or in a series of steps to $n = 1$) and release the energy the atom originally absorbed. The energy emitted is observed as light. *This is the source of the lines observed in the emission spectrum of H atoms*, and the same basic explanation holds for the spectra of other elements and for the colors of fireworks.

For hydrogen, a series of emission lines having energies in the ultraviolet region (called the **Lyman series**; Figure 7.12) arises from electrons moving from states with $n > 1$ to the $n = 1$ state. The series of lines that have energies in the visible region—the **Balmer series**—arises from electrons moving from states with $n > 2$ to the lower state with $n = 2$.

In summary, we now recognize that *the origin of atomic spectra is the movement of electrons between quantized energy states*. If an electron is excited from a lower energy state to a higher one, energy is absorbed. Conversely, if an electron moves from a higher energy state to a lower one, energy is emitted. If the energy is emitted as electromagnetic radiation, an emission line is observed. The energy of a specific emission line for excited hydrogen atoms is

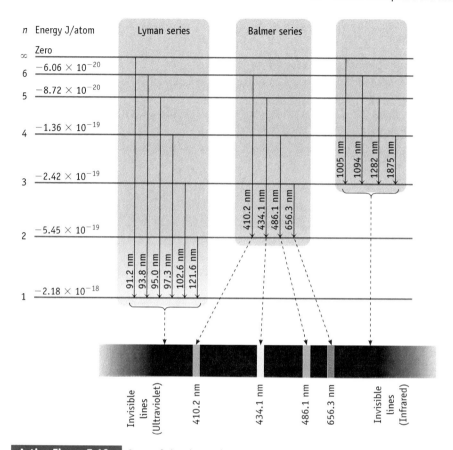

Active Figure 7.12 **Some of the electronic transitions that can occur in an excited H atom.** The Lyman series of lines in the ultraviolet region results from transitions to the $n = 1$ level. Transitions from levels with values of $n > 2$ to $n = 2$ occur in the visible region (Balmer series; see Figure 7.8). Lines in the infrared region result from transitions from levels with $n > 3$ or 4 to the $n = 3$ or 4 levels. (Only the series ending at $n = 3$ is illustrated.)

GENERAL
Chemistry·Now™ See the General ChemistryNow CD-ROM or website to explore an interactive version of this figure accompanied by an exercise.

$$\Delta E = E_{final} - E_{initial} = -Rhc\left(\frac{1}{n_{final}^2} - \frac{1}{n_{initial}^2}\right) \qquad (7.5)$$

where *Rhc* is 2.179×10^{-18} J/atom, or 1312 kJ/mol.

Bohr was able to use his model of the atom to calculate the wavelengths of the lines in the hydrogen spectrum. He had tied the unseen (the interior of the atom) to the seen (the observable lines in the hydrogen spectrum)—a fantastic achievement! In addition, he introduced the concept of energy quantization in describing atomic structure, a concept that remains an important part of modern science.

As mentioned previously, agreement between theory and experiment is taken as evidence that the theoretical model is valid. It soon became apparent, however, that a flaw existed in Bohr's theory. Bohr's model of the atom explained only the spectrum of H atoms and of other systems having one electron (such as He⁺). Furthermore, *the idea that electrons are particles moving about the nucleus with a path of fixed radius, like that of the planets about the sun, is no longer the accepted model for the atom.*

GENERAL
Chemistry⋅♣⋅Now™

See the General ChemistryNow CD-ROM or website:

- **Screen 7.6 Atomic Line Spectrum,** for an exercise on the light emitted by excited hydrogen atoms

  (a) for a simulation and exercise examining the radiation emitted when the electrons of excited hydrogen atoms return to the ground state

  (b) for a tutorial on calculating the wavelength of radiation emitted when electrons change energy levels

Example 7.3—Energies of Emission Lines for Excited Atoms

Problem Calculate the wavelength of the green line in the visible spectrum of excited H atoms using the Bohr theory.

Strategy First locate the green line in Figure 7.12 and determine the quantum states involved. That is, decide on $n_{initial}$ and n_{final}. Then use Equation 7.5 to calculate the difference in energy, ΔE, between these states, a difference that appears as visible light in the green region of the spectrum. Finally, express ΔE in terms of a wavelength.

Solution The green line is the second most energetic line in the visible spectrum of hydrogen and arises from electrons moving from $n = 4$ to $n = 2$. Using Equation 7.5 where $n_{final} = 2$ and $n_{initial} = 4$, we have

$$\Delta E = E_{final} - E_{initial} = \left(-\frac{Rhc}{2^2}\right) - \left(-\frac{Rhc}{4^2}\right)$$

$$\Delta E = -Rhc\left(\frac{1}{4} - \frac{1}{16}\right) = -Rhc(0.1875)$$

Earlier, we found that Rhc is 1312 kJ/mol, so the $n = 4$ to $n = 2$ transition involves an energy change of

$$\Delta E = -(1312 \text{ kJ}/\text{mol})(0.1875) = -246.0 \text{ kJ}/\text{mol}$$

The wavelength can now be calculated. First the photon energy, E_{photon}, is expressed as J/photon.

$$E_{photon} = \frac{\left(246.0 \frac{\text{kJ}}{\text{mol}}\right)\left(1 \times 10^3 \frac{\text{J}}{\text{kJ}}\right)}{6.022 \times 10^{23} \frac{\text{photons}}{\text{mol}}} = 4.085 \times 10^{-19} \frac{\text{J}}{\text{photon}}$$

Now apply Planck's equation where $E_{photon} = h\nu = hc/\lambda$, and so $\lambda = hc/E_{photon}$.

$$\lambda = \frac{hc}{E_{photon}} = \frac{\left(6.626 \times 10^{-34} \frac{\text{J} \cdot \text{s}}{\text{photon}}\right)(2.998 \times 10^8 \text{ m} \cdot \text{s}^{-1})}{4.085 \times 10^{-19} \text{ J/photon}}$$

$$= 4.863 \times 10^{-7} \text{ m}$$

$$= (4.863 \times 10^{-7} \text{ m})(1 \times 10^9 \text{ nm/m})$$

$$= 486.3 \text{ nm}$$

The experimental value is 486.1 nm (see Figure 7.12). This represents excellent agreement between experiment and theory.

Experimental Evidence for Bohr's Theory

Niels Bohr's model of the hydrogen atom was powerful because it could reproduce experimentally observed line spectra. But additional experimental confirmation of the model soon arose.

If the electron in the hydrogen atom is moved from the ground state, where $n = 1$, to the energy level where $n =$ infinity, the electron is considered to have been removed from the atom. That is, the atom has been ionized.

$$H(g) \longrightarrow H^+(g) + e^-$$

We can calculate the energy for this process from Equation 7.5 where $n_{final} = \infty$ and $n_{initial} = 1$.

$$\Delta E = -Rhc\left(\frac{1}{n_{final}^2} - \frac{1}{n_{initial}^2}\right)$$

$$\Delta E = -Rhc\left(\frac{1}{\infty^2} - \frac{1}{1^2}\right) = Rhc$$

Because $Rhc = 1312$ kJ/mol, the energy to move an electron from $n = 1$ to $n = \infty$ is 1312 kJ/mol of H atoms. We now call this the *ionization energy* of the atom [▶ Section 8.6] and can measure it in the laboratory. The experimental value is 1312 kJ/mol, in exact agreement with the result calculated from Bohr's theory!

$H^+(g)$ $n = \infty$ $E = 0$ kJ/mol

$\Delta E = +Rhc = +1312$ kJ/mol

$H(g)$ $n = 1$ $E = -1312$ kJ/mol

Exercise 7.5—Energy of an Atomic Spectral Line

The Lyman series of spectral lines for the H atom occurs in the ultraviolet region. They arise from transitions from higher levels to $n = 1$. Calculate the frequency and wavelength of the least energetic line in this series.

7.4—The Wave Properties of the Electron

Einstein used the photoelectric effect to demonstrate that light, usually thought of as having wave properties, can also have the properties of particles, albeit without mass [◀ page 303]. This fact was pondered by Louis Victor de Broglie (1892–1987). If light can be considered as having both wave and particle properties, would matter behave similarly? That is, could a tiny object such as an electron, normally considered a particle, also exhibit wave properties in some circumstances? In 1925, de Broglie proposed that a free electron of mass m moving with a velocity v should have an associated wavelength given by the equation

$$\lambda = \frac{h}{mv} \tag{7.6}$$

This idea was revolutionary because it linked the particle properties of the electron (m and v) with a wave property (λ). Experimental proof was soon produced. In 1927, C. J. Davisson (1881–1958) and L. H. Germer (1896–1971), working at Bell Telephone Laboratories in New Jersey, found that a beam of electrons was diffracted like light waves by the atoms of a thin sheet of metal foil (Figure 7.13) and that de Broglie's relation was followed quantitatively. Because diffraction is an effect best explained based on the wave properties of radiation, it follows that *electrons can be described as having wave properties under some circumstances.*

De Broglie's equation suggests that any moving particle has an associated wavelength. For λ to be measurable, however, the product of m and v must be very small

R. K. Bohn, Department of Chemistry, University of Connecticut

Figure 7.13 **Electron diffraction pattern obtained for magnesium oxide (MgO).**

because h is so small. For example, a 114-g baseball traveling at 110 mph has a large mv product (5.6 kg · m/s) and therefore the incredibly small wavelength of 1.2×10^{-34} m! This tiny value cannot be measured with any instrument now available. As consequence, we will never assign wave properties to a baseball or any other massive object. It is possible to observe wave-like properties only for particles of extremely small mass, such as protons, electrons, and neutrons.

GENERAL
Chemistry⚛Now™

See the General ChemistryNow CD-ROM or website:
- **Screen 7.8 Wave Properties of the Electron,** for a tutorial on calculating the wavelength of a moving electron

---

Example 7.4—Using de Broglie's Equation

Problem Calculate the wavelength associated with an electron of mass $m = 9.109 \times 10^{-28}$ g that travels at 40.0% of the speed of light.

Strategy First, consider the units involved. Wavelength is calculated from h/mv, where h is Planck's constant expressed in units of joule seconds (J · s). As discussed in Chapter 6, 1 J = 1 kg · m^2/s^2. Therefore, the mass must be in kilograms and speed in meters per second.

Solution

Electron mass $= 9.109 \times 10^{-31}$ kg

Electron speed (40.0% of light speed) $= (0.400)(2.998 \times 10^8 \text{ m} \cdot \text{s}^{-1}) = 1.20 \times 10^8 \text{ m} \cdot \text{s}^{-1}$

Substituting these values into de Broglie's equation, we have

$$\lambda = \frac{h}{mv}$$

$$= \frac{6.626 \times 10^{-34}(\text{kg} \cdot \text{m}^2/\text{s}^2)(\text{s})}{(9.109 \times 10^{-31} \text{ kg})(1.20 \times 10^8 \text{ m/s})}$$

$$= 6.07 \times 10^{-12} \text{ m}$$

In nanometers, the wavelength is

$$\lambda = (6.07 \times 10^{-12} \text{ m})(1.00 \times 10^9 \text{ nm/m}) = 6.07 \times 10^{-3} \text{ nm}$$

Comment The calculated wavelength is about $\frac{1}{12}$ of the diameter of the H atom.

Exercise 7.6—De Broglie's Equation

Calculate the wavelength associated with a neutron having a mass of 1.675×10^{-24} g and a kinetic energy of 6.21×10^{-21} J. (Recall that the kinetic energy of a moving particle is $E = \frac{1}{2}mv^2$.)

7.5—Quantum Mechanical View of the Atom

After World War I, Erwin Schrödinger (1887–1961), an Austrian, worked toward a comprehensive theory of the behavior of electrons in atoms. Starting with de Broglie's hypothesis that an electron in an atom could be described by equations for

Historical Perspectives

20th-Century Giants of Science

Many of the advances in science described in this chapter occurred during the early part of the 20th century, as the result of theoretical studies by some of the greatest minds in the history of science.

Max Karl Ernst Ludwig Planck

(1858–1947) was raised in Germany, where his father was a profes-

Max Planck

sor at a university. While still in his teens Planck decided to become a physicist, against the advice of the head of the physics department at Munich, who told him, "the important discoveries [in physics] have been made. It is hardly worth entering physics anymore." Fortunately, Planck did not take this advice and went on to study thermodynamics. This interest led him eventually to consider the ultraviolet catastrophe and to develop his revolutionary hypothesis, which was announced two weeks before Christmas in 1900. He was awarded the Nobel Prize in physics in 1918 for this work. Einstein later said it was a longing to find harmony and order in nature, a "hunger in his soul," that spurred Planck on.

Erwin Schrödinger

(1887–1961) was born in Vienna, Austria. Following his service as an artillery officer in World War I, he became a professor of physics. In 1928, he succeeded Planck as pro-

Erwin Schrödinger

fessor of physics at the University of Berlin. He shared the Nobel Prize in physics in 1933.

Niels Bohr

(1885–1962) was born in Copenhagen, Denmark. He earned a Ph.D. in physics in Copenhagen in 1911 and then went to work first with J. J. Thomson and later with Ernest Rutherford in

Niels Bohr

England. It was there that he began to develop his theory of atomic structure and his explanation of atomic spectra. (He received the Nobel Prize in physics in 1922 for this work.) Bohr returned to Copenhagen, where he eventually became director of the Institute for Theoretical Physics. Many young physicists worked with him at the Institute, seven of whom eventually received Nobel Prizes in chemistry and physics. Among these scientists were Werner Heisenberg, Wolfgang Pauli, and Linus Pauling. Element 107 was recently named bohrium in Bohr's honor.

Werner Heisenberg

(1901–1976) studied with Max Born and later with Bohr. He received the Nobel Prize in physics in 1932. The recent play *Copenhagen*, which has been staged in London and New York,

Werner Heisenberg

centers on the relationship between Bohr and Heisenberg and their involvement in the development of atomic weapons in World War II.

wave motion, Schrödinger developed the concept that has come to be called **quantum mechanics** or **wave mechanics**.

The Uncertainty Principle

De Broglie's suggestion that an electron can be described as having wave properties was confirmed by experiment [◀ Section 7.4]. J. J. Thomson's experiments were interpreted on the basis of the particle-like nature of the electron (see page 62). But how can an electron be both a particle and a wave? No single experiment can be done to show the electron behaves *simultaneously* as a wave *and* a particle. Scientists now accept **wave-particle duality**—that is, the idea that the electron indeed has the properties of both.

What does wave-particle duality have to do with electrons in atoms? Werner Heisenberg (1901–1976) and Max Born (1882–1970) provided the answer. Heisenberg concluded, in what is now known as the **uncertainty principle**, that it is impossible to fix both the position of an electron in an atom and its energy with any degree of certainty. Attempting to determine accurately either the location or the energy leaves the other uncertain. (Contrast this principle with the world around you: For objects larger than those on the atomic level—say, an automobile—you can determine, with considerable accuracy, both their energy and location at a given time.)

Based on Heisenberg's idea, Born proposed that the results of quantum mechanics should be interpreted as follows: If we choose to know the energy of an electron in an atom with only a small uncertainty, then we must accept a correspondingly large uncertainty in its position in the space about the atom's nucleus. In practical terms, this means we can assess only the likelihood, or **probability**, of finding an electron with a given energy within a given region of space. In the next section you will see that the result of this viewpoint is the definition of the regions around an atom's nucleus in which there is the highest probability of finding a given electron.

Schrödinger's Model of the Hydrogen Atom and Wave Functions

■ **Wave Functions and Energy**
In Bohr's theory, the electron energy for the H atom is given by $E_n = -Rhc/n^2$. This same result came from Schrödinger's electron wave model.

Schrödinger's model of the hydrogen atom is based on the premise that the electron can be described as a wave and not as a particle. Unlike Bohr's model, Schrödinger's approach resulted in mathematical equations that are complex and difficult to solve except in simple cases. We need not be concerned here with the mathematics, but the solutions to these equations—called **wave functions** and symbolized by the Greek letter ψ (psi)—are important. Understanding the implications of these wave functions is essential to understanding the modern view of the atom. The following important points can be made concerning wave functions:

1. The behavior of the electron in the atom is best described as a standing wave. In a vibrating string, only certain vibrations or standing waves (see Figure 7.2) can be observed. Similarly, *only certain wave functions are allowed for the electron in the atom.*

2. Each wave function ψ is associated with an allowed energy value, E_n, for the electron.

3. Taken together, points 1 and 2 say that *the energy of the electron is quantized*; that is, the electron can have only certain values of energy.

4. The concept of energy quantization enters Schrödinger's theory naturally with the basic assumption that an electron is a standing wave. This is in contrast with Bohr's theory, in which quantization was imposed as a postulate at the start.

5. The square of the wave function (ψ^2) is related to the probability of finding the electron within a given region of space. Scientists refer to this probability as the **electron density**.

6. Schrödinger's theory defines the energy of the electron precisely. The uncertainty principle, however, tells us there must be a large uncertainty in the electron's position. Thus, we can describe only the *probability* of the electron being within a certain region in space when in a given energy state. The region of space in which an electron of a given energy is most probably located is called its **orbital**.

7. To solve Schrödinger's equation for an electron in three-dimensional space, three integer numbers—the **quantum numbers n, ℓ, and m_ℓ**—are an integral part of the mathematical solution. These quantum numbers may have only certain combinations of values, as outlined below.

Quantum numbers are used to define the energy states and orbitals available to the electron. Let us first describe the quantum numbers and the information they provide. We will then turn to the connection between quantum numbers and the energies and shapes of atomic orbitals.

Quantum Numbers

Before looking into the meanings of the three quantum numbers n, ℓ, and m_ℓ, it is important to note two points:

- The quantum numbers are all integers, but their values cannot be selected randomly.
- The three quantum numbers (and their values) are not parameters that scientists dreamed up. Instead, when the behavior of the electron in the hydrogen atom is described mathematically as a wave, the quantum numbers are a natural consequence.

n, the Principal Quantum Number = 1, 2, 3, . . .

The principal quantum number *n* can have any integer value from 1 to infinity. The value of *n* is the primary factor in determining the energy of an electron. Indeed, for the hydrogen atom (with its single electron), the energy of the electron varies *only* with the value of *n* and is given by the same equation derived by Bohr for the H atom: $E_n = -Rhc/n^2$.

The value of *n* also defines the size of an orbital: The greater the value of *n*, the greater the electron's average distance from the nucleus.

Each electron is labeled according to its value of *n*. In atoms having more than one electron, two or more electrons may have the same *n* value. These electrons are then said to be in the same **electron shell** or same **electron level**.

ℓ, the Angular Momentum Quantum Number = 0, 1, 2, 3, . . . , *n* − 1

The electrons of a given shell can be grouped into **subshells**, where each subshell is characterized by a different value of the quantum number ℓ and by a characteristic shape. The quantum number ℓ can have any integer value from 0 to *n* − 1. *Each value of ℓ corresponds to a different orbital shape or orbital type.*

Because ℓ can be no larger than *n* − 1, the value of *n* limits the number of subshells possible for the *n*th shell. Thus, for *n* = 1, ℓ must equal 0. Because ℓ has only one value when *n* = 1, only one subshell is possible for an electron assigned to *n* = 1. When *n* = 2, ℓ can be either 0 or 1. Because two values of ℓ are now possible, there are two subshells in the *n* = 2 electron shell.

The values of ℓ are usually coded by letters according to the following scheme:

Value of ℓ	Corresponding Subshell Label
0	s
1	p
2	d
3	f

For example, an ℓ = 1 subshell is called a "*p* subshell," and an orbital found in that subshell is called a "*p* orbital." Conversely, an electron assigned to a *p* subshell has an ℓ value of 1.

m_ℓ, the Magnetic Quantum Number = 0, ±1, ±2, ±3, . . . , ±ℓ

The magnetic quantum number, m_ℓ, is related to the orientation in space of the orbitals within a subshell. *Orbitals in a given subshell differ only in their orientation in space, not in their energy.*

The value of ℓ limits the integer values assigned to m_ℓ: m_ℓ can range from +ℓ to −ℓ with 0 included. For example, when ℓ = 2, m_ℓ has five values: −2, −1, 0, +1, and +2. The number of values of m_ℓ for a given subshell (= 2ℓ + 1) specifies the number of orientations that exist for the orbitals of that subshell and thus the number of orbitals in the subshell.

■ **Electron Energy and Quantum Numbers**
The electron energy in the H atom depends *only* on the value of *n*. In atoms with more electrons, the energy depends on *both n* and ℓ. This is discussed in more detail in Section 8.3.

■ **Orbital Symbols**
Early studies of the emission spectra of elements classified lines into four groups on the basis of their appearance. These groups were labeled *sharp, principal, diffuse,* and *fundamental*. From these names came the labels we now apply to orbitals: *s, p, d,* and *f*.

Useful Information from Quantum Numbers

The three quantum numbers introduced thus far are a kind of ZIP code for electrons. For example, suppose you live in an apartment building. You could specify your location as being on a particular floor (n), in a particular apartment on that floor (ℓ), and in a particular room in the apartment (m_ℓ). Analogously, n describes the shell to which an electron is assigned in an atom, ℓ describes the subshell within that shell, and m_ℓ is related to the orientation of the orbital within that subshell.

Allowed values of the three quantum numbers are summarized in Table 7.1. Before describing the composition of the first four electron shells ($n = 1, 2, 3,$ and 4), let us summarize some useful points:

- Electrons in atoms are assigned to orbitals, which are grouped into subshells. One or more subshells with the same value of n constitute an electron shell.

- Electron subshells are labeled by first giving the value of n and then the value of ℓ in the form of its letter code. For $n = 1$ and $\ell = 0$, for example, the label is $1s$.

If you describe sets of quantum numbers, starting with a given value of n and then deciding the values of ℓ and then m_ℓ that follow (see Table 7.1), you would discover the following:

- n = the number of subshells in a shell
- $2\ell + 1$ = the number of orbitals in a subshell = the number of values of m_ℓ
- n^2 = the number of orbitals in a shell

First Electron Shell, $n = 1$

When $n = 1$ the value of ℓ can only be 0, and so m_ℓ must also have a value of 0. This means that, in the electron shell closest to the nucleus, only one subshell exists, and that subshell consists of only a single orbital, the $1s$ orbital.

For the Second Shell, $n = 2$

When $n = 2$, ℓ can have two values (0 and 1), so two subshells or two types of orbitals occur in the second shell. One of these is the $2s$ subshell ($n = 2$ and $\ell = 0$), and the other is the $2p$ subshell ($n = 2$ and $\ell = 1$). Because the values of m_ℓ can be $-1, 0,$ and $+1$ when $\ell = 1$, three p orbitals exist. All three orbitals have $\ell = 1$ so they all have the same shape. However, because each has a different m_ℓ value, the three orbitals differ in their orientation in space.

For the Third Shell, $n = 3$

When $n = 3$, three subshells, or orbital types, are possible for an electron because ℓ has the values 0, 1, and 2. Because you see ℓ values of 0 and 1 again, you know that two of the subshells within the $n = 3$ shell are $3s$ ($\ell = 0$, one orbital) and $3p$ ($\ell = 1$, three orbitals). The third subshell is d, indicated by $\ell = 2$. Because m_ℓ has five values ($-2, -1, 0, +1,$ and $+2$) when $\ell = 2$, five d orbitals (no more and no less) occur in the $\ell = 2$ subshell.

For the Fourth Shell, $n = 4$, and Beyond

Table 7.1 shows that there are four subshells in the $n = 4$ shell. In addition to s, p, and d orbitals, there is an f subshell; that is, there are orbitals for which $\ell = 3$. Seven such orbitals exist because there are seven values of m_ℓ when $\ell = 3$ ($-3, -2, -1, 0, +1, +2,$ and $+3$).

■ **Subshells and Orbitals**

Subshell	Number of Orbitals in Subshell
s	1
p	3
d	5
f	7

Table 7.1   Summary of the Quantum Numbers, Their Interrelationships, and the Orbital Information Conveyed

Principal Quantum Number	Angular Momentum Quantum Number	Magnetic Quantum Number	Number and Type of Orbitals in the Subshell
Symbol = n Values = 1, 2, 3, . . . n = number of subshells	Symbol = ℓ Values = 0 . . . $n-1$	Symbol = m_ℓ Values = $-\ell$ . . . 0 . . . $+\ell$	Number of orbitals in shell = n^2 and number of orbitals in subshell = $2\ell + 1$
1	0	0	one 1s orbital (one orbital of one type in the $n = 1$ shell)
2	0 1	0 + 1, 0, −1	one 2s orbital three 2p orbitals (four orbitals of two types in the $n = 2$ shell)
3	0 1 2	0 + 1, 0, −1 + 2, +1, 0, −1, −2	one 3s orbital three 3p orbitals five 3d orbitals (nine orbitals of three types in the $n = 3$ shell)
4	0 1 2 3	0 + 1, 0, −1 + 2, +1, 0, −1, −2 + 3, +2, +1, 0, −1, −2, −3	one 4s orbital three 4p orbitals five 4d orbitals seven 4f orbitals (16 orbitals of four types in the $n = 4$ shell)

GENERAL
Chemistry·⚛·Now™

See the General ChemistryNow CD-ROM or website:

- **Screen 7.9 Heisenberg's Uncertainty Principle,** to view an animation on the quantum mechanical view of the atom

- **Screen 7.12 Quantum Numbers and Orbitals,** for a tutorial on determining values for the quantum numbers for an orbital

Exercise 7.7—Using Quantum Numbers

Complete the following statements:

(a) When $n = 2$, the values of ℓ can be _____ and _____.
(b) When $\ell = 1$, the values of m_ℓ can be _____ , _____ , and _____ , and the subshell has the letter label _____.
(c) When $\ell = 2$, the subshell is called a _____ subshell.
(d) When a subshell is labeled s, the value of ℓ is _____ and m_ℓ has the value _____.
(e) When a subshell is labeled p, _____ orbitals occur within the subshell.
(f) When a subshell is labeled f, there are _____ values of m_ℓ, and _____ orbitals occur within the subshell.

7.6—The Shapes of Atomic Orbitals

The chemistry of an element and its compounds is determined by the atom's electrons, and particularly by the electrons with the highest value of n, which are often called *valence electrons* [▶ Section 9.1]. The types of orbitals to which these electrons are assigned are also important, so we turn now to the question of orbital shape and orientation.

s Orbitals

When an electron has $\ell = 0$, we often say the electron is assigned to, or "occupies," an s orbital. But what does this mean? What is an s orbital? What does it look like? To answer these questions, we begin with the wave function for an electron with $n = 1$ and $\ell = 0$, that is, with a $1s$ orbital. If we assume the electron is a tiny particle and not a wave, and if we could photograph the $1s$ electron at 1-second intervals for a few thousand seconds, the composite picture would look like the drawing in Figure 7.14a. It resembles a cloud of dots, so chemists refer to such representations of electron orbitals as **electron cloud pictures**.

The fact that the density of dots is greater close to the nucleus (the electron cloud is denser close to the nucleus) indicates that the electron is most often found near the nucleus (or, conversely, it is less likely to be found farther away). Putting this statement in the language of quantum mechanics, we say the greatest probability of finding the electron is in a tiny volume of space close to the nucleus. Conversely, the probability of finding the electron declines upon moving away from the nucleus; it is less probable that the electron is farther away. The thinning of the electron cloud at increasing distance, shown by the decreasing density of dots in Figure 7.14a, is illustrated in a different way in Figure 7.14b. Here we plotted the square of the wave function for the electron in a $1s$ orbital (ψ^2), times 4π and the distance squared ($4\pi r^2$), as a function of the distance of the electron from the nucleus. The units of $4\pi r^2 \psi^2$ at each point are $1/$distance, so the vertical axis of this plot represents the probability of finding the electron in a thin spherical shell a distance r from the nucleus. For this reason, $4\pi r^2 \psi^2$ is sometimes called a **surface density plot** or a **radial distribution plot**. For the $1s$ orbital, $4\pi r^2 \psi^2$ is zero at the nucleus—there is no probability the electron will be *at* the nucleus—but the probability is very high a short distance from the nucleus and decreases rapidly as the distance from the nucleus increases. Notice that the probability of finding the electron approaches but never quite reaches zero, even at very large distances.

For the $1s$ orbital, Figure 7.14a shows that the electron is most likely found within a sphere with the nucleus at the center. No matter in which direction you proceed from the nucleus, the probability of finding an electron is the same at the same distance from the nucleus (Figure 7.14b). *The $1s$ orbital is spherical in shape.*

The visual image in Figure 7.14a is that of a cloud whose density is small at large distances from the center; there is no sharp boundary beyond which the electron is never found. The s and other orbitals, however, are often depicted as having a sharp boundary surface (Figure 7.14c), largely because it is easier to draw such pictures. To arrive at the diagram in Figure 7.14c, we drew a sphere about the nucleus in such a way that the chance of finding the electron somewhere inside is 90%.

Misconceptions exist about pictures such as Figure 7.14c. First, there is not an impenetrable surface within which the electron is "contained." Second, the proba-

■ **Surface Density Plot for 1s**
The wave nature of the electron is evident from Figure 7.14b. The maximum amplitude of the electron wave occurs at 0.0529 nm. It is interesting to note that this maximum is at exactly the same distance from the nucleus as Niels Bohr calculated for the radius of the orbit occupied by the $n = 1$ electron.

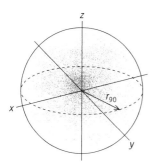

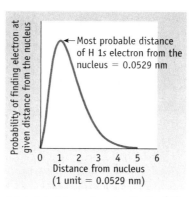

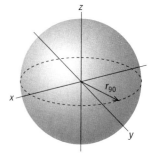

(a) Dot picture of an electron in a 1s orbital. Each dot represents the position of the electron at a different instant in time. Note that the dots cluster closest to the nucleus. r_{90} is the radius of a sphere within which the electron is found 90% of the time.

(b) A plot of the surface density ($4\pi r^2\psi^2$) as a function of distance for a hydrogen atom 1s orbital. This gives the probability of finding the electron at a given distance from the nucleus.

(c) The surface of the sphere within which the electron is found 90% of the time for a 1s orbital. This surface is often called a "boundary surface." (A 90% surface was chosen arbitrarily. If the choice was the surface within which the electron is found 50% of the time, the sphere would be considerably smaller.)

Active Figure 7.14   **Different views of a 1s ($n = 1$ and $\ell = 0$) orbital.**

GENERAL
Chemistry ⚛ Now™ See the General ChemistryNow CD-ROM or website to explore an interactive version of this figure accompanied by an exercise.

bility of finding the electron is not the same throughout the volume enclosed by the surface. For example, the electron in the H atom 1s orbital has a greater probability of being at 0.0529 nm from the nucleus than closer or farther away. Third, the terms "electron cloud" and "electron distribution" seem to imply that the electron is a particle, but quantum mechanics treats the electron as having wave properties.

Finally, an important feature of all *s* orbitals (1*s*, 2*s*, 3*s*, and so on) is that they are *spherical in shape*. One important difference between *s* orbitals with different *n* values, however, is that *the size of s orbitals increases as n increases* (Figure 7.15). Thus, the 1*s* orbital is more compact than the 2*s* orbital, which is in turn more compact than the 3*s* orbital.

p Orbitals

Atomic orbitals for which $\ell = 1$, *p* orbitals, all have the same basic shape. *All p orbitals have one imaginary plane that slices through the nucleus and that divides the region of electron density in half* (Figures 7.15 and 7.16). This imaginary plane is called a **nodal surface**, a planar surface on which there is zero probability of finding the electron. The electron can never be found on the nodal surface; the regions of electron density lie on either side of the nucleus. A plot of electron probability ($4\pi r^2\psi^2$) versus distance would start at zero at the nucleus, rise to a maximum, and then drop off at still greater distances.

If you enclose 90% of the electron density within a surface, the views in Figure 7.16 are appropriate. The electron cloud has a shape that resembles a weight lifter's "dumbbell," so chemists often describe *p* orbitals as having dumbbell shapes.

According to Table 7.1, when $\ell = 1$, then m_ℓ can only be -1, 0, or $+1$. That is, three orientations are possible for $\ell = 1$ or *p* orbitals. There are three mutually perpendicular directions in space (*x, y,* and *z*), and the *p* orbitals are commonly visualized as lying along those directions (with the nodal surface perpendicular to the axis). Each orbital is labeled according to the axis along which it lies (p_x, p_y, or p_z).

■ **Standing Waves and Nodal Surfaces** Recall that standing waves have nodes (Figure 7.2). Similarly, the electron waves in an atom have nodes.

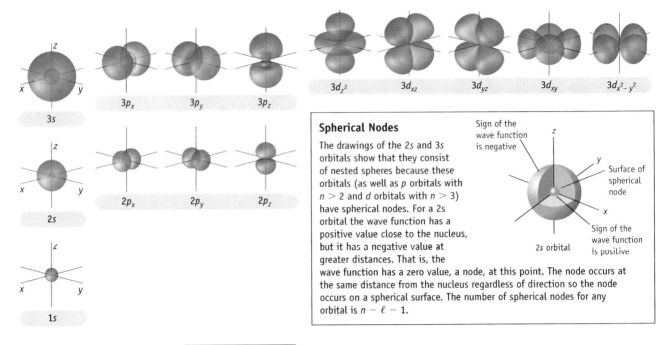

Spherical Nodes

The drawings of the 2s and 3s orbitals show that they consist of nested spheres because these orbitals (as well as p orbitals with $n > 2$ and d orbitals with $n > 3$) have spherical nodes. For a 2s orbital the wave function has a positive value close to the nucleus, but it has a negative value at greater distances. That is, the wave function has a zero value, a node, at this point. The node occurs at the same distance from the nucleus regardless of direction so the node occurs on a spherical surface. The number of spherical nodes for any orbital is $n - \ell - 1$.

Active Figure 7.15   **Atomic Orbitals.** Boundary surface diagrams for electron densities of 1s, 2s, 2p, 3s, 3p, and 3d orbitals for a hydrogen atom. For the p orbitals, the subscript letter on the orbital notation indicates the cartesian axis along which the orbital lies.

GENERAL Chemistry ·Now™ See the General ChemistryNow CD-ROM or website to explore an interactive version of this figure accompanied by an exercise.

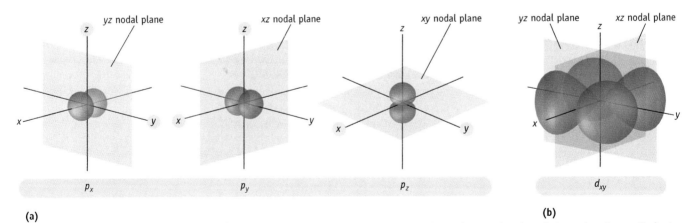

Figure 7.16   Nodal surfaces in p and d orbitals. A plane passing through the nucleus (perpendicular to this axis) is called a nodal surface. (a) The three p orbitals each have one nodal surface ($\ell = 1$). (b) The d_{xy} orbital. All five d orbitals have two nodal surfaces ($\ell = 2$). Here the nodal surfaces are the xz- and yz-planes, so the regions of electron density lie in the xy-plane and between the x- and y-axes.

d Orbitals

The value of ℓ is equal to the number of nodal surfaces that slice through the nucleus. Thus, *s* orbitals, for which $\ell = 0$, have no nodal surfaces, and *p* orbitals, for which $\ell = 1$, have one planar nodal surface. It follows that the five *d* orbitals, for which $\ell = 2$, have two nodal surfaces, which results in four regions of electron density. The d_{xy} orbital, for example, lies in the *xy*-plane and the two nodal surfaces are the *xz*- and *yz*-planes (see Figure 7.16). Two other orbitals, d_{xz} and d_{yz}, lie in planes defined by the *xz*- and *yz*-axes, respectively; they also have two, mutually perpendicular nodal surfaces (Figure 7.15).

Of the two remaining *d* orbitals, the $d_{x^2-y^2}$ orbital is easier to visualize. Like the d_{xy} orbital, the $d_{x^2-y^2}$ orbital results from two vertical planes slicing the electron density into quarters. Now, however, the planes bisect the *x*- and *y*-axes, so the regions of electron density lie along the *x*- and *y*-axes.

The final *d* orbital, d_{z^2} (Figure 7.15), has two main regions of electron density along the *z*-axis, but a "doughnut" of electron density also occurs in the *xy*-plane. This orbital has two nodal surfaces, but the surfaces are not flat.

f Orbitals

The seven *f* orbitals all have $\ell = 3$. The three nodal surfaces cause the electron density to lie in eight regions of space. These orbitals are less easily visualized, but one *f* orbital is illustrated in Figure 7.17.

GENERAL
Chemistry ⚛ Now ™

See the General ChemistryNow CD-ROM or website:
- **Screen 7.13 Shapes of Atomic Orbitals,** for exercises on orbital shapes, quantum numbers, and nodes

---

Exercise 7.8—Orbital Shapes

(a) What are the *n* and ℓ values for each of the following orbitals: *6s, 4p, 5d,* and *4f*?
(b) How many nodal planes exist for a *4p* orbital? For a *6d* orbital?

7.7—Atomic Orbitals and Chemistry

We close with some questions to ponder: When an element is part of a molecule, are the orbitals the same? Do they have the same shapes? What do the shapes of orbitals have to do with the chemistry of an element? We will take up these questions in the rest of the book, but a few answers are in order here.

Schrödinger's wave equation can be solved exactly for the hydrogen atom but not for heavier atoms or their ions. Nonetheless, chemists make the assumption that orbitals in other atoms are hydrogen-like, even when those atoms are part of a molecule. This approach has allowed chemists to make predictions, using computer-based simulations, about the behavior of molecules. Such predictions are often remarkably accurate, as confirmed by experimental observations.

■ **ℓ and Nodal Surfaces**

Orbital	ℓ	Number of Nodal Surfaces
s	0	0
p	1	1
d	2	2
f	3	3

■ **Nodal surfaces**

Nodal surfaces occur for all *p, d,* and *f* orbitals. These surfaces are usually flat, so they are referred to as nodal planes. In some cases (for example, d_{z^2}), however, the "plane" is not flat and so is better referred to as a "surface."

Charles D. Winters

Figure 7.17   One of the seven possible *f* orbitals. Notice the presence of three nodal planes as required by an orbital with $\ell = 3$.

Chemistry is the study of molecules and their transformations. By thinking about the orbitals of the atoms in molecules, and by making the simple assumption that they resemble those of the hydrogen atom, we can understand much of the chemistry of even complex systems such as those in plants and animals.

Chapter Goals Revisited

When you have finished studying this chapter, you should ask whether you have met the chapter goals. In particular, you should be able to

Describe the properties of electromagnetic radiation

a. Use the terms *wavelength, frequency, amplitude,* and *node* (Section 7.1). General ChemistryNow homework: Study Question(s) 3

b. Use Equation 7.1 ($c = \lambda\nu$), the relationship between the wavelength (λ) and frequency (ν) of electromagnetic radiation and the speed of light (c).

c. Recognize the relative wavelength (or frequency) of the various types of electromagnetic radiation (Figure 7.3). General ChemistryNow homework: SQ(s) 1

d. Understand that the energy of a photon, a massless particle of radiation, is proportional to its frequency (Planck's equation, Equation 7.2). This is an extension of Planck's idea that energy at the atomic level is quantized (Section 7.2). General ChemistryNow homework: SQ(s) 5, 12, 14, 54, 56, 61, 62, 76c

Understand the origin of light from excited atoms and its relationship to atomic structure

a. Describe the Bohr model of the atom, its ability to account for the emission line spectra of excited hydrogen atoms, and the limitations of the model (Section 7.3).

b. Understand that, in the Bohr model of the H atom, the electron can occupy only certain energy levels, each with an energy proportional to $1/n^2$ ($E = -Rhc/n^2$), where n is the principal quantum number (Equation 7.4, Section 7.3). If an electron moves from one energy state to another, the amount of energy absorbed or emitted in the process is equal to the difference in energy between the two states (Equation 7.5, Section 7.3). General ChemistryNow homework: SQ(s) 18, 22, 58

Describe the experimental evidence for wave-particle duality

a. Understand that in the modern view of the atom, electrons are described by the physics of waves (Section 7.4). The wavelength of an electron or any subatomic particle is given by de Broglie's equation (Equation 7.6). General ChemistryNow homework: SQ(s) 24

Describe the basic ideas of quantum mechanics

a. Recognize the significance of quantum mechanics in describing the modern view of atomic structure (Section 7.5).

b. Understand that an orbital for an electron in an atom corresponds to the allowed energy of that electron.

c. Understand that the position of the electron is not known with certainty; only the probability of the electron being at a given point of space can be calculated. This is the interpretation of the quantum mechanical model and embodies the postulate called the Heisenberg uncertainty principle.

Define the three quantum numbers (n, ℓ, and m_ℓ) and their relationship to atomic structure

    a. Describe the allowed energy states of the electron in an atom using three quantum numbers n, ℓ, and m_ℓ (Section 7.5). General ChemistryNow homework: SQ(s) 28, 30, 36, 38, 40, 74

    b. Describe the shapes of the orbitals (Section 7.6). General ChemistryNow homework: SQ(s) 44, 51, 65f

Key Equations

Equation 7.1 (page 297)
This equation states that the product of the wavelength (λ) and frequency (ν) of electromagnetic radiation is equal to the speed of light (c).

$$c = \lambda \times \nu$$

Equation 7.2 (page 302)
Planck's equation states that the energy of a photon, a massless particle of radiation, is proportional to its frequency (ν).

$$E = h\nu$$

where h is Planck's constant ($6.626 \times 10^{-34}\,\text{J}\cdot\text{s}$).

Equation 7.4 (page 308)
In Bohr's theory, the potential energy of the electron, E_n, in the nth quantum level of the H atom is proportional to $1/n^2$.

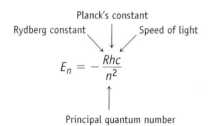

$$E_n = -\frac{Rhc}{n^2}$$

Planck's constant · Rydberg constant · Speed of light · Principal quantum number

where n is an integer equal to or greater than 1 and $Rhc = 2.179 \times 10^{-18}\,\text{J/atom}$ or $1312\,\text{kJ/mol}$.

Equation 7.5 (page 311)
The change in energy for an electron moving between two quantum levels (n_{final} and n_{initial}) in the H atom.

$$\Delta E = E_{\text{final}} - E_{\text{initial}} = -Rhc\left(\frac{1}{n_{\text{final}}^2} - \frac{1}{n_{\text{initial}}^2}\right)$$

Equation 7.6 (page 313)
De Broglie's equation relates the wavelength of the electron (λ) to its mass (m) and speed (v). h is Planck's constant.

$$\lambda = \frac{h}{mv}$$

Study Questions

▲ denotes more challenging questions.

■ denotes questions available in the Homework and Goals section of the General ChemistryNow CD-ROM or website.

Blue numbered questions have answers in Appendix O and fully worked solutions in the *Student Solutions Manual.*

Structures of many of the compounds used in these questions are found on the General ChemistryNow CD-ROM or website in the Models folder.

GENERAL Chemistry‑‑Now™ Assess your understanding of this chapter's topics with additional quizzing and conceptual questions at http://now.brookscole.com/kotz6e

Practicing Skills

Electromagnetic Radiation

(See Example 7.1, Exercise 7.1, Figure 7.3, and General ChemistryNow Screen 7.3.)

1. ■ Answer the following questions based on Figure 7.3:
   (a) Which type of radiation involves less energy, x-rays or microwaves?
   (b) Which radiation has the higher frequency, radar or red light?
   (c) Which radiation has the longer wavelength, ultraviolet or infrared light?

2. Consider the colors of the visible spectrum.
   (a) Which colors of light involve less energy than green light?
   (b) Which color of light has photons of greater energy, yellow or blue?
   (c) Which color of light has the greater frequency, blue or green?

3. ■ Traffic signals are often now made of LEDs (light-emitting diodes). Amber and green ones are pictured here.
   (a) The light from an amber signal has a wavelength of 595 nm, and that from a green signal has wavelength of 500 nm. Which has the higher frequency?
   (b) Calculate the frequency of amber light.

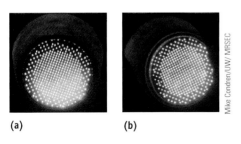

(a)   (b)

Mike Condren/UW/MRSEC

4. Suppose you are standing 225 m from a radio transmitter. What is your distance from the transmitter in terms of the number of wavelengths if
   (a) The station is broadcasting at 1150 kHz (on the AM radio band)? (1 kHZ = 1×10^3 Hz or 1000 cycles per second.)
   (b) The station is broadcasting at 98.1 MHz (on the FM radio band)? (1 MHz = 10^6 Hz, or cycles per second.)

Electromagnetic Radiation and Planck's Equation

(See page 304, Exercise 7.2, and General ChemistryNow Screens 7.4 and 7.5.)

5. ■ Green light has a wavelength of 5.0×10^2 nm. What is the energy, in joules, of one photon of green light? What is the energy, in joules, of 1.0 mol of photons of green light?

6. Violet light has a wavelength of about 410 nm. What is its frequency? Calculate the energy of one photon of violet light. What is the energy of 1.0 mol of violet photons? Compare the energy of photons of violet light with those of red light. Which is more energetic?

7. The most prominent line in the spectrum of aluminum is at 396.15 nm. What is the frequency of this line? What is the energy of one photon with this wavelength? Of 1.00 mol of these photons?

8. The most prominent line in the spectrum of magnesium is 285.2 nm. Other lines are found at 383.8 and 518.4 nm. In what region of the electromagnetic spectrum are these lines found? Which is the most energetic line? What is the energy of 1 mol of photons with the wavelength of the most energetic line?

9. Place the following types of radiation in order of increasing energy per photon:
   (a) yellow light from a sodium lamp
   (b) x-rays from an instrument in a dentist's office
   (c) microwaves in a microwave oven
   (d) your favorite FM music station at 91.7 MHz

10. Place the following types of radiation in order of increasing energy per photon.
    (a) radar signals
    (b) radiation within a microwave oven
    (c) gamma rays from a nuclear reaction
    (d) red light from a neon sign
    (e) ultraviolet radiation from a sun lamp

Photoelectric Effect

(See page 303 and Figure 7.6.)

11. An energy of 2.0×10^2 kJ/mol is required to cause a cesium atom on a metal surface to lose an electron. Calculate the longest possible wavelength of light that can ionize a cesium atom. In what region of the electromagnetic spectrum is this radiation found?

12. ■ You are an engineer designing a switch that works by the photoelectric effect. The metal you wish to use in your device requires 6.7×10^{-19} J/atom to remove an electron. Will the switch work if the light falling on the metal has a wavelength of 540 nm or greater? Why or why not?

Atomic Spectra and the Bohr Atom

(See Examples 7.2 and 7.3, Figures 7.9 –7.12, and General ChemistryNow Screens 7.6 and 7.7.)

13. The most prominent line in the spectrum of mercury is at 253.652 nm. Other lines are located at 365.015 nm, 404.656 nm, 435.833 nm, and 1013.975 nm.

    (a) Which of these lines represents the most energetic light?

    (b) What is the frequency of the most prominent line? What is the energy of one photon with this wavelength?

    (c) Are any of these lines found in the spectrum of mercury shown in Figure 7.9? What color or colors are these lines?

14. ■ The most prominent line in the spectrum of neon is found at 865.438 nm. Other lines are located at 837.761 nm, 878.062 nm, 878.375 nm, and 1885.387 nm.

    (a) In what region of the electromagnetic spectrum are these lines found?

    (b) Are any of these lines found in the spectrum of neon shown in Figure 7.9?

    (c) Which of these lines represents the most energetic light?

    (d) What is the frequency of the most prominent line? What is the energy of one photon with this wavelength?

15. A line in the Balmer series of emission lines of excited H atoms has a wavelength of 410.2 nm (Figure 7.12). What color is the light emitted in this transition? What quantum levels are involved in this emission line? What are the values of $n_{initial}$ and n_{final}?

16. What are the wavelength and frequency of the radiation involved in the least energetic emission line in the Lyman series? What quantum levels are involved in this emission line? What are the values of $n_{initial}$ and n_{final}?

17. Consider only transitions involving the $n = 1$ through $n = 5$ energy levels for the H atom (where the energy level spacings below are not to scale).

    _____ $n = 5$

    _____ $n = 4$

    _____ $n = 3$

    _____ $n = 2$

    _____ $n = 1$

    (a) How many emission lines are possible, considering only the five quantum levels?

    (b) Photons of the highest frequency are emitted in a transition from the level with $n =$ _____ to a level with $n =$ _____.

    (c) The emission line having the longest wavelength corresponds to a transition from the level with $n =$ _____ to the level with $n =$ _____.

18. ■ Consider only transitions involving the $n = 1$ through $n = 4$ energy levels for the hydrogen atom (using the diagram in Study Question 17).

    (a) How many emission lines are possible, considering only the four quantum levels?

    (b) Photons of the lowest energy are emitted in a transition from the level with $n =$ _____ to a level with $n =$ _____.

    (c) The emission line having the shortest wavelength corresponds to a transition from the level with $n =$ _____ to the level with $n =$ _____.

19. The energy emitted when an electron moves from a higher energy state to a lower energy state in any atom can be observed as electromagnetic radiation.

    (a) Which involves the emission of less energy in the H atom, an electron moving from $n = 4$ to $n = 2$ or an electron moving from $n = 3$ to $n = 2$?

    (b) Which involves the emission of more energy in the H atom, an electron moving from $n = 4$ to $n = 1$ or an electron moving from $n = 5$ to $n = 2$? Explain fully.

20. If energy is absorbed by a hydrogen atom in its ground state, the atom is excited to a higher energy state. For example, the excitation of an electron from the level with $n = 1$ to the level with $n = 3$ requires radiation with a wavelength of 102.6 nm. Which of the following transitions would require radiation of *longer wavelength* than this?

    (a) $n = 2$ to $n = 4$ (c) $n = 1$ to $n = 5$

    (b) $n = 1$ to $n = 4$ (d) $n = 3$ to $n = 5$

21. Calculate the wavelength and frequency of light emitted when an electron changes from $n = 3$ to $n = 1$ in the H atom. In what region of the spectrum is this radiation found?

22. ■ Calculate the wavelength and frequency of light emitted when an electron changes from $n = 4$ to $n = 3$ in the H atom. In what region of the spectrum is this radiation found?

DeBroglie and Matter Waves

(See Example 7.4 and General ChemistryNow Screen 7.8.)

23. An electron moves with a velocity of 2.5×10^8 cm $\cdot$ s^{-1}. What is its wavelength?

24. ■ A beam of electrons ($m = 9.11 \times 10^{-31}$ kg/electron) has an average speed of 1.3×10^8 m $\cdot$ s^{-1}. What is the wavelength of electrons having this average speed?

25. Calculate the wavelength, in nanometers, associated with a 1.0×10^2-g golf ball moving at 30. m $\cdot$ s^{-1} (about 67 mph). How fast must the ball travel to have a wavelength of 5.6×10^{-3} nm?

26. A rifle bullet (mass = 1.50 g) has a velocity of 7.00×10^2 mph. What is the wavelength associated with this bullet?

Quantum Mechanics

(See Sections 7.5 and 7.6 and General ChemistryNow Screens 7.9–7.14.)

27. (a) When $n = 4$, what are the possible values of ℓ?
    (b) When ℓ is 2, what are the possible values of m_ℓ?
    (c) For a 4s orbital, what are the possible values of n, ℓ, and m_ℓ?
    (d) For a 4f orbital, what are the possible values of n, ℓ, and m_ℓ?

28. ■ (a) When $n = 4$, $\ell = 2$, and $m_\ell = -1$, to what orbital type does this refer? (Give the orbital label, such as 1s.)
    (b) How many orbitals occur in the $n = 5$ electron shell? How many subshells? What are the letter labels of the subshells?
    (c) If a subshell is labeled f, how many orbitals occur in the subshell? What are the values of m_ℓ?

29. A possible excited state of the H atom has the electron in a 4p orbital. List all possible sets of quantum numbers n, ℓ, and m_ℓ for this electron.

30. ■ A possible excited state for the H atom has an electron in a 5d orbital. List all possible sets of quantum numbers n, ℓ, and m_ℓ for this electron.

31. How many subshells occur in the electron shell with the principal quantum number $n = 4$?

32. How many subshells occur in the electron shell with the principal quantum number $n = 5$?

33. Explain briefly why each of the following is not a possible set of quantum numbers for an electron in an atom.
    (a) $n = 2$, $\ell = 2$, $m_\ell = 0$
    (b) $n = 3$, $\ell = 0$, $m_\ell = -2$
    (c) $n = 6$, $\ell = 0$, $m_\ell = 1$

34. Which of the following represent valid sets of quantum numbers? For a set that is invalid, explain briefly why it is not correct.
    (a) $n = 3$, $\ell = 3$, $m_\ell = 0$ (c) $n = 6$, $\ell = 5$, $m_\ell = -1$
    (b) $n = 2$, $\ell = 1$, $m_\ell = 0$ (d) $n = 4$, $\ell = 3$, $m_\ell = -4$

35. What is the maximum number of orbitals that can be identified by each of the following sets of quantum numbers? When "none" is the correct answer, explain your reasoning.
    (a) $n = 3$, $\ell = 0$, $m_\ell = +1$ (c) $n = 7$, $\ell = 5$
    (b) $n = 5$, $\ell = 1$ (d) $n = 4$, $\ell = 2$, $m_\ell = -2$

36. ■ What is the maximum number of orbitals that can be identified by each of the following sets of quantum numbers? When "none" is the correct answer, explain your reasoning.
    (a) $n = 4$, $\ell = 3$ (c) $n = 2$, $\ell = 2$
    (b) $n = 5$ (d) $n = 3$, $\ell = 1$, $m_\ell = -1$

37. State which of the following orbitals cannot exist according to the quantum theory: 2s, 2d, 3p, 3f, 4f, and 5s. Briefly explain your answers.

38. ■ State which of the following are incorrect designations for orbitals according to the quantum theory: 3p, 4s, 2f, and 1p. Briefly explain your answers.

39. Write a complete set of quantum numbers (n, ℓ, and m_ℓ) that quantum theory allows for each of the following orbitals: (a) 2p, (b) 3d, and (c) 4f.

40. ■ Write a complete set of quantum numbers (n, ℓ, and m_ℓ) for each of the following orbitals: (a) 5f, (b) 4d, and (c) 2s.

41. A particular orbital has $n = 4$ and $\ell = 2$. What must this orbital be: (a) 3p, (b) 4p, (c) 5d, or (d) 4d?

42. A given orbital has a magnetic quantum number of $m_\ell = -1$. This could *not* be a (an)
    (a) f orbital   (c) p orbital
    (b) d orbital   (d) s orbital

43. How many nodal surfaces are associated with each of the following orbitals?
    (a) 2s   (b) 5d   (c) 5f

44. ■ How many nodal surfaces are associated with each of the following atomic orbitals?
    (a) 4f   (b) 2p   (c) 6s

General Questions on Atomic Structure

These questions are not designated as to type or location in the chapter. They may combine several concepts. More challenging questions are indicated by ▲.

45. Which of the following are applicable when explaining the photoelectric effect? Correct any statements that are wrong.
    (a) Light is electromagnetic radiation.
    (b) The intensity of a light beam is related to its frequency.
    (c) Light can be thought of as consisting of massless particles whose energy is given by Planck's equation, $E = h\nu$.

46. In what region of the electromagnetic spectrum for hydrogen is the Lyman series of lines found? The Balmer series?

47. Give the number of nodal surfaces for each orbital type: s, p, d, and f.

48. What is the maximum number of s orbitals found in a given electron shell? The maximum number of p orbitals? Of d orbitals? Of f orbitals?

49. Match the values of ℓ shown in the table with orbital type (s, p, d, or f).

ℓ Value	Orbital Type
3	_____
0	_____
1	_____
2	_____

---

50. Sketch a picture of the 90% boundary surface of an s orbital and the p_x orbital. Be sure the latter drawing shows why the p orbital is labeled p_x and not p_y, for example.

51. ■ Complete the following table.

Orbital Type	Number of Orbitals in a Given Subshell	Number of Nodal Surfaces
s	_____	_____
p	_____	_____
d	_____	_____
f	_____	_____

52. Excited H atoms have many emission lines. One series of lines, called the Pfund series, occurs in the infrared region. It results when an electron changes from higher energy levels to a level with $n = 5$. Calculate the wavelength and frequency of the lowest energy line of this series.

53. An advertising sign gives off red light and green light.
    (a) Which light has the higher-energy photons?
    (b) One of the colors has a wavelength of 680 nm and the other has a wavelength of 500 nm. Which color has which wavelength?
    (c) Which light has the higher frequency?

54. ■ Radiation in the ultraviolet region of the electromagnetic spectrum is quite energetic. It is this radiation that causes dyes to fade and your skin to develop a sunburn. If you are bombarded with 1.00 mol of photons with a wavelength of 375 nm, what amount of energy, in kilojoules per mole of photons, are you being subjected to?

55. A cell phone sends signals at about 850 MHz (1 MHz = 1×10^6 Hz or cycles per second).
    (a) What is the wavelength of this radiation?
    (b) What is the energy of 1.0 mol of photons with a frequency of 850 MHz?
    (c) Compare the energy in part (b) with the energy of a mole of photons of blue light (420 nm).
    (d) Comment on the difference in energy between 850 MHz radiation and blue light.

56. ■ Assume your eyes receive a signal consisting of blue light, $\lambda = 470$ nm. The energy of the signal is 2.50×10^{-14} J. How many photons reach your eyes?

57. If sufficient energy is absorbed by an atom, an electron can be lost by the atom and a positive ion formed. The amount of energy required is called the ionization energy. In the H atom, the ionization energy is that required to change the electron from $n = 1$ to $n = $ infinity. (See "A Closer Look: Experimental Evidence for Bohr's Theory," page 313.) Calculate the ionization energy for He$^+$ ion. Is the ionization energy of the He$^+$ more or less than that of H? (Bohr's theory applies to He$^+$ because it, like the H atom, has a single electron. The electron energy, however, is now given by $E = -Z^2 Rhc/n^2$, where Z is the atomic number of helium.)

58. ■ Suppose hydrogen atoms absorb energy so that electrons are excited to the $n = 7$ energy level. Electrons then undergo these transitions, among others: (a) $n = 7 \longrightarrow n = 1$; (b) $n = 7 \longrightarrow n = 6$; and (c) $n = 2 \longrightarrow n = 1$. Which transition produces a photon with (i) the smallest energy, (ii) the highest frequency, and (iii) the shortest wavelength?

59. Rank the following orbitals in the H atom in order of increasing energy: $3s$, $2s$, $2p$, $4s$, $3p$, $1s$, and $3d$.

60. How many orbitals correspond to each of the following designations?
    (a) $3p$ (d) $6d$ (g) $n = 5$
    (b) $4p$ (e) $5d$ (h) $7s$
    (c) $4p_x$ (f) $5f$

61. ■ Cobalt-60 is a radioactive isotope used in medicine for the treatment of certain cancers. It produces β particles and γ rays, the latter having energies of 1.173 and 1.332 MeV. (1 MeV = 1 million electron-volts and 1 eV = 9.6485×10^4 J/mol.) What are the wavelength and frequency of a γ-ray photon with an energy of 1.173 MeV?

62. ▲ ■ Exposure to high doses of microwaves can cause damage. Estimate how many photons, with $\lambda = 12$ cm, must be absorbed to raise the temperature of your eye by 3.0 °C. Assume the mass of an eye is 11 g and its specific heat capacity is 4.0 J/g · K.

63. When the *Sojourner* spacecraft landed on Mars in 1997, the planet was approximately 7.8×10^7 km from the earth. How long did it take for the television picture signal to reach earth from Mars?

64. The most prominent line in the emission spectrum of chromium is found at 425.4 nm. Other lines in the chromium spectrum are found at 357.9 nm, 359.3 nm, 360.5 nm, 427.5 nm, 429.0 nm, and 520.8 nm.
    (a) Which of these lines represents the most energetic light?
    (b) What color is light of wavelength 425.4 nm?

65. Answer the following questions as a summary quiz on the chapter.
    (a) The quantum number n describes the _____ of an atomic orbital.
    (b) The shape of an atomic orbital is given by the quantum number _____.
    (c) A photon of green light has _____ (less or more) energy than a photon of orange light.
    (d) The maximum number of orbitals that may be associated with the set of quantum numbers $n = 4$ and $\ell = 3$ is _____.
    (e) The maximum number of orbitals that may be associated with the quantum number set $n = 3$, $\ell = 2$, and $m_\ell = -2$ is _____.

(f) ■ Label each of the following orbital pictures with the appropriate letter:

(g) When $n = 5$, the possible values of ℓ are _____.

(h) The number of orbitals in the $n = 4$ shell is _____.

66. Answer the following questions as a summary quiz on this chapter.

(a) The quantum number n describes the _____ of an atomic orbital and the quantum number ℓ describes its _____.

(b) When $n = 3$, the possible values of ℓ are _____.

(c) What type of orbital corresponds to $\ell = 3$? _____

(d) For a 4d orbital, the value of n is _____ , the value of ℓ is _____, and a possible value of m_ℓ is _____.

(e) Each of the following drawings represents a type of atomic orbital. Give the letter designation for the orbital, give its value of ℓ, and specify the number of nodal surfaces.

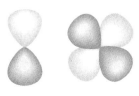

Letter =   _____   _____

ℓ value =   _____   _____

Nodal surfaces =   _____   _____

(f) An atomic orbital with three nodal surfaces is _____.

(g) Which of the following orbitals *cannot* exist according to modern quantum theory: 2s, 3p, 2d, 3f, 5p, 6p?

(h) Which of the following is *not* a valid set of quantum numbers?

n	ℓ	m_ℓ
3	2	1
2	1	2
4	3	0

(i) What is the maximum number of orbitals that can be associated with each of the following sets of quantum numbers? (One possible answer is "none.")

(i) $n = 2$ and $\ell = 1$

(ii) $n = 3$

(iii) $n = 3$ and $\ell = 3$

(iv) $n = 2$, $\ell = 1$, and $m_\ell = 0$

Summary and Conceptual Questions

The following questions use concepts from the previous chapters.

67. What are two major assumptions of Bohr's theory of atomic structure?

68. Bohr pictured the electrons of the atom as being located in definite orbits about the nucleus, just as the planets orbit the sun. Criticize this model.

69. Light is given off by a sodium- or mercury-containing streetlight when the atoms are excited. The light you see arises for which of the following reasons?

(a) Electrons are moving from a given energy level to one of higher n.

(b) Electrons are being removed from the atom, thereby creating a metal cation.

(c) Electrons are moving from a given energy level to one of lower n.

70. How do we interpret the physical meaning of the square of the wave function? What are the units of $4\pi r^2 \psi^2$?

71. What does "wave-particle duality" mean? What are its implications in our modern view of atomic structure?

72. Which of these are observable?

(a) position of an electron in an H atom

(b) frequency of radiation emitted by H atoms

(c) path of an electron in an H atom

(d) wave motion of electrons

(e) diffraction patterns produced by electrons

(f) diffraction patterns produced by light

(g) energy required to remove electrons from H atoms

(h) an atom

(i) a molecule

(j) a water wave

73. In principle, which of the following can be determined?

(a) the energy of an electron in the H atom with high precision and accuracy

(b) the position of a high-speed electron with high precision and accuracy

(c) at the same time, both the position and the energy of a high-speed electron with high precision and accuracy

74. ▲ ■ Suppose you live in a different universe where a different set of quantum numbers is required to describe the atoms of that universe. These quantum numbers have the following rules:

N, principal $1, 2, 3, \ldots, \infty$

L, orbital $= N$

M, magnetic $-1, 0, +1$

How many orbitals are there altogether in the first three electron shells?

75. A photon with a wavelength of 93.8 nm strikes a hydrogen atom, and light is emitted by the atom. How many emission lines would be observed? At what wavelengths? Explain briefly. (See Figure 7.12.)

76. ▲ Technetium is not found naturally on earth; it must be synthesized in the laboratory. Nonetheless, because it is radioactive it has valuable medical uses. For example, the element in the form of sodium pertechnetate ($NaTcO_4$) is used in imaging studies of the brain, thyroid, and salivary glands and in renal blood flow studies, among other things.

    (a) In what group and period of the periodic table is the element found?

    (b) The valence electrons of technetium are found in the $5s$ and $4d$ subshells. What is a set of quantum numbers (n, ℓ, and m_ℓ) for one of the electrons of the $5s$ subshell?

    (c) ■ Technetium emits a γ-ray with an energy of 0.141 MeV. (1 MeV = 1 million electron-volts, where 1 eV = 9.6485×10^4 J/mol.) What are the wavelength and frequency of a γ-ray photon with an energy of 0.141 MeV?

    (d) To make $NaTcO_4$, the metal is dissolved in nitric acid.

    $$7\,HNO_3(aq)\ +\ Tc(s) \longrightarrow$$
    $$HTcO_4(aq)\ +\ 7\,NO_2(g) +\ 3\,H_2O(\ell)$$

    and the product, $HTcO_4$, is treated with NaOH to make $NaTcO_4$.

    (i) Write a balanced equation for the reaction of $HTcO_4$ with NaOH.

    (ii) If you begin with 4.5 mg of Tc metal, how much $NaTcO_4$ can be made? What mass of NaOH, in grams, is required to convert all of the $HTcO_4$ into $NaTcO_4$?

77. Explain why you could or could not measure the wavelength of a golf ball in flight.

78. See the General ChemistryNow CD-ROM or website, Screen 7.1 Chemical Puzzler. This screen shows that light of different colors can come from a "neon" sign or from certain salts when they are placed in a burning organic liquid. ("Neon" signs are glass tubes filled with neon, argon, and other gases, and the gases are excited by an electric current. They are very similar in this regard to common fluorescent lights, although the light in fluorescent tubes comes from the phosphor that coats the inside of the tube.) What do these two sources of light have in common? How is the light generated in each case?

79. A large pickle is attached to two electrodes, which are then attached to a 110-V power supply (see the problem on Screen 7.7 of the General ChemistryNow CD-ROM or website). As the voltage is increased across the pickle, it begins to glow with a yellow color. Knowing that pickles are made by soaking the vegetable in a concentrated salt solution, describe why the pickle might emit light when electrical energy is added.

The "electric pickle."

80. See the General ChemistryNow CD-ROM or website, Screen 7.7 Bohr's Model of the Hydrogen Atom, Simulation. A photon with a wavelength of 97.3 nm is fired at a hydrogen atom and leads to the emission of light. How many emission lines are emitted? Explain why more than one line is emitted.

Do you need a live tutor for homework problems?
Access vMentor at General ChemistryNow at
http://now.brookscole.com/kotz6e
for one-on-one tutoring from a chemistry expert

▲ More challenging   ■ In General ChemistryNow   Blue-numbered questions answered in Appendix O

8— Atomic Electron Configurations and Chemical Periodicity

Courtesy of Chemical and Engineering News

Oliver Sacks. Born in London in 1933 to two physicians, Sacks now lives in New York City, where he is a practicing neurologist. He is a member of the American Academy of Arts and Letters and is the author of books such as *The Man Who Mistook His Wife for a Hat* and *Awakenings*. His most recent book, *Uncle Tungsten* (New York, Alfred Knopf, 2001), describes his lifelong fascination with chemistry.

Everything in Its Place

The periodic table of elements has put "everything in its place," according to Oliver Sacks. Sacks is a well-known neurologist, but he is also a writer of books such as *The Man Who Mistook His Wife for a Hat* and *Awakenings*. Less well known is the fact that he has had a love affair with chemistry since he was a boy growing up in London during World War II. On a trip to the London Science Museum, he saw a wall-sized periodic table that displayed samples of many of the 92 chemical elements known at that time. Said Sacks, "Seeing the table, with its actual samples of the elements, was one of the formative experiences of my boyhood and showed me, with the force of revelation, the beauty of science. The periodic table seemed so economical and simple: everything, the whole 92-ishness, reduced to two axes, and yet along each axis an ordered progression of different properties."

Dmitri Mendeleev, one of two people responsible for the creation of the periodic table, was born in Tobolsk in western Siberia on February 8, 1834. He was the youngest of 14 or 17 children (the number is not certain). His father became incapacitated shortly after Dmitri's birth, so, to support the large family, his mother took over a glass manufacturing business begun by her father.

Catastrophe struck the family in 1848 and 1849, when Mendeleev's father died and the glass factory burned. Young Mendeleev's mother was determined to ensure that he be schooled properly, so they journeyed 1300 miles to Moscow so that the boy could enroll in the university. Once in Moscow they found that students from Siberia were not permitted at the university, so they went another 400 miles to St. Petersburg. There Mendeleev's mother was able to secure a place for him at the Central Pedagogical Institute. She died shortly thereafter.

Mendeleev was an extraordinary student of science and published original work before he was 20, even though he was afflicted

See Chapter Goals Revisited (page 365). Test your knowledge of these goals by taking the exam-prep quiz on the General ChemistryNow CD-ROM or website.

- Understand the role magnetism plays in determining and revealing atomic structure.
- Understand effective nuclear charge and its role in determining atomic properties.
- Write the electron configuration for elements and monatomic ions.
- Understand the fundamental physical properties of the elements and their periodic trends.

with tuberculosis and had to do much of his writing in bed. He took the gold medal as the top student at the Institute in 1855 and shortly thereafter was sent to the Crimea as a teacher. The climate in the Crimea was hospitable, much suited to recovering from his illness. However, one reason he was sent far away from St. Petersburg was because he had a terrible temper and was less than beloved by his former teachers and colleagues.

Within a few years Mendeleev returned to St. Petersburg as a lecturer at the university. Soon thereafter, he went to study and do research in Paris, France, and Heidelberg, Germany. In Heidelberg he worked briefly with Robert Bunsen, the inventor of a burner used for spectroscopic studies. There also Mendeleev's temper got the better of him, and he was forced to retreat to a small room where he worked in isolation.

A defining moment for Mendeleev came in 1860 at a conference in Karlsruhe, Germany, where leading chemists from all over Europe came to settle on a system for determining atomic weights. This system, once in place, was crucial to Mendeleev's discovery a few years later of the periodic law and his publication of the first periodic table.

Novosti/Science Photo Library/Photo Researchers, Inc.

Dmitri Mendeleev seated at his desk. Every picture of him shows his long hair and beard. He was in the habit of having it cut only once a year. For more on the story of Mendeleev and the periodic table, see *Mendeleyev's Dream* by P. Strathern: New York, St. Martin's Press, 2001.

In 1861 Mendeleev returned to St. Petersburg and joined the faculty of the Technical Institute. His love of chemistry, as well as his intense blue eyes and flowing beard and hair, made him a popular teacher. He also realized that the teaching of chemistry in Russia was in a sorry state. To remedy this situation, he wrote a 500-page textbook of organic chemistry in only 60 days!

At the age of 32 Mendeleev was appointed professor of general chemistry at the University of St. Petersburg. By 1869 he had completed the first volume of a new textbook, *The Principles of Chemistry*, which was subsequently translated into all the major languages of the world. As he began the second volume, Mendeleev was searching for an organizing principle underlying chemistry. To look for patterns in the chemical and physical behaviors of the elements, he wrote lists of those properties on small cards, one for each element. After four days of pondering the problem for hours on end, he was so exhausted he fell asleep at his desk. In his words, "I saw in a dream a table where all the elements fell into place as required. Awakening, I immediately wrote it down on a piece of paper." This was the beginning of the periodic table chemists use today.

To Review Before You Begin
- Review Chapter 2 on atoms and atomic structure
- Review Chapter 7 on quantum numbers

The wave mechanical model of the atom accurately describes atoms or ions that have a single electron, such as H and He$^+$. It is obvious, however, that a truly useful model must be applicable to atoms with more than one electron—that is, to all the other known elements. One objective of this chapter, therefore, is to develop a workable model for the electronic structure of elements other than hydrogen.

A second objective is to explore some of the physical properties of elements, among them the ease with which atoms lose or gain electrons to form ions and the sizes of atoms and ions. These properties are directly related to the arrangement of electrons in atoms and thus to the chemistry of the elements and their compounds.

8.1—Electron Spin

Around 1920 it was demonstrated experimentally that the electron behaves as though it has a spin, just as the earth has a spin. To understand this property and its relationship to atomic structure requires understanding some aspects of the general phenomenon of magnetism. You will see that electron spin must be represented by a fourth quantum number, the **electron spin magnetic quantum number, m_s.** That is, the complete description of an electron in an atom requires *four* quantum numbers (n, ℓ, m_ℓ, and m_s).

Magnetism

In 1600, William Gilbert (1544–1603) concluded that the earth is a large spherical magnet giving rise to a magnetic field that surrounds the planet (Figure 8.1). The needle of a compass, itself a small magnet, lines up with earth's magnetic field, with

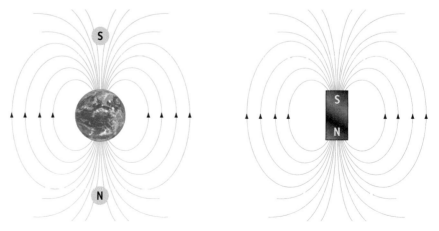

Figure 8.1 The magnetic fields of the earth and of a bar magnet. The lines of magnetic force of the earth come from one pole, arbitrarily called the "north magnetic pole" (N) and loop toward the "south magnetic pole" (S). (The geographic North Pole of the earth, named before the introduction of the term "magnetic pole," is actually the magnetic south pole.)

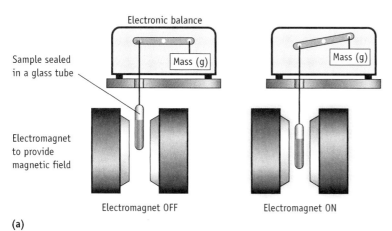

Charles D. Winters

(a)   (b)

Active Figure 8.2 **Observing and measuring paramagnetism.** (a) A magnetic balance is used to measure the magnetism of a sample. The sample is first weighed with the electromagnet turned off. The magnet is then turned on and the sample reweighed. If the substance is paramagnetic, the sample is drawn into the magnetic field and the apparent weight increases. (b) Liquid oxygen (boiling point 90.2 K) clings to the poles of a strong magnet. Elemental oxygen is paramagnetic because it has unpaired electrons. (See Chapter 10.)

GENERAL
Chemistry ⚛ Now™ See the General ChemistryNow CD-ROM or website to explore an interactive version of this figure accompanied by an exercise.

one end of the needle pointing approximately to the earth's geographic North Pole. We say the end of the compass needle pointing north is the magnet's "magnetic north pole" or simply its "north pole" (N). The other end of the needle is its "south pole" (S). Because opposite poles (N–S) attract, this means that earth's geographic north pole is its magnetic south pole.

Paramagnetism and Unpaired Electrons

Most substances are slightly repelled by a strong magnet; that is, they are **diamagnetic**. In contrast, some metals and compounds are attracted to a magnetic field. Such substances are called **paramagnetic**, and the magnitude of the effect can be determined with an apparatus such as that illustrated in Figure 8.2a.

The magnetism of most paramagnetic materials is so weak that you can observe the effect only in the presence of a strong magnetic field. For example, the oxygen we breathe is paramagnetic; it sticks to the poles of a strong magnet (Figure 8.2b).

Paramagnetism arises from electron spins. An electron in an atom has the magnetic properties expected for a spinning, charged particle (Figure 8.3). Experiments have shown that, if an atom with a single unpaired electron is placed in a magnetic field, only two orientations are possible for the electron spin: aligned with the field or opposed to the field. One orientation is associated with an electron spin quantum number value of $m_s = +\frac{1}{2}$ and the other with an m_s value of $-\frac{1}{2}$. *Electron spin is quantized.*

When one electron is assigned to an orbital in an atom, the electron's spin orientation can take either value of m_s. We observe experimentally that hydrogen atoms, each of which has a single electron, are paramagnetic; when an external magnetic field is applied, the electron magnets align with the field—like the needle of a compass—and experience an attractive force. Helium, with two electrons, is diamagnetic. To account for this observation, we assume that *the two electrons have*

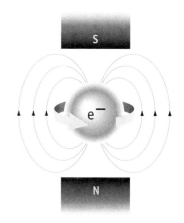

Figure 8.3 **Electron spin and magnetism.** The electron, with its spin and negative electric charge, acts as a "micromagnet." Relative to a magnetic field, only two spin directions are possible: clockwise or counterclockwise. The north pole of the spinning electron can therefore be either aligned with an external magnetic field or opposed to that field. (*See General ChemistryNow Screen 8.3 Spinning Electrons and Magnetism, to view an animation of concepts in this figure.*)

A Closer Look

Paramagnetism and Ferromagnetism

Magnetic materials are relatively common and many are important in our economy. For example, a large magnet is at the heart of the magnetic resonance imaging (MRI) used in medicine, and tiny magnets are found in stereo speakers and in telephone handsets. Magnetic oxides are used in recording tapes and computer disks.

The magnetic materials we use are *ferromagnetic*. The magnetic effect of ferro-

magnetic materials is much larger than that of paramagnetic ones. Ferromagnetism occurs when the spins of unpaired electrons in a cluster of atoms (called a *domain*) in the solid align themselves in the same direction. Only the metals of the iron, cobalt, and nickel subgroups, as well as a few other metals such as neodymium, exhibit this property. They are also unique in that, once the domains are aligned in a magnetic field, the metal is permanently magnetized.

Many alloys exhibit greater ferromagnetism than do the pure metals themselves. One example of such a material is Alnico,

and another is an alloy of neodymium, iron, and boron.

Audio and video tapes are plastics coated with crystals of ferromagnetic oxides such as Fe_2O_3 or CrO_2. The recording head uses an electromagnetic field to create a varying magnetic field based on signals from a microphone. This magnetizes the tape as it passes through the head, with the strength and direction of magnetization varying with the frequency of the sound to be recorded. When the tape is played back, the magnetic field of the moving tape induces a current, which is amplified and sent to the speakers.

Magnets. Many common consumer products contain magnetic materials.

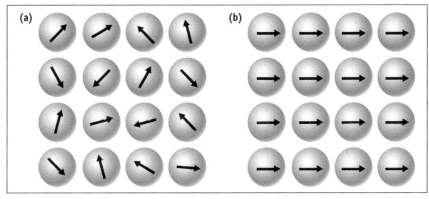

Magnetism. (a) Paramagnetism: In the absence of an external magnetic field, the unpaired electrons in the atoms or ions of the substance are randomly oriented. If a magnetic field is imposed, however, these spins will tend to become aligned with the field. (b) Ferromagnetism: The spins of the unpaired electrons in a cluster of atoms or ions align themselves in the same direction.

opposite spin orientations. We say their *spins are paired*, which means that the magnetic field of one electron is canceled out by the magnetic field of the second electron with opposite spin.

In summary, *paramagnetism is the attraction to a magnetic field of substances in which the constituent ions or atoms contain unpaired electrons.* Substances in which all electrons are paired with partners of opposite spin are diamagnetic. This explanation opens the way to understanding the arrangement of electrons in atoms with more than one electron.

GENERAL
Chemistry ⚛ Now™

See the General ChemistryNow CD-ROM or website:

- **Screen 8.3 Spinning Electrons and Magnetism,** to watch a video demonstrating the paramagnetism of liquid oxygen

Chemical Perspectives

Quantized Spins and MRI

Just as electrons have a spin, so do atomic nuclei. In the hydrogen atom, the single proton of the nucleus spins on its axis. For most heavier atoms, such as carbon, the atomic nucleus includes both protons and neutrons, and the entire entity has a spin. This property is important, because nuclear spin allows scientists to detect these atoms in molecules and to learn something about their chemical environments.

The technique used to detect the spins of atomic nuclei is **nuclear magnetic resonance (NMR)**. It is one of the most powerful methods currently available to determine molecular structures. About 20 years ago it was adapted as a diagnostic technique in medicine, where it is known as **magnetic resonance imaging (MRI)**.

Just as electron spin is quantized, so too is nuclear spin. The H atom nucleus can spin in either of two directions. If the H atom is placed in a strong, external magnetic field, however, the spinning nuclear magnet can align itself with the external field or against. If a sample of ethanol (CH_3CH_2OH), for example, is placed in a strong magnetic field, a slight excess of the H atom nuclei (and C atom nuclei) is aligned with the lines of force of the field.

The nuclei aligned with the field have a slightly lower energy than those not aligned. The NMR and MRI technologies depend on the fact that energy in the

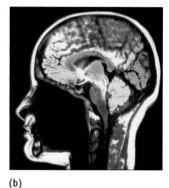

Magnetic resonance imaging. (a) MRI instrument. The patient is placed inside a large magnet, and the tissues to be examined are irradiated with radio-frequency radiation. (b) An MRI image of the human brain.

radio-frequency region can be absorbed by the sample and can cause the nuclear spins to go out of alignment—that is, to move to a higher energy state. This absorption of energy is detected by the instrument.

The most important aspect of the magnetic resonance technique is that the difference in energy between two different spin states depends on the locations of atoms in the molecule. In the case of ethanol, the three CH_3 protons are different from the two CH_2 protons, and both sets are different from the OH proton. These three different sets of H atoms absorb radiation of slightly different energies. The instrument measures the frequencies absorbed, and a scientist familiar with the technique can quickly distinguish the three different environments in the molecule.

The MRI technique closely resembles the NMR method. Hydrogen is abundant in the human body as water and in numerous organic molecules. In the MRI device, the patient is placed in a strong magnetic field, and the tissues being examined are irradiated with pulses of radio-frequency radiation.

The MRI image is produced by detecting how fast the excited nuclei "relax" from the higher energy state to the lower energy state. The "relaxation time" depends on the type of tissue. When the tissue is scanned, the H atoms in different regions of the body show different relaxation times, and an accurate "image" is built up.

MRI gives information on soft tissue— muscle, cartilage, and internal organs— which is unavailable from x-ray scans. This technology is also noninvasive, and the magnetic fields and radio-frequency radiation used are not harmful to the body.

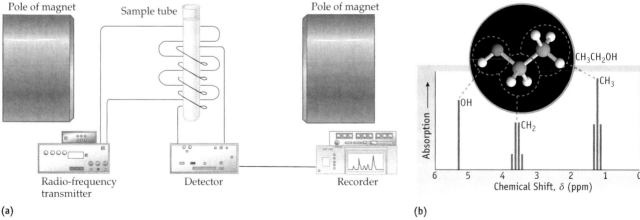

Nuclear magnetic resonance. (a) A schematic diagram of an NMR spectrometer. (b) The NMR spectrum of ethanol, showing that the three different types of protons appear in distinctly different regions of the spectrum. The pattern observed for the CH_2 and CH_3 protons, for example, is characteristic of these groups of atoms and signals the chemist that they are present in the molecule.

Table 8.3 Electron Configurations of Atoms in the Ground State

Z	Element	Configuration	Z	Element	Configuration	Z	Element	Configuration
1	H	$1s^1$	37	Rb	$[Kr]5s^1$	74	W	$[Xe]4f^{14}5d^46s^2$
2	He	$1s^2$	38	Sr	$[Kr]5s^2$	75	Re	$[Xe]4f^{14}5d^56s^2$
3	Li	$[He]2s^1$	39	Y	$[Kr]4d^15s^2$	76	Os	$[Xe]4f^{14}5d^66s^2$
4	Be	$[He]2s^2$	40	Zr	$[Kr]4d^25s^2$	77	Ir	$[Xe]4f^{14}5d^76s^2$
5	B	$[He]2s^22p^1$	41	Nb	$[Kr]4d^45s^1$	78	Pt	$[Xe]4f^{14}5d^96s^1$
6	C	$[He]2s^22p^2$	42	Mo	$[Kr]4d^55s^1$	79	Au	$[Xe]4f^{14}5d^{10}6s^1$
7	N	$[He]2s^22p^3$	43	Tc	$[Kr]4d^55s^2$	80	Hg	$[Xe]4f^{14}5d^{10}6s^2$
8	O	$[He]2s^22p^4$	44	Ru	$[Kr]4d^75s^1$	81	Tl	$[Xe]4f^{14}5d^{10}6s^26p^1$
9	F	$[He]2s^22p^5$	45	Rh	$[Kr]4d^85s^1$	82	Pb	$[Xe]4f^{14}5d^{10}6s^26p^2$
10	Ne	$[He]2s^22p^6$	46	Pd	$[Kr]4d^{10}$	83	Bi	$[Xe]4f^{14}5d^{10}6s^26p^3$
11	Na	$[Ne]3s^1$	47	Ag	$[Kr]4d^{10}5s^1$	84	Po	$[Xe]4f^{14}5d^{10}6s^26p^4$
12	Mg	$[Ne]3s^2$	48	Cd	$[Kr]4d^{10}5s^2$	85	At	$[Xe]4f^{14}5d^{10}6s^26p^5$
13	Al	$[Ne]3s^23p^1$	49	In	$[Kr]4d^{10}5s^25p^1$	86	Rn	$[Xe]4f^{14}5d^{10}6s^26p^6$
14	Si	$[Ne]3s^23p^2$	50	Sn	$[Kr]4d^{10}5s^25p^2$	87	Fr	$[Rn]7s^1$
15	P	$[Ne]3s^23p^3$	51	Sb	$[Kr]4d^{10}5s^25p^3$	88	Ra	$[Rn]7s^2$
16	S	$[Ne]3s^23p^4$	52	Te	$[Kr]4d^{10}5s^25p^4$	89	Ac	$[Rn]6d^17s^2$
17	Cl	$[Ne]3s^23p^5$	53	I	$[Kr]4d^{10}5s^25p^5$	90	Th	$[Rn]6d^27s^2$
18	Ar	$[Ne]3s^23p^6$	54	Xe	$[Kr]4d^{10}5s^25p^6$	91	Pa	$[Rn]5f^26d^17s^2$
19	K	$[Ar]4s^1$	55	Cs	$[Xe]6s^1$	92	U	$[Rn]5f^36d^17s^2$
20	Ca	$[Ar]4s^2$	56	Ba	$[Xe]6s^2$	93	Np	$[Rn]5f^46d^17s^2$
21	Sc	$[Ar]3d^14s^2$	57	La	$[Xe]5d^16s^2$	94	Pu	$[Rn]5f^67s^2$
22	Ti	$[Ar]3d^24s^2$	58	Ce	$[Xe]4f^15d^16s^2$	95	Am	$[Rn]5f^77s^2$
23	V	$[Ar]3d^34s^2$	59	Pr	$[Xe]4f^36s^2$	96	Cm	$[Rn]5f^76d^17s^2$
24	Cr	$[Ar]3d^54s^1$	60	Nd	$[Xe]4f^46s^2$	97	Bk	$[Rn]5f^97s^2$
25	Mn	$[Ar]3d^54s^2$	61	Pm	$[Xe]4f^56s^2$	98	Cf	$[Rn]5f^{10}7s^2$
26	Fe	$[Ar]3d^64s^2$	62	Sm	$[Xe]4f^66s^2$	99	Es	$[Rn]5f^{11}7s^2$
27	Co	$[Ar]3d^74s^2$	63	Eu	$[Xe]4f^76s^2$	100	Fm	$[Rn]5f^{12}7s^2$
28	Ni	$[Ar]3d^84s^2$	64	Gd	$[Xe]4f^75d^16s^2$	101	Md	$[Rn]5f^{13}7s^2$
29	Cu	$[Ar]3d^{10}4s^1$	65	Tb	$[Xe]4f^96s^2$	102	No	$[Rn]5f^{14}7s^2$
30	Zn	$[Ar]3d^{10}4s^2$	66	Dy	$[Xe]4f^{10}6s^2$	103	Lr	$[Rn]5f^{14}6d^17s^2$
31	Ga	$[Ar]3d^{10}4s^24p^1$	67	Ho	$[Xe]4f^{11}6s^2$	104	Rf	$[Rn]5f^{14}6d^27s^2$
32	Ge	$[Ar]3d^{10}4s^24p^2$	68	Er	$[Xe]4f^{12}6s^2$	105	Db	$[Rn]5f^{14}6d^37s^2$
33	As	$[Ar]3d^{10}4s^24p^3$	69	Tm	$[Xe]4f^{13}6s^2$	106	Sg	$[Rn]5f^{14}6d^47s^2$
34	Se	$[Ar]3d^{10}4s^24p^4$	70	Yb	$[Xe]4f^{14}6s^2$	107	Bh	$[Rn]5f^{14}6d^57s^2$
35	Br	$[Ar]3d^{10}4s^24p^5$	71	Lu	$[Xe]4f^{14}5d^16s^2$	108	Hs	$[Rn]5f^{14}6d^67s^2$
36	Kr	$[Ar]3d^{10}4s^24p^6$	72	Hf	$[Xe]4f^{14}5d^26s^2$	109	Mt	$[Rn]5f^{14}6d^77s^2$
			73	Ta	$[Xe]4f^{14}5d^36s^2$			

Electron configurations are often written in abbreviated form by combining the **noble gas notation** with the *spdf* or orbital box notation. The arrangement preceding the $2s$ electron is that of the noble gas helium so, instead of writing out $1s^22s^1$, the completed electron shells are represented by placing the symbol of the corresponding noble gas in brackets. Thus, lithium's configuration would be written as $[He]2s^1$.

The electrons included in the noble gas notation are often referred to as the **core electrons** of the atom. Not only is it a time-saving way to write electron config-

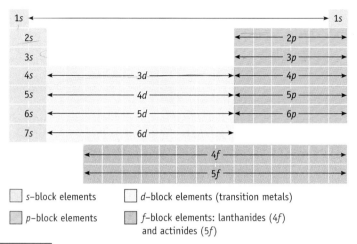

s–block elements   d–block elements (transition metals)

p–block elements   f–block elements: lanthanides ($4f$) and actinides ($5f$)

Active Figure 8.7   **Electron configurations and the periodic table.** The outermost electrons of an element are assigned to the indicated orbitals. See Table 8.3.

GENERAL

Chemistry · Now™ See the General ChemistryNow CD-ROM or website to explore an interactive version of this figure accompanied by an exercise.

urations, but the noble gas notation also conveys the idea that the core electrons can generally be ignored when considering the chemistry of an element. The electrons beyond the core electrons—the $2s^1$ electron in the case of lithium—are the **valence electrons**, the electrons that determine the chemical properties of an element.

The position of lithium in the periodic table tells you its configuration immediately. All the elements of Group 1A have one electron assigned to an s orbital of the nth shell, for which n is the number of the period in which the element is found (Figure 8.7). For example, potassium is the first element in the $n = 4$ row (the fourth period), so potassium has the electron configuration of the element preceding it in the table (Ar) plus a final electron assigned to the $4s$ orbital: $[Ar]4s^1$.

Beryllium (Be) and Other Elements of Group 2A

Beryllium, in Group 2A, has two electrons in the 1s orbital plus two additional electrons.

Beryllium: *spdf* notation $1s^2 2s^2$ or $[He]2s^2$

          Box notation   ⇅   ⇅   ☐☐☐
                          $1s$ $2s$ $2p$

All elements of Group 2A have electron configurations of [*electrons of preceding noble gas*] ns^2, where n is the period in which the element is found in the periodic table. Because all the elements of Group 1A have the valence electron configuration ns^1, and those in Group 2A have ns^2, these elements are called *s-block elements.*

Boron (B) and Other Elements of Group 3A

Boron (Group 3A) is the first element in the block of elements on the right side of the periodic table. Because the 1s and 2s orbitals are filled in a boron atom, the fifth electron must be assigned to a $2p$ orbital.

Boron: *spdf* notation $1s^2 2s^2 2p^1$ or $[He]2s^2 2p^1$

          Box notation   ⇅   ⇅   ↑ ☐☐
                          $1s$ $2s$ $2p$

Elements from Group 3A through Group 8A are often called the *p*-block elements. All have the general configuration ns^2np^x, where *x* varies from 1 to 6 (and is equal to the group number minus 2) (plus filled *d* orbitals for heavier elements as outlined below).

Carbon (C) and Other Elements of Group 4A

Carbon (Group 4A) is the second element in the *p* block, so a second electron is assigned to the 2*p* orbitals. For carbon to be in its lowest energy (or ground state) this electron must be assigned to either of the remaining *p* orbitals, and it will have the same spin direction as the first *p* electron.

Carbon: *spdf* notation $1s^22s^22p^2$ or $[He]2s^22p^2$

Box notation   ⊡ ⊡ ⊡⊡⊡
                   1s   2s   2p

In general, when electrons are assigned to *p*, *d*, or *f* orbitals, each successive electron is assigned to a different orbital of the subshell, and each electron has the same spin as the previous one; this pattern continues until the subshell is half full. Additional electrons must then be assigned to half-filled orbitals. This procedure follows **Hund's rule**, which states that the most stable arrangement of electrons is that with the maximum number of unpaired electrons, all with the same spin direction. This arrangement makes the total energy of an atom as low as possible.

Carbon is the second element in the *p* block of elements, so it has two electrons in *p* orbitals. Because carbon is a second-period element, the *p* orbitals involved are 2*p*. Thus, you can immediately write the carbon electron configuration by referring to the periodic table: Starting at H and moving from left to right across the successive periods, you write $1s^2$ to reach the end of period 1, and then $2s^2$ and finally $2p^2$ to bring the electron count to six. Carbon, the lightest element of Group 4A, has four electrons in the $n = 2$ shell.

Nitrogen (N) and Oxygen (O) and Elements of Groups 5A and 6A

Nitrogen (Group 5A) has five valence electrons. Besides the two 2*s* electrons, it has three electrons, all with the same spin, in three different 2*p* orbitals.

Nitrogen: *spdf* notation $1s^22s^22p^3$ or $[He]2s^22p^3$

Box notation   ⊡ ⊡ ⊡⊡⊡
                   1s   2s   2p

Oxygen (Group 6A) has six valence electrons. Two of these six electrons are assigned to the 2*s* orbital, and, as oxygen is the fourth element in the *p* block, the other four electrons are assigned to 2*p* orbitals.

Oxygen: *spdf* notation $1s^22s^22p^4$ or $[He]2s^22p^4$

Box notation   ⊡ ⊡ ⊡⊡⊡
                   1s   2s   2p

This means the fourth 2*p* electron must pair up with one already present. It makes no difference to which orbital this electron is assigned (the 2*p* orbitals all have the same energy), but it must have a spin opposite to the other electron already assigned to that orbital so that each electron has a different set of quantum numbers (the Pauli exclusion principle).

Fluorine (F) and Neon (Ne) and Elements of Groups 7A and 8A

Fluorine (Group 7A) has seven electrons in the $n = 2$ shell. Two of these electrons occupy the $2s$ subshell, and the remaining five electrons occupy the $2p$ subshell.

Fluorine: *spdf* notation $\qquad 1s^2 2s^2 2p^5 \qquad$ or $\qquad$ [He]$2s^2 2p^5$

Box notation

$\uparrow\downarrow$	$\uparrow\downarrow$	$\uparrow\downarrow$	$\uparrow\downarrow$	$\uparrow$
$1s$	$2s$		$2p$	

All halogen atoms have a similar configuration, $ns^2 np^5$, where n is the period in which the element is located.

Like the other elements in Group 8A, neon is a noble gas. All Group 8A elements (except helium) have eight electrons in the shell of highest n value, so all have the configuration $ns^2 np^6$, where n is the period in which the element is found. That is, all the noble gases have filled ns and np subshells. As you will see, the nearly complete chemical inertness of the noble gases correlates with this electron configuration.

Neon: *spdf* notation $\qquad 1s^2 2s^2 2p^6 \qquad$ or $\qquad$ [He]$2s^2 2p^6$

Box notation

$\uparrow\downarrow$	$\uparrow\downarrow$	$\uparrow\downarrow$	$\uparrow\downarrow$	$\uparrow\downarrow$
$1s$	$2s$		$2p$	

Elements of Period 3

The first element of the third period, sodium, is in Group 1A. The electron configuration of the element is that of a neon core plus one $3s$ electron.

Sodium: *spdf* notation $\qquad 1s^2 2s^2 2p^6 3s^1 \qquad$ or $\qquad$ [Ne]$3s^1$

Box notation

$\uparrow\downarrow$	$\uparrow\downarrow$	$\uparrow\downarrow$	$\uparrow\downarrow$	$\uparrow\downarrow$	$\uparrow$
$1s$	$2s$		$2p$		$3s$

Moving across the third period, we come to silicon. This element is in Group 4A and so has four electrons beyond the neon core. Because it is the second element in the p block, it has two electrons in $3p$ orbitals. Thus, its electron configuration is

Silicon: *spdf* notation $\qquad 1s^2 2s^2 2p^6 3s^2 3p^2 \qquad$ or $\qquad$ [Ne]$3s^2 3p^2$

Box notation

$\uparrow\downarrow$	$\uparrow\downarrow$	$\uparrow\downarrow$	$\uparrow\downarrow$	$\uparrow\downarrow$	$\uparrow\downarrow$	$\uparrow$	$\uparrow$	
$1s$	$2s$		$2p$		$3s$		$3p$	

From silicon to the end of the third period, electrons are added to the $3p$ orbitals in the same manner as the elements in the second period. Finally, at argon the $3p$ subshell is completed with six electrons.

GENERAL
Chemistry ⚛ Now™

See the General ChemistryNow CD-ROM or website:

• **Screen 8.7 Atomic Electron Configurations**

(a) for a simulation exploring the relationship between an element's electron configuration and its position in the periodic table

(b) for a tutorial on determining an element's box notation

(c) for a tutorial on detemining an element's *spdf* notation

(d) for a tutorial on determining whether an element is diamagnetic or paramagnetic

---

Example 8.1—Electron Configurations

Problem Give the electron configuration of sulfur, using the *spdf*, noble gas, and orbital box notations.

Strategy Sulfur, atomic number 16, is the sixth element in the third period ($n = 3$), and is in the *p* block. The last six electrons assigned to the atom, therefore, have the configuration $3s^2 3p^4$. These are preceded by the completed shells $n = 1$ and $n = 2$, the electron arrangement for Ne.

Solution The electron configuration of sulfur is

Complete *spdf* notation:	$1s^2 2s^2 2p^6 3s^2 3p^4$
spdf with noble gas notation:	$[Ne]3s^2 3p^4$
Orbital box notation:	$[Ne]$ [↑↓] [↑↓][↑][↑]
	$\quad\quad\quad 3s \quad\quad 3p$

Example 8.2—Electron Configurations and Quantum Numbers

Problem Write the electron configuration for Al using the noble gas notation, and give a set of quantum numbers for each of the electrons with $n = 3$ (the valence electrons).

Strategy Aluminum is the third element in the third period. It therefore has three electrons with $n = 3$. Because Al is in the *p* block of elements, two of the electrons are assigned to 3s and the remaining electron is assigned to 3p.

Solution The element is preceded by the noble gas neon, so the electron configuration is $[Ne]3s^2 3p^1$. Using box notation, the configuration is

Aluminum configuration:	$[Ne]$ [↑↓] [↑][ ][ ]
	$\quad\quad\quad 3s \quad\quad 3p$

The possible sets of quantum numbers for the two 3s electrons are

	n	ℓ	m_ℓ	m_s
For ↑	3	0	0	$+\frac{1}{2}$
For ↓	3	0	0	$-\frac{1}{2}$

For the single 3p electron, one of six possible sets is $n = 3$, $\ell = 1$, $m_\ell = +1$, and $m_s = +\frac{1}{2}$.

Exercise 8.2—*spdf* Notation, Orbital Box Diagrams, and Quantum Numbers

(a) Which element has the configuration $1s^2 2s^2 2p^6 3s^2 3p^5$?

(b) Using *spdf* notation and a box diagram, show the electron configuration of phosphorus.

(c) Write one possible set of quantum numbers for the valence electrons of calcium.

Electron Configurations of the Transition Elements

The elements of the fourth through the seventh periods use d or f subshells, in addition to s and p subshells, to accommodate electrons (see Figure 8.7 and Table 8.4). Elements whose atoms are filling d subshells are described as *transition elements*. Those for which f subshells are filling are sometimes called the inner transition elements or, more usually, the **lanthanides** (filling $4f$ orbitals) and **actinides** (filling $5f$ orbitals).

The transition elements are always preceded in the periodic table by two s-block elements (Figure 8.7). Accordingly, scandium, the first transition element, has the configuration $[Ar]3d^1 4s^2$, and titanium follows with $[Ar]3d^2 4s^2$ (Table 8.4).

The general procedure for assigning electrons would suggest that the configuration of the chromium atom is $[Ar]3d^4 4s^2$. The actual configuration, however, has one electron assigned to each of the six available $3d$ and $4s$ orbitals: $[Ar]3d^5 4s^1$. This phenomenon is explained by assuming that the $4s$ and $3d$ orbitals have approximately the same energy in Cr, and each of the six valence electrons of chromium is assigned to one of these orbitals. This element illustrates the fact that occasionally minor differences crop up between the predicted and actual configurations. These discrepancies have little or no effect on the chemistry of the element, however.

Following chromium, atoms of manganese, iron, and nickel have the configurations that would be expected from the order of orbital filling in Figure 8.5. The Group 1B element copper, however, has a single electron in the $4s$ orbital, and the remaining ten electrons beyond the argon core are assigned to the $3d$ orbitals. Zinc ends the first transition series. This Group 2B element has two electrons assigned to the $4s$ orbital, and the $3d$ orbitals are completely filled with ten electrons.

Lanthanides and Actinides

The fifth period ($n = 5$) follows the pattern of the fourth period with minor variations. The sixth period, however, includes the lanthanide series beginning with lanthanum, La. As the first element in the d block, lanthanum has the configuration $[Xe]5d^1 6s^2$. The next element, cerium (Ce), is set out in a separate row at the bottom of the periodic table, and it is with the elements in this row (Ce through Lu) that electrons are first assigned to f orbitals. Thus, the configuration of cerium is $[Xe]4f^1 5d^1 6s^2$. Moving across the lanthanide series, the pattern continues with

■ **Writing Electron Configurations**
Although it does not necessarily reflect the filling order, we follow the convention of writing the orbitals in order of increasing n when writing electron configurations. For a given n, the subshells are listed in order of increasing ℓ.

Table 8.4   Orbital Box Diagrams for the Elements Ca Through Zn

		3d	4s
Ca	$[Ar]4s^2$	☐☐☐☐☐	↑↓
Sc	$[Ar]3d^14s^2$	↑ ☐☐☐☐	↑↓
Ti	$[Ar]3d^24s^2$	↑ ↑ ☐☐☐	↑↓
V	$[Ar]3d^34s^2$	↑ ↑ ↑ ☐☐	↑↓
Cr*	$[Ar]3d^54s^1$	↑ ↑ ↑ ↑ ↑	↑
Mn	$[Ar]3d^54s^2$	↑ ↑ ↑ ↑ ↑	↑↓
Fe	$[Ar]3d^64s^2$	↑↓ ↑ ↑ ↑ ↑	↑↓
Co	$[Ar]3d^74s^2$	↑↓ ↑↓ ↑ ↑ ↑	↑↓
Ni	$[Ar]3d^84s^2$	↑↓ ↑↓ ↑↓ ↑ ↑	↑↓
Cu*	$[Ar]3d^{10}4s^1$	↑↓ ↑↓ ↑↓ ↑↓ ↑↓	↑
Zn	$[Ar]3d^{10}4s^2$	↑↓ ↑↓ ↑↓ ↑↓ ↑↓	↑↓

*These configurations do not follow the "$n + \ell$" rule.

some variation, with 14 electrons being assigned to the seven $5f$ orbitals in the last element, lutetium (Lu, $[Xe]4f^{14}5d^16s^2$) (see Table 8.3).

The seventh period also includes an extended series of elements utilizing f orbitals, the actinides, which begins with actinium (Ac, $[Rn]6d^17s^2$). The next element is thorium (Th), which is followed by protactinium (Pa) and uranium (U). The electron configuration of uranium is $[Rn]5f^36d^17s^2$. The third element in the actinide series, it has three $5f$ electrons.

When you have completed this section, you should be able to depict accurately the electron configuration of any element in the s and p blocks using the periodic table as a guide. Prediction of the electron configurations for atoms of elements in the d and f blocks (Table 8.3) is somewhat less precise, but you are reminded that these small "anomalies" have little effect on the chemical behavior of the elements.

Example 8.3—Electron Configurations of the Transition Elements

Problem Using the *spdf* and noble gas notations, give electron configurations for (a) technetium, Tc, and (b) osmium, Os.

Strategy Base your answer on the positions of the elements in the periodic table. That is, for each element, find the preceding noble gas and then note the number of s, p, d, and f electrons that lead from the noble gas to the element.

Solution

(a) *Technetium, Tc:* The noble gas that precedes Tc is krypton, Kr, at the end of the $n = 4$ row. After the 36 electrons of Kr are assigned, seven electrons remain. Two of these electrons are in the 5s orbital, and the remaining five are in 4d orbitals. Therefore, the technetium configuration is $[Kr]4d^5 5s^2$.

(b) *Osmium, Os:* Osmium is a sixth-period element and the twenty-second element following the noble gas xenon. Of the 22 electrons to be added after the Xe core, 2 are assigned to the 6s orbital and 14 to 4f orbitals. The remaining 6 are assigned to 5d orbitals. Thus, the osmium configuration is $[Xe]4f^{14}5d^6 6s^2$.

Exercise 8.3—Electron Configurations

Using the periodic table and without looking at Table 8.3, write electron configurations for the following elements:

(a) P **(c)** Zr **(e)** Pb
(b) Zn **(d)** In **(f)** U

Use the *spdf* and noble gas notations. When you have finished, check your answers with Table 8.3.

8.5—Electron Configurations of Ions

Much of the chemistry of the elements involves the formation of ions, and we can write their electron configurations as well as those of the elements. To form a cation from a neutral atom, one or more of the valence electrons is removed; that is, electrons are removed from the electron shell of highest n. If several subshells are present within the nth shell, the electron or electrons of maximum ℓ are removed. Thus, a sodium ion is formed by removing the $3s^1$ electron from the Na atom,

$$Na: [1s^2 2s^2 2p^6 3s^1] \longrightarrow Na^+: [1s^2 2s^2 2p^6] + e^-$$

and Ge^{2+} is formed by removing two 4p electrons from a germanium atom,

$$Ge: [Ar]3d^{10}4s^2 4p^2 \longrightarrow Ge^{2+}: [Ar]3d^{10}4s^2 + 2\ e^-$$

The same general rule applies to transition metal atoms. This means the titanium(II) cation has the configuration $[Ar]3d^2$, for example:

$$Ti: [Ar]3d^2 4s^2 \longrightarrow Ti^{2+}: [Ar]3d^2 + 2e^-$$

The iron(II) and iron(III) cations have the configurations $[Ar]3d^6$ and $[Ar]3d^5$, respectively:

$$Fe: [Ar]3d^6 4s^2 \longrightarrow Fe^{2+}: [Ar]3d^6 + 2\ e^-$$
$$Fe^{2+}: [Ar]3d^6 \longrightarrow Fe^{3+}: [Ar]3d^5 + e^-$$

All common transition metal cations have electron configurations of the general type [noble gas core]$(n - 1)d^x$. That is, in the process of ionization the ns electrons are lost before $(n - 1)d$ electrons. It is important to remember this point because the chemical and physical properties of transition metal cations are determined by the presence of electrons in d orbitals.

Atoms and ions with unpaired electrons are paramagnetic; that is, they are capable of being attracted to a magnetic field [◀ Section 8.1]. Paramagnetism is

Charles D. Winters

Formation of iron(III) chloride. When iron reacts with chlorine (Cl_2) to produce $FeCl_3$, each iron atom loses three electrons to give a paramagnetic Fe^{3+} ion with the configuration $[Ar]3d^5$.

Figure 8.8   Paramagnetism. (a) A sample of iron(III) oxide is packed into a plastic tube and suspended from a thin nylon filament. (b) When a powerful magnet is brought near it, the paramagnetic iron(III) ions in Fe_2O_3 cause the sample to be attracted to the magnet. [The magnet is made of neodymium, iron, and boron ($Nd_2Fe_{14}B$). These magnets are powerful enough to attract a U.S. $1 bill, which is printed with ink containing a small quantity of an iron-based compound.]

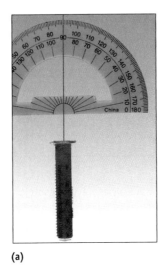

(a)   (b)

Charles D. Winters

important here because it provides experimental evidence that transition metal ions with charges of 2+ or greater have no ns electrons. For example, the Fe^{3+} ion is paramagnetic to the extent of five unpaired electrons (Figure 8.8). If three $3d$ electrons had been removed instead to form Fe^{3+}, the ion would still be paramagnetic but only to the extent of three unpaired electrons.

GENERAL
Chemistry $\cdot$ Now™

See the General ChemistryNow CD-ROM or website:

● **Screen 8.8 Electron Configuration in Ions**

    (a) for a simulation exploring the changes to an element's electron configuration when it ionizes

    (b) for a tutorial on determining an ion's box notation

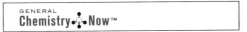

Example 8.4—Configurations of Transition Metal Ions

Problem Give the electron configurations for copper, Cu, and for its 1+ and 2+ ions. Are either of these ions paramagnetic? How many unpaired electrons does each have?

Strategy Observe the configuration of copper in Table 8.4. Recall that s and then d electrons are removed to form a transition metal ion.

Solution Copper has only one electron in the $4s$ orbital and ten electrons in $3d$ orbitals:

Cu: [Ar]$3d^{10}4s^1$ ⇅ ⇅ ⇅ ⇅ ⇅   ↑
                                    3d   4s

When copper is oxidized to Cu^+, the $4s$ electron is lost.

Cu⁺:   [Ar]$3d^{10}$ ⇅ ⇅ ⇅ ⇅ ⇅   ☐
                                  3d   4s

The copper(II) ion is formed from copper(I) by removal of one of the $3d$ electrons.

$$Cu^{2+}: \quad [Ar]3d^9 \quad \boxed{\uparrow\downarrow}\boxed{\uparrow\downarrow}\boxed{\uparrow\downarrow}\boxed{\uparrow\downarrow}\boxed{\uparrow} \quad \boxed{}$$
$$\qquad\qquad\qquad\qquad\quad 3d \qquad\qquad\quad 4s$$

Copper(II) ions (Cu^{2+}) have one unpaired electron, so they should be paramagnetic. In contrast, Cu^+ has no unpaired electrons, so the ion and its compounds are diamagnetic.

Exercise 8.4—Metal Ion Configurations

Depict the electron configurations for V^{2+}, V^{3+}, and Co^{3+}. Use orbital box diagrams and noble gas notation. Are any of the ions paramagnetic? If so, give the number of unpaired electrons.

8.6—Atomic Properties and Periodic Trends

Once electron configurations were understood, chemists realized that *similarities in properties of the elements are the result of similar valence shell electron configurations.* An objective of this section is to describe how atomic electron configurations are related to some of the physical and chemical properties of the elements and why those properties change in a reasonably predictable manner when moving down groups and across periods (Figure 8.9). This background should make the periodic table an even more useful tool in your study of chemistry. With an understanding of electron configurations and their relation to properties, you should be able to organize and predict chemical and physical properties of the elements and their compounds. We will concentrate on physical properties in this section and then look briefly at chemical behavior in Section 8.7.

Atomic Size

An orbital has no sharp boundary [◀ Figure 7.14, page 321], so how can we define the size of an atom? There are actually several ways, and they can give slightly different results.

One of the simplest and most useful ways to define atomic size is to say that it is the distance between atoms in a sample of the element. Let us take a diatomic molecule such as Cl_2 (Figure 8.10a). The radius of a Cl atom is assumed to be one half the experimentally determined distance between the centers of the two atoms. This distance is 198 pm, so the radius of one Cl atom is 99 pm. Similarly, the C—C distance in diamond is 154 pm, so a radius of 77 pm can be assigned to carbon. To test these estimates, we can add them together to estimate the distance between Cl and C in CCl_4. The predicted distance of 176 pm agrees with the experimentally measured C—Cl distance of 176 pm.

This approach to determining atomic radii will apply only if molecular compounds of the element exist. For metals, atomic radius can be estimated from measurements of the atom-to-atom distance in a crystal of the element (Figure 8.10b).

A set of atomic radii has been assembled (Figure 8.11), and some interesting periodic trends are seen immediately. *For the main group elements, atomic radii generally increase going down a group in the periodic table and decrease going across a period.* These trends reflect two important effects:

■ **Atomic Radii—Caution**
Numerous tabulations of atomic and covalent radii exist, and the values quoted in them may differ. The variation comes about because several methods are used to determine the radii of atoms, and the different methods can give slightly different values.

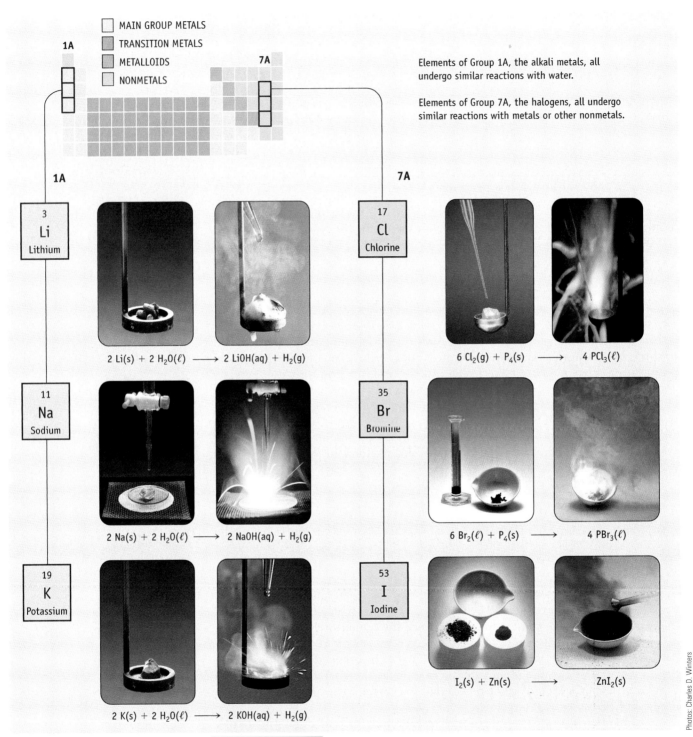

MAIN GROUP METALS
TRANSITION METALS
METALLOIDS
NONMETALS

Elements of Group 1A, the alkali metals, all undergo similar reactions with water.

Elements of Group 7A, the halogens, all undergo similar reactions with metals or other nonmetals.

1A

3 Li Lithium

$2 Li(s) + 2 H_2O(\ell) \longrightarrow 2 LiOH(aq) + H_2(g)$

11 Na Sodium

$2 Na(s) + 2 H_2O(\ell) \longrightarrow 2 NaOH(aq) + H_2(g)$

19 K Potassium

$2 K(s) + 2 H_2O(\ell) \longrightarrow 2 KOH(aq) + H_2(g)$

7A

17 Cl Chlorine

$6 Cl_2(g) + P_4(s) \longrightarrow 4 PCl_3(\ell)$

35 Br Bromine

$6 Br_2(\ell) + P_4(s) \longrightarrow 4 PBr_3(\ell)$

53 I Iodine

$I_2(s) + Zn(s) \longrightarrow ZnI_2(s)$

Photos: Charles D. Winters

Active Figure 8.9   Examples of the Periodicity of Group 1A and Group 7A Elements. Dimitri Mendeleev developed the first periodic table by listing elements in order of increasing atomic weight. Every so often an element had properties similar to those of a lighter element, and these elements were placed in vertical columns or groups. We now recognize that the elements should be listed in order of increasing atomic number and that the periodic occurrence of similar properties is related to the electron configurations of the elements.

GENERAL
Chemistry·Now™ See the General ChemistryNow CD-ROM or website to explore an interactive version of this figure accompanied by an exercise.

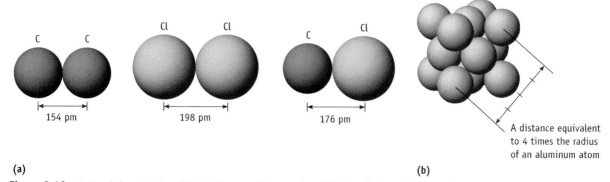

(a) **(b)**

Figure 8.10 **Determining atomic radii.** (a) The sum of the atomic radii of C and Cl provides a good estimate of the C — Cl distance in a molecule having such a bond. (b) Each sphere in this tiny piece of an aluminum crystal represents an aluminum atom. Measuring the distance shown, for example, allows a scientist to estimate the radius of an aluminum atom.

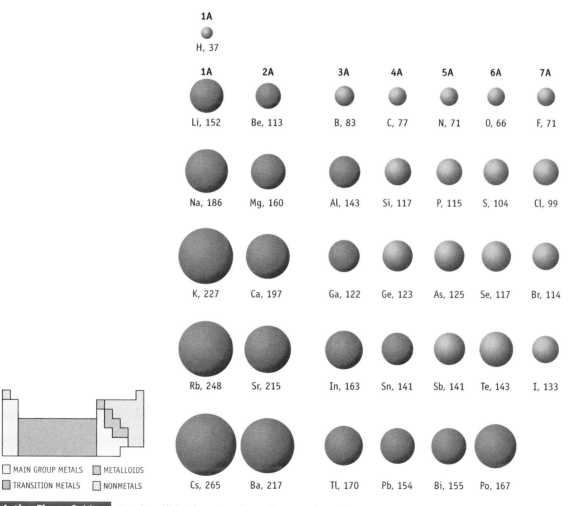

Active Figure 8.11 **Atomic radii in picometers for main group elements.**
1 pm = 1×10^{-12} m. Data taken from J. Emsley: *The Elements,* 3rd ed., Oxford, Clarendon Press, 1998.

GENERAL
Chemistry ❖ Now™ See the General ChemistryNow CD-ROM or website to explore an interactive version of this figure accompanied by an exercise.

- The size of an atom is determined by the outermost electrons. In going from the top to the bottom of a group in the periodic table, the outermost electrons are assigned to orbitals with increasingly higher values of the principal quantum number, n. The underlying electrons require some space, so the electrons of the outer shell must be farther from the nucleus.

- For main group elements of a given period, the principal quantum number, n, of the valence electron orbitals is the same. In going from one element to the next across a period, a proton is added to each nucleus and an electron is added to each outer shell. In each step, the effective nuclear charge, Z^* [◀ Table 8.2] increases slightly because the effect of each additional proton is more important than the effect of an additional electron. The result is that attraction between the nucleus and electrons increases, and atomic radius decreases.

The periodic trend in the atomic radii of transition metal atoms (Figure 8.12) is somewhat different from that for main group elements. Going from left to right across a given period, the radii initially decrease across the first few elements. The sizes of the elements in the middle of a transition series then change very little until a small increase in size occurs at the end of the series. The size of the atom is determined largely by electrons in the outermost shell—that is, by the electrons of the ns subshell. In the first transition series, for example, the outer shell contains the $4s$ electrons, but electrons are being added to the $3d$ orbitals across the series. The increased nuclear charge on the atoms as one moves from left to right should cause the radius to decrease. This effect, however, is mostly cancelled out by increased electron–electron repulsion among the electrons. On reaching the Groups 1B and 2B elements at the end of the series, the size increases slightly because the d subshell is filled, and electron–electron repulsions cause the size to increase.

■ **Trends in Atomic Radii**
General trends in atomic radii of s- and p-block elements with position in the periodic table.

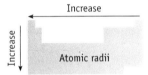

Figure 8.12   Trends in atomic radii for the transition elements. Atomic radii of the Group 1A and 2A metals and the transition metals of the fourth, fifth, and sixth periods.

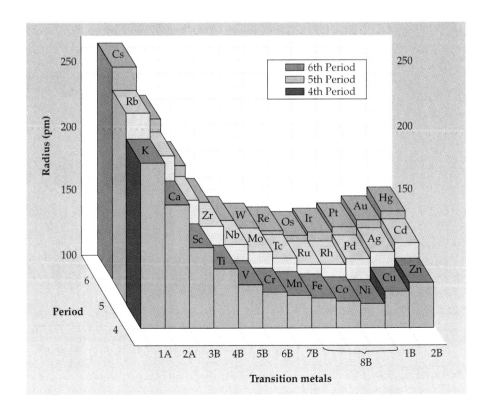

See the General ChemistryNow CD-ROM or website:

- **Screen 8.9 Atomic Properties and Periodic Trends,** for a simulation exploring energy levels of orbitals and the ability to retain electrons
- **Screen 8.10 Atomic Sizes,** for a simulation exploring the trends in atomic size moving across and down the periodic table

Exercise 8.5—Periodic Trends in Atomic Radii

Place the three elements Al, C, and Si in order of increasing atomic radius.

Exercise 8.6—Estimating Atom–Atom Distances

(a) Using Figure 8.11, estimate the H — O and H — S distances in H_2O and H_2S, respectively.

(b) If the interatomic distance in Br_2 is 228 pm, what is the radius of Br? Using this value, and that for Cl (99 pm), estimate the distance between atoms in BrCl.

Ionization Energy

Ionization energy is the energy required to remove an electron from an atom in the gas phase.

$$\text{Atom in ground state(g)} \longrightarrow \text{Atom}^+ \text{(g)} + e^-$$

$$\Delta E \equiv \text{ionization energy, } IE$$

To separate an electron from an atom, energy must be supplied to overcome the attraction of the nuclear charge. Because energy must be supplied (an endothermic process), ionization energies always have positive values.

Atoms other than hydrogen have a series of ionization energies, because more than one electron can always be removed [◀ page 351]. For example, the first three ionization energies of magnesium are

First ionization energy, $IE_1 = 738$ kJ/mol
$$\text{Mg(g)} \longrightarrow \text{Mg}^+\text{(g)} + e^-$$
$$1s^2 2s^2 2p^6 3s^2 \qquad 1s^2 2s^2 2p^6 3s^1$$

Second ionization energy, $IE_2 = 1451$ kJ/mol
$$\text{Mg}^+\text{(g)} \longrightarrow \text{Mg}^{2+}\text{(g)} + e^-$$
$$1s^2 2s^2 2p^6 3s^1 \qquad 1s^2 2s^2 2p^6 3s^0$$

Third ionization energy, $IE_3 = 7733$ kJ/mol
$$\text{Mg}^{2+}\text{(g)} \longrightarrow \text{Mg}^{3+}\text{(g)} + e^-$$
$$1s^2 2s^2 2p^6 \qquad 1s^2 2s^2 2p^5$$

■ **Valence and Core Electrons**
Removal of core electrons requires much more energy than removal of a valence electron. Core electrons are not lost in chemical reactions.

Notice that removing each subsequent electron requires more energy because the electron is being removed from an increasingly positive ion. Most importantly, notice the large increase in ionization energy for removing the third electron to give Mg^{3+}. *This large increase is experimental evidence for the electron shell structure of atoms.* The first two ionization steps are for the removal of electrons from the outermost or valence shell of electrons. The third electron, however, must come from the $2p$ subshell. This subshell is significantly lower in energy than the 3s subshell (page 340), and considerably more energy is required to remove the $n = 2$ electron than the $n = 3$ electrons.

■ **Trends in Ionization Energy**
General trends in first ionization energies of *s*- and *p*-block elements with position in the periodic table.

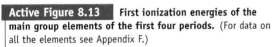

■ **Factors Controlling Trends in Ionization Energies**
The ionization energy of an atom is always a balance between electron–nuclear attraction (which depends on Z) and electron–electron repulsion.

As another example, consider the first two ionization energies for lithium.

First ionization energy, IE_1 = 513.3 kJ/mol
$$\text{Li(g)} \longrightarrow \text{Li}^+\text{(g)} + e^-$$
$$1s^2 2s^1 \qquad\qquad 1s^2$$

Second ionization energy, IE_2 = 7298 kJ/mol
$$\text{Li}^+\text{(g)} \longrightarrow \text{Li}^{2+}\text{(g)} + e^-$$
$$1s^2 \qquad\qquad 1s^1$$

The ionization energy for the removal of the second electron is large because the second electron is removed from a much lower energy (inner) subshell.

For main group (*s*- and *p*-block) elements, *first ionization energies generally increase across a period and decrease down a group* (Figure 8.13 and Appendix F). The trend *across a period* is rationalized by the increase in effective nuclear charge, Z*, with increasing atomic number. Not only does this mean that the atomic radius decreases, but the energy required to remove an electron also increases. The general decrease in ionization energy *down a group* occurs because the electron removed is increasingly farther from the nucleus, thus reducing the nucleus-electron attractive force.

A closer look at ionization energies reveals that the trend across a given period is not smooth, particularly in the second period. Variations are seen on going from *s*-block to *p*-block elements—from beryllium to boron, for example. This occurs because the 2*p* electrons are slightly higher in energy than the 2*s* electrons (see Figure 8.4, page 340), so the ionization energy for boron is lower than that for beryllium.

Moving from boron to carbon and then to nitrogen, the effective nuclear charge increases (see Table 8.2), which again means an increase in ionization energy. Another dip to lower ionization energy occurs on passing from Group 5A to Group 6A. This is especially noticeable in the second period (N and O). No change occurs in either *n* or *ℓ*, but electron–electron repulsions increase for the following

Active Figure 8.13   **First ionization energies of the main group elements of the first four periods.** (For data on all the elements see Appendix F.)

GENERAL
Chemistry⚛Now™ See the General ChemistryNow CD-ROM or website to explore an interactive version of this figure accompanied by an exercise.

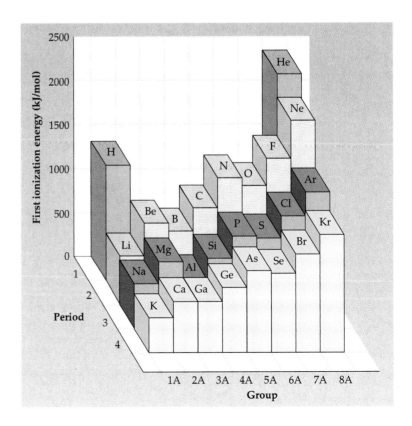

reason. In Groups 3A–5A, electrons are assigned to separate p orbitals (p_x, p_y, and p_z). Beginning in Group 6A, however, two electrons are assigned to the same p orbital. The fourth p electron shares an orbital with another electron and thus experiences greater repulsion than it would if it had been assigned to an orbital of its own:

$$\text{O (oxygen atom)} \xrightarrow{+1314 \text{ kJ/mol}} \text{O}^+ \text{ (oxygen cation)} + e^-$$

[Ne] $\boxed{\uparrow\downarrow}$ $\boxed{\uparrow\downarrow\;\uparrow\;\uparrow}$ [Ne] $\boxed{\uparrow\downarrow}$ $\boxed{\uparrow\;\uparrow\;\uparrow}$
    $2s$ $2p$ $2s$ $2p$

The greater repulsion experienced by the fourth $2p$ electron makes it easier to remove, and each of the remaining p electrons has an orbital of its own. The usual trend resumes on going from oxygen to fluorine to neon, however, reflecting the increase in Z^*.

Electron Affinity

Some atoms have an affinity, or "liking," for electrons and can acquire one or more electrons to form a negative ion. The **electron affinity**, *EA*, of an atom is defined as the energy of a process in which an electron is acquired by the atom in the gas phase (Figure 8.14 and Appendix F).

$$A(g) + e^-(g) \rightarrow A^-(g) \qquad \Delta E \equiv \text{electron affinity, } EA$$

The greater the affinity an atom has for an electron, the more negative the value of *EA* will be. For example, the electron affinity of fluorine is -328 kJ/mol, a large value indicating an exothermic, product-favored reaction to form the anion, F^-. Boron has a much lower electron affinity for an electron, as indicated by a less negative *EA* value of -26.7 kJ/mol.

■ **Electron Affinity and Sign Conventions** For a useful discussion of electron affinity, see J. C. Wheeler: "Electron affinities of the alkaline earth metals and the sign convention for electron affinity." *Journal of Chemical Education*, Vol. 74, pp. 123–127, 1997. Numerical values for *EA* are given in Appendix F.

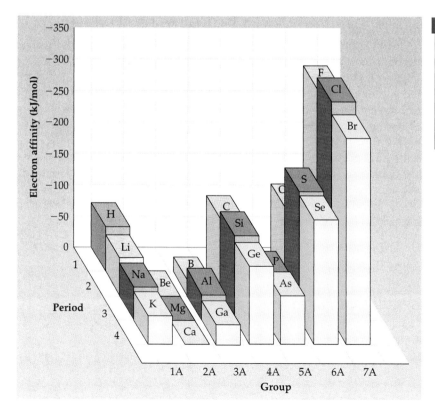

Active Figure 8.14   **Electron affinity.** The larger the affinity (*EA*) of an atom for an electron, the more negative the value. For numerical data, see Appendix F. (Data were taken from H. Hotop and W. C. Lineberger: "Binding energies of atomic negative ions," *Journal of Physical Chemistry*, Reference Data, Vol. 14, p. 731, 1985.)

GENERAL
Chemistry ⚛ Now™ See the General ChemistryNow CD-ROM or website to explore an interactive version of this figure accompanied by an exercise.

■ *EA* **Values of Zero**
The value of *EA* for Be is not measurable because the Be⁻ anion does not exist. Most tables assign a value of 0 to the *EA* for this element and similar cases (in particular the Group 2A elements).

■ **Trends In *EA***
General trends in electron affinities of A-group elements. Exceptions occur at Groups 2A and 5A.

Increase in
affinity for electron
(*EA* becomes more negative)

Electron affinity

Increase in affinity
for electron
(*EA* becomes
more negative)

Electron affinity and ionization energy represent the energy involved in the gain or loss of an electron by an atom, respectively. It is therefore not surprising that periodic trends in electron affinity are related to the periodic trends in ionization energy. The effective nuclear charge of atoms increases across a period (Table 8.2), not only making it more difficult to ionize the atom but also increasing the attraction of the atom for an additional electron. Thus, *an element with a high ionization energy generally has a high affinity for an electron.* As seen in Figure 8.14, the values of *EA* generally become more negative on moving across a period as the affinity for electrons increases.

One result of increasing Z* across a period is that nonmetals generally have much more negative values of *EA* than do metals, reflecting the greater affinity of nonmetals for electrons. This prediction agrees with chemical experience, which tells us that metals generally do not form negative ions and that nonmetals have an increasing tendency to form anions as we proceed across a period.

The trend to more negative electron affinities across a period is not smooth, however. For example, beryllium has no affinity for an electron. A beryllium anion, Be⁻, is not stable because the added electron must be assigned to a higher energy subshell (2*p*) than the valence electrons (2*s*) (see page 340). Nitrogen atoms also have no affinity for electrons. Here an electron pair must be formed when an N atom acquires an electron. Significant electron–electron repulsions occur in an N⁻ ion, making the ion much less stable. The increase in Z* on going from carbon to nitrogen cannot overcome the effect of these electron–electron repulsions.

The noble gases are not included in a discussion of electron affinity. They have no affinity for electrons, because any additional electron must be added to the next higher electron shell. The higher Z* of the noble gases is not sufficient to overcome this effect.

The affinity for an electron generally declines on descending a group of the periodic table (Figure 8.14). Electrons are added increasingly farther from the nucleus, so the attractive force between the nucleus and electrons decreases. However, this general trend does not extend to the elements in period 2. For example, the affinity of the fluorine atom for an electron is lower than that of chlorine (*EA* for F is less negative than *EA* for Cl), and the same phenomenon is observed in Groups 3A through 6A. One explanation is that significant electron–electron repulsions occur in small anions such as the F⁻ ion. That is, adding an electron to the seven electrons already present in the n = 2 shell of the small F atom leads to considerable repulsion between electrons. Chlorine has a larger atomic volume than fluorine, so adding an electron does not result in such significant electron–electron repulsions in the Cl⁻ anion.

No atom has a negative electron affinity for a *second* electron. Attaching a second electron to an ion that already has a negative charge leads to severe repulsions. So how can you account for ions such as O^{2-}, which is present in so many naturally occurring substances (for example, CaO)? The answer is that doubly charged anions can sometimes be stabilized in crystalline environments by electrostatic attraction to neighboring positive ions [▶ Chapters 9 and 13].

GENERAL
Chemistry┅Now™

See the General ChemistryNow CD-ROM or website:
- **Screen 8.11 Ionization Energy,** for a simulation exploring the trends in ionization energy moving across and down the periodic table
- **Screen 8.12 Electron Affinity,** for a simulation exploring the trend in electron affinity moving across the periodic table

Example 8.5—Periodic Trends

Problem Compare the three elements C, O, and Si.

(a) Place them in order of increasing atomic radius.

(b) Which has the largest ionization energy?

(c) Which has the more negative electron affinity, O or C?

Strategy Review the trends in atomic properties in Figures 8.11–8.14 and Appendix F.

Solution

(a) *Atomic size:* Atomic radius declines on moving across a period, so oxygen must have a smaller radius than carbon. However, radius increases on moving down a periodic group. Because C and Si are in the same group (Group 4A), Si must be larger than C. In order of increasing size, the trend is $O < C < Si$.

(b) *Ionization energy:* Ionization energy generally increases across a period and decreases down a group; a large decrease in IE occurs from the second- to the third-period elements. Thus, the trend in ionization energies should be $Si < C < O$.

(c) *Electron affinity:* Electron affinity values generally become more negative across a period and less negative down a group. Therefore, the EA for O should be more negative than the EA for C. That is, O ($EA = -141.0$ kJ/mol) has a greater affinity for an electron than does C ($EA = -121.9$ kJ/mol).

Comment EA for third-period elements (Si, $EA = -133.6$ kJ/mol) is generally slightly more negative than EA for second-period elements (C, $EA = -121.85$ kJ/mol). This trend occurs because of electron–electron repulsions; such repulsions are larger in the small C^- ion than in the larger Si^- ion.

Exercise 8.7—Periodic Trends

Compare the three elements B, Al, and C.

(a) Place the three elements in order of increasing atomic radius.
(b) Rank the elements in order of increasing ionization energy. (Try to do this without looking at Figure 8.13; then compare your estimates with the graph.)
(c) Which element is expected to have the most negative electron affinity value?

Ion Sizes

Having considered the energies involved in forming positive and negative ions, let us now look at periodic trends in ion radii.

Periodic trends in the sizes of ions in the same group are the same as those for neutral atoms: Positive or negative ions increase in size when descending the group (Figure 8.15). Pause for a moment, however, and compare the ionic radii in Figure 8.15 with the atomic radii in Figure 8.11. When an electron is removed from an atom to form a cation, the size shrinks considerably; *the radius of a cation is always smaller than that of the atom from which it is derived.* For example, the radius of Li is 152 pm, whereas the radius of Li^+ is only 78 pm. When an electron is removed from a Li atom, the attractive force of three protons is now exerted on only two electrons, so the remaining electrons contract toward the nucleus. The decrease in ion size is especially great when the last electron of a particular shell is removed, as is the case

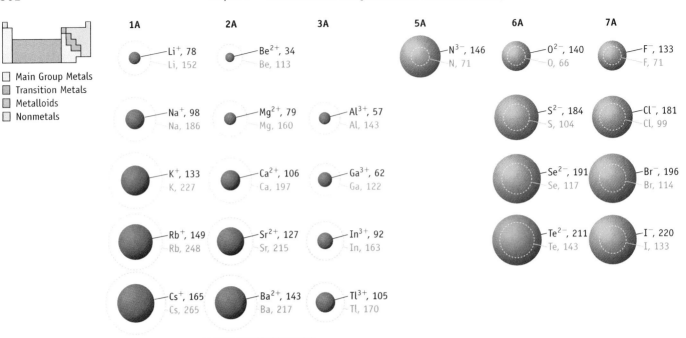

□ Main Group Metals
■ Transition Metals
■ Metalloids
□ Nonmetals

Active Figure 8.15 **Relative sizes of some common ions compared with neutral atom size.** Radii are given in picometers (1 pm = 1×10^{-12} m). (Data taken from J. Emsley: *The Elements*, 3rd ed., Oxford, Clarendon Press, 1998.)

GENERAL
Chemistry⚛Now™ See the General ChemistryNow CD-ROM or website to explore an interactive version of this figure accompanied by an exercise.

for Li. The loss of the $2s$ electron from Li leaves Li$^+$ with no electrons in the $n = 2$ shell.

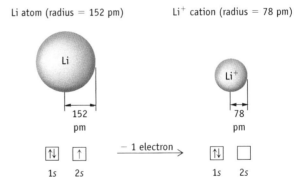

The shrinkage will also be great when two or more electrons are removed, as for Al^{3+} in which it exceeds 50%:

Al atom (radius = 143 pm)   Al^{3+} cation (radius = 57 pm)

[Ne]⇅ ↑□□ $\xrightarrow{-3 \text{ electrons}}$ [Ne]□ □□□
     3s   3p   3s   3p

You can also see by comparing Figures 8.11 and 8.15 that *anions are always larger than the atoms from which they are derived.* Here the argument is the opposite of that used to explain positive ion radii. The F atom, for example, has nine protons and nine electrons. On forming the anion, the nuclear charge is still +9, but now ten electrons are in the anion. The F^- ion is much larger than the F atom because of increased electron–electron repulsions.

F atom (radius = 71 pm) F^- anion (radius = 133 pm)

Finally, it is useful to compare the sizes of isoelectronic ions across the periodic table. **Isoelectronic** ions have the same number of electrons (but a different number of protons). One such series of commonly occurring ions is O^{2-}, F^-, Na^+, and Mg^{2+}:

Ion	O^{2-}	F^-	Na^+	Mg^{2+}
Number of electrons	10	10	10	10
Number of nuclear protons	8	9	11	12
Ionic radius (pm)	140	133	98	79

All these ions have a total of ten electrons. The O^{2-} ion, however, has only 8 protons in its nucleus to attract these electrons, whereas F^- has 9, Na^+ has 11, and Mg^{2+} has 12. As the number of protons increases in a series of isoelectronic ions, the balance in electron–proton attraction and electron–electron repulsion shifts in favor of attraction, and the radius decreases. As you can see in Figure 8.15, this is true for all isoelectronic series of ions.

GENERAL
Chemistry ⚛ Now™

See the General ChemistryNow CD-ROM or website:
- **Screen 8.14 Ion Sizes**

   (a) for a simulation exploring the relationship between ion formation and orbital energies in main group elements

   (b) for a simulation exploring the relationship between orbital energies and electron configurations on the size of the main group element ions

Exercise 8.8—Ion Sizes

What is the trend in sizes of the ions N^{3-}, O^{2-}, and F^-? Briefly explain why this trend exists.

8.7—Periodic Trends and Chemical Properties

Atomic and ionic radii, ionization energies, and electron affinities are properties associated with atoms and their ions. It is reasonable to expect that knowledge of these properties will be useful as we explore the chemistry of the elements. Let us consider just one example here, the formation of ionic compounds.

Reaction of sodium metal and chlorine gas. This reaction produces the ionic compound NaCl, which consists of Na^+ and Cl^- ions. Because of their electron configurations, and their values of ionization energy and electron affinity, the reaction does not produce ions such as Na^{2+}, Cl^+, or Cl^{2-}. (See also Figure 1.7 on page 19.)

As described in Section 2.6, the periodic table was created by grouping together elements having similar chemical properties. Alkali metals, for example, characteristically form compounds in which the metal is in the form of a 1+ ion, such as Li^+, Na^+, or K^+. Thus, the reaction between sodium and chlorine gives the ionic compound, NaCl (composed of Na^+ and Cl^- ions) [◀ Figure 1.7, page 19], and potassium and water react to form an aqueous solution of KOH, a solution containing the hydrated ions $K^+(aq)$ and $OH^-(aq)$ (Figure 8.9).

$$2\ Na(s) + Cl_2(g) \longrightarrow 2\ NaCl(s)$$
$$2\ K(s) + 2\ H_2O(\ell) \longrightarrow 2\ K^+(aq) + 2\ OH^-(aq) + H_2(g)$$

Both of these observations agree with the fact that alkali metals have electron configurations of the type [noble gas core]ns^1 and have low ionization energies.

Ionization energies also account in part for the fact that these reactions of sodium and potassium do not produce compounds such as $NaCl_2$ or $K(OH)_2$. The formation of a Na^{2+} or K^{2+} ion is clearly a very unfavorable process. Removing a second electron from these metals requires a great deal of energy because this electron must come from the atom's core electrons. Indeed, removal of core electrons from any atom is exceedingly unfavorable. This is the underlying reason that *main group metals generally form cations with an electron configuration equivalent to that of the nearest noble gas.*

Why isn't Na_2Cl another possible product from the sodium and chlorine reaction? This formula would imply that the compound contains Na^+ and Cl^{2-} ions. Chlorine atoms have a relatively high electron affinity, but only for the addition of one electron. Adding two electrons per atom means that the second electron must enter the next higher shell at much higher energy. An anion such as Cl^{2-} is simply not stable. This example leads us to a general statement. *Nonmetals generally acquire enough electrons to form an anion with the electron configuration of the next, higher noble gas.*

We can use similar logic to rationalize results of other reactions. Ionization energies increase on going from left to right across a period. We have seen that elements from Groups 1A and 2A form ionic compounds, an observation directly related to the low ionization energies for these elements. Ionization energies for elements toward the middle and right side of a period, however, are sufficiently large that cation formation is unfavorable. Thus, we generally do not expect to encounter ionic compounds containing carbon; instead, we find carbon *sharing* electrons with other elements in compounds such as CO_2 and CCl_4. On the right side of the second period, oxygen and fluorine much prefer taking on electrons to giving them up; these elements have high ionization energies and relatively large, negative electron affinities. Thus, oxygen and fluorine form anions and not cations when they react.

GENERAL
Chemistry Now™

See the General ChemistryNow CD-ROM or website:
- **Screen 8.15 Chemical Reactions and Periodic Properties,** to watch videos on the relationship of atomic electron configurations and orbital energies on periodic trends

---

Exercise 8.9—Energies and Compound Formation

Give a plausible explanation for the observation that magnesium and chlorine react to form $MgCl_2$ and not $MgCl_3$.

Chapter Goals Revisited

Now that you have studied this chapter, you should ask whether you have met the chapter goals. In particular, you should be able to

Understand the role magnetism plays in determining and revealing atomic structure

a. Classify substances as paramagnetic (attracted to a magnetic field; characterized by unpaired electron spins) or diamagnetic (repelled by a magnetic field) (Section 8.1).

b. Recognize that each electron in an atom has a different set of the four quantum numbers, n, ℓ, m_ℓ, and m_s, where m_s, the spin quantum number, has values of $+\frac{1}{2}$ or $-\frac{1}{2}$ (Section 8.2). General ChemistryNow homework: Study Question(s) 18, 19

c. Understand that the Pauli exclusion principle leads to the conclusion that no atomic orbital can be assigned more than two electrons and that the two electrons in an orbital must have opposite spins (different values of m_s) (Section 8.2).

Understand effective nuclear charge and its role in determining atomic properties

a. Understand effective nuclear charge, Z^*, and its ability to explain why different subshells in the same shell have different energies. Also, understand the role of Z^* in determining the properties of atoms (Sections 8.3 and 8.6).

Write the electron configuration for elements and monatomic ions

a. Using the periodic table as a guide, depict electron configurations of the elements and monatomic ions using orbital box or the *spdf* notation. In both cases, configurations can be abbreviated with the noble gas notation (Sections 8.3 and 8.4). General ChemistryNow homework: SQ(s) 2, 3, 6, 12, 14, 21, 37, 38, 39, 48

b. Recognize that electrons are assigned to the subshells of an atom in order of increasing subshell energy. In the H atom the subshell energies increase with increasing n, but, in a many-electron atom, the energies depend on both n and ℓ (see Figure 8.3).

c. When assigning electrons to atomic orbitals, apply the Pauli exclusion principle and Hund's rule (Sections 8.3 and 8.4).

Understand the fundamental physical properties of the elements and their periodic trends

a. Predict how properties of atoms—size, ionization energy (*IE*), and electron affinity (*EA*)—change on moving down a group or across a period of the periodic table (Section 8.6). The general periodic trends for these properties are as follows:

(i) Atomic size decreases across a period and increases down a group.

(ii) *IE* increases across a period and decreases down a group.

(iii) The affinity for an electron generally increases across a period (the value of *EA* becomes more negative) and decreases down a group. General ChemistryNow homework: SQ(s) 26, 28, 30, 45, 46, 50, 53

b. Recognize the role that ionization energy and electron affinity play in the chemistry of the elements (Section 8.7). General ChemistryNow homework: SQ(s) 62

Study Questions

▲ denotes more challenging questions.

■ denotes questions available in the Homework and Goals section of the General ChemistryNow CD-ROM or website.

Blue numbered questions have answers in Appendix O and fully worked solutions in the *Student Solutions Manual.*

Structures of many of the compounds used in these questions are found on the General ChemistryNow CD-ROM or website in the Models folder.

GENERAL
Chemistry⋅☆⋅Now™ Assess your understanding of this chapter's topics with additional quizzing and conceptual questions at http://now.brookscole.com/kotz6e

Practicing Skills

Writing Electron Configurations of Atoms

(See Examples 8.1–8.3; Tables 8.1, 8.3, and 8.4; and the Toolbox on the General ChemistryNow.)

1. Write the electron configurations for P and Cl using both *spdf* notation and orbital box diagrams. Describe the relationship between each atom's electron configuration and its position in the periodic table.

2. ■ Write the electron configurations for Mg and Ar using both *spdf* notation and orbital box diagrams. Describe the relation of the atom's electron configuration to its position in the periodic table.

3. ■ Using *spdf* notation, write the electron configurations for atoms of chromium and iron, two of the major components of stainless steel.

4. Using *spdf* notation, give the electron configuration of vanadium, V, an element found in some brown and red algae and some toadstools.

5. Depict the electron configuration for each of the following atoms using *spdf* and noble gas notations.
   (a) Arsenic, As. A deficiency of As can impair growth in animals even though larger amounts are poisonous.
   (b) Krypton, Kr. It ranks seventh in abundance of the gases in the earth's atmosphere.

6. ■ Using *spdf* and noble gas notations, write electron configurations for atoms of the following elements and then check your answers with Table 8.3.
   (a) Strontium, Sr. This element is named for a town in Scotland.
   (b) Zirconium, Zr. The metal is exceptionally resistant to corrosion and so has important industrial applications. Moon rocks show a surprisingly high zirconium content compared with rocks on earth.
   (c) Rhodium, Rh. This metal is used in jewelry and in catalysts in industry.

   (d) Tin, Sn. The metal was used in the ancient world. Alloys of tin (solder, bronze, and pewter) are important.

7. Use noble gas and *spdf* notations to depict electron configurations for the following metals of the third transition series.
   (a) Tantalum, Ta. The metal and its alloys resist corrosion and are often used in surgical and dental tools.
   (b) Platinum, Pt. This metal was used by pre-Columbian Indians in jewelry. It is used now in jewelry and for anticancer drugs and industrial catalysts.

8. The lanthanides, once called the rare earth elements, are really only "medium rare." Using noble gas and *spdf* notations, depict reasonable electron configurations for the following elements.
   (a) Samarium, Sm. This lanthanide is used in magnetic materials.
   (b) Ytterbium, Yb. This element was named for the village of Ytterby in Sweden, where a mineral source of the element was found.

9. The actinide americium, Am, is a radioactive element that has found use in home smoke detectors. Depict its electron configuration using noble gas and *spdf* notations.

10. Predict reasonable electron configurations for the following elements of the actinide series of elements. Use noble gas and *spdf* notations.
    (a) Plutonium, Pu. The element is best known as a byproduct of nuclear power plant operations.
    (b) Curium, Cm. This actinide was named for Madame Curie (page 57).

Electron Configurations of Atoms and Ions and Magnetic Behavior

(See Example 8.4 and General ChemistryNow Screens 8.3, 8.7, and 8.8.)

11. Using orbital box diagrams, depict an electron configuration for each of the following ions: (a) Mg^{2+}, (b) K^+, (c) Cl^-, and (d) O^{2-}.

12. ■ Using orbital box diagrams, depict an electron configuration for each of the following ions: (a) Na^+, (b) Al^{3+}, (c) Ge^{2+}, and (d) F^-.

13. Using orbital box diagrams and noble gas notation, depict the electron configurations of (a) V, (b) V^{2+}, and (c) V^{5+}. Are any of the ions paramagnetic?

14. ■ Using orbital box diagrams and noble gas notation, depict the electron configurations of (a) Ti, (b) Ti^{2+}, and (c) Ti^{4+}. Are any of the ions paramagnetic?

15. Manganese is found as MnO_2 in deep ocean deposits.
    (a) Depict the electron configuration of this element using the noble gas notation and an orbital box diagram.
    (b) Using an orbital box diagram, show the electrons beyond those of the preceding noble gas for the 2+ ion.
    (c) Is the 2+ ion paramagnetic?
    (d) How many unpaired electrons does the Mn^{2+} ion have?

16. Nickel generally forms 2+ ions but alkaline batteries have Ni^{3+} ions in NiOOH. Using orbital box diagrams and the noble gas notation, show electron configurations of these ions. Are either of these ions paramagnetic?

Quantum Numbers and Electron Configurations

(See Example 8.2 and General ChemistryNow Screens 7.12, 8.4, and 8.7.)

17. Explain briefly why each of the following is not a possible set of quantum numbers for an electron in an atom. In each case, change the incorrect value (or values) to make the set valid.
    (a) $n = 4$, $\ell = 2$, $m_\ell = 0$, $m_s = 0$
    (b) $n = 3$, $\ell = 1$, $m_\ell = -3$, $m_s = -\frac{1}{2}$
    (c) $n = 3$, $\ell = 3$, $m_\ell = -1$, $m_s = +\frac{1}{2}$

18. ■ Explain briefly why each of the following is not a possible set of quantum numbers for an electron in an atom. In each case, change the incorrect value (or values) to make the set valid.
    (a) $n = 2$, $\ell = 2$, $m_\ell = 0$, $m_s = +\frac{1}{2}$
    (b) $n = 2$, $\ell = 1$, $m_\ell = -1$, $m_s = 0$
    (c) $n = 3$, $\ell = 1$, $m_\ell = +2$, $m_s = +\frac{1}{2}$

19. ■ What is the maximum number of electrons that can be identified with each of the following sets of quantum numbers? In one case, the answer is "none." Explain why this is true.
    (a) $n = 4$, $\ell = 3$
    (b) $n = 6$, $\ell = 1$, $m_\ell = -1$
    (c) $n = 3$, $\ell = 3$, $m_\ell = -3$

20. What is the maximum number of electrons that can be identified with each of the following sets of quantum numbers? In some cases, the answer may be "none." In such cases, explain why "none" is the correct answer.
    (a) $n = 3$
    (b) $n = 3$ and $\ell = 2$
    (c) $n = 4$, $\ell = 1$, $m_\ell = -1$, and $m_s = -\frac{1}{2}$
    (d) $n = 5$, $\ell = 0$, $m_\ell = +1$

21. ■ Depict the electron configuration for magnesium using an orbital box diagram and noble gas notation. Give a complete set of four quantum numbers for each of the electrons beyond those of the preceding noble gas.

22. Depict the electron configuration for phosphorus using an orbital box diagram and noble gas notation. Give one possible set of four quantum numbers for each of the electrons beyond those of the preceding noble gas.

23. Using an orbital box diagram and noble gas notation, show the electron configuration of gallium, Ga. Give a set of quantum numbers for the highest-energy electron.

24. Using an orbital box diagram and noble gas notation, show the electron configuration of titanium. Give one possible set of four quantum numbers for each of the electrons beyond those of the preceding noble gas.

Periodic Properties

(See Section 8.6, Example 8.5, and General ChemistryNow Screens 8.9–8.12.)

25. Arrange the following elements in order of increasing size: Al, B, C, K, and Na. (Try doing it without looking at Figure 8.11, and then check yourself by looking up the necessary atomic radii.)

26. ■ Arrange the following elements in order of increasing size: Ca, Rb, P, Ge, and Sr. (Try doing it without looking at Figure 8.11, then check yourself by looking up the necessary atomic radii.)

27. Select the atom or ion in each pair that has the larger radius.
    (a) Cl or Cl^- (c) In or I
    (b) Al or O

28. ■ Select the atom or ion in each pair that has the larger radius.
    (a) Cs or Rb   (c) Br or As
    (b) O^{2-} or O

29. Which of the following groups of elements is arranged correctly in order of increasing ionization energy?
    (a) C < Si < Li < Ne   (c) Li < Si < C < Ne
    (b) Ne < Si < C < Li   (d) Ne < C < Si < Li

30. ■ Arrange the following atoms in order of increasing ionization energy: Li, K, C, and N.

31. Compare the elements Na, Mg, O, and P.
    (a) Which has the largest atomic radius?
    (b) Which has the most negative electron affinity?
    (c) Place the elements in order of increasing ionization energy.

32. Compare the elements B, Al, C, and Si.
    (a) Which has the most metallic character?
    (b) Which has the largest atomic radius?
    (c) Which has the most negative electron affinity?
    (d) Place the three elements B, Al, and C in order of increasing first ionization energy.

33. Explain each answer briefly.
    (a) Place the following elements in order of increasing ionization energy: F, O, and S.
    (b) Which has the largest ionization energy: O, S, or Se?
    (c) Which has the most negative electron affinity: Se, Cl, or Br?
    (d) Which has the largest radius: O^{2-}, F^-, or F?

34. Explain each answer briefly.
    (a) Rank the following in order of increasing atomic radius: O, S, and F.
    (b) Which has the largest ionization energy: P, Si, S, or Se?
    (c) Place the following in order of increasing radius: O^{2-}, N^{3-}, and F^-.
    (d) Place the following in order of increasing ionization energy: Cs, Sr, and Ba.

General Questions

These questions are not designated as to type or location in the chapter. They may combine several concepts. More challenging questions are indicated by ▲.

35. The diagrams below represent a small section of a solid. Each circle represents an atom and an arrow represents an electron.

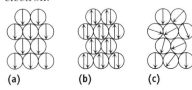

    (a)   (b)   (c)

    (a) Which represents a diamagnetic solid, which a paramagnetic solid, and which a ferromagnetic solid?
    (b) Which is most strongly attracted to a magnetic field? Which is least strongly attracted?

36. The name rutherfordium, Rf, has been given to element 104 to honor the physicist Ernest Rutherford (page 65). Depict its electron configuration using *spdf* and noble gas notations.

37. ■ Using an orbital box diagram and noble gas notation, show the electron configurations of uranium and of the uranium(IV) ion. Is either of these paramagnetic?

38. ■ The rare earth elements, or lanthanides, commonly exist as 3+ ions. Using an orbital box diagram and noble gas notation, show the electron configurations of the following elements and ions.
    (a) Ce and Ce^{3+} (cerium)   (b) Ho and Ho^{3+} (holmium)

39. ■ A neutral atom has two electrons with $n = 1$, eight electrons with $n = 2$, eight electrons with $n = 3$, and two electrons with $n = 4$. Assuming this element is in its ground state, supply the following information:
    (a) atomic number
    (b) total number of *s* electrons
    (c) total number of *p* electrons
    (d) total number of *d* electrons
    (e) is the element a metal, metalloid, or nonmetal?

40. Element 109, now named meitnerium (in honor of the Austrian–Swedish physicist, Lise Meitner [1878–1968]), was produced in August 1982 by a team at Germany's Institute for Heavy Ion Research. Depict its electron configuration using *spdf* and noble gas notations. Name another element found in the same group as meitnerium.

41. Which of the following is *not* an allowable set of quantum numbers? Explain your answer briefly.

	n	ℓ	m_ℓ	m_s
(a)	2	0	0	$-\frac{1}{2}$
(b)	1	1	0	$+\frac{1}{2}$
(c)	2	1	-1	$-\frac{1}{2}$
(d)	4	3	$+2$	$-\frac{1}{2}$

42. A possible excited state for the H atom has an electron in a $4p$ orbital. List all possible sets of quantum numbers (n, ℓ, m_ℓ, and m_s) for this electron.

43. The magnet in the photo is made from neodymium, iron, and boron.

A magnet made of an alloy containing the elements Nd, Fe, and B.

    (a) Write the electron configuration of each of these elements using an orbital box diagram and noble gas notation.
    (b) Are these elements paramagnetic or diamagnetic?
    (c) Write the electron configurations of Nd^{3+} and Fe^{3+} using orbital box diagrams and noble gas notation. Are these ions paramagnetic or diamagnetic?

44. Name the element corresponding to each characteristic below.
    (a) the element with the electron configuration $1s^2 2s^2 2p^6 3s^2 3p^3$
    (b) the alkaline earth element with the smallest atomic radius
    (c) the element with the largest ionization energy in Group 5A
    (d) the element whose 2+ ion has the configuration $[Kr]4d^5$
    (e) the element with the most negative electron affinity in Group 7A
    (f) the element whose electron configuration is $[Ar]3d^{10}4s^2$

45. ■ Arrange the following atoms in the order of increasing ionization energy: Si, K, P, and Ca.

46. ■ Rank the following in order of increasing ionization energy: Cl, Ca^{2+}, and Cl^-. Briefly explain your answer.

47. Answer the questions below about the elements A and B, which have the electron configurations shown.

$$A = [Kr]5s^1 \qquad B = [Ar]3d^{10}4s^2 4p^4$$

    (a) Is element A a metal, nonmetal, or metalloid?
    (b) Which element has the greater ionization energy?
    (c) Which element has the less negative electron affinity?
    (d) Which element has the larger atomic radius?

48. ■ Answer the following questions about the elements with the electron configurations shown here:

$$A = [Ar]4s^2 \qquad B = [Ar]3d^{10}4s^24p^5$$

  (a) Is element A a metal, metalloid, or nonmetal?

  (b) Is element B a metal, metalloid, or nonmetal?

  (c) Which element is expected to have the larger ionization energy?

  (d) Which element has the smaller atomic radius?

49. Which of the following ions are unlikely to be found in a chemical compound: Cs^+, In^{4+}, Fe^{6+}, Te^{2-}, Sn^{5+}, and I^-? Explain briefly.

50. ■ Place the following elements and ions in order of decreasing size: K^+, Cl^-, S^{2-}, and Ca^{2+}.

51. Answer each of the following questions:

  (a) Of the elements S, Se, and Cl, which has the largest atomic radius?

  (b) Which has the larger radius, Br or Br^-?

  (c) Which should have the largest difference between the first and second ionization energy: Si, Na, P, or Mg?

  (d) Which has the largest ionization energy: N, P, or As?

  (e) Which of the following has the largest radius: O^{2-}, N^{3-}, or F^-?

52. The following are isoelectronic species: Cl^-, K^+, and Ca^{2+}. Rank them in order of increasing (a) size, (b) ionization energy, and (c) electron affinity.

53. ■ Compare the elements Na, B, Al, and C with regard to the following properties:

  (a) Which has the largest atomic radius?

  (b) Which has the most negative electron affinity?

  (c) Place the elements in order of increasing ionization energy.

54. ▲ Two elements in the second transition series (Y through Cd) have four unpaired electrons in their 3+ ions. What elements fit this description?

55. The configuration for an element is given here.

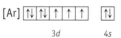

  (a) What is the identity of the element with this configuration?

  (b) Is a sample of the element paramagnetic or diamagnetic?

  (c) How many unpaired electrons does a 3+ ion of this element have?

56. The configuration of an element is given here.

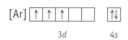

  (a) What is the identity of the element?

  (b) In what group and period is the element found?

  (c) Is the element a nonmetal, a main group element, a transition metal, a lanthanide, or an actinide?

  (d) Is the element diamagnetic or paramagnetic? If paramagnetic, how many unpaired electrons are there?

  (e) Write a complete set of quantum numbers (n, ℓ, m_ℓ, m_s) for each of the valence electrons.

  (f) What is the configuration of the 2+ ion formed from this element? Is the ion diamagnetic or paramagnetic?

Summary and Conceptual Questions

The following questions use concepts from the previous chapters.

57. Why is the radius of Li^+ so much smaller than the radius of Li? Why is the radius of F^- so much larger than the radius of F?

58. Which ions in the following list are not likely to be found in chemical compounds: K^{2+}, Cs^+, Al^{4+}, F^{2-}, and Se^{2-}? Explain briefly.

59. Write electron configurations to show the first two ionization processes for potassium. Explain why the second ionization energy is much greater than the first.

60. Explain how the ionization energy of atoms changes and why the change occurs when proceeding down a group of the periodic table.

61. (a) Explain why the sizes of atoms change when proceeding across a period of the periodic table.

  (b) Explain why the sizes of transition metal atoms change very little across a period.

62. ■ Which of the following elements has the greatest difference between its first and second ionization energies: C, Li, N, Be? Explain your answer.

63. What arguments would you use to convince another student in general chemistry that MgO consists of the ions Mg^{2+} and O^{2-} and not the ions Mg^+ and O^-? What experiments could be done to provide some evidence that the correct formulation of magnesium oxide is $Mg^{2+}O^{2-}$?

64. Explain why the first ionization energy of Ca is greater than that of K, whereas the second ionization energy of Ca is lower than the second ionization energy of K.

65. The energies of the orbitals in many elements have been determined. For the first two periods they have the following values:

Element	1s (kJ/mol)	2s (kJ/mol)	2p (kJ/mol)
H	−1313		
He	−2373		
Li		−520.0	
Be		−899.3	
B		−1356	−800.8
C		−1875	−1029
N		−2466	−1272
O		−3124	−1526
F		−3876	−1799
Ne		−4677	−2083

(a) Why does the energy generally become more negative on proceeding across the second period?

(b) How are these values related to the ionization energy and electron affinity of the elements?

(c) Use these energy values to explain the observation that the ionization energies of the first four second-period elements are in the order Li < Be > B < C.

Note that these energy values are the basis for the discussion in the Simulation on the General ChemistryNow CD-ROM or website Screen 8.9. Data from J. B. Mann, T. L. Meek, and L. C. Allen: *Journal of the American Chemical Society*, Vol. 122, p. 2780, 2000.

66. ▲ The ionization energies for the removal of the first electron in Si, P, S, and Cl are as listed in the table below. Briefly rationalize this trend.

Element	First Ionization Energy (kJ/mol)
Si	780
P	1060
S	1005
Cl	1255

67. Using your knowledge of the trends in element sizes on going across the periodic table, explain briefly why the density of the elements increases from K through V.

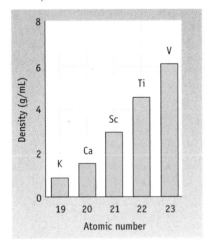

68. The densities (in g/cm^3) of elements in Groups 6B, 8B, and 1B are given in the table below.

Period 4	Cr, 7.19	Co, 8.90	Cu, 8.96
Period 5	Mo, 10.22	Rh, 12.41	Ag, 10.50
Period 6	W, 19.30	Ir, 22.56	Au, 19.32

Transition metals in the sixth period all have much greater densities than the elements in the same groups in the fourth and fifth periods. Refer to Figure 8.12 and explain this observation.

69. The discovery of two new elements (atomic numbers 113 and 115) was announced in February 2004.

Some members of the team that discovered elements 113 and 115 at the Lawrence Livermore National Laboratory (left to right): Jerry Landrum, Dawn Shaughnessy, Joshua Patin, Philip Wilk, and Kenton Moody.

(a) Use *spdf* and noble gas notations to give the electron configurations of these two elements.

(b) Name an element in the same periodic group as the two elements.

(c) Element 113 was made by firing a light atom at a heavy americium atom. The two combine to give a nucleus with 113 protons. What light atom was used as a projectile?

70. Explain why the reaction of calcium and fluorine does *not* form CaF_3.

71. ▲ Thionyl chloride, $SOCl_2$, is an important chlorinating and oxidizing agent in organic chemistry. It is prepared industrially by oxygen atom transfer from SO_3 to SCl_2.

$$SO_3(g) + SCl_2(g) \longrightarrow SO_2(g) + SOCl_2(g)$$

(a) Give the electron configuration for an atom of sulfur using an orbital box diagram. Do not use the noble gas notation.

(b) Using the configuration given in part (a), write a set of quantum numbers for the highest-energy electron in a sulfur atom.

(c) What element involved in this reaction (O, S, Cl) should have the smallest ionization energy? The smallest radius?

(d) Which should be smaller: the sulfide ion, S^{2-}, or a sulfur atom, S?

(e) If you want to make 675 g of $SOCl_2$, what mass of SCl_2 is required?

(f) If you use 10.0 g of SO_3 and 10.0 g of SCl_2, what is the theoretical yield of $SOCl_2$?

(g) $\Delta H°_{rxn}$ for the reaction of SO_3 and SCl_2 is -96.0 kJ/mol $SOCl_2$ produced. Using data in Appendix L, calculate the standard molar enthalpy of formation of SCl_2.

72. Sodium metal reacts readily with chlorine gas to give sodium chloride. (*See General ChemistryNow CD-ROM or website Screen 8.16 Chemical Puzzler.*)

$$Na(s) + \tfrac{1}{2}Cl_2(g) \longrightarrow NaCl(s)$$

(a) What is the reducing agent in this reaction? What property of the element contributes to its ability as a reducing agent?

(b) What is the oxidizing agent in this reaction? What property of the element contributes to its ability as an oxidizing agent?

(c) Why does the reaction produce NaCl and not a compound such as Na_2Cl or $NaCl_2$?

73. If a C atom is attached or "bonded" to a Cl atom, the calculated distance between the atoms is the sum of their radii. Calculate the expected distance between the pairs of atoms in the following table. Then use model molecules in the Molecular Models folder on the General ChemistryNow CD-ROM or website to examine the appropriate distance in the designated molecules. Is there reasonably good agreement between the calculated and measured distances?

Molecule	Atom Distance	Calculated (pm)	Measured (pm)
BF_3	B — F	_____	_____
PF_3	P — F	_____	_____
CH_4	C — H	_____	_____
H_3COH	C — O	_____	_____

(Note that BF_3 and PF_3 are in the Inorganic folder. CH_4 and CH_3OH are in the Organic folder. The latter (CH_3OH) is called methanol and is in the Alcohol folder. The distances given on these models are in angstrom units, where $1\ \text{Å} = 100$ pm.)

Do you need a live tutor for homework problems?
Access vMentor at General ChemistryNow at
http://now.brookscole.com/kotz6e
for one-on-one tutoring from a chemistry expert

9—Bonding and Molecular Structure: Fundamental Concepts

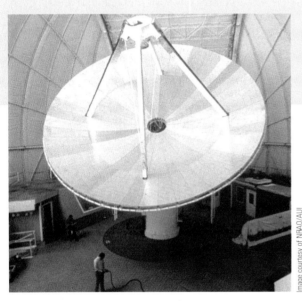

The 12-meter radio telescope at the National Radio Astronomy Observatory. It is used to search for molecules in deep space.

Comets deliver many complex molecules to earth. This is the Hale-Bopp comet in 1997.

Molecules in Space

Life is based on simple molecules like water and ammonia, slightly more complex ones like sugars, and very complex ones like DNA and hemoglobin. Where do they come from? How are they formed? What do they look like? Are their properties connected to how they look— that is, to their structures?

The origins of molecules is a topic eagerly studied by astronomers. Since the 1960s some space scientists have surmised that comets bring water, ammonia, and even more complex molecules of all kinds to earth from outer space. Every day an average of about 30 tons of organic material arrive on earth from space.

More than 120 molecules have been identified by radio astronomers in the far reaches of our galaxy. These range from hydrogen molecules to other simple molecules such as CO, H_2O, NH_3, and HCl.

| CO | H_2O | NH_3 | HCl |

Some molecules from deep space.

Recently, more complex molecules have been observed, including a simple sugar, glycolaldehyde, $C_2H_4O_2$, discovered in 2001 in a cloud of gas and dust about 26,000 light-years from earth. According to a researcher at the NASA Goddard Space Center, "The discovery of this sugar molecule in a cloud from which new stars are forming means it is increasingly likely that the chemical precursors to life are formed in such clouds long before planets develop around the stars."

Glycolaldehyde is a member of the carbohydrate family, all of which have the general formula $C_a(H_2O)_b$. The molecule has two C atoms in its "backbone." One C atom is attached to an H atom and an O atom. The other C atom has two H atoms and one OH group

Chapter Goals

See Chapter Goals Revisited (page 425). Test your knowledge of these goals by taking the exam-prep quiz on the General ChemistryNow CD-ROM or website.

- Understand the difference between ionic and covalent bonds.
- Draw Lewis electron dot structures for small molecules and ions.
- Use the valence shell electron-pair repulsion theory (VSEPR) to predict the shapes of simple molecules and ions and to understand the structures of more complex molecules.
- Use electronegativity to predict the charge distribution in molecules and ions and to define the polarity of bonds.
- Predict the polarity of molecules.
- Understand the properties of covalent bonds and their influence on molecular structure.

Chapter Outline

J. Hester and P. Scowan, of Arizona State University, and NASA.

The Eagle Nebula. These pillar-like structures are vast columns of gas and dust, within which new stars have recently formed. It is here that many molecules are created. The tallest of the pillars (at left) is about one light-year in length from base to tip. The Eagle Nebula is a star-forming region 7000 light-years away in the constellation Serpens.

attached. Indeed, its structure is quite predictable, and recognizing that pattern is one objective of this chapter. We want to know, for example, why the angles made by the atom attachments are not 90°, and why there is a difference in the C — O links. We would also like to know how this structure influences its chemical and physical properties, so that we might predict how it would interact with other molecules.

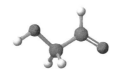

$HOCH_2CHO$

Glycolaldehyde.

How are some of these complex molecules formed? Temperatures in deep space hover near absolute zero, and astronomers believe that simple molecules such as water, CO, CO_2, and CH_3OH (methanol) freeze onto the surface of minute pieces of interstellar dust. These dust particles are subjected to intense radiation from nearby stars, causing the molecules to fragment (much as you saw in the mass spectrum in Figure 3.15). The fragments rearrange and combine in new ways, forming larger molecules such as glycolaldehyde.

Other molecules found recently in space include hydrocarbons (compounds composed only of C and H) such as anthracene. Anthracene is a member of a large class of compounds called polycyclic aromatic hydrocarbons. You may be aware of them because they are carcinogenic pollutants on earth, and you produce minute quantities when you cook a hamburger on a charcoal grill.

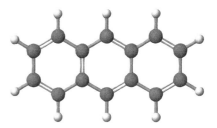

Anthracene, a polycyclic aromatic hydrocarbon.

Why are these compounds flat? What relation do they have to other carbon based compounds? These are just a few of the subjects we begin to explore in this and subsequent chapters.

To Review Before You Begin

- Know the names of common compounds and ions (Chapter 3)
- Review Coulomb's law (page 112)
- Understand energy changes in chemical reactions (Chapter 6)

Scientists have long known that the key to interpreting the properties of a chemical substance is first to recognize and understand its structure and bonding. **Structure** refers to the way atoms are arranged in space, and **bonding** describes the forces that hold adjacent atoms together. In Chapter 3, we told the story of how the basic structure of DNA was uncovered. This structure raises many interesting questions, such as why DNA chains have a helical shape. The answer is related to the geometry of the chemical bonds around each of the carbon, phosphorus, and oxygen atoms of the chain. Just how this relationship works will become more evident as you learn more about the topics of structure and bonding.

The goal of this and the next two chapters is to explain how atoms are arranged in chemical compounds and what holds them together. At the same time, we want to begin to relate the structure and bonding in a molecule to its chemical and physical properties.

Our discussion of structure and bonding begins with small molecules and ions, and then progresses to larger molecules. From compound to compound, atoms of the same element participate in bonding and structure in a predictable manner. This consistency allows us to develop a group of principles that apply to many different chemical compounds, including such complex structures as DNA.

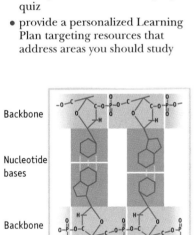

The arrangement of atoms in DNA.

9.1—Valence Electrons

The electrons in an atom can be divided into two groups: *valence electrons* and *core electrons*. Valence electrons determine the chemical properties of the atom because chemical reactions result in the loss, gain, or rearrangement of these electrons [◄ page 345]. The remaining electrons, the core electrons, are not involved in chemical behavior.

For main group elements (elements of the A groups in the periodic table), the valence electrons are the s and p electrons in the outermost shell (Table 9.1). All electrons in inner shells (such as those in filled d subshells) are core electrons. A useful guideline for *main group elements* is that *the number of valence electrons is equal to the group number*. The fact that all elements in a periodic group have the same number of valence electrons accounts for the similarity of chemical properties among members of the group.

Valence electrons for *transition elements* include the electrons in the ns and $(n-1)d$ orbitals (see Table 9.1). The remaining electrons are core electrons. As with main group elements, the valence electrons for transition metals determine the chemical properties of these elements.

See General ChemistryNow CD-ROM or website:

- **Screen 9.2 Valence Electrons,** for the correlation of the periodic table and valence electrons

Table 9.1 Core and Valence Electrons for Several Common Elements

Element	Periodic Group	Core Electrons	Valence Electrons	Total Configuration
Main Group Elements				
Na	1A	$1s^2 2s^2 2p^6 = $ [Ne]	$3s^1$	[Ne]$3s^1$
Si	4A	$1s^2 2s^2 2p^6 = $ [Ne]	$3s^2 3p^2$	[Ne]$3s^2 3p^2$
As	5A	$1s^2 2s^2 2p^6 3s^2 3p^6 3d^{10} = $ [Ar]$3d^{10}$	$4s^2 4p^3$	[Ar]$3d^{10} 4s^2 4p^3$
Transition Elements				
Ti	4B	$1s^2 2s^2 2p^6 3s^2 3p^6 = $ [Ar]	$3d^2 4s^2$	[Ar]$3d^2 4s^2$
Co	8B	[Ar]	$3d^7 4s^2$	[Ar]$3d^7 4s^2$
Mo	6B	[Kr]	$4d^5 5s^1$	[Kr]$4d^5 5s^1$

Lewis Symbols for Atoms

G. N. Lewis (1875–1946) introduced a useful way to describe electrons in the valence shell of an atom. In the system he developed, the element's symbol represents the atomic nucleus together with the core electrons. Up to four valence electrons, represented by dots, are placed one at a time around the symbol; then, if any valence electrons remain, they are placed next to ones already there. Chemists now refer to these pictures as **Lewis electron dot symbols**. Lewis dot symbols for the main group elements of the second and third periods are shown in Table 9.2.

Arranging the valence electrons of a main group element around an atom in four groups suggests that the valence shell can accommodate a maximum of four pairs of electrons. Because this arrangement represents eight electrons, it is referred to as an *octet* of electrons. An octet of electrons surrounding an atom is regarded as a stable configuration. The noble gases, with the exception of helium, have eight valence electrons and demonstrate a notable lack of reactivity. (Helium, neon, and argon do not undergo any chemical reactions, and the other noble gases have very limited chemical reactivity.) Because chemical reactions involve changes in the valence electron shell, the limited reactivity of the noble gases is taken as evidence of the stability of their noble gas ($ns^2 np^6$) electron configuration. Hydrogen, which in its compounds has two electrons in its valence shell, obeys the spirit of this rule by matching the electron configuration of He.

■ **H Atoms and Electron Octets**
Hydrogen cannot be surrounded by an octet of electrons. An atom of H, which has only a $1s$ valence electron orbital, can accommodate only a pair of electrons.

Table 9.2 Lewis Dot Symbols for Main Group Atoms

1A ns^1	2A ns^2	3A $ns^2 np^1$	4A $ns^2 np^2$	5A $ns^2 np^3$	6A $ns^2 np^4$	7A $ns^2 np^5$	8A $ns^2 np^6$
Li·	·Be·	·B·	·C·	·N·	:O·	:F·	:Ne:
Na·	·Mg·	·Al·	·Si·	·P·	:S·	:Cl·	:Ar:

Example 9.1—Valence Electrons

Problem Give the number of valence electrons for Ca and Se. Draw the Lewis electron dot symbol for each element.

Strategy Locate the elements in the periodic table. Note that, for main group elements, the number of valence electrons equals the group number.

Solution Calcium, in Group 2A, has two valence electrons, and selenium, in Group 6A, has six. Dots representing electrons are placed around the element symbol one at a time until there are four electrons. Subsequent electrons are paired with those already present:

$$\cdot \text{Ca} \cdot \qquad \cdot \overset{\cdot\cdot}{\underset{\cdot\cdot}{\text{Se}}} \cdot$$

calcium   selenium

Exercise 9.1—Electrons

Give the number of valence electrons for Ba, As, and Br. Draw the Lewis dot symbol for each of these elements.

9.2—Chemical Bond Formation

When a chemical reaction occurs between two atoms, their valence electrons are reorganized so that a net attractive force—a **chemical bond**—occurs between atoms. There are two general types of bonds: ionic and covalent. Their formation can be depicted using Lewis symbols.

An **ionic bond** forms when *one or more valence electrons are transferred from one atom to another*, creating positive and negative ions. When sodium and chlorine react (Figure 9.1a), an electron is transferred from a sodium atom to a chlorine atom to form Na^+ and Cl^-.

Figure 9.1 Formation of ionic compounds. Both reactions shown here are quite exothermic, as reflected by the very negative molar enthalpies of formation for the reaction products. (*See General ChemistryNow Screen 9.5 Chemical Reactions and Periodic Properties, to watch a video of the sodium chlorine reaction.*)

(a) The reaction of elemental sodium and chlorine.
ΔH_f° [NaCl(s)] = −411.12 kJ/mol

(b) The reaction of elemental calcium and oxygen to give calcium oxide.
ΔH_f° [CaO(s)] = −635.09 kJ/mol

$$Na \cdot \; + \; \cdot \overset{..}{\underset{..}{Cl}} : \quad \longrightarrow \quad \left[Na \overset{\frown}{} \cdot \overset{..}{\underset{..}{Cl}} : \right] \quad \longrightarrow \quad \left[Na^{+} \quad : \overset{..}{\underset{..}{Cl}} : ^{-} \right]$$

Metal atom	Nonmetal atom	Electron transfer from reducing agent to oxidizing agent.	Ionic compound. Ions have noble gas electron configurations.

The "bond" is the attractive force between the positive and negative ions.

Covalent bonding, in contrast, *involves sharing of valence electrons between atoms.* Two chlorine atoms, for example, share a pair of electrons, one electron from each atom, to form a covalent bond.

$$: \overset{..}{\underset{..}{Cl}} \cdot \; + \; \cdot \overset{..}{\underset{..}{Cl}} : \quad \longrightarrow \quad : \overset{..}{\underset{..}{Cl}} : \overset{..}{\underset{..}{Cl}} :$$

It is useful to reflect on the differences in the Lewis electron dot structure representations of ionic and covalent bonding. In both processes, unpaired electrons in the reactants are paired up. Both processes give products in which each atom is surrounded by eight electrons (an octet). The position of the electron pair between the two bonded atoms differs significantly, however. In a chlorine molecule (Cl_2), the electron pair is shared equally by the two atoms. In contrast, the electron pair in sodium chloride has become part of the valence shell of chlorine.

As bonding is described in greater detail, you will discover that the two types of bonding—complete electron transfer and the equal sharing of electrons—are extreme cases. In most chemical compounds electrons are shared unequally, with the extent of sharing varying widely from very little sharing (largely ionic) to considerable sharing (largely covalent).

GENERAL
Chemistry • ⚛ • **Now**™

See General ChemistryNOW CD-ROM or website:

- **Screen 9.5 Chemical Reactions and Periodic Properties,** to watch a video of the formation of sodium chloride from the elements

9.3—Bonding in Ionic Compounds

Metallic sodium reacts vigorously with gaseous chlorine to give sodium chloride, (Figure 9.1a), and calcium metal and oxygen react to give calcium oxide (Figure 9.1b). In each case, the product is an ionic compound: NaCl contains Na^+ and Cl^- ions, whereas CaO is composed of Ca^{2+} and O^{2-} ions.

$$Na(s) + \tfrac{1}{2} Cl_2(g) \longrightarrow NaCl(s) \qquad \Delta H° = -411.12 \text{ kJ}$$
$$Ca(s) + \tfrac{1}{2} O_2(g) \longrightarrow CaO(s) \qquad \Delta H° = -635.09 \text{ kJ}$$

These exothermic reactions, which are examples of the chemical behavior demonstrated by elements in these periodic groups, can be understood based on the atomic properties described in Chapter 8. The alkali and alkaline earth metals have relatively low ionization energies. Loss of the ns valence electrons from these elements leads to cations with a noble gas configuration $(n-1)s^2(n-1)p^6$. In

■ **Valence Electron Configurations and Ionic Compound Formation**
For the formation of NaCl:

Na changes from $1s^2 2s^2 2p^6 3s^1$ to Na^+ with $1s^2 2s^2 2p^6$, equivalent to the Ne configuration.

Cl changes from $[Ne]3s^2 3p^5$ to Cl^- with $[Ne]3s^2 3p^6$, equivalent to the Ar configuration.

contrast, elements immediately preceding Group 8A (the halogens and the Group 6A elements) have high affinities for electrons. These elements typically form anions by adding electrons, giving an ion with an electron configuration equivalent to that of the next noble gas. The tendency to achieve a noble gas configuration by gain or loss of electrons is an important aspect in the chemistry of main group elements.

Ion Attraction and Lattice Energy

To understand bonding in ionic compounds, it is useful to think about the energy involved in their formation. We begin by asking what the energy change is for the formation of the ion pair [Na^+,Cl^-] in the gas phase starting with sodium and chlorine atoms, also in the gas phase. The overall energy of this reaction can be thought of as the sum of three individual steps: (1) the ionization of a sodium atom to form Na^+ (the energy of this process is the *ionization energy* of the element); (2) the addition of an electron to a chlorine atom to form the Cl^- ion (the energy here is the *electron affinity* of the element); and (3) the formation of the [Na^+,Cl^-](g) ion pair from Na^+(g) and Cl^-(g).

1. Formation of Na^+(g) and an electron

   $Na(g) \longrightarrow Na^+(g) + e^-$ ΔE_{ion} = ionization energy of Na = +496 kJ/mol

2. Formation of Cl^-(g) from a Cl atom and an electron

   $Cl(g) + e^- \longrightarrow Cl^-(g)$ ΔE_{EA} = electron affinity of Cl = −349 kJ/mol

3. Formation of the ion pair

   $Na^+(g) + Cl^-(g) \longrightarrow [Na^+,Cl^-](g)$ $\Delta E_{ion\ pair}$ = −498 kJ/mol

The energy for the last step in this process can be calculated from an equation related to Coulomb's law [◀ Equation 3.1, page 112].

$$E_{ion\ pair} = C(N)\left(\frac{(n^+e)(n^-e)}{d}\right)$$

The symbol C represents a constant, d is the distance between the ion centers, n is the number of positive (n^+) and negative (n^-) charges on an ion, and e is the charge on an electron. Including Avogadro's number, N, allows us to calculate the energy change for 1 mol of ion pairs. Because the charges are opposite in sign, the energy value is negative. Inspection of this equation reveals that the energy of attraction between ions of opposite charge depends on two factors:

- *The magnitude of the ion charges.* The higher the ion charges, the greater the attraction, so ΔE for ion-pair formation has a larger negative value. For example, the attraction between Ca^{2+} and O^{2-} ions will be about four times larger [$(2+) \times (2-)$] than the attraction between Na^+ and Cl^- ions, and the energy will be more negative by a factor of about 4.

- *The distance between the ions.* This is an inverse relationship because, as the distance between ions becomes greater (d becomes larger), the attractive force between the ions declines and the energy is less negative. The distance is determined by the sizes of the ions [◀ Figure 8.15].

Ionic compounds exist as solids under normal conditions. Their structures contain not ion pairs but rather positive and negative ions arranged in a three-dimensional lattice [▶ Chapter 13]. Models of a small segment of a NaCl lattice are pictured in Figure 9.2. In crystalline NaCl, each Na^+ cation is surrounded by six Cl^-

Photo: Charles D. Winters

Figure 9.2 Crystalline NaCl and models of the sodium chloride crystal lattice. (*left*) A ball-and-stick model. The "sticks" in the model are there to help identify the locations of the atoms. (*right*) A space-filling model. These models represent only a small portion of the lattice. Ideally, it extends infinitely in all directions. Sodium ions are colored silver and chloride ions are colored yellow to distinguish them in this illustration.

anions, and six Na^+ ions are nearest neighbors to each Cl^-. The equation for $E_{ion\ pair}$ can be modified to take into account the extensive attractions between ions of opposite charge and repulsions between ions of like charge in a crystal lattice. That is, we can calculate the **lattice energy**, $\Delta E_{lattice}$, which is the energy of formation of one mole of a solid crystalline ionic compound when ions in the gas phase combine (see Table 9.3).

$$Na^+(g) + Cl^-(g) \longrightarrow NaCl(s) \qquad \Delta E_{lattice} = -786\ kJ/mol$$

Lattice energy is a measure of the strength of ionic bonding in solid compounds. As we shall see, these energy values are closely related to the temperatures required to melt ionic compounds [▶ Section 13.6].

What is important about lattice energy here is the dependence of $\Delta E_{lattice}$ on ion charges and sizes. The effect of ion charge is illustrated by the lattice energies of MgO and NaF. The value of $\Delta E_{lattice}$ for MgO ($-4050\ kJ/mol$) is about four times more negative than the value for NaF ($-926\ kJ/mol$) because the charges on the Mg^{2+} and O^{2-} ions are each twice as large as those on Na^+ and F^- ions.

The effect of ion size on lattice energy is also predictable: A lattice built from smaller ions generally leads to a more negative value for the lattice energy (Table 9.3 and Figure 9.3). For alkali metal halides, for example, the lattice energy for lithium compounds is generally more negative than that for potassium compounds because the Li^+ ion is much smaller than the K^+ cation. Similarly, fluorides are more strongly bonded than are iodides with the same cation.

Calculating a Lattice Energy

The lattice energies in Table 9.3 were calculated using a thermodynamic energy level diagram known as a Born-Haber cycle. This calculation is an application of Hess's law, which says that the energy involved in one pathway from reactants to products (the heat of formation of the compound, ΔH_f°) is the sum of the energies involved in another pathway (Steps 1–3). Such a cycle is illustrated in Figure 9.4 for solid sodium chloride.

Steps 1 and 2 in Figure 9.4 involve formation of $Na^+(g)$ and $Cl^-(g)$ ions from the elements, and the enthalpy change of each step is available from experiments (Appendices F and L). Step 3 in Figure 9.4 gives the lattice enthalpy, $\Delta H_{lattice}$, and

Table 9.3 Lattice Energies of Some Ionic Compounds

Compound	$\Delta E_{lattice}$ (kJ/mol)
LiF	−1037
LiCl	−852
LiBr	−815
LiI	−761
NaF	−926
NaCl	−786
NaBr	−752
NaI	−702
KF	−821
KCl	−717
KBr	−689
KI	−649

Source: D. Cubicciotti: "Lattice energies of the alkali halides and electron affinities of the halogens." *Journal of Chemical Physics*, Vol. 31, p. 1646, 1959.

■ Born-Haber Cycles

These energy level diagrams are named for Max Born (1882–1970) and Fritz Haber (1868–1934), German scientists who played prominent roles in thermodynamic research.

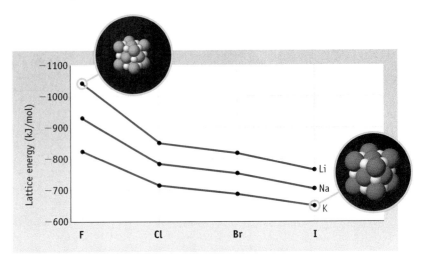

Figure 9.3   Lattice energy. $\Delta E_{lattice}$ is illustrated for the formation of the alkali metal halides, MX(s), from the ions $M^+(g) + X^-(g)$.

ΔH_f° is the standard molar enthalpy of formation of NaCl(s) (Appendix L). The enthalpy values for each step are related by the following equation:

$$\Delta H_f^\circ \, [\text{NaCl(s)}] = \Delta H_{\text{Step 1a}} + \Delta H_{\text{Step 1b}} + \Delta H_{\text{Step 2a}} + \Delta H_{\text{Step 2b}} + \Delta H_{\text{Step 3}}$$

Because the values for all of these quantities are known except for $\Delta H_{\text{Step 3}}$ ($\Delta H_{\text{lattice}}$), the value for this step can be calculated.

Step 1a. Enthalpy of formation of Cl(g) = $+121.3$ kJ/mol (Appendix L)

Step 1b. ΔH for Cl(g) + e$^-$ $\longrightarrow$ Cl$^-$(g) = -349 kJ/mol (Appendix F)

Step 2a. Enthalpy of formation of Na(g) = $+107.3$ kJ/mol (Appendix L) ⌐

Figure 9.4   Born-Haber cycle for the formation of NaCl(s) from the elements. The calculation here uses enthalpy values, and the value obtained is the lattice enthalpy, $\Delta H_{\text{lattice}}$. The difference between $\Delta H_{\text{lattice}}$ and $\Delta E_{\text{lattice}}$ is generally not significant and can be corrected for, if desired.

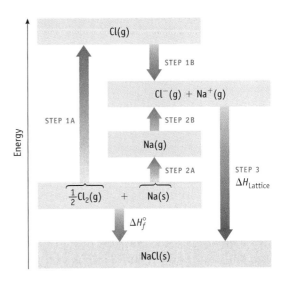

Step 2b. ΔH for $Na(g) \longrightarrow Na^+(g) + e^- = +496$ kJ/mol (Appendix F)

Given that ΔH_f°, the standard heat of formation of NaCl(s) is -411.12 kJ/mol, we can calculate ΔH_3, which is the lattice enthalpy, $\Delta H_{lattice}$.

Step 3. Formation of NaCl(s) from the ions in the gas phase = ΔH_3 = $\Delta H_{lattice} = -787$ kJ/mol

GENERAL
Chemistry ⚛ Now™

See General ChemistryNOW CD-ROM or website:
- **Screen 9.4 Lattice Energy,** for an illustration of lattice and lattice energy

---

Exercise 9.2—Using Lattice Energies

Calculate the molar enthalpy of formation, ΔH_f°, of solid sodium iodide using the approach outlined in Figure 9.4. The required data can be found in Appendices F and L and in Table 9.3.

Why Don't Compounds Such as NaCl$_2$ and NaNe Exist?

The Born-Haber energy cycle in Figure 9.4 can help us answer some interesting questions. For example, why is it unlikely that a compound such as $NaCl_2$ will exist? Here the sodium is present as the Na^{2+} ion, and the formation of this ion (with the electron configuration $1s^2 2s^2 2p^5$) would require the loss of *two electrons* from a sodium atom. Because the second electron must be removed from the $n = 2$ shell, formation of Na^{2+} requires a substantial amount of energy. That is, the total energy for Step 2b in the Born-Haber cycle in Figure 9.4 would be the sum of the first *and* second ionization energies for Na (496 kJ + 4562 kJ). We can also make the reasonable assumption that the lattice energy for $NaCl_2$ (Step 3) is negative and has a value at least double that of NaCl (the increase reflects the facts that the cation charge has doubled and the size of Na^{2+} is less than that of Na^+). If we add up the energies for Steps 1 and 2, and then use a lattice energy in Step 3 of about 1500 kJ, we would estimate a very positive value for ΔH_f° of $NaCl_2$ (about $+3400$ kJ/mol). The formation of $NaCl_2$ from Na(s) and $Cl_2(g)$ is quite unfavorable because the lattice energy is not enough to offset the high energy of formation of Na^{2+}.

Because sodium is a good reducing agent, why doesn't it reduce neon to Ne^- and form NaNe? Again, we can think about this question in terms of the Born-Haber cycle for NaCl. Put Ne in place of Cl_2. Because neon exists in the form of atoms, $\Delta H_{Step\ 1a}$ is not required. More important to the outcome is the fact that neon's affinity for an electron should be extremely low, and $\Delta H_{Step\ 1b}$ will be quite positive. (The additional electron has to be placed in the next higher electron shell, and the $n = 3$ shell is much higher in energy than the $n = 2$ shell.)

$$Ne\ (1s^2 2s^2 2p^6) + e^- \longrightarrow Ne^-\ (1s^2 2s^2 2p^6 3s^1) \qquad \Delta H \gg 0$$

The lattice energy of NaNe is not expected to be exothermic enough to overcome this and other endothermic steps, so an overall positive enthalpy change is again expected. The formation of NaNe is energetically unfavorable.

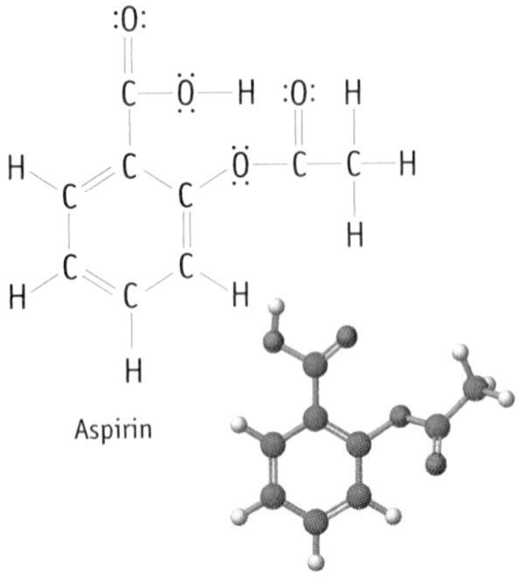

Aspirin

One goal in this chapter is to understand why a molecule such as aspirin has the shape that it exhibits.

9.4—Covalent Bonding and Lewis Structures

The remainder of this chapter is concerned with covalent bonding, in which electron pairs are shared between bonded atoms. Examples of compounds having covalent bonds include gases in our atmosphere (O_2, N_2, H_2O, and CO_2), common fuels (CH_4), and most of the compounds in your body. Covalent bonding is also responsible for the atom-to-atom connections in common ions such as CO_3^{2-}, CN^-, NH_4^+, NO_3^-, and PO_4^{3-}. We will develop the basic principles of structure and bonding using as examples molecules and ions made up of only a few atoms, but the same principles apply to larger molecules from aspirin to proteins and DNA with thousands of atoms.

The molecules and ions just mentioned are composed entirely of *nonmetal* atoms. A point that needs special emphasis is that, in molecules or ions made up *only* of nonmetal atoms, the atoms are attached by covalent bonds. Conversely, the presence of a metal in a formula is usually a signal that the compound is likely to be ionic.

Lewis Electron Dot Structures

In a simple description of covalent bonding, a bond results when one or more electron pairs are shared between two atoms. The electron-pair bond between the two atoms of an H_2 molecule is represented by a pair of dots or, alternatively, a line.

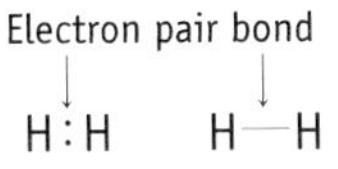

Electron pair bond

H:H   H—H

The representation of molecules such as H_2 in this fashion is called a **Lewis electron dot structure** or just a **Lewis structure** in honor of the American chemist Gilbert Newton Lewis (1875–1946) (page 415).

Simple Lewis structures can be drawn by starting with Lewis dot symbols for atoms and arranging the valence electrons to form bonds. To create the Lewis structure for F_2, for example, we could start with the Lewis dot symbol for a fluorine atom. Fluorine, an element in Group 7A, has seven valence electrons. The Lewis symbol shows that an F atom has a single unpaired electron along with three electron pairs. In F_2, the single electrons, one on each F atom, pair up in the covalent bond.

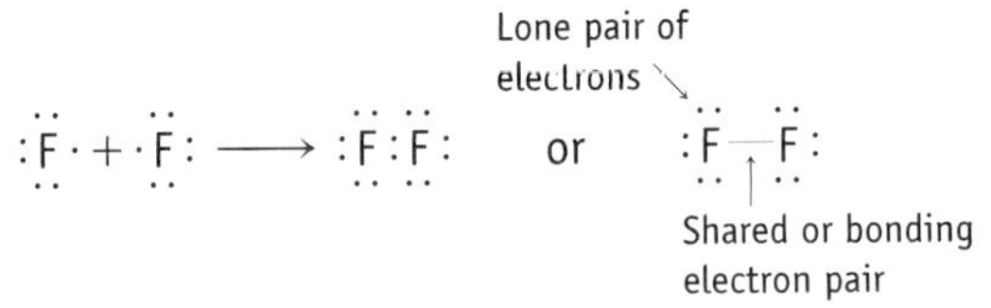

Lone pair of electrons

:F· + ·F: ⟶ :F:F: or   :F—F:

Shared or bonding electron pair

In the Lewis dot structure for F_2 the pair of electrons in the F — F bond is the bonding pair, or **bond pair**. The other six pairs reside on single atoms and are called **lone pairs**. Because they are not involved in bonding; they are also called **nonbonding electrons**.

Carbon dioxide, CO_2, and dinitrogen, N_2, are examples of molecules in which two atoms are multiply bonded, that is, they share more than one electron pair.

O=C=O   :N≡N:

In carbon dioxide, the carbon atom shares two pairs of electrons with each oxygen and so is linked to each O atom by a **double bond**. The valence shell of each oxygen atom in CO_2 has two bonding pairs and two lone pairs. In dinitrogen, the two nitrogen atoms share three pairs of electrons, so they are linked by a **triple bond**. In addition, each N atom has a single lone pair.

The Octet Rule

An important observation can be made about the molecules you have seen so far: Each atom (except H) has a share in four pairs of electrons, so each has achieved a *noble gas configuration. Each atom is surrounded by an octet of eight electrons.* (Hydrogen typically forms a bond to only one other atom, resulting in two electrons in its valence shell.) *The tendency of molecules and polyatomic ions to have structures in which eight electrons surround each atom* is known as the **octet rule**. As an example, a triple bond is necessary in dinitrogen to have an octet around each nitrogen atom. The carbon atom and both oxygen atoms in CO_2 achieve the octet configuration by forming double bonds.

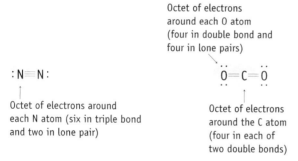

Octet of electrons around each O atom (four in double bond and four in lone pairs)

Octet of electrons around each N atom (six in triple bond and two in lone pair)

Octet of electrons around the C atom (four in each of two double bonds)

The octet rule is extremely useful, but keep in mind that it is more a *guideline* than a rule. It directs you to seek a Lewis structure in which each atom has eight electrons in its valence shell (or two in the case of hydrogen). Particularly for the second-period elements C, N, O, and F, a Lewis structure in which each atom achieves an octet is likely to be correct. Although some exceptions exist, if an atom such as C, N, O, or F in a Lewis structure does not follow the octet rule, you should probably doubt the structure's validity. If a structure obeying the octet rule cannot be written, then it is possible an incorrect formula has been assigned to the compound.

There is a systematic approach to constructing Lewis structures of molecules and ions. Let us take formaldehyde, CH_2O, as an example.

1. *Determine the arrangement of atoms within a molecule.* The central atom is *usually* the one with the least negative electron affinity. In CH_2O the central atom is C. You will come to recognize that certain elements often appear as the center atom, among them C, N, P, and S. Halogens are often terminal atoms forming a single bond to one other atom, but they can be the central atom when combined

with O in oxoacids (such as $HClO_4$). Oxygen is the central atom in water, but in conjunction with carbon, nitrogen, phosphorus, and the halogens it is usually a terminal atom. Hydrogen is a terminal atom because it typically bonds to only one other atom.

2. *Determine the total number of valence electrons in the molecule or ion.* In a neutral molecule this number will be the sum of the valence electrons for each atom. For an anion, *add* a number of electrons equal to the negative charge; for a cation, *subtract* the number of electrons equal to the positive charge. The number of valence electron pairs will be half the total number of valence electrons. For CH_2O,

Valence electrons = 12 electrons (or 6 electron pairs)
= 4 for C + (2 × 1) for two H atoms + 6 for O

3. *Place one pair of electrons between each pair of bonded atoms to form a single bond.*

Here three electron pairs are used to make three single bonds (which are represented by single lines). Three pairs of electrons remain to be used.

4. *Use any remaining pairs as lone pairs around each terminal atom (except H) so that each terminal atom is surrounded by eight electrons.* If there are electrons left over after this step, assign them to the central atom. If the central atom is an element in the third or higher period, it can have more than eight electrons.

Here all six pairs have been assigned, but notice that the C atom has a share in only three pairs.

5. *If the central atom has fewer than eight electrons at this point, move one or more of the lone pairs on the terminal atoms into a position intermediate between the center and the terminal atom to form multiple bonds.*

As a general rule, double or triple bonds are formed *only* when both atoms are from the following list: C, N, O, or S. That is, bonds such as C=C, C=N, and C=O, and C=S will be encountered.

GENERAL
Chemistry ⋅Now™

See General ChemistryNOW CD-ROM or website:
- **Screen 9.8 Drawing Lewis Structures,** for a tutorial and exercise on drawing Lewis structures

Example 9.2—Drawing Lewis Structures

Problem Draw Lewis structures for ammonia (NH_3), the hypochlorite ion (ClO^-), and the nitronium ion (NO_2^+).

Strategy Follow the five steps outlined for CH_2O in the text.

Solution for NH_3

1. *Decide on the central atom.* Hydrogen atoms are always terminal atoms, so nitrogen must be the central atom in the molecule.

2. *Count the number of valence electrons.* The total is eight (four valence pairs).

$$\text{Valence electrons} = 5 \text{ (for N)} + 3 \text{ (1 for each H)}$$

3. *Form single covalent bonds between each pair of atoms.* This uses three of the four pairs available.

$$\begin{array}{ccc} \text{H} & \!\!\!-\text{N}-\!\!\! & \text{H} \\ & | & \\ & \text{H} & \end{array}$$

4. *Place the remaining pair of electrons on the central atom.*

$$\begin{array}{ccc} \text{H} & \!\!\!-\overset{..}{\text{N}}-\!\!\! & \text{H} \\ & | & \\ & \text{H} & \end{array}$$

Each H atom has a share in one pair of electrons as required, and the central N atom has achieved an octet configuration with four electron pairs. No additional steps are required; this is the correct Lewis structure.

Solution for ClO^- Ion

1. With two atoms, there is no "central" atom.

2. Valence electrons = 14 (7 valence pairs)

$$= 7 \text{ (for Cl)} + 6 \text{ (for O)} + 1 \text{ (for the negative charge on the ion)}$$

3. One electron pair is used in the Cl—O bond: Cl—O.

4. Distribute the six remaining electron pairs around the "terminal" atoms.

$$\left[:\!\overset{..}{\underset{..}{\text{Cl}}}\!-\!\overset{..}{\underset{..}{\text{O}}}\!: \right]^-$$

5. As no electrons remain to be assigned and both atoms have an octet of electrons, this is the correct Lewis structure.

Solution for NO_2^+ Ion

1. Nitrogen is the center atom, because its electron affinity is less negative than that of oxygen.

2. Valence electrons = 16 (8 valence pairs)

$$= 5 \text{ (for N)} + 12 \text{ (six for each O)} - 1 \text{ (for the positive charge)}$$

3. Two electron pairs form the single bonds from the nitrogen to each oxygen:

$$\text{O}-\text{N}-\text{O}$$

4. Distribute the remaining six pairs of electrons on the terminal O atoms:

$$\left[:\!\overset{..}{\underset{..}{\text{O}}}\!-\!\text{N}-\!\overset{..}{\underset{..}{\text{O}}}\!: \right]^+$$

5. The central nitrogen atom is two electron pairs short of an octet. Thus, a lone pair of electrons on each oxygen atom is converted to a bonding electron pair to give two $N=O$ double bonds. Each atom in the ion now has four electron pairs. Nitrogen has four bonding pairs, and each oxygen atom has two lone pairs and shares two bond pairs.

Move lone pairs to create
double bonds and satisfy
the octet for N.

$$\left[:\ddot{O} \; N \; \ddot{O}: \right]^{+} \longrightarrow \left[\ddot{O}=N=\ddot{O} \right]^{+}$$

Exercise 9.3—Drawing Lewis Structures

Draw Lewis structures for NH_4^+, CO, NO^+, and SO_4^{2-}.

Predicting Lewis Structures

Lewis structures are useful in gaining a perspective on the structure and chemistry of a molecule or ion. The guidelines for drawing Lewis structures are helpful, but chemists also rely on patterns of bonding in related molecules.

Hydrogen Compounds

Some common compounds and ions formed from second-period nonmetal elements and hydrogen are shown in Table 9.4. Their Lewis structures illustrate the fact that the Lewis symbol for an element is a useful guide in determining the number of bonds formed by that element. For example, if there is no charge, nitrogen has five valence electrons. Two electrons occur as a lone pair; the other three occur as unpaired electrons. To achieve an octet, it is necessary to pair each of the unpaired electrons with an electron from another atom. Thus, N is predicted to form three bonds in uncharged molecules, which is indeed the case. Similarly, carbon is expected to form four bonds, oxygen two, and fluorine one.

Group 4A   Group 5A   Group 6A   Group 7A

$-\overset{|}{\underset{|}{C}}-$ $-\overset{..}{\underset{|}{N}}-$ $-\overset{..}{\underset{..}{O}}-$ $:\overset{..}{\underset{..}{F}}-$

Hydrocarbons are compounds formed from carbon and hydrogen, and the first two members of the series called the *alkanes* are CH_4 and C_2H_6 (see Table 9.4). What is the Lewis structure of the third member of the series, propane (C_3H_8)? We can rely on the idea that the atoms in this species all bond in predictable ways. Carbon is expected to form four bonds, and hydrogen can bond to only one other atom. The only arrangement of atoms that meets these criteria has three atoms of carbon linked together by carbon–carbon single bonds. The remaining positions around the carbon atoms are filled in with hydrogen—three hydrogen atoms on the end carbons and two on the middle carbon:

$$\begin{array}{ccccccc} & & H & & H & & H \\ & & | & & | & & | \\ H & - & C & - & C & - & C & - & H \\ & & | & & | & & | \\ & & H & & H & & H \end{array}$$

propane, C_3H_8

Table 9.4   Common Hydrogen-Containing Compounds and Ions of the Second-Period Elements

Group 4A	Group 5A	Group 6A	Group 7A
CH_4 methane	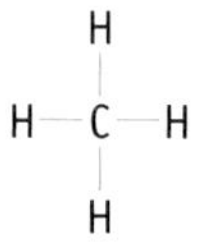 NH_3 ammonia	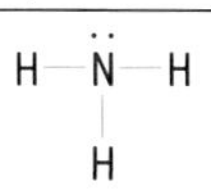 H_2O water	 HF hydrogen fluoride 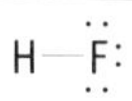
C_2H_6 ethane	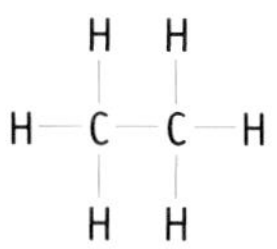 N_2H_4 hydrazine	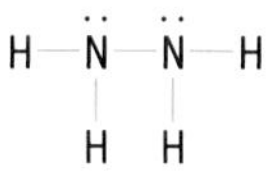 H_2O_2 hydrogen peroxide	
C_2H_4 ethylene	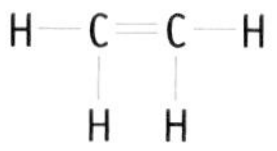 NH_4^+ ammonium ion	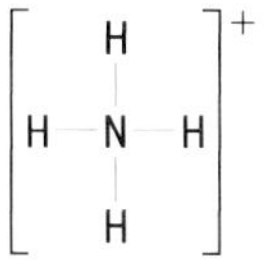 H_3O^+ hydronium ion	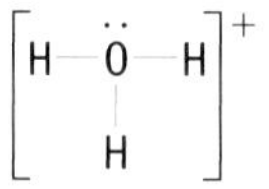
C_2H_2 acetylene	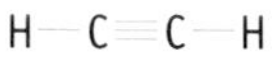 NH_2^- amide ion	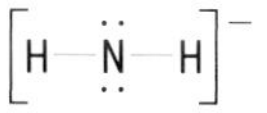 OH^- hydroxide ion	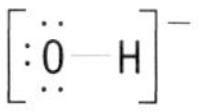

Example 9.3—Predicting Lewis Structures

Problem Draw Lewis electron dot structures for CCl_4 and NF_3.

Strategy One way to solve this problem is to recognize that CCl_4 and NF_3 are similar to CH_4 and NH_3, respectively, except that H atoms have been replaced by halogen atoms.

Solution Recall that carbon is expected to form four bonds and nitrogen three bonds to give an octet of electrons. In addition, halogen atoms have seven valence electrons, so both Cl and F can attain an octet by forming one covalent bond, just as hydrogen does.

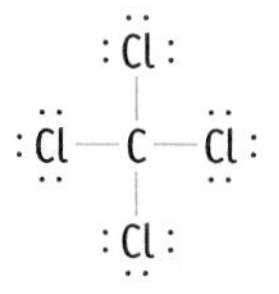

carbon tetrachloride 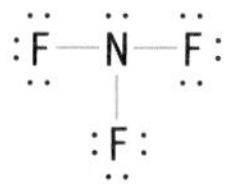   nitrogen trifluoride

As a check, count the number of valence electrons for each molecule and verify that all are present.

CCl_4: Valence electrons = 4 for C + 4 × 7 (for Cl) = 32 electrons (16 pairs)

The structure shows 8 electrons in single bonds and 24 electrons as lone-pair electrons, for a total of 32 electrons. The structure is correct.

NF_3: Valence electrons = 5 for N + 3 × 7 (for F) = 26 electrons (13 pairs)

The structure shows 6 electrons in single bonds and 20 electrons as lone-pair electrons, for a total of 26 electrons. The structure is correct.

Exercise 9.4—Predicting Lewis Structures

Predict Lewis structures for methanol, CH_3OH and hydroxylamine, NH_2OH. (*Hint:* The formulas of these compounds are written to guide you in choosing the correct arrangement of atoms.)

> **Problem-Solving Tip 9.1**
>
> **Useful Ideas to Consider When Drawing Lewis Electron Dot Structures**
>
> - The octet rule is a useful guideline when drawing Lewis structures.
> - Carbon can form four bonds (4 single bonds; 2 double bonds; 2 single bonds and 1 double bond; or one single bond and 1 triple bond). In uncharged species, nitrogen forms three bonds and oxygen forms two bonds. Hydrogen typically forms only one bond to another atom.
> - When multiple bonds are formed, both of the atoms involved are usually one of the following: C, N, O and S. Oxygen has the ability to form multiple bonds with a variety of elements. Carbon forms many compounds having multiple bonds to another carbon or to N or O.
> - Nonmetals may form single, double, and triple bonds but never quadruple bonds.
> - Always account for single bonds and lone pairs before determining whether multiple bonds are present.
> - Be alert for the possibility that the molecule or ion you are considering is isoelectronic (page 389) with a species you have seen before.

Oxoacids and Their Anions

Lewis structures of common acids and their anions are illustrated in Table 9.5. In the absence of water, these acids are covalently bonded molecular compounds, a conclusion that we should draw because all elements in the formula are nonmetals. (Nitric acid, for example, has properties that we associate with a covalent molecule: It is a colorless liquid with a boiling point of 83 °C.) In aqueous solution, however, HNO_3, H_2SO_4, and $HClO_4$ are ionized to give a hydrogen ion and the appropriate anion. A Lewis structure for the nitrate ion, for example, can be created using the guidelines on page 383, and the result is a structure with two N—O single bonds and one N═O double bond. To form nitric acid from the nitrate ion, a hydrogen ion is attached to one of the O atoms that has a single bond to the central N.

Table 9.5 Lewis Structures of Common Oxoacids and Their Anions

HNO_3 nitric acid	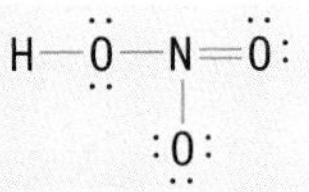	H_3PO_4 phosphoric acid	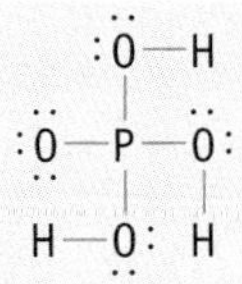	H_2SO_4 sulfuric acid	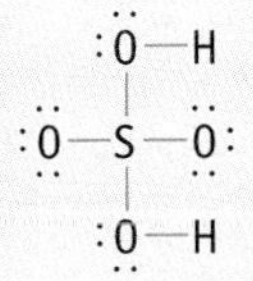
NO_3^- nitrate ion	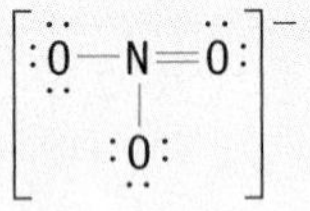	PO_4^{3-} phosphate ion	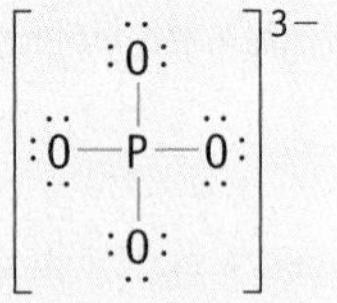	HSO_4^- hydrogen sulfate ion	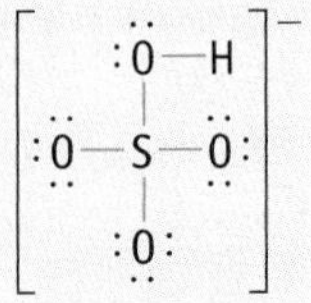
$HClO_4$ perchloric acid	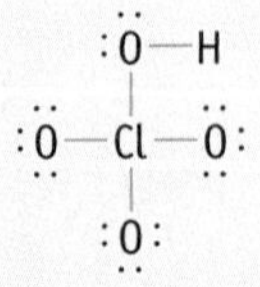	HOCl hypochlorous acid	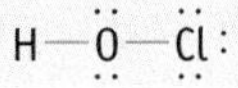	SO_4^{2-} sulfate ion	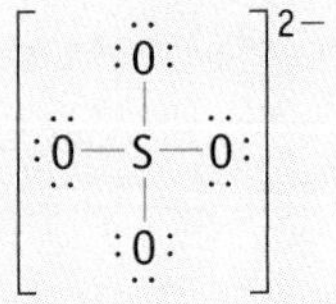
ClO_4^- perchlorate ion	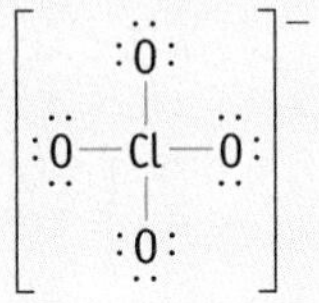	OCl^- hypochlorite ion	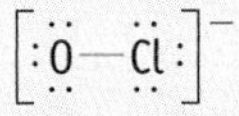		

nitrate ion   nitric acid

A characteristic property of acids in aqueous solution is their ability to donate a hydrogen ion (H^+). The NO_3^- anion is formed when the acid, HNO_3, loses a hydrogen ion. The H^+ ion separates from the acid by breaking the $H-O$ bond, the electrons of the bond staying with the O atom. As a result, HNO_3 and NO_3^- have the same number of electrons, 24, and their structures are closely related.

Exercise 9.5—Lewis Structures of Acids and Their Anions

Draw a Lewis structure for the anion $H_2PO_4^-$, which is derived from phosphoric acid.

Isoelectronic Species

In what way are NO^+, N_2, CO, and CN^- similar? Most important, all of them have two atoms and the same total number of valence electrons, 10, which leads to the same Lewis structure for each molecule or ion. The two atoms in each are linked with a triple bond. With three bonding pairs and one lone pair, each atom thus has an octet of electrons.

Molecules and ions having the *same number of valence electrons and the same Lewis structures* are said to be **isoelectronic** (Table 9.6). You will find it helpful to think in terms of isoelectronic molecules and ions because this perspective offers another way to see relationships in bonding among common chemical substances.

There are both similarities and important differences in chemical properties of isoelectronic species. For example, both carbon monoxide, CO, and cyanide ion, CN^-, are very toxic, which results from the fact that they can bind to the iron of hemoglobin in blood and block the uptake of oxygen. They differ, though, in terms of their acid–base chemistry. In aqueous solution, cyanide ion readily adds H^+ to form hydrogen cyanide, whereas CO does not. The isoelectronic species Cl_2 and OCl^- provide a similar example. Attachment of H^+ to OCl^- forms hypochlorous acid, HOCl. In contrast, Cl_2 does not add a proton.

■ **Isoelectronic and Isostructural**
The term **isostructural** is often used in conjunction with isoelectronic species. Species that are isostructural have the same structure. For example, the PO_4^{3-}, SO_4^{2-}, and ClO_4^- ions in Table 9.6 all have four oxygens bonded to the central atom. In addition, they are isoelectronic in that all have 32 valence electrons.

Table 9.6   Some Common Isoelectronic Molecules and Ions

Formulas	Representative Lewis Structure	Formulas	Representative Lewis Structure
BH_4^-, CH_4, NH_4^+		CO_3^{2-}, NO_3^-	
NH_3, H_3O^+		PO_4^{3-}, SO_4^{2-}, ClO_4^-	
CO_2, OCN^-, SCN^-, N_2O NO_2^+, OCS, CS_2			

Exercise 9.6—Identifying Isoelectronic Species

(a) Is the acetylide ion, C_2^{2-}, isoelectronic with N_2?

(b) Identify a common molecular (uncharged) species that is isoelectronic with nitrite ion, NO_2^-. Identify a common ion that is isoelectronic with HF.

9.5—Resonance

Ozone, O_3, an unstable, blue, diamagnetic gas with a characteristic pungent odor, protects the earth and its inhabitants from intense ultraviolet radiation from the sun. An important feature of its structure is that the two oxygen–oxygen bonds are the same length. This suggests that the two oxygen–oxygen bonds are equivalent. That is, equal O — O bond lengths imply an equal number of bond pairs in each O — O bond. Using the guidelines for drawing Lewis structures, however, you might come to a different conclusion. There are two possible ways of writing the Lewis structure for the molecule:

Alternative ways of creating the Lewis structure of ozone

Double bond on the right: :O — O — O: ⟶ :O — O = O:

Double bond on the left: :O — O — O: ⟶ :O = O — O:

These structures are equivalent in that each has a double bond on one side of the central oxygen atom and a single bond on the other side. If either were the actual structure of ozone, one bond (O=O) should be shorter than the other (O — O). The actual structure of ozone shows this is not the case. The inescapable conclusion is that these Lewis structures do not correctly represent the bonding in ozone.

Linus Pauling proposed the theory of **resonance** to reconcile this discrepancy. *Resonance structures are used to represent bonding in a molecule or ion when a single Lewis structure fails to describe accurately the actual electronic structure.* The alternative structures shown for ozone are called **resonance structures**. They have identical patterns of bonding and equal energy. The actual structure of this molecule is a *composite*, or **resonance hybrid**, of the equivalent resonance structures. This conclusion is reasonable because we see that the O — O bonds both have a length of 127.8 pm, intermediate between the average length of an O = O double bond (121 pm) and an O — O single bond (132 pm).

Benzene is the classic example of the use of resonance to represent a structure. The benzene molecule is a six-member ring of carbon atoms with six equivalent carbon–carbon bonds (and a hydrogen atom attached to each carbon atom). The carbon–carbon bonds are 139 pm long, intermediate between the average length of a C=C double bond (134 pm) and a C — C single bond (154 pm).

■ **Linus Pauling (1901–1994)**
Linus Pauling was born in Portland, Oregon, earned a B.S. degree in chemical engineering from Oregon State College in 1922, and completed his Ph.D. in chemistry at the California Institute of Technology in 1925. In chemistry he is widely known for his book *The Nature of the Chemical Bond*. For more on Pauling, see page 436.

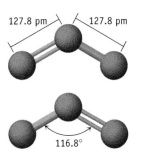

127.8 pm   127.8 pm

116.8°

Ozone, O_3. This bent molecule has oxygen–oxygen bonds of the same length.

Resonance structures of benzene, C_6H_6

Abbreviated representation of resonance structures

Problem-Solving Tip 9.2

Resonance Structures

- Resonance is a means of representing the bonding when a single Lewis structure fails to give an accurate picture.

- The atoms must have the same structural arrangement in each resonance structure. Attaching the atoms in a different fashion creates a different compound.

- Resonance structures differ only in the assignment of electron-pair positions, never in their atom positions.

- Resonance structures differ in the number of bond pairs between a given pair of atoms.

- Even though the formal process of converting one resonance structure to another seems to move electrons about, resonance is not meant to indicate the motion of electrons.

- The actual structure of a molecule is a composite or hybrid of the resonance structures.

- There will always be at least one multiple bond (double or triple) in each resonance structure.

Two resonance structures can be written for the molecule that differ only in double bond placement. A hybrid of these two structures, however, will lead to a molecule with six equivalent carbon–carbon bonds.

Let us apply the concept of resonance to describe bonding in the carbonate ion, CO_3^{2-}. This anion has 24 valence electrons (12 pairs).

Three equivalent structures can be drawn for this ion, differing only in the location of the $C=O$ double bond. This fits the classical situation for resonance, so it is appropriate to conclude that no single structure correctly describes this ion. Instead, the actual structure is a hybrid of the three structures, in good agreement with experimental results. In the CO_3^{2-} ion, all three carbon–oxygen bond distances are 129 pm, intermediate between the $C-O$ single bond (143 pm) and $C=O$ double bond (122 pm) distances.

In aqueous solution, a hydrogen ion can be attached to the carbonate ion to give the hydrogen carbonate, or bicarbonate, ion. This ion can be described as a resonance hybrid of two Lewis structures.

■ **Depicting Resonance Structures**
The use of an arrow (↔) as a symbol to link resonance structures and the name "resonance" are somewhat unfortunate. An arrow might seem to imply that a change is occurring, and the term *resonance* has the connotation of vibrating or alternating back and forth between different forms. Neither view is correct. Resonance is simply a way of representing a structure. Electron pairs are not actually moving from one place to another.

A Closer Look

Resonance Structures, Lewis Structures, and Molecular Models

When drawing structures of molecules or ions that have resonance structures, or when illustrating their structures with computer-based molecular models, we generally show only one resonance structure. Thus, a model of benzene, C_6H_6, would have alternating double bonds. This is one of the two possible resonance structures. A model of the nitrate ion would have one double bond and two single bonds in each of the three possible resonance structures.

$$\left[\begin{array}{c} :\!O\!=\!C\!-\!\overset{..}{\underset{..}{O}}: \\ | \\ :\!\overset{..}{O}\!-\!H \end{array} \right]^{-} \longleftrightarrow \left[\begin{array}{c} :\!\overset{..}{\underset{..}{O}}\!-\!C\!=\!O: \\ | \\ :\!\overset{..}{O}\!-\!H \end{array} \right]^{-}$$

Finally, notice that in each of the examples of resonance structures, the structures have been "linked" by a double-headed arrow ($\longleftrightarrow$). This convention is followed throughout chemistry.

GENERAL
Chemistry•⚛•Now™

See General ChemistryNOW CD-ROM or website:

● **Screen 9.9 Resonance Structures,** for a tutorial on drawing resonance structures

---

Example 9.4—Drawing Resonance Structures

Problem Draw resonance structures for the nitrite ion, NO_2^-. Are the N — O bonds single, double, or intermediate in value?

Strategy Draw the dot structure in the usual manner. If multiple bonds are required, resonance structures may exist. This will be the case if the octet of an atom can be completed by using an electron pair from more than one terminal atom to form a multiple bond. Bonds to the central atom cannot then be "pure" single or double bonds but rather are somewhere between the two.

Solution Nitrogen is the center atom in the nitrite ion, which has a total of 18 valence electrons (9 pairs).

Valence electrons = 5 (for the N atom) + 12 (6 for each O atom) + 1 (for negative charge)

After forming N — O single bonds, and distributing lone pairs on the terminal O atoms, a pair remains, which is placed on the central N atom.

$$\left[:\!\overset{..}{\underset{..}{O}}\!-\!\overset{..}{N}\!-\!\overset{..}{\underset{..}{O}}: \right]^{-}$$

To complete the octet of electrons about the N atom, form an N $=$ O double bond.

$$\left[:\!O\!=\!\overset{..}{N}\!-\!\overset{..}{\underset{..}{O}}: \right]^{-} \longleftrightarrow \left[:\!\overset{..}{\underset{..}{O}}\!-\!\overset{..}{N}\!=\!O: \right]^{-}$$

Because there are two ways to do this, two equivalent structures can be drawn, and the actual structure must be a resonance hybrid of these two structures. The nitrogen–oxygen bonds are neither single nor double bonds, but rather have an intermediate value.

Exercise 9.7—Drawing Resonance Structures

Draw resonance structures for the nitrate ion, NO_3^-. Sketch a plausible Lewis dot structure for nitric acid, HNO_3.

9.6—Exceptions to the Octet Rule

Although the vast majority of molecular compounds and ions obey the octet rule, there are exceptions. These include molecules and ions that have fewer than four pairs of electrons on a central atom, those that have more than four pairs, and those that have an odd number of electrons.

Compounds in Which an Atom Has Fewer Than Eight Valence Electrons

Boron, a nonmetal in Group 3A, has three valence electrons and so is expected to form three covalent bonds with other nonmetallic elements. This behavior results in a valence shell for boron in its compounds with only six electrons, two short of an octet. Many boron compounds of this type are known, including such common compounds as boric acid [$B(OH)_3$], borax [$Na_2B_4O_5(OH)_4 \cdot 8 H_2O$] (Figure 9.5), and the boron trihalides (BF_3, BCl_3, BBr_3, and BI_3).

boron trifluoride   boric acid

Boron compounds such as BF_3 that are two electrons short of an octet can be quite reactive. The boron atom can accommodate a fourth electron pair when that pair is provided by another atom. In general, molecules or ions with lone pairs can fulfill this role. Ammonia, for example, reacts with BF_3 to form $H_3N \rightarrow BF_3$. The bond between the B and N atoms in this compound uses an electron pair that originated on the N atom. The reaction of an F^- ion with BF_3 to form BF_4^- is another example.

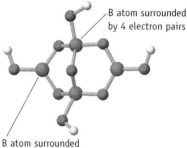
B atom surrounded by 4 electron pairs

B atom surrounded by 3 electron pairs

Figure 9.5 **Borax.** This common material, which is used in soaps, contains an interesting anion, $B_4O_5(OH)_4^{2-}$. This boron–oxygen ion has two B atoms surrounded by four electron pairs, and two B atoms surrounded by only three electron pairs.

coordinate covalent bond

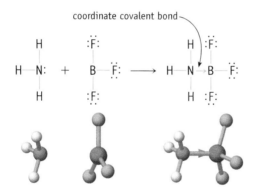

If a bonding pair of electrons originates on one of the bonded atoms, the bond is called a **coordinate covalent bond**. In Lewis structures, a coordinate covalent bond is often designated by an arrow that points away from the atom donating the electron pair.

Compounds in Which an Atom Has More Than Eight Valence Electrons

Elements in the third and higher periods often form compounds and ions in which the central element is surrounded by more than four valence electron pairs (Table 9.7). With most compounds and ions in this category, the central atom is bonded to fluorine, chlorine, or oxygen.

It is often obvious from the formula of a compound that an octet around an atom has been exceeded. As an example, consider sulfur hexafluoride, SF_6, a gas formed by the reaction of sulfur and excess fluorine. Sulfur is the central atom in this compound, and fluorine typically bonds to only one other atom with a single electron-pair bond (as in HF and CF_4). Six S—F bonds are required in SF_6, meaning there will be six electron pairs in the valence shell of the sulfur atom.

Table 9.7   Lewis Structures in Which the Central Atom Exceeds an Octet

Group 4A	Group 5A	Group 6A	Group 7A	Group 8
SiF_5^-	PF_5	SF_4	ClF_3	XeF_2
SiF_6^{2-}	PF_6^-	SF_6	BrF_5	XeF_4

More than four groups bonded to a central atom is a reliable signal that there are more than eight electrons around a central atom. But be careful—the central atom octet can also be exceeded with four or fewer atoms bonded to the central atom. Consider three examples from Table 9.7: The central atom in SF_4, ClF_3, and XeF_2 has five electron pairs in its valence shell.

A useful observation is that *only elements of the third and higher periods in the periodic table form compounds and ions in which an octet is exceeded.* Second-period elements (B, C, N, O, and F) are restricted to a maximum of eight electrons in their compounds. For example, nitrogen forms compounds and ions such as NH_3, NH_4^+, and NF_3, but NF_5 is unknown. Phosphorus, the third-period element just below nitrogen in the periodic table, forms many compounds similar to nitrogen (PH_3, PH_4^+, PF_3), but it also readily accommodates five or six valence electron pairs in compounds such as PF_5 or in ions such as PF_6^-. Arsenic, antimony, and bismuth—the elements below phosphorus in Group 5A—resemble phosphorus in their behavior.

The usual explanation for the contrasting behavior of second- and third-period elements centers on the number of orbitals in the valence shell of an atom. Second-period elements have four valence orbitals (one $2s$ and three $2p$ orbitals). Two electrons per orbital result in a total of eight electrons being accommodated around an atom. For elements in the third and higher periods, the *d* orbitals in the outer shell are traditionally included among valence orbitals for the elements. Thus, for phosphorus, the $3d$ orbitals are included with the $3s$ and $3p$ orbitals as valence orbitals. The extra orbitals provide the element with an opportunity to accommodate up to 12 electrons.

See General ChemistryNow CD-ROM or website:

- **Screen 9.10 Exceptions to the Octet Rule,** for a tutorial on identifying compounds that do not follow the octet rule

Example 9.5—Lewis Structures in Which the Central Atom Has More Than Eight Electrons

Problem Sketch the Lewis structure of the $[ClF_4]^-$ ion.

Strategy Use the guidelines on page 384.

Solution

1. The Cl atom is the central atom.

2. This ion has 36 valence electrons [= 7 for Cl + 4 × (7 for F) + 1 for ion charge] or 18 pairs.

3. Draw the ion with four single covalent Cl—F bonds.

$$
\left[
\begin{array}{c}
F \\
| \\
F-Cl-F \\
| \\
F
\end{array}
\right]^-
$$

4. Place lone pairs on the terminal atoms. Because two electron pairs remain after placing lone pairs on the four F atoms, and because we know that Cl can accommodate more than four pairs, these two pairs are placed on the central Cl atom.

The last two electron pairs are added to the central Cl atom. ⟶

Exercise 9.8—Lewis Structures in Which the Central Atom Has More Than Eight Electrons

Sketch the Lewis structures for $[ClF_2]^+$ and $[ClF_2]^-$. How many lone pairs and bond pairs surround the Cl atom in each ion?

Molecules with an Odd Number of Electrons

Two nitrogen oxides—NO, with 11 valence electrons, and NO_2, with 17 valence electrons—are among a very small group of stable molecules with an odd number of electrons. Because these compounds have an odd number of electrons, it is impossible to draw a structure obeying the octet rule; at least one electron must be unpaired.

Even though NO_2 does not obey the octet rule, an electron dot structure can be written that approximates the bonding in the molecule. This Lewis structure places the unpaired electron on nitrogen. Two resonance structures show that the nitrogen–oxygen bonds are equivalent, as observed experimentally.

Experimental evidence for NO indicates that the bonding between N and O is intermediate between a double bond and a triple bond. It is not possible to write a

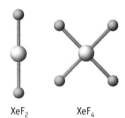

Xenon Compounds
Compounds of xenon are among the more interesting entries in Table 9.7. Noble gas compounds were not discovered until the early 1960s. One of the more intriguing compounds is XeF_2, in part because of the simplicity of its synthesis. Xenon difluoride can be made by placing a flask containing xenon gas and fluorine gas in sunlight. After several weeks, crystals of colorless XeF_2 are found in the flask.

XeF_2 XeF_4

Chemical Perspectives

The Importance of Odd-Electron Molecules

Small molecules such as H_2, O_2, H_2O, CO, and CO_2 are among the most important molecules commercially, environmentally, and biologically. Imagine the surprise of chemists and biologists when it was discovered a few years ago that nitrogen monoxide (nitric oxide, NO), which was widely considered to be toxic, plays an important biological role.

Nitric oxide is a colorless, paramagnetic gas that is moderately soluble in water. In the laboratory, it can be synthesized by the reduction of nitrite ion with iodide ion:

$$KNO_2(aq) + KI(aq) + H_2SO_4(aq) \longrightarrow$$
$$NO(g) + K_2SO_4(aq) + H_2O(\ell) + \tfrac{1}{2} I_2(aq)$$

The formation of NO from the elements was an unfavorable, energetically uphill reaction ($\Delta H_f^\circ = 90.2$ kJ/mol). Nevertheless, small quantities of this compound form from nitrogen and oxygen at high temperatures. For example, conditions in an internal combustion engine favor this reaction.

Nitric oxide reacts rapidly with O_2 to form the reddish brown gas NO_2.

$$2\,NO(\text{colorless, g}) + O_2(g) \longrightarrow$$
$$2\,NO_2(\text{brown, g})$$

The result is that NO (and compounds such as NO_2 and HNO_3 arising from reactions of NO with O_2 and H_2O) are among the air pollutants produced by automobiles.

A few years ago chemists learned that NO is synthesized in a biological process by animals as diverse as barnacles, fruit flies, horseshoe crabs, chickens, trout, and humans. Even more recently they have found that NO is important in an astonishing range of physiological processes in humans and other animals. For example, it has a role in neurotransmission, blood clotting, and blood pressure control as well as in the immune system's ability to kill tumor cells and intracellular parasites.

On September 11, 2001, the United States was struck by terrorists. These attacks were followed in October when an unknown person or persons mailed letters containing anthrax to various people, including the leaders of the U.S. Senate. No one in the Senate was taken ill, but the building had to be decontaminated.

A hazardous materials worker is sprayed down on Capitol Hill on October 24, 2001, as buildings are checked for anthrax contamination.

One way to kill anthrax spores is to fumigate the contaminated area with chlorine dioxide, ClO_2. Chlorine dioxide, an odd-electron molecule with 19 valence electrons, was the first oxide of chlorine discovered. It was prepared by Humphry Davy in 1811. It is now made in several ways, all involving the reduction of sodium chlorate, $NaClO_3$. The compound is very reactive. Despite this tendency, thousands of tons are made annually, primarily for water-treatment and bleaching wood pulp to make paper.

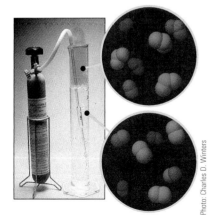

The colorless gas NO is bubbled into water from a high-pressure tank. When the gas emerges into the air, the NO reacts rapidly with O_2 to give brown NO_2 gas.

Lewis structure for NO that is in accord with the properties of this substance, so a different theory is needed to understand bonding in this molecule. We shall return to compounds of this type when molecular orbital theory is introduced in Chapter 10.

The two nitrogen oxides, NO and NO_2, are members of a class of chemical substances called free radicals. **Free radicals** are chemical species—both atomic and molecular—with an unpaired electron. Free radicals are generally quite reactive. Free atoms such as H and Cl, for example, readily combine with other atoms to give molecules such as H_2, Cl_2, and HCl.

Free radicals are involved in many reactions in the environment. For example, small amounts of NO are released from vehicle exhausts. The NO rapidly forms NO_2, which decomposes in the presence of sunlight and oxygen to give more NO as well as ozone, O_3, an air pollutant that affects the respiratory system.

$$NO_2(g) + O_2(g) \longrightarrow NO(g) + O_3(g)$$

The two nitrogen oxides, NO and NO_2, are unique in that they can be isolated, and neither has the extreme reactivity of most free radicals. When cooled, however, two NO_2 molecules join or "dimerize" to form colorless N_2O_4; the unpaired electrons combine to form an N—N bond in N_2O_4. Even though this bond is weak, the reaction is easily observed in the laboratory (Figure 9.6).

When cooled, NO_2 free
radicals couple to form
N_2O_4 molecules.

N_2O_4 gas is colorless.

A flask of brown NO_2 gas in warm water

A flask of NO_2 gas in ice water

Figure 9.6 **Free radical chemistry.** When cooled, the brown gas NO_2, a free radical, forms colorless N_2O_4, a molecule with an N—N single bond. The coupling of two free radicals is a common type of chemical reactivity. Because two identical free radicals come together, the product is called a dimer, and the process is called a dimerization. (*See the General ChemistryNow Screen 9.11 Free Radicals, to watch a video of this process.*)

9.7—Molecular Shapes

One reason for drawing Lewis electron dot structures is to be able to predict the three-dimensional geometry of molecules and ions. Because the physical and chemical properties of compounds are tied to their structures, the importance of this subject cannot be overstated.

The **valence shell electron-pair repulsion (VSEPR)** model provides a reliable method for predicting the shapes of covalent molecules and polyatomic ions. The VSEPR model is based on the idea that *bond and lone electron pairs in the valence shell of an element repel each other and seek to be as far apart as possible.* The positions assumed by the valence electrons of an atom thus define the angles between bonds to surrounding atoms. The VSEPR theory is remarkably successful in predicting structures of molecules and ions of main group elements. However, it is less effective (and seldom used) to predict structures of compounds containing transition metals.

To get a sense of how valence shell electron pairs repel one another and determine structure, blow up several balloons to a similar size. Imagine that each balloon represents an electron cloud. A repulsive force prevents other balloons from occupying the same space. When two, three, four, five, or six balloons are tied together at a central point (representing the nucleus and core electrons of a central atom), the balloons naturally form the shapes shown in Figure 9.7. These geometric arrangements minimize interactions between the balloons.

■ **VSEPR Theory**
The VSEPR theory was devised by Ronald J. Gillespie (1924–) and Ronald S. Nyholm (1917–1971).

GENERAL
Chemistry••Now™

See General ChemistryNow CD-ROM or website:

- **Screen 9.13 Ideal Electron Repulsion Shapes,** to view an animation of the electron-pair geometries and a tutorial on identifying geometries
- **Screen 9.14 Determining Molecular Shape,** for an exercise and a tutorial on predicting molecular geometry

Charles D. Winters

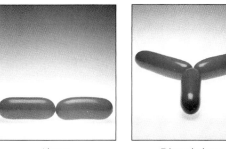

Linear

Trigonal planar

Tetrahedral

Trigonal bipyramidal

Octahedral

Charles D. Winters

Figure 9.7   Balloon models of electron-pair geometries for two to six electron-pairs. If two to six balloons of similar size and shape are tied together, they will naturally assume the arrangements shown. These pictures illustrate the predictions of the VSEPR.

Central Atoms Surrounded Only by Single-Bond Pairs

The simplest application of VSEPR theory is to molecules and ions in which all of the electron pairs around the central atom are involved in single covalent bonds. Figure 9.8 illustrates the geometries predicted for molecules or ions with the general formula AX_n, where A is the central atom and n is the number of X groups bonded to it.

The **linear** geometry for two bond pairs and the **trigonal-planar** geometry for three bond pairs involve a central atom that does not have an octet of electrons (see Section 9.6). The central atom in a **tetrahedral** molecule obeys the octet rule with four bond pairs. The central atoms in **trigonal-bipyramidal** and **octahedral** molecules have five and six bonding pairs, respectively, and are expected only when the central atom is an element in Period 3 or higher of the periodic table [▶ page 401].

Example 9.6—Predicting Molecular Shapes

Problem Predict the shape of silicon tetrachloride, $SiCl_4$.

Strategy The first step is to draw the Lewis structure. The Lewis structure does not need to be drawn in any particular way because its purpose is merely to describe the number of bonds around an atom and to indicate whether there are any lone pairs. The number of bond and lone pairs of electrons around the central atom determines the molecular shape (Figure 9.8).

Solution The Lewis structure of $SiCl_4$ has four electron pairs, all bond pairs, around the central Si atom. Therefore, a tetrahedral structure is predicted for the $SiCl_4$ molecule, with Cl—Si—Cl bond angles of 109.5°. This agrees with the actual structure for $SiCl_4$.

Lewis structure   Molecular geometry

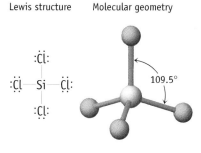

109.5°

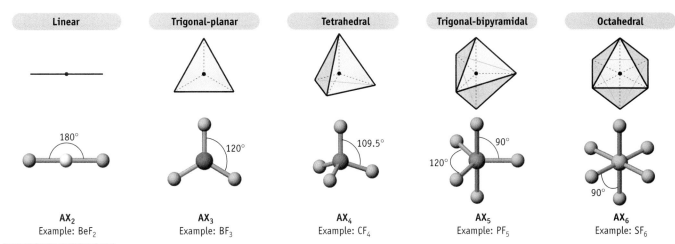

| **Linear** | **Trigonal-planar** | **Tetrahedral** | **Trigonal-bipyramidal** | **Octahedral** |

180°　　　120°　　　109.5°　　　90°　120°　　　90°

| AX_2 | AX_3 | AX_4 | AX_5 | AX_6 |
| Example: BeF_2 | Example: BF_3 | Example: CF_4 | Example: PF_5 | Example: SF_6 |

Active Figure 9.8　　**Various geometries predicted by VSEPR.** Geometries predicted by VSEPR for molecules that contain only single covalent bonds around the central atom.

GENERAL
Chemistry Now™ See the General ChemistryNow CD-ROM or website to explore an interactive version of this figure accompanied by an exercise.

Exercise 9.9—Predicting Molecular Shapes

What is the shape of the dichloromethane (CH_2Cl_2) molecule? Predict the Cl — C — Cl bond angle.

Central Atoms with Single-Bond Pairs and Lone Pairs

To see how *lone pairs* affect the geometry of a molecule or polyatomic ion, return to the balloon models in Figure 9.7. Recall that the balloons represented *all* of the electron pairs in the valence shell. The balloon model therefore predicts the "electron-pair geometry" rather than the "molecular geometry." The **electron-pair geometry** is the geometry taken up by *all* valence electron pairs around a central atom, whereas the **molecular geometry** describes the arrangement in space of the central atom and the atoms directly attached to it. It is important to recognize that *lone pairs of electrons on the central atom occupy spatial positions even though their locations are not included in the verbal description of the shape of the molecule or ion.*

Let us use the VSEPR model to predict the molecular geometry and bond angles in the NH_3 molecule, which has a lone pair on the central atom. First, draw the Lewis structure and count the total number of electron pairs around the central nitrogen atom. There are four pairs of electrons in the nitrogen valence shell, so the *electron-pair geometry* is predicted to be *tetrahedral*. We have drawn a tetrahedron with nitrogen as the central atom and the three bond pairs represented by lines. The lone pair is included here to indicate its spatial position in the tetrahedron. The *molecular geometry* is described as a *trigonal pyramid*. The nitrogen atom is at the apex of the pyramid, and the three hydrogen atoms form the trigonal base.

H — N — H
　　|
　　H

Lewis structure

N
H　　H
　H

Electron-pair
geometry, tetrahedral

Actual H–N–H
angle = 107.5°

Molecular geometry,
trigonal pyramidal

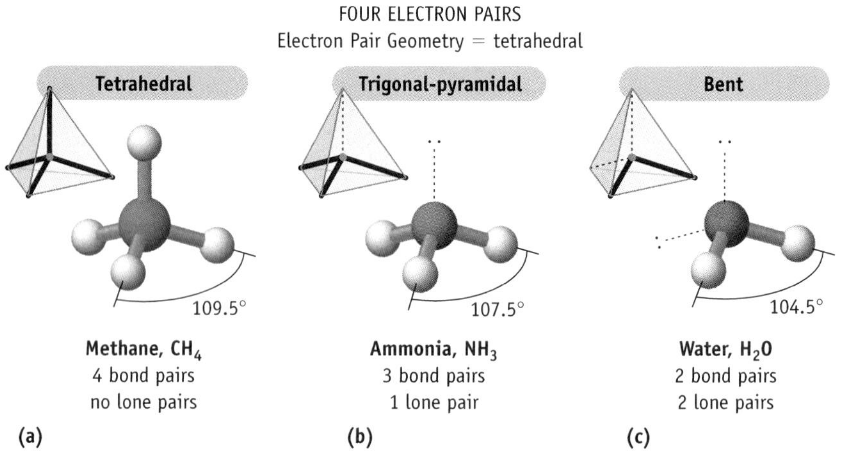

FOUR ELECTRON PAIRS
Electron Pair Geometry = tetrahedral

Figure 9.9   **The molecular geometries of methane, ammonia, and water.** All have four electron pairs around the central atom, so all have a tetrahedral electron-pair geometry. (a) Methane has four bond pairs, so it has a tetrahedral molecular shape. (b) Ammonia has three bond pairs and one lone pair, so it has a trigonal-pyramidal molecular shape. (c) Water has two bond pairs and two lone pairs, so it has a bent, or angular, molecular shape. The decrease in bond angles in the series can be explained by the fact that the lone pairs have a larger spatial requirement than the bond pairs.

Effect of Lone Pairs on Bond Angles

Because the electron-pair geometry in NH_3 is tetrahedral, we would expect the H—N—H bond angle to be 109.5°. In fact, the experimentally determined bond angles in NH_3 are 107.5°, and the H—O—H angle in water is smaller still (104.5°) (Figure 9.9). These angles are close to the tetrahedral angle but not exactly that value. This discrepancy highlights the fact that VSEPR is not a precise model; it can only predict the approximate geometry. Small variations in geometry (e.g., bond angles that are a few degrees different from those predicted) are quite common and often arise because of differences between the spatial requirements of lone pairs and bond pairs. Lone pairs of electrons seem to occupy a larger volume than bonding pairs, and the increased volume of lone pairs causes bond pairs to squeeze closer together. In general, the relative strengths of repulsions are in the order

Lone pair–lone pair > lone pair–bond pair > bond pair–bond pair

The different spatial requirements of lone pairs and bond pairs are important and are included as part of the VSEPR model. For example, the VSEPR model can be used to predict variations in the bond angles in the series of molecules CH_4, NH_3, and H_2O. The bond angles decrease in this series as the number of lone pairs on the central atom increases (Figure 9.9).

Example 9.7—Finding the Shapes of Molecules and Ions

Problem What are the shapes of the ions (a) H_3O^+ and (b) ClF_2^+?

Strategy Draw the Lewis structures for each ion. Count the number of lone and bond pairs around each central atom. Use Figure 9.8 to decide on the electron-pair geometry. Finally, the location of the atoms in the ion—which are determined by the bond and lone pairs—gives the geometry of the ion.

Solution

(a) The Lewis structure of the hydronium ion, H_3O^+, shows that the oxygen atom is surrounded by four electron pairs, so the electron-pair geometry is tetrahedral.

Lewis structure   Electron-pair geometry,   Molecular geometry,
                        tetrahedral   trigonal pyramid

Because three of the four pairs are used to bond terminal atoms, the central O atom and the three H atoms form a **trigonal-pyramidal molecular shape** like that of NH_3.

(b) Chlorine is the central atom in ClF_2^+. It is surrounded by four electron pairs, so the electron-pair geometry around chlorine is tetrahedral. Because only two of the four pairs are bonding pairs, the ion has a **bent geometry**.

Lewis structure   Electron-pair geometry,   Molecular geometry,
                        tetrahedral   bent or angular

Exercise 9.10—VSEPR and Molecular Geometry

Give the electron-pair geometry and molecular geometry for BF_3 and BF_4^-. What is the effect on the molecular geometry of adding an F^- ion to BF_3 to give BF_4^-?

Central Atoms with More Than Four Valence Electron Pairs

The situation becomes more complicated if the central atom has five or six electron pairs, some of which are lone pairs. A trigonal-bipyramidal structure (Figures 9.8 and 9.10) has two sets of positions that are not equivalent. The positions in the trigonal plane lie in the equator of an imaginary sphere around the central atom and are called the *equatorial* positions. The north and south poles in this representation are called the *axial* positions. Each equatorial atom has two neighboring groups (the axial atoms) at 90°, and each axial atom has three groups (the equatorial atoms) at 90°. The result is that the lone pairs, which require more space than bonding pairs, prefer to occupy equatorial positions rather than axial positions.

The entries in the top line of Figure 9.11 show species having a total of five valence electron pairs, with zero, one, two, and three lone pairs. In SF_4, with one lone pair, the molecule assumes a seesaw shape with the lone pair in one of the equatorial positions. The ClF_3 molecule has three bond pairs and two lone pairs. The two lone pairs in ClF_3 are in equatorial positions; two bond pairs are axial and the third is in the equatorial plane, so the molecular geometry is T-shaped. The third molecule shown is XeF_2. Here, all three equatorial positions are occupied by lone pairs, so the molecular geometry is linear.

The geometry assumed by six electron pairs is octahedral (see Figure 9.11), and all the angles at adjacent positions are 90°. Unlike the trigonal bipyramid, the

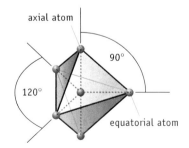

Figure 9.10 The trigonal bipyramid showing the axial and equatorial atoms. The angles between atoms in the equatorial position are 120°. The angles between equatorial and axial atoms are 90°.

The techniques just outlined can be used to find the geometries of the atoms in more complicated molecules. Consider, for example, cysteine, one of the natural amino acids.

H—S̈—C₃—C₂—C₁—Ö—H

Cysteine, $HSCH_2CH(NH_2)CO_2H$

Four pairs of electrons occur around the S, N, C_2, and C_3 atoms, so the electron-pair geometry around each is tetrahedral. Thus, the H—S—C and H—N—H angles are predicted to be approximately 109°. The O atom in the grouping C_1—O—H also is surrounded by four pairs, so this angle is likewise approximately 109°. Finally, the angle made by O—C_1—O is 120° because the electron-pair geometry around C_1 is planar and trigonal.

Example 9.9—Finding the Shapes of Molecules and Ions

Problem What are the shapes of the (a) nitrate ion, NO_3^-, and (b) $XeOF_4$?

Strategy Draw the Lewis structure and then decide on the electron-pair geometry. The positions of the atoms give the molecular geometry of the ion. Follow the procedure used in Examples 9.7 and 9.8.

Solution

(a) The NO_3^- and CO_3^{2-} ions are isoelectronic. Thus, like the carbonate ion, the electron-pair geometry and molecular shape of NO_3^- are trigonal-planar .

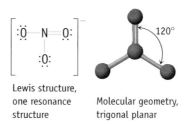

Lewis structure,
one resonance
structure

Molecular geometry,
trigonal planar

(b) The $XeOF_4$ molecule has a Lewis structure with a total of six electron pairs about the central Xe atom, one of which is a lone pair. It has a square-pyramidal molecular structure . Two structures are possible based on the position occupied by the oxygen atom, but there is no way to predict which one is correct. The actual structure is the one shown, with the oxygen in the apex of the square pyramid.

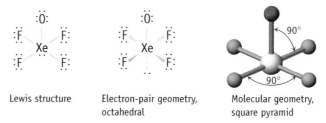

Lewis structure

Electron-pair geometry,
octahedral

Molecular geometry,
square pyramid

Exercise 9.12—Determining Molecular Shapes

Use Lewis structures and the VSEPR model to determine the electron-pair and molecular geometries for the following:

(a) phosphate ion, PO_4^{3-}
(b) sulfite ion, SO_3^{2-}
(c) IF_5

9.8—Charge Distribution In Covalent Bonds and Molecules

Lewis structures generally provide a fairly good picture of bonding in a covalently bonded molecule or ion. It is possible to "fine-tune" this picture, however, to get a more precise description of the distribution of the electrons. This effort will provide further insight into the chemical and physical properties of covalent molecules.

Closer analysis of covalently bonded molecules reveals that the valence electrons are not distributed among the atoms as evenly as Lewis structures might suggest. Some atoms may have a slight negative charge; others may have a slight positive charge. This situation occurs because the electron pair or pairs in a given bond may be drawn more strongly toward one atom than the other. The way the electrons are distributed in the molecule is called its *charge distribution*.

Charge distribution affects the properties of the molecule. Consider a diatomic (two-atom) molecule in which one atom is partially positive and the other is partially negative. In the solid or liquid state, for example, the molecules could be expected to line up with the positive end of one molecule near the negative end of another. The intermolecular or "between molecule" force of attraction would be enhanced by the attraction of opposite charges, affecting properties of the substance that are related to intermolecular forces, such as boiling point.

Positive or negative charges in a molecule or ion will influence, among other things, the site at which reactions occur. For example, does a positive H^+ ion attach itself to the O or the Cl of OCl^-? Is the product HOCl or HClO? It is reasonable to expect H^+ to attach to the more negatively charged atom. We can predict this outcome by evaluating atom formal charges in molecules and ions.

Formal Charges on Atoms

The **formal charge** for an atom in a molecule or ion is the charge calculated for that atom based on the Lewis structure of the molecule or ion, using Equation 9.1.

$$\text{Formal charge of an atom in a molecule or ion} =$$
$$\text{group number of the atom} - \left[\text{LPE} + \tfrac{1}{2}(\text{BE})\right] \quad (9.1)$$

In this equation,

- The group number gives the number of valence electrons brought by a particular atom to the molecule or ion.
- LPE = number of lone-pair electrons on an atom.
- BE = number of bonding electrons around an atom.

The term in square brackets is the number of electrons assigned by the Lewis structure to an atom in a molecule or ion. The difference between this term and the

(b) *Formal charges for the $CO_3{}^{2-}$ ion*

$$Formal\ charge = 0$$
$$= 6 - [4 + \tfrac{1}{2}(4)]$$

$$\begin{bmatrix} :\ddot{O}: \\ \| \\ :\ddot{O} - C - \ddot{O}: \end{bmatrix}^{2-}$$

Formal charge = −1
= 6 − [6 + ½(2)]

Formal charge = 0
= 4 − [0 + ½(8)]

Formal charge = −1
= 6 − [6 + ½(2)]

In each case notice that the sum of the atom's formal charges is the charge on the ion. In the carbonate ion, which has three resonance structures, the average charge on the O atoms is $-\left(\tfrac{2}{3}\right)$.

Exercise 9.13—Calculating Formal Charges

Calculate the formal charge on each atom in the following:

(a) CN^- **(b)** SO_3

Bond Polarity and Electronegativity

The models used to represent covalent and ionic bonding are the extreme situations in bonding. Pure covalent bonding, in which atoms share an electron pair equally, occurs *only* when two identical atoms are bonded. When two dissimilar atoms form a covalent bond, the electron pair will be unequally shared. The result is a **polar covalent bond,** a bond in which the two atoms have residual or partial charges (Figure 9.12).

Bonds are polar because not all atoms hold onto their valence electrons with the same force, nor do atoms take on additional electrons with equal ease. Recall from the discussion of atomic properties that different elements have different values of ionization energy and electron affinity (Section 8.6). These differences in behavior for free atoms carry over to atoms in molecules.

If a bond pair is not equally shared between atoms, the bonding electrons are nearer to one of the atoms. The atom toward which the pair is displaced has a larger share of the electron pair and thus acquires a partial negative charge. At the same time, the atom at the other end of the bond is depleted in electrons and acquires a partial positive charge. The bond between the two atoms has a positive end and a negative end; that is, it has negative and positive poles. The bond is called a **polar bond**, and the molecule is said to be **dipolar** (having two poles).

In ionic compounds, displacement of the bonding pair to one of the two atoms is essentially complete, and + and − symbols are written alongside the atom symbols in the Lewis drawings. For a polar covalent bond, the polarity is indicated by writing the symbols δ^+ and δ^- alongside the atom symbols, where δ (the Greek letter "delta") stands for a *partial* charge. Hydrogen fluoride, water, and ammonia are three simple molecules having polar, covalent bonds (Figure 9.13).

With so many atoms to use in covalent bond formation, it is not surprising that bonds between atoms can fall along in a continuum from pure ionic to pure covalent. There is no sharp dividing line between an ionic bond and a covalent bond.

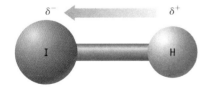

Figure 9.12 A polar covalent bond in HI. Iodine has a larger share of the bonding electrons and hydrogen has the smaller share. The result is that I has a partial negative charge (δ^-), and H has a partial positive charge (δ^+).

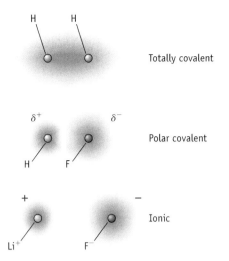

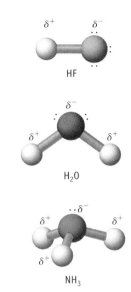

Figure 9.13 Three simple molecules with polar covalent bonds. In each case F, O, and N are more electronegative than H. See Figure 9.14.

In the 1930s, Linus Pauling proposed a parameter called atom electronegativity that allows us to decide whether a bond is polar, which atom of the bond is negative and which is positive, and whether one bond is more polar than another. The **electronegativity**, χ, of an atom is defined as a measure of *the ability of an atom in a molecule to attract electrons to itself.*

Values of electronegativity are given in Figure 9.14. Several features and periodic trends are apparent. The element with the largest electronegativity is fluorine; it is assigned a value of $\chi = 4.0$. The element with the smallest value is the alkali metal cesium. Electronegativities generally increase from left to right across a period and decrease down a group—the opposite of the trend observed for metallic character. Metals typically have low values of electronegativity, ranging from slightly less than 1 to about 2. Electronegativity values for the metalloids are around 2,

1A	2A											3A	4A	5A	6A	7A
														H 2.2		
Li 1.0	Be 1.6											B 2.0	C 2.5	N 3.0	O 3.5	F 4.0
Na 0.9	Mg 1.3	3B	4B	5B	6B	7B		8B		1B	2B	Al 1.6	Si 1.9	P 2.2	S 2.6	Cl 3.2
K 0.8	Ca 1.0	Sc 1.4	Ti 1.5	V 1.6	Cr 1.7	Mn 1.5	Fe 1.8	Co 1.9	Ni 1.9	Cu 1.9	Zn 1.6	Ga 1.8	Ge 2.0	As 2.2	Se 2.6	Br 3.0
Rb 0.8	Sr 1.0	Y 1.2	Zr 1.3	Nb 1.6	Mo 2.2	Tc 1.9	Ru 2.2	Rh 2.3	Pd 2.2	Ag 1.9	Cd 1.7	In 1.8	Sn 2.0	Sb 1.9	Te 2.1	I 2.7
Cs 0.8	Ba 0.9	La 1.1	Hf 1.3	Ta 1.5	W 2.4	Re 1.9	Os 2.2	Ir 2.2	Pt 2.3	Au 2.5	Hg 2.0	Tl 1.6	Pb 2.3	Bi 2.0	Po 2.0	At 2.2

■ <1.0   □ 1.5–1.9   ▨ 2.5–2.9
□ 1.0–1.4   ▨ 2.0–2.4   ■ 3.0–4.0

Figure 9.14 Electronegativity values for the elements according to Pauling. Trends for electronegativities are the opposite of the trends defining metallic character. Nonmetals have high values of electronegativity, the metalloids have intermediate values, and the metals have low values. (Values to one decimal place. J. Emsley: *The Elements*, 3rd ed., Clarendon Press, Oxford, 1998.)

A Closer Look

Electronegativity

Electronegativity is a useful, if somewhat vague, concept. It is, however, related to the ionic character of bonds. Chemists have found, as illustrated in the figure, that a correlation exists between the difference in electronegativity of bonded atoms and the degree of ionicity expressed as "% ionic character."

As the difference in electronegativity increases, ionic character increases. Does this trend allow us to say that one compound is ionic and another is covalent? No, we can say only that one bond is more ionic or more covalent than another.

Electron affinity was introduced in Section 8.6. At first glance it may appear that electronegativity and electron affinity measure the same property, but they do not. Electronegativity is a parameter that applies only to atoms in molecules, whereas electron affinity is a measurable energy quantity that refers to isolated atoms.

Although electron affinity was introduced earlier as a criterion with which to predict the central atom in a molecule, experience indicates that electronegativity is a better choice. That is, *the central atom is generally the atom of lowest electronegativity.*

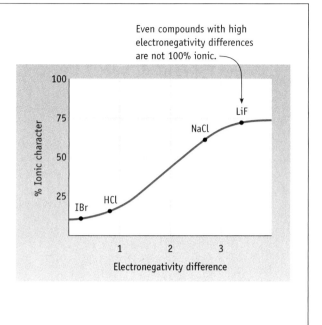

Even compounds with high electronegativity differences are not 100% ionic.

whereas nonmetals have values greater than 2. No values are given for He, Ne, and Ar because these elements do not form chemical compounds.

A large *difference* in electronegativities for atoms is observed when moving from the left- and right-hand side of the periodic table. For cesium fluoride, for example, the difference in electronegativity values, $\Delta\chi$, is 3.2 [= 4.0 (for F) − 0.8 (for Cs)]. The bond is ionic in CsF, therefore, with Cs as the cation (Cs$^+$) and F as the anion (F$^-$). In contrast, the electronegativity difference between H and Cl in HCl is only 1.0 [= 3.2 (for Cl) − 2.2 (for H)]. We conclude that bonding in HCl must be more covalent, as expected for a compound formed from two nonmetals. The H—Cl bond is polar, however, with hydrogen being the positive end of the molecule and chlorine the negative end (H$^{\delta+}$—Cl$^{\delta-}$).

Predicting trends in bond polarity in groups of related compounds is possible using values of electronegativity. Among the hydrogen halides, for example, the trend in polarity is HF ($\Delta\chi = 1.8$) > HCl ($\Delta\chi = 1.0$) > HBr ($\Delta\chi = 0.8$) > HI ($\Delta\chi = 0.5$). The HX bonds become less polar on going down the elements of Group 7A from F to I.

GENERAL
Chemistry•⚛•Now™

See General ChemistryNOW CD-ROM or Website:
- **Screen 9.17 Bond Polarity and Electronegativity,** for relative electronegativity values

Example 9.11—Estimating Bond Polarities

Problem For each of the following bond pairs, decide which is the more polar and indicate the negative and positive poles.

(a) B—F and B—Cl   **(b)** Si—O and P—P   **(c)** C＝O and C＝S

Strategy Locate the elements in the periodic table. Recall that electronegativity generally increases across a period and up a group.

Solution

(a) B and F lie relatively far apart in the periodic table. B is a metalloid and F is a nonmetal. Here χ for B = 2.0 and χ for F = 4.0. Similarly, B and Cl are relatively far apart in the periodic table, but Cl is below F in the periodic table (χ for Cl = 3.2) and is therefore less electronegative than F. The difference in electronegativity for B — F is 2.0: for B — Cl it is 1.2. Both bonds are expected to be polar, with B positive and the halide atom negative, but a B—F bond is more polar than a B—Cl bond .

(b) Because the bond is between two atoms of the same kind, the P — P bond is nonpolar. Silicon is in Group 4A and the third period, whereas O is in Group 6A and the second period. Consequently, O has a greater electronegativity (3.5) than Si (1.9), so the Si—O bond is highly polar ($\Delta\chi = 1.6$), with O the more negative atom.

(c) Oxygen lies above sulfur in the periodic table, so oxygen is more electronegative than S. This means the C—O bond is more polar than the C—S bond . For the C — O bond, O is the more negative atom. The value of $\Delta\chi$ (1.0) for CO indicates a moderately polar bond.

Exercise 9.14—Bond Polarity

For each of the following pairs of bonds, decide which is the more polar. For each polar bond, indicate the positive and negative poles. First make your prediction from the relative positions of the atoms in the periodic table; then check your prediction by calculating $\Delta\chi$.

(a) H — F and H — I   (c) C — Si and C — S
(b) B — C and B — F

Combining Formal Charge and Bond Polarity

Using formal charge calculations alone to locate the site of a charge in an ion can sometimes lead to results that seem incorrect. The ion BF_4^- illustrates this point. Boron has a formal charge of −1 in this ion, whereas the formal charge calculated for the fluorine atoms is 0. This is not logical: Fluorine is the more electronegative atom, so the negative charge should reside on F and not on B.

To resolve this dilemma, we must consider electronegativity in conjunction with formal charge. Based on the electronegativity difference between fluorine and boron ($\Delta\chi = 2.0$) the B — F bonds are expected to be polar, with fluorine being the negative end of the bond, $B^{\delta+}$ — $F^{\delta-}$. So, in this instance, predictions based on electronegativity and formal charge work in opposite directions. The formal charge calculation places the negative charge on boron, but the electronegativity difference says that the negative charge on boron is distributed onto the fluorine atoms. In effect, the charge is "spread out" over the molecule.

Linus Pauling pointed out two basic guidelines to use when describing charge distributions in molecules and ions. First, the **electroneutrality principle** declares that electrons will be distributed in such a way that the charges on all atoms are as close to zero as possible. Second, if a negative charge is present, it should reside on the most electronegative atoms. Similarly, positive charges would be expected on the least electronegative atoms. The effect of these principles is clearly seen in the case of BF_4^-, where the negative charge is distributed over the four fluorine atoms rather than residing on the boron atom.

Considering the concepts of electronegativity and formal charge together can also help to decide which of several resonance structures is the more important. For example, Lewis structure A for CO_2 is the logical one to draw. But what is wrong with B, in which each atom also has an octet of electrons?

Formal charges   0   0   0   +1   0   −1

Resonance structures $\ddot{O}=C=\ddot{O}$ $:\ddot{O}=C—\ddot{O}:$

            A   B

In structure A, each atom has a formal charge of 0, a favorable situation. In structure B, one oxygen atom has a formal charge of $+1$ and the other has $−1$. This is contrary to the principle of electroneutrality. In addition, B places a positive charge on the very electronegative O atom. Thus, we can conclude that structure B is a less satisfactory structure than A.

Now use what you have learned with CO_2 to decide which of the three possible resonance structures for the OCN^- ion is the most reasonable. Formal charges for each atom are given above the element's symbol.

Formal charges   −1   0   0   0   0   −1   +1   0   −2

Resonance structures $\left[:\ddot{O}—C\equiv N:\right]^-$ $\longleftrightarrow$ $\left[:\ddot{O}=C=\ddot{N}:\right]^-$ $\longleftrightarrow$ $\left[:O\equiv C—\ddot{N}:\right]^-$

            A   B   C

■ **Formal Charges in OCN^-**
Example of formal charge calculation: For resonance form C for OCN^-, we have

O: $6 − [2 + \frac{1}{2}(6)] = +1$

C: $4 − [0 + \frac{1}{2}(8)] = 0$

N: $5 − [6 + \frac{1}{2}(2)] = −2$

Sum of formal charges $= −1 =$ charge on the ion

Structure C will not contribute significantly to the overall electronic structure of the ion. It has a $−2$ formal charge on the N atom and a $+1$ formal charge on the O atom, whereas the formal charges in the other structures are 0 or $−1$. Structure A is more significant than structure B because the negative charge in A is placed on the most electronegative atom (O). We predict, therefore, that structure A is the best representation for this ion and that the carbon–nitrogen bond will resemble a triple bond.

The result for OCN^- also allows us to predict that protonation of the ion will lead to HOCN and not HNCO. That is, an H^+ ion will add to the more negative oxygen atom.

Example 9.12—Calculating Formal Charges

Problem Boron-containing compounds often have a boron atom with only three bonds (and no lone pairs). Why not form a double bond with a terminal atom to complete the boron octet? To answer this question, consider possible resonance structures of BF_3 and calculate the atom formal charges. Are the bonds polar in BF_3? If so, which is the more negative atom?

Strategy Calculate the formal charges on each atom in the resonance structures. The preferred structure will have atoms with low formal charges. Negative formal charges should be on the most electronegative atoms.

Solution The two possible structures for BF_3 are illustrated here with the calculated formal charges on the B and F atoms.

Formal charge $= 0$ Formal charge $= +1$
   $= 7 − [6 + \frac{1}{2}(2)]$ $= 7 − [4 + \frac{1}{2}(4)]$

      $:\ddot{F}:$ $:\ddot{F}:$

   $:\ddot{F}—B—\ddot{F}:$ $:\ddot{F}—B—\ddot{F}:$

Formal charge $= 0$ Formal charge $= −1$
   $= 3 − [0 + \frac{1}{2}(6)]$ $= 3 − [0 + \frac{1}{2}(8)]$

The structure on the left is preferred because all atoms have a zero formal charge and the very electronegative F atom does not have a charge of $1+$.

F ($\chi = 4.0$) is more electronegative than B ($\chi = 2.0$), so the B—F bond is polar, with the F atom being partially negative and the B atom being partially positive.

Exercise 9.15—Formal Charge, Bond Polarity, and Electronegativity

Consider all possible resonance structures for SO_2. What is the formal charge on each atom in each resonance structure? What are the bond polarities? Do they agree with the formal charges?

9.9—Molecular Polarity

The term "polar" was used in Section 9.8 to describe a bond in which one atom has a partial positive charge and the other has a partial negative charge. Because most molecules have polar bonds, molecules as a whole can also be polar. In a polar molecule, electron density accumulates toward one side of the molecule, giving that side a negative charge, $-\delta$, and leaving the other side with a positive charge of equal value, $+\delta$ (Figure 9.15).

Before describing the factors that determine whether a molecule is polar, let us look at the experimental measurement of the polarity of a molecule. When placed in an electric field, polar molecules experience a force that tends to align them with the field (Figure 9.15). When the electric field is created by a pair of oppositely charged plates, the positive end of each molecule is attracted to the negative plate, and the negative end is attracted to the positive plate. The extent to which the molecules line up with the field depends on their **dipole moment**, μ, which is defined as the product of the magnitude of the partial charges ($+\delta$ and $-\delta$) on the molecule and the distance by which they are separated. The SI unit of the dipole moment is the coulomb-meter, but dipole moments have traditionally been given using a derived unit called the debye (D; $1\ D = 3.34 \times 10^{-30}\ C \cdot m$). Experimental values of some dipole moments are listed in Table 9.8.

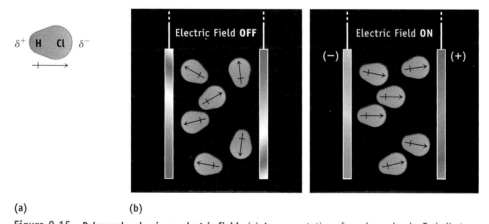

(a)   (b)

Figure 9.15 **Polar molecules in an electric field.** (a) A representation of a polar molecule. To indicate the direction of molecular polarity, an arrow is drawn with the head pointing to the negative side and a plus sign placed at the positive end. (b) When placed in an electric field (between charged plates), polar molecules experience a force that tends to align them with the field. The negative ends of the molecules are drawn to the positive plate, and vice versa. The orientation of the polar molecules affects the electrical capacitance of the plates (their ability to hold a charge), which provides a way to measure experimentally the magnitude of the dipole.

Table 9.8   Dipole Moments of Selected Molecules

Molecule (AB)	Moment (μ, D)	Geometry	Molecule (AB$_2$)	Moment (μ, D)	Geometry
HF	1.78	linear	H$_2$O	1.85	bent
HCl	1.07	linear	H$_2$S	0.95	bent
HBr	0.79	linear	SO$_2$	1.62	bent
HI	0.38	linear	CO$_2$	0	linear
H$_2$	0	linear			

Molecule (AB$_3$)	Moment (μ, D)	Geometry	Molecule (AB$_4$)	Moment (μ, D)	Geometry
NH$_3$	1.47	trigonal-pyramidal	CH$_4$	0	tetrahedral
NF$_3$	0.23	trigonal-pyramidal	CH$_3$Cl	1.92	tetrahedral
BF$_3$	0	trigonal-planar	CH$_2$Cl$_2$	1.60	tetrahedral
			CHCl$_3$	1.04	tetrahedral
			CCl$_4$	0	tetrahedral

The force of attraction between the negative end of one polar molecule and the positive end of another (called a *dipole–dipole force* and discussed in Section 13.2) affects the properties of polar compounds. Intermolecular forces (forces between molecules) influence the temperature at which a liquid freezes or boils, for example. These forces will also help determine whether a liquid dissolves certain gases or solids or whether it mixes with other liquids, and whether it adheres to glass or other solids.

To predict whether a molecule is polar, we need to consider whether the molecule has polar bonds and how these bonds are positioned relative to one another. Diatomic molecules composed of two atoms with different electronegativities are always polar (see Table 9.8); there is one bond, and the molecule has a positive and a negative end. But what happens with a molecule composed of three or more atoms, in which there are two or more polar bonds? Let us look at a series of molecules with stoichiometry AX_2, AX_3, and AX_4, evaluating how the choice of substituent or "terminal" groups (X) and molecular geometry influence the molecular polarity.

Consider first a linear triatomic molecule such as carbon dioxide, CO$_2$ (Figure 9.16). Here each C — O bond is polar, with the oxygen atom being the negative end of the bond dipole. The terminal atoms are at the same distance from the C atom, both have the same δ^- charge, and they are symmetrically arranged around the cen-

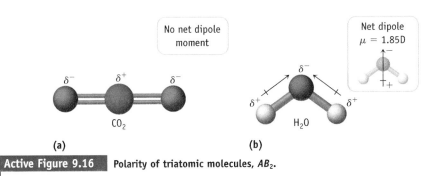

No net dipole moment

Net dipole
μ = 1.85D

CO$_2$ H$_2$O

(a)   (b)

Active Figure 9.16   Polarity of triatomic molecules, *AB$_2$*.

GENERAL
Chemistry Now™ See the General ChemistryNow CD-ROM or website to explore an interactive version of this figure accompanied by an exercise.

Developing Concepts of Bonding and Structure

Gilbert Newton Lewis (1875–1946) introduced the theory of the shared electron-pair chemical bond in a paper published in the *Journal of the American Chemical Society* in 1916. This theory revolutionized chemistry, and it is to honor his contribution that electron dot structures are now known as Lewis structures. Lewis also made major contributions in thermodynamics, and his research included studies on isotopes and on the interaction of light with substances. We will encounter his work again in Chapters 17 and 18, where the important Lewis theory of acids and bases is described.

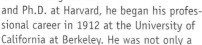

Lewis was born in Massachusetts but raised in Nebraska. After earning his B.A. and Ph.D. at Harvard, he began his professional career in 1912 at the University of California at Berkeley. He was not only a productive researcher, but also a teacher who profoundly influenced his students. Among his ideas was the use of problem sets in teaching, an idea still in use today.

The commonly used unit of dipole moments is named in honor of **Peter Debye (1884–1966)**. Because he had not studied Greek and Latin in high school, Debye could not gain entry to a university in the Netherlands, where he was born. Instead, he attended university in Aachen, Germany. Later he studied for his Ph.D. in physics in Munich and then had a long career in Swiss and German universities. During that period Debye developed a theory on the diffraction of x-rays by solids, a new concept for magnetic cooling, and (with E. Hückel) a model for interionic attractions in aqueous solution. As his interests turned more to chemistry, he worked on methods of determining the shapes of polar molecules.

Debye received the Nobel Prize in chemistry in 1936, a prize for which he had been nominated every year from 1927 to 1936. (He was also nominated for the physics prize in 15 of the years from 1916 to 1936). In addition, he was elected to 22 academies of science and received 12

medals and 18 honorary degrees.

In 1939 Debye was invited to lecture at Cornell University in New York. Because the Nazi government was increasingly interfering in his home institution in Germany, he decided to stay permanently in the United States.

A coworker of Debye, H. Sack, said that Debye "was not only endowed with a most powerful and penetrating intellect and an unmatched ability for presenting his ideas in a most lucid way, but he also knew the art of living a full life. He greatly enjoyed his scientific endeavors, he had a deep love for his family and home life, and he had an eye for the beauties of nature and a taste for the pleasure of the out-of-doors . . ."

One of the authors of this book knew Professor Debye at Cornell and remembers him as a true gentleman.

tral C atom. Therefore, CO_2 has no molecular dipole, even though each bond is polar. This is analogous to a tug-of-war in which the people at opposite ends of the rope are pulling with equal force.

In contrast, water is a bent triatomic molecule. Because O has a larger electronegativity ($\chi = 3.5$) than H ($\chi = 2.2$), each of the O—H bonds is polar, with the H atoms having the same δ^+ charge and oxygen having a negative charge (δ^-) (Figure 9.16). Electron density accumulates on the O side of the molecule, making the molecule electrically "lopsided" and therefore polar ($\mu = 1.85$ D).

In trigonal-planar BF_3, the B—F bonds are highly polar because F is much more electronegative than B (χ of B = 2.0 and χ of F = 4.0) (Figure 9.17). The molecule is nonpolar, however, because the three terminal F atoms have the same δ^- charge, are located the same distance from the boron atom, and are arranged symmetrically and in the same plane as the central boron atom. In contrast, the planar-trigonal molecule phosgene, Cl_2CO, is polar ($\mu = 1.17$ D) (Figure 9.17). Here the angles are all approximately 120°, so the O and Cl atoms are symmetrically arranged around the C atom. The electronegativities of the three atoms in the molecule differ, however: $\chi(O) > \chi(Cl) > \chi(C)$. As a consequence, there is a net displacement of electron density away from the center of the molecule, mostly toward the O atom.

Ammonia, like BF_3, has AX_3 stoichiometry and polar bonds. In contrast to BF_3, however, NH_3 is a trigonal-pyramidal molecule. The slightly positive H atoms are located in the base of the pyramid, and the slightly negative N atom is at the apex of the pyramid. As a consequence, NH_3 is polar (Figure 9.17). Indeed, trigonal-pyramidal molecules are generally polar.

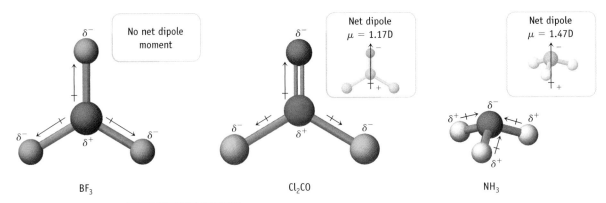

BF₃ — No net dipole moment

Cl₂CO — Net dipole $\mu = 1.17D$

NH₃ — Net dipole $\mu = 1.47D$

Active Figure 9.17   Polar and nonpolar molecules of the type AB_3.

GENERAL
Chemistry✦Now™ See the General ChemistryNow CD-ROM or website to explore an interactive version of this figure accompanied by an exercise.

Molecules like carbon tetrachloride, CCl_4, and methane, CH_4, are nonpolar owing to their symmetrical, tetrahedral structures. The four atoms bonded to C have the same partial charge and are located the same distance from the C atom. In contrast, tetrahedral molecules with both Cl and H atoms ($CHCl_3$, CH_2Cl_2, and CH_3Cl) are polar (Table 9.8 and Figure 9.8). The electronegativity for H atoms (2.2) is less than that of Cl atoms (3.2), and the carbon–hydrogen distance is different from the carbon–chlorine distances. Because Cl is more electronegative than H, the Cl atoms are on the more negative side of the molecule. Thus, the positive end of the molecular dipole is toward the H atom.

To summarize this discussion of molecular polarity, look again at Figure 9.8 (page 399). These sketches show molecules of the type AX_n where A is the central atom and X is a terminal atom. You can predict that a molecule AX_n will *not* be polar, regardless of whether the A—X bonds are polar, if

• All of the terminal atoms (or groups), X, are identical, and
• All of the X atoms (or groups) are arranged symmetrically around the central atom, A, in the geometries shown.

Chemical Perspectives

Cooking with Microwaves

Microwave ovens are common appliances in homes, dorm rooms, and offices. They work by capitalizing on the polarity of water.

Microwaves are generated in a magnetron, a device invented during World War II for antiaircraft radar. The microwaves (frequency = 2.45×10^9 s^{-1}) bounce off the metal walls of the oven and strike the food from many angles. They pass through glass or plastic dishes with little effect.

Because electromagnetic radiation consists of oscillating electric (and magnetic) fields (Figure 7.1), however, microwaves can affect mobile, charged particles such as dissolved ions or the polar water molecules commonly found in food. As each wave crest approaches a water molecule, the molecule turns to align itself with the wave and continues turning over or rotating as the trough of the wave passes. Thus, the molecules absorb energy through increased molecular motions, which translates to a higher temperature.

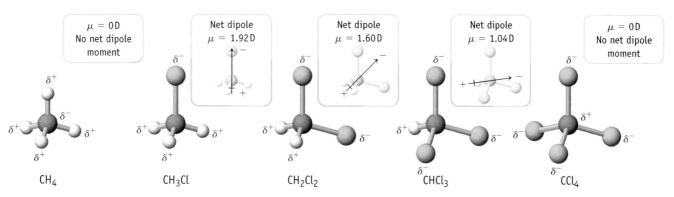

Figure 9.18 **Polarity of tetrahedral molecules.** The electronegativities of the atoms involved are in the order Cl (3.2) > C (2.5) > H (2.2). This means the C—H and C—Cl bonds are polar with a net displacement of electron density away from the H atoms and toward the Cl atoms [H(δ^+)—C(δ^-) and C(δ^+)—Cl(δ^-)]. Although the electron-pair geometry around the C atom in each molecule is tetrahedral, only in CH$_4$ and CCl$_4$ are the polar bonds totally symmetrical in their arrangement. Therefore, CH$_3$Cl, CH$_2$Cl$_2$, and CHCl$_3$ are polar molecules, with the negative end being toward the Cl atoms and the positive end being toward the H atoms.

On the other hand, if one of the *X* atoms (or groups) is different in the structures in Figure 9.8 (as in Figures 9.17 and 9.18), or if one of the *X* positions is occupied by a lone pair, the molecule will be polar.

GENERAL
Chemistry⚛Now™

See the General ChemistryNOW CD-ROM or website:
• **Screen 9.18 Molecular Polarity,** for practice in determining polarity

Example 9.13—Molecular Polarity

Problem Are (a) nitrogen trifluoride, NF$_3$, and (b) sulfur tetrafluoride, SF$_4$, polar or nonpolar? If polar, indicate the negative and positive sides of the molecule.

Strategy You cannot decide whether a molecule is polar without determining its structure. Therefore, start with the Lewis structure, decide on the electron-pair geometry, and then decide on the molecular geometry. If the molecular geometry is one of the highly symmetrical geometries in Figure 9.8 on page 399, the molecule is *not* polar. If it does not fit one of these categories, it will be polar.

Solution

(a) NF$_3$ has the same pyramidal structure as NH$_3$. Because F is more electronegative than N, each bond is polar, with the more negative end being the F atom. This means that the NF$_3$ molecule as a whole is polar.

(b) Sulfur tetrafluoride, SF$_4$, has an electron-pair geometry of a trigonal bipyramid (see Figure 9.11). Because the lone pair occupies one of the positions, the S—F bonds are not arranged symmetrically. Furthermore, the S—F bonds are highly polar, with the bond dipole having F as the negative end (χ for S is 2.6 and χ for F is 4.0). SF$_4$ is therefore a polar molecule. The axial S—F bond dipoles cancel each other because they point in opposite directions. The equatorial S—F bonds, however, both point to one side of the molecule.

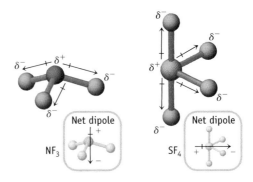

NF₃   SF₄

Example 9.14—Molecular Polarity

Problem 1,2-Dichloroethylene can exist in two forms. Is either of these molecules polar?

$$
\begin{array}{cc}
\text{H} \quad\quad \text{H} & \text{Cl} \quad\quad \text{H} \\
\text{C} = \text{C} & \text{C} = \text{C} \\
\text{Cl} \quad\quad \text{Cl} & \text{H} \quad\quad \text{Cl} \\
\text{A} & \text{B}
\end{array}
$$

Strategy To decide whether a molecule is polar we first sketch the structure. Then, using electronegativity values, we decide on the bond polarity. Finally, we decide whether the electron density in the bonds is distributed symmetrically or is shifted to one side of the molecule.

Solution Here the H and Cl atoms are arranged around the C=C double bonds with all bond angles being 120°. The electronegativities of the atoms involved are in the order Cl (3.2) > C (2.5) > H (2.2) (Figures 9.14 and 9.18). This means the C—H and C—Cl bonds are polar with a net displacement of electron density away from the H atoms and toward the Cl atoms [$H^{\delta+}$—$C^{\delta-}$ and $C^{\delta+}$—$Cl^{\delta-}$]. In structure A, the Cl atoms are located on one side of the molecule, so electrons in the H—C and C—Cl bonds move toward the side of the molecule with Cl atoms and away from the side with the H atoms. **Molecule A is polar**.

In molecule B, the movement of electron density toward the Cl atom on one end of the molecule is counterbalanced by an opposing movement on the other end. **Molecule B is not polar**.

Overall displacement of bonding electrons

$$
\begin{array}{cc}
\overset{\delta+}{\text{H}} \quad\quad \overset{\delta+}{\text{H}} \\
\text{C} = \text{C} \\
\underset{\delta-}{\text{Cl}} \quad\quad \underset{\delta-}{\text{Cl}}
\end{array}
$$

A, polar, displacement of bonding electrons to one side of the molecule.

Displacement of bonding electrons

$$
\begin{array}{cc}
\overset{\delta-}{\text{Cl}} \quad\quad \overset{\delta+}{\text{H}} \\
\text{C} = \text{C} \\
\underset{\delta+}{\text{H}} \quad\quad \underset{\delta-}{\text{Cl}}
\end{array}
$$

(*Displacement of bonding electrons*)

B, not polar, no net displacement of bonding electrons to one side of the molecule.

Exercise 9.16—Molecular Polarity

For each of the following molecules, decide whether the molecule is polar and which side is positive and which negative: $BFCl_2$, NH_2Cl, and SCl_2.

9.10—Bond Properties: Order, Length, and Energy

Bond Order

The **order of a bond** is the number of bonding electron pairs shared by two atoms in a molecule (Figure 9.19). You will encounter bond orders of 1, 2, and 3, as well as fractional bond orders.

When the bond order is 1, only a single covalent bond exists between a pair of atoms. Examples are the bonds in molecules such as H_2, NH_3, and CH_4. The bond order is 2 when two electron pairs are shared between atoms, such as the $C=O$ bonds in CO_2 and the $C=C$ bond in ethylene, $H_2C=CH_2$. The bond order is 3 when two atoms are connected by three bonds. Examples include the carbon–oxygen bond in carbon monoxide, CO, and the nitrogen–nitrogen bond in N_2.

Fractional bond orders occur in molecules and ions having resonance structures. For example, what is the bond order for each oxygen–oxygen bond in O_3? Each resonance structure of O_3 has one $O—O$ single bond and one $O=O$ double bond, for a total of three shared bonding pairs accounting for two oxygen–oxygen links.

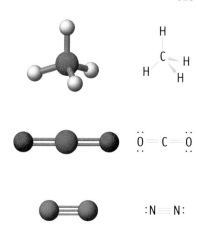

Figure 9.19   Bond order. The four $C—H$ bonds in methane each have a bond order of 1. The two $C=O$ bonds of CO_2 each have a bond order of two, whereas the nitrogen–nitrogen bond in N_2 has a bond order of 3.

Bond order = 1

Bond order = 2

Bond order for each oxygen–oxygen bond = $\frac{3}{2}$, or 1.5

One resonance structure

The bond order between any bonded pair of atoms X and Y is defined as

$$\text{Bond order} = \frac{\text{number of shared pairs linking X and Y}}{\text{number of X—Y links in the molecule or ion}} \qquad (9.2)$$

For ozone there are three bond pairs involved in two oxygen–oxygen bonds, so the bond order is $\frac{3}{2}$, or 1.5.

Bond Length

Bond length is the distance between the nuclei of two bonded atoms. Bond lengths are therefore related to the sizes of the atoms (Section 8.6), but, for a given pair of atoms, the order of the bond determines the final value of the distance.

Table 9.9 lists average bond lengths for a number of common chemical bonds. It is important to recognize that these are *average* values. Neighboring parts of a molecule can affect the length of a particular bond. For example, Table 9.9 specifies that the average $C—H$ bond has a length of 110 pm. In methane, CH_4, the measured bond length is 109.4 pm, whereas the $C—H$ bond is only 105.9 pm long in acetylene, $H—C\equiv C—H$. Variations as great as 10% from the average values listed in Table 9.9 are possible.

Because atom sizes vary in a regular fashion with the position of the element in the periodic table (Figure 8.11), predictions of trends in bond length can be made quickly. For example, the $H—X$ distance in the hydrogen halides increases in the order predicted by the relative sizes of the halogens: $H—F < H—Cl < H—Br < H—I$. Likewise, bonds between carbon and another element in a given period decrease going from left to right, in a predictable fashion; for example, $C—C > C—N > C—O > C—F$. Trends for multiple bonds are similar. A $C=O$ bond is shorter than a $C=S$ bond, and a $C=N$ bond is shorter than a $C=C$ bond.

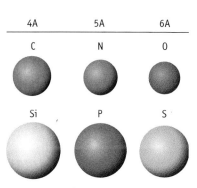

Relative sizes of some atoms of Groups 4A, 5A, and 6A.

Table 9.9   Some Average Single- and Multiple-Bond Lengths in Picometers (pm)*

Single Bond Lengths

	Group										
	1A	**4A**	**5A**	**6A**	**7A**	**4A**	**5A**	**6A**	**7A**	**7A**	**7A**
	H	C	N	O	F	Si	P	S	Cl	Br	I
H	74	110	98	94	92	145	138	132	127	142	161
C		154	147	143	141	194	187	181	176	191	210
N			140	136	134	187	180	174	169	184	203
O				132	130	183	176	170	165	180	199
F					128	181	174	168	163	178	197
Si						234	227	221	216	231	250
P							220	214	209	224	243
S								208	203	218	237
Cl									200	213	232
Br										228	247
I											266

Multiple Bond Lengths

C=C	134	C≡C	121	
C=N	127	C≡N	115	
C=O	122	C≡O	113	
N=O	115	N≡O	108	

*1 pm = 10^{-12} m.

The effect of bond order is evident when bonds between the same two atoms are compared. For example, the bonds become shorter as the bond order increases in the series C—O, C=O, and C≡O:

Bond	C—O	C=O	C≡O
Bond order	1	2	3
Bond length (pm)	143	122	113

Double bonds are shorter than single bonds between the same set of atoms, and triple bonds between those same atoms are shorter still.

The carbonate ion, CO_3^{2-}, has three equivalent resonance structures. It has a CO bond order of 1.33 (or $\frac{4}{3}$) because four electron pairs are used to form three carbon–oxygen links. The CO bond distance (129 pm) is intermediate between a C—O single bond (143 pm) and a C=O double bond (122 pm).

See General ChemistryNOW CD-ROM or website:
- **Screen 9.19 Bond Properties,** to see how bond order, bond length, and bond energy are related

---

Exercise 9.17—Bond Order and Bond Length

(a) Give the bond order of each of the following bonds and arrange them in order of decreasing bond distance: $C=N$, $C\equiv N$, and $C-N$.

(b) Draw resonance structures for NO_2^-. What is the NO bond order in this ion? Consult Table 9.9 for $N-O$ and $N=O$ bond lengths. Compare these with the NO bond length in NO_2^- (124 pm). Account for any differences you observe.

Bond Energy

The **bond dissociation energy,** symbolized by D, is the enthalpy change for breaking a bond in a molecule with the reactants and products in the gas phase.

$$\text{Molecule (g)} \xrightleftharpoons[\text{Energy released} = -D]{\text{Energy supplied} = D} \text{Molecular fragments (g)}$$

Suppose you wish to break the carbon–carbon bonds in ethane (H_3C-CH_3), ethylene ($H_2C=CH_2$), and acetylene ($HC\equiv CH$), for which the bond orders are 1, 2, and 3, respectively. For the same reason that the ethane $C-C$ bond is the longest of the series, and the acetylene $C\equiv C$ bond is the shortest, bond breaking requires the least energy for ethane and the most energy for acetylene.

$$
\begin{array}{lll}
H_3C-CH_3(g) & \longrightarrow H_3C(g) + CH_3(g) & D = +346 \text{ kJ} \\
H_2C=CH_2(g) & \longrightarrow H_2C(g) + CH_2(g) & D = +610 \text{ kJ} \\
HC\equiv CH(g) & \longrightarrow HC(g) + CH(g) & D = +835 \text{ kJ}
\end{array}
$$

Because D represents the energy transferred to the molecule from its surroundings, D has a positive value; that is, *the process of breaking bonds in a molecule is always endothermic.*

The energy supplied to break carbon–carbon bonds must be the same as the energy released when the same bonds form. *The formation of bonds from atoms or radicals in the gas phase is always exothermic.* This means, for example, that ΔH for the formation of H_3C-CH_3 from two $CH_3(g)$ radicals is -346 kJ/mol.

$$H_3C\cdot(g) + \cdot CH_3(g) \rightarrow H_3C-CH_3(g) \qquad \Delta H = -D = -346 \text{ kJ}$$

Generally, the bond energy for a given type of bond (a $C-C$ bond, for example) varies somewhat depending on the compound, just as bond lengths vary from one molecule to another. They are sufficiently similar, however, that it is possible to create a table of *average bond energies* (Table 9.10). The values in such tables may be used to *estimate* enthalpies of reactions, as described below.

In reactions between molecules, bonds in reactants are broken and new bonds are formed as products form. If the total energy released when new bonds form exceeds the energy required to break the original bonds, the overall reaction is

---

■ **Bond Energy and Electronegativity**
Linus Pauling derived electronegativity values from a consideration of bond energies. He recognized that the energy of a bond between two different atoms is often greater than expected if the bond electrons are shared equally. He postulated that the "extra energy" arises from the fact that the atoms do not share electrons equally. One atom is slightly positive and the other is slightly negative. Thus, there is a small coulombic force of attraction involving oppositely charged ions in addition to the force of attraction arising from the sharing of electrons. This coulombic force enhances the overall force of attraction.

Draw Lewis electron dot structures for small molecules and ions

a. Draw Lewis structures for molecular compounds and ions (Section 9.4). General ChemistryNow homework: SQ(s) 12, 14

b. Understand and apply the octet rule; recognize exceptions to the octet rule (Sections 9.4 and 9.6). General ChemistryNow homework: SQ(s) 18

c. Write resonance structures, understand what resonance means, and know how and when to use this means of representing bonding (Section 9.5). General ChemistryNow homework: SQ(s) 16

Use the valence shell electron-pair repulsion theory (VSEPR) to predict the shapes of simple molecules and ions and to understand the structures of more complex molecules

a. Predict the shape or geometry of molecules and ions of main group elements using VSEPR theory (Section 9.7). Table 9.11 summarizes the relation between valence electron pairs, electron-pair and molecular geometry, and molecular polarity. General ChemistryNow homework: SQ(s) 20, 24, 28, 87a, 90b, 97a

Use electronegativity and formal charge to predict the charge distribution in molecules and ions and to define the polarity of bonds

a. Calculate formal charges for atoms in a molecule based on the Lewis structure (Section 9.8). General ChemistryNow homework: SQ(s) 30

b. Define electronegativity and understand how it is used to describe the unequal sharing of electrons between atoms in a bond (Section 9.8).

c. Combine formal charge and electronegativity to gain a perspective on the charge distribution in covalent molecules and ions (Section 9.8). General ChemistryNow homework: SQ(s) 38, 40

Predict the polarity of molecules

a. Understand why some molecules are polar whereas others are nonpolar (Section 9.9). See Table 9.8. General ChemistryNow homework: SQ(s) 34, 44

b. Predict the polarity of a molecule (Section 9.9 and Examples 9.13 and 9.14). General ChemistryNow homework: SQ(s) 35

Table 9.11   Summary of Molecular Shapes and Molecular Polarity

Valence Electron Pairs	Electron-Pair Geometry	Number of Bond Pairs	Number of Lone Pairs	Molecular Geometry	Molecular Dipole?*	Examples
2	linear	2	0	linear	no	$BeCl_2$
3	trigonal planar	3	0	trigonal planar	no	BF_3, BCl_3
		2	1	bent (V-shaped)	yes	$SnCl_2(g)$
4	tetrahedral	4	0	tetrahedral	no	CH_4, BF_4^-
		3	1	trigonal-bipyramidal	yes	NH_3, PF_3
		2	2	bent (V-shaped)	yes	H_2O, SCl_2
5	trigonal bipyramid	5	0	trigonal bipyramidal	no	PF_5
		4	1	seesaw	yes	SF_4
		3	2	T-shaped	yes	ClF_3
		2	3	linear	no	XeF_2, I_3^-
6	octahedral	6	0	octahedral	no	SF_6, PF_6^-
		5	1	square-pyramidal	yes	ClF_5
		4	2	square-planar	no	XeF_4

*For molecules form of the form AX_n, where the X atoms are identical.

Understand the properties of covalent bonds and their influence on molecular structure

a. Define and predict trends in bond order, bond length, and bond dissociation energy (Section 9.10). General ChemistryNow homework: SQ(s) 48, 50, 54

b. Use bond dissociation energies, *D*, in calculations (Section 9.10 and Example 9.15). General ChemistryNow homework: SQ(s) 56, 57, 58

Key Equations

Equation 9.1 (page 405)
Calculating the formal charge on an atom in a molecule

$$\text{Formal charge of an atom in a molecule or ion} = \text{group number} - \left[\text{LPE} + \tfrac{1}{2}(\text{BE})\right]$$

Equation 9.2 (page 419)
Calculating bond order

$$\text{Bond order} = \frac{\text{number of shared pairs linking X and Y}}{\text{number of X—Y links in the molecule or ion}}$$

Equation 9.3 (page 423)
Calculating the enthalpy change for a reaction using bond dissociation energies (*D*)

$$\Delta H^{\circ}_{\text{rxn}} = \sum D(\text{bonds broken}) - \sum D(\text{bonds formed})$$

Study Questions

▲ denotes more challenging questions.

■ denotes questions available in the Homework and Goals section of the General ChemistryNow CD-ROM or website.

Blue numbered questions have answers in Appendix O and fully worked solutions in the *Student Solutions Manual*.

Structures of many of the compounds used in these questions are found on the General ChemistryNow CD-ROM or website in the Models folder.

GENERAL
Chemistry•♦•Now™ Assess your understanding of this chapter's topics with additional quizzing and conceptual questions at http://now.brookscole.com/kotz6e

Practicing Skills

Valence Electrons and the Octet Rule
(See Example 9.1 and General ChemistryNow Screen 9.2.)

1. Give the periodic group number and number of valence electrons for each of the following atoms.

(a) O   (d) Mg
(b) B   (e) F
(c) Na   (f) S

2. Give the periodic group number and number of valence electrons for each of the following atoms.

(a) C   (d) Si
(b) Cl   (e) Se
(c) Ne   (f) Al

3. For elements in Groups 3A–7A of the periodic table, give the number of bonds an element is expected to form if it obeys the octet rule.

4. Which of the following elements are capable of forming compounds in which the indicated atom has more than four valence electron pairs?

(a) C   (d) F   (g) Se
(b) P   (e) Cl   (h) Sn
(c) O   (f) B

Ionic Compounds
(See Section 9.3 and General ChemistryNow Screens 9.3 and 9.4.)

5. Which compound has the most negative energy of ion pair formation? Which has the least negative value?

(a) NaCl   (b) MgS   (c) KI

---

6. Which of the following ionic compounds are *not* likely to exist: $MgCl$, $ScCl_3$, BaF_3, $CsKr$, Na_2O? Explain your choices.

7. List the following compounds in order of increasing lattice energy (from least negative to most negative): LiI, LiF, CaO, RbI.

8. Calculate the molar enthalpy of formation, ΔH_f°, of solid lithium fluoride using the approach outlined on pages 378–381. $\Delta H_f^\circ[Li(g)] = 159.37$ kJ/mol, and other required data can be found in Appendices F and L. (See also Exercise 9.2.)

9. To melt an ionic solid, energy must be supplied to disrupt the forces between ions so the regular array of ions collapses. If the distance between the anion and the cation in a crystalline solid decreases (but ion charges remain the same), should the melting point decrease or increase? Explain.

10. ■ Which compound in each of the following pairs should require the higher temperature to melt? (See Study Question 9.)
    (a) $NaCl$ or $RbCl$
    (b) BaO or MgO
    (c) $NaCl$ or MgS

Lewis Electron Dot Structures

(See Examples 9.2–9.5 and General ChemistryNow Screens 9.7–9.8.)

11. Draw a Lewis structure for each of the following molecules or ions.
    (a) NF_3
    (b) ClO_3^-
    (c) $HOBr$
    (d) SO_3^{2-}

12. ■ Draw a Lewis structure for each of the following molecules or ions:
    (a) CS_2
    (b) BF_4^-
    (c) NO_2^-
    (d) $SOCl_2$

13. Draw a Lewis structure for each of the following molecules:
    (a) Chlorodifluoromethane, $CHClF_2$ (C is the central atom)
    (b) Acetic acid, CH_3CO_2H. Its basic structure is pictured.

$$\begin{array}{cccc} & H & O & \\ & | & \| & \\ H - & C - & C - & O - H \\ & | & & \\ & H & & \end{array}$$

    (c) Acetonitrile, CH_3CN (the framework is $H_3C - C - N$)
    (d) Allene, H_2CCCH_2

14. ■ Draw a Lewis structure for each of the following molecules.
    (a) Methanol, CH_3OH (C is the central atom)
    (b) Vinyl chloride, $H_2C = CHCl$, the molecule from which PVC plastics are made.
    (c) Acrylonitrile, $H_2C = CHCN$, the molecule from which materials such as Orlon are made

$$\begin{array}{ccccc} & H & H & & \\ & | & | & & \\ H - & C - & C - & C - & N \end{array}$$

15. Show all possible resonance structures for each of the following molecules or ions.
    (a) SO_2
    (b) NO_2^-
    (c) SCN^-

16. ■ Show all possible resonance structures for each of the following molecules or ions:
    (a) Nitrate ion, NO_3^-
    (b) Nitric acid, HNO_3
    (c) Nitrous oxide (laughing gas), N_2O

17. Draw a Lewis structure for each of the following molecules or ions.
    (a) BrF_3
    (b) I_3^-
    (c) XeO_2F_2
    (d) XeF_3^+

18. ■ Draw a Lewis structure for each of the following molecules or ions:
    (a) BrF_5
    (b) IF_3
    (c) IBr_2^-
    (d) BrF_2^+

Molecular Geometry

(See Examples 9.6–9.9 and General ChemistryNow Screens 9.12–9.14. Note that many of these molecular structures are available on the General ChemistryNow CD-ROM or website.)

19. Draw a Lewis structure for each of the following molecules or ions. Describe the electron-pair geometry and the molecular geometry around the central atom.
    (a) NH_2Cl
    (b) Cl_2O (O is the central atom)
    (c) SCN^-
    (d) HOF

20. ■ Draw a Lewis structure for each of the following molecules or ions. Describe the electron-pair geometry and the molecular geometry around the central atom.
    (a) ClF_2^+
    (b) $SnCl_3^-$
    (c) PO_4^{3-}
    (d) CS_2

21. The following molecules or ions all have two oxygen atoms attached to a central atom. Draw a Lewis structure for each one and then describe the electron-pair geometry and the molecular geometry around the central atom. Comment on similarities and differences in the series.
    (a) CO_2
    (b) NO_2^-
    (c) O_3
    (d) ClO_2^-

22. The following molecules or ions all have three oxygen atoms attached to a central atom. Draw a Lewis structure for each one and then describe the electron-pair geometry and the molecular geometry around the central atom. Comment on similarities and differences in the series.
    (a) CO_3^{2-}
    (b) NO_3^-
    (c) SO_3^{2-}
    (d) ClO_3^-

23. Draw a Lewis structure for each of the following molecules or ions. Describe the electron-pair geometry and the molecular geometry around the central atom.
    (a) ClF_2^-
    (b) ClF_3
    (c) ClF_4^-
    (d) ClF_5

24. ■ Draw a Lewis structure of each of the following molecules or ions. Describe the electron-pair geometry and the molecular geometry around the central atom.
    (a) SiF_6^{2-}
    (b) PF_5
    (c) SF_4
    (d) XeF_4

25. Give approximate values for the indicated bond angles.
    (a) O—S—O in SO_2
    (b) F—B—F angle in BF_3
    (c) Cl—C—Cl angle in Cl_2CO
    (c) H—C—H (angle 1) and C—C≡N (angle 2) in acetonitrile

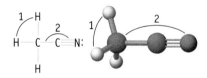

26. Give approximate values for the indicated bond angles.
    (a) Cl—S—Cl in SCl_2
    (b) N—N—O in N_2O
    (c) Bond angles in vinyl alcohol (a component of polymers and another molecule found in space).

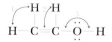

27. Phenylalanine is one of the natural amino acids and is a "breakdown" product of aspartame. Estimate the values of the indicated angles in the amino acid. Explain why the —CH_2—$CH(NH_2)$—CO_2H chain is not linear.

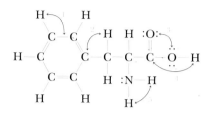

28. ■ Acetylacetone has the structure shown here. Estimate the values of the indicated angles.

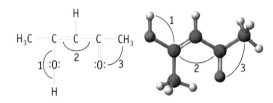

Formal Charge
(See Example 9.10 and General ChemistryNow Screen 9.16.)

29. Determine the formal charge on each atom in the following molecules or ions:
    (a) N_2H_4 (c) BH_4^-
    (b) PO_4^{3-} (d) NH_2OH

30. ■ Determine the formal charge on each atom in the following molecules or ions.
    (a) SCO
    (b) HCO_2^- (formate ion)
    (c) O_3
    (d) HCO_2H (formic acid)

31. Determine the formal charge on each atom in the following molecules and ions.
    (a) NO_2^+ (c) NF_3
    (b) NO_2^- (d) HNO_3

32. Determine the formal charge on each atom in the following molecules and ions.
    (a) SO_2 (c) SO_2Cl_2
    (b) $SOCl_2$ (d) FSO_3^-

Bond Polarity and Electronegativity
(See Example 9.11 and General ChemistryNow Screen 9.17.)

33. For each pair of bonds, indicate the more polar bond and use an arrow to show the direction of polarity in each bond.
    (a) C—O and C—N   (c) B—O and B—S
    (b) P—Br and P—Cl   (d) B—F and B—I

▲ More challenging   ■ In General ChemistryNow   Blue-numbered questions answered in Appendix O

34. ■ For each of the bonds listed below, tell which atom is the more negatively charged.
    (a) C — N   (c) C — Br
    (b) C — H   (d) S — O

35. ■ Acrolein, C_3H_4O, is the starting material for certain plastics.

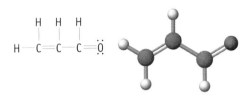

    (a) Which bonds in the molecule are polar and which are nonpolar?
    (b) Which is the most polar bond in the molecule? Which is the more negative atom of this bond?

36. Urea, $(NH_2)_2CO$, is used in plastics and fertilizers. It is also the primary nitrogen-containing substance excreted by humans.
    (a) Which bonds in the molecule are polar and which are nonpolar?
    (b) Which is the most polar bond in the molecule? Which atom is the negative end of the bond dipole?

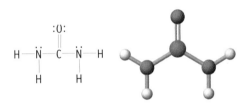

Bond Polarity and Formal Charge
(See Example 9.12 and General ChemistryNow Screens 9.16 and 9.17.)

37. Considering both formal charges and bond polarities, predict on which atom or atoms the negative charge resides in the following anions:
    (a) OH^- (b) BH_4^- (c) $CH_3CO_2^-$

38. ■ Considering both formal charge and bond polarities, predict on which atom or atoms the positive charge resides in the following cations.
    (a) H_3O^+ (c) NO_2^+
    (b) NH_4^+ (d) NF_4^+

39. Three resonance structures are possible for dinitrogen oxide, N_2O.
    (a) Draw the three resonance structures.
    (b) Calculate the formal charge on each atom in each resonance structure.
    (c) Based on formal charges and electronegativity, predict which resonance structure is the most reasonable.

40. ■ Compare the electron dot structures of the carbonate (CO_3^{2-}) and borate (BO_3^{3-}) ions.
    (a) Are these ions isoelectronic?
    (b) How many resonance structures does each ion have?
    (c) What are the formal charges of each atom in these ions?
    (d) If an H^+ ion attaches to CO_3^{2-} to form the bicarbonate ion, HCO_3^-, does it attach to an O atom or to the C atom?

41. Two resonance structures are possible for NO_2^-. Draw these structures and then find the formal charge on each atom in each resonance structure. If an H^+ ion is attached to NO_2^- (to form the acid HNO_2), does it attach to O or N?

42. Draw the resonance structures for the formate ion, HCO_2^- and find the formal charge on each atom. If an H^+ ion is attached to HCO_2^- (to form formic acid), does it attach to C or O?

Molecular Polarity
(See Examples 9.13 and 9.14 and General ChemistryNow Screen 9.18.)

43. Consider the following molecules:
    (a) H_2O (c) CO_2 (e) CCl_4
    (b) NH_3 (d) ClF
    (i)   In which compound are the bonds most polar?
    (ii) Which compounds in the list are *not* polar?
    (iii) Which atom in ClF is more negatively charged?

44. ■ Consider the following molecules:
    (a) CH_4 (c) BF_3
    (b) NH_2Cl (d) CS_2
    (i) Which compound has the most polar bonds?
    (ii) Which compounds in the list are *not* polar?

45. Which of the following molecules is (are) polar? For each polar molecule, indicate the direction of polarity—that is, which is the negative end and which is the positive end of the molecule.
    (a) $BeCl_2$ (c) CH_3Cl
    (b) HBF_2 (d) SO_3

46. Which of the following molecules is (are) not polar? Which molecule has the most polar bonds?
    (a) CO   (d) PCl_3
    (b) BCl_3 (e) GeH_4
    (c) CF_4

Bond Order and Bond Length
(See Exercise 9.17 and General ChemistryNow Screen 9.19.)

47. Give the bond order for each bond in the following molecules or ions.
    (a) CH_2O (c) NO_2^+
    (b) SO_3^{2-} (d) NOCl

48. ■ Give the bond order for each bond in the following molecules or ions.
    (a) CN^-
    (b) CH_3CN
    (c) SO_3
    (d) $CH_3CH = CH_2$

49. In each pair of bonds, predict which is shorter.
    (a) $B — Cl$ or $Ga — Cl$
    (b) $Sn — O$ or $C — O$
    (c) $P — S$ or $P — O$
    (d) $C = O$ or $C = N$

50. ■ In each pair of bonds, predict which is shorter.
    (a) $Si — N$ or $Si — O$
    (b) $Si — O$ or $C — O$
    (c) $C — F$ or $C — Br$
    (d) The $C — N$ bond or the $C ≡ N$ bond in $H_2NCH_2C ≡ N$

51. Consider the nitrogen–oxygen bond lengths in NO_2^+, NO_2^-, and NO_3^-. In which ion is the bond predicted to be longest? In which is it predicted to be the shortest? Explain briefly.

52. Compare the carbon–oxygen bond lengths in the formate ion (HCO_2^-), in methanol (CH_3OH), and in the carbonate ion (CO_3^{2-}). In which species is the carbon–oxygen bond predicted to be longest? In which is it predicted to be shortest? Explain briefly.

Bond Energy

(See Table 9.10, Example 9.9, and General ChemistryNow Screen 9.20.)

53. Consider the carbon–oxygen bond in formaldehyde (CH_2O) and carbon monoxide (CO). In which molecule is the CO bond shorter? In which molecule is the CO bond stronger?

54. ■ Compare the nitrogen–nitrogen bond in hydrazine, H_2NNH_2, with that in "laughing gas," N_2O. In which molecule is the nitrogen–nitrogen bond shorter? In which is the bond stronger?

55. Hydrogenation reactions, which involve the addition of H_2 to a molecule, are widely used in industry to transform one compound into another. For example, 1-butene (C_4H_8) is converted to butane (C_4H_{10}) by addition of H_2.

Use the bond energies of Table 9.10 to estimate the enthalpy change for this hydrogenation reaction.

56. ■ Phosgene, Cl_2CO, is a highly toxic gas that was used as a weapon in World War I. Using the bond energies of Table 9.10, estimate the enthalpy change for the reaction of carbon monoxide and chlorine to produce phosgene. (*Hint:* First draw the electron dot structures of the reactants and products so you know the types of bonds involved.)

$$CO(g) + Cl_2(g) \longrightarrow Cl_2CO(g)$$

57. ■ The compound oxygen difluoride is quite reactive, giving oxygen and HF when treated with water:

$$OF_2(g) + H_2O(g) \longrightarrow O_2(g) + 2\ HF(g)$$
$$\Delta H^\circ_{rxn} = -318\ kJ$$

Using bond energies, calculate the bond dissociation energy of the $O — F$ bond in OF_2.

58. ■ Oxygen atoms can combine with ozone to form oxygen:

$$O_3(g) + O(g) \longrightarrow 2\ O_2(g) \qquad \Delta H^\circ_{rxn} = -394\ kJ$$

Using ΔH°_{rxn} and the bond energy data in Table 9.10, estimate the bond energy for the oxygen–oxygen bond in ozone, O_3. How does your estimate compare with the energies of an $O — O$ single bond and an $O = O$ double bond? Does the oxygen–oxygen bond energy in ozone correlate with its bond order?

General Questions on Bonding and Molecular Structure

These questions are not designated as to type or location in the chapter. They may combine several concepts. More challenging questions are indicated by ▲.

59. Specify the number of valence electrons for Li, Ti, Zn, Si, and Cl.

60. Describe the formation of KF from K and F atoms using Lewis symbols. Is bonding in KF ionic or covalent?

61. Predict whether the following compounds are ionic or covalent: KI, MgS, CS_2, P_4O_{10}.

62. Define lattice energy. Which should have the more negative lattice energy, LiF or CsF? Explain.

63. Which compound is not likely to exist: $CaCl_2$ or $CaCl_4$? Explain.

64. In boron compounds the B atom often is not surrounded by four valence electron pairs. Illustrate this with BCl_3. Show how the molecule can achieve an octet configuration by forming a coordinate covalent bond with ammonia (NH_3).

65. Which of the following compounds or ions do *not* have an octet of electrons surrounding the central atom: BF_4^-, SiF_4, SeF_4, BrF_4^-, XeF_4?

66. In which of the following does the central atom obey the octet rule: NO_2, SF_4, NH_3, SO_3, O_2^-? Are any of these species odd-electron molecules or ions?

67. Give the bond order of each bond in acetylene, $H — C ≡ C — H$, and phosgene, Cl_2CO.

---

68. Draw resonance structures for the formate ion, HCO_2^- and then determine the C—O bond order in the ion.

69. Determine the N—O bond order in the nitrate ion, NO_3^-.

70. Consider a series of molecules in which carbon is bonded by single bonds to atoms of second-period elements: C—O, C—F, C—N, C—C, and C—B. Place these bonds in order of increasing bond length.

71. To estimate the enthalpy change for the reaction

$$O_2(g) + 2\ H_2(g) \longrightarrow 2\ H_2O(g)$$

what bond energies do you need? Outline the calculation, being careful to show correct algebraic signs.

72. What is the principle of electroneutrality? Use this rule to exclude a possible resonance structure of CO_2.

73. Draw Lewis structures (and resonance structures where appropriate) for the following molecules and ions. What similarities and differences are there in this series?
    (a) CO_2 (b) N_3^- (c) OCN^-

74. Does SO_2 have a dipole moment? If so, what is the direction of the net dipole in SO_2?

75. What are the orders of the N—O bonds in NO_2^- and NO_2^+? The nitrogen–oxygen bond length in one of these ions is 110 pm and 124 pm in the other. Which bond length corresponds to which ion? Explain briefly.

76. Which has the greater O—N—O bond angle, NO_2^- or NO_2^+? Explain briefly.

77. Compare the F—Cl—F angles in ClF_2^+ and ClF_2^-. Using Lewis structures, determine the approximate bond angle in each ion. Decide which ion has the greater bond angle and explain your reasoning.

78. Draw an electron dot structure for the cyanide ion, CN^-. In aqueous solution this ion interacts with H^+ to form the acid. Should the acid formula be written as HCN or CNH?

79. Draw the electron dot structure for the sulfite ion, SO_3^{2-}. In aqueous solution the ion interacts with H^+. Does H^+ attach itself to the S atom or the O atom of SO_3^{2-}?

80. Dinitrogen monoxide, N_2O, can decompose to nitrogen and oxygen gas:

$$2\ N_2O(g) \longrightarrow 2\ N_2(g) + O_2(g)$$

Use bond energies to estimate the enthalpy change for this reaction.

81. ▲ The equation for the combustion of gaseous methanol is

$$2\ CH_3OH(g) + 3\ O_2(g) \longrightarrow 2\ CO_2(g) + 4\ H_2O(g)$$

   (a) Using the bond energies in Table 9.10, estimate the enthalpy change for this reaction. What is the heat of combustion of one mole of gaseous methanol?
   (b) Compare your answer in part (a) with a calculation of ΔH°_{rxn} using thermochemical data and the methods of Chapter 6 (see Equation 6.6).

82. Acrylonitrile, C_3H_3N, is the building block of the synthetic fiber Orlon.

   (a) Give the approximate values of angles 1, 2, and 3.
   (b) Which is the shorter carbon–carbon bond?
   (c) Which is the stronger carbon–carbon bond?
   (d) Which is the most polar bond?

83. ▲ The cyanate ion, NCO^-, has the least electronegative atom, C, in the center. The very unstable fulminate ion, CNO^-, has the same formula, but the N atom is in the center.
   (a) Draw the three possible resonance structures of CNO^-.
   (b) On the basis of formal charges, decide on the resonance structure with the most reasonable distribution of charge.
   (c) Mercury fulminate is so unstable it is used in blasting caps. Can you offer an explanation for this instability? (*Hint:* Are the formal charges in any resonance structure reasonable in view of the relative electronegativities of the atoms?)

84. Vanillin is the flavoring agent in vanilla extract and in vanilla ice cream. Its structure is shown here:

   (a) Give values for the three bond angles indicated.
   (b) Indicate the shortest carbon–oxygen bond in the molecule.
   (c) Indicate the most polar bond in the molecule.

85. ▲ Given that the spatial requirement of a lone pair is much greater than that of a bond pair, explain why
   (a) XeF_2 has a linear molecular structure and not a bent one.
   (b) ClF_3 has a T-shaped structure and not a trigonal-planar one.

86. The formula for nitryl chloride is $ClNO_2$. Draw the Lewis structure for the molecule, including all resonance structures. Describe the electron-pair and molecular geometries, and give values for all bond angles.

▲ More challenging ■ In General ChemistryNow Blue-numbered questions answered in Appendix O

87. Hydroxyproline is a less common amino acid.

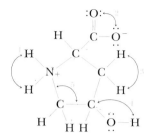

(a) ■ Give approximate values for the indicated bond angles.
(b) Which are the most polar bonds in the molecule?

88. Amides are an important class of organic molecules. They are usually drawn as sketched here, but another resonance structure is possible.

$$\text{H} \quad :\text{O}:$$
$$\text{H}-\text{C}-\text{C}-\text{N}-\text{H}$$
$$\qquad \text{H} \qquad \text{H}$$

(a) Draw that structure, and then suggest why it is usually not pictured.
(b) Suggest a reason for the fact that the H — N — H angle is close to 120°.

89. Use the bond energies in Table 9.10 to calculate the enthalpy change for the decomposition of urea (Study Question 36) to hydrazine, $H_2N — NH_2$, and carbon monoxide. (Assume all compounds are in the gas phase.)

90. The molecule shown here, 2-furylmethanethiol, is responsible for the aroma of coffee:

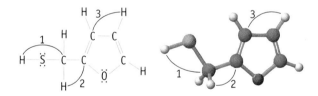

(a) What are the formal charges on the S and O atoms?
(b) ■ Give approximate values of angles 1, 2, and 3.
(c) Which are the shorter carbon–carbon bonds in the molecule?
(d) Which bond in this molecule is the most polar?
(e) Is the molecule as a whole polar or nonpolar?
(f) The molecular model makes it clear that the four C atoms of the ring are all in a plane. Is the O atom in that same plane (making the five-member ring planar), or is the O atom bent above or below the plane?

91. ▲ Dihydroxyacetone is a component of quick-tanning lotions. (It reacts with the amino acids in the upper layer of skin and colors them brown in a reaction similar to that occurring when food is browned as it cooks.)

(a) Supposing you can make this compound by treating acetone with oxygen, use bond energies to estimate the enthalpy change for the following reaction (which is assumed to occur in the gas phase). Is the reaction exothermic or endothermic?

$$\text{H} \quad :\text{O}: \quad \text{H} \qquad\qquad \text{H} \quad :\text{O}: \quad \text{H}$$
$$\text{H}-\text{C}-\text{C}-\text{C}-\text{H} + \text{O}_2 \longrightarrow \text{H}-\text{O}-\text{C}-\text{C}-\text{C}-\text{O}-\text{H}$$
$$\quad \text{H} \qquad\quad \text{H} \qquad\qquad\qquad\quad \text{H} \qquad\quad \text{H}$$

acetone   dihydroxyacetone

(b) Is acetone polar?
(c) Positive H atoms can sometimes be removed (as H^+) from molecules with strong bases (which is in part what happens in the tanning reaction). Which H atoms are the most positive in dihydroxyacetone?

92. Nitric acid, HNO_3, has three resonance structures. One of them, however, contributes much less to the resonance hybrid than the other two. Sketch the three resonance structures and assign a formal charge to each atom. Which one of your structures is the least important?

93. ▲ Acrolein is used to make plastics. Suppose this compound can be prepared by inserting a carbon monoxide molecule into the C — H bond of ethylene.

$$\qquad\qquad\qquad\qquad\qquad\qquad\qquad \text{H}$$
$$\qquad\qquad\qquad\qquad\qquad\qquad\qquad \text{H} \quad \text{C}=\ddot{\text{O}}$$
$$\text{H} \quad \text{H} \qquad\qquad\qquad\qquad \text{H} \quad |$$
$$\text{C}=\text{C} + :\text{C}=\text{O}: \longrightarrow \text{C}=\text{C}$$
$$\text{H} \quad \text{H} \qquad\qquad\qquad\qquad \text{H} \quad \text{H}$$

ethylene   acrolein

(a) Which is the stronger carbon–carbon bond in acrolein?
(b) Which is the longer carbon–carbon bond in acrolein?
(c) Is ethylene or acrolein polar?
(d) Is the reaction of CO with C_2H_4 to give acrolein endothermic or exothermic?

94. (a) Glycolaldehyde was featured in the story "Molecules in Space" (page 372). Indicate the unique bond angles in this molecule.
(b) One molecule found in the 1995 Hale-Bopp comet is HC_3N. Suggest a structure for this molecule. (*Hint:* it is based on a chain of atoms.)

95. 1,2-Dichloroethylene can be synthesized by adding Cl_2 to the carbon–carbon triple bond of acetylene.

$$\qquad\qquad\qquad\qquad\qquad \text{H} \qquad\quad \text{Cl}$$
$$\text{H}-\text{C}=\text{C}-\text{H} + \text{Cl}_2 \longrightarrow \quad \text{C}=\text{C}$$
$$\qquad\qquad\qquad\qquad\qquad \text{Cl} \qquad\quad \text{H}$$

Using bond energies, estimate the enthalpy change for this reaction in the gas phase.

96. The following molecules or ions have fluorine atoms attached to a central atom from Groups 3A through 7A. Draw the Lewis structure for each one and then describe the electron-pair geometry and the molecular geometry. Comment on similarities and differences in the series.

  (a) BF_3

  (b) CF_4

  (c) PF_3

  (d) OF_2

  (e) HF

97. The molecule pictured below is epinephrine, a compound used as a bronchodilator and antiglaucoma agent.

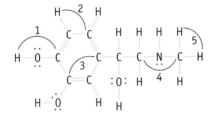

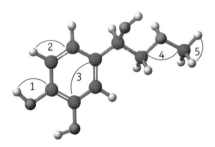

  (a) ■ Give a value for each of the indicated bond angles.

  (b) What are the most polar bonds in the molecule?

Summary and Conceptual Questions

The following questions use concepts from the previous chapters.

98. Define "bond dissociation energy." Does the enthalpy change for a bond-breaking reaction [e.g., $C-H(g) \longrightarrow C(g) + H(g)$] always have a positive sign, always have a negative sign, or vary? Explain briefly.

99. A molecule has four electron pairs around a central atom. Explain how the molecule can have a pyramidal structure. How can the molecule have a bent structure? What bond angles are predicted in each case?

100. What is the difference between the electron-pair geometry and the molecular geometry of a molecule? Use the water molecule as an example in your discussion.

101. Bromine plays a role in environmental chemistry. It is, for example, evolved in volcanic eruptions.

  (a) The following molecules are important in bromine environmental chemistry: HBr, BrO, HOBr, and OH. Which are odd-electron molecules?

  (b) Use bond energies to estimate the energies of three reactions of bromine:

$$Br_2(g) \longrightarrow 2\ Br(g)$$
$$2\ Br(g) + O_2(g) \longrightarrow 2\ BrO(g)$$
$$BrO(g) + H_2O(g) \longrightarrow HOBr(g) + OH(g)$$

  (c) Using bond energies, estimate the standard heat of formation of $HOBr(g)$ from $H_2(g)$, $O_2(g)$, and $Br_2(g)$.

  (d) Are the reactions in parts (b) and (c) exothermic or endothermic?

102. The simple molecule acrylamide, $H_2C=CHC(=O)NH_2$, is a known neurotoxin and possible carcinogen. It was a shock to all consumers of potato chips and french fries a few years ago when was found to occur in those products. (Acrylamide arises during the cooking process from a reaction of the sugar glucose and the amino acid asparagine, both naturally found in many foods.)

  (a) Draw an electron dot structure for acrylamide, showing any possible resonance structures.

  (b) Sketch the molecular structure of acrylamide, showing all unique bond angles.

  (c) Indicate which carbon–carbon bond is the stronger of the two.

  (d) Is the molecule polar or nonpolar?

  (e) The amount of acrylamide found in potato chips is 1.7 mg/kg. If a serving of potato chips is 28 g, how many moles of acrylamide are you consuming?

103. Examine the trends in lattice energy in Table 9.3. The value of the lattice energy becomes somewhat more negative on going from NaI to NaBr to NaCl, and all are in the range of −700 to −800 kJ/mol. Suggest a reason for the observation that the lattice energy of NaF ($\Delta E_{lattice}$ = −926 kJ/mol) is much more negative than those of the other sodium halides.

104. Locate the molecules in the table shown here in the Molecular Models folder on the General ChemistryNow CD-ROM or website. Measure the carbon–carbon bond length in each and complete the table. (Note that the bond lengths are given in angstrom units, where 1 Å = 0.1 nm.)

Measured Bond Formula	Bond Distance (Å)	Order
ethane, C_2H_6	_____	_____
butane, C_4H_{10}	_____	_____
ethylene, C_2H_4	_____	_____
acetylene, C_2H_2	_____	_____
benzene, C_6H_6	_____	_____

What relationship between bond order and carbon–carbon bond length do you observe?

105. See General ChemistryNow CD-ROM or website Screen 9.18 Molecular Polarity. Use the Molecular Polarity tool on this screen to explore the polarity of molecules.

    (a) Is BF_3 a polar molecule? Does the molecular polarity change as F is replaced by H on BF_3? Does the polarity change as F is replaced by H? What happens when two F atoms are replaced by H?

    (b) Is $BeCl_2$ a polar molecule? Does the polarity change when Cl is replaced by Br?

106. Locate the following molecules in the Molecular Models folder on the General ChemistryNow CD-ROM or website. In each case, measure unique bond angles and bond lengths and use them to label a sketch of the molecule.

    (a) Tylenol (Drugs folder)

    (b) ClF_3 (Inorganic folder)

    (c) Ethylene glycol (Organic Alcohols folder)

Do you need a live tutor for homework problems?
Access vMentor at General ChemistryNow at
http://now.brookscole.com/kotz6e
for one-on-one tutoring from a chemistry expert

10—Bonding and Molecular Structure: Orbital Hybridization and Molecular Orbitals

Linus Pauling (1901–1994).

Linus Pauling: A Life of Chemical Thought

Linus Pauling received the Nobel Prize for chemistry in 1954, an honorary high school diploma in 1962, and the Nobel Peace Prize in 1962. That is an extraordinary sequence—but then Pauling was an extraordinary man. He was born on February 28, 1901, in Portland, Oregon. His father, an itinerant pharmaceutical salesman, died when he was nine, and Pauling soon became a partial provider for his mother and two younger sisters. Although an excellent high school student, he refused to wait around to complete a civics requirement and so did not graduate. It was merely the first of many civil disobediences.

Against the wishes of his mother in 1917, Pauling enrolled as a chemical engineering major at Oregon Agricultural College. His interests soon turned to chemistry and, after taking a year off to help support his family, he graduated in 1922. He decided to embark on graduate work at the California Institute of Technology, then a fledgling institution, unlike today's research powerhouse. Pauling's research involved the use of the relatively new technique of x-ray crystallography to determine the atomic-level structure of crystals. Experiment alone, however, was not sufficient; he also needed to master the relevant theory. Consequently, Pauling followed his Ph.D. studies with a tour of European centers of the emerging discipline of quantum mechanics. He was superbly—perhaps uniquely—equipped for his life's work.

Returning to Caltech, Pauling began an intensive program of structural determination, using x-ray crystallography for solids and electron diffraction for vapors. Interatomic distances and angles were digested and analyzed, and quantum mechanical calculations were made. Prediction became possible. This endeavor was superbly summarized in his 1939 book, *The Nature of the Chemical Bond, and*

Thomas Hollyman/Photo Researchers, Inc.

436

the Structure of Molecules and Crystals, which was to prove the most influential chemistry text of the 20th century.

Having largely solved the structural chemistry of inorganic and simple organic substances, Pauling then turned his attention to biochemical materials. He was to become, in Francis Crick's words, "one of the founders of molecular biology." Pauling, along with coworker Robert Corey, systematically tackled the basic structural chemistry of proteins. On his fiftieth birthday, he communicated his landmark paper on the α-helix to the Proceedings of the National Academy of Sciences. It was his work on proteins, together with his studies of the nature of the chemical bond, that was cited in the award of the 1954 Nobel Prize for chemistry.

But Pauling's scientific career was not yet half over. He met with both disappointments (his failure to solve the structure of DNA and the nonacceptance of his spheron model of nuclear stability) and successes (a diagnosis of sickle cell anemia as a "molecular disease" and the introduction of the molecular evolutionary clock). However, from about 1950, science was to be merely one part, and at times even a minor part, of his active life.

Born into a relatively conservative family, Pauling became, under the urgent prompting of his wife, Ava Helen, an active political propagandist and agitator.

In particular, he played a major role in bringing about the nuclear test ban treaty of 1962. For this effort, he received the 1962 Nobel Peace prize. While his chemistry prize was universally praised, his peace prize was widely denounced in the conservative press. Only much later would his contributions be given their due.

The last years of Pauling's life were mainly spent in the advocacy of what he called "ortho-molecular medicine"—the optimization of levels of various minerals and vitamins in the human body. The most familiar was the prescription of megadoses of vitamin C to treat various ailments, especially the common cold. The nutritional establishment was outraged. The RDA value was vastly smaller than the 1 to 10 grams per day recommended by Pauling, who argued that the amount of vitamin C necessary to prevent scurvy was not necessarily sufficient to contribute maximally to bodily health. While the jury is still out on some of Pauling's more extreme claims, the medical establishment has suddenly developed a fondness for antioxidants such as vitamin C.

Active and optimistic to the end, Linus Pauling died at his ranch on the Big Sur coast of California on August 19, 1994.

Essay by Derek Davenport, Professor Emeritus of Chemistry, Purdue University

From the Ava and Linus Pauling Papers, Special Collections, Oregon State University.

Linus and Ava Helen Pauling. Dr. Pauling received the Nobel Peace prize in 1962, but his wife also played a major role in the effort to bring about a nuclear test ban treaty.

To Review Before You Begin

- Review the theory of chemical bonding (Sections 9.1 and 9.2)
- Review drawing Lewis structures (Sections 9.4–9.6)
- Review bond properties (Section 9.10)
- Review the principles of molecular structure and VSEPR theory (Section 9.7)

Just how are molecules held together? How can two distinctly different molecules have the same formula? Why is oxygen paramagnetic, and how is this property connected with bonding in the molecule? These are just a few of the fundamental and interesting questions that are raised in this chapter and that require us to take a more advanced look at bonding.

10.1—Orbitals and Bonding Theories

Orbitals, both atomic and molecular, are the focus of this chapter. The quantum mechanical model for the atom, which is the most successful way to explain the properties of atoms that scientists have devised, describes electrons in atoms as waves. An atomic orbital has a specific energy related to electrostatic forces: an attractive force due to the positively charged atomic nucleus acting on an electron in that orbital, and a repulsive force acting on the electron due to the other electrons in the atom. If the energy of the orbital is known accurately, an electron's position is known less well (the Heisenberg uncertainty principle). For this reason, we think of orbitals as regions in space in which there is a high probability of finding the electron (Figures 7.14 and 7.15).

From Chapters 7 and 8 you know that the locations of the valence electrons in atoms are described by an orbital model. It seems reasonable that an orbital model could also be used to describe electrons in molecules.

Two common approaches to rationalizing chemical bonding based on orbitals are the **valence bond (VB) theory** and the **molecular orbital (MO) theory**. The former was developed largely by Linus Pauling (page 436) and the latter by Robert S. Mulliken, another American chemist. The valence bond approach is closely tied to Lewis's idea of bonding electron pairs between atoms and lone pairs of electrons localized on a particular atom. In contrast, Mulliken's approach was to derive molecular orbitals that are "spread out," or *delocalized*, over the molecule. One way to do so is to combine atomic orbitals to form a set of orbitals that are the property of the molecule, and then distribute the electrons of the molecule within these orbitals.

Why are two theories used? Isn't one more correct than the other? Actually, both give good descriptions of the bonding in molecules and polyatomic ions, but they are used for different purposes. Valence bond theory is generally the method of choice to provide a qualitative, visual picture of molecular structure and bonding. This theory is particularly useful for molecules made up of many atoms. In contrast, molecular orbital theory is used when a more quantitative picture of bonding is needed. Furthermore, VB theory provides a good description of bonding for molecules in their ground, or lowest, energy state. In contrast, MO theory is essential if we want to describe molecules in higher-energy excited states. Among other things, it is important in explaining the colors of compounds. Finally, for a few molecules such as NO and O_2, MO theory is the only theory to describe their bonding accurately.

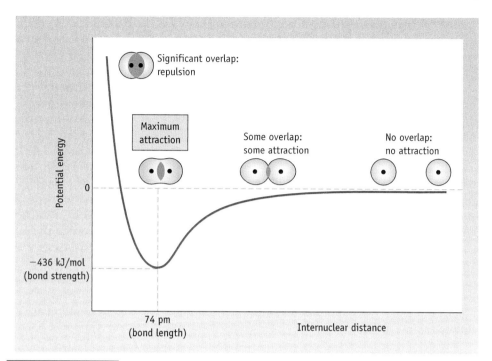

Active Figure 10.1 **Potential energy change during H — H bond formation from isolated hydrogen atoms.** The lowest energy is reached at an H — H separation of 74 pm, where there is overlap of 1s orbitals. At greater distances the overlap is less, and the bond is weaker. At H — H distances less than 74 pm, repulsions between the nuclei and between the electrons of the two atoms increase rapidly, and the potential energy curve rises steeply. Thus, an H_2 molecule is expected to be less stable when the distance between the atoms is very small.

GENERAL Chemistry ⚛ Now™ See the General ChemistryNow CD-ROM or website to explore an interactive version of this figure accompanied by an exercise.

10.2—Valence Bond Theory

Orbital Overlap Model of Bonding

What happens if two atoms at an infinite distance apart are brought together to form a bond? This process is often illustrated with H_2 because, with just two electrons and two nuclei, it is the simplest molecular compound known (Figure 10.1). Initially, when two hydrogen atoms are widely separated, they do not interact. If the atoms move closer together, however, the electron on one atom begins to experience an attraction to the positive charge of the nucleus of the other atom. Because of the attractive forces, the electron clouds on the atoms become distorted as the electron of one atom is drawn toward the nucleus of the second atom, and the potential energy of the system is lowered. Calculations show that when the distance between the H atoms is 74 pm, the potential energy reaches a minimum and the H_2 molecule is most stable. Significantly, 74 pm corresponds to the experimentally measured bond distance in the H_2 molecule.

Each individual hydrogen atom has a single electron. In H_2 the two electrons pair up to form the bond. There is a net stabilization, representing the extent to which the energies of the two electrons are lowered from their values in the free atoms. The net stabilization (the extent by which the potential energy is lowered)

■ **Bonds in Valence Bond Theory**
In the language of valence bond theory, a pair of electrons of opposite spin located between a pair of atoms constitutes a bond.

can be calculated, and the calculated value approximates the experimentally determined bond energy [◀ Section 9.10]. Agreement between theory and experiment on both bond distance and energy constitutes evidence that this theoretical approach has merit.

Bond formation is depicted in Figures 10.1 and 10.2 as occurring when the electron clouds on the two atoms interpenetrate, or overlap. This **orbital overlap** increases the probability of finding the bonding electrons in the region of space between the two nuclei. *The idea that bonds are formed by overlap of atomic orbitals is the basis for valence bond theory.*

When the single covalent bond is formed in H_2, the electron cloud of each atom becomes distorted in a way that gives the electrons a higher probability of being in the region between the two hydrogen atoms. This outcome makes sense, because the distortion results in the electrons being situated so that they can be attracted equally to the two positively charged nuclei. Placing the electrons between the nuclei also matches the Lewis electron dot model.

The covalent bond that arises from the overlap of two s orbitals, one from each of two atoms as in H_2, is called a **sigma (σ) bond**. *The electron density of a sigma bond is greatest along the axis of the bond.*

In summary, the main points of the valence bond approach to bonding are as follows:

- Orbitals overlap to form a bond between two atoms (see Figure 10.2).

- Two electrons, of opposite spin, can be accommodated in the overlapping orbitals. Usually one electron is supplied by each of the two bonded atoms.

- Because of orbital overlap, the bonding electrons have a higher probability of being found within a region of space influenced by both nuclei. Both electrons are simultaneously attracted to both nuclei.

What happens with elements beyond hydrogen? In the Lewis structure of HF, for example, a bonding electron pair is placed between H and F, and three lone pairs of electrons are depicted as localized on the F atom (Figure 10.2b). To use an orbital approach, look at the valence shell electrons and orbitals for each atom that will overlap. The hydrogen atom will use its $1s$ orbital in bond formation. The electron configuration of fluorine is $1s^2 2s^2 2p^5$, and the unpaired electron for this atom is assigned to one of the $2p$ orbitals. A sigma bond results from overlap of the hydrogen $1s$ and the fluorine $2p$ orbital.

Formation of the H — F bond is similar to formation of an H — H bond. A hydrogen atom approaches a fluorine atom along the axis containing the $2p$ orbital with a single electron. The orbitals ($1s$ on H and $2p$ on F) become distorted as each atomic nucleus influences the electron and orbital of the other atom. Still closer together, the $1s$ and $2p$ orbitals overlap, and the two electrons pair up to give a σ bond (see Figure 10.2b). There is an optimal distance (92 pm) at which the energy is lowest, which corresponds to the bond distance in HF. The net stabilization achieved in this process is the energy for the H — F bond.

The remaining electrons on the fluorine atom (two electrons in the $2s$ orbital and four electrons in the other two $2p$ orbitals) are not involved in bonding. The lone pairs associated with this element in the Lewis structure are nonbonding electrons.

Extension of this model gives a description of bonding in F_2. The $2p$ orbitals on the two atoms overlap, and the single electron from each atom is paired in the resulting σ bond (Figure 10.2c). The $2s$ and $2p$ electrons not involved in the bond are the lone pairs on each atom.

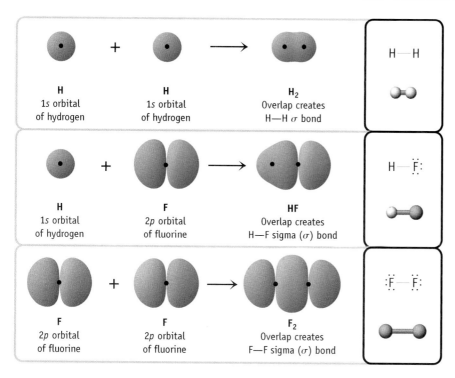

Figure 10.2   Covalent bond formation in H₂, HF, and F₂.

(a) Overlap of hydrogen 1s orbitals to form the H — H sigma bond.

(b) Overlap of hydrogen 1s and fluorine 2p orbitals to form the sigma (σ) bond in HF.

(c) Overlap of 2p orbitals on two fluorine atoms to form the sigma (σ) bond in F₂.

GENERAL
Chemistry•Now™

See the General ChemistryNow CD-ROM or website:
• **Screen 10.3 Valence Bond Theory,** for an animation of bond formation

Hybridization of Atomic Orbitals

The simple picture using orbital overlap to describe bonding in H₂, HF, and F₂ works well, but we run into difficulty when molecules with more atoms are considered. For example, a Lewis dot structure of methane, CH₄, shows four C — H covalent bonds. VSEPR theory predicts, and experiments confirm, that the electron-pair geometry of the C atom in CH₄ is tetrahedral, with an angle of 109.5° between the bond pairs. The hydrogens are identical in this structure. Thus four equivalent bonding electron pairs occur around the C atom. An orbital picture of the bonds should convey both the geometry and the fact that all C — H bonds are the same.

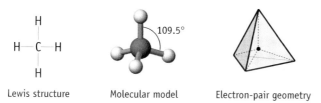

Lewis structure   Molecular model   Electron-pair geometry

If we apply the orbital overlap model used for H₂ and F₂ without modification to describe the bonding in CH₄, a problem arises. The three orbitals for the 2p valence electrons of carbon are at right angles, 90° (Figure 10.3), and do not match the tetrahedral angle of 109.5°. The spherical 2s orbital could bond in any direction.

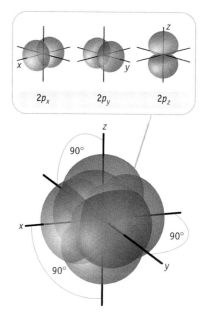

Figure 10.3   The 2p orbitals of an atom. The $2p_x$, $2p_y$, and $2p_z$ orbitals lie long the x-, y-, and z-axes, 90° to each other.

Charles D. Winters

Figure 10.4   Hybridization: an analogy. Atomic orbitals can mix, or hybridize, to form hybrid orbitals. When two atomic orbitals on an atom combine, two new orbitals are produced on that atom. The new orbitals have a different direction in space than the original orbitals. An analogy is mixing two different colors (*left*) to produce a third color, which is a "hybrid" of the original colors (*center*). After mixing there are still two beakers (*right*), each containing the same volume of solution as before, but the color is a "hybrid" color. (*See General ChemistryNow Screens 10.4 Hybrid Orbitals, and 10.6 Determining Hybrid Orbitals.*)

Furthermore, a carbon atom in its ground state $(1s^2 2s^2 2p^2)$ has only two unpaired electrons (in the $2p$ orbitals), not the four that are needed to allow formation of four bonds.

To describe the bonding in methane and other molecules, Linus Pauling proposed the theory of **orbital hybridization** (Figure 10.4). He suggested that a new set of orbitals, called **hybrid orbitals**, could be created by mixing the *s*, *p*, and (when required) *d* atomic orbitals on an atom. Two important principles govern the outcome. First, *the number of hybrid orbitals is always the same as the number of atomic orbitals that are mixed to create the hybrid orbital set.* Second, *the hybrid orbitals are more directed from the central atom toward the terminal atoms than are the unhybridized atomic orbitals, leading to better orbital overlap and a stronger bond between the central and terminal atoms.*

The sets of hybrid orbitals that arise from mixing *s*, *p*, and *d* atomic orbitals are illustrated in Figure 10.5. The following features are important:

- The hybrid orbitals required by an atom in a molecule or ion are determined by the electron-pair geometry around that atom. A hybrid orbital is required for each sigma bond or lone electron pair on a central atom.

- If the valence shell *s* orbital on the central atom in a molecule or ion is mixed with a valence shell *p* orbital on that same atom, two hybrid orbitals are created. They are separated by 180°. The set of two orbitals is labeled *sp*.

- If an *s* orbital is combined with two *p* orbitals, all in the same valence shell, three hybrid orbitals are created. They are separated by 120°, and the set of three orbitals is labeled sp^2.

- When the *s* orbital in a valence shell is combined with three *p* orbitals, the result is four hybrid orbitals, each labeled sp^3. The hybrid orbitals are separated by 109.5°, the tetrahedral angle.

- If one or two *d* orbitals are added to the sp^3 set, then two other hybrid orbital sets are created. They are utilized by the central atom of a molecule or ion with a trigonal-bipyramidal or octahedral electron-pair geometry.

Let us examine a case of each type of hybridization in simple molecules, returning first to the case of methane. Keep in mind, however, that these principles apply to atoms in even the most complex molecules, such as DNA.

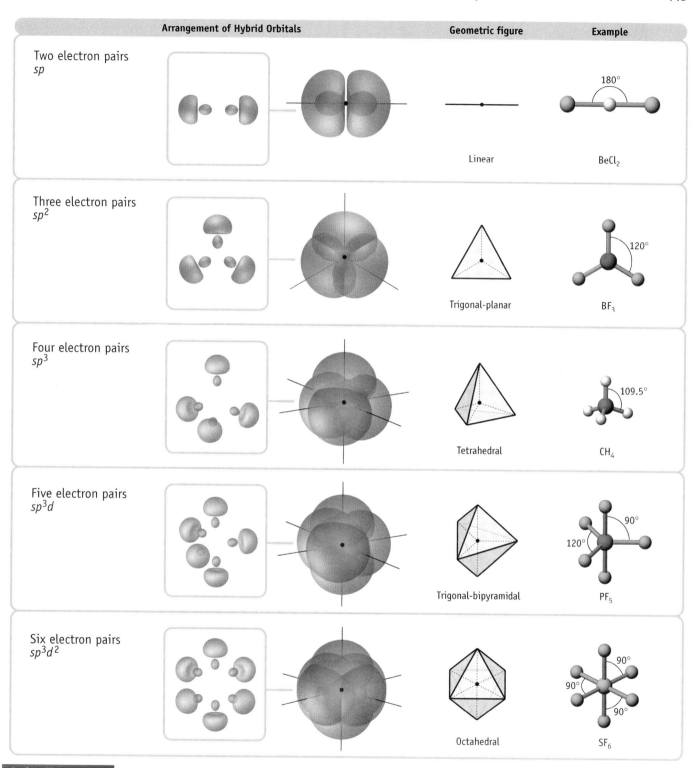

Arrangement of Hybrid Orbitals			Geometric figure	Example
Two electron pairs *sp*			Linear	BeCl₂ 180°
Three electron pairs *sp²*			Trigonal-planar	BF₃ 120°
Four electron pairs *sp³*			Tetrahedral	CH₄ 109.5°
Five electron pairs *sp³d*			Trigonal-bipyramidal	PF₅ 90° 120°
Six electron pairs *sp³d²*			Octahedral	SF₆ 90° 90° 90°

Active Figure 10.5 **Hybrid orbitals for two to six electron pairs.** The geometry of the hybrid orbital sets for two to six valence shell electron pairs is given in the right column. In forming a hybrid orbital set, the *s* orbital is always used, plus as many *p* orbitals (and *d* orbitals) as are required to give the necessary number of σ-bonding and lone-pair orbitals.

GENERAL
Chemistry ⚛ Now™ See the General ChemistryNow CD-ROM or website to explore an interactive version of this figure accompanied by an exercise.

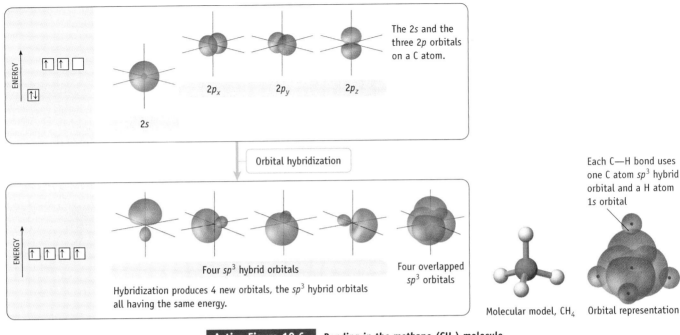

The 2s and the three 2p orbitals on a C atom.

2s   2p$_x$ 2p$_y$ 2p$_z$

Orbital hybridization

Four sp^3 hybrid orbitals

Hybridization produces 4 new orbitals, the sp^3 hybrid orbitals all having the same energy.

Four overlapped sp^3 orbitals

Each C—H bond uses one C atom sp^3 hybrid orbital and a H atom 1s orbital

Molecular model, CH$_4$ Orbital representation

Active Figure 10.6   Bonding in the methane (CH$_4$) molecule.

GENERAL
Chemistry·❀·Now™ See the General ChemistryNow CD-ROM or website to explore an interactive version of this figure accompanied by an exercise.

GENERAL
Chemistry·❀·Now™

See the General ChemistryNow CD-ROM or website:
• **Screen 10.4 Hybrid Orbitals,** for an animation of the formation of sp^3 hybrid orbitals

Valence Bond Theory for Methane, CH$_4$

In methane, four orbitals directed to the corners of a tetrahedron are needed to match the electron-pair geometry on the central carbon atom. By mixing the four valence shell orbitals (the 2s and all three 2p orbitals on carbon), a new set of four hybrid orbitals is created that has tetrahedral geometry (Figures 10.5 and 10.6). Each of the four hybrid orbitals is labeled sp^3 to indicate the atomic orbital combination (an s orbital and three p orbitals) from which they are derived. The four sp^3 orbitals have an identical shape, and the angle between them is 109.5°, the tetrahedral angle. Because the orbitals have the same energy, one electron can be assigned to each according to Hund's rule [◀ Section 8.4]. Then, each C—H bond is formed by overlap of one of the carbon sp^3 hybrid orbitals with the 1s orbital of a hydrogen atom; one electron from the C atom is paired with an electron from an H atom.

Valence Bond Theory for Ammonia, NH$_3$

The Lewis structure for ammonia shows four electron pairs in the valence shell of nitrogen: three bond pairs and a lone pair (Figure 10.7). VSEPR theory predicts a tetrahedral electron-pair geometry and a trigonal-pyramidal molecular geometry. Structure evidence is a close match to prediction; the H—N—H bond angles are 107.5° in this molecule.

■ **Hybrid Orbitals & Atomic Orbitals**
Note that *four* atomic orbitals produce *four* hybrid orbitals. The number of hybrid orbitals produced is always the same as the number of atomic orbitals used.

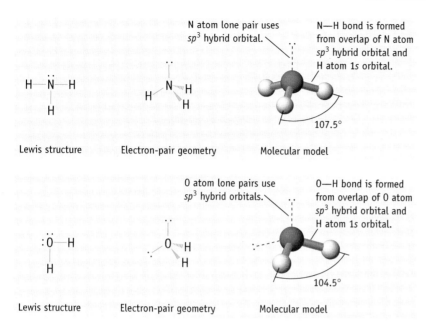

Figure 10.7 Bonding in ammonia, NH₃, and water, H₂O.

N atom lone pair uses sp^3 hybrid orbital.

N—H bond is formed from overlap of N atom sp^3 hybrid orbital and H atom 1s orbital.

107.5°

Lewis structure   Electron-pair geometry   Molecular model

O atom lone pairs use sp^3 hybrid orbitals.

O—H bond is formed from overlap of O atom sp^3 hybrid orbital and H atom 1s orbital.

104.5°

Lewis structure   Electron-pair geometry   Molecular model

Based on the electron-pair geometry of NH_3, we predict sp^3 hybridization to accommodate the four electron pairs on the N atom. The lone pair is assigned to one of the hybrid orbitals, and each of the other three hybrid orbitals is occupied by a single electron. Overlap of each of the singly occupied, sp^3 hybrid orbitals with a 1s orbital for hydrogen, and pairing of the electrons in these orbitals, creates the N—H bonds about 109° apart.

Valence Bond Theory for Water, H₂O

The oxygen atom of water has two bonding pairs and two lone pairs in its valence shell, and the H—O—H angle is 104.5° (Figure 10.7). Four sp^3 hybrid orbitals are created from the 2s and 2p atomic orbitals of oxygen. Two of these sp^3 orbitals are occupied by unpaired electrons and are used to form O—H bonds. Lone pairs occupy the other two hybrid orbitals.

GENERAL
Chemistry ·ᐧ·Now™

See the General ChemistryNow CD-ROM or website:
• **Screen 10.5 Sigma Bonding,** for a tutorial on sigma bond formation
• **Screen 10.6 Determining Hybrid Orbitals,** for a tutorial on determining hybrid orbitals

■ **Hybridization and Geometry**
Hybridization reconciles the electron-pair geometry with the orbital overlap criterion of bonding. A statement such as "the atom is tetrahedral because it is sp^3 hybridized" is backward. That the electron-pair geometry around the atom is tetrahedral is a fact. Hybridization is one way to rationalize that fact.

Example 10.1—Valence Bond Description of Bonding in Ethane

Problem Describe the bonding in ethane, C_2H_6, using valence bond theory.

Strategy First, draw the Lewis structure and predict the electron-pair geometry at both carbon atoms. Next, assign a hybridization to these atoms. Finally, describe covalent bonds that arise based on orbital overlap, and place electron pairs in their proper locations.

Solution Each carbon atom has an octet configuration, sharing electron pairs with three hydrogen atoms and with the other carbon atom. The electron pairs around carbon have tetrahedral geometry, so carbon is assigned sp^3 hybridization. The C—C bond is formed by overlap of sp^3 orbitals on each C atom, and each of the C—H bonds is formed by overlap of an sp^3 orbital on carbon with a hydrogen $1s$ orbital.

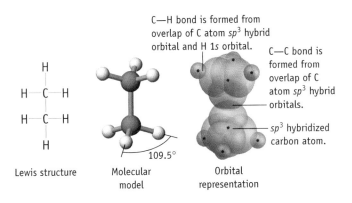

C—H bond is formed from overlap of C atom sp^3 hybrid orbital and H $1s$ orbital.

C—C bond is formed from overlap of C atom sp^3 hybrid orbitals.

sp^3 hybridized carbon atom.

109.5°

Lewis structure   Molecular model   Orbital representation

Example 10.2—Valence Bond Description of Bonding in Methanol

Problem Describe the bonding in the methanol molecule, CH₃OH, using valence bond theory.

Strategy Construct the Lewis structure for the molecule. The electron-pair geometry around each atom determines the hybrid orbital set used by that atom.

Solution The electron-pair geometry around both the C and O atoms in CH₃OH is tetrahedral. Thus, we may assign sp^3 hybridization to each atom, and the C—O bond is formed by overlap of sp^3 orbitals on these atoms. Each C—H bond is formed by overlap of a carbon sp^3 orbital with a hydrogen $1s$ orbital, and the O—H bond is formed by overlap of an oxygen sp^3 orbital with the hydrogen $1s$ orbital. Two lone pairs on oxygen occupy the remaining sp^3 orbitals.

Comment Notice that one end of the CH₃OH molecule (the CH₃ or methyl group) is just like the CH₃ group in the ethane molecule (Example 10.1), and the OH group resembles the OH group in water. It is helpful to recognize pieces of molecules and their bonding descriptions.

This example also shows how to predict the structure and bonding in a complicated molecule by looking at each atom separately. This important principle is essential when dealing with molecules made up of many atoms.

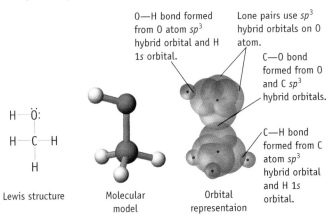

O—H bond formed from O atom sp^3 hybrid orbital and H $1s$ orbital.

Lone pairs use sp^3 hybrid orbitals on O atom.

C—O bond formed from O and C sp^3 hybrid orbitals.

C—H bond formed from C atom sp^3 hybrid orbital and H $1s$ orbital.

Lewis structure   Molecular model   Orbital representaion

Exercise 10.1—Valence Bond Description of Bonding

Use valence bond theory to describe the bonding in the hydronium ion, H_3O^+, and methylamine, CH_3NH_2.

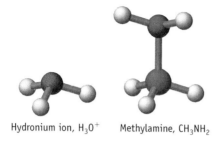

Hydronium ion, H_3O^+ Methylamine, CH_3NH_2

Hybrid Orbitals for Molecules and Ions with Trigonal-Planar Electron-Pair Geometries

Atoms having trigonal-planar geometries are commonly encountered in molecules and ions. For example, BF_3 and other boron halides are trigonal-planar, as are a number of other species, such as NO_3^- and CO_3^{2-}. The carbon atoms in ethylene, $CH_2=CH_2$, are also trigonal-planar, and the electron-pair geometry of O_3 and NO_2^- is trigonal-planar.

A trigonal-planar electron-pair geometry requires a central atom with three hybrid orbitals in a plane, 120° apart. Three hybrid orbitals mean three atomic orbitals must be combined, and the combination of an s orbital with two p orbitals is appropriate (Figure 10.5). If p_x and p_y orbitals are used in hybrid orbital formation, the three hybrid sp^2 orbitals will lie in the xy-plane. The p_z orbital not used to form these hybrid orbitals is perpendicular to the plane containing the three sp^2 orbitals (Figure 10.8).

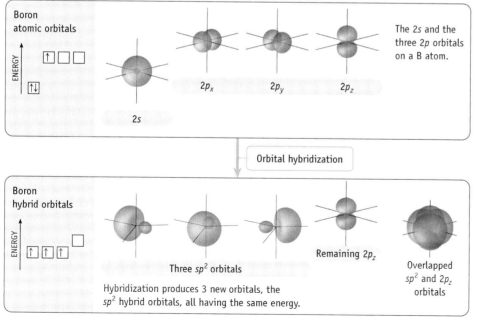

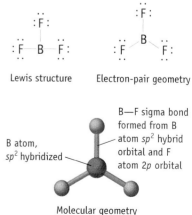

Figure 10.8 Bonding in a trigonal-planar molecule.

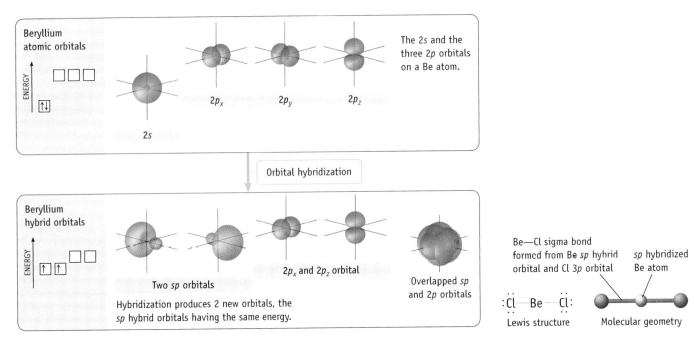

Figure 10.9   Bonding in a linear molecule. Because only one *p* orbital is incorporated in the hybrid orbital, two *p* orbitals remain. These orbitals are perpendicular to each other and to the axis along which the two *sp* hybrid orbitals lie.

Boron trifluoride has trigonal-planar electron-pair and molecular geometries. Each boron–fluorine bond in this compound results from overlap of an sp^2 orbital on boron with a *p* orbital on fluorine. Notice that the p_z orbital on boron, which is not used to form the sp^2 hybrid orbitals, is not occupied by electrons.

Hybrid Orbitals for Molecules and Ions with Linear Electron-Pair Geometries

For molecules in which the central atom has a linear electron-pair geometry, two hybrid orbitals, 180° apart, are required. One *s* and one *p* orbital can be hybridized to form two *sp* hybrid orbitals (Figure 10.9). If the p_x orbital is used, then the *sp* orbitals are oriented along the *x*-axis. The p_y and p_z orbitals are perpendicular to this axis.

Beryllium dichloride, $BeCl_2$, is a solid under ordinary conditions. When it is heated to more than 520 °C, however, it vaporizes to give $BeCl_2$ vapor. In the gas phase, $BeCl_2$ is a linear molecule, so *sp* hybridization is appropriate for the beryllium atom in this species. Combining beryllium's 2*s* and $2p_x$ orbitals gives the two *sp* hybrid orbitals that lie along the *x*-axis. Each Be — Cl bond arises by overlap of an *sp* hybrid orbital on beryllium with a 3*p* orbital on chlorine. In this molecule, there are only two electron pairs around the beryllium atom, so the p_y and p_z orbitals are not occupied (Figure 10.9).

Hybrid Orbitals Involving *s, p,* and *d* Atomic Orbitals

A basic assumption of Pauling's valence bond theory is that *the number of hybrid orbitals equals the number of valence orbitals used in their creation.* As a consequence, the maximum number of hybrid orbitals that can be created from the *s* and *p* orbitals for an atom is four.

How, then, should we deal with compounds like PF_5 and SF_6, which have more than four electron pairs in their valence shells? To describe the bonding in compounds having five or six electron pairs on a central atom requires the atom to have five or six hybrid orbitals, which must be created from five or six atomic orbitals. This is possible if additional atomic orbitals from the d subshell are used in hybrid orbital formation. The d orbitals are considered to be valence shell orbitals for main group elements of the third and higher periods.

To accommodate six electron pairs in the valence shell of an element, six sp^3d^2 hybrid orbitals can be created from the one s, three p, and two d orbitals. The six sp^3d^2 hybrid orbitals are directed to the corners of an octahedron (Figure 10.5). Thus, they are oriented to accommodate the valence electron pairs for a compound that has an octahedral electron-pair geometry. Five coordination and trigonal-bipyramidal geometry are matched to sp^3d hybridization. One s, three p, and one d orbital combine to produce five sp^3d hybrid orbitals.

Example 10.3—Hybridization Involving d Orbitals

Problem Describe the bonding in PF_5 using valence bond theory.

Strategy The first step is to establish the electron pair and molecular geometries of PF_5. The electron-pair geometry around the P atom gives the number of hybrid orbitals required. If five hybrid orbitals are required, the combination of atomic orbitals is sp^3d.

Solution Here the P atom is surrounded by five electron pairs, so PF_5 has trigonal-bipyramidal electron-pair and molecular geometries. The hybridization scheme is therefore sp^3d.

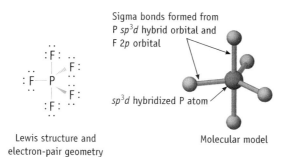

Sigma bonds formed from
P sp^3d hybrid orbital and
F $2p$ orbital

sp^3d hybridized P atom

Lewis structure and
electron-pair geometry

Molecular model

Example 10.4—Recognizing Hybridization

Problem Identify the hybridization of the central atom in the following compounds and ions:

(a) SF_3^+ **(b)** SO_4^{2-} **(c)** SF_4 **(d)** I_3^-

Strategy The hybrid orbitals used by a central atom are determined by the electron-pair geometry (see Figure 10.5). To answer this question, first write the Lewis structure and then predict the electron-pair geometry.

Solution Following the procedures in Chapter 9, the Lewis structures for SF_3^+ and SO_4^{2-} can be written as follows:

Four electron-pairs surround the center atom in each of these ions, and the electron-pair geometry for these atoms is tetrahedral. Thus, sp^3 hybridization for the central atom is used to describe the bonding.

For SF_4 and I_3^-, five pairs of electrons are in the valence shell of the center atom. For these, sp^3d hybridization is appropriate for the central S or I atom.

Exercise 10.2—Hybridization Involving d Orbitals

Describe the bonding in XeF_4 using hybrid orbitals. Remember to consider first the Lewis structure, then the electron-pair geometry (based on VSEPR theory), and finally the molecular shape.

Exercise 10.3—Recognizing Hybridization

Identify the hybridization of the central atom in the following compounds and ions:

(a) BH_4^- (c) OSF_4 (e) BCl_3
(b) SF_5^- (d) ClF_3 (f) XeO_6^{4-}

Multiple Bonds

According to valence bond theory, bond formation requires that two orbitals on adjacent atoms overlap. Many molecules have two or three bonds between pairs of atoms. Therefore, according to valence bond theory, a double bond requires *two* sets of overlapping orbitals and *two* electron pairs. For a triple bond, *three* sets of atomic orbitals are required, with each set accommodating a pair of electrons.

Double Bonds

Consider ethylene, $H_2C=CH_2$, one of the more common molecules with a double bond. The molecular structure of ethylene places all six atoms in a plane, with H—C—H and H—C—C angles of approximately 120°. Each carbon atom has trigonal-planar geometry, so sp^2 hybridization is assumed for these atoms. Thus, a description of bonding in ethylene starts with each carbon atom having three sp^2 hybrid orbitals in the molecular plane and an unhybridized p orbital perpendicular to that plane (see Figure 10.8). Because each carbon atom is involved in four bonds, a single unpaired electron is placed in each of these orbitals.

> ⬜↑ Unhybridized p orbital. Used for π bonding in C_2H_4.
>
> ⬜↑⬜↑⬜↑ Three sp^2 hybrid orbitals. Used for C—H and C—C σ bonding in C_2H_4.

Now we can visualize the C—H bonds, which arise from overlap of sp^2 orbitals on carbon with hydrogen $1s$ orbitals. After accounting for the C—H bonds, one sp^2 orbital on each carbon atom remains. These orbitals point toward each other and over-

---

■ Multiple Bonds

C=C
⟋
Double bond requires two sets of overlapping orbitals and two pairs of electrons.

C≡C
⟋
Triple bond requires three sets of overlapping orbitals and three pairs of electrons.

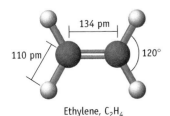

134 pm
110 pm
120°

Ethylene, C_2H_4

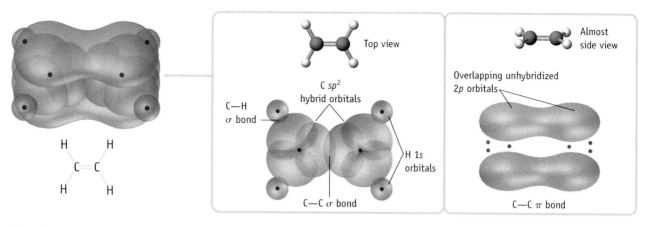

(a) Lewis structure and bonding of ethylene, C_2H_4.

(b) The C—H σ bonds are formed by overlap of C atom sp^2 hybrid orbitals with H atom $1s$ orbitals. The σ bond between C atoms arises from overlap of sp^2 orbitals.

(c) The carbon-carbon π bond is formed by overlap of an unhybridized $2p$ orbital on each atom. Note the lack of electron density along the C—C bond axis.

Active Figure 10.10 **The valence bond model of bonding in ethylene, C_2H_4.** Each C atom is assumed to be sp^2 hybridized.

GENERAL
Chemistry ⚛ Now ™ See the General ChemistryNow CD-ROM or website to explore an interactive version of this figure accompanied by an exercise.

lap to form one of the bonds linking the carbon atoms (Figure 10.10a). This leaves only one other orbital unaccounted for on each carbon, an unhybridized p orbital (see Figure 10.8). These orbitals can be used to create the second bond between carbon atoms in C_2H_4. If they are aligned correctly, the unhybridized p orbitals on the two carbons can overlap, allowing the electrons in these orbitals to be paired. The overlap does not occur directly along the C — C axis, however. Instead, the arrangement compels these orbitals to overlap sideways, and the electron pair occupies an orbital with electron density above and below the plane containing the six atoms (Figure 10.10c).

This description results in two types of bonds in C_2H_4. One type is the C — H and C — C bonds that arise from the overlap of atomic orbitals so that the bonding electrons that lie along the bond axis form sigma (σ) bonds. The other is the bond formed by sideways overlap of p atomic orbitals, called a **pi (π) bond**. In a π bond, the overlap region is above and below the internuclear axis, and the electron density of the π bond is above and below the σ bond axis (Figures 10.10b and 10c).

Notice that a π bond can form *only* if (1) there are unhybridized p orbitals on adjacent atoms and (2) the p orbitals are perpendicular to the plane of the molecule and parallel to one another. This happens only if the sp^2 orbitals of both carbon atoms are in the same plane. Thus, *the π bond requires that all six atoms of the molecule lie in one plane.*

Double bonds between carbon and oxygen, sulfur, or nitrogen are quite common. Consider formaldehyde, CH_2O, in which a carbon–oxygen π bond occurs (Figure 10.11). A trigonal-planar electron-pair geometry indicates sp^2 hybridization for the C atom. The σ bonds from the C atom to the O atom and the two H atoms form by overlap of sp^2 hybrid orbitals with half-filled orbitals from the oxygen and two hydrogen atoms. An unhybridized p orbital on carbon is oriented perpendicular to the molecular plane (just as for the carbon atoms of C_2H_4). This p orbital is available for π bonding, this time with an oxygen orbital.

What orbitals on oxygen are used in this model? The approach in Figure 10.11 assumes sp^2 hybridization for oxygen. This uses one O atom sp^2 orbital in σ bond

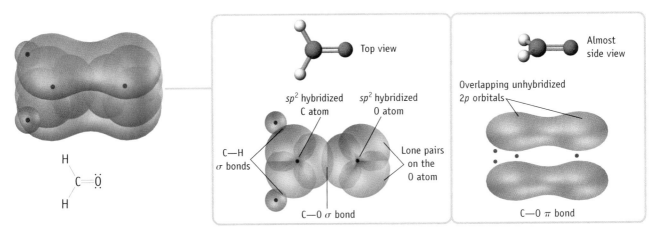

(a) Lewis structure and bonding of formaldehyde, CH_2O.

(b) The C—H σ bonds are formed by overlap of C atom sp^2 hybrid orbitals with H atom $1s$ orbitals. The σ bond between C and O atoms arises from overlap of sp^2 orbitals.

(c) The C—O π bond comes from the side-by-side overlap of p orbitals on the two atoms.

Figure 10.11   Valence bond description of bonding in formaldehyde, CH_2O.

formation, leaving two sp^2 orbitals to accommodate lone pairs. The remaining p orbital on the O atom participates in the π bond.*

GENERAL
Chemistry⋅᛭⋅Now™

See the General ChemistryNow CD-ROM or website:

- **Screen 10.7 Multiple Bonding,** for a tutorial on hybrid orbitals and σ and π bonding

Example 10.5—Bonding in Acetic Acid

Problem Using valence bond theory, describe the bonding in acetic acid, CH_3CO_2H, the important ingredient in vinegar.

Strategy Write a Lewis electron dot structure and determine the electron-pair geometry around each atom using VSEPR. Use this geometry to decide on the hybrid orbitals used in σ bonding. If unhybridized p orbitals are available, then C═O π bonding can occur.

Solution The carbon atom of the CH_3 group has tetrahedral electron-pair geometry, which means that it is sp^3 hybridized. Three sp^3 orbitals are used to form the C—H bonds. The fourth sp^3 orbital is used to bond to the adjacent carbon atom. This carbon atom has a trigonal-planar electron-pair geometry, so it must be sp^2 hybridized. The C—C bond is formed using one of these orbitals, and the other two sp^2 orbitals are used to form the σ bonds to the two oxygens. The oxygen of the O—H group has four electron pairs, so it must be tetrahedral and sp^3 hybridized. Thus, this O atom uses two sp^3 orbitals to bond to the adjacent carbon and the hydrogen, and two sp^3 orbitals accommodate the two lone pairs.

* A second approach is to use unhybridized orbitals on oxygen in bonding. If unhybridized oxygen is assumed, the two p orbitals that are oriented at right angles, and that each contain a single electron, are used to create σ and π bonds. The argument favoring hybridization for oxygen is that it adds consistency to the valence bond approach; because hybridization is required for some atoms, it makes sense to use it for all of them. The objection is that hybridization was introduced simply to explain molecular geometry. The O atom is bonded to only one other atom, so there is no geometry to explain; that is, hybridization does not add anything to the explanation, and it could be regarded as an additional complication.

Finally, the carbon–oxygen double bond can be described exactly as in the CH_2O molecule (Figure 10.11). Both the C and O atoms are assumed to be sp^2 hybridized, and the unhybridized p orbital remaining on each atom is used to form the carbon–oxygen π bond.

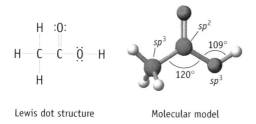

Lewis dot structure   Molecular model

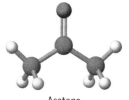

Acetone

Exercise 10.4—Bonding in Acetone

Use valence bond theory to describe the bonding in acetone, CH_3COCH_3.

Triple Bonds

Acetylene, $H-C\equiv C-H$, is an example of a molecule with a triple bond. VSEPR allows us to predict that the four atoms lie in a straight line with $H-C-C$ angles of 180°. This arrangement implies that the carbon atom is sp hybridized (Figure 10.9). For each carbon atom, there are two sp orbitals: one directed toward hydrogen and used to create the $C-H$ σ bond, and one directed toward the other carbon and used to create a σ bond between the two carbon atoms. Two unhybridized p orbitals remain on each carbon, and they are oriented so that it is possible to form *two* π bonds in $H-C\equiv C-H$ (Figures 10.9 and 10.12).

| ↑ | ↑ | Two unhybridized p orbitals. Used for π bonding in C_2H_2.

| ↑ | ↑ | Two sp hybrid orbitals. Used for $C-H$ and $C-C$ σ bonding in C_2H_2.

These π bonds are perpendicular to the molecular axis and perpendicular to each other. Three electrons on each carbon atom are paired to form the triple bond consisting of a σ bond and two π bonds (Figure 10.12).

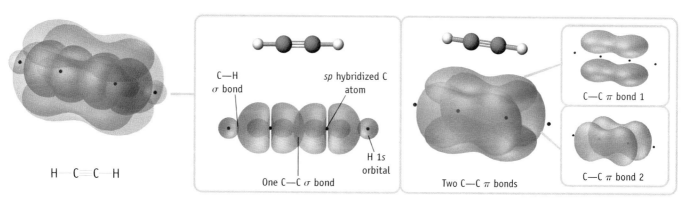

Figure 10.12 **Bonding in acetylene.**

Now that we have examined two cases of multiple bonds, let us summarize several important points:

- A double bond always consists of a σ bond and a π bond. Similarly, a triple bond always consists of a σ bond and *two* π bonds.
- A π bond may form only if unhybridized p orbitals remain on the bonded atoms.
- If a Lewis structure shows multiple bonds, the atoms involved must be either sp^2 or sp hybridized. Only in this manner will unhybridized p orbitals be available to form a π bond.

Exercise 10.5—Triple Bonds Between Atoms

Describe the bonding in a nitrogen molecule, N_2.

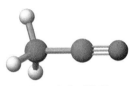

Acetonitrile, CH_3CN

Exercise 10.6—Bonding and Hybridization

Estimate values for the H—C—H, H—C—C, and C—C—N angles in acetonitrile, $CH_3C \equiv N$. Indicate the hybridization of both carbon atoms and the nitrogen atom, and analyze the bonding using valence bond theory.

Cis-Trans Isomerism: A Consequence of π Bonding

Ethylene, C_2H_4, is a planar molecule. This geometry allows the unhybridized p orbitals on the two carbon atoms to line up and form a π bond (see Figure 10.10). Let us speculate on what would happen if one end of the ethylene molecule is twisted relative to the other end (Figure 10.13). This action would distort the molecule away from planarity, and the p orbitals would rotate out of alignment. Rotation

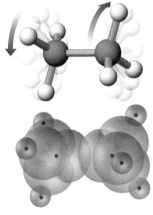

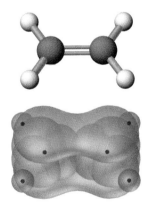

(a) Free rotation can occur around the axis of a single (σ) bond.

(b) Rotation is severely restricted around double bonds because doing so would break the π bond, a process generally requiring a great deal of energy.

Active Figure 10.13   **Rotation around bonds.**

GENERAL
Chemistry･ﾞ･Now™ See the General ChemistryNow CD-ROM or website to explore an interactive version of this figure accompanied by an exercise.

would diminish the extent of overlap of these orbitals. If a twist of 90° was achieved, the two p orbitals would no longer overlap; the π bond would be broken. However, so much energy is required to break this bond (about 260 kJ/mol) that rotation around a $C{=}C$ bond is not expected to occur at room temperature.

A consequence of restricted rotation is that isomers occur for many compounds containing a $C{=}C$ bond. **Isomers** are compounds that have the same formula but different structures. In this case, the two isomeric compounds differ with respect to the orientation of the groups attached to the carbons of the double bond. Two isomeric compounds with the formula $C_2H_2Cl_2$ are *cis*- and *trans*-1,2-dichloroethylene. Their structures resemble ethylene, except that two hydrogen atoms have been replaced by chlorine atoms. Because a large amount of energy is required to break the π bond, the *cis* compound cannot rearrange to form the *trans* compound under ordinary conditions. Each compound can be obtained separately, and each has its own identity. *Cis*-1,2-dichloroethylene boils at 60.3 °C, whereas *trans*-1,2-dichloroethylene boils at 47.5 °C.

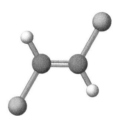

cis-1,2-dichloroethylene   trans-1,2-dichloroethylene

Although *cis* and *trans* isomers do not interconvert at ordinary temperatures, they will do so at higher temperatures. According to the kinetic theory of matter [◀ Section 1.1], molecules in the gas and liquid phases move rapidly and often collide with one another. Molecules also constantly flex or vibrate along or around the bonds holding them together. If the temperature is sufficiently high, the molecular motions can become sufficiently energetic that rotation around the $C{=}C$ bond can occur. It may also occur under other special conditions, such as when the molecule absorbs light energy. Indeed, this specific situation is found to occur in the physiological process that allows us to see (Figure 10.14).

<small>GENERAL</small>
Chemistry∙⁺∙Now™

See the General ChemistryNow CD-ROM or website:
- **Screen 10.8 Molecular Fluxionality,** for an exercise on isomers and multiple bonds

Benzene: A Special Case of π Bonding

Benzene, C_6H_6, is the simplest member of a large group of substances known as *aromatic* compounds, a historical reference to their odor. It occupies a pivotal place in the history and practice of chemistry.

To 19th-century chemists, benzene was a perplexing substance with an unknown structure. Based on its chemical reactions, however, August Kekulé (1829–1896) suggested that the molecule has a planar, symmetrical ring structure. We know now that he was correct. The ring is flat, and all of the carbon–carbon bonds are the same length (139 pm) a distance intermediate between the average single bond (154 pm) and double bond (134 pm) lengths. Assuming the molecule

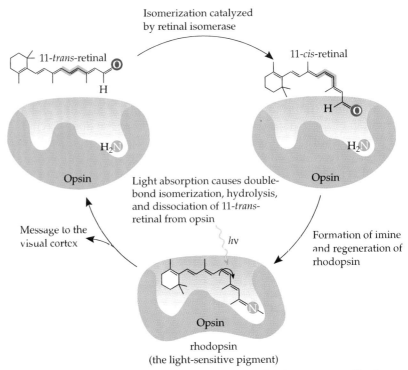

The primary chemical reaction of vision, occurring in the photoreceptor cells of the eyes, is absorption of light by rhodopsin, followed by isomerization of a carbon–carbon double bond from a *cis* configuration to a *trans* configuration.

Figure 10.14 The chemistry of vision. Rotation around a double bond occurs in the reactions that allow you to see. A yellow-orange compound, β-carotene, which is the natural coloring agent in carrots, breaks down in your liver to produce vitamin A, also called retinol. Retinol is oxidized to 11-*trans*-retinal, which isomerizes to 11-*cis*-retinal. The *cis* isomer reacts with the protein opsin in the eye to give the pigment rhodopsin. This light-sensitive combination absorbs light in the blue-green region of the visible spectrum. Light striking the pigment triggers rotation around a carbon–carbon double bond, transforming rhodopsin into meta-rhodopsin. This change in molecular shape causes a nerve impulse to be sent to your brain, and you perceive a visual image.

Eventually meta-rhodopsin reacts chemically to produce 11-*trans*-retinal, and the cycle of chemical changes begins again. Conversion of meta-rhodopsin back to 11-*trans*-retinal is not as rapid as its formation, however, and an image formed on the retina persists for a tenth of a second or so. This persistence of vision allows you to perceive movies and videos as continuously moving images, even though they actually consist of separate pictures, each captured on a piece of film or tape for a thirtieth of a second. (*See General ChemistryNow Screen 10.13 Molecular Orbitals and Vision.*)

has two resonance structures with alternating double bonds, the observed structure is rationalized [◀ Section 9.5]. The C—C bond order in C_6H_6 (1.5) is the average of a single bond and a double bond.

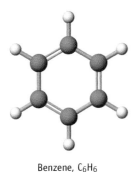

Benzene, C_6H_6

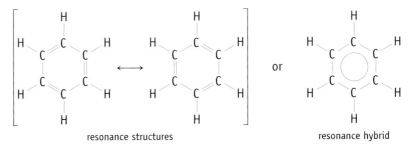

resonance structures   resonance hybrid

Understanding the bonding in benzene (Figure 10.15) is important because the benzene ring structure occurs in an enormous number of chemical compounds.

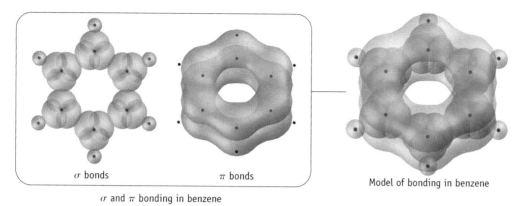

σ bonds   π bonds

σ and π bonding in benzene

Model of bonding in benzene

Figure 10.15 Bonding in benzene, C_6H_6. (*left*) The C atoms of the ring are bonded to each other through σ bonds using sp^2 hybrid orbitals of the C atom. The C — H bonds also use C atom sp^2 hybrid orbitals. The π bonding framework of the molecule arises from overlap of C atom p orbitals not used in hybrid orbital formation. As these orbitals are perpendicular to the ring, π electron density is above and below the plane of the ring. (*right*) A composite of σ and π bonding in benzene.

We assume that the trigonal-planar carbon atoms have sp^2 hybridization. Each C — H bond is formed by overlap of an sp^2 orbital of a carbon atom with a $1s$ orbital of hydrogen, and the C — C σ bonds arise by overlap of sp^2 orbitals on adjacent carbon atoms. After accounting for the σ bonding, an unhybridized p orbital remains on each C atom, and each is occupied by a single electron. These six orbitals and six electrons form three π bonds. Because all carbon–carbon bond lengths are the same, each p orbital overlaps equally well with the p orbitals of both adjacent carbons, and the π interaction is unbroken around the six-member ring.

The orbital picture of benzene underscores an important point. The basis of valence bond theory, which states that a bond is described as a pair of electrons between two atoms, does not work well for the π electrons in benzene—nor does it work whenever resonance is needed to describe a structure. However, molecular orbital theory does give us a better view, and that is the subject of the next section.

10.3—Molecular Orbital Theory

Molecular orbital (MO) theory is an alternative way to view orbitals in molecules. In contrast to the localized bond and lone-pair electrons of valence bond theory, MO theory assumes that pure s and p atomic orbitals of the atoms in the molecule combine to produce orbitals that are spread out, or delocalized, over several atoms or even over an entire molecule. These orbitals are called **molecular orbitals**.

One reason for learning about the MO concept is that it correctly predicts the electronic structures of molecules such as O_2 that do not follow the electron-pairing assumptions of the Lewis approach. The rules of Chapter 9 would guide you to draw the electron dot structure of O_2 with all the electrons paired, which fails to explain its paramagnetism (Figure 10.16). The molecular orbital approach can account for this property, but valence bond theory cannot. To see how MO theory can be used to describe the bonding in O_2 and other diatomic molecules, we shall first describe four principles used to develop the theory.

Principles of Molecular Orbital Theory

In MO theory we begin with a given arrangement of atoms in the molecule at the known bond distances. We then determine the *sets* of molecular orbitals. One way to do so is to combine available valence orbitals on all the constituent atoms. These

■ **A Failure of the Valence Bond Theory** Lewis electron dot structures fail to describe the bonding correctly in a well-known diatomic molecule, O_2. The O_2 molecule is paramagnetic, which requires the presence of unpaired electrons. The obvious Lewis structure, however, has all electrons paired. The molecular orbital approach shows that the molecule has two unpaired electrons.

■ **Diatomic Molecules** Molecules such as H_2, Li_2, and N_2, in which two identical atoms are bonded, are often called *homonuclear* diatomic molecules.

Figure 10.16   Liquid oxygen. Oxygen gas condenses (*left*) to a pale blue liquid at −183 °C (*middle*). Oxygen in the liquid state is paramagnetic and clings to the poles of a magnet (*right*). (*See General ChemistryNow Screen 10.12 Paramagnetism, to watch a video of this figure.*)

molecular orbitals more or less encompass all the atoms of the molecule, and the valence electrons for all the atoms in the molecule are assigned to the molecular orbitals. Just as with orbitals in atoms, electrons are assigned according to the Pauli exclusion principle and Hund's rule [◀ Sections 8.2 and 8.4].

The **first principle of molecular orbital theory** is that *the total number of molecular orbitals is always equal to the total number of atomic orbitals contributed by the atoms that have combined.* To illustrate this orbital conservation principle, let us consider the H_2 molecule.

Molecular Orbitals for H_2

Molecular orbital theory specifies that when the $1s$ orbitals of two hydrogen atoms overlap, *two* molecular orbitals result. In the molecular orbital resulting from *addition* of the atomic orbitals, the $1s$ regions of electron density add together, leading to an increased probability that electrons will reside in the bond region between the two nuclei (Figure 10.17). This **bonding molecular orbital** is the same as the chemical bond described by valence bond theory. It is also a σ orbital because the region of electron probability lies directly along the bond axis. This molecular orbital is labeled σ_{1s}, where the subscript $1s$ indicates that $1s$ atomic orbitals were used to create the molecular orbital.

The other molecular orbital is constructed by *subtracting* one atomic orbital from the other (see Figure 10.17). When this happens, the probability of finding an electron between the nuclei in the molecular orbital is reduced, and the probability of finding the electron in other regions is higher. Without significant electron density between them, the nuclei repel one another. This type of orbital is called an **antibonding molecular orbital**. Because it is also a σ orbital, it is labeled σ_{1s}^{*}. The asterisk signifies that it is antibonding. *Antibonding orbitals have no counterpart in valence bond theory.*

The **second principle of molecular orbital theory** is that *the bonding molecular orbital is lower in energy than the parent orbitals, and the antibonding orbital is higher in energy* (Figure 10.17). As a result, the energy of a group of atoms is lower than the energy of the separated atoms when electrons are assigned to bonding molecular orbitals. Chemists say the system is "stabilized" by chemical bond formation. Conversely, the system is "destabilized" when electrons are assigned to antibonding orbitals because the energy of the system is higher than that of the atoms themselves.

■ **Orbitals and Electron Waves**
Orbitals are characterized as electron waves; therefore, a way to view molecular orbital formation is to assume that two electron waves, one from each atom, interfere with each other. The interference can be constructive, giving a bonding MO, or destructive, giving an antibonding MO.

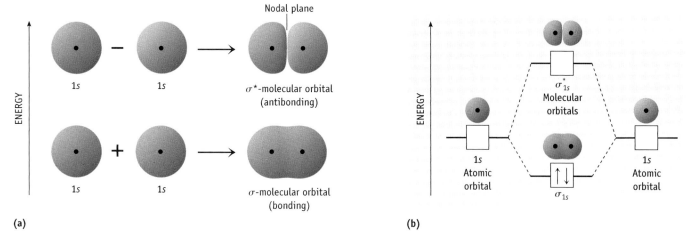

(a)   **(b)**

Figure 10.17   Molecular orbitals. (a) Bonding and antibonding σ molecular orbitals are formed from two 1s atomic orbitals on adjacent atoms. Notice the presence of a node in the antibonding orbital. (The node is a plane on which there is zero probability of finding an electron.) (b) A molecular orbital diagram for H_2. The two electrons are placed in the σ_{1s} orbital, the molecular orbital lower in energy. (*See General ChemistryNow Screen 10.9 Molecular Orbital Theory, to view animations based on this figure.*)

The **third principle of molecular orbital theory** is that the *electrons of the molecule are assigned to orbitals of successively higher energy* according to the Pauli exclusion principle and Hund's rule. This is analogous to the procedure for building up electronic structures of atoms. Thus, electrons occupy the lowest energy orbitals available: when two electrons are assigned to an orbital, their spins must be paired. Because the energy of the electrons in the bonding orbital of H_2 is lower than that of either parent 1s electron (see Figure 10.17b), the H_2 molecule is stable. We write the electron configuration of H_2 as $(\sigma_{1s})^2$.

What would happen if we try to combine two helium atoms to form dihelium, He_2? Both He atoms have a 1s valence orbital that can be added and subtracted to produce the same kind of molecular orbitals as in H_2. Unlike in H_2, however, four electrons need to be assigned to these orbitals (Figure 10.18). The pair of electrons in the σ_{1s} orbital stabilizes He_2. The two electrons in the σ_{1s}^* orbital, however, destabilize the He_2 molecule. The energy decrease from the electrons in the σ_{1s} bonding molecular orbital is offset by the energy increase due to the electrons in the σ_{1s}^* antibonding molecular orbital. Thus, molecular orbital theory predicts that He_2 has no net stability; two He atoms have no tendency to combine. This confirms what we already know—elemental helium exists in the form of single atoms and not as a diatomic molecule.

Bond Order

Bond order was defined in Chapter 9 as the net number of bonding electron pairs linking a pair of atoms. This same concept can be applied directly to molecular orbital theory, but now bond order is defined as

$$\text{Bond order} = \tfrac{1}{2}(\text{number of electrons in bonding MOs} \\ - \text{number of electrons in antibonding MOs}) \qquad (10.1)$$

In the H_2 molecule, there are two electrons in a bonding orbital and none in an antibonding orbital, so H_2 has a bond order of 1. In contrast, in He_2 the stabilizing

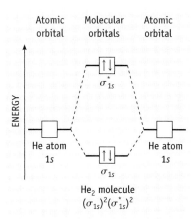

Figure 10.18   A molecular orbital energy level diagram for the dihelium molecule, He_2. This diagram provides a rationalization for the nonexistence of the molecule. In He_2 both the bonding (σ_{1s}) and antibonding orbitals (σ_{1s}^*) would be fully occupied. Note that occupation of antibonding orbitals leads to a greater destabilization than occupation of bonding orbitals leads to stabilization.

effect of the σ_{1s} pair is canceled by the destabilizing effect of the σ_{1s}^* pair, so the bond order is 0.

Fractional bond orders are possible. Consider the ion He_2^+. Its molecular orbital electron configuration is $(\sigma_{1s})^2(\sigma_{1s}^*)^1$. In this ion, there are two electrons in a bonding molecular orbital, but only one in an antibonding orbital. MO theory predicts that He_2^+ should have a bond order of 0.5; that is, a weak bond should exist between helium atoms in such a species. Interestingly, this ion has been identified in the gas phase using special experimental techniques.

GENERAL
Chemistry⋅**Now**™

See the General ChemistryNow CD-ROM or website:
- **Screen 10.9 Molecular Orbital Theory,** for an exercise on identifying molecular orbitals
- **Screen 10.10 Molecular Orbital Configurations,** for a description of the dihydrogen molecular orbital diagram

---

Example 10.6—Molecular Orbitals and Bond Order

Problem Write the electron configuration of the H_2^- ion in molecular orbital terms. What is the bond order of the ion?

Strategy Count the number of valence electrons in the ion and then place those electrons in the MO diagram for the H_2 molecule. Find the bond order from Equation 10.1.

Solution This ion has three electrons (one each from the H atoms plus one for the negative charge). Therefore, its electronic configuration is $(\sigma_{1s})^2(\sigma_{1s}^*)^1$, identical with the configuration for He_2^+. This means H_2^- also has a net bond order of 0.5. The H_2^- ion is thus predicted to exist under special circumstances.

Exercise 10.7—Molecular Orbitals and Bond Order

What is the electron configuration of the H_2^+ ion? Compare the bond order of this ion with those of He_2^+ and H_2^-. Do you expect H_2^+ to exist?

Molecular Orbitals of Li₂ and Be₂

The **fourth principle of molecular orbital theory** is that *atomic orbitals combine to form molecular orbitals most effectively when the atomic orbitals are of similar energy.* This principle becomes important when we move past He_2 to Li_2 (dilithium) and to even heavier molecules.

A lithium atom has electrons in two orbitals of the s type ($1s$ and $2s$), so a $1s \pm 2s$ combination is theoretically possible. Because the $1s$ and $2s$ orbitals are quite different in energy, however, this interaction can be disregarded. Thus, the molecular orbitals come only from $1s \pm 1s$ and $2s \pm 2s$ combinations (Figure 10.19). This means the molecular orbital electron configuration of dilithium, Li_2, is

$$Li_2 \text{ MO configuration: } (\sigma_{1s})^2(\sigma_{1s}^*)^2(\sigma_{2s})^2$$

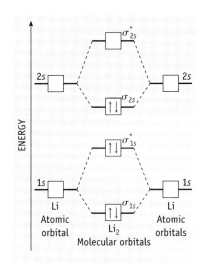

Figure 10.19 Energy level diagram for the combination of two Li atoms. Notice that the molecular orbitals are created by combining orbitals of similar energies. The electron configuration is shown for Li_2.

The bonding effect of the σ_{1s} electrons is canceled by the antibonding effect of the σ_{1s}^* electrons, so these pairs make no net contribution to bonding in Li_2. Bonding in Li_2 is due to the electron pair assigned to the σ_{2s} orbital, and the bond order is 1.

The fact that the σ_{1s} and σ_{1s}^* electron pairs of Li_2 make no net contribution to bonding is exactly what you observed in drawing electron dot structures in Chapter 9: *Core electrons are ignored.* In molecular orbital terms, core electrons are assigned to bonding and antibonding molecular orbitals that offset one another.

A diberyllium molecule, Be_2, is not expected to exist. Its electron configuration is

$$Be_2 \text{ MO configuration: } [\text{core electrons}](\sigma_{2s})^2(\sigma_{2s}^*)^2$$

The effects of σ_{2s} and σ_{2s}^* electrons cancel, and there is no net bonding. The bond order is 0, so the molecule does not exist.

Example 10.7—Molecular Orbitals in Diatomic Molecules

Problem Be_2 does not exist. But what about the Be_2^+ ion? Describe its electron configuration in molecular orbital terms and give the net bond order. Do you expect the ion to exist?

Strategy Count the number of electrons in the ion and place them in the MO diagram in Figure 10.19. Write the electron configuration and calculate the bond order from Equation 10.1.

Solution The Be_2^+ ion has seven electrons (Be_2 has eight), of which four are core electrons. (The core electrons are assigned to σ_{1s} and σ_{1s}^* molecular orbitals.) The remaining three electrons are assigned to the σ_{2s} and σ_{2s}^* molecular orbitals, so the MO electron configuration is $[\text{core electrons}](\sigma_{2s})^2(\sigma_{2s}^*)^1$. This means the net bond order is 0.5, and so Be_2^+ is predicted to exist under special circumstances.

Exercise 10.8—Molecular Orbitals in Diatomic Molecules

Could the anion Li_2^- exist? What is the ion's bond order?

Molecular Orbitals from Atomic *p* Orbitals

With the principles of molecular orbital theory in place, we are ready to account for bonding in such important homonuclear diatomic molecules as N_2, O_2, and F_2. To describe the bonding in these molecules we will use both *s* and *p* valence orbitals in forming molecular orbitals.

Sigma-bonding and antibonding molecular orbitals are formed by *s* orbitals interacting as illustrated in Figure 10.19. Similarly, it is possible for a *p* orbital on one atom to interact with a *p* orbital on the other atom to produce a pair of σ-bonding and σ^*-antibonding molecular orbitals (Figure 10.20).

In addition, each atom has *two p* orbitals in planes perpendicular to the σ bond connecting the two atoms. These *p* orbitals can interact sideways to give π-bonding and π-antibonding molecular orbitals (Figure 10.21). Combining these two *p* orbitals on each atom produces *two* π-bonding molecular orbitals (π_p) and *two* pi-antibonding molecular orbitals (π_p^*).

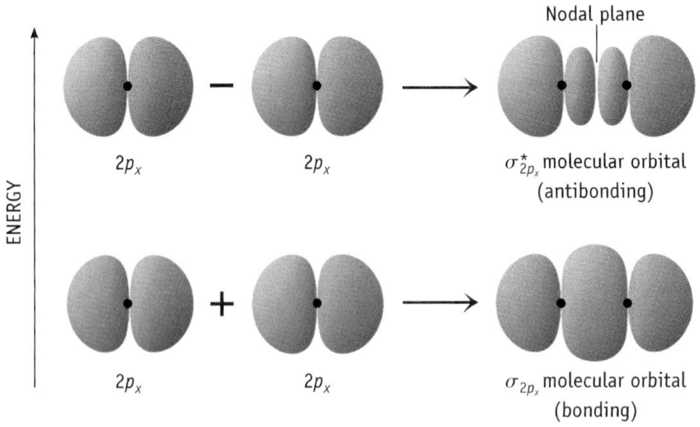

Figure 10.20   Sigma molecular orbitals from *p* atomic orbitals. Sigma-bonding (σ_{2p}) and antibonding (σ^{*}_{2p}) molecular orbitals arise from overlap of 2*p* orbitals. Each orbital can accommodate two electrons. The *p* orbitals in electron shells of higher *n* give molecular orbitals of the same basic shape.

Figure 10.21   Formation of π molecular orbitals. Sideways overlap of atomic 2*p* orbitals that lie in the same direction in space gives rise to pi-bonding (π_{2p}) and pi-antibonding (π^{*}_{2p}) molecular orbitals. The *p* orbitals in shells of higher *n* give molecular orbitals of the same basic shape.

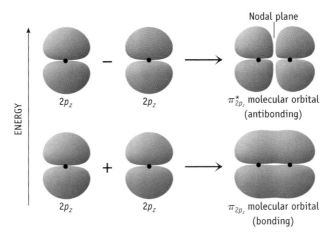

Electron Configurations for Homonuclear Molecules for Boron Through Fluorine

Orbital interactions in a second-period, homonuclear, diatomic molecule lead to the energy level diagram shown in Figure 10.22. Electron assignments can be made using this diagram, and the results for the diatomic molecules B_2 through F_2 are tabulated in Table 10.1, which has two noteworthy features.

First, notice the correlation between the electron configurations and the bond orders, bond lengths, and bond energies at the bottom of Table 10.1. As the bond order between a pair of atoms increases, the energy required to break the bond increases, and the bond distance decreases. Dinitrogen, N_2, with a bond order of 3, has the largest bond energy and the shortest bond distance.

Second, notice the configuration for dioxygen, O_2. Dioxygen has 12 valence electrons (6 from each atom), so it has the molecular orbital configuration

$$O_2 \text{ MO configuration: } [\text{core electrons}](\sigma_{2s})^2(\sigma^{*}_{2s})^2(\pi_{2p})^4(\sigma_{2p})^2(\pi^{*}_{2p})^2$$

This configuration leads to a bond order of 2 in agreement with experiment, and it specifies two unpaired electrons (in π^{*}_{2p} molecular orbitals). Thus, molecular orbital theory succeeds where valence bond theory fails. MO theory explains both the observed bond order and the paramagnetic behavior of O_2.

■ **Highest Occupied Molecular Orbital (HOMO)**
Chemists often refer to the highest energy MO that contains electrons as the HOMO. For O_2 this is the π^{*}_{2p} orbital. Chemists also use the term LUMO, for the lowest unoccupied molecular orbital. For O_2, it would be σ^{*}_{2p}.

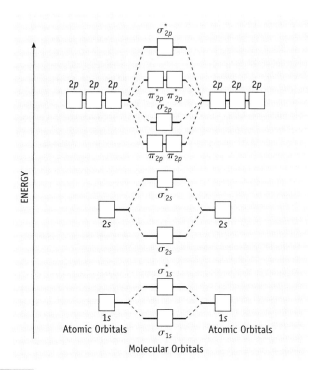

Active Figure 10.22 **Molecular orbital energy level diagram for homonuclear diatomic molecules of second period elements.** Although the diagram leads to the correct conclusions regarding bond order and magnetic behavior for O_2, N_2, and F_2, the energy ordering of the MOs is correct only for N_2 and F_2. For O_2, the σ_{2p} MO is lower in energy than the π_{2p} MOs. See "A Closer Look," page 464.

GENERAL
Chemistry ·+· Now™ See the General ChemistryNow CD-ROM or website to explore an interactive version of this figure accompanied by an exercise.

Table 10.1 Molecular Orbital Occupations and Physical Data for Homonuclear Diatomic Molecules of Second-Period Elements

	B_2	C_2	N_2	O_2	F_2
σ_{2p}^*	☐	☐	☐	☐	☐
π_{2p}^*	☐☐	☐☐	☐☐	↑ ↑	↿⇂ ↿⇂
σ_{2p}	☐	☐	↿⇂	↿⇂	↿⇂
π_{2p}	↑ ↑	↿⇂ ↿⇂	↿⇂ ↿⇂	↿⇂ ↿⇂	↿⇂ ↿⇂
σ_{2s}^*	↿⇂	↿⇂	↿⇂	↿⇂	↿⇂
σ_{2s}	↿⇂	↿⇂	↿⇂	↿⇂	↿⇂
Bond order	One	Two	Three	Two	One
Bond-dissociation energy (kJ/mol)	290	620	945	498	155
Bond distance (pm)	159	131	110	121	143
Observed magnetic behavior (paramagnetic or diamagnetic)	Para	Dia	Dia	Para	Dia

A Closer Look

Molecular Orbitals for Compounds Formed from *p*-Block Elements

Several features of the molecular orbital energy level diagram in Figure 10.22 might be described in more detail.

- The bonding and antibonding σ orbitals from 2*s* interactions are lower in energy than the σ and π MOs from 2*p* interactions. The reason is that 2*s* orbitals have a lower energy than 2*p* orbitals in the separated atoms.

- The energy separation of the bonding and antibonding orbitals is greater for

σ_{2p} than for π_{2p}. This happens because *p* orbitals overlap to a greater extent when they are oriented head to head (to give σ_{2p} MOs) than when they are side by side (to give π_{2p} MOs). The greater the orbital overlap, the greater the stabilization of the bonding MO and the greater the destabilization of the antibonding MO.

Figure 10.22 shows an energy ordering of molecular orbitals that you might not have expected, but there are reasons for this order. A more sophisticated approach takes into account the "mixing" of *s* and *p* atomic orbitals, which have similar energies. This causes the σ_{2s} and σ_{2s}^{*} molecular

orbitals to be lower in energy than expected, and the σ_{2p} and σ_{2p}^{*} orbitals to be higher in energy than expected. This is the reason the energy lowering and raising for the σ_{2s} and σ_{2s}^{*} orbitals (and for the σ_{2p} and σ_{2p}^{*} orbitals) in Figure 10.22 is not symmetrical with respect to the 2*s* and 2*p* atomic orbital energies.

The mixing of *s* and *p* orbitals is important only for B_2, C_2, and N_2, so the figure applies just to these molecules. For O_2 and F_2, σ_{2p} is lower in energy than π_{2p}. Nonetheless, Figure 10.22 gives the correct bond order and magnetic behavior for these two molecules.

GENERAL
Chemistry • Now™

See the General ChemistryNow CD-ROM or website:

- **Screen 10.11 Homonuclear Diatomic Molecules,** for an exercise on molecular orbital configurations

---

Example 10.8—Electron Configuration for a Homonuclear Diatomic Ion

Problem When potassium reacts with O_2, potassium superoxide, KO_2, is one of the products. This is an ionic compound, in which the anion is the superoxide ion, O_2^{-}. Write the molecular orbital electron configuration for the ion. Predict its bond order and magnetic behavior.

Strategy Use the energy level diagram of Figure 10.22 to generate the configuration of this ion. Use Equation 10.1 to determine the bond order.

Solution The MO configuration for O_2^{-} is

$$O_2^{-} \text{ MO configuration: } [\text{core electrons}](\sigma_{2s})^2(\sigma_{2s}^{*})^2(\pi_{2p})^4(\sigma_{2p})^2(\pi_{2p}^{*})^3$$

The ion is predicted to be paramagnetic to the extent of one unpaired electron, a prediction confirmed by experiment. The bond order is 1.5, because there are eight bonding electrons and five antibonding electrons. The bond order for O_2^{-} is lower than that for O_2, so we predict that the $O—O$ bond length in O_2^{-} will be longer than the oxygen–oxygen bond length in O_2. In fact, the superoxide ion has an $O—O$ bond length of 134 pm, whereas the bond length in O_2 is 121 pm.

Comment You should quickly spot the fact that the superoxide ion (O_2^{-}) contains an odd number of electrons. It is another diatomic species (in addition to NO and O_2) for which it is not possible to write a Lewis structure that accurately represents the bonding.

Exercise 10.9—Molecular Electron Configurations

The cations O_2^{+} and N_2^{+} are important components of the earth's upper atmosphere. Write the electron configuration of O_2^{+}. Predict its bond order and magnetic behavior.

Electron Configurations for Heteronuclear Diatomic Molecules

The compounds NO, CO, and ClF—all molecules containing two different elements—are examples of **heteronuclear diatomic molecules**. MO descriptions for heteronuclear diatomic molecules generally resemble those for homonuclear diatomic molecules. As a consequence, an energy level diagram like Figure 10.22 can be used to judge the bond order and magnetic behavior for heteronuclear diatomics.

Consider nitrogen monoxide, NO. Nitrogen monoxide has 11 molecular valence electrons. If they are assigned to the MOs for a homonuclear diatomic molecule, the molecular electron configuration is

NO MO configuration: $[\text{core electrons}](\sigma_{2s})^2(\sigma_{2s}^*)^2(\pi_{2p})^4(\sigma_{2p})^2(\pi_{2p}^*)^1$

The net bond order is 2.5, in accordance with the bond length information. The single unpaired electron is assigned to the π_{2p}^* molecular orbital. The molecule is paramagnetic, as predicted for a molecule with an odd number of electrons.

Resonance and MO Theory

Ozone, O_3, is a simple triatomic molecule with equal oxygen–oxygen bond lengths. Equal X — O bond lengths are also observed in other triatomic molecules and ions, such as SO_2, NO_2^-, and HCO_2^-. Valence bond theory introduced resonance to rationalize the equivalent bonding to the oxygen atoms in these structures. MO theory provides another view of this problem.

O_3 SO_2 NO_2^- HCO_2^-

To visualize the bonding in ozone, begin by assuming that all three O atoms are sp^2 hybridized. The central atom uses its sp^2 hybrid orbitals to form two σ bonds and to accommodate a lone pair. The terminal atoms use their sp^2 hybrid orbitals to form one σ bond and to accommodate two lone pairs. In total, the lone pairs and bonding pairs in the σ framework of O_3 account for seven of the nine valence electron pairs in O_3.

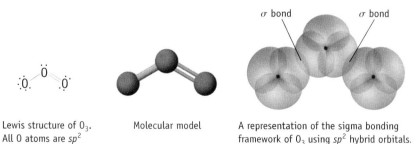

Lewis structure of O_3.
All O atoms are sp^2
hybridized.

Molecular model

A representation of the sigma bonding
framework of O_3 using sp^2 hybrid orbitals.

The π bond in ozone arises from the two remaining pairs (Figure 10.23). Because we have assumed that each oxygen atom in O_3 is sp^2 hybridized, an unhybridized p orbital perpendicular to the O_3 plane remains on each of the three oxygen atoms. The orbitals are in the correct orientation to form π bonds. *A principle of MO theory is that the number of molecular orbitals must equal the number of atomic orbitals.* Thus, the three $2p$ atomic orbitals must be combined in a way that forms three molecular orbitals.

One π_p MO for ozone is a bonding orbital because the three p orbitals are "in phase" across the molecule. Another π_p MO is an antibonding orbital because the

Figure 10.23   Pi-bonding in ozone, O_3. Each O atom in O_3 is sp^2 hybridized. The three $2p$ orbitals, one on each atom, are used to create the three π molecular orbitals. Two pairs of electrons are assigned to the orbitals: one pair in the bonding orbital and one pair in the nonbonding orbital. The π bond order is 0.5, as one bonding pair is spread across two bonds.

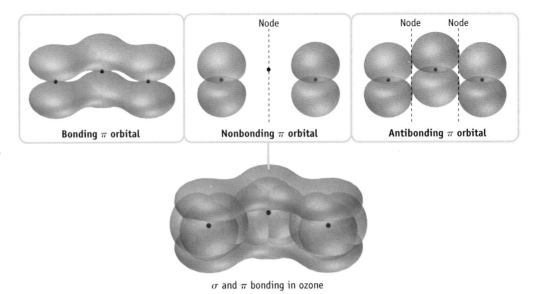

σ and π bonding in ozone

■ **Metals and Molecular Orbitals**
The bonding in metals can be described best using molecular orbital theory. See "The Chemistry of Materials," page 642.

atomic orbital on the central atom is "out of phase" with the terminal atom p orbitals. The third π_p MO is a nonbonding orbital because the middle p orbital does not participate in the MO. The bonding π_p MO is filled by a pair of electrons that is delocalized, or "spread over," the molecule, just as the resonance hybrid implies. The nonbonding orbital is also occupied, but the electrons in this orbital are concentrated near the two terminal oxygens. As the name implies, electrons in this molecular orbital neither help nor hinder the bonding in the molecule. The π bond order of O_3 is 0.5, since one bond pair is spread over two O — O linkages. Because the σ bond order is 1.0 and the π bond order is 0.5, the net oxygen–oxygen bond order is 1.5—the same value given by valence bond theory.

The observation that two of the π molecular orbitals for ozone extend over three atoms illustrates an important point regarding molecular orbital theory: Orbitals can extend beyond two atoms. In valence bond theory, in contrast, all representations for bonding were based on being able to localize pairs of electrons in bonds between two atoms. To further illustrate the MO approach, look again at benzene (Figure 10.24). On page 457 we noted that the π electrons in this molecule were spread out over all six carbon atoms. We can now see how the same case can be made with MO theory. Six p orbitals contribute to the π system. Based on the premise that the number of molecular orbitals must equal the number of atomic orbitals, there must be six π molecular orbitals in benzene. An energy level diagram for benzene shows that the six p electrons reside in the three lowest-energy (bonding) molecular orbitals.

Figure 10.24   Molecular orbital energy level diagram for benzene. Because there are six unhybridized p orbitals, six π molecular orbitals can be formed—three bonding and three antibonding. The three bonding molecular orbitals accommodate the six π electrons.

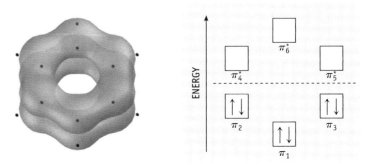

Chapter Goals Revisited

When you have finished studying this chapter, you should ask whether you have met the chapter goals. In particular, you should be able to

Understand the differences between valence bond theory and molecular orbital theory

a. Describe the main features of valence bond theory and molecular orbital theory, the two commonly used theories for covalent bonding (Section 10.1).

b. Recognize that the premise for valence bond theory is that bonding results from the overlap of atomic orbitals. By virtue of the overlap of orbitals, electrons are concentrated (or localized) between two atoms (Section 10.2).

c. Distinguish how sigma (σ) and pi (π) bonds arise. For σ bonding, orbitals overlap in a head-to-head fashion, concentrating electrons along the bond axis. Sideways overlap of p atomic orbitals results in π bond formation, with electrons above and below the molecular plane (Section 10.2).

d. Understand how molecules having double bonds can have isomeric forms.
General ChemistryNow homework: Study Question(s) 14

Identify the hybridization of an atom in a molecule or ion

a. Use the concept of hybridization to rationalize molecular structure (Section 10.2). General ChemistryNow homework: SQ(s) 3, 4, 6, 8, 11, 22, 27, 32, 34, 38, 51

Hybrid Orbitals	Atomic Orbitals Used	Number of Hybrid Orbitals	Electron-Pair Geometry
sp	$s + p$	2	Linear
sp^2	$s + p + p$	3	Trigonal-planar
sp^3	$s + p + p + p$	4	Tetrahedral
sp^3d	$s + p + p + p + d$	5	Trigonal-bipyramidal
sp^3d^2	$s + p + p + p + d + d$	6	Octahedral

Understand the differences between bonding and antibonding molecular orbitals

a. Understand molecular orbital theory (Section 10.3), in which atomic orbitals are combined to form bonding orbitals, nonbonding orbitals, or antibonding orbitals that are delocalized over several atoms. In this description, the electrons of the molecule or ion are assigned to the orbitals beginning with the one at lowest energy, according to the Pauli exclusion principle and Hund's rule.

b. Use molecular orbital theory to explain the properties of O_2 and other diatomic molecules. General ChemistryNow homework: SQ(s) 15, 16, 18, 20, 42, 44

GENERAL
Chemistry⋅⋅⋅Now™

See the General ChemistryNow CD-ROM or website to:

- Assess your understanding with homework questions keyed to each goal
- Check your readiness for an exam by taking the exam-prep quiz and exploring the resources in the personalized Learning Plan it provides

Key Equations

Equation 10.1 (page 459)

Calculating the order of a bond from the molecular orbital electron configuration

Bond order $= \frac{1}{2}$(number of electrons in bonding MOs
$-$ number of electrons in antibonding MOs)

Study Questions

▲ denotes more challenging questions.

■ denotes questions available in the Homework and Goals section of the General ChemistryNow CD-ROM or website.

Blue numbered questions have answers in Appendix O and fully worked solutions in the *Student Solutions Manual.*

Structures of many of the compounds used in these questions are found on the General ChemistryNow CD-ROM or website in the Models folder.

GENERAL
Chemistry ⚛ Now™ Assess your understanding of this chapter's topics with additional quizzing and conceptual questions at http://now.brookscole.com/kotz6e

Practicing Skills

Valence Bond Theory

(See Examples 10.1–10.5 and General ChemistryNow Screens 10.2–10.7)

1. Draw the Lewis structure for chloroform, $CHCl_3$. What are its electron-pair and molecular geometries? What orbitals on C, H, and Cl overlap to form bonds involving these elements?

2. ■ Draw the Lewis structure for NF_3. What are its electron-pair and molecular geometries? What is the hybridization of the nitrogen atom? What orbitals on N and F overlap to form bonds between these elements?

3. Specify the electron-pair and molecular geometry for each of the following. Describe the hybrid orbital set used by the underlined atom in each molecule or ion.
   (a) $\underline{B}Br_3$ (b) $\underline{C}O_2$ (c) $\underline{C}H_2Cl_2$ (d) $\underline{C}O_3{}^{2-}$

4. ■ Specify the electron-pair and molecular geometry for each of the following. Describe the hybrid orbital set used by the underlined atom in each molecule or ion.
   (a) $\underline{C}Se_2$ (b) $\underline{S}O_2$ (c) $\underline{C}H_2O$ (d) $\underline{N}H_4{}^+$

5. ■ Describe the hybrid orbital set used by each of the indicated atoms in the molecules below:
   (a) the carbon atoms and the oxygen atom in dimethyl ether, H_3COCH_3
   (b) each carbon atom in propene

$$H_3C - \overset{\overset{\textstyle H}{|}}{C} = CH_2$$

   (c) the two carbon atoms and the nitrogen atom in the amino acid glycine

$$H - \overset{\overset{\textstyle H}{|}}{\underset{\underset{\textstyle H}{|}}{N}} - \overset{\overset{\textstyle H}{|}}{C} - \overset{\overset{\textstyle :O:}{\|}}{C} - \overset{\cdot\cdot}{\underset{\cdot\cdot}{O}} - H$$

6. ■ Give the hybrid orbital set used by each of the underlined atoms in the following molecules.

   (a)
$$H - \underset{\cdot\cdot}{\underline{N}} - \overset{:\overset{\cdot\cdot}{O}:}{\underline{C}} - \underset{\cdot\cdot}{N} - H$$
   (with H atoms on the N's)

   (c)
$$H - \underline{C} = C - \underline{C} \equiv N:$$
   (with H H on first C)

   (b)
$$H_3\underline{C} - \overset{\overset{\textstyle H}{|}}{C} = \overset{\overset{\textstyle H}{|}}{C} - \overset{\overset{\textstyle H}{|}}{\underline{C}} = \overset{\cdot\cdot}{\underset{\cdot\cdot}{O}}$$

7. Draw the Lewis structure and then specify the electron-pair and molecular geometries for each of the following molecules or ions. Identify the hybridization of the central atom.
   (a) $SiF_6{}^{2-}$ (b) SeF_4 (c) $ICl_2{}^-$ (d) XeF_4

8. ■ Draw the Lewis structure and then specify the electron-pair and molecular geometries for each of the following molecules or ions. Identify the hybridization of the central atom.
   (a) $XeOF_4$ (c) OSF_4
   (b) BrF_5 (d) central Br in $Br_3{}^-$

9. Draw the Lewis structures of the acid HPO_2F_2 and its anion $PO_2F_2{}^-$. What is the molecular geometry and hybridization for the phosphorus atom in each species? (H is bonded to the O atom in the acid.)

10. Draw the Lewis structures of HSO_3F and SO_3F^-. What is the molecular geometry and hybridization for the sulfur atom in each species? (H is bonded to the O atom in the acid.)

11. ■ What is the hybridization of the carbon atom in phosgene, Cl_2CO? Give a complete description of the σ and π bonding in this molecule.

12. What is the hybridization of the sulfur atom in sulfuryl fluoride, SO_2F_2?

13. The arrangement of groups attached to the C atoms involved in a $C = C$ double bond leads to *cis* and *trans* isomers. For each compound below, draw the other isomer.

   (a)
$$\begin{array}{ccc} H_3C & & H \\ & C = C & \\ H & & CH_3 \end{array}$$

   (b)
$$\begin{array}{ccc} Cl & & CH_3 \\ & C = C & \\ H & & H \end{array}$$

14. ■ For each compound below decide whether *cis* and *trans* isomers are possible. If isomerism is possible, draw the other isomer.

   (a)
$$\begin{array}{ccc} H_3C & & H \\ & C = C & \\ H & & CH_2CH_3 \end{array}$$

   (c)
$$\begin{array}{ccc} Cl & & CH_2OH \\ & C = C & \\ H & & H \end{array}$$

   (b)
$$\begin{array}{ccc} H & & CH_3 \\ & C = C & \\ H & & H \end{array}$$

Molecular Orbital Theory

(See Examples 10.6–10.8 and General ChemistryNow Screens 10.9–10.12.)

15. ■ The hydrogen molecular ion, H_2^+, can be detected spectroscopically. Write the electron configuration of the ion in molecular orbital terms. What is the bond order of the ion? Is the hydrogen–hydrogen bond stronger or weaker in H_2^+ than in H_2?

16. ■ Give the electron configurations for the ions Li_2^+ and Li_2^- in molecular orbital terms. Compare the Li—Li bond order in these ions with the bond order in Li_2.

17. Calcium carbide, CaC_2, contains the acetylide ion, C_2^{2-}. Sketch the molecular orbital energy level diagram for the ion. How many net σ and π bonds does the ion have? What is the carbon–carbon bond order? How has the bond order changed on adding electrons to C_2 to obtain C_2^{2-}? Is the C_2^{2-} ion paramagnetic?

18. ■ Oxygen, O_2, can acquire one or two electrons to give O_2^- (superoxide ion) or O_2^{2-} (peroxide ion). Write the electron configuration for the ions in molecular orbital terms, and then compare them with the O_2 molecule on the following bases.
    (a) magnetic character
    (b) net number of σ and π bonds
    (c) bond order
    (d) oxygen–oxygen bond length

19. Assume the energy level diagram for homonuclear diatomic molecules (Figure 10.22) can be applied to heteronuclear diatomics such as CO.
    (a) Write the electron configuration for carbon monoxide, CO.
    (b) What is the highest-energy, occupied molecular orbital? (Chemists call this the HOMO.)
    (c) Is the molecule diamagnetic or paramagnetic?
    (d) What is the net number of σ and π bonds? What is the CO bond order?

20. ■ The nitrosyl ion, NO^+, has an interesting chemistry.
    (a) Is NO^+ diamagnetic or paramagnetic? If paramagnetic, how many unpaired electrons does it have?
    (b) Assume the molecular orbital diagram for a homonuclear diatomic molecule (Figure 10.22) applies to NO^+. What is the highest-energy molecular orbital occupied by electrons?
    (c) What is the nitrogen–oxygen bond order?
    (d) Is the N—O bond in NO^+ stronger or weaker than the bond in NO?

General Questions on Valence Bond and Molecular Orbital Theory

These questions are not designated as to type or location in the chapter. They may combine several concepts.

21. Draw the Lewis structure for AlF_4^-. What are its electron-pair and molecular geometries? What orbitals on Al and F overlap to form bonds between these elements?

22. ■ Draw the Lewis structure for ClF_3. What are its electron-pair and molecular geometries? What is the hybridization of the chlorine atom? What orbitals on Cl and F overlap to form bonds between these elements?

23. Describe the O—S—O angle and the hybrid orbital set used by sulfur in each of the following molecules or ions:
    (a) SO_2 (c) SO_3^{2-}
    (b) SO_3 (d) SO_4^{2-}

    Do all have the same value for the O—S—O angle? Does the S atom in all these species use the same hybrid orbitals?

24. Sketch the Lewis structures of ClF_2^+ and ClF_2^-. What are the electron-pair and molecular geometries of each ion? Do both have the same F—Cl—F angle? What hybrid orbital set is used by Cl in each ion?

25. Sketch the resonance structures for the nitrite ion, NO_2^-. Describe the electron-pair and molecular geometries of the ion. From these geometries, decide on the O—N—O bond angle, the average NO bond order, and the N atom hybridization.

26. Sketch the resonance structures for the nitrate ion, NO_3^-. Is the hybridization of the N atom the same or different in each structure? Describe the orbitals involved in bond formation by the central N atom.

27. ■ Sketch the resonance structures for the N_2O molecule. Is the hybridization of the N atoms the same or different in each structure? Describe the orbitals involved in bond formation by the central N atom.

28. Compare the structure and bonding in CO_2 and CO_3^{2-} with regard to the O—C—O bond angles, the CO bond order, and the C atom hybridization.

29. Numerous molecules are detected in deep space (page 372). Three of them are illustrated here.

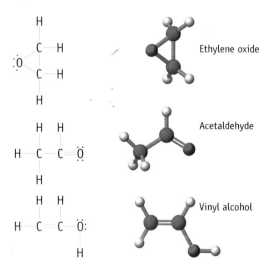

Ethylene oxide

Acetaldehyde

Vinyl alcohol

    (a) Comment on the similarities or differences in the formulas of these compounds. Are they isomers?
    (b) Indicate the hybridization of each C atom in each molecule.

(c) Indicate the value of the H—C—H angle in each of the three molecules.

(d) Are any of these molecules polar?

(e) Which molecule should have the strongest carbon–carbon bond? The strongest carbon–oxygen bond?

30. Acrolein, a component of photochemical smog, has a pungent odor and irritates eyes and mucous membranes.

$$H—C=C—C—H$$

(a) What are the hybridizations of carbon atoms 1 and 2?

(b) What are the approximate values of angles A, B, and C?

(c) Is *cis-trans* isomerism possible here?

31. The organic compound below is a member of a class known as oximes.

$$H—C—C=N:$$

(a) What are the hybridizations of the two C atoms and of the N atom?

(b) What is the approximate C—N—O angle?

32. ■ The compound sketched below is acetylsalicylic acid, commonly known as aspirin:

(a) What are the approximate values of the angles marked A, B, C, and D?

(b) What hybrid orbitals are used by carbon atoms 1, 2, and 3?

33. Phosphoserine is a less common amino acid.

(a) Describe the hybridizations of atoms 1 through 5.

(b) What are the approximate values of the bond angles A, B, C, and D?

(c) What are the most polar bonds in the molecule?

34. ■ Lactic acid is a natural compound found in sour milk.

(a) How many π bonds occur in lactic acid? How many σ bonds?

(b) Describe the hybridization of atoms 1, 2, and 3.

(c) Which CO bond is the shortest in the molecule? Which CO bond is the strongest?

(d) What are the approximate values of the bond angles A, B, and C?

35. ■ Boron trifluoride, BF_3, can accept a pair of electrons from another molecule to form a coordinate covalent bond, as in the following reaction with ammonia:

$$H—N: + B—F \longrightarrow H—N\rightarrow B—F$$

(a) What is the geometry about the boron atom in BF_3? In $H_3N \longrightarrow BF_3$?

(b) What is the hybridization of the boron atom in the two compounds?

(c) Does the boron atom's hybridization change on formation of the coordinate covalent bond?

36. The sulfamate ion, $H_2NSO_3^-$, can be thought of as having been formed from the amide ion, NH_2^-, and sulfur trioxide, SO_3.

(a) Sketch a structure for the sulfamate ion and estimate the bond angles.

(b) What changes in hybridization do you expect for N and S in the course of the reaction $NH_2^- + SO_3 \longrightarrow H_2N—SO_3^-$?

37. Cinnamaldehyde occurs naturally in cinnamon oil.

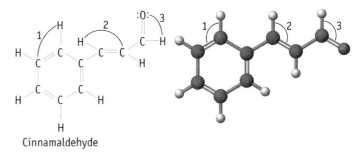

Cinnamaldehyde

(a) What is the most polar bond in the molecule?

▲ More challenging   ■ In General ChemistryNow   Blue-numbered questions answered in Appendix O

(b) How many sigma (σ) bonds and how many pi (π) bonds are there?

(c) Is *cis-trans* isomerism possible? If so, draw the isomers of the molecule.

(d) Give the hybridization of the C atoms in the molecule.

(e) What are the values of the bond angles 1, 2, and 3?

38. ■ Iodine and oxygen form a complex series of ions, among them IO_4^- and IO_5^{3-}. Draw the Lewis structures for these ions, and specify their electron-pair geometries and the shapes of the ions. What is the hybridization of the I atom in these ions?

39. Antimony pentafluoride reacts with HF according to the equation

$$2\ HF + SbF_5 \longrightarrow [H_2F]^+[SbF_6]^-$$

(a) What is the hybridization of the Sb atom in the reactant and product?

(b) Draw a Lewis structure for H_2F^+. What is the geometry of H_2F^+? What is the hybridization of F in H_2F^+?

40. Xenon forms well-characterized compounds. Two xenon–oxygen compounds are XeO_3 and XeO_4. Draw the Lewis structures of these compounds, and give their electron-pair and molecular geometries. What are the hybrid orbital sets used by xenon in these two oxides?

41. The simple valence bond picture of O_2 does not agree with the molecular orbital view. Compare these two theories with regard to the peroxide ion, O_2^{2-}.

(a) Draw an electron dot structure for O_2^{2-}. What is the bond order of the ion?

(b) Write the molecular orbital electron configuration for O_2^{2-}. What is the bond order based on this approach?

(c) Do the two theories of bonding lead to the same magnetic character and bond order for O_2^{2-}?

42. ■ Nitrogen, N_2, can ionize to form N_2^+ or add an electron to give N_2^-. Using molecular orbital theory, compare these species with regard to (a) their magnetic character, (b) net number of π bonds, (c) bond order, (d) bond length, and (e) bond strength.

43. Which of the homonuclear, diatomic molecules of the second-period elements (from Li_2 to Ne_2) are paramagnetic? Which have a bond order of 1? Which have a bond order of 2? Which diatomic molecule has the highest bond order?

44. ■ Which of the following molecules or molecule ions should be paramagnetic? What is the highest occupied molecular orbital (HOMO) in each one? Assume the molecular orbital diagram in Figure 10.22 applies to all of them.

(a) NO   (c) O_2^{2-} (e) CN

(b) OF^- (d) Ne_2^+

45. The CN molecule has been found in interstellar space. Assuming the electronic structure of the molecule can be described using the molecular orbital energy level diagram in Figure 10.22, answer the following questions.

(a) What is the highest energy occupied molecular orbital (HOMO) to which an electron (or electrons) is (are) assigned?

(b) What is the bond order of the molecule?

(c) How many net σ bonds are there? How many net π bonds?

(d) Is the molecule paramagnetic or diamagnetic?

46. Amphetamine is a stimulant. Replacing one H atom on the NH_2, or amino, group with CH_3 gives methamphetamine, a particularly dangerous drug commonly known as "speed."

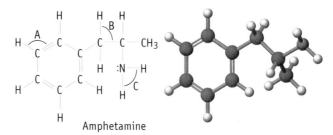

Amphetamine

(a) What are the hybrid orbitals used by the C atoms of the C_6 ring, by the C atoms of the side chain, and by the N atom?

(b) Give approximate values for the bond angles *A*, *B*, and *C*.

(c) How many σ bonds and π bonds are in the molecule?

(d) Is the molecule polar or nonpolar?

(e) Amphetamine reacts readily with a proton (H^+) in aqueous solution. Where does this proton attach to the molecule?

47. Menthol is used in soaps, perfumes, and foods. It is present in the common herb mint, and it can be prepared from turpentine.

(a) What are the hybridizations used by the C atoms in the molecule?

(b) What is the approximate C—O—H bond angle?

(c) Is the molecule polar or nonpolar?

(d) Is the six-member carbon ring planar or nonplanar? Explain why or why not.

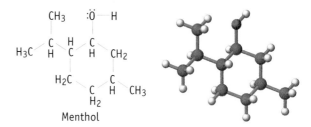

Menthol

48. The elements of the second period from boron to oxygen form compounds of the type $X_nE—EX_n$, where X can be H or a halogen. Sketch possible molecular structures for B_2F_4, C_2H_4, N_2H_4, and O_2H_2. Give the hybridizations of E in each molecule and specify approximate X—E—E bond angles.

---

49. ▲ The compound whose structure is shown here is acetylacetone. It exists in two forms: the *enol* form and the *keto* form.

$$H_3C-C=C-C-CH_3 \rightleftharpoons H_3C-C-C-C-CH_3$$

enol form   *keto* form

The molecule reacts with OH⁻ to form an anion, $[CH_3COCHCOCH_3]^-$ (often abbreviated acac⁻ for acetylacetonate ion). One of the most interesting aspects of this anion is that one or more of them can react with a transition metal cations to give very stable, highly colored compounds.

(a) Are the *keto* and *enol* forms of acetylacetone contributing resonance forms? Explain your answer.

(b) What is the hybridization of each atom (except H) in the *enol* form? What changes in hybridization occur when it is transformed into the *keto* form?

(c) What is the electron-pair geometry and molecular geometry around each C atom in the *keto* and *enol* forms? What changes in geometry occur when the *keto* form changes to the *enol* form?

(d) Draw two possible resonance structures for the acac⁻ ion.

(e) Is *cis-trans* isomerism possible in either the *enol* or the *keto* form?

50. ▲ Ethylene oxide has a three-member ring of two C atoms and an O atom.

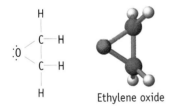

Ethylene oxide

(a) What are the expected bond angles in the ring?

(b) What is the hybridization of each atom in the ring?

(c) Comment on the relation between the bond angles expected based on hybridization and the bond angles expected for a three-member ring.

Summary and Conceptual Questions

The following questions may use concepts from the previous chapters.

51. ■ What is the maximum number of hybrid orbitals that a carbon atom may form? What is the minimum number? Explain briefly.

52. Consider the three fluorides BF_4^-, SiF_4, and SF_4.

(a) Identify a molecule that is isoelectronic with BF_4^-.

(b) Are SiF_4 and SF_4 isoelectronic?

(c) What is the hybridization of the central atom in each of these species?

53. ▲ When two amino acids react with each other, they form a linkage called an amide group, or a peptide link. (If more linkages are added, a protein or polypeptide is formed.)

(a) What are the hybridizations of the C and N atoms in the peptide linkage?

(b) Is the structure illustrated the only resonance structure possible for the peptide linkage? If another resonance structure is possible, compare it with the one shown. Decide which is the more important structure.

(c) The computer-generated structure shown here, which contains a peptide linkage, shows that the linkage is flat. This is an important feature of proteins. Speculate on reasons that the CO — NH linkage is planar.

$-H_2O$

Peptide linkage

54. What is the connection between bond order, bond length, and bond energy? Use ethane (C_2H_6), ethylene (C_2H_4), and acetylene (C_2H_2) as examples.

55. When is it desirable to use MO theory rather than valence bond theory?

56. How do valence bond theory and molecular orbital theory differ in their explanation of the bond order of 1.5 for ozone?

57. Examine the Hybrid Orbitals tool on Screen 10.6 of the General ChemistryNow CD-ROM or website. Use this tool to systematically combine atomic orbitals to form hybrid atomic orbitals.

(a) What is the relationship between the number of hybrid orbitals produced and the number of atomic orbitals used to create them?

(b) Do hybrid atomic orbitals form between different *p* orbitals without involving *s* orbitals?

(c) What is the relationship between the energy of hybrid atomic orbitals and the atomic orbitals from which they are formed?

(d) Compare the shapes of the hybrid orbitals formed from an *s* orbital and a p_x orbital with the hybrid atomic orbitals formed from an *s* orbital and a p_z orbital.

(e) Compare the shape of the hybrid orbitals formed from *s*, p_x, and p_y orbitals with the hybrid atomic orbitals formed from *s*, p_x, and p_z orbitals.

58. Screen 10.2 of the General ChemistryNow CD-ROM or website shows the change in energy as a function of the H—H distance when H_2 forms from separated H atoms.

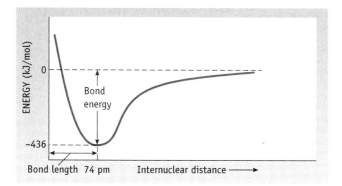

(a) Screen 10.3 describes the attractive and repulsive forces that occur when two atoms approach each other. What must be true about the relative strengths of those attractive and repulsive forces if a covalent bond is to form?

(b) When two atoms are widely separated, the energy of the system is defined as zero. As the atoms approach each other, the energy drops, reaches a minimum, and then increases as they approach still more closely. Explain these observations.

(c) For a bond to form, orbitals on adjacent atoms must overlap, and each pair of overlapping orbitals will contain two electrons. Explain why neon does not form a diatomic molecule, Ne_2, whereas fluorine forms F_2.

59. Examine the bonding in ethylene, C_2H_4, on Screen 10.7 of the General ChemistryNow CD-ROM or website and then go to the "A Closer Look" Auxiliary screen.

(a) Explain why the allene molecule is not flat. That is, explain why the CH_2 groups at opposite ends do not lie in the same plane.

(b) Based on the theory of orbital hybridization, explain why benzene is a planar, symmetrical molecule.

(c) What are the hybrid orbitals used by the three C atoms of allyl alcohol?

60. Screen 10.8 of the General ChemistryNow CD-ROM or website describes the motions of molecules.

(a) Observe the animations of the rotations of *trans*-2-butene and butane about their carbon–carbon bonds.

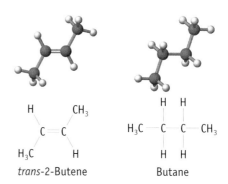

trans-2-Butene   Butane

As one end of *trans*-2-butene rotates relative to the other end, the energy increases about 200 kJ/mol and then drops as the rotation produces *cis*-2-butene. In contrast, the rotation of the butane molecule requires much less energy (only 60 kJ/mol). When butane has reached the halfway point in its rotation, the energy has reached a maximum. Why does *trans*-2-butene require so much more energy to rotate about the central carbon–carbon bond than does butane?

(b) The structure of propene, C_3H_6, is pictured here. Which carbon–hydrogen group (CH_3 or CH_2) can rotate freely with respect to the rest of the molecule?

$$H_3C - \overset{\overset{\displaystyle H}{|}}{C} = CH_2$$

(c) Can the two CH_2 fragments of allene (see Screen 10.7SB) rotate with respect to each other? Briefly explain why or why not.

Do you need a live tutor for homework problems?
Access vMentor at General ChemistryNow at
http://now.brookscole.com/kotz6e
for one-on-one tutoring from a chemistry expert

11—Carbon: More Than Just Another Element

Original, stoppered bottle of mauveine prepared by Perkin. The structure of the mauveine cation is shown here.

A Colorful Beginning

The color purple was once associated with royalty because the dyes were rare and expensive. William Henry Perkin changed everything.

Among the roots of modern organic chemistry one finds the discovery, in 1856, of the compound mauveine (or mauve) by William Henry Perkin (1838–1907). This discovery dates from before the creation of the first periodic table; before the discovery of electrons, protons, and neutrons; before chemists knew anything about bonding; and even before the tetrahedral geometry of carbon was recognized. Perkin's work led to a flourishing dye industry in the latter part of the 19th century, which represented one of the first chemical industries to gain major importance. By 1900, more than 1000 synthetic dyes were known and in use.

Before the discovery of mauve, almost all dyes came from natural sources. Because the dye used for the color purple, Tyrian purple, was the rarest and most expensive, it became the exclusive color of royalty. Tyrian purple was the origin of both fame and fortune for the ancient empire of Tyre, because the dye was obtained only from a small mollusk found in the Mediterranean Sea in that region. More than 9000 mollusks were needed to obtain 1 gram of dye!

William Henry Perkin (1838–1907). See the book on Perkin's life, *Mauve*, S. Garfield: New York, W. W. Norton, 2001.

Chapter Goals

See Chapter Goals Revisited (page 520). Test your
knowledge of these goals by taking the exam-prep quiz
on the General ChemistryNow CD-ROM or website.

- Classify organic compounds based on formula and
  structure.
- Recognize and draw structures of structural isomers and
  stereoisomers for carbon compounds.
- Name and draw structures of common organic
  compounds.
- Know the common reactions of organic functional groups.
- Relate properties to molecular structure.
- Identify common polymers.

The discovery of mauve by Perkin is an interesting tale of
serendipity. At the age of 13, Perkin enrolled at the City of
London School. His father paid an extra fee for him to attend a lunchtime chemistry course and set up a lab at home
for him to do experiments. Hooked on chemistry, Perkin
attended the public lectures that Michael Faraday gave on
Saturdays at the Royal Institution. At 15, he enrolled in
the Royal College of Science in London to study chemistry under the famous chemist August Wilhelm von
Hofmann. Perkin completed his studies at age 17 (the
field of chemistry was a lot smaller then than it is
today) and took a position at the college as Hofmann's assistant, rather a great honor.

Perkin's first chemistry project was to try to
synthesize quinine ($C_{20}H_{24}N_2O_2$), an antimalarial
drug. The route he proposed involved oxidizing
anilinium sulfate [$(C_6H_5NH_3)_2SO_4$]. Instead of
quinine, he obtained a black solid that dissolved in a water–ethanol mixture to give a
purple solution. Using a cloth to mop up a spill
on the lab bench, he noticed that the substance
stained the cloth a beautiful purple color. Furthermore, the color didn't wash out, an essential feature for
a useful dye. Later it was learned that the anilinium
sulfate used in the original reaction was impure and
that the impurity was essential in the synthesis. Had Perkin used a
pure sample as his starting reagent, the discovery of mauve would
not have happened, at least not in this way. A study in 1994 on

Science & Society Picture Library/Science Museum/London

**A silk dress dyed with Perkin's orig-
inal sample of mauve in 1862, at
the dawning of the synthetic dye
industry.**

samples of mauve preserved in museums
determined that Perkin's mauve was a mixture
of primarily two compounds, which have
closely related structures, along with traces
of several others.

At the age of 18, Perkin quit his assis-
tantship to exploit this new discovery. It
was not an easy decision because it in-
curred the great displeasure of his mentor,
Professor Hofmann. With financial help
from his family, Perkin set up a factory
outside of London. Although the road to
success was not smooth, Perkin perse-
vered and by the age of 36 he was a very
wealthy man. He then retired from the
dye business and devoted the rest of his
life to chemical research on various
topics including the synthesis of fragrances. He also studied optical activity, the ability of certain compounds to
rotate polarized light. During his life he received numerous honors for his research. One
honor, however, came many years after his
death. In 1972, when the Chemical Society of
London renamed its research journals after
famous society members, it chose Perkin's name for the journals in
which organic chemists publish their research.

To Review Before You Begin

- Review writing Lewis structures and predicting molecular structures (Section 9.4)
- Recall how to draw structures of molecules (Section 9.7)
- Review covalent bonding: valence bond and molecular orbital theory (Chapter 10)

The vast majority of the 20 million chemical compounds currently known are organic; that is, they are compounds built on a carbon framework. Organic compounds vary greatly in size and complexity, from the simplest hydrocarbon, methane, to molecules made up of many thousands of atoms. As you read this chapter, you will see that the range of possible materials is huge.

11.1—Why Carbon?

We begin this discussion of organic chemistry with a question: What features of carbon lead to both the abundance and the complexity of organic compounds? Answers fall into two categories: structural diversity and stability.

Structural Diversity

With four electrons in its outer shell, carbon will form four bonds to reach an octet configuration. In contrast, the elements boron and nitrogen form three bonds in molecular compounds, oxygen forms two bonds, and hydrogen and the halogens form one bond. With a larger number of bonds comes the opportunity to create more complex structures. This will become increasingly evident in this brief tour of organic chemistry.

A carbon atom can reach an octet of electrons in various ways (Figure 11.1):

- *By forming four single bonds.* A carbon atom can bond to four other atoms, which can be either atoms of other elements (often H, N, or O) or other carbon atoms.

- *By forming a double bond and two single bonds.* The carbon atoms in ethylene, $H_2C=CH_2$, are linked to other atoms in this way.

- *By forming two double bonds,* as in carbon dioxide ($O=C=O$).

- *By forming a triple bond and a single bond,* an arrangement seen in acetylene, $HC\equiv CH$.

Recognize, with each of these arrangements, the various possible geometries around carbon: tetrahedral, trigonal planar, and linear. Carbon's tetrahedral geometry is of special significance because it leads to three-dimensional chains and rings of carbon atoms, as in propane and cyclopentane. The ability to form multiple bonds leads to whole families of compounds based on structures such as ethylene, acetylene, and benzene.

propane, C_3H_8 cyclopentane, C_5H_{10} benzene, C_6H_6

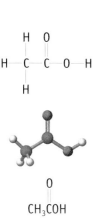

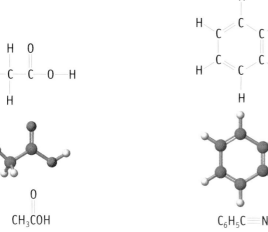

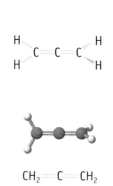

$$CH_3COH$$

$$C_6H_5C \equiv N$$

$$CH_2 = C = CH_2$$

(a) Acetic acid. One carbon atom in this compound is attached to 4 other atoms by single bonds and has tetrahedral geometry. The second carbon atom, connected by a double bond to one oxygen, and by single bonds to the other oxygen and to carbon, has trigonal planar geometry.

(b) Benzonitrile. Six trigonal planar carbon atoms make up the benzene ring. The seventh C atom, bonded by a single bond to carbon and a triple bond to nitrogen, has a linear geometry.

(c) Carbon is linked by double bonds to two other carbon atoms in C_3H_4, a linear molecule commonly called allene.

Figure 11.1 Ways that carbon atoms bond.

Isomers

A hallmark of carbon chemistry is the remarkable array of isomers that can exist. **Isomers** are compounds that have identical composition but different structures. Two broad categories of isomers exist: structural isomers and stereoisomers.

Structural isomers are compounds having the same elemental composition, but in which the atoms are linked together in different ways. Ethanol and dimethyl ether are structural isomers, as are 1-butene and 2-methylpropene.

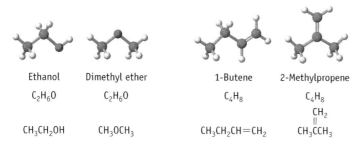

Ethanol	Dimethyl ether	1-Butene	2-Methylpropene
C_2H_6O	C_2H_6O	C_4H_8	C_4H_8
			CH_2
			$\parallel$
CH_3CH_2OH	CH_3OCH_3	$CH_3CH_2CH=CH_2$	CH_3CCH_3

Stereoisomers are compounds with the same formula and in which there is a similar attachment of atoms. However, the atoms have different orientations in space. Two types of stereoisomers exist: geometric isomers and optical isomers.

Cis- and *trans*-2-butene are **geometric isomers.** Geometric isomerism in these compounds occurs as a result of the C=C double bond. Recall that the carbon atom and the attached groups cannot rotate around a double bond (page 455). Thus, the geometry around the C=C double bond is fixed in space. If two groups occur on the adjacent carbon atoms and on the same side of the double bond, a *cis* isomer is produced. If groups appear on opposite sides, a *trans* isomer is produced.

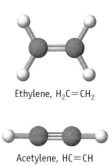

Ethylene, $H_2C=CH_2$

Acetylene, $HC\equiv CH$

Ethylene and acetylene. These two-carbon hydrocarbons can be the building blocks of more complex molecules. These are their common names, but their systematic names are ethene and ethyne.

A Closer Look

Writing Formulas and Drawing Structures

In Chapter 3 you learned that there are various ways of presenting structures (page 101). It is appropriate to return to that point as we look at organic compounds. Consider methane and ethane, for example. We can represent these molecules in several ways:

1. *Molecular formula:* CH_4 or C_2H_6. This type of formula gives information only on composition.

2. *Condensed formula:* For ethane this would be written CH_3CH_3 (or as H_3CCH_3). This method of writing the formula gives some information on the way atoms are connected.

3. *Structural formula:* You will recognize this formula as the Lewis structure. An elaboration on the condensed formula in (2), this representation defines more clearly how each atom is connected, but it fails to describe the shapes of molecules.

```
        H                H   H
        |                |   |
   H — C — H        H — C — C — H
        |                |   |
        H                H   H
  Methane, CH₄      Ethane, C₂H₆
```

4. *Perspective drawings:* These drawings are used to convey the three-dimensional nature of molecules. Bonds extending out of the plane of the paper are drawn with wedges, and bonds behind the plane of the paper are represented as dashed wedges (page 101). Using these guidelines, the structures of methane and ethane could be drawn as follows:

5. *Computer-drawn ball-and-stick and space-filling models.*

Ball-and stick

Space-filling

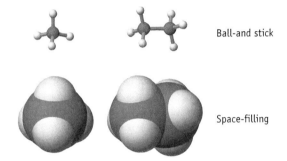

Cis-2-butene, C_4H_8 *Trans*-2-butene, C_4H_8

Optical isomerism is a second type of stereoisomerism. Optical isomers are molecules that have nonsuperimposable mirror images (Figure 11.2). Molecules (and other objects) that have nonsuperimposable mirror images are termed **chiral**. Pairs of non-superimposable molecules are called **enantiomers**.

Pure samples of enantiomers have the same physical properties, such as melting point, boiling point, density, and solubility in common solvents. They differ in one significant way, however: When a beam of plane-polarized light passes through a solution of a pure enantiomer, the plane of polarization rotates. The two enantiomers rotate polarized light to an equal extent, but in opposite directions (Figure 11.3). The term "optical isomerism" is used because this effect involves light (see "A Closer Look: Optical Isomers").

The most common examples of chiral compounds are those in which four different atoms (or groups of atoms) are attached to a tetrahedral carbon atom. Lactic acid, found in milk and a product of normal human metabolism, is an example of one such chiral compound (Figure 11.2). Optical isomerism is particularly important in the amino acids and other biologically important molecules.

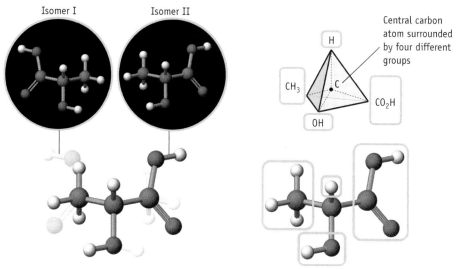

(a) Lactic acid isomers are nonsuperimposable

(b) Lactic acid, $CH_3CH(OH)CO_2H$

Active Figure 11.2 **Optical isomers.** (a) Optical isomerism occurs if a molecule and its mirror image cannot be superimposed. The situation is seen if four different groups are attached to carbon. (b) Lactic acid, a chiral molecule. Four different groups (H, OH, CH_3, and CO_2H) are attached to the central carbon atom.

Lactic acid is produced from milk when milk is fermented to make cheese. It is also found in other sour foods such as sauerkraut and is a preservative in pickled foods such as onions and olives. In our bodies it is produced by muscle activity and normal metabolism.

GENERAL
Chemistry ⚛ Now™ See the General ChemistryNow CD-ROM or website to explore an interactive version of this figure accompanied by an exercise.

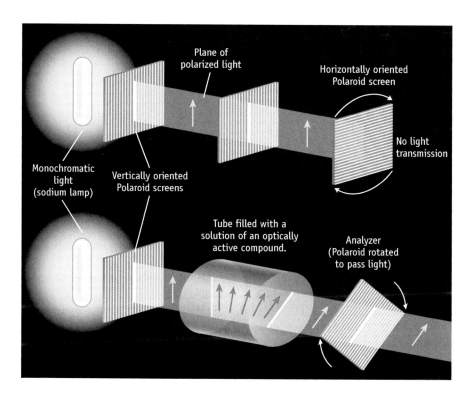

Figure 11.3 **Rotation of plane-polarized light by an optical isomer.** (*Top*) Monochromatic light (light of only one wavelength) is produced by a sodium lamp. After it passes through a polarizing filter, the light vibrates in only one direction—it is polarized. Polarized light will pass through a second polarizing filter if this filter is positioned parallel to the first filter, but not if the second filter is perpendicular.

(*Bottom*) A solution of an optical isomer placed between the first and second polarizing filters causes rotation of the plane of polarized light. The angle of rotation can be determined by rotating the second filter until maximum light transmission occurs. The magnitude and direction of rotation are unique physical properties of the optical isomer being tested.

A Closer Look

Optical Isomers and Chirality

Everyone has accidentally put a left shoe on a right foot, or a left-handed glove on a right hand. It doesn't work very well. Even though our two hands and two feet appear generally similar, a very important distinction separates them. Left hands and feet are mirror images of right hands and feet. Most importantly, these mirror images cannot be superimposed. We describe them as *chiral*.

Many common objects have this property. Some seashells are chiral, for example. Wood screws and machine bolts are also chiral, being distinguished by left-handed or right-handed threads.

Certain molecules have the same characteristic as gloves and hands: A given structure and its mirror image—its enantiomer—cannot be superimposed. There are various ways to visualize that two enantiomers are different. Imagine a tetrahedral carbon atom attached to four other atoms or groups, all different. For simplicity the atoms bonded to the central C atom in the amino acid alanine are shown as 1, 2, 3, and 4 in the drawing. Sight down one of the bonds to carbon (say, the bond from atom 1 to C) in one enantiomer. The other three atoms (2, 3, and 4) will then appear in a clockwise order. In the second enantiomer, atoms 2, 3, and 4 will appear in counterclockwise order.

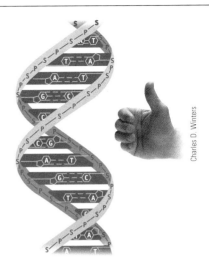

The helical chain of DNA is like the threads of a screw. It twists to the left or it twists to the right. Here it twists to the right. If you curl your right hand around the chain, with your thumb extended, your fingers will show the direction of the twist and your thumb will point along the chain.

The handedness of seashells. Seashells are almost all right-handed. This photo shows the egg cases for whelk shells. Each egg case is about 3 cm in diameter and about 2–3 mm thick. Each egg case is attached to a spine, and the arrangement of egg cases around the spine is right-handed.

Enantiomers of alanine.

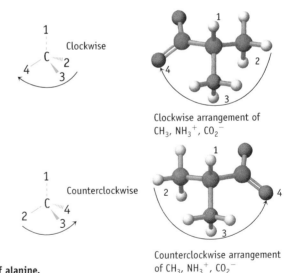

Clockwise arrangement of CH_3, NH_3^+, CO_2^-

Counterclockwise arrangement of CH_3, NH_3^+, CO_2^-

Stability of Carbon Compounds

Carbon compounds are notable for their resistance to chemical change. Were this not so, far fewer compounds of carbon would be known.

Strong bonds are needed for molecules to survive in their environment. Molecular collisions in gases, liquids, and solutions often provide enough energy to break some chemical bonds, and bonds can be broken if the energy associated with photons of visible and ultraviolet light exceeds the bond energy. Carbon–carbon bonds are relatively strong, however, as are the bonds between carbon and most other atoms. The average C—C bond energy is 346 kJ/mol, the C—H bond energy is

413 kJ/mol, and carbon–carbon double and triple bond energies are even higher [◀ Section 9.10]. Contrast these values with bond energies for the Si—H bond (328 kJ/mol) and the Si—Si bond (222 kJ/mol). The consequence of high bond energies for bonds to carbon is that, for the most part, organic compounds do not degrade under normal conditions.

Oxidation of most organic compounds is strongly product-favored, but most organic compounds can survive lengthy contact with O_2. The reason is that these reactions occur slowly. Most organic compounds burn only if their combustion is initiated by heat or by a spark. As a consequence, oxidative degradation is not a barrier to the existence of organic compounds.

11.2—Hydrocarbons

Hydrocarbons, compounds made of carbon and hydrogen only, are classified into several subgroups: alkanes, cycloalkanes, alkenes, alkynes, and aromatic compounds (Table 11.1). We begin our discussion by considering compounds that have carbon atoms with four single bonds, the alkanes and cycloalkanes.

GENERAL
Chemistry ⋅🔧⋅Now™

See the General ChemistryNow CD-ROM or website:
- **Screen 11.3 Hydrocarbons,** for a description of the classes of hydrocarbons

Table 11.1   Some Types of Hydrocarbons

Type of Hydrocarbon	Characteristic Features	General Formula	Example
alkanes	C—C single bonds and all C atoms have four single bonds	C_nH_{2n+2}	CH_4, methane C_2H_6, ethane
cyclic alkanes	C—C single bonds and all C atoms have four single bonds	C_nH_{2n}	C_6H_{12}, cyclohexane
alkenes	C=C double bond	C_nH_{2n}	$H_2C=CH_2$, ethylene
alkynes	C≡C triple bond	C_nH_{2n-2}	HC≡CH, acetylene
aromatics	rings with π bonding extending over several C atoms	—	benzene, C_6H_6

Alkanes

Alkanes have the general formula C_nH_{2n+2}, with n having integer values (Table 11.2). Formulas of specific compounds can be generated from this general formula, the first four of which are CH_4 (methane), C_2H_6 (ethane), C_3H_8 (propane), and C_4H_{10} (butane) (Figure 11.4). Methane has four hydrogen atoms arranged tetrahedrally around a single carbon atom. Replacing a hydrogen atom in methane by a —CH_3 group gives ethane. If an H atom of ethane is replaced by yet another —CH_3 group, propane results. Butane is derived from propane by replacing an H atom of one of the chain-ending carbon atoms with a —CH_3 group. In all of these compounds each C atom is attached to four other atoms, either C or H, so alkanes are often called **saturated compounds**.

Table 11.2   Selected Hydrocarbons of the Alkane Family, C_nH_{2n+2}*

Name	Molecular Formula	State at Room Temperature
methane	CH_4	
ethane	C_2H_6	gas
propane	C_3H_8	
butane	C_4H_{10}	
pentane	C_5H_{12} (pent- = 5)	
hexane	C_6H_{14} (hex- = 6)	
heptane	C_7H_{16} (hept- = 7)	liquid
octane	C_8H_{18} (oct- = 8)	
nonane	C_9H_{20} (non- = 9)	
decane	$C_{10}H_{22}$ (dec- = 10)	
octadecane	$C_{18}H_{38}$ (octadec- = 18)	solid
eicosane	$C_{20}H_{42}$ (eicos- = 20)	

* This table lists only selected alkanes. Liquid compounds with 11 to 16 carbon atoms are also known. Many solid alkanes with more than 20 carbon atoms also exist.

Structural Isomers

The formulas for alkanes do not hint at their structural diversity. Structural isomers are possible for all alkanes larger than propane. For example, there are two structural isomers for C_4H_{10} and three for C_5H_{12}. As the number of carbon atoms in an alkane increases, the number of possible structural isomers greatly increases; there are 5 isomers possible for C_6H_{14}, 9 isomers for C_7H_{16}, 18 for C_8H_{18}, 75 for $C_{10}H_{22}$, and 366,319 for $C_{20}H_{42}$.

To recognize the isomers corresponding to a given formula, keep in mind the following points:

- Each alkane is built upon a framework of tetrahedral carbon atoms, and each carbon must have four single bonds.

- An effective approach is to create a framework of carbon atoms and then fill the remaining positions around carbon with H atoms so that each C atom has four bonds.

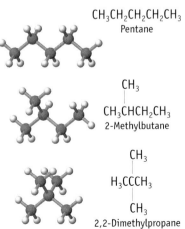

Structural isomers of butane, C_4H_{10}.

$CH_3CH_2CH_2CH_3$

CH_3
CH_3CHCH_3

Butane   2-Methylpropane

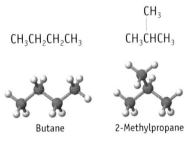

$CH_3CH_2CH_2CH_2CH_3$
Pentane

CH_3
$CH_3CHCH_2CH_3$
2-Methylbutane

CH_3
H_3CCCH_3
CH_3
2,2-Dimethylpropane

Structural isomers of pentane, C_5H_{12}.

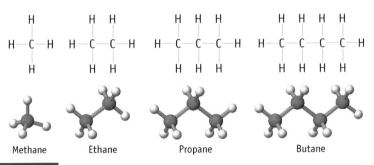

Methane   Ethane   Propane   Butane

Active Figure 11.4   Alkanes. The lowest-molecular-weight alkanes, all gases under normal conditions, are methane, ethane, propane, and butane.

GENERAL
Chemistry⚛Now™ See the General ChemistryNow CD-ROM or website to explore an interactive version of this figure accompanied by an exercise.

• Free rotation occurs around carbon–carbon single bonds. Therefore, when atoms are assembled to form the skeleton of an alkane, the emphasis is on how carbon atoms are attached to one another and not on how they might lie relative to one another in the plane of the paper.

Example 11.1—Drawing Structural Isomers of Alkanes

Problem Draw structures of the five isomers of C_6H_{14}. Are any of these isomers chiral?

Strategy Focus first on the different frameworks that can be built from six carbon atoms. Having created a carbon framework, fill hydrogen atoms into the structure so that each carbon has four bonds.

Solution

Step 1. Placing six carbon atoms in a chain gives the framework for the first isomer. Now fill in hydrogen atoms: three on the carbons on the ends of the chain, two on each of the carbons in the middle. You have created the first isomer, hexane.

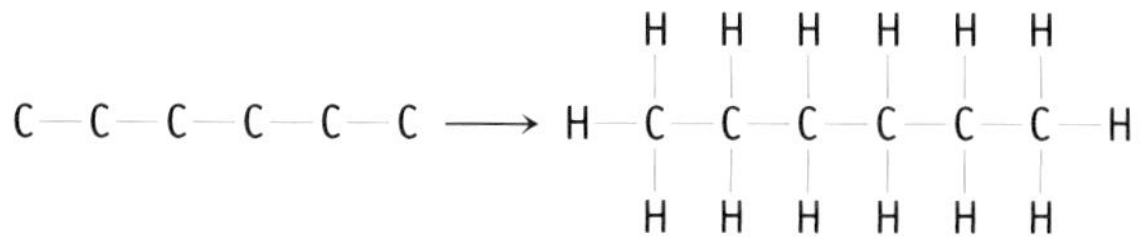

carbon framework of hexane   hexane

Step 2. Draw a chain of five carbon atoms, then add the sixth carbon atom to one of the carbons in the middle of this chain. (Adding it to a carbon at the end of the chain gives a six-carbon chain, the same framework drawn in Step 1.) Two different carbon frameworks can be built from the five-carbon chain, depending on whether the sixth carbon is linked to the 2 or 3 position. For each of these frameworks, fill in the hydrogens.

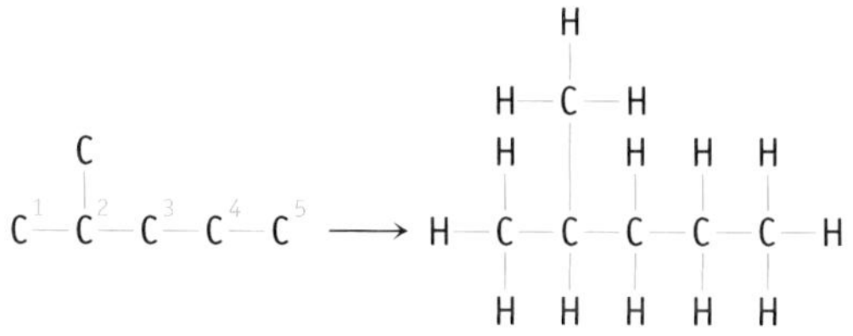

carbon framework
of methylpentane isomers   2-methylpentane

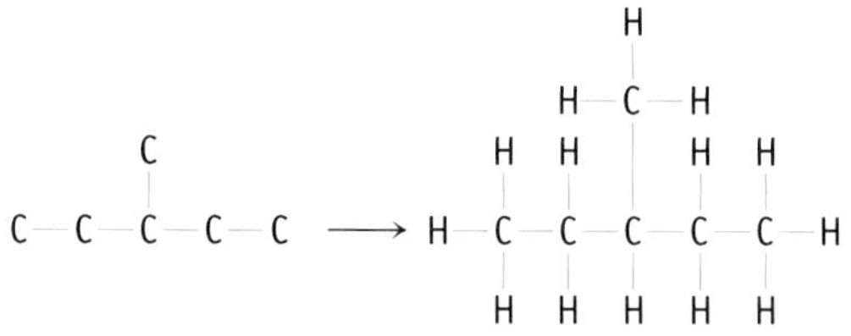

3-methylpentane

Step 3. Draw a chain of four carbon atoms. Add in the two remaining carbons, again being careful not to extend the chain length. Two different structures are possible: one with the remaining carbon atoms each in the 2 and 3 positions, and another with both extra carbon atoms attached at the 2 position. Fill in the 14 hydrogens. You have now drawn the fourth and fifth isomers.

■ **Chirality in Alkanes**
To be chiral, a compound must have at least one C atom attached to four different groups. Thus, the C_7H_{16} isomer here is chiral.

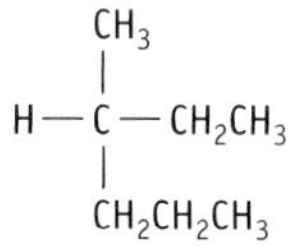

carbon atom frameworks
for dimethylbutane isomers

2,3-dimethylbutane

2,2-dimethylbutane

None of the isomers of C_6H_{14} is chiral. To be chiral, a compound must have at least one C atom with four different groups attached. This condition is not met in any of these isomers.

Comment Should we look for structures in which the longest chain is three carbon atoms? Try it, but you will see that it is not possible to add the three remaining carbons to a three-carbon chain without creating one of the carbon chains already drawn in a previous step. Thus, we have completed the analysis, with five isomers of this compound being identified.

Names have been given to each of these compounds. See the text that follows this Example and see Appendix E for guidelines on nomenclature.

Exercise 11.1—Drawing Structural Isomers of Alkanes

(a) Draw the nine isomers having the formula C_7H_{16}. [*Hint*: There is one structure with a seven-carbon chain, two structures with six-carbon chains, five structures in which the longest chain has five carbons (one is illustrated in the margin), and one structure with a four-carbon chain.]

(b) Identify the isomers of C_7H_{16} that are chiral.

One possible isomer of an alkane with the formula C_7H_{16}.

■ **Naming Guidelines**
For more details on naming organic compounds, see Appendix E.

Naming Alkanes

With so many possible isomers for a given alkane, chemists need a systematic way of naming them. The rules for naming alkanes and their derivatives follow:

- The names of alkanes end in "-ane."
- The names of alkanes with chains of one to ten carbon atoms are given in Table 11.2. After the first four compounds, the names are derived from Latin numbers—pentane, hexane, heptane, octane, nonane, decane—and this regular naming continues for higher alkanes.
- When naming a specific alkane, the root of the name corresponds to the longest carbon chain in the compound. One isomer of C_5H_{12} has a three-carbon chain with two —CH_3 groups on the second C atom of the chain. Thus, its name is based on propane.

Problem Solving Tip 11.1

Drawing Structural Formulas

An error students sometimes make is to suggest that the three carbon skeletons drawn here are different. They are, in fact, the same. All are five-carbon chains with another C atom in the 2 position.

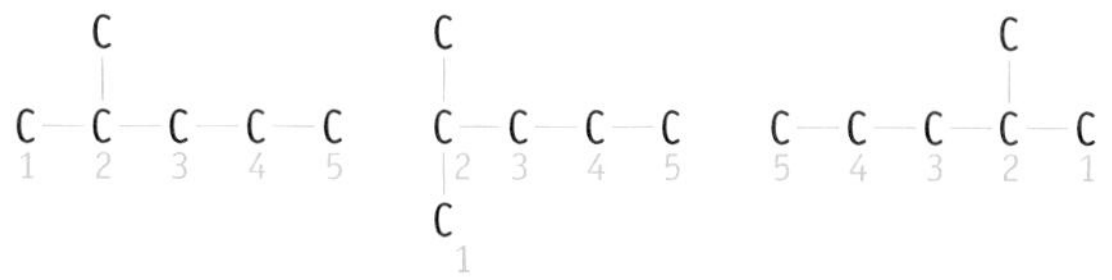

Remember that Lewis structures do not indicate the geometry of molecules.

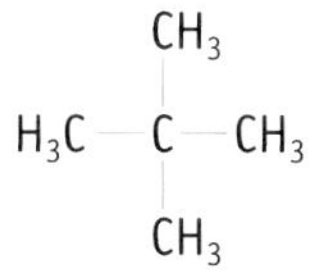

$$
\begin{array}{c}
CH_3 \\
| \\
H_3C - C - CH_3 \\
| \\
CH_3
\end{array}
$$

2,2-dimethylpropane

- Substituent groups on a hydrocarbon chain are identified by a name and the position of substitution in the carbon chain; this information precedes the root of the name. The position is indicated by a number that refers to the carbon atom to which the substituent is attached. (Numbering of the carbon atoms in a chain should begin at the end of the carbon chain that allows the substituent groups to have the lowest numbers.) Both —CH_3 groups in 2,2-dimethylpropane are located at the 2 position.

- Names of hydrocarbon substituents, called **alkyl groups**, are derived from the name of the hydrocarbon. The group —CH_3, derived by taking a hydrogen from methane, is called the methyl group; the C_2H_5 group is the ethyl group.

- If two or more of the same substituent groups occur, the prefixes di-, tri-, and tetra- are added. When different substituent groups are present, they are generally listed in alphabetical order.

■ **Systematic and Common Names**
Many organic compounds are known by common names. For example, 2,2-dimethyl-propane is also called neopentane. However, the IUPAC (International Union of Pure and Applied Chemistry) has formulated rules for systematic names, which are generally used in this book. See Appendix E.

Example 11.2—Naming Alkanes

Problem Give the systematic name for

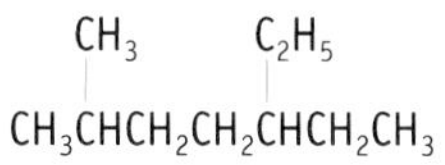

$$
\begin{array}{cc}
CH_3 & C_2H_5 \\
| & | \\
\end{array}
$$
$$CH_3CHCH_2CH_2CHCH_2CH_3$$

Strategy Identify the longest carbon chain and base the name of the compound on that alkane. Identify the substituent groups on the chain and their locations. When there are two or more substituents (the groups attached to the chain), number the parent chain from the end that gives the lower number to the substituent encountered first. If the substituents are different, list them in alphabetical order. (For more on naming compounds, see Appendix E.)

Solution Here the longest chain has seven C atoms, so the root of the name is *heptane*. There is a methyl group on C-2 and an ethyl group on C-5. Giving the substituents in alphabetic order, and numbering the chain from the end having the methyl group, the systematic name is 5-ethyl-2-methylheptane.

Figure 11.5 **Paraffin wax and mineral oil.** These common consumer products are mixtures of alkanes.

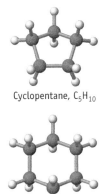

Cyclopentane, C_5H_{10}

Cyclohexane, C_6H_{12}

The structures of cyclopentane, C_5H_{10}, and cyclohexane, C_6H_{12}. The C_5 ring is nearly planar. In contrast, the tetrahedral geometry around carbon means that the C_6 ring is decidedly puckered.

Exercise 11.2—Naming Alkanes

Name the nine isomers of C_7H_{16} in Exercise 11.1.

Properties of Alkanes

Methane, ethane, propane, and butane are gases at room temperature and pressure, whereas the higher-molecular-weight compounds are liquids or solids (Table 11.2). An increase in melting point and boiling point with molecular weight is a general phenomenon that reflects the increased forces of attraction between molecules [▶ Section 13.2].

You already know about alkanes in a nonscientific context because several are common fuels. Natural gas, gasoline, kerosene, fuel oils, and lubricating oils are all mixtures of various alkanes. White mineral oil is also a mixture of alkanes, as is paraffin wax (Figure 11.5).

Pure alkanes are colorless. (The colors seen in gasoline and other petroleum products are due to additives.) The gases and liquids have noticeable but not unpleasant odors. All of these substances are insoluble in water, which is typical of compounds that are nonpolar or nearly so. Low polarity is expected for alkanes because the electronegativity of carbon ($\chi = 2.5$) and hydrogen ($\chi = 2.2$) are not greatly different [◀ Section 9.9].

All alkanes burn readily in air to give CO_2 and H_2O in very exothermic reactions. This is, of course, the reason they are widely used as fuels.

$$CH_4(g) + 2\ O_2(g) \longrightarrow CO_2(g) + 2\ H_2O(\ell) \qquad \Delta H^\circ_{rxn} = -890.3\ kJ$$

Other than in combustion reactions, alkanes exhibit relatively low chemical reactivity. One reaction that does occur, however, is the replacement of the hydrogen atoms of an alkane by chlorine atoms on reaction with Cl_2. It is formally an oxidation because Cl_2, like O_2, is a strong oxidizing agent. These reactions, which can be initiated by ultraviolet radiation, are free radical reactions. Highly reactive Cl atoms are formed from Cl_2 under UV radiation. Reaction of methane with Cl_2 under these conditions proceeds in a series of steps, eventually yielding CCl_4, commonly known as carbon tetrachloride. (HCl is the other product of these reactions.)

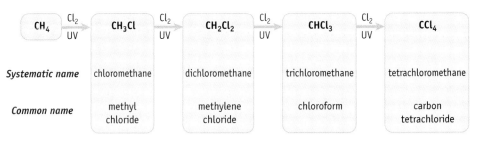

	CH_4	$\xrightarrow[UV]{Cl_2}$	CH_3Cl	$\xrightarrow[UV]{Cl_2}$	CH_2Cl_2	$\xrightarrow[UV]{Cl_2}$	$CHCl_3$	$\xrightarrow[UV]{Cl_2}$	CCl_4
Systematic name			chloromethane		dichloromethane		trichloromethane		tetrachloromethane
Common name			methyl chloride		methylene chloride		chloroform		carbon tetrachloride

The last three compounds are used as solvents, albeit less frequently today because of their toxicity. Carbon tetrachloride was also once widely used as a dry cleaning fluid and, because it does not burn, in fire extinguishers.

Cycloalkanes, C_nH_{2n}

Cycloalkanes are constructed with tetrahedral carbon atoms joined together to form a ring. For example, cyclopentane, C_5H_{10}, consists of a ring of five carbon atoms. Each carbon atom is bonded to two adjacent carbon atoms and to two

A Closer Look

Flexible Molecules

Most organic molecules are flexible; that is, they can twist and bend in various ways. Few molecules better illustrate this behavior than cyclohexane. Two structures are possible, "chair" and "boat" forms.

These forms can interconvert by partial rotation of several bonds.

The more stable structure is the chair form which allows the hydrogen atoms to remain as far apart as possible. A side view of this form of cyclohexane reveals two sets of hydrogen atoms in this molecule. Six hydrogen atoms, called the equatorial

hydrogens, lie in a plane around the carbon ring. The other six hydrogens are positioned above and below the plane and are called axial hydrogens. Flexing the ring (a rotation around the C—C single bonds) moves the hydrogen atoms between axial and equatorial environments.

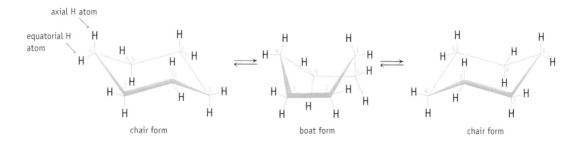

chair form   boat form   chair form

hydrogen atoms. Notice that the five carbon atoms fall very nearly in a plane. This is because the internal angles of a pentagon, 110°, closely match the tetrahedral angle of 109.5°. The small distortion from planarity allows hydrogen atoms on adjacent carbon atoms to be a little farther apart.

Cyclohexane has a nonpolar ring with six —CH$_2$ groups. If the carbon atoms were in the form of a regular hexagon with all carbon atoms in one plane, the C—C—C bond angles would be 120°. To have tetrahedral bond angles of 109.5° around each C atom, the ring has to pucker. The C$_6$ ring is flexible, however, and exists in two interconverting forms (see "A Closer Look: Flexible Molecules").

Interestingly, cyclobutane and cyclopropane are also known, although the bond angles in these species are much less than 109.5°. These compounds are examples of **strained hydrocarbons**, so named because an unfavorable geometry is imposed around carbon. One of the features of strained hydrocarbons is that the C—C bonds are weaker and the molecules readily undergo ring-opening reactions that relieve the bond angle strain.

Alkenes and Alkynes

The abundance and diversity of alkanes are repeated with **alkenes**, hydrocarbons with one or more C==C double bonds. The presence of the double bond adds two features missing in alkanes: the possibility of geometric isomerism and increased reactivity.

The general formula for alkenes is C$_n$H$_{2n}$. The first two members of the series of alkenes are ethene, C$_2$H$_4$ (common name, ethylene), and propene, C$_3$H$_6$ (common name, propylene). Only a single structure can be drawn for these compounds. As with alkanes, the occurrence of isomers begins with species containing four carbon atoms. Four alkene isomers have the formula C$_4$H$_8$, and each has distinct chemical and physical properties (Table 11.3).

Cyclopropane, C$_3$H$_6$ Cyclobutane, C$_4$H$_8$

Cyclopropane and cyclobutane. Cyclopropane was at one time used as a general anesthetic in surgery. However, its explosive nature when mixed with oxygen soon eliminated this application. The *Columbia Encyclopedia* states that "cyclopropane allowed the transport of more oxygen to the tissues than did other common anesthetics and also produced greater skeletal muscle relaxation. It is not irritating to the respiratory tract. Because of the low solubility of cyclopropane in the blood, postoperative recovery was usually rapid but nausea and vomiting were common."

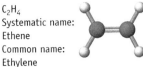

C_2H_4
Systematic name:
Ethene
Common name:
Ethylene

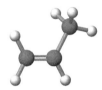

C_3H_6
Systematic name:
Propene
Common name:
Propylene

Table 11.3   Properties of Butene Isomers

Name	Boiling Point	Melting Point	Dipole Moment (D)	ΔH_f° (gas) (kJ/mol)
1-butene	−6.26 °C	−185.4 °C	—	−20.5
2-methylpropene	−6.95 °C	−140.4 °C	0.503	−37.5
cis-2-butene	3.71 °C	−138.9 °C	0.253	−29.7
trans-2-butene	0.88 °C	−105.5 °C	0	−33.0

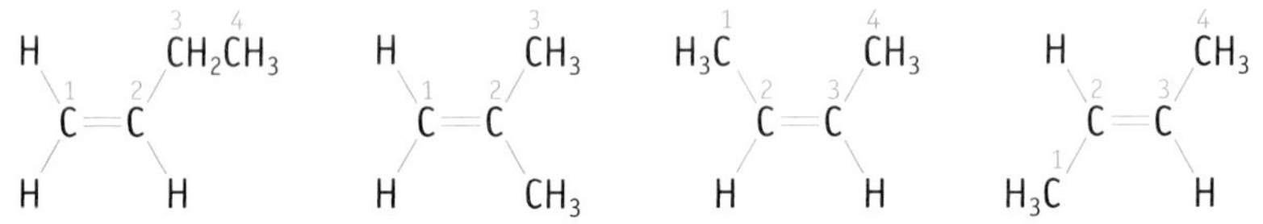

1-butene   2-methylpropene   cis-2-butene   trans-2-butene

Alkene names end in "-ene." As with alkanes, the root name for alkenes is that of the longest carbon chain. The position of the double bond is indicated with a number, and, when appropriate, the prefix *cis* or *trans* is added. Three of the C_4H_8 isomers have four-carbon chains and so are butenes. One has a three-carbon chain and is a propene. Notice that the carbon chain is numbered from the end that gives the double bond the lowest number. In the first isomer at the left, the double bond is between C atoms 1 and 2, so the name is 1-butene and not 3-butene.

Example 11.3—Determining Isomers of Alkenes from a Formula

Problem Draw structures for the six possible alkene isomers with the formula C_5H_{10}. Give the systematic name of each.

Strategy A procedure that involved drawing the carbon skeleton and then adding hydrogen atoms served well when drawing structures of alkanes (Example 11.1), and a similar approach can be used here. It will be necessary to put one double bond into the framework and to be alert for *cis-trans* isomerism.

Solution

1. A five-carbon chain with one double bond can be constructed in two ways. One gives rise to *cis–trans* isomers.

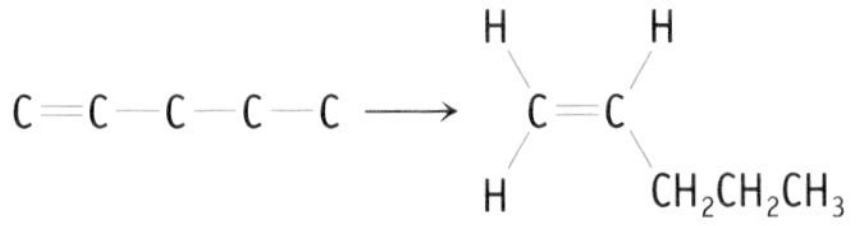

$$C=C-C-C-C \longrightarrow$$

1-pentene

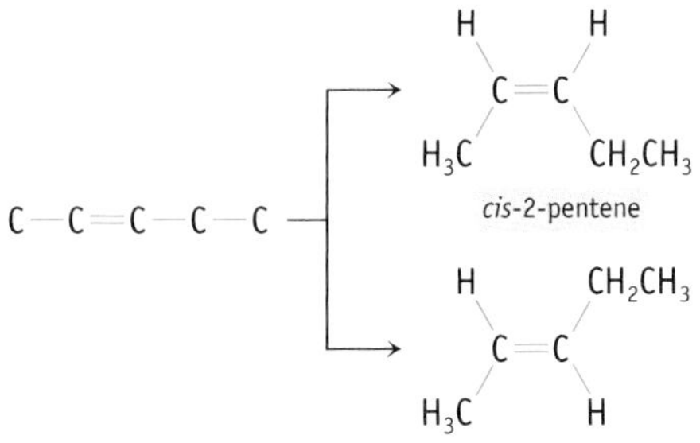

cis-2-pentene

trans-2-pentene

2. Draw the possible four-carbon chains containing a double bond. Add the fifth carbon atom to either the 2 or 3 position. When all three possible combinations are found, fill in the hydrogen atoms. This results in three more structures:

$$C-C-C-C \longrightarrow \begin{array}{c} H \quad\quad CH_3 \\ C=C \\ H \quad\quad CH_2CH_3 \end{array}$$
$$\text{1 2 3 4}$$

2-methyl-1-butene

$$C=C-C-C \longrightarrow \begin{array}{c} H \quad\quad H \\ C=C \\ H \quad\quad CHCH_3 \\ \quad\quad\quad CH_3 \end{array}$$
$$\text{1 2 3 4}$$

3-methyl-1-butene

$$C-C=C-C \longrightarrow \begin{array}{c} H \quad\quad CH_3 \\ C=C \\ H_3C \quad\quad CH_3 \end{array}$$
$$\text{4 3 2 1}$$

2-methyl-2-butene

Exercise 11.3—Determining Structural Isomers of Alkenes from a Formula

There are 17 possible alkene isomers with the formula C_6H_{12}. Draw structures of the five isomers in which the longest chain has six carbon atoms and give the name of each. Which of these isomers is chiral? (There are also eight isomers in which the longest chain has five carbon atoms, and four isomers in which the longest chain has four carbon atoms. How many can you find?)

More than one double bond can be present in a hydrocarbon. Butadiene, for example, has two double bonds and is known as a *diene*. Many natural products have numerous double bonds (Figure 11.6). There are also cyclic hydrocarbons, such as cyclohexene, with double bonds.

$$\begin{array}{c} \quad\quad\quad H_2 \\ \quad\quad\quad C \\ H_2C \quad\quad CH_2 \\ H_2C \quad\quad CH \\ \quad\quad\quad C \\ \quad\quad\quad H \end{array}$$

Cyclohexene, C_6H_{10}

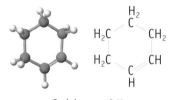

$$H_2C=CHCH=CH_2$$

1,3-Butadiene, C_4H_6

Cycloalkenes and dienes. Cyclohexene, C_6H_{10} (*top*), and 1,3-butadiene (C_4H_6) (*bottom*).

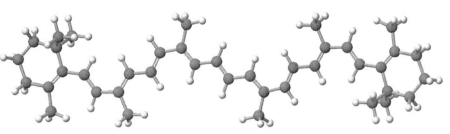

Figure 11.6   Carotene, a naturally occurring compound with 11 C=C bonds. The π electrons can be excited by visible light in the blue-violet region of the spectrum. As a result, carotene appears orange-yellow to the observer. Carotene or carotene-like molecules are partnered with chlorophyll in nature in the role of assisting in the harvesting of sunlight. Green leaves have a high concentration of carotene. In autumn, green chlorophyll molecules are destroyed and the yellows and reds of carotene and related molecules are seen. The red color of tomatoes, for example, comes from a molecule very closely related to carotene. As a tomato ripens, its chlorophyll disintegrates and the green color is replaced by the red of the carotene-like molecule.

An oxy-acetylene torch. The reaction of ethyne (acetylene) with oxygen produces a very high temperature. Oxy-acetylene torches, used in welding, take advantage of this fact.

Alkynes, compounds with a carbon–carbon triple bond, have the general formula (C_nH_{2n-2}). Table 11.4 lists alkynes that have four or fewer carbon atoms. The first member of this family is ethyne (common name, acetylene), a gas used as a fuel in metal cutting torches.

Table 11.4   Some Simple Alkynes C_nH_{2n-2}

Structure	Systematic Name	Common Name	BP (°C)
$HC \equiv CH$	ethyne	acetylene	−85
$CH_3C \equiv CH$	propyne	methylacetylene	−23
$CH_3CH_2C \equiv CH$	1-butyne	ethylacetylene	9
$CH_3C \equiv CCH_3$	2-butyne	dimethylacetylene	27

Properties of Alkenes and Alkynes

Like alkanes, alkenes and alkynes are colorless. Low-molecular-weight compounds are gases, whereas compounds with higher molecular weights are liquids or solids. Alkanes, alkenes, and alkynes are also oxidized by O_2 to give CO_2 and H_2O.

In contrast to alkanes, alkenes and alkynes have an elaborate chemistry. We gain an insight into their chemical behavior by noting that they are called **unsaturated compounds**. Carbon atoms are capable of bonding to a maximum of four other atoms, and they do so in alkanes and cycloalkanes. In alkenes, however, the carbon atoms linked by a double bond are bonded to only three atoms; in alkynes, they bond to two atoms. It is possible to increase the number of bonds to carbon by **addition reactions** in which molecules with the general formula X—Y (such as hydrogen, halogens, hydrogen halides, and water) add across the carbon–carbon double bond (Figure 11.7). The result is a compound with four atoms bonded to each carbon.

$$
\begin{array}{c}
H \qquad\qquad H \qquad\qquad\qquad\qquad\qquad X \quad Y \\
\diagdown \qquad\quad \diagup \qquad\qquad\qquad\qquad\qquad\quad | \quad\; | \\
C = C \;\; + X - Y \longrightarrow H - C - C - H \\
\diagup \qquad\quad \diagdown \qquad\qquad\qquad\qquad\qquad\quad | \quad\; | \\
H \qquad\qquad H \qquad\qquad\qquad\qquad\qquad H \quad H
\end{array}
$$

X — Y = H_2, Cl_2, Br_2; H — Cl, H — Br, H — OH, HO — Cl

The products of addition reactions are substituted alkanes. For example, the addition of bromine to ethylene forms 1,2-dibromoethane.

$$
\begin{array}{c}
H \qquad\qquad H \qquad\qquad\qquad\qquad\qquad Br \quad Br \\
\diagdown \qquad\quad \diagup \qquad\qquad\qquad\qquad\qquad\quad | \quad\;\; | \\
C = C \;\; + Br_2 \longrightarrow H - C - C - H \\
\diagup \qquad\quad \diagdown \qquad\qquad\qquad\qquad\qquad\quad | \quad\;\; | \\
H \qquad\qquad H \qquad\qquad\qquad\qquad\qquad H \quad\; H
\end{array}
$$

1,2-dibromoethane

The addition of 2 mol of chlorine to acetylene gives 1,1,2,2-tetrachloroethane.

$$
\begin{array}{c}
\qquad\qquad\qquad\qquad\qquad\qquad\qquad\qquad Cl \quad Cl \\
\qquad\qquad\qquad\qquad\qquad\qquad\qquad\qquad | \quad\;\; | \\
HC \equiv CH + 2\,Cl_2 \longrightarrow Cl - C - C - Cl \\
\qquad\qquad\qquad\qquad\qquad\qquad\qquad\qquad | \quad\;\; | \\
\qquad\qquad\qquad\qquad\qquad\qquad\qquad\qquad H \quad\; H
\end{array}
$$

1,1,2,2-tetrachloroethane

Charles D. Winters

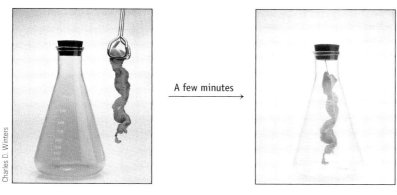

Figure 11.7   Bacon fat and addition reactions. The fat in bacon is partially unsaturated. Like other unsaturated compounds, bacon fat reacts with Br_2 in an addition reaction. Here you see the color of Br_2 vapor fade when a strip of bacon is introduced.

If the reagent added to a double bond is hydrogen ($X—Y = H_2$), the reaction is called **hydrogenation** and the product is an alkane. Hydrogenation is usually a very slow reaction, but it can be speeded up in the presence of a catalyst, often a specially prepared form of a metal, such as platinum, palladium, and rhodium. You may have heard the term hydrogenation because certain foods contain "hydrogenated" or "partially hydrogenated" ingredients. One brand of crackers has a label that says, "Made with 100% pure vegetable shortening . . . (partially hydrogenated soybean oil with hydrogenated cottonseed oil)." One reason for hydrogenating an oil is to make it less susceptible to spoilage; another is to convert it from a liquid to a solid.

■ **Catalysts**
A substance that causes a reaction to occur at a faster rate is called a catalyst. We will describe catalysts in more detail in Chapter 15.

GENERAL
Chemistry Now™

See the General ChemistryNow CD-ROM or website:

- **Screen 11.4 Hydrocarbons and Addition Reactions,** for a simulation and tutorial on alkene addition reactions

Example 11.4—Reaction of an Alkene

Problem Draw the structure of the compound obtained from the reaction of Br_2 with propene and name the compound.

Strategy Bromine will add across the $C{=}C$ double bond. The name will include the name of the carbon chain and indicate the positions of the Br atoms.

Solution

$$
\begin{array}{c}
\underset{\text{propene}}{
\begin{array}{c}
\text{H} \qquad \text{H} \\
\text{C}{=}\text{C} \\
\text{H} \qquad \text{CH}_3
\end{array}}
\; + Br_2 \longrightarrow
\underset{\text{1,2-dibromopropane}}{
\begin{array}{c}
\text{Br} \quad \text{Br} \\
\text{H}-\text{C}-\text{C}-\text{CH}_3 \\
\text{H} \quad \text{H}
\end{array}}
$$

Exercise 11.4—Reactions of Alkenes

(a) Draw the structure of the compound obtained from the reaction of HBr with ethylene and name the compound.

(b) Draw the structure of the product of the reaction of Br_2 with *cis*-2-butene and name this compound.

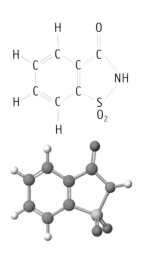

Saccharin ($C_7H_5NO_3S$). This compound, an artificial sweetener, is a benzene derivative.

Aromatic Compounds

Benzene, C_6H_6, is a key molecule in chemistry. It is the simplest **aromatic compound**, one of a class of compounds so named because they have significant, and usually not unpleasant, odors. Other members of this class, which are all based on benzene, include toluene and naphthalene. A source of many aromatic compounds is coal and the volatile substances that are released when coal is heated to a high temperature in the absence of air (Table 11.5).

benzene   toluene   naphthalene

Benzene occupies a pivotal place in the history and practice of chemistry. Michael Faraday discovered this compound in 1825 as a byproduct of illuminating gas, itself produced by heating coal. Today, benzene is an important industrial chemical, usually ranking among the top 25 chemicals in production annually in the United States. It is used as a solvent and is also the starting point for making thousands of different compounds by replacing the H atoms of the ring.

Toluene was originally obtained from Tolu balsam, the pleasant-smelling gum of a South American tree, *Toluifera balsamum*. This balsam has been used in cough syrups and perfumes. Naphthalene is an ingredient in "moth balls," although 1,4-dichlorobenzene is now more commonly used. Aspartame and another artificial sweetener, saccharin, are also benzene derivatives.

Table 11.5   Some Aromatic Compounds from Coal Tar

Common Name	Formula	Boiling Point (°C)	Melting Point (°C)
benzene	C_6H_6	80	+6
toluene	$C_6H_5CH_3$	111	−95
o-xylene	$1,2\text{-}C_6H_4(CH_3)_2$	144	−25
m-xylene	$1,3\text{-}C_6H_4(CH_3)_2$	139	−48
p-xylene	$1,4\text{-}C_6H_4(CH_3)_2$	138	+13
naphthalene	$C_{10}H_8$	218	+80

The Structure of Benzene

The formula of benzene suggested to 19th-century chemists that this compound should be unsaturated, but, if viewed this way, its chemistry was perplexing. Whereas alkenes readily add Br_2, for example, benzene does not do so under similar conditions. The benzene structural question was finally solved by August Kekulé (1829–1896). We now recognize that benzene's different reactivity relates to its structure and bonding, both of which are quite different from the structure and bonding in alkenes. Benzene has equivalent carbon–carbon bonds, 139 pm in length, intermediate between a C—C single bond (154 pm) and a C=C double bond (134 pm). The π bonds are formed by the continuous overlap of the p orbitals on the six carbon atoms (page 455). Using valence bond terminology, the structure is a hybrid of two resonance structures.

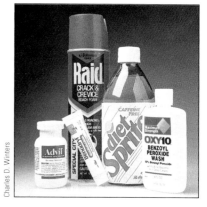

Some products containing compounds based on benzene. Examples include sodium benzoate in soft drinks, ibuprofen in Advil, and benzoyl peroxide in Oxy-10.

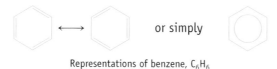

Representations of benzene, C_6H_6

Benzene Derivatives

Toluene, chlorobenzene, styrene, benzoic acid, aniline, and phenol are common examples of benzene derivatives.

chlorobenzene   benzoic acid   aniline   styrene   phenol

If more than one H atom of benzene is replaced, isomers can arise. Thus, the systematic nomenclature for benzene derivatives involves naming substituent groups and identifying their positions on the ring by numbering the six carbon atoms [▶ Appendix E]. Some common names, which are based on an older naming scheme, are also regularly used. This scheme identified isomers of disubstituted benzenes with the prefixes ***ortho*** (*o-*, substituent groups on adjacent carbons in the benzene ring), ***meta*** (*m-*, substituents separated by one carbon atom), and ***para*** (*p-*, substituent groups on carbons on opposite sides of the ring).

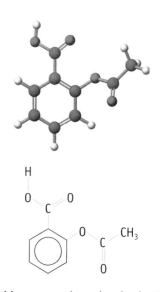

Aspirin, a commonly used analgesic. It is based on benzoic acid with an acetate group, —O_2CCH_3, in the *ortho* position.

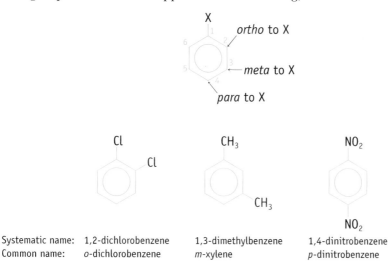

Systematic name: 1,2-dichlorobenzene   1,3-dimethylbenzene   1,4-dinitrobenzene
Common name: *o*-dichlorobenzene   *m*-xylene   *p*-dinitrobenzene

Example 11.5—Isomers of Substituted Benzenes

Problem Draw and name the isomers of $C_6H_3Cl_3$.

Strategy Begin by drawing the structure of C_6H_5Cl. Place a second Cl atom on the ring in the *ortho*, *meta*, and *para* positions. Add the third Cl in one of the remaining positions.

Solution The three isomers of $C_6H_3Cl_3$ are shown here. They are named as derivatives of benzene by specifying the number of substituent groups with the prefix "tri-," the name of the substituent, and the positions of the three groups around the six-member ring.

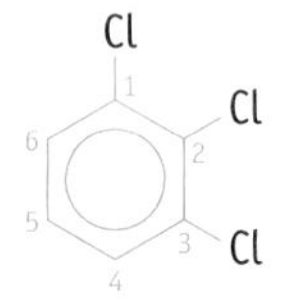

1,2,3-trichlorobenzene

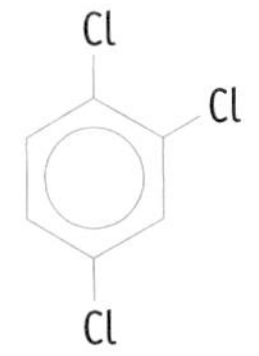

1,2,4-trichlorobenzene

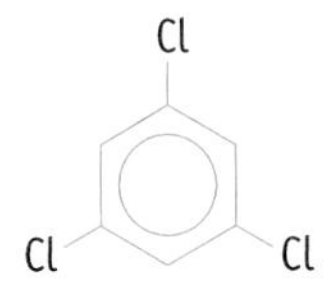

1,3,5-trichlorobenzene

Exercise 11.5—Isomers of Substituted Benzenes

Aniline, $C_6H_5NH_2$, is the common name for aminobenzene. Draw a structure for *p*-diaminobenzene, a compound used in dye manufacture. What is the systematic name for *p*-diaminobenzene?

Properties of Aromatic Compounds

Benzene is a colorless liquid, and simple substituted benzenes are liquids or solids under normal conditions. The properties of aromatic compounds are typical of hydrocarbons in general: They are insoluble in water, soluble in nonpolar solvents, and oxidized by O_2 to form CO_2 and H_2O.

One of the most important properties of benzene and other aromatic compounds is an unusual stability that is associated with the unique π bonding in this molecule. Because the π bonding in benzene is typically described using resonance structures, the extra stability is termed **resonance stabilization**. The extent of resonance stabilization in benzene is evaluated by comparing the energy evolved in the hydrogenation of benzene to form cyclohexane

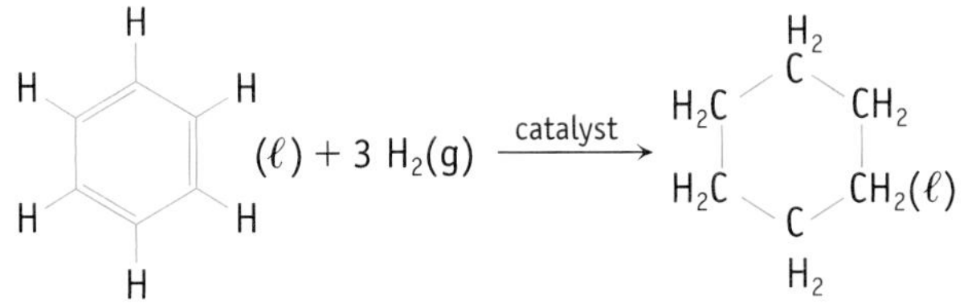

$(\ell) + 3\ H_2(g) \xrightarrow{\text{catalyst}}$

$\Delta H^\circ_{rxn} = -206.7$ kJ

A Closer Look

Petroleum Chemistry

Much of the world's current technology relies on petroleum. Burning fuels derived from petroleum provides by far the largest amount of energy in the industrial world (see "The Chemistry of Fuels and Energy Sources", page 282). Petroleum and natural gas are also the chemical raw materials used in the manufacture of plastics, rubber, pharmaceuticals, and a vast array of other compounds.

The petroleum that is pumped out of the ground is a complex mixture whose composition varies greatly depending on its source. The primary components of petroleum are always alkanes, but, to varying degrees, nitrogen and sulfur-containing compounds are also present. Aromatic compounds are present as well, but alkenes and alkynes are not.

A modern petrochemical plant.

An early step in the petroleum refining process is distillation (Chapter 14), in which the crude mixture is separated into a series of fractions based on boiling point: first a gaseous fraction (mostly alkanes with one to four carbon atoms; this fraction is often burned off), and then gasoline, kerosene, and fuel oils. After distillation, considerable material, in the form of a semisolid, tar-like residue, remains.

The petrochemical industry seeks to maximize the amounts of the higher-valued fractions of petroleum produced and to make specific compounds for which a particular need exists. This means carrying out chemical reactions involving the raw materials on a huge scale. One process to which petroleum is subjected is known as *cracking*. At very high temperatures, bond breaking or "cracking" can occur, and longer-chain hydrocarbons will fragment into smaller molecular units. These reactions are carried out in the presence of a wide array of catalysts, materials that speed up reactions and direct them toward specific products. Among the important products of cracking are ethylene and other alkenes, which serve as the raw materials for the formation of materials such as polyethylene. Cracking also produces gaseous hydrogen, a widely used raw material in the chemical industry.

Other important reactions involving petroleum are run at elevated temperatures and in the presence of specific catalysts.

Such reactions include *isomerization* reactions, in which the carbon skeleton of an alkane rearranges to form a new isomeric species, and *reformation* processes, in which smaller molecules combine to form new molecules. Each process is directed toward achieving a specific goal, such as increasing the proportion of branched-chain hydrocarbons in gasoline to obtain higher octane ratings. A great amount of chemical research has gone into developing and understanding these highly specialized processes.

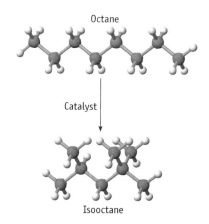

Producing gasoline. Branched hydrocarbons have a higher octane rating in gasoline. Therefore, an important process in producing gasoline is the isomerization of octane to a branched hydrocarbon such as isooctane, 2,2,4-trimethylpentane.

---

with the energy evolved in hydrogenation of three isolated double bonds.

$$3\ H_2C{=}CH_2(g) + 3\ H_2(g) \longrightarrow 3\ C_2H_6(g) \qquad \Delta H^\circ = -410.8\ kJ$$

The hydrogenation of benzene is about 200 kJ less exothermic than the hydrogenation of three moles of ethylene. The difference is attributable to the added stability associated with π bonding in benzene.

Although aromatic compounds are unsaturated hydrocarbons, they do not undergo the addition reactions typical of alkenes and alkynes. Instead, *substitution reactions* occur, in which one or more hydrogen atoms are replaced by other groups. Such reactions require a second reagent, such as H_2SO_4, $AlCl_3$, or $FeBr_3$.

Nitration: $\quad C_6H_6(\ell) + HNO_3(\ell) \xrightarrow{H_2SO_4} C_6H_5NO_2(\ell) + H_2O(\ell)$

Alkylation: $\quad C_6H_6(\ell) + CH_3Cl(\ell) \xrightarrow{AlCl_3} C_6H_5CH_3(\ell) + HCl(g)$

Halogenation: $\quad C_6H_6(\ell) + Br_2(\ell) \xrightarrow{FeBr_3} C_6H_5Br(\ell) + HBr(g)$

Table 11.6 Common Functional Groups and Derivatives of Alkanes

Functional Group*	General Formula*	Class of Compound	Examples
F, Cl, Br, I	RF, RCl, RBr, RI	haloalkane	CH_3CH_2Cl, chloroethane
OH	ROH	alcohol	CH_3CH_2OH, ethanol
OR'	ROR'	ether	$(CH_3CH_2)_2O$, diethyl ether
NH$_2$†	RNH$_2$	(primary) amine	$CH_3CH_2NH_2$, ethylamine
$-\overset{\displaystyle O \atop \parallel}{C}H$	RCHO	aldehyde	CH_3CHO, ethanal (acetaldehyde)
$-\overset{\displaystyle O \atop \parallel}{C}-R'$	RCOR'	ketone	CH_3COCH_3, propanone (acetone)
$-\overset{\displaystyle O \atop \parallel}{C}-OH$	RCO$_2$H	carboxylic acid	CH_3CO_2H, ethanoic acid (acetic acid)
$-\overset{\displaystyle O \atop \parallel}{C}-OR'$	RCO$_2$R'	ester	$CH_3CO_2CH_3$, methyl acetate
$-\overset{\displaystyle O \atop \parallel}{C}-NH_2$	RCONH$_2$	amide	CH_3CONH_2, acetamide

* R and R' can be the same or different hydrocarbon groups.
† Secondary amines (R$_2$NH) and tertiary amines (R$_3$N) are also possible, see discussion in the text.

Alcohol racing fuel. Methanol, CH_3OH, is used as the fuel in cars of the type that race in Indianapolis.

11.3—Alcohols, Ethers, and Amines

Other types of organic compounds arise as elements other than carbon and hydrogen are included in the compound. Two elements in particular, oxygen and nitrogen, add a rich dimension to carbon chemistry.

Organic chemistry organizes compounds containing elements other than carbon and hydrogen as derivatives of hydrocarbons. Formulas (and structures) are represented by substituting one or more hydrogens in a hydrocarbon molecule by a **functional group**. A functional group is an atom or group of atoms attached to a carbon atom in the hydrocarbon. Formulas of hydrocarbon derivatives are then written as R—X, in which R is a hydrocarbon lacking a hydrogen atom, and X is the functional group that has replaced the hydrogen in the structure. The chemical and physical properties of the hydrocarbon derivatives are a blend of the properties associated with hydrocarbons and the group that has been substituted for hydrogen.

Table 11.6 identifies some common functional groups and the families of organic compounds resulting from their attachment to a hydrocarbon.

GENERAL
Chemistry•┇•Now™

See the General ChemistryNow CD-ROM or website:

• **Screen 11.5 Functional Groups,** for a description of the types of organic functional groups and for tutorials on their structures, bonding, and chemistry

Alcohols and Ethers

If one of the hydrogen atoms of an alkane is replaced by a hydroxyl (—OH) group, the result is an **alcohol**, ROH. Methanol, CH_3OH, and ethanol, CH_3CH_2OH, are the most important alcohols, but others are also commercially important (Table 11.7). Notice that several have more than one OH functional group.

Table 11.7 Some Important Alcohols

Condensed Formula	BP (°C)	Systematic Name	Common Name	Use
CH_3OH	65.0	methanol	methyl alcohol	fuel, gasoline additive, making formaldehyde
CH_3CH_2OH	78.5	ethanol	ethyl alcohol	beverages, gasoline additive, solvent
$CH_3CH_2CH_2OH$	97.4	1-propanol	propyl alcohol	industrial solvent
$CH_3CH(OH)CH_3$	82.4	2-propanol	isopropyl alcohol	rubbing alcohol
$HOCH_2CH_2OH$	198	1,2-ethanediol	ethylene glycol	antifreeze
$HOCH_2CH(OH)CH_2OH$	290	1,2,3-propanetriol	glycerol (glycerin)	moisturizer in consumer products

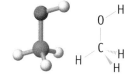

Methanol, CH_3OH, is the simplest alcohol. Methanol is often called "wood alcohol" because it was originally produced by heating wood in the absence of air.

More than 5×10^8 kg of methanol is produced in the United States annually. Most of this production is used to make formaldehyde (CH_2O) and acetic acid (CH_3CO_2H), both important chemicals in their own right. Methanol is also used as a solvent, as a de-icer in gasoline, and as a fuel in high-powered racing cars. It is found in low concentration in new wine, where it contributes to the odor, or "bouquet." Like ethanol, methanol causes intoxication, but methanol differs in being more poisonous, largely because the human body converts it to formic acid (HCO_2H) and formaldehyde (CH_2O). These compounds attack the cells of the retina in the eye, leading to permanent blindness.

Ethanol is the "alcohol" of alcoholic beverages, in which it is formed by the anaerobic (without air) fermentation of sugar. For many years, industrial alcohol, which is used as a solvent and as a starting material for the synthesis of other compounds, was made by fermentation. In the last several decades, however, it has become cheaper to make ethanol from petroleum byproducts—specifically, by the addition of water to ethylene.

■ **Aerobic Fermentation**
Aerobic fermentation (in the presence of O_2) of ethanol leads to the formation of acetic acid. This is how wine vinegar is made.

$$\underset{\text{ethylene}}{\begin{array}{c}H\\|\\C\\|\\H\end{array}=\begin{array}{c}H\\|\\C\\|\\H\end{array}}(g) + H_2O(g) \xrightarrow{\text{catalyst}} \underset{\text{ethanol}}{H-\begin{array}{c}H\\|\\C\\|\\H\end{array}-\begin{array}{c}H\\|\\C\\|\\H\end{array}-OH(\ell)}$$

Beginning with three-carbon alcohols, structural isomers are possible. For example, 1-propanol and 2-propanol (common name, isopropyl alcohol) are different compounds (Table 11.7).

Ethylene glycol and glycerol are common alcohols having two and three —OH groups, respectively. Ethylene glycol is used as antifreeze in automobiles. Glycerol's most common use is as a softener in soaps and lotions. It is also a raw material for the preparation of nitroglycerin (Figure 11.8).

Charles D. Winters

Rubbing alcohol. Common rubbing alcohol is 2-propanol, also called isopropyl alcohol.

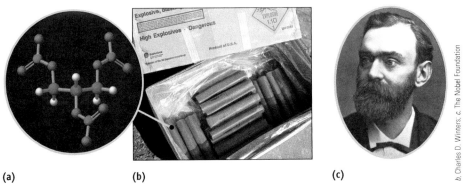

Figure 11.8 Nitroglycerin. (a) Concentrated nitric acid and glycerin react to form an oily, highly unstable compound called nitroglycerin, $C_3H_5(ONO_2)_3$. (b) Nitroglycerin is more stable if absorbed onto an inert solid, a combination called dynamite. (c) The fortune of Alfred Nobel (1833–1896), built on the manufacture of dynamite, now funds the Nobel Prizes.

b, Charles D. Winters; c, The Nobel Foundation

Systematic name:	1,2-ethanediol	1,2,3-propanetriol
Common name:	ethylene glycol	glycerol or glycerin

Example 11.6—Structural Isomers of Alcohols

Problem How many different alcohols are derivatives of pentane? Draw structures and name each alcohol.

Strategy Pentane, C_5H_{12}, has a five-carbon chain. An —OH group can replace a hydrogen atom on one of the carbon atoms. Alcohols are named as derivatives of the alkane (pentane) by replacing the "-e" at the end with "-ol" and indicating the position of the —OH group by a numerical prefix (Appendix E.).

Solution Three different alcohols are possible, depending on whether the —OH group is placed on the first, second, or third carbon atom in the chain. (The fourth and fifth positions are identical to the second and first positions in the chain, respectively.)

1-pentanol

2-pentanol

3-pentanol

Comment Additional structural isomers with the formula $C_5H_{11}OH$ are possible in which the longest carbon chain has three C atoms (one isomer) or four C atoms (four isomers).

Exercise 11.6—Structures of Alcohols

Draw the structure of 1-butanol and alcohols that are structural isomers of the compound.

Properties of Alcohols and Ethers

Methane, CH_4, is a gas (boiling point, -161 °C) with low solubility in water. Methanol, CH_3OH, by contrast, is a liquid that is *miscible* with water in all proportions. The boiling point of methanol, 65 °C, is 226 °C higher than the boiling point of methane. What a difference the addition of a single atom into the structure can make in the properties of simple molecules!

Alcohols are related to water, with one of the H atoms of H_2O being replaced by an organic group. If a methyl group is substituted for one of the hydrogens of water, methanol results. Ethanol has a $—C_2H_5$ (ethyl) group, and propanol has a $—C_3H_7$ (propyl) group in place of one of the hydrogens of water. Viewing alcohols as related to water also helps in understanding the properties of alcohols.

The two parts of methanol, the $—CH_3$ group and the $—OH$ group, contribute to its properties. For example, methanol will burn, a property associated with hydrocarbons. On the other hand, its boiling point is more like that of water. The temperature at which a substance boils is related to the forces of attraction between molecules, called *intermolecular forces:* The stronger the attractive, intermolecular forces in a sample, the higher the boiling point [▶ Section 13.5]. These forces are particularly strong in water, a result of the polarity of the $—OH$ group in this molecule [◀ Section 9.9]. Methanol is also a polar molecule, and it is the polar $—OH$ group that leads to methanol's high boiling point. In contrast, methane is nonpolar and its low boiling point is the result of weak intermolecular forces.

It is also possible to explain the differences in the solubility of methane and methanol in water. The solubility of methanol is conferred by the polar $—OH$ portion of the molecule. Methane, which is nonpolar, has low water solubility.

■ **Hydrogen Bonding**
The intermolecular forces of attraction of compounds with hydrogen attached to a highly electronegative atom, like O, N, or F, are so exceptional that they are accorded a special name: hydrogen bonding. We will discuss hydrogen bonding in Section 13.3.

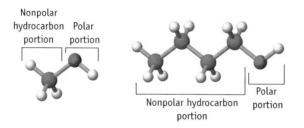

Nonpolar hydrocarbon portion   Polar portion

Nonpolar hydrocarbon portion   Polar portion

As the size of the alkyl group in an alcohol increases, the alcohol's boiling point rises, a general trend seen in families of similar compounds (see Table 11.7). The solubility in water in this series decreases. Methanol and ethanol are completely miscible in water, whereas 1-propanol is moderately water-soluble, and 1-butanol is less soluble than 1-propanol. With an increase in the size of the hydrocarbon group, the organic group (the nonpolar part of the molecule) has become a larger fraction of the molecule, and properties associated with nonpolarity begin to dominate. Space-filling models show that in methanol, the polar and nonpolar parts of the molecule are approximately similar in size, but in 1-butanol the $—OH$ group is less than 20% of the molecule. The molecule is less like water and more "organic."

Attaching more than one $—OH$ group to a hydrocarbon framework has an effect that is opposite to the effect of increased hydrocarbon size. Two $—OH$ groups on a three-carbon framework, as found in propylene glycol, convey complete miscibility with water, in contrast to the limited solubility of 1-propanol and 2-propanol (Figure 11.9).

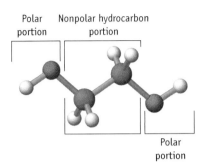

Polar portion   Nonpolar hydrocarbon portion

Polar portion

Photos: Charles D. Winters

Methanol is often added to automobile gasoline tanks in the winter to prevent fuel lines from freezing. It is soluble in water and lowers the water's freezing point.

Ethylene glycol is used in automobile radiators. It is soluble in water, and lowers the freezing point and raises the boiling point of the water in the cooling system. (See Section 14.4.)

Ethylene glycol, a major component of automobile antifreeze, is completely miscible with water.

Figure 11.9   Properties and uses of methanol and ethylene glycol.

Charles D. Winters

Safe antifreeze—propylene glycol, $CH_3CHOHCH_2OH$. Most antifreeze sold today consists of about 95% ethylene glycol. Cats and dogs are attracted by the smell and taste of the compound, but it is toxic. In fact, only a few milliliters can prove fatal to a small dog or cat. In the first stage of poisoning, an animal may appear drunk, but within 12–36 hours the kidneys stop functioning and the animal slips into a coma. To avoid accidental poisoning of domestic and wild animals, you can use propylene glycol antifreeze. This compound affords the same antifreeze protection but is much less toxic.

Ethers have the general formula ROR′. The best known ether is diethyl ether, $CH_3CH_2OCH_2CH_3$. Lacking an —OH group, the properties of ethers are in sharp contrast to those of alcohols. Diethyl ether, for example, has a lower boiling point (34.5 °C) than ethanol, CH_3CH_2OH (78.3 °C), and is only slightly soluble in water.

GENERAL
Chemistry⚛Now™

See the General ChemistryNow CD-ROM or website:
- **Screen 11.6 Functional Groups (1): Reactions of Alcohols**, for an exercise on substitution and elimination reactions of alcohols

Amines

It is often convenient to think about water and ammonia as being similar molecules: They are the simplest hydrogen compounds of adjacent second-period elements. Both are polar, and they exhibit some similar chemistry, such as protonation (to give H_3O^+ and NH_4^+) and deprotonation (to give OH^- and NH_2^-).

This comparison of water and ammonia can be extended to alcohols and amines. Alcohols have formulas related to water in which one hydrogen in H_2O is replaced with an organic group (R—OH). In organic **amines**, one or more hydrogen atoms of NH_3 are replaced with an organic group. Amine structures are similar to ammonia's structure; that is, the geometry about the N atom is trigonal-pyramidal.

Amines are categorized based on the number of organic substituents as primary (one organic group), secondary (two organic groups), or tertiary (three organic groups). As examples, consider the three amines with methyl groups: CH_3NH_2, $(CH_3)_2NH$, and $(CH_3)_3N$.

CH_3NH_2	$(CH_3)_2NH$	$(CH_3)_3N$
Primary amine	Secondary amine	Tertiary amine
Methylamine	Dimethylamine	Trimethylamine

Properties of Amines

Amines usually have offensive odors. You know what the odor is if you have ever smelled decaying fish. Two appropriately named amines, putrescine and cadaverine, add to the odor of urine, rotten meat, and bad breath.

$H_2NCH_2CH_2CH_2CH_2NH_2$
putrescine
1,4-butanediamine

$H_2NCH_2CH_2CH_2CH_2CH_2NH_2$
cadaverine
1,5-pentanediamine

The smallest amines are water-soluble, but most amines are not. All amines are bases, however, and they react with acids to give salts, many of which are water-soluble. As with ammonia, the reactions involve adding H^+ to the lone pair of electrons on the N atom. This is illustrated by the reaction of aniline (aminobenzene) with H_2SO_4 to give anilinium sulfate.

$$2\ C_6H_5NH_2(aq) + H_2SO_4(aq) \longrightarrow 2\ C_6H_5NH_3^+(aq) + SO_4^{2-}(aq)$$

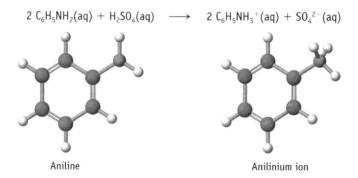

Aniline	Anilinium ion

Recall that Perkin started with this salt in his serendipitous discovery of the dye mauve [◀ page 474].

The facts that an amine can be protonated, and that the proton can be removed again by treating the compound with a base have practical and physiological importance. Nicotine in cigarettes is normally found in the protonated form. (This water-soluble form is often used in insecticides.) Adding a base such as ammonia removes the H^+ ion to leave nicotine in its "free-base" form.

$$NicH_2^{2+}\ (aq) + 2\ NH_3(aq) \longrightarrow Nic(aq) + 2\ NH_4^+(aq)$$

In this form, nicotine is much more readily absorbed by the skin and mucous membranes, so the compound is a much more potent poison.

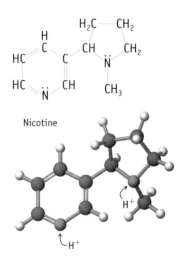

Nicotine

Nicotine Two nitrogen atoms in the nicotine molecule can be protonated, which is the form in which nicotine is normally found. The protons can be removed, however, by treating it with a base. This "free-base" form is much more poisonous and addictive. See J. F. Pankow: *Environmental Science & Technology,* Vol 31, p. 2428, August 1997.

11.4—Compounds with a Carbonyl Group

Formaldehyde, acetic acid, and acetone are among the organic compounds referred to in previous examples. These compounds have a common structural feature: Each contains a trigonal-planar carbon atom doubly bonded to an oxygen. The C=O group is called the **carbonyl group**, and all of these compounds are members of a large class of compounds called **carbonyl compounds**.

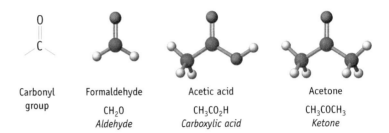

Carbonyl group	Formaldehyde	Acetic acid	Acetone
	CH_2O	CH_3CO_2H	CH_3COCH_3
	Aldehyde	*Carboxylic acid*	*Ketone*

In this section, we will examine five groups of carbonyl compounds (Table 11.6, page 496):

- *Aldehydes* (RCHO) have an organic group (—R) and an H atom attached to a carbonyl group.
- *Ketones* (RCOR′) have two —R groups attached to the carbonyl carbon; they may be the same groups, as in acetone, or different groups.
- *Carboxylic acids* (RCO$_2$H) have an —R group and an —OH group attached to the carbonyl carbon.
- *Esters* (RCO$_2$R′) have —R and —OR′ groups attached to the carbonyl carbon.
- *Amides* (RCONR$_2$′, RCONHR′, and RCONH$_2$) have an —R group and an amino group (—NH$_2$, —NHR, —NR$_2$) bonded to the carbonyl carbon.

Aldehydes, ketones, and carboxylic acids are oxidation products of alcohols and, indeed, are commonly made by this route. The product obtained through oxidation of an alcohol depends on the alcohol's structure, which is classified according to the number of carbon atoms bonded to the C atom bearing the —OH group. *Primary alcohols* have one carbon and two hydrogen atoms attached, whereas *secondary alcohols* have two carbon atoms and one hydrogen atom attached. *Tertiary alcohols* have three carbon atoms attached to the C atom bearing the —OH group.

A primary alcohol is oxidized in two steps. It is first oxidized to an aldehyde and then in a second step to a carboxylic acid:

For example, the air oxidation of ethanol in wine produces wine vinegar, the most important ingredient of which is acetic acid.

Primary alcohol: ethanol

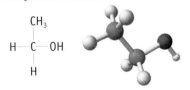

Secondary alcohol: 2-propanol

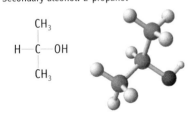

Tertiary alcohol: 2-methyl-2-propanol

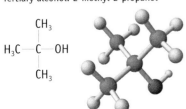

Acids have a sour taste. The word "vinegar" (from the French *vin aigre*) means sour wine. A device to test one's breath for alcohol relies on a similar oxidation of ethanol (Figures 5.16 and 11.10).

In contrast to primary alcohols, oxidation of a secondary alcohol produces a ketone:

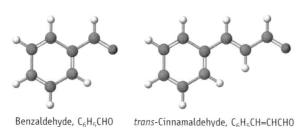

secondary alcohol   ketone
(—R and —R′ are organic groups. They may be the same or different.)

Common oxidizing agents used for these reactions are reagents such as $KMnO_4$ and $K_2Cr_2O_7$ (Table 5.4).

Finally, tertiary alcohols do *not* react with the usual oxidizing agents.

$$(CH_3)_3COH \xrightarrow{\text{oxidizing agent}} \text{no reaction}$$

Aldehydes and Ketones

Aldehydes and **ketones** have pleasant odors and are often used in fragrances. Benzaldehyde is responsible for the odor of almonds and cherries, cinnamaldehyde is found in the bark of the cinnamon tree, and the ketone *p*-hydroxyphenyl-2-butanone is responsible for the odor of ripe raspberries (a favorite of the authors of this book). Table 11.8 lists several simple aldehydes and ketones.

Benzaldehyde, C_6H_5CHO *trans*-Cinnamaldehyde, $C_6H_5CH=CHCHO$

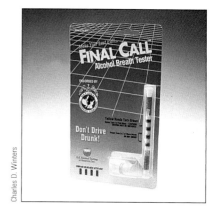

Figure 11.10   Alcohol tester. This device for testing a person's breath for the presence of ethanol relies on the oxidation of the alcohol. If present, ethanol is oxidized by potassium dichromate, $K_2Cr_2O_7$, to acetaldehyde, and then to acetic acid. The yellow-orange dichromate ion is reduced to green $Cr^{3+}(aq)$, the color change indicating that ethanol was present.

Table 11.8   Simple Aldehydes and Ketones

Structure	Common Name	Systematic Name	BP (°C)
$\overset{O}{\overset{\|\|}{HCH}}$	formaldehyde	methanal	−19
$\overset{O}{\overset{\|\|}{CH_3CH}}$	acetaldehyde	ethanal	20
$\overset{O}{\overset{\|\|}{CH_3CCH_3}}$	acetone	propanone	56
$\overset{O}{\overset{\|\|}{CH_3CCH_2CH_3}}$	methyl ethyl ketone	butanone	80
$\overset{O}{\overset{\|\|}{CH_3CH_2CCH_2CH_3}}$	diethyl ketone	3-pentanone	102

Aldehydes and odors. The odors of almonds and cinnamon are due to aldehydes, but the odor of fresh raspberries comes from a ketone.

Aldehydes and ketones are the oxidation products of primary and secondary alcohols, respectively. The reverse reactions—reduction of aldehydes to primary alcohols, and reduction of ketones to secondary alcohols—are also known. Commonly used reagents for such reductions are $NaBH_4$ and $LiBH_4$, although H_2 is used on an industrial scale.

$$R-\overset{\overset{\displaystyle O}{\|}}{C}-H \xrightarrow{NaBH_4 \text{ or } LiAlH_4} R-\overset{\overset{\displaystyle OH}{|}}{\underset{\underset{\displaystyle H}{|}}{C}}-H$$

aldehyde   primary alcohol

$$R-\overset{\overset{\displaystyle O}{\|}}{C}-R \xrightarrow{NaBH_4 \text{ or } LiAlH_4} R-\overset{\overset{\displaystyle OH}{|}}{\underset{\underset{\displaystyle H}{|}}{C}}-R$$

ketone   secondary alcohol

Exercise 11.7—Aldehydes and Ketones

(a) Draw the structural formula for 2-pentanone. Draw structures for a ketone and two aldehydes that are isomers of 2-pentanone, and name each of these compounds.

(b) What is the product of the reduction of 2-pentanone with $LiBH_4$?

Exercise 11.8—Aldehydes and Ketones

Draw the structures and name the aldehyde or ketone formed upon oxidation of the following alcohols: (a) 1-butanol, (b) 2-butanol, (c) 2-methyl-1-propanol. Are these three alcohols structural isomers?

Figure 11.11   Acetic acid in bread. Acetic acid is produced in bread when leavened with the yeast *Saccharomyces exigus*. Another group of bacteria, *Lacto-bacillus sanfrancisco*, contribute to the flavor of sourdough bread. These bacteria metabolize the sugar maltose, excreting acetic acid and lactic acid, $CH_3CH(OH)CO_2H$, thereby giving the bread its unique sour taste.

Carboxylic Acids

Acetic acid is the most common and most important **carboxylic acid**. For many years, acetic acid was made by oxidizing ethanol produced by fermentation. Now, however, acetic acid is generally made by combining carbon monoxide and methanol in the presence of a catalyst:

$$\underset{\text{methanol}}{CH_3OH(\ell)} + CO(g) \xrightarrow{\text{catalyst}} \underset{\text{acetic acid}}{CH_3CO_2H(\ell)}$$

About 1 billion kilograms of acetic acid is produced annually in the United States for use in plastics, synthetic fibers, and fungicides.

Many organic acids are found naturally (Table 11.9). Acids are recognizable by their sour taste (Figure 11.11) and are found in common foods: Citric acid in fruits, acetic acid in vinegar, and tartaric acid in grapes are just three examples.

Some carboxylic acids have common names derived from the source of the acid (Table 11.9). Because formic acid is found in ants, its name comes from the Latin

Table 11.9 Some Naturally Occurring Carboxylic Acids

Name	Structure	Natural Source
benzoic acid	⬡—CO₂H	berries
citric acid	HO₂C—CH₂—C(OH)(CO₂H)—CH₂—CO₂H	citrus fruits
lactic acid	H₃C—CH(OH)—CO₂H	sour milk
malic acid	HO₂C—CH₂—CH(OH)—CO₂H	apples
oleic acid	CH₃(CH₂)₇—CH=CH—(CH₂)₇—CO₂H	vegetable oils
oxalic acid	HO₂C—CO₂H	rhubarb, spinach cabbage, tomatoes
stearic acid	CH₃(CH₂)₁₆—CO₂H	animal fats
tartaric acid	HO₂C—CH(OH)—CH(OH)—CO₂H	grape juice, wine

Formic acid, HCO₂H. This acid puts the sting in ant bites.

word for ant (*formica*). Butyric acid gives rancid butter its unpleasant odor, and the name is related to the Latin word for butter (*butyrum*). The systematic names of acids (Table 11.10) are formed by dropping the "-e" on the name of the corresponding alkane and adding "-oic" (and the word "acid").

Because of the substantial electronegativity of oxygen, we expect the two O atoms of the carboxylic acid group to be slightly negatively charged, and the H atom of the —OH group to be positively charged. This distribution of charges has several important implications:

Table 11.10 Some Simple Carboxylic Acids

Structure	Common Name	Systematic Name	BP (°C)
HCOH (=O)	formic acid	methanoic acid	101
CH₃COH (=O)	acetic acid	ethanoic acid	118
CH₃CH₂COH (=O)	propionic acid	propanoic acid	141
CH₃(CH₂)₂COH (=O)	butyric acid	butanoic acid	163
CH₃(CH₂)₃COH (=O)	valeric acid	pentanoic acid	187

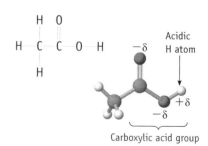

Acetic Acid. The H atom of the carboxylic acid group (—CO₂H) is the acidic proton of this and other carboxylic acids.

A Closer Look

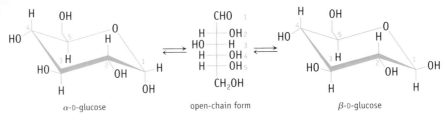

Glucose and Sugars

Having described alcohols and carbonyl compounds, we now pause to look at glucose, the most common, naturally occurring carbohydrate.

As their name implies, formulas of carbohydrates can be written as though they are a combination of carbon and water, $C_x(H_2O)_y$. Thus, the formula of glucose, $C_6H_{12}O_6$, is equivalent to $C_6(H_2O)_6$. This compound is a sugar, or, more accurately, a **monosaccharide.**

Carbohydrates are polyhydroxy aldehydes or ketones. Glucose is an interesting molecule that exists in three different isomeric forms. Two of the isomers contain six-member rings; the third isomer features a chain structure. In solution, the three forms rapidly interconvert.

Notice that glucose is a chiral molecule. In the chain structure, four of the carbon atoms are bonded to four different groups. In nature, glucose occurs in just one of its enantiomeric forms; thus, a solution of glucose rotates polarized light.

Knowing glucose's structure allows one to predict some of its properties. With five polar —OH groups in the molecule, glucose is, not surprisingly, soluble in water.

The aldehyde group is susceptible to chemical oxidation to form a carboxylic acid. Detection of glucose (in urine or blood) takes advantage of this fact; diagnostic tests for glucose involve oxidation with subsequent detection of the products.

Glucose is in a class of sugar molecules called hexoses, molecules having six carbon atoms. 2-Deoxyribose, the sugar in the backbone of the DNA molecule, is a pentose, a molecule with five carbon atoms.

Glucose and other monosaccharides serve as the building blocks for larger carbohydrates. Sucrose, a disaccharide, is formed from a molecule of glucose and a molecule of fructose, another monosaccharide. Starch is a polymer composed of many monosaccharide units.

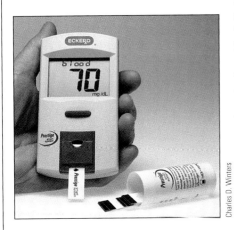

Laboratory test for glucose.

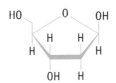

deoxyribose, a pentose in the DNA backbone

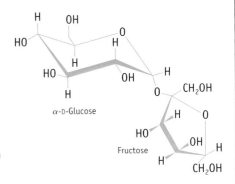

The structure of sucrose. Sucrose is formed from the hexoses α-D-glucose and fructose. An ether linkage is formed by loss of H_2O from two —OH groups.

- The polar acetic acid molecule dissolves readily in water, which you already know because vinegar is an aqueous solution of acetic acid. (Acids with larger organic groups are less soluble, however.)

- The hydrogen of the —OH group is the acidic hydrogen. As noted in Chapter 5, acetic acid is a weak acid in water, as are all other organic acids.

Carboxylic acids undergo a number of reactions. Among these is the reduction of the acid (with reagents such as $LiAlH_4$ or $NaBH_4$) first to an aldehyde and then to an alcohol. For example, acetic acid is reduced first to acetaldehyde and then to ethanol.

$$CH_3CO_2H \xrightarrow{\text{LiAlH}_4} CH_3CHO \xrightarrow{\text{LiAlH}_4} CH_3CH_2OH$$

     acetic acid   acetaldehyde   ethanol

Chemical Perspectives

Aspirin Is More Than 100 Years Old!

Aspirin is one of the most successful non-prescription drugs ever made. Americans swallow more than 50 million aspirin tablets a day, mostly for the pain-relieving (analgesic) effects of the drug. Aspirin also wards off heart disease and thrombosis (blood clots), and it has even been suggested as a possible treatment for certain cancers and for senile dementia.

Hippocrates (460–370 BC), the ancient Greek physician, recommended an infusion of willow bark to ease the pain of childbirth. It was not until the 19th century that an Italian chemist, Raffaele Piria, isolated salicylic acid, the active compound in the bark. Soon thereafter, it was found that the acid could be extracted

from a wildflower, *Spiraea ulmaria*. It is from the name of this plant that the name "aspirin" (a + spiraea) is derived.

Hippocrates's willow bark extract, salicylic acid, is an analgesic, but it is also very irritating to the stomach lining. It was therefore an important advance when Felix Hoffmann and Henrich Dreser of Bayer Chemicals in Germany found, in 1897, that a derivative of salicylic acid, acetylsalicylic acid, was also a useful drug and had fewer side effects. This derivative is the compound we now call "aspirin."

Acetylsalicylic acid slowly reverts to salicylic acid and acetic acid in the presence of moisture. Indeed, if you smell the characteristic odor of acetic acid in a bottle of aspirin tablets, they are too old and should be discarded.

Aspirin is a component of various over-the-counter medicines, such as Anacin,

Ecotrin, Excedrin, and Alka-Seltzer. The latter is a combination of aspirin with citric acid and sodium bicarbonate. Sodium bicarbonate is a base, and it reacts with the acid to produce the sodium salt of acetylsalicylic acid, a form of aspirin that is water-soluble and quicker-acting.

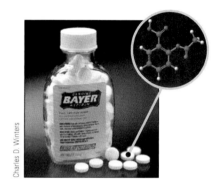

Charles D. Winters

Acetylsalicylic acid, aspirin.

Yet another important aspect of carboxylic acid chemistry is these acids' reaction with bases to give carboxylate anions. For example, acetic acid reacts with sodium hydroxide to give sodium acetate (sodium ethanoate).

$$CH_3CO_2H(aq) + OH^-(aq) \longrightarrow CH_3CO_2^-(aq) + H_2O(\ell)$$

Esters

Carboxylic acids (RCO_2H) react with alcohols ($R'OH$) to form esters (RCO_2R') in an **esterification** reaction. (These reactions are generally run in the presence of strong acids because acids accelerate the reaction.)

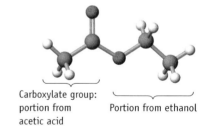

Carboxylate group: portion from acetic acid · Portion from ethanol

Ethyl acetate, an ester
$CH_3CO_2CH_2CH_3$

Table 11.11 lists a few common esters and the acid and alcohol from which they are formed. The two-part name of an ester is given by (1) the name of the hydrocarbon group from the alcohol and (2) the name of the carboxylate group derived from the acid name by replacing "-ic" with "-ate." For example, ethanol (commonly called ethyl alcohol) and acetic acid combine to give the ester ethyl acetate.

94. One of the resonance structures for pyridine is illustrated here. Draw another resonance structure for the molecule. Comment on the similarity between this compound and benzene.

pyridine

95. Write balanced equations for the combustion of ethane and ethanol.

(a) Calculate the heat of combustion for each compound. Which has the more negative enthalpy change for combustion per gram?

(b) If ethanol is assumed to be partially oxidized ethane, what effect does this have on the heat of combustion?

96. Describe a simple chemical test to tell the difference between $CH_3CH_2CH_2CH=CH_2$ and its isomer cyclopentane.

97. Describe a simple chemical test to tell the difference between 2-propanol and its isomer methyl ethyl ether.

98. Plastics make up about 20% of the volume of landfills. There is, therefore, considerable interest in reusing or recycling these materials. To identify common plastics, a set of universal symbols is now used, five of which are illustrated here. They symbolize low- and high-density polyethylene, poly(vinyl chloride), polypropylene, and polyethylene terephthalate.

    3

PETE   HDPE   V

LDPE   PP

(a) Tell which symbol belongs to which type of plastic.

(b) Find an item in the grocery or drug store made from each of these plastics.

(c) Properties of several plastics are listed in the table. Based on this information, describe how to separate samples of these plastics from one another.

Plastic	Density (g/cm³)	Melting Point (°C)
Polypropylene	0.92	170
High-density polyethylene	0.97	135
Polyethylene terephthalate	1.34–1.39	245

99. ▲ Maleic acid is prepared by the catalytic oxidation of benzene. It is a dicarboxylic acid; that is, it has two carboxylic acid groups.

(a) Combustion of 0.125 g of the acid gives 0.190 g of CO_2 and 0.0388 g of H_2O. What is the empirical formula of the acid?

(b) A 0.261-g sample of the acid requires 34.60 mL of 0.130 M NaOH for complete titration (so that the H^+ ions from both carboxylic acid groups are used). What is the molecular formula of the acid?

(c) Draw a Lewis structure for the acid.

(d) Describe the hybridization used by the C atoms.

(e) What are the bond angles around each C atom?

100. Benzene, C_6H_6, is a planar molecule. As General ChemistryNow CD-ROM or website Screen 11.2 shows, another six-carbon cyclic molecule, cyclohexane (C_6H_{12}), is not planar.

(a) Contrast the carbon atom hybridization in these two molecules.

(b) Why is π electron delocalization possible in benzene?

(c) Why is cyclohexane not planar?

101. ▲ Addition reactions of hydrocarbons are described on the General ChemistryNow CD-ROM or website Screen 11.4. In the Simulation you learn that the product of an addition reaction of an alkene is controlled by Markovnikov's rule.

(a) Draw the structure of the product obtained by adding HBr to propene, and give its name.

(b) Draw the structure and give the name of the compound that results from adding H_2O to 2-methyl-1-butene.

(c) If you add H_2O to 2-methyl-2-butene, is the product the same or different than the product from the reaction in part (b)?

102. Refer to the General ChemistryNow CD-ROM or website Screen 11.5, and then describe the hybrid orbitals used by the indicated atoms:

(a) the O atom in an alcohol

(b) the $C=O$ carbon in an aldehyde

(c) the $C=O$ carbon in a carboxylic acid

(d) the $C-O-C$ oxygen atom in an ester

(e) the N atom in an amine

103. Addition and substitution reactions are described in Chapter 11. Another type of reaction of organic compounds, elimination, is described on the General ChemistryNow CD-ROM or website Screen 11.6.

(a) What is the difference between a substitution reaction and an elimination reaction?

(b) Compare the elimination reaction shown on this screen with the hydrogenation reaction shown on Screen 11.4. In what ways are they similar or dissimilar?

104. Properties of fats and oils are described on the General ChemistryNow CD-ROM or website Screen 11.7, and on page 510.

    (a) What type of reaction is used to make a fat or oil from glycerol and a fatty acid: addition, substitution, or elimination?

    (b) What is the primary structural difference between fats and oils? What types of functional groups do each contain?

    (c) What structural feature of oil molecules prevents them from coiling up on themselves as fat molecules do?

105. Addition polymerization is described on Screen 11.9 of the General ChemistryNow CD-ROM and website.

    (a) What is the primary structural feature of the molecules used to form addition polymers?

    (b) Consider the animation of a polymerization reaction shown on this screen. The polymer made here has a chain of 14 carbon atoms. Could the chain have been shorter or longer? Explain briefly.

    (c) What controls the length of the polymer chains formed?

    (d) Can the addition polymerization reaction be classified as one of the reaction types studied earlier: addition, substitution, or elimination?

106. Condensation polymerization is described on Screen 11.10 of the General ChemistryNow CD-ROM or website.

    (a) What is the primary structural feature necessary for a molecule to be useful in a condensation polymerization reaction?

    (b) Describe the appearance of the nylon being made in this video.

    (c) What does the designation "6,6" mean in nylon-6,6?

Do you need a live tutor for homework problems?
Access vMentor at General ChemistryNow at
http://now.brookscole.com/kotz6e
for one-on-one tutoring from a chemistry expert

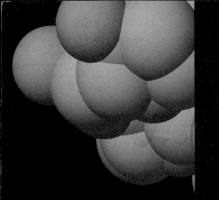

The Chemistry of Life: Biochemistry

John Townsend

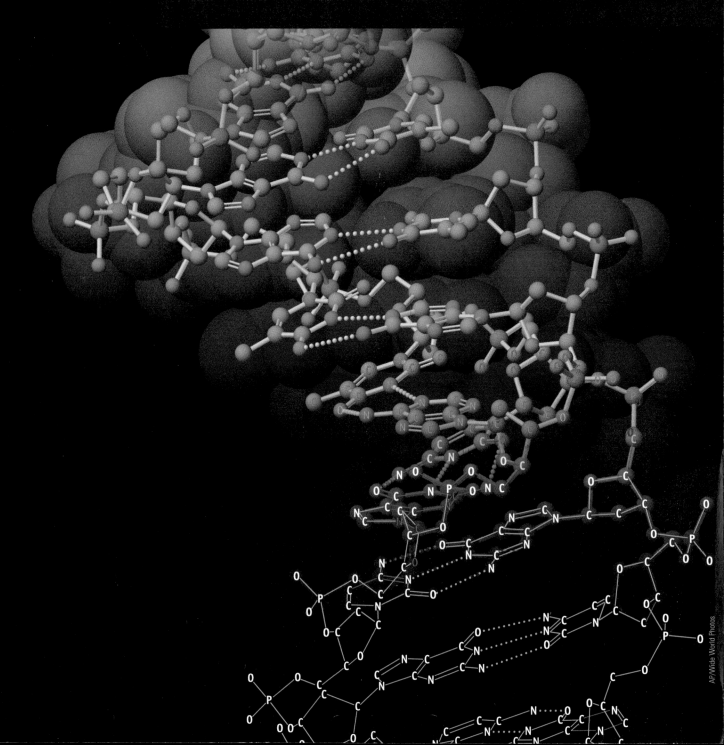

You are a marvelously complicated biological organism. So is every other organism on earth. What molecules are present in you, and what are their properties? How is genetic information passed from generation to generation? How does your body carry out the numerous reactions that are needed for life?

These questions and many others fall into the realm of biochemistry, one of the most rapidly expanding areas of science. As the name implies, biochemistry exists at the interface of two scientific disciplines: biology and chemistry.

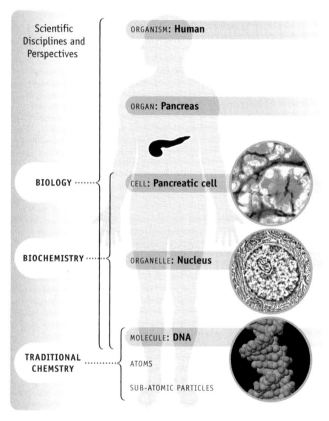

The human body with areas of interest to biologists, biochemists, and chemists.

What separates a biochemist's perspective of biological phenomena from a biologist's perspective? The difference is becoming less distinct, but biochemists tend to concentrate more on the specific molecules involved in biological processes and on how chemical reactions occur in an organism. In other words, biochemists use the strategies of chemists to understand processes in living things.

The goal in this supplementary chapter is to consider how chemistry is involved in answering important biological ques-

◀ **Different representations of the framework of double helical DNA.** *(bottom to top)* Structural formula, ball-and-stick model, and space-filling model. (Note that the hydrogen atoms are omitted.)

tions. To do so, we will begin by examining two major classes of biological compounds: proteins and nucleic acids. We will also discuss some chemical reactions that occur in living things, including some reactions involved in obtaining energy from food.

Proteins

Your body contains thousands of different proteins, and about 50% of the dry weight of your body consists of proteins. Proteins provide structural support (muscle, collagen), help organisms move (muscle), store and transport chemicals from one area to another (hemoglobin), regulate when certain chemical reactions will occur (hormones), and catalyze a host of chemical reactions (enzymes). All of these different functions and others are accomplished using this one type of molecule.

Amino Acids Are the Building Blocks of Proteins

Proteins are condensation polymers (Section 11.5) formed from amino acids. **Amino acids** are organic compounds that contain two functional groups: an amino group ($-NH_2$) and a carboxylic acid group ($-CO_2H$) (Figure 1). Each of these functional groups can exist in two different states: an ionized form

(a) Generic alpha-amino acid.

(b) Zwitterionic form of an alpha-amino acid.

(c) Alanine

Figure 1 α-Amino acids. (a) α-Amino acids have a C atom to which is attached an amino group ($-NH_2$), a carboxylic acid group ($-CO_2H$), an organic group (R), and an H atom. (b) The zwitterionic form of an α-amino acid. (c) Alanine, one of the naturally occurring amino acids.

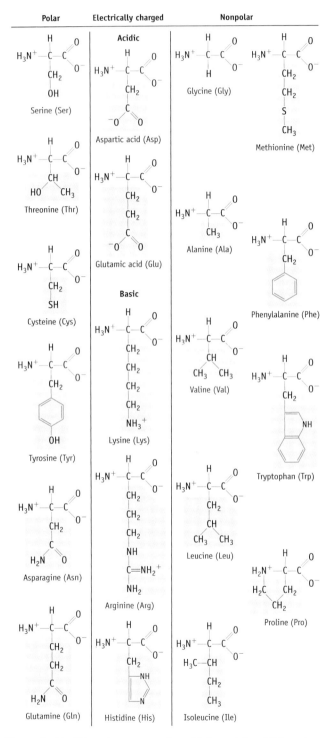

Figure 2 **The 20 most common amino acids in proteins.** All (except proline and glycine) share the characteristic that there is an NH_3^+ group, a CO_2^- group, an H atom, and an organic group attached to a chiral C atom, called the alpha (α) carbon. The organic groups may be polar, nonpolar, or electrically charged. (Histidine is shown in the electrically charged column because the unprotonated N in the organic group can easily be protonated.)

($-NH_3^+$ and $-CO_2^-$) and an unionized form ($-NH_2$ and $-CO_2H$). If both groups are in their ionized forms, the resulting species contains both a positive and a negative charge and is called a **zwitterion**. In an aqueous (polar) environment at physiological pH (about 7.4), amino acids are predominantly in the zwitterionic form.

Almost all amino acids that make up proteins are α-amino acids. In an α-amino acid, the amino group is at one end of the molecule, and the acid group is at the other end. In between these two groups, a single carbon atom (the α-carbon) has attached to it a hydrogen atom and either another hydrogen atom or an organic group, denoted R (Figure 1). Naturally occurring proteins are predominantly built using 20 amino acids, which differ only in terms of the identity of the organic group, R. These organic groups can be nonpolar (groups derived from alkanes or aromatic hydrocarbons) or polar (with alcohol, acidic, basic, or other polar functional groups) (Figure 2). Depending on which amino acids are present, a region in a protein may be nonpolar, very polar, or anything in between.

All α-amino acids, except glycine, have four different groups attached to the α-carbon. The α-carbon is thus a *chiral* center (page 479), and two enantiomers exist. Interestingly, all of these amino acids occur in nature in a single enantiomeric form.

Condensation reactions between two amino acids result in the elimination of water and the formation of an amide linkage (Figure 3). The amide linkage in proteins is often referred to as a **peptide bond**, and the polymer (the protein) that results from a series of these reactions is called a **polypeptide**. The amide linkage is planar (page 509), and both the carbon and the nitrogen atoms are sp^2 hybridized. There is partial double bond

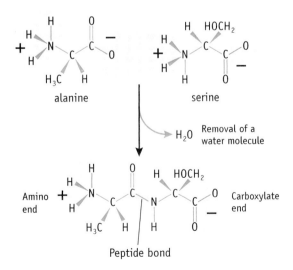

Figure 3 **Formation of a peptide.** Two α-amino acids condense to form an amide linkage, which is often called a peptide bond. Proteins are polypeptides, polymers consisting of many amino acid units linked through peptide bonds.

character in the C—O and C—N bonds, leading to restricted rotation about the carbon-nitrogen bond. As a consequence, each peptide bond in a protein possesses a rigid, planar section, which plays a role in determining its structure.

Naturally occurring proteins typically have molar masses of 5000 g/mol or greater and consist of one or more polypeptide chains. For example, insulin is a small protein produced in the pancreas that is involved in controlling the amount of sugar in the blood; bovine (cow) insulin has a molar mass of 5733 g/mol. It consists of two peptide chains, one having 21 amino acids linked together and the other having 30 linked amino acids.

Protein Structure and Hemoglobin

With this basic understanding of amino acids and peptide bonds, let us examine some larger issues related to protein structure. One of the central tenets of biochemistry is that "structure determines function." In other words, what a molecule can do is determined by which atoms or groups of atoms are present and how they are arranged in space. It is not surprising, therefore, that much effort has been devoted to determining the structures of proteins.

To simplify their discussions, biochemists describe proteins as having different structural levels. Each level of structure can be illustrated using hemoglobin.

Hemoglobin is the molecule in red blood cells that carries oxygen from the lungs to all of the body's other cells. It is a large iron-containing protein, made up of more than 10,000 atoms and having a molar mass of 64,500 g/mol. Hemoglobin includes four polypeptide segments: two identical segments called the α-subunits containing 141 amino acids each and two other segments called the β-subunits containing 146 amino acids each.

The β-subunits are identical to each other but different from the α-subunits. Each subunit contains an iron(II) ion locked inside an organic ion called a **heme** unit. (Figure 4). The oxygen molecules transported by hemoglobin bind to these iron(II) ions. (For more information about the heme group, see "A Closer Look" on page 1084.)

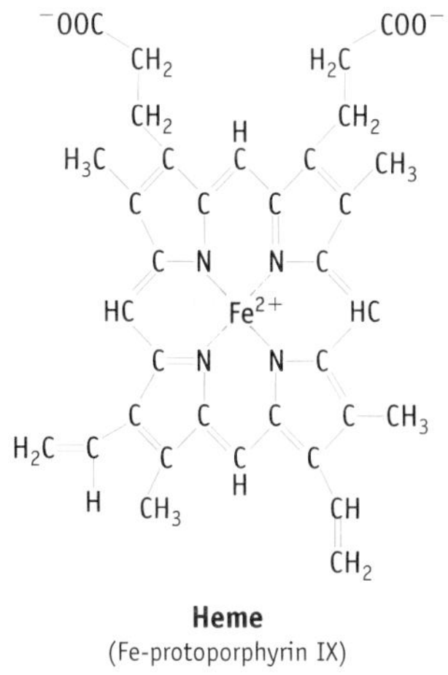

Let us focus on the polypeptide part of hemoglobin (Figure 5). The first step in describing a structure is to identify how the atoms are linked together. This is the **primary structure** of a protein, which is simply the sequence of amino acids linked together by peptide bonds. For example, a glycine unit can be fol-

Heme
(Fe-protoporphyrin IX)

Figure 4 Heme. The heme unit in hemoglobin (and in myoglobin, a related protein) consists of an iron ion in the center of a porphyrin ring system.

lowed by an alanine, followed by a valine, and so on. A particular sequence will lead to a particular structure.

The remaining levels of structure all deal with noncovalent interactions between amino acids in the protein. The **secondary structure** of a protein refers to how amino acids near one another in the sequence arrange themselves. Some regular patterns often emerge, such as helices, sheets, and turns. In hemoglobin, it was discovered that the amino acids in large portions of the polypeptide chains arrange themselves into many *helical* regions, a commonly observed polypeptide secondary structure.

The **tertiary structure** of a protein refers to how the chain is folded, including how amino acids that are far apart in the sequence interact with each other. In other words, this structure deals with how the regions of the polypeptide chain fold into the overall three-dimensional structure.

For proteins consisting of only one chain, the tertiary structure is the highest level of structure present. In proteins consisting of more than one polypeptide chain, such as hemoglobin, there is a fourth level of structure, called the **quaternary structure**. It deals with how the different chains interact. The quaternary structure of hemoglobin shows how the four subunits are related to one another in the overall protein.

Sickle Cell Anemia

The subtleties of sequence, structure, and function are dramatically illustrated in the case of hemoglobin. Seemingly small structural features in hemoglobin and other molecules can be important in determining function, as is clearly illustrated by the disease called *sickle cell anemia*. This disease, which is sometimes fatal, affects some individuals of African descent. Persons affected by this disease are anemic; that is, they have low red blood cell counts. In addition, many of their red blood cells are elongated and curved like a sickle instead of being round disks (Figure 6a). These elongated red blood cells are more fragile than normal blood cells, leading to the anemia that is observed. They also restrict the flow of blood within the capillaries, thereby decreasing the amount of oxygen that the individual's cells receive.

The cause of sickle cell anemia has been traced to a small structural change in hemoglobin. In the β subunits of the hemoglobin in individuals carrying the sickle cell trait, a valine has been substituted for a glutamic acid at position 6. An amino acid in this position ends up on the surface of the protein, where it is exposed to the aqueous environment of the cell. The problem arises from the fact that glutamic acid and valine are quite different from each other. The side chain in glutamic acid is ionic, whereas that in valine is nonpolar. The nonpolar side chain on valine causes a nonpolar region to stick out from the molecule where one should not appear. When hemoglobin (normal or sickle cell) is in the deoxygenated state, it has a nonpolar cavity in another region. The nonpolar region around the valine on one sickle cell hemoglobin molecule fits nicely into this nonpolar cavity on another hemoglobin. The sickle cell hemoglobins thus link together, forming long chainlike structures that lead to the symptoms described (Figure 6b).

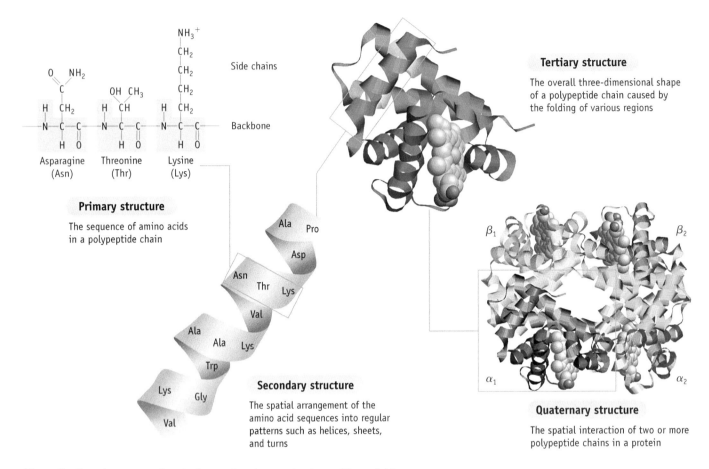

Primary structure

The sequence of amino acids in a polypeptide chain

Secondary structure

The spatial arrangement of the amino acid sequences into regular patterns such as helices, sheets, and turns

Tertiary structure

The overall three-dimensional shape of a polypeptide chain caused by the folding of various regions

Quaternary structure

The spatial interaction of two or more polypeptide chains in a protein

Figure 5   The primary, secondary, tertiary, and quaternary structures of hemoglobin.

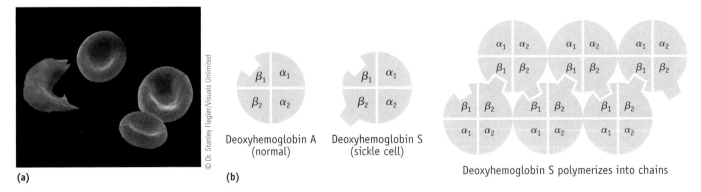

© Dr. Stanley Flegler/Visuals Unlimited

(a)

(b)

Figure 6   Normal and sickled red blood cells. (a) Red blood cells are normally rounded in shape, but people afflicted with sickle cell anemia have cells with a characteristic "sickle" shape. (b) Sickle cell hemoglobin has a nonpolar region that can fit into a nonpolar cavity on another hemoglobin. Sickle cell hemoglobins can link together to form long chainlike structures.

Just one amino acid substitution in each β-subunit causes sickle cell anemia! While other amino acid substitutions may not lead to such severe consequences, sequence, structure, and function are intimately linked and of crucial importance throughout biochemistry.

Enzymes, Active Sites, and Lysozyme

Many reactions necessary for life occur too slowly on their own, so organisms speed them up to the appropriate level using biological catalysts called **enzymes**. Almost every metabolic reaction in a living organism requires an enzyme, and most of these enzymes are proteins. Enzymes are often able to speed up reactions by tremendous amounts; catalyzed rates are typically 10^7 to 10^{14} times faster than uncatalyzed rates.

For an enzyme to catalyze a reaction, several key steps must occur:

1. A reactant (often called the **substrate**) must bind to the enzyme.
2. The chemical reaction must take place.
3. The product(s) of the reaction must leave the enzyme so that more substrate can bind and the process can be repeated.

Typically, enzymes are very specific; that is, only a limited number of compounds (often only one) serve as substrates for a given enzyme, and the enzyme catalyzes only one type of reaction. The place in the enzyme where the substrate binds and the reaction occurs is called the **active site**. The active site often consists of a cavity or cleft in the structure into which the substrate or part of the substrate can fit. The R groups of amino acids or the presence of metal ions in an active site, for example, are often important factors in binding a substrate and catalyzing a reaction.

Lysozyme is an enzyme that can be obtained from human mucus and tears and from other sources, such as egg whites. Alexander Fleming (1881–1955) (who later discovered penicillin) is said to have discovered its presence in mucus when he had a cold. He purposely allowed some of the mucus from his nose to drip onto a dish containing a bacteria culture and found that some of the bacteria died. The chemical in the mucus responsible for this effect was a protein. Fleming called it lysozyme because it is an enzyme that causes some bacteria to undergo *lysis* (rupture).

Lysozyme's antibiotic activity has been traced to its ability to catalyze a reaction that breaks down the cell walls of some bacteria. These cell walls contain a polysaccharide, a polymer of sugar molecules. This polysaccharide is composed of two alternating sugars: *N*-acetylmuramic

Crystals of lysozyme. These crystals were grown on the Space Shuttle in zero gravity.

NASA

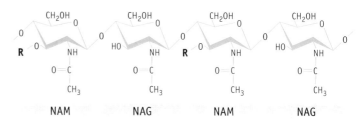

N-acetylmuramic acid (NAM)   *N*-acetylglucosamine (NAG)

Polysaccharide chain of alternating NAM and NAG

Figure 7 **The structures of *N*-acetylmuramic acid (NAM) and *N*-acetylglucosamine (NAG).** The cell walls of some bacteria contain a polysaccharide chain of alternating NAM and NAG units.

acid (NAM) and *N*-acetylglucosamine (NAG) (Figure 7). Lysozyme speeds up the reaction that breaks the bond between C-1 of NAM and C-4 of NAG (Figure 8). Lysozyme has also been shown to catalyze the breakdown of polysaccharides containing only NAG.

Lysozyme (Figure 9) is a protein containing 129 amino acids linked together in a single polypeptide chain. Its molar mass is 14,000 g/mol. As was true in the determination of the double-helical structure of DNA [◀ page 96], x-ray crystallography and model building were key techniques used in determining its three-dimensional structure and method of action.

The structure of lysozyme does not reveal the location of the active site in the enzyme, however. If the enzyme and the substrate could be observed bound together, then the active site would be revealed. In reality, the enzyme–substrate complex lasts for too short a time to be observed by a technique such as x-ray crystallography. Another method had to be used to identify the active site.

Lysozyme is not very effective in cleaving molecules consisting of only two or three NAG units [$(NAG)_3$]. In fact, these molecules act as inhibitors of the enzyme. Researchers surmised that the

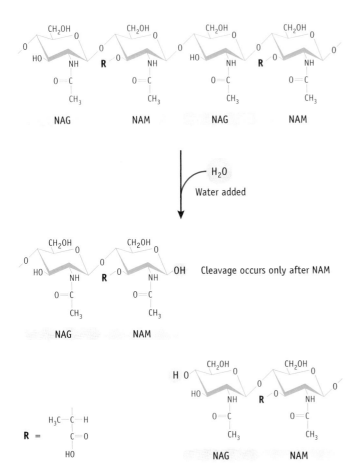

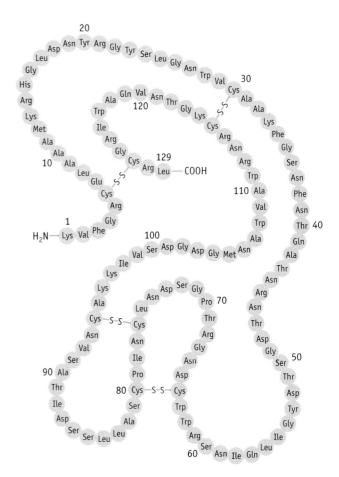

Figure 8   Cleavage of a bond between _N_-acetylmuramic acid (NAM) and _N_-acetylglucosamine (NAG). This reaction is accelerated by the enzyme lysozyme.

Figure 9   The primary structure of lysozyme. The cross-chain disulfide links (—S—S—) are links between cysteine amino acid residues.

inhibition resulted from these small molecules binding to the active site in the enzyme. Therefore, x-ray crystallography was performed on crystals of lysozyme that had been treated with $(NAG)_3$. It revealed that $(NAG)_3$ binds to a cleft in lysozyme (Figure 10).

The cleft in lysozyme where $(NAG)_3$ binds has room for a total of six NAG units. Molecular models of the enzyme and $(NAG)_6$ showed that five of the six sugars fit nicely into the cleft but that the fourth sugar in the sequence did not fit well. To get this sugar into the active site, its structure has to be distorted in the direction that the sugar must move during the cleavage reaction (assuming the bond cleaved is the one connecting it and the next sugar). Amino acids immediately around this location could also assist in the cleavage reaction. In addition, models showed that if an alternating sequence of NAM and NAG binds to the enzyme in this cleft, NAM must bind to this location in the active site: NAM cannot fit into the sugar-binding site immediately before this one, whereas NAG can. For this reason, cleavage must occur only between C-1 of NAM and C-4 of the following NAG, not the other way around—and this is exactly what occurs.

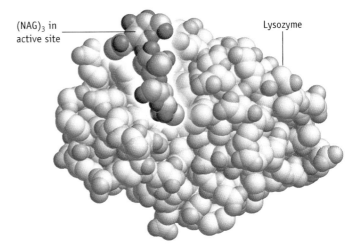

(NAG)₃ in active site   Lysozyme

Figure 10   Lysozyme with $(NAG)_3$.

Nucleic Acids

In the first half of the 20th century, researchers identified deoxyribonucleic acid (DNA) as the genetic material in cells. Also found in cells was a close relative of DNA called ribonucleic acid (RNA). Once it was known that DNA was the molecule involved in heredity, scientists set about determining how it accomplishes this task. Because structure determines function, to understand how a molecule works, you must first know its structure.

Nucleic Acid Structure

RNA and DNA are both polymers (Figure 11). They consist of sugars having five carbons (β-D-ribose in RNA and β-D-2-deoxyribose in DNA) that are connected by phosphodiester groups. A phosphodiester group links the 3' (pronounced "three prime") position of one sugar to the 5' position of the next sugar. Attached at the 1' position of each sugar is an aromatic, nitrogen-containing (nitrogenous) base. The bases in DNA are adenine (A), cytosine (C), guanine (G), and thymine (T); in RNA, the nitrogenous bases are the same as in DNA except that uracil (U) is used rather than thymine (Figure 12). A single ribose (or 2-deoxyribose) with a nitrogenous base attached is called a **nucleoside**. If a phosphate group is also attached, then the combination is called a **nucleotide** (Figure 12).

The principal chemical difference between RNA and DNA is the identity of the sugar (Figure 13). Ribose has a hydroxyl group (—OH) at the 2 position, whereas 2-deoxyribose has only a hydrogen atom at this position. This seemingly small difference turns out to have profound effects. A chain of RNA is cleaved many times faster than a corresponding chain of DNA under similar conditions due to the involvement of this hydroxyl group in the cleavage reaction. The greater stability of DNA contributes to it being a better repository for genetic material.

How does DNA store genetic information? DNA consists of a double helix; one strand of DNA is paired with another strand

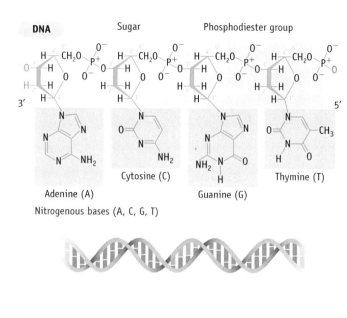

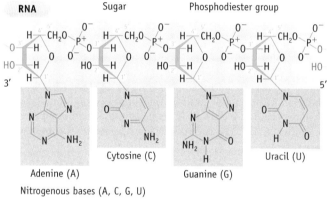

Figure 11   DNA and RNA.

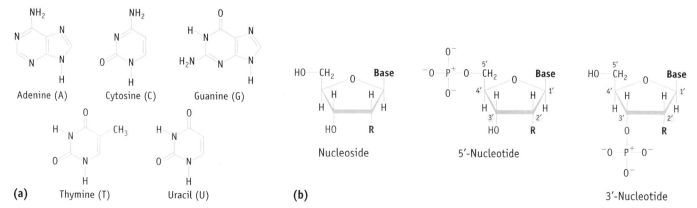

Figure 12   Bases, nucleotides, and nucleosides. (a) The five bases present in DNA and RNA. (b) A nucleoside, a 5'-nucleotide, and a 3'-nucleotide.

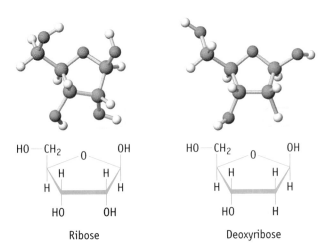

HO—CH₂ ⟍ O ⟋ OH

Ribose

HO—CH₂ ⟍ O ⟋ OH

Deoxyribose

Figure 13   Ribose and Deoxyribose. The sugars found in RNA and DNA, respectively.

running in the opposite direction. The key parts of the structure of DNA for this function are the nitrogenous bases. James Watson and Francis Crick (page 96) noticed that A can form two hydrogen bonds with T and that C can form three hydrogen bonds with G. The spacing in the double helix is just right for either an A-T pair or for a C-G pair to fit, but other combinations (such as A-G) do not fit properly (Figure 14). Thus, if we know the identity of a nucleotide on one strand of the double helix, then we can figure out which nucleotide must be bound to it on the other strand. The two strands are referred to as **complementary strands**.

If the two strands are separated from each other, as they are in the cell division process called *mitosis*, then the cell could construct a new complementary strand for each of the original strands by placing a G wherever there is a C, a T wherever there is an A, and so forth. Through this process, called **replication**, the cell would end up with two identical double-stranded DNA molecules. When the cell divides, each of the two resulting cells will get one molecule of DNA (Figure 15). In this way, the genetic information is passed along from one generation to the next.

Protein Synthesis

The sequence of nucleotides in a cell's DNA contains the instructions to make all of the various proteins the cell needs. DNA is the information storage molecule. To use this information, the cell first makes a complementary copy of the required portion of the DNA using RNA. This step is called **transcription**. The molecule of RNA that results is called **messenger RNA (mRNA)** because it takes this message to where protein synthesis occurs in the cell. The cell uses the less stable RNA rather than DNA to carry out this function.

It makes sense to use DNA, the more stable molecule, to store the genetic information because the cell wants this information to be passed from generation to generation intact. Conversely, it makes sense to use RNA to send the message to make a particular protein. By using the less stable RNA, the message will not be permanent but rather will be destroyed after a certain time, thus allowing the cell to turn off its synthesis.

The mRNA goes to the **ribosomes**, complex bodies in a cell consisting of a mixture of proteins and RNA. Protein synthesis

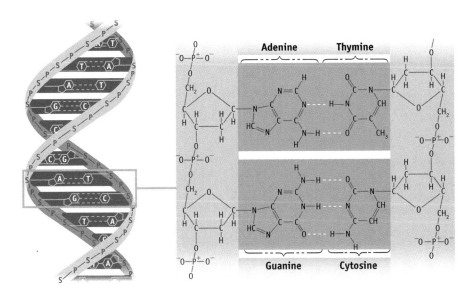

Figure 14   Base pairs and complementary strands in DNA. With the four bases in DNA, the usual pairings are adenine with thymine and cytosine with guanine. The pairing is promoted by hydrogen bonding, the interaction of an H atom bound to an O or N atom with an O or N atom in a neighboring molecule.

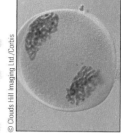

Two strands of DNA. Each base is paired with its partner: adenine (A) with thymine (T), guanine (G) with cytosine (C).

The two DNA strands are separated from each other.

Two new complementary strands are built using the original strands.

Replication results in two identical double-stranded DNA molecules.

At this stage during cell division, the chromosomes containing the DNA have been duplicated, and the two sets have been separated.

Figure 15 **The main steps in DNA replication.** The products of this replication are two identical double helical DNA molecules. When a cell divides, each resulting cell gets one of these.

actually occurs in the ribosomes. The protein is made as the ribosome moves along the strand of mRNA. The sequence of nucleotides in mRNA contains information about the order of amino acids in the desired protein. Following the signal in mRNA to start protein synthesis, every sequence of three nucleotides provides the code for an amino acid until the ribosome reaches the signal to stop (Table 1). These three-nucleotide sequences in mRNA are referred to as **codons**, and the correspondence between each codon and its message (start, a particular amino acid, or stop) is referred to as the **genetic code**.

How is the genetic code used to make a protein? In the ribosome–mRNA complex, there are two neighboring binding sites, the P site and the A site. Each cycle that seeks to add an amino acid to a growing protein begins with that part of the protein already constructed being located in the P site. The A site is where the next amino acid is brought in. Yet another type of RNA becomes involved at this point. This **transfer RNA (tRNA)** consists of a strand of RNA to which an amino acid can be attached (Figure 16). A strand of tRNA has a particular region that contains a sequence of three nucleotides that can attempt to base pair to a codon in the mRNA at the ribosome's A site. This three-nucleotide sequence in the tRNA is called the **anticodon**. Only if the base pairing between the codon and anticodon is complementary (for example, A with U) will the tRNA be able to bind to the mRNA–ribo-

Table 1 Examples of the 64 Codons in the Genetic Code

Codon Base Sequence*	Amino Acid to be Added
AAA	Lysine
AAC	Asparagine
AUG	Start
CAA	Glutamine
CAU	Histidine
GAA	Glutamic acid
GCA	Alanine
UAA	Stop
UAC	Tyrosine

* A = adenine, C = cytosine, G = guanine, U = uracil.

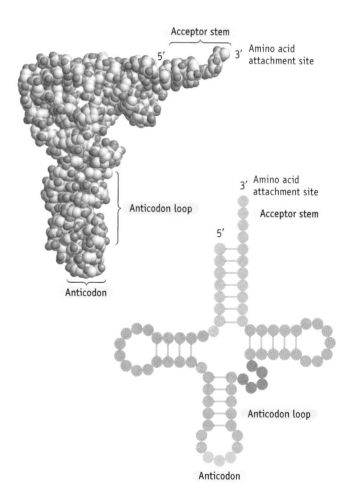

Figure 16 **tRNA structure.**

some complex. Not only does the anticodon determine to which codon a particular strand of tRNA can bind, but it also determines which amino acid will be attached to the end of the tRNA molecule. Thus, a codon in the mRNA selects for a particular tRNA anticodon, which in turn selects for the correct amino acid.

The growing protein chain in the P site reacts with the amino acid in the A site, resulting in the protein chain being elongated by one amino acid and moving the chain into the A site. The ribosome then moves down the mRNA chain, placing the tRNA just used into the P site along with the protein strand and exposing a new codon in the A site. The process is then repeated (Figure 17).

Converting the information from a nucleotide sequence in mRNA into an amino acid sequence in a protein is called **translation**. Protein synthesis thus consists of two main processes: transcription of the DNA's information into RNA, followed by translation of the RNA's message into the amino acid sequence of the protein.

The RNA World and the Origin of Life

One of the most fascinating and persistent questions scientists pursue is how life arose on earth. Plaguing those trying to answer this question is a molecular chicken-and-egg problem: Which came first, DNA or proteins? DNA is good at storing genetic information, but it is not good at catalyzing reactions. Proteins are good

at catalyzing reactions, but they are not good at storing genetic information. In trying to picture an early self-replicating molecule, deciding whether it should be based on DNA or proteins seemed hopeless. Ultimately, both functions are important. These problems have caused some scientists to turn away from considering either DNA or proteins as candidates for the first molecule of life. One hypothesis that has gained support in recent years suggests that the first life on earth may have been based on RNA instead.

Like DNA, RNA is a nucleic acid and can serve as a genetic storage molecule. We have already seen how it serves as an information molecule in the process of protein synthesis. In addition, scientists have discovered that retroviruses, like the human immunodeficiency virus (HIV) that causes AIDS, use RNA as the repository of genetic information instead of DNA. Perhaps the first organisms on earth also used RNA to store genetic information.

In the 1980s, researchers discovered that particular strands of RNA catalyze some reactions involving cutting and joining together strands of RNA. Thomas Cech and Sidney Altman shared the 1989 Nobel Prize in chemistry for their independent discoveries of systems that utilize "catalytic RNA." One might imagine that an organism could use RNA both as the genetic material and as a catalyst. Information and action are thus combined in this one molecule.

According to proponents of the "RNA World" hypothesis, the first organism used RNA for both information and catalysis. At

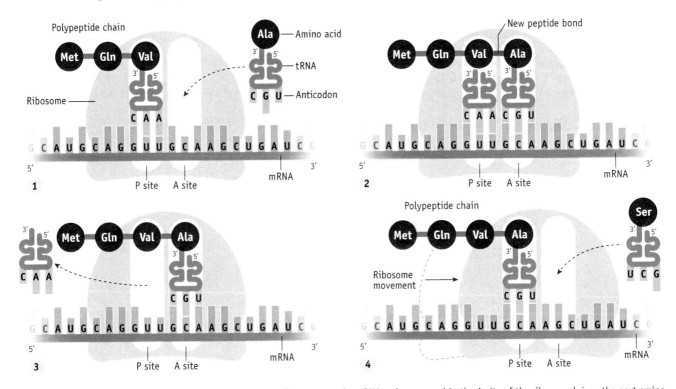

Figure 17 **Protein synthesis.** The tRNA with an anticodon complementary to the mRNA codon exposed in the A site of the ribosome brings the next amino acid to be added to the growing protein chain. After the new peptide bond is formed, the ribosome moves down the mRNA exposing a new codon in the A site and transferring the previous tRNA and the protein chain to the P site.

some later date, DNA evolved and had better information storage capabilities, so it took over the genetic information storage functions from RNA. Likewise, proteins eventually evolved and proved better at catalysis than RNA, so they took over this role for most reactions in a cell. RNA still plays a central role in the flow of genetic information, however. Genetic information does not go directly from DNA to proteins; it must pass through RNA along the way. Those favoring the RNA World hypothesis also point out that many **enzyme cofactors**, molecules that must be present for an enzyme to work, are RNA nucleotides or are based on RNA nucleotides. As we shall see, one of the most important molecules in metabolism is an RNA nucleotide, adenosine 5′-triphosphate (ATP). The importance of these nucleotides might date back to an earlier time when organisms were based on RNA alone.

The RNA World hypothesis is interesting and can answer some of the questions that arise in research on the origins of life, but it is not the only current hypothesis dealing with the origin of a self-replicating system. Much research remains to be done before we truly understand how life could have arisen on earth.

Metabolism

Why do we eat? Some components of our food, such as water, are used directly in our bodies. We break down other chemicals to obtain the molecular building blocks we need to make the many chemicals in our bodies. Oxidation of foods also provides the energy we need to perform the activities of life. The many different chemical reactions that foods undergo in the body to provide energy and chemical building blocks fall into the area of biochemistry called **metabolism**. We have already studied some aspects of energy changes in chemical reactions in Chapter 6 and some aspects of oxidation–reduction reactions in Chapter 5. We shall now examine some of these same considerations in biochemical reactions.

Energy and ATP

Substances in food, such as *carbohydrates*, are oxidized in part of the metabolic process. These oxidations are energetically favorable reactions, releasing large quantities of energy. For example, the thermochemical equation for the oxidation of the sugar glucose ($C_6H_{12}O_6$) to form carbon dioxide and water is

$$C_6H_{12}O_6 \text{ (s)} + 6 \text{ } O_2(g) \longrightarrow 6 \text{ } CO_2(g) + 6 \text{ } H_2O(\ell) \quad \Delta H° = -2803 \text{ kJ}$$

Figure 18 shows the oxidation of table sugar, sucrose. Of course, we do not want to have such a rapid, flame-producing oxidation occur in our bodies. Instead, we want to carry out a more controlled oxidation. The body does so in a way that allows it to obtain the energy in small increments.

It would be inefficient if every part of the cell needed to have all the mechanisms necessary to carry out the oxidation of every type of molecule used for energy. Instead, the cell carries out the oxidation of compounds such as glucose in one location

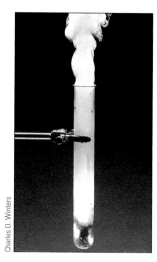

Charles D. Winters

Figure 18 **Oxidation of sucrose.** The oxidation of sucrose (here with $KClO_3$) is carried out in a more controlled manner in your body.

and stores the energy in a small set of compounds that can be used almost anywhere in the cell.

The principal compound used to perform this function is **adenosine 5′-triphosphate (ATP)**. This ribonucleotide consists of a ribose to which the nitrogenous base adenine is connected at the 1′ position and a triphosphate group is connected at the 5′ position (Figure 19). In aerobic respiration the equivalent of 30–32 moles of ATP is produced per mole of glucose that reacts. (Some bacteria can produce up to 38 moles.) Based on the ΔH values for the processes a greater production of ATP might be expected, but the process is not completely efficient.

The hydrolysis of ATP to adenosine 5′-diphosphate (ADP) and inorganic phosphate (P_i) is an exothermic process (Figure 20).

$$ATP + H_2O \longrightarrow ADP + P_i \quad \Delta H \approx -24 \text{ kJ}$$

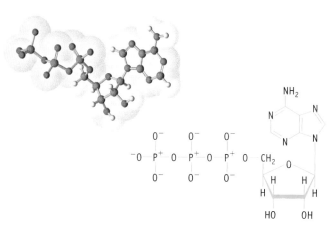

Figure 19 **Adenosine-5′-triphosphate (ATP).**

Why is this reaction exothermic? We can assess this by evaluating bond energies (page 421). In this reaction, we must break two bonds, a P—O bond in ATP and a H—O bond in water. But we also form two new bonds: a P—O bond between the phosphate group being cleaved off the ATP and the OH of the original water and a H—O bond between the hydrogen from the water and the portion of the ATP that forms ADP. In the overall process, more energy is released in forming these new bonds in the products than is required to break the necessary bonds in the reactants. Thus, the overall reaction is exothermic.

In cells many chemical processes that would be endothermic on their own are linked with the hydrolysis of ATP. The combination of an energetically unfavorable process with the energetically favorable hydrolysis of ATP can yield a process that is energetically favorable. For example, most cells have a greater concentration of potassium ions and a smaller concentration of sodium ions inside them than are present outside them. The natural tendency, therefore, is for sodium ions to flow into the cell and for potassium ions to flow out. To maintain the correct concentrations, the cell must counteract this movement and actively pump sodium ions out of the cell and potassium ions into the cell.

Figure 20   The exothermic conversion of adenosine-5'-triphosphate (ATP) to adenosine-5'-diphosphate (ADP).

Chemical Perspectives

AIDS and Reverse Transcriptase

One of the major health crises in modern times is the epidemic associated with the disease called acquired immune deficiency syndrome (AIDS). A person develops AIDS in the final stages of infection with the human immunodeficiency virus (HIV). At the time of this writing, an estimated 40 million people worldwide were infected with HIV. HIV is a retrovirus. Unlike all organisms and most viruses, a retrovirus does not use DNA as its genetic material, but rather single-stranded RNA.

During the course of infection, the viral RNA is transcribed into DNA by means of an enzyme called reverse transcriptase. It is so named because the direction of information flow is in the opposite direction (RNA ⟶ DNA) than that usually found in cells. The resulting DNA is inserted into the cell's DNA. The infected cell then produces the proteins and RNA to make new virus particles.

Reverse transcriptase consists of two subunits (see the accompanying figure). One subunit has a molar mass of approximately 6.6×10^4 g/mol, and the other has a molar mass of roughly 5.1×10^4 g/mol. Reverse transcriptase is not a very accurate enzyme, however. It makes an error in transcription for every 2000 to 4000 nucleotides copied. This is a much larger error rate than that for most cellular enzymes that copy DNA, which typically make one error for every 10^9 to 10^{10} nucleotides copied. The high error rate for reverse transcriptase contributes to the challenge scientists face in trying to combat HIV because these replication errors lead to frequent mutations in the virus. That is, the virus keeps changing, which means that developing a treatment that works and will continue to work is very difficult. Some treatments have been successful in significantly delaying the onset of AIDS, but none has yet proven to be a cure. More research is needed to combat this deadly disease.

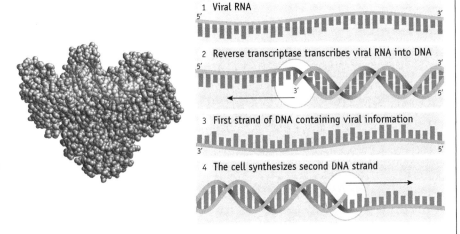

Reverse transcriptase. The reverse transcriptase enzyme consists of two subunits (shown in red and purple). Reverse transcriptase catalyzes the transcription of viral RNA into DNA. The cell then constructs a complementary strand of DNA. The resulting double stranded DNA is inserted into the cell's DNA.

This activity goes in the direction that is not favored, so it requires energy. To accomplish this feat, the cell links this pumping process to the hydrolysis of ATP to ADP. The energy released from the hydrolysis reaction provides the energy to run a molecular pump (an enzyme) that moves the ions in the direction the cell needs.

Oxidation–Reduction and NADH

Cells also need compounds that can be used to carry out oxidation–reduction reactions. Just as ATP is a compound used in many biochemical reactions when energy is needed, so nature uses another small set of compounds to run many redox reactions. An important example is **nicotinamide adenine dinucleotide (NADH)**. This compound consists of two ribonucleotides joined at their 5′ positions via a diphosphate linkage. One of the nucleotides has adenine as its nitrogenous base, whereas the other has a nicotinamide ring (Figure 21). When NADH is oxidized, changes occur in the nicotinamide ring, such that the equivalent of a hydride ion (H⁻) is lost. Because this hydride ion has two electrons associated with it, the nicotinamide ring loses two electrons in this process. The resulting species, referred to as NAD⁺, is shown on the right in Figure 21.

In many biochemical reactions, when a particular species needs to be reduced, it reacts with NADH. The NADH is oxidized to NAD⁺, losing two electrons in the process, and the species of interest is reduced by gaining these electrons. If a species must be oxidized, the opposite process often occurs; that is, it reacts with NAD⁺. The NAD⁺ is reduced to NADH, and the species of interest is oxidized.

Respiration and Photosynthesis

In the process of **respiration**, a cell breaks down glucose, which is oxidized to CO_2 and H_2O.

$$C_6H_{12}O_6(s) + 6\ O_2(g) \longrightarrow 6\ CO_2(g) + 6\ H_2O(\ell)$$

The energy released in this reaction is used to generate the ATP needed by the cell. The sugars employed in this process can be traced back to green plants, where sugars are made via the process of photosynthesis. In **photosynthesis**, plants carry out the reverse of glucose oxidation—that is, the synthesis of glucose.

$$6\ CO_2(g) + 6\ H_2O(\ell) \longrightarrow C_6H_{12}O_6(s) + 6\ O_2(g)$$

Green plants have found a way to use light to provide the energy needed to run this endothermic reaction.

The key molecule involved in trapping the energy from light in photosynthesis is chlorophyll. Green plants contain two types of chlorophyll: chlorophyll a and chlorophyll b (Figure 22). The absorbance spectra of chlorophyll a and chlorophyll b are also shown in Figure 22. Notice that these molecules absorb best in the blue-violet and red-orange regions. Not much light is absorbed in the green region. When white light shines on chlorophyll, therefore, red-orange and blue-violet light are absorbed by the chlorophyll; green light is not absorbed but rather is reflected. We see the reflected light, so plants appear green to us. The light energy absorbed by the chlorophyll is used to drive the process of photosynthesis.

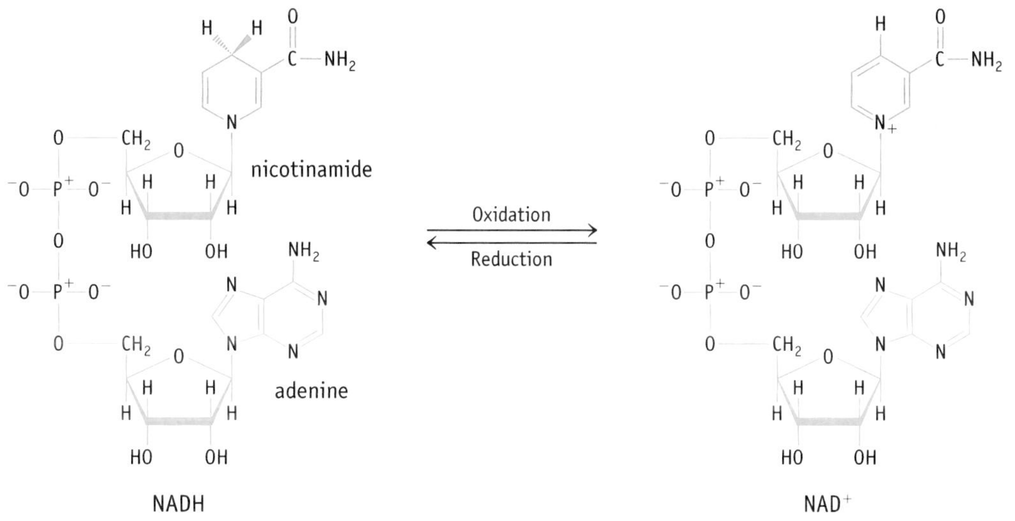

Figure 21　The structures of NADH and NAD.

12—Gases & Their Properties

Modern hot air balloons. A hot-air balloon rises because the heated air in the balloon has a lower density than the surrounding atmosphere.

Up, Up, and Away!

The invention of the balloon appears...to be a discovery of great importance.

Benjamin Franklin, about 1784

The first flight of a manned hot-air balloon took place in Paris, France, on November 21, 1783. The balloon was designed by Joseph and Étienne Montgolfier and piloted by Pilatre de Rozier and the Marquis d'Arlandes. The two pilots traveled 12 kilometers in less than half an hour at about 900 meters in altitude.

How does a hot-air balloon fly? As you will learn in this chapter, it can ascend because heating affects the density of the air in the balloon. When the air inside a hot-air balloon is heated (usually with a propane heater in a modern balloon), the gas expands. Initially, this expansion is used to inflate the balloon. At a certain point, however, the balloon no longer increases in volume. Air is then forced from the inside of the balloon as the air continues to expand on heating. As a result, less air remains inside. The smaller mass of air in the same volume means that the gas inside the balloon has a lower density than the surrounding atmosphere, so the balloon ascends.

Jacques Charles and M. S. Roberts ascended over Paris on December 1, 1783, in a hydrogen-filled balloon.

See Chapter Goals Revisited (page 578). Test your knowledge of these goals by taking the exam-prep quiz on the General ChemistryNow CD-ROM or website.

- Understand the basis of the gas laws and know how to use those laws (Boyle's law, Charles's law, Avogadro's hypothesis, Dalton's law).
- Use the ideal gas law.
- Apply the gas laws to stoichiometric calculations.
- Understand kinetic-molecular theory as it is applied to gases, especially the distribution of molecular speeds (energies).
- Recognize why real gases do not behave like ideal gases.

The typical modern hot-air balloon is about 18 meters tall (60 feet) and has a volume of about 2250 cubic meters of air heated with a propane burner. These balloons can carry enough propane fuel to fly for about 2 hours.

Historians have speculated that the Montgolfier brothers got their idea for a hot-air balloon after reading about the experiments on gases made by Joseph Black (1728–1799) of Scotland. Indeed, the 18th century was a time of great discoveries about the nature of chemistry. Experiments on gases by scientists such as Black, Henry Cavendish (1731–1810), Joseph Priestley (1733–1804), and Antoine Lavoisier (1743–1794) gave birth to modern chemistry.

Among the chemists working on gases was Jacques Charles (1746–1823). In August 1783, Charles exploited his recent studies on hydrogen gas by inflating a balloon with this gas. Because hydrogen would escape easily from a paper bag, Charles made a silk bag coated with rubber. Inflating the bag took several days and required nearly 225 kg of sulfuric acid and 450 kg of iron to produce the H_2 gas. The balloon stayed aloft for almost 45 minutes and traveled about 15 miles. When it landed in a village, however, the people were so terrified that they tore it to shreds. Several months later, Charles and a passenger flew a new hydrogen-filled balloon some distance across the French countryside and ascended to the then-incredible altitude of 2 miles.

Balloons that remain aloft for long periods of time typically need to use a lighter-than-air gas such as helium or hydrogen to produce lift. This approach was first tried in June 1785 by Montgolfier's first pilot, de Rozier. He and a friend, Pierre Romain, tried to fly from Paris to London in a balloon containing a hydrogen-filled cell and an air-filled cell heated by a flame. Unfortunately, at an altitude of about 900 feet, the hydrogen gas exploded, killing the two pilots. They were the first people to die in manned flight.

The latest balloons designed for long-distance flight are now called Rozier balloons. They have one or more cells filled with nonflammable helium as well as a cell in which the air can be heated. The helium provides much of the lift for the balloon, while the cell containing the hot air allows for adjustments to be made in amount of lift—that is, to move to a different altitude or to compensate for the cooling of the helium at night (and the consequent contraction in the volume of the helium). For example, at night when the air is colder, the pilot heats the air in an unsealed cell with a propane burner, which in turn transfers heat to upper helium cells. Such a device was used in the first successful circumnavigation of the globe by a balloon in March 1999.

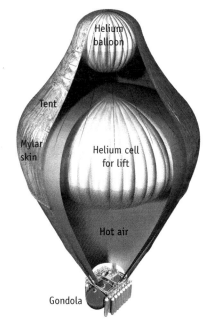

The Rozier balloon. A balloon of this design set a ballooning distance record by flying around the globe in 1999. The upper helium cell "stakes out the tent" around the larger helium cell, which helps to insulate the latter cell. Lift is provided by both the helium cells and heated air. The gondola below the balloon is insulated and sealed with life-support equipment for the crew. The balloon flies at an altitude of 6000 to 15,000 meters, a height at which the outside temperature can be about −60 °C.

To Review Before You Begin

- Review chemical stoichiometry in Chapters 4 and 5

Hot-air balloons, SCUBA diving equipment, and automobile air bags (Figure 12.1) depend on the properties of gases. Aside from understanding how these devices work, there are at least three reasons for studying gases. First, some common elements and compounds (such as helium, hydrogen, oxygen, nitrogen, and methane) exist in the gaseous state under normal conditions of pressure and temperature. Furthermore, many common liquids such as water can be vaporized, and the physical properties of these vapors are important. Second, our gaseous atmosphere provides one means of transferring energy and material throughout the globe, and it is the source of life-giving chemicals.

The third reason for studying gases is also compelling. Of the three states of matter, gases are reasonably simple when viewed at the molecular level and, as a result, gas behavior is well understood. It is possible to describe the properties of gases *qualitatively* in terms of the behavior of the molecules that make up the gas. Even more impressive, it is possible to describe the properties of gases *quantitatively* using simple mathematical models. One objective of scientists is to develop precise mathematical and conceptual models of natural phenomena, and a study of gas behavior will introduce you to this approach.

12.1—The Properties of Gases

To describe gases, chemists have learned that only four quantities are needed: the pressure (P), volume (V), and temperature (T, kelvins) of the gas, and its amount (n, mol). Let us examine the first of these parameters, gas pressure, and its units.

Gas Pressure

You are already familiar with pressure. Meteorologists tell us that the pressure of the atmosphere is rising when nice weather approaches and that it is falling when a storm approaches. They also often speak of rising or falling barometric pressure—a barometer is the device used to measure atmospheric pressure.

Figure 12.1 Automobile air bags. Most automobiles are now equipped with air bags to protect the driver and the front-seat passenger in the event of a head-on or side crash. Such bags are inflated with nitrogen gas, which is generated by the explosive decomposition of sodium azide in the event of a crash.

$$2 \text{ NaN}_3(s) \longrightarrow 2 \text{ Na}(s) + 3 \text{ N}_2(g)$$

The air bag is fully inflated in about 0.050 second. This is important because the typical automobile collision lasts about 0.125 second. (*See General ChemistryNow Screen 12.1 Puzzler: Air Bags, for questions about automobile air bags.*)

SAAB Car USA, Inc.

A barometer can be made by filling a tube with a liquid, often mercury, and inverting the tube in a dish containing the same liquid (Figure 12.2). If the air has been removed completely from the vertical tube, the liquid in the tube assumes a level such that the pressure exerted by the mass of the column of liquid in the tube is balanced by the pressure of the atmosphere pressing down on the surface of the liquid in the dish.

At sea level, the mercury in a mercury-filled barometer will rise about 760 mm above the surface of the mercury in the dish. Thus, pressure is often reported in units of **millimeters of mercury (mm Hg).** Pressures are also reported as **standard atmospheres (atm)**, a unit defined as follows:

1 standard atmosphere (1 atm) = 760 mm Hg (exactly)

Though it is not the SI unit, *the atmosphere is the pressure unit used in this book.*

The SI unit of pressure is the **pascal (Pa)**, named for the French mathematician and philosopher Blaise Pascal (1623–1662). **Pressure** is defined as the force exerted on an object divided by the area over which it is exerted, and the pascal is the only pressure unit that is expressed in these terms.

1 pascal (Pa) = 1 newton/meter^2

(The newton is the SI unit of force.) Because the pascal is a very small unit compared with ordinary pressures, the unit kilopascal (kPa) is used more frequently.

The pascal has a simple relationship to another unit of pressure called the **bar**, where 1 bar = 100,000 Pa. The thermodynamic data in Chapters 6 and 19 and in Appendix L are given for gas pressures of 1 bar. To summarize, the units used in science for pressure are

1 atm = 760 mm Hg (exactly) = 101.3 kilopascals (kPa) = 1.013 bar

or

1 bar = 1 × 10^5 Pa (exactly) = 1 × 10^2 kPa = 0.9872 atm

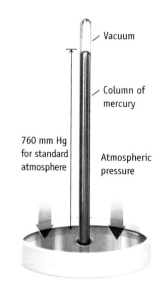

Figure 12.2 A barometer. The pressure of the atmosphere on the surface of the mercury in the dish is balanced by the downward pressure exerted by the column of mercury. The barometer was invented in 1643 by Evangelista Torricelli (1608–1647). A unit of pressure called the *torr* in his honor is equivalent to 1 mm Hg.

Example 12.1—Pressure Unit Conversions

Problem Convert a pressure of 635 mm Hg into its corresponding value in units of atmospheres (atm), bars, and kilopascals (kPa).

Strategy Use the relationships between millimeters of mercury, atmospheres, bars, and pascals described earlier in the text.

Solution The relationship between millimeters of mercury and atmospheres is 1 atm = 760 mm Hg. Notice that the given pressure is less than 760 mm Hg—that is, less than 1 atm:

$$635 \text{ mm Hg} \left(\frac{1 \text{ atm}}{760. \text{ mm Hg}} \right) = 0.836 \text{ atm}$$

The relationship between atmospheres and bars is 1 atm = 1.013 bar. We have

$$0.836 \text{ atm} \left(\frac{1.013 \text{ bar}}{1 \text{ atm}} \right) = 0.847 \text{ bar}$$

The factor relating units of millimeters of mercury and kilopascals is 101.325 kPa = 760 mm Hg. Therefore,

$$635 \text{ mm Hg} \left(\frac{101.3 \text{ kPa}}{760. \text{ mm Hg}} \right) = 84.6 \text{ kPa}$$

A Closer Look

Measuring Gas Pressure

Pressure is the force exerted on an object divided by the area over which the force is exerted:

$$\text{Pressure} = \frac{\text{force}}{\text{area}}$$

This book, for example, weighs more than 4 lb and has an area of 82 in.², so it exerts a pressure of about 0.05 lb/in.² when it lies flat on a surface. (In metric units, the pressure is about 3 g/cm².)

Now consider the pressure that the column of mercury exerts on the mercury in the dish in the barometer shown in Figure 12.2. This pressure exactly balances the pressure of the atmosphere. Thus the pressure of the atmosphere (or of any other gas) can be measured by relating it to the height of the column of mercury (or any other liquid) that the gas can support.

Mercury is the liquid of choice for barometers because of its high density. The height of a barometer filled with water would exceed 10 m. [A column of water is about 13.6 times as high as a column of mercury because mercury's density (13.53 g/cm³) is about 13.6 times that of water (density = 0.997 g/cm³, at 25 °C).]

In the laboratory we often use a U-tube manometer, which is a mercury-filled, U-shaped glass tube. The closed side of the tube is evacuated so that no gas remains to exert pressure on the mercury on that side. The other side is open to the gas whose pressure we want to measure. When the gas presses on the mercury in the open side, the gas pressure is read directly (in mm Hg) as the difference in mercury levels on the closed and open sides.

You may have used a tire gauge to check the pressure in your car or bike tires. In the United States, such gauges usually indicate the pressure in pounds per square inch (psi) where 1 atm = 14.7 psi. Some newer gauges give the pressure in kilopascals as well. The reading on the scale refers to the pressure *in excess of atmospheric pressure*. (A flat tire is not a vacuum; it contains air at atmospheric pressure.) For example, if the gauge reads 35 psi (2.4 atm), the pressure in the tire is actually about 50 psi (3.4 atm). (*See General ChemistryNow Screen 12.2.*)

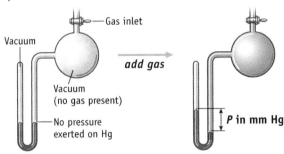

Exercise 12.1—Pressure Unit Conversions

Rank the following pressures in decreasing order of magnitude (from largest to smallest): 75 kPa, 250 mm Hg, 0.83 bar, 0.63 atm.

12.2—Gas Laws: The Experimental Basis

Experimentation in the 17th and 18th centuries led to three gas laws that provide the basis for understanding gas behavior.

The Compressibility of Gases: Boyle's Law

When you pump up the tires of your bicycle, the pump squeezes the air into a smaller volume (Figure 12.3). This property of a gas is called its **compressibility.** While studying the compressibility of gases, Robert Boyle (1627–1691) observed that the volume of a fixed amount of gas at a given temperature is inversely proportional to the pressure exerted by the gas. All gases behave in this manner, and we now refer to this relationship as **Boyle's law.**

Boyle's law can be demonstrated in many ways. In Figure 12.4 a hypodermic syringe is filled with air and sealed. When pressure is applied to the movable plunger of the syringe, the air inside is compressed. As the pressure (P) increases on the syringe, the gas volume in the syringe (V) decreases. When the pressure of the gas in the syringe is plotted as a function of $1/V$, a straight line results. This type of plot

Figure 12.3 A bicycle pump—Boyle's law in action. The pump compresses air into a smaller volume. You experience Boyle's law because you can feel the increasing pressure of the gas as you press down on the plunger.

Charles D. Winters

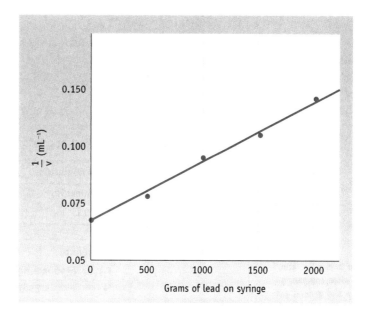

Active Figure 12.4   **An experiment to demonstrate Boyle's law.** A syringe filled with air is sealed. Pressure is applied by adding lead shot to the beaker on top of the syringe. As the mass of lead increases, the pressure on the air in the sealed syringe increases and the gas is compressed. A plot of (1/volume of air in the syringe) versus P (as measured by the mass of lead) is a straight line.

GENERAL
Chemistry ⚛ Now™ See the General ChemistryNow CD-ROM or website to explore an interactive version of this figure accompanied by an exercise.

demonstrates that the pressure and volume of the gas are inversely proportional; that is, they change in opposite directions. Mathematically, we can write this as:

$$P \propto \frac{1}{V} \quad \text{when } n \text{ and } T \text{ are constant}$$

Where the symbol $\propto$ means "proportional to." For a given amount of gas (n) at a fixed temperature (T), the gas volume decreases if the pressure increases. Conversely, if the pressure is lowered, then the gas volume increases.

Boyle's experimentally determined relationship can be put into a useful mathematical form. When two quantities are proportional to each other, they can be equated if a *proportionality constant*, here called C_B, is introduced. Thus,

$$P = C_B \times \frac{1}{V} \quad \text{or} \quad PV = C_B \quad \text{when } n \text{ and } T \text{ are constant}$$

This form of Boyle's law expresses the fact that *the product of the pressure and volume of a gas sample is a constant at a given temperature*, where the constant C_B is determined by the amount of gas (in moles) and its temperature (in kelvins). It follows from this that, if the pressure–volume product is known for a gas sample under one set of conditions (P_1 and V_1), then it is known for another set of conditions (P_2 and V_2). Under either set of conditions, the PV product is equal to C_B, so

$$P_1V_1 = P_2V_2 \quad \text{at constant } n \text{ and } T \tag{12.1}$$

This form of Boyle's law is useful when we want to know, for example, what happens to the volume of a given quantity of gas when the pressure changes at a constant temperature.

Example 12.3—Charles's Law

Problem Suppose you have a sample of CO_2 in a gas-tight syringe (as in Figure 12.4). The gas volume is 25.0 mL at room temperature (20.0 °C). What is the final volume of the gas if you hold the syringe in your hand to raise its temperature to 37 °C?

Strategy Because a given quantity of gas is heated (at a constant pressure), Charles's law applies. Because we know the original V and T, and we want to calculate a new volume at a new, but known, temperature, we can use Equation 12.2.

Solution Organize the information in a table. Notice that the temperature *must* be converted to kelvins.

Original Conditions	Final Conditions
$V_1 = 25.0$ mL	$V_2 = ?$
$T_1 = 20.0 + 273 = 293$ K	$T_2 = 37 + 273 = 310.$ K

Substitute the known quantities into Equation 12.2 and solve for V_2:

$$V_2 = T_2 \times \frac{V_1}{T_1} = (310.\ \text{K}) \times \frac{25.0\ \text{mL}}{293\ \text{K}} = 26.5\ \text{mL}$$

Comment As expected, the volume of the gas increased with a temperature increase. The new volume (V_2) must equal the original volume (V_1) multiplied by a temperature fraction that is greater than 1 to reflect the effect of the temperature increase. That is,

$$V_2 = V_1 \times \frac{310.\ \text{K}}{293\ \text{K}} = 26.5\ \text{mL}$$

Exercise 12.3—Charles's Law

A balloon is inflated with helium to a volume of 45 L at room temperature (25 °C). If the balloon is cooled to −10 °C, what is the new volume of the balloon? Assume that the pressure does not change.

Combining Boyle's and Charles's Laws: The General Gas Law

The volume of a given amount of gas is inversely proportional to its pressure at constant temperature (Boyle's law) and directly proportional to the Kelvin temperature at constant pressure (Charles's law). But what if we need to know what happens to the gas when two of the three parameters (*P*, *V*, and *T*) change? For example, what would happen to the pressure of a sample of nitrogen in an automobile air bag if the same amount of gas were placed in a smaller bag and heated to a higher temperature? You can deal with this situation by combining the two equations that express Boyle's and Charles's laws:

$$\frac{P_1 V_1}{T_1} = \frac{P_2 V_2}{T_2} \quad \text{for a given amount of gas, } n \qquad (12.3)$$

This equation is sometimes called the **general gas law** or **combined gas law.** It applies specifically to situations in which the *amount of gas does not change.*

Example 12.4—General Gas Law

Problem Helium-filled balloons are used to carry scientific instruments high into the atmosphere. Suppose a balloon is launched when the temperature is 22.5 °C and the barometric pressure is 754 mm Hg. If the balloon's volume is 4.19×10^3 L (and no helium escapes from the balloon), what will the volume be at a height of 20 miles, where the pressure is 76.0 mm Hg and the temperature is −33.0 °C?

Strategy Here we know the initial volume, temperature, and pressure of the gas. We want to know the volume of the same amount of gas at a new pressure and temperature. It is most convenient to use Equation 12.3, the general gas law.

Solution Begin by setting out the information given in a table.

A weather balloon is filled with helium. As it ascends into the troposphere, does the volume increase or decrease?

Initial Conditions	Final Conditions
$V_1 = 4.19 \times 10^3$ L	$V_2 = ?$ L
$P_1 = 754$ mm Hg	$P_2 = 76.0$ mm Hg
$T_1 = 22.5$ °C (295.7 K)	$T_2 = −33.0$ °C (240.2 K)

We can rearrange the general gas law to calculate the new volume, V_2:

$$V_2 = \left(\frac{T_2}{P_2}\right) \times \left(\frac{P_1 V_1}{T_1}\right) = V_1 \times \frac{P_1}{P_2} \times \frac{T_2}{T_1}$$

$$= 4.19 \times 10^3 \text{ L} \times \frac{754 \text{ mm Hg}}{76.0 \text{ mm Hg}} \times \frac{240.2 \text{ K}}{295.7 \text{ K}}$$

$$= 3.38 \times 10^4 \text{ L}$$

Comment The pressure decreased by almost a factor of 10, which should lead to about a 10-fold volume increase. This increase is partly offset by a drop in temperature, which leads to a volume decrease. On balance, the volume increases because the pressure has dropped so substantially.

Notice that the solution was to multiply the original volume (V_1) by a pressure factor (larger than 1 because the volume increases with a lower pressure) and a temperature factor (smaller than 1 because volume decreases with a decrease in temperature).

Exercise 12.4—The General Gas Law

You have a 22.-L cylinder of helium at a pressure of 150 atm and a temperature of 31 °C. How many balloons can you fill, each with a volume of 5.0 L, on a day when the atmospheric pressure is 755 mm Hg and the temperature is 22 °C?

The general gas law leads to other useful predictions of gas behavior. For example, if a given amount of gas is held in a closed container, the pressure of the gas will increase with increasing temperature.

$$\frac{P_1}{T_1} = \frac{P_2}{T_2} \quad \text{when } V_1 = V_2 \text{ and so } P_2 = P_1 \times \frac{T_2}{T_1}$$

That is, when T_2 is greater than T_1, P_2 will be greater than P_1. In fact, this is the reason tire manufacturers recommend checking tire pressures when the tires are cold. After driving for some distance, friction warms a tire and increases the internal pressure. Filling a warm tire to the recommended pressure may lead to a dangerously underinflated tire.

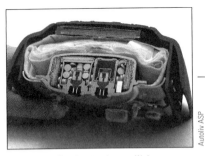

When a car decelerates in a collision, an electrical contact is made in the sensor unit. The propellant (green solid) detonates, releasing nitrogen gas, and the folded nylon bag explodes out of the plastic housing.

Driver side air bags inflate with 35-70 L of N_2 gas, whereas passenger air bags hold about 60-160 L.

The bag deflates within 0.2 s, the gas escaping through holes in the bottom of the bag.

Figure 12.7   Automobile air bags. *(See Figure 12.1 and General ChemistryNow Screen 12.13 Puzzler: Air Bags.)*

Avogadro's Hypothesis

The air bag is a safety device found in most of today's automobiles. In the event of an accident, it is rapidly inflated with nitrogen gas generated by a chemical reaction. The air bag unit has a sensor that is sensitive to sudden deceleration of the vehicle and will send an electrical signal that will trigger the reaction (Figures 12.1 and 12.7). The explosion of sodium azide generates nitrogen gas.

$$2\ NaN_3(s) \longrightarrow 2\ Na(s) + 3\ N_2(g)$$

Driver-side air bags inflate to a volume of about 35–70 L, and passenger-side air bags inflate to about 60–160 L. The final volume of the bag will depend on the amount of nitrogen gas generated.

The relationship between volume and amount of gas was first noted by Amadeo Avogadro. In 1811 he used work on gases by the chemist (and early experimenter with hot-air balloons) Joseph Guy-Lussac (1778–1850) to propose that *equal volumes of gases under the same conditions of temperature and pressure have equal numbers of particles* (either molecules or atoms, depending on the composition of the gas). This idea came to be known as **Avogadro's hypothesis**. Stated another way, the volume of a gas at a given temperature and pressure is directly proportional to the amount of gas in moles:

$$V \propto n \text{ at constant } T \text{ and } P$$

GENERAL
Chemistry ·ϟ· Now™

See the General ChemistryNow CD-ROM or website:
- **Screen 12.3 Gas Laws,** for interactive versions the the three gas laws

Example 12.5—Avogadro's Hypothesis

Problem Ammonia can be made directly from the elements:

$$N_2(g) + 3\ H_2(g) \longrightarrow 2\ NH_3(g)$$

If you begin with 15.0 L of $H_2(g)$, what volume of $N_2(g)$ is required for complete reaction (both gases being at the same T and P)? What is the theoretical yield of NH_3, in liters, under the same conditions?

Strategy From Avogadro's hypothesis we know that gas volume is proportional to the amount of gas. Therefore, we can substitute gas volumes for moles in this stoichiometry problem.

Solution Calculate the volumes of N_2 required and of NH_3 produced (in liters) by multiplying the volume of H_2 available by a stoichiometric factor (also in liters) obtained from the chemical equation:

$$V(N_2 \text{ required}) = 15.0 \text{ L } H_2 \text{ available} \times \frac{1 \text{ L } N_2}{3 \text{ L } H_2} = 5.00 \text{ L } N_2$$

$$V(NH_3 \text{ produced}) = 15.0 \text{ L } H_2 \text{ available} \times \frac{2 \text{ L } NH_3}{3 \text{ L } H_2} = 10.0 \text{ L } NH_3$$

Exercise 12.5—Avogadro's Hypothesis

Methane burns in oxygen to give CO_2 and H_2O, according to the equation

$$CH_4(g) + 2 O_2(g) \longrightarrow CO_2(g) + 2 H_2O(g)$$

If 22.4 L of gaseous CH_4 is burned, what volume of O_2 is required for complete combustion? What volumes of CO_2 and H_2O are produced? Assume all gases are at the same temperature and pressure.

12.3—The Ideal Gas Law

Four interrelated quantities can be used to describe a gas: pressure, volume, temperature, and amount (moles). We know from experiments that three gas laws can be used to describe the relationship of these properties (Section 12.2).

Boyle's Law	Charles's Law	Avogadro's Hypothesis
$V \propto (1/P)$	$V \propto T$	$V \propto n$
(constant T, n)	(constant P, n)	(constant T, P)

If all three relationships are combined, the result is

$$V \propto \frac{nT}{P}$$

This can be made into a mathematical equation by introducing a proportionality constant, R, called the **gas constant**. It is a *universal constant*—a number that you can use to interrelate the properties of any gas:

$$V = R\left(\frac{nT}{P}\right)$$

or

$$PV = nRT \qquad (12.4)$$

The equation $PV = nRT$ is called the **ideal gas law**. It describes the behavior of an "ideal" gas. As you will learn in Section 12.9, there is no such thing as an "ideal" gas.

■ **Properties of an ideal gas**
For ideal gases it is assumed that there are no forces of attraction between molecules and the molecules themselves occupy no volume.

However, real gases at pressures around one atmosphere or less and temperatures around room temperature usually behave close enough to ideality that $PV = nRT$ adequately describes their behavior.

To use the equation $PV = nRT$, we need a value for R. It is readily determined experimentally. By carefully measuring P, V, n, and T for a sample of gas, we can calculate the value of R from these values using the ideal gas law equation. For example, under conditions of **standard temperature and pressure (STP)**, a gas temperature of 0 °C or 273.15 K and a pressure of 1 atm, 1 mol of gas occupies 22.414 L, a quantity called the **standard molar volume**. Substituting these values into the ideal gas law equation gives a value for R:

$$R = \frac{PV}{nT} = \frac{(1.0000 \text{ atm})(22.414 \text{ L})}{(1.0000 \text{ mol})(273.15 \text{ K})} = 0.082057 \frac{\text{L} \cdot \text{atm}}{\text{K} \cdot \text{mol}}$$

With a value for R, we can now use the ideal gas law in calculations.

■ **STP—What Is It?**
A gas is at STP, or standard temperature and pressure, when its temperature is 0 °C or 273.15 K and its pressure is 1 atm. Under these conditions, exactly 1 mol of a gas occupies 22.414 L.

GENERAL
Chemistry•⦙•Now™

See the General ChemistryNow CD-ROM or website:
• **Screen 12.4 The Ideal Gas Law,** for a simulation of the ideal gas law and a tutorial

---

Example 12.6—Ideal Gas Law

Problem The nitrogen gas in an automobile air bag, with a volume of 65 L, exerts a pressure of 829 mm Hg at 25 °C. What amount of N_2 gas (in moles) is in the air bag?

Strategy You are given P, V, and T and want to calculate the amount of gas (n). Use the ideal gas law, Equation 12.4.

Solution First list the information provided.

$$P = 829 \text{ mm Hg} \quad V = 65 \text{ L} \quad T = 25 \text{ °C} \quad n = ?$$

To use the ideal gas law with R having units of (L · atm/K · mol), the pressure must be expressed in atmospheres and the temperature in kelvins. Therefore,

$$P = 829 \text{ mm Hg} \times \frac{1 \text{ atm}}{760 \text{ mm Hg}} = 1.09 \text{ atm}$$

$$T = 25 + 273 = 298 \text{ K}$$

Now substitute the values of P, V, T, and R into the ideal gas law and solve for the amount of gas, n:

$$n = \frac{PV}{RT} = \frac{(1.09 \text{ atm})(65 \text{ L})}{(0.082057 \text{ L} \cdot \text{atm}/\text{K} \cdot \text{mol})(298 \text{ K})} = 2.9 \text{ mol } N_2$$

Notice that the units of atmospheres, liters, and kelvins cancel to leave the answer in units of moles.

Exercise 12.6—Ideal Gas Law

The balloon used by Jacques Charles in his historic flight in 1783 was filled with about 1300 mol of H_2. If the temperature of the gas was 23 °C and its pressure was 750 mm Hg, what was the volume of the balloon?

(a)

(b)

Figure 12.8 Gas density. (a) The balloons are filled with nearly equal amounts of gas at the same temperature and pressure. One yellow balloon contains helium, a low-density gas ($d = 0.090$ g/L). The other balloons contain air, a higher-density gas ($d = 1.2$ g/L). (b) A hot-air balloon rises because the heated air has a lower density.

The Density of Gases

The density of a gas at a given temperature and pressure (Figure 12.8) is a useful quantity. Let us see how density is related to the ideal gas law. Because the amount (n, mol) of any compound is given by its mass (m) divided by its molar mass (M), we can substitute m/M for n in the ideal gas equation.

$$PV = \left(\frac{m}{M}\right)RT$$

Density (d) is defined as mass divided by volume (m/V). We can rearrange this form of the gas law to give the following equation, which has the term (m/V) on the left. This is the density of the gas.

$$d = \frac{m}{V} = \frac{PM}{RT} \tag{12.5}$$

Gas density is directly proportional to the pressure and molar mass and inversely proportional to the temperature. Equation 12.5 is useful because gas density can be calculated from the molar mass, or the molar mass can be found from a measurement of gas density of a gas at a given pressure and temperature.

Example 12.7—Density and Molar Mass

Problem Calculate the density of CO_2 at STP. Is CO_2 more or less dense than air (1.2 g/L)?

Strategy Use Equation 12.5, the equation relating gas density and molar mass. Here we know the molar mass (44.0 g/mol), pressure ($P = 1.00$ atm), temperature ($T = 273.15$ K), and the gas constant (R). Only the density (d) is unknown.

Solution The known values are substituted into Equation 12.5, which is then solved for density:

$$d = \frac{PM}{RT} = \frac{(1.00 \text{ atm})(44.0 \text{ g/mol})}{(0.082057 \text{ L} \cdot \text{atm/K} \cdot \text{mol})(273 \text{ K})} = 1.96 \text{ g/L}$$

The density of CO_2 is considerably greater than that of dry air at STP (1.3 g/L).

Figure 12.9   **Gas density.** Because carbon dioxide from fire extinguishers is denser than air, it settles on top of a fire and smothers it. (When CO_2 gas is released from the tank, it expands and cools significantly. The white cloud is solid CO_2 and condensed moisture from the air.)

Exercise 12.7—Gas Density Calculation

The density of an unknown gas is 5.02 g/L at 15.0 °C and 745 mm Hg. Calculate its molar mass.

Gas density has practical implications. From the equation $d = PM/RT$ we recognize that the density of a gas is directly proportional to its molar mass. Dry air, which has an average molar mass of about 29 g/mol, has a density of about 1.2 g/L at 1 atm and 25 °C. Gases or vapors with molar masses greater than 29 g/mol have densities larger than 1.2 g/L under these same conditions (1 atm and 25 °C). As a consequence, gases such as CO_2, SO_2, and gasoline vapor settle along the ground if released into the atmosphere (Figure 12.9). Conversely, gases such as H_2, He, CO, CH_4 (methane), and NH_3 rise if released into the atmosphere.

The significance of gas density was tragically revealed in several recent events. One occurred in the African country of Cameroon in 1986, when Lake Nyos expelled a huge bubble of CO_2 into the atmosphere. Because CO_2 is denser than air, the CO_2 cloud hugged the ground, killing 1700 people living in a nearby village.

Calculating the Molar Mass of a Gas from *P, V,* and *T* Data

When a new compound is isolated in the laboratory, one of the first things to be done is to determine its molar mass. If the compound is in the gas phase, a classical method of determining the molar mass is to measure the pressure and volume exerted by a given mass of the gas at a given temperature.

GENERAL
Chemistry·Now™

See the General ChemistryNow CD-ROM or website:

- **Screen 12.5 Gas Density,** to watch a video of a hot-air balloon and to work an exercise on gas density and molar mass
- **Screen 12.6 Using Gas Laws: Determining Molar Mass,** for a tutorial on gas density

---

Example 12.8—Calculating the Molar Mass of a Gas from *P, V,* and *T* Data

Problem You are trying to determine, by experiment, the empirical formula of a gaseous compound to replace chlorofluorocarbons in air conditioners. Your results give an empirical formula of CHF_2. Now you need the molar mass of the compound to find the molecular formula. You conduct another experiment and find that a 0.100-g sample of the compound exerts a pressure of 70.5 mm Hg in a 256-mL container at 22.3 °C. What is the molar mass of the compound? What is its molecular formula?

Strategy Here you know the mass of a gas in a given volume (*V*), so you can calculate its density (*d*). Then, knowing the gas pressure and temperature, you can use Equation 12.5 to calculate the molar mass.

Solution Begin by organizing the data:

$$m = \text{mass of gas} = 0.100 \text{ g}$$
$$P = 70.5 \text{ mm Hg, or } 0.0928 \text{ atm}$$
$$V = 256 \text{ mL, or } 0.256 \text{ L}$$
$$T = 22.3 \text{ °C, or } 295.5 \text{ K}$$

The density of the gas is the mass of the gas divided by the volume:

$$d = \frac{0.100 \text{ g}}{0.256 \text{ L}} = 0.391 \text{ g/L}$$

Use this value of density, along with the values of pressure and temperature in Equation 12.5 ($d = PM/RT$), and solve for the molar mass (M).

$$M = \frac{dRT}{P} = \frac{(0.391 \text{ g/L})(0.082057 \text{ L} \cdot \text{atm/K} \cdot \text{mol})(295.5 \text{ K})}{0.0928 \text{ atm}} = 102 \text{ g/mol}$$

With this result, you can compare the experimentally determined molar mass with the mass of a mole of gas having the empirical formula CHF_2.

$$\frac{\text{Experimental molar mass}}{\text{Mass of 1 mol } CHF_2} = \frac{102 \text{ g/mol}}{51.0 \text{ g/formula unit}} = 2 \text{ formula units of } CHF_2 \text{ per mol}$$

Therefore, the formula of the compound is $C_2H_2F_4$.

Comment Alternatively, you can use the ideal gas law. Here you know P and T for a gas in a given volume (V), so you can calculate the amount of gas (n).

$$n = \frac{PV}{RT} = \frac{(0.0928 \text{ atm})(0.256 \text{ L})}{(0.082057 \text{ L} \cdot \text{atm/K} \cdot \text{mol})(295.5 \text{ K})} = 9.80 \times 10^{-4} \text{ mol}$$

You now know that 0.100 g of gas is equivalent to 9.80×10^{-4} mol. Therefore,

$$\text{Molar mass} = \frac{0.100 \text{ g}}{9.80 \times 10^{-4} \text{ mol}} = 102 \text{ g/mol}$$

Exercise 12.8—Molar Mass from P, V, and T Data

A 0.105-g sample of a gaseous compound has a pressure of 561 mm Hg in a volume of 125 mL at 23.0 °C. What is its molar mass?

12.4—Gas Laws and Chemical Reactions

Many industrially important reactions involve gases. Two examples are the combination of nitrogen and hydrogen to produce ammonia,

$$N_2(g) + 3 \text{ } H_2(g) \longrightarrow 2 \text{ } NH_3(g)$$

and the electrolysis of aqueous NaCl to produce hydrogen and chlorine,

$$2 \text{ } NaCl(aq) + 2 \text{ } H_2O(\ell) \longrightarrow 2 \text{ } NaOH(aq) + H_2(g) + Cl_2(g)$$

If we want to understand the quantitative aspects of such reactions, we need to carry out stoichiometry calculations. The scheme in Figure 12.10 connects these calculations for gas reactions with the stoichiometry calculations done in Chapters 4 and 5.

GENERAL
Chemistry Now™

See the General ChemistryNow CD-ROM or website:

- **Screen 12.7 Gas Laws and Chemical Reactions: Stoichiometry,** for a tutorial on gas laws and chemical reactions

Figure 12.10   A scheme for performing stoichiometry calculations. Here A and B may be either reactants or products. The amount of A (mol) can be calculated from its mass in grams, from the concentration and volume of a solution, or from P, V, and T data by using the ideal gas law. Once the amount of B is determined, this value can be converted to a mass or solution concentration or volume, or to a volume of gas at a given pressure and temperature.

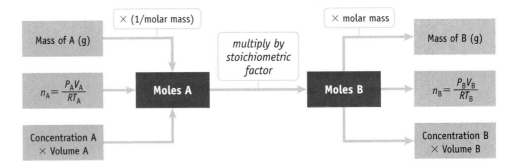

Example 12.9—Gas Laws and Stoichiometry

Problem You are asked to design an air bag for a car. You know that the bag should be filled with gas having a pressure higher than atmospheric pressure, say 829 mm Hg, at a temperature of 22.0 °C. The bag has a volume of 45.5 L. What quantity of sodium azide, NaN_3, should be used to generate the required quantity of gas? The gas-producing reaction is

$$2\ NaN_3(s) \longrightarrow 2\ Na(s) + 3\ N_2(g)$$

Strategy The general logic to be used here follows one of the pathways in Figure 12.10 (middle left to upper right).

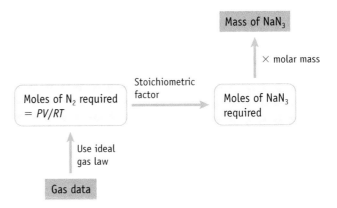

Solution The first step is to find the amount (mol) of gas required so that it can be related in the next step to the amount of sodium azide required:

$$P = 829 \text{ mm Hg } (1 \text{ atm}/760 \text{ mm Hg}) = 1.09 \text{ atm}$$

$$V = 45.5 \text{ L}$$

$$T = 22.0 \text{ °C, or } 295.2 \text{ K}$$

$$n = N_2 \text{ required (mol)} = \frac{PV}{RT}$$

$$n = \frac{(1.09 \text{ atm})(45.5 \text{ L})}{(0.082057 \text{ L} \cdot \text{atm/K} \cdot \text{mol})(295.2 \text{ K})} = 2.05 \text{ mol } N_2$$

Now that the required amount of nitrogen has been calculated, we can calculate the quantity of sodium azide that will produce 2.05 mol of N_2 gas.

$$\text{Mass of } NaN_3 = 2.05 \text{ mol } N_2 \left(\frac{2 \text{ mol } NaN_3}{3 \text{ mol } N_2}\right)\left(\frac{65.01 \text{ g}}{1 \text{ mol } NaN_3}\right) = 88.8 \text{ g } NaN_3$$

Example 12.10—Gas Laws and Stoichiometry

Problem You wish to prepare some deuterium gas, D_2, for use in an experiment. One technique is to react heavy water, D_2O, with an active metal such as lithium.

$$2\ Li(s) + 2\ D_2O(\ell) \longrightarrow 2\ LiOD(aq) + D_2(g)$$

Suppose you place 0.125 g of Li metal in 15.0 mL of D_2O ($d = 1.11$ g/mL). What amount of D_2 (in moles) can be prepared? If dry D_2 gas is captured in a 1450-mL flask at 22.0 °C, what is the pressure of the gas (in atm)? (Deuterium has an atomic weight of 2.0147 g/mol.)

Strategy You are combining two reactants with no guarantee that they are in the correct stoichiometric ratio. This reaction must therefore be approached as a *limiting reactant problem*. You have to find the amount of each substance and then see if one of them is present in a limited amount. Once the limiting reactant is known, the amount of D_2 produced and its pressure under the conditions given can be calculated.

Lithium metal (in the spoon) reacts with drops of water, H_2O, to produce LiOH and hydrogen gas, H_2. If heavy water, D_2O, is used, deuterium gas, D_2, can be produced.

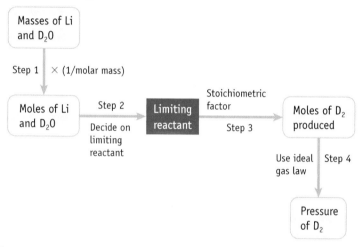

Solution

Step 1. *Calculate the amount (mol) of Li and of D_2O:*

$$0.125\ g\ Li \times \frac{1\ mol\ Li}{6.941\ g\ Li} = 0.0180\ mol\ Li$$

$$15.0\ mL\ D_2O \times \frac{1.11\ g\ D_2O}{1\ mL\ D_2O} \times \frac{1\ mol\ D_2O}{20.03\ g\ D_2O} = 0.831\ mol\ D_2O$$

Step 2. *Decide which reactant is the limiting reactant:*

$$\text{Ratio of moles of reactants available} = \frac{0.831\ mol\ D_2O}{0.0180\ mol\ Li} = \frac{46.2\ mol\ D_2O}{1\ mol\ Li}$$

The balanced equation shows that the ratio should be 1 mol of D_2O to 1 mol of Li. From the calculated values, we can see that D_2O is in large excess and Li is the limiting reactant. Therefore, further calculations are based on the amount of Li available.

Step 3. *Use the limiting reactant to calculate the amount of D_2 produced:*

$$0.0180\ mol\ Li \left(\frac{1\ mol\ D_2\ produced}{2\ mol\ Li}\right) = 0.00900\ mol\ D_2\ produced$$

Step 4. *Calculate the pressure of D_2:*

$P = ?$ $\qquad\qquad T - 22.0$ °C, or 295.2 K

$V = 1450$ mL, or 1.45 L $\qquad n = 0.00900$ mol D_2

$$P = \frac{nRT}{V} = \frac{(0.00900\ mol)(0.082057\ L \cdot atm/K \cdot mol)(295.2\ K)}{1.45\ L} = 0.150\ atm$$

Exercise 12.9—Gas Laws and Stoichiometry

Gaseous ammonia is synthesized by the reaction

$$N_2(g) + 3\ H_2(g) \xrightarrow[\substack{500\ °C}]{\text{iron catalyst}} 2\ NH_3(g)$$

Assume that 355 L of H_2 gas at 25.0 °C and 542 mm Hg is combined with excess N_2 gas. What amount of NH_3 gas, in moles, can be produced? If this amount of NH_3 gas is stored in a 125-L tank at 25.0 °C, what is the pressure of the gas?

12.5—Gas Mixtures and Partial Pressures

The air you breathe is a mixture of nitrogen, oxygen, carbon dioxide, water vapor, and small amounts of other gases (Table 12.1). Each of these gases exerts its own pressure, and atmospheric pressure is the sum of the pressures exerted by each gas. The pressure of each gas in the mixture is called its **partial pressure**.

■ **John Dalton (1766–1844)**
For a short biography of John Dalton, see page 65.

John Dalton (1766–1844) was the first to observe that the pressure of a mixture of gases is the sum of the pressures of the various gases in the mixture. This observation is now known as **Dalton's law of partial pressures** (Figure 12.11). Mathematically, we can write Dalton's law of partial pressures as

$$P_{\text{total}} = P_1 + P_2 + P_3 + \cdots \tag{12.6}$$

where P_1, P_2, and P_3 are the pressures of the different gases in a mixture and P_{total} is the total pressure.

In a mixture of gases, each gas behaves independently of all others in the mixture. Therefore, we can consider the behavior of each gas in a mixture separately. As an example let us take a mixture of three ideal gases, labeled A, B, and C. There are n_A moles of A, n_B moles of B, and n_C moles of C. Assume that the mixture ($n_{\text{total}} = n_A + n_B + n_C$) is contained in a given volume (V) at a given temperature (T). We can calculate the pressure exerted by each gas from the ideal gas law equation:

$$P_A V = n_A RT \quad P_B V = n_B RT \quad P_C V = n_C RT$$

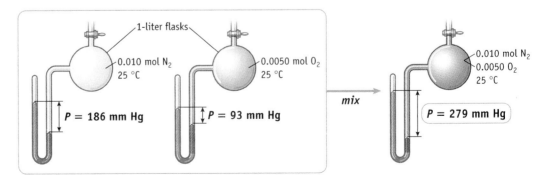

Figure 12.11 **Dalton's law.** In a 1.0-L flask at 25 °C, 0.010 mol of N_2 exerts a pressure of 186 mm Hg, and 0.0050 mol of O_2 in a 1.0-L flask at 25 °C exerts a pressure of 93 mm Hg (*left and middle*). The N_2 and O_2 samples are mixed in a 1.0-L flask at 25 °C (*right*). The total pressure, 279 mm Hg, is the sum of the partial pressures that each gas alone exerts in the flask.

Table 12.1 Components of Atmospheric Dry Air

Constituent	Molar Mass*	Mole Percent	Partial Pressure at STP (atm)
N_2	28.01	78.08	0.7808
O_2	32.00	20.95	0.2095
CO_2	44.01	0.033	0.00033
Ar	39.95	0.934	0.00934

*The average molar mass of dry air = 28.960 g/mol.

where each gas (A, B, and C) is in the same volume V and is at the same temperature T. According to Dalton's law, the total pressure exerted by the mixture is the sum of the pressures exerted by each component:

$$P_{total} = P_A + P_B + P_C = n_A\left(\frac{RT}{V}\right) + n_B\left(\frac{RT}{V}\right) + n_C\left(\frac{RT}{V}\right)$$

$$P_{total} = (n_A + n_B + n_C)\left(\frac{RT}{V}\right)$$

$$P_{total} = n_{total}\left(\frac{RT}{V}\right) \tag{12.7}$$

For mixtures of gases, it is convenient to introduce a quantity called the **mole fraction, X**, which is defined as the number of moles of a particular substance in a mixture divided by the total number of moles of all substances present. Mathematically, the mole fraction of a substance A in a mixture with B and C is expressed as

$$X_A = \frac{n_A}{n_A + n_B + n_C} = \frac{n_A}{n_{total}}$$

Now we can combine this equation (written as $n_{total} = n_A/X_A$) with the equations for P_A and P_{total}, and derive the equation

$$P_A = X_A P_{total} \tag{12.8}$$

This equation is useful because it tells us that the pressure of a gas in a mixture of gases is the product of its mole fraction and the total pressure of the mixture. For example, the mole fraction of N_2 in air is 0.78, so, at STP, its partial pressure is 0.78 atm or 590 mm Hg.

GENERAL
Chemistry ⚛ Now™

See the General ChemistryNow CD-ROM or website:
- **Screen 12.8 Gas Mixtures and Partial Pressures,** for two tutorials on Dalton's Law

Example 12.11—Partial Pressures of Gases

Problem Halothane, $C_2HBrClF_3$, is a nonflammable, nonexplosive, and nonirritating gas that is commonly used as an inhalation anesthetic.

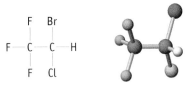

$$F \quad Br$$
$$| \quad |$$
$$F - C - C - H$$
$$| \quad |$$
$$F \quad Cl$$

1,1,1-trifluorobromochloroethane, halothane

Suppose you mix 15.0 g of halothane vapor with 23.5 g of oxygen gas. If the total pressure of the mixture is 855 mm Hg, what is the partial pressure of each gas?

Strategy One way to solve this problem is to recognize that the partial pressure of a gas is given by the total pressure of the mixture multiplied by the mole fraction of the gas.

Solution Let us first calculate the mole fractions of halothane and of O_2.

Step 1. *Calculate mole fractions:*

$$\text{Amount } C_2HBrClF_3 = 15.0 \text{ g} \left(\frac{1 \text{ mol}}{197.4 \text{ g}} \right) = 0.0760 \text{ mol}$$

$$\text{Amount } O_2 = 23.5 \text{ g} \left(\frac{1 \text{ mol}}{32.00 \text{ g}} \right) = 0.734 \text{ mol}$$

$$\text{Mole fraction } C_2HBrClF_3 = \frac{0.0760 \text{ mol } C_2HBrClF_3}{0.810 \text{ total moles}} = 0.0938$$

Because the sum of the mole fractions of halothane and of O_2 must equal 1.000, this means that the mole fraction of oxygen is 0.906.

$$X_{halothane} + X_{oxygen} = 1.000$$

$$0.0938 + X_{oxygen} = 1.000$$

$$X_{oxygen} = 0.906$$

Step 2. *Calculate partial pressures:*

$$\text{Partial pressure of halothane} = P_{halothane} = X_{halothane} \times P_{total}$$

$$P_{halothane} = 0.0938 \times P_{total} = 0.0938 \, (855 \text{ mm Hg})$$

$$P_{halothane} = 80.2 \text{ mm Hg}$$

The total pressure of the mixture is the sum of the partial pressures of the gases in the mixture.

$$P_{halothane} + P_{oxygen} = 855 \text{ mm Hg}$$

and so

$$P_{oxygen} = 855 \text{ mm Hg} - P_{halothane}$$

$$P_{oxygen} = 855 \text{ mm Hg} - 80.2 \text{ mm Hg} = 775 \text{ mm Hg}$$

Exercise 12.10—Partial Pressures of Gases

The halothane–oxygen mixture described in Example 12.11 is placed in a 5.00-L tank at 25.0 °C. What is the total pressure (in mm Hg) of the gas mixture in the tank? What are the partial pressures (in mm Hg) of the gases?

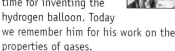
12.6—The Kinetic-Molecular Theory of Gases

So far we have discussed the macroscopic properties of gases, properties such as pressure and volume that result from the behavior of a system with a large number of particles. Now we turn to the kinetic-molecular theory [◀ Section 1.5] for a description of the behavior of matter at the molecular or atomic level. Hundreds of experimental observations have led to the following postulates regarding the behavior of gases:

- Gases consist of particles (molecules or atoms), whose separation is much greater than the size of the particles themselves (see Figure 12.12).

- The particles of a gas are in continual, random, and rapid motion. As they move, they collide with one another and with the walls of their container, but they do so without loss of energy.

- The average kinetic energy of gas particles is proportional to the gas temperature. *All gases, regardless of their molecular mass, have the same average kinetic energy at the same temperature.*

Let us discuss the behavior of gases from this point of view.

Molecular Speed and Kinetic Energy

If your friend walks into your room carrying a pizza, how do you know it? In scientific terms, we know that the odor-causing molecules of food enter the gas phase and drift through space until they reach the cells of your body that react to odors. The same thing happens in the laboratory when bottles of aqueous ammonia (NH_3) and hydrochloric acid (HCl) sit side by side (Figure 12.13). Molecules of the two compounds enter the gas phase and drift along until they encounter one another, at which time they react and form a cloud of tiny particles of solid ammonium chloride (NH_4Cl).

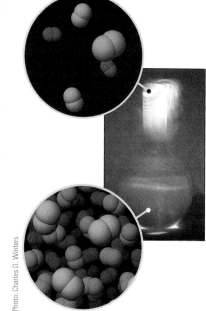

Photo: Charles D. Winters

Figure 12.12   A molecular view of gases and liquids. The fact that a large volume of N_2 gas can be condensed to a small volume of liquid indicates that the distances between molecules in the gas phase are very large as compared with the distances between molecules in liquids. (Liquid N_2 boils at −196 °C.)

Figure 12.13 The movement of gas molecules. Open dishes of aqueous ammonia and hydrochloric acid were placed side by side. When molecules of NH_3 and HCl escape from solution to the atmosphere and encounter one another, we observe a cloud of solid ammonium chloride, NH_4Cl.

If you change the temperature of the environment of the containers in Figure 12.13 and measure the time needed for the cloud of ammonium chloride to form, you would find that this time is longer at lower temperatures. The reason is that the speed at which molecules move depends on the temperature. Let us expand on this idea.

The molecules in a gas sample do not all move at the same speed. Rather, as illustrated in Figure 12.14 for O_2 molecules, there is a distribution of speeds. Figure 12.14 shows the number of particles in a gas sample that are moving at certain speeds at a given temperature. We can make two important observations. First, at a given temperature, some molecules have high speeds and others have low speeds. Most of the molecules, however, have some intermediate speed, and their most probable speed corresponds to the maximum in the curve. For oxygen gas at 25 °C, for example, most molecules have speeds in the range of 200 m/s to 700 m/s, and their most probable speed is about 400 m/s. (These are very high speeds, indeed. A speed of 400 m/s corresponds to about 900 miles per hour!)

A second observation regarding the distribution of speeds is that as the temperature increases, the most probable speed increases, and the number of molecules traveling at very high speeds increases greatly.

The kinetic energy of a single molecule of mass m in a gas sample is given by the equation

$$KE = \frac{1}{2}(\text{mass})(\text{speed})^2 = \frac{1}{2}mu^2$$

where u is the speed of that molecule. We can calculate the kinetic energy of a single gas molecule from this equation but not the kinetic energy of a collection of molecules, because not all of the molecules in a gas sample are moving at the same speed. However, we can calculate the *average* kinetic energy of a collection of molecules by relating it to other averaged quantities of the system. In particular, the average kinetic energy is related to the average speed.

$$\overline{KE} = \frac{1}{2}m\overline{u^2}$$

Figure 12.14 The distribution of molecular speeds. A graph of the number of molecules with a given speed versus that speed shows the distribution of molecular speeds. The red curve shows the effect of increased temperature. Even though the curve for the higher temperature is "flatter" and broader than the curve for the lower temperature, the areas under the curves are the same because the number of molecules in the sample is fixed.

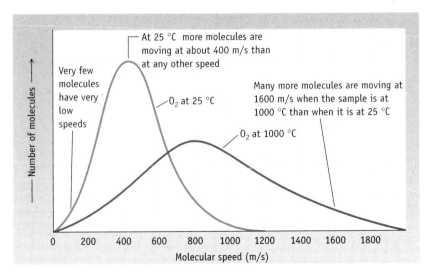

(The horizontal bar over the symbols KE and u indicate an average value.) This equation states that the average kinetic energy of the molecules in a gas sample, $\overline{KE}$, is related to $\overline{u^2}$, the average of the squares of their speeds (called the "mean square speed").

Experiments also show that the average kinetic energy, $\overline{KE}$, of a sample of gas molecules is directly proportional to temperature with a proportionality constant of $\frac{3}{2}R$:

$$\overline{KE} = \frac{3}{2}RT$$

where R is the gas constant expressed in SI units ($8.314472\ \text{J/K} \cdot \text{mol}$).

Because $\overline{KE}$ is proportional to both $\frac{1}{2}m\overline{u^2}$ and T, temperature and $\frac{1}{2}m\overline{u^2}$ must also be proportional; that is, $\frac{1}{2}m\overline{u^2} \propto T$. This relationship among mass, average speed, and temperature is expressed in Equation 12.9. Here the square root of the mean square speed ($\sqrt{\overline{u^2}}$, called the **root-mean-square** or **rms speed**), the temperature (T, in kelvins), and the molar mass (M) are related.

$$\sqrt{\overline{u^2}} = \sqrt{\frac{3RT}{M}} \tag{12.9}$$

This equation, sometimes called *Maxwell's equation* after James Clerk Maxwell [◀ page 296], shows that the speeds of gas molecules are indeed related directly to the temperature (Figure 12.14). The rms speed is a useful quantity because of its direct relationship to the average kinetic energy and because it is very close to the true average speed for a sample. (The average speed is 92% of the rms speed.)

All gases have the same average kinetic energy at the same temperature. However, if you compare a sample of one gas with another—say, compare O_2 and N_2—the molecules do not necessarily have the same average speed (Figure 12.15). Instead, Maxwell's equation shows that the smaller the molar mass of the gas, the greater the rms speed.

■ **Maxwell-Boltzmann Curves**
Plots showing the relationship between the number of molecules and their speed or energy (Figure 12.14) are often called Maxwell–Boltzmann distribution curves. They are named after James Clerk Maxwell (1831–1879) and Ludwig Boltzmann (1844–1906). The distribution of speeds (or kinetic energies) of molecules (as illustrated in Figures 12.14 and 12.15) is often used when explaining chemical phenomena.

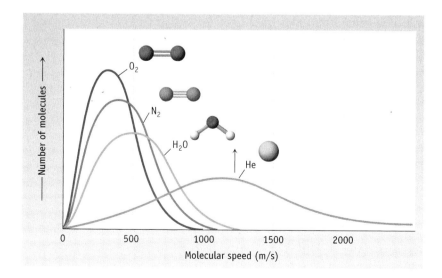

Figure 12.15 The effect of molecular mass on the distribution of speeds. At a given temperature, molecules with higher masses have lower speeds.

See the General ChemistryNow CD-ROM or website:
- **Screen 12.9 The Kinetic-Molecular Theory of Gases: Gases on the Molecular Scale,** to view an animation of gases at different temperatures
- **Screen 12.11 Distribution of Molecular Speeds: Maxwell-Boltzmann Curves,** to view an animation of a Boltzmann distribution and to see a simulation in which distribution curves are calculated

Example 12.12—Molecular Speed

Problem Calculate the rms speed of oxygen molecules at 25 °C.

Strategy We must use Equation 12.9 with M in units of kg/mol. The reason is that R is in units of J/K $\cdot$ mol, and 1 J $= 1$ kg $\cdot$ m^2/s^2.

Solution The molar mass of O_2 is 32.0×10^{-3} kg/mol.

$$\sqrt{\overline{u^2}} = \sqrt{\frac{3(8.3145 \text{ J/K} \cdot \text{mol})(298 \text{ K})}{32.0 \times 10^{-3} \text{ kg/mol}}} = \sqrt{2.32 \times 10^5 \text{ J/kg}}$$

To obtain the answer in meters per second, we use the relation 1 J $= 1$ kg $\cdot$ m^2/s^2. This means we have

$$\sqrt{\overline{u^2}} = \sqrt{2.32 \times 10^5 \text{ kg} \cdot \text{m}^2/(\text{kg} \cdot \text{s}^2)} = \sqrt{2.32 \times 10^5 \text{ m}^2/\text{s}^2} = 482 \text{ m/s}$$

This speed is equivalent to about 1100 miles per hour!

Exercise 12.11—Molecular Speed

Calculate the rms speeds of helium atoms and N_2 molecules at 25 °C.

Kinetic-Molecular Theory and the Gas Laws

The gas laws, which come from experiment, can be explained by the kinetic-molecular theory. The starting place is to describe how pressure arises from collisions of gas molecules with the walls of the container holding the gas (Figure 12.16). Recall that pressure is related to the force of the collisions (see Section 12.1).

$$\text{Gas pressure from collisions} = \frac{\text{force of collisions}}{\text{area}}$$

The force exerted by the collisions depends on the number of collisions and the average force per collision. When the temperature of a gas increases, the average kinetic energy of the molecules increases as well. In turn, the average force of the collisions with the walls increases. (This is akin to the difference in the force exerted by a car traveling at high speed versus one moving at only a few kilometers per hour.) Also, because the speed of gas molecules increases with temperature, more collisions occur per second. Thus, the collective force per square centimeter is greater, and the pressure increases. Mathematically, this is related to the direct proportionality between P and T when n and V are fixed; that is, $P = (nR/V)T$.

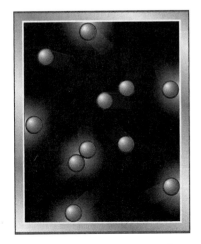

Figure 12.16 Gas pressure. According to the kinetic-molecular theory, gas pressure is caused by gas molecules bombarding the container walls.

Increasing the number of molecules of a gas at a fixed temperature and volume does not change the average collision force, but it does increase the number of collisions occurring per second. Thus, the pressure increases, and we can say that P is proportional to n when V and T are constant; that is, $P = n(RT/V)$.

If the pressure is to remain constant when either the number of molecules of gas or the temperature is increased, then the volume of the container (and the area over which the collisions can take place) must increase. This is expressed by stating that V is proportional to nT when P is constant $[V = nT(R/P)]$, a statement that is a *combination of Avogadro's hypothesis and Charles's law.*

Finally, if the temperature is constant, the average impact force of molecules of a given mass with the container walls must be constant. If n is kept constant while the volume of the container becomes smaller, the number of collisions with the container walls per second must increase. This means the pressure increases, so P is proportional to $1/V$ when n and T are constant, as stated by *Boyle's law*; that is, $P = (1/V)(nRT)$.

GENERAL
Chemistry⚛Now™

See the General ChemistryNow CD-ROM or website:

• **Screen 12.10 Gas Laws and Kinetic-Molecular Theory,** to view an animation and to see a simulation of the gas laws at the molecular level.

12.7—Diffusion and Effusion

When a pizza is brought into a room, the volatile aroma-causing molecules vaporize into the atmosphere, where they mix with the oxygen, nitrogen, carbon dioxide, water vapor, and other gases present. Even if there were no movement of the air in the room caused by fans or people moving about, the odor would eventually reach everywhere in the room. This mixing of molecules of two or more gases due to their random molecular motions is called **diffusion**. Given time, the molecules of one component in a gas mixture will thoroughly and completely mix with all other components of the mixture (Figure 12.17).

time →

(a)   (b)

Figure 12.17 Diffusion. (a) Liquid bromine, Br_2, was placed in a small flask inside a larger container. (b) The cork was removed from the flask and, with time, bromine vapor diffused into the larger container. Bromine vapor is now distributed evenly in the containers.

Charles D. Winters

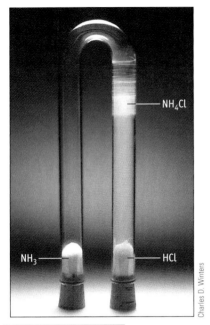

Active Figure 12.18   **Gaseous diffusion.** HCl gas (from hydrochloric acid) and NH₃ gas (from aqueous ammonia) diffuse from opposite ends of a glass U-tube. When they meet, they react to form white, solid NH₄Cl. The NH₄Cl is formed closer to the end from which the HCl gas begins because HCl molecules are heavier than NH₃ molecules and diffuse slower. See also Figure 12.13.

GENERAL
Chemistry•⚛•Now™ See the General ChemistryNow CD-ROM or website to explore an interactive version of this figure accompanied by an exercise.

Diffusion is also illustrated by the experiment shown in Figure 12.18. Here cotton moistened with hydrochloric acid is placed at one end of a U-tube, and cotton moistened with aqueous ammonia is placed at the other end. Molecules of HCl and NH₃ diffuse into the tube. When they meet, they produce white, solid NH₄Cl (just as in Figure 12.13).

$$HCl(g) + NH_3(g) \longrightarrow NH_4Cl(s)$$

Notice that the gases do not meet in the middle. Rather, because the heavier HCl molecules diffuse less rapidly than the lighter NH₃ molecules, the molecules meet closer to the HCl end of the U-tube.

Closely related to diffusion is **effusion**, which is the movement of gas through a tiny opening in a container into another container where the pressure is very low (Figure 12.19). Thomas Graham (1805–1869), a Scottish chemist, studied the effusion of gases and found that the rate of effusion of a gas—the amount of gas moving from one place to another in a given amount of time—is inversely proportional to the square root of its molar mass. Based on these experimental results, the rates of effusion of two gases can be compared:

$$\frac{\text{Rate of effusion of gas 1}}{\text{Rate of effusion of gas 2}} = \sqrt{\frac{\text{molar mass of gas 2}}{\text{molar mass of gas 1}}} \qquad (12.10)$$

The relationship in Equation 12.10—now known as **Graham's law**—is readily derived from Maxwell's equation by recognizing that the rate of effusion depends on the speed of the molecules. The ratio of the rms speeds is the same as the ratio of the effusion rates:

$$\frac{\text{Rate of effusion of gas 1}}{\text{Rate of effusion of gas 2}} = \frac{\sqrt{\overline{u^2} \text{ of gas 1}}}{\sqrt{\overline{u^2} \text{ of gas 2}}} = \frac{\sqrt{3RT/(M \text{ of gas 1})}}{\sqrt{3RT/(M \text{ of gas 2})}}$$

Canceling out like terms gives the expression in Equation 12.10.

GENERAL
Chemistry•⚛•Now™

See the General ChemistryNow CD-ROM or website:
• **Screen 12.12 Application of the Kinetic-Molecular Theory: Diffusion,** to watch a video of diffusion and for an interactive exercise

Figure 12.19   Effusion. H₂ and N₂ gas molecules effuse through the pores of a porous barrier. Lighter molecules (H₂) with higher average speeds strike the barrier more often and pass more often through it than heavier, slower molecules (N₂) at the same temperature. According to Graham's law, H₂ molecules effuse 3.73 times faster than N₂ molecules.

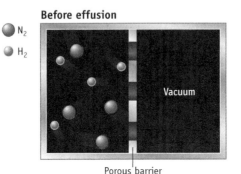

Before effusion

N₂

H₂

Vacuum

Porous barrier

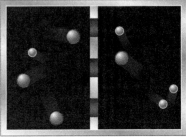

During effusion

Example 12.13—Using Graham's Law of Effusion to Calculate Molar Mass

Problem Tetrafluoroethylene, C_2F_4, effuses through a barrier at a rate of 4.6×10^{-6} mol/h. An unknown gas, consisting of only boron and hydrogen, effuses at a rate of 5.8×10^{-6} mol/h under the same conditions. What is the molar mass of the unknown gas?

Strategy From Graham's law we know that a light molecule will effuse more rapidly than a heavier one. Because the unknown gas effuses more rapidly than C_2F_4 ($M = 100.0$ g/mol), the unknown must have a molar mass less than 100 g/mol. Substitute the experimental data into Graham's law (Equation 12.10).

Solution

$$\frac{5.8 \times 10^{-6} \text{ mol/h}}{4.6 \times 10^{-6} \text{ mol/h}} = 1.3 = \sqrt{\frac{100.0 \text{ g/mol}}{M \text{ of unknown}}}$$

To solve for the unknown molar mass, square both sides of the equation and rearrange to find M for the unknown.

$$1.6 = \frac{100.0 \text{ g/mol}}{M \text{ of unknown}}$$

$$M = 63 \text{ g/mol}$$

Comment A boron–hydrogen compound corresponding to this molar mass is B_5H_9, called pentaborane.

Exercise 12.12—Graham's Law

A sample of pure methane, CH_4, is found to effuse through a porous barrier in 1.50 min. Under the same conditions, an equal number of molecules of an unknown gas effuse through the barrier in 4.73 min. What is the molar mass of the unknown gas?

12.8—Some Applications of the Gas Laws and Kinetic-Molecular Theory

Separating Isotopes

The effusion process played a central role in the development of the atomic bomb in World War II and is still used today to prepare fissionable uranium for nuclear power plants. Naturally occurring uranium exists primarily as two isotopes: ^{235}U (0.720% abundant) and ^{238}U (99.275% abundant). Because only the lighter isotope, ^{235}U, is suitable as a fuel in reactors, uranium ore must be enriched in this isotope.

Gas effusion is one way to separate the ^{235}U and ^{238}U isotopes. A uranium oxide sample is first converted to uranium hexafluoride, UF_6. This solid fluoride sublimes readily; it has a vapor pressure of 760 mm Hg at 55.6 °C. When UF_6 vapor is placed in a chamber with porous walls, the lighter, more rapidly moving $^{235}UF_6$ molecules effuse through the walls to a greater extent than the heavier $^{238}UF_6$ molecules.

To assess the separation of uranium isotopes, let us compare the rates of effusion of $^{235}UF_6$ and $^{238}UF_6$. Using Graham's law,

■ **Vapor pressure**
The vapor pressure of volatile liquids and solids is described in detail in Section 13.5.

$$\frac{\text{Rate of }^{235}UF_6}{\text{Rate of }^{238}UF_6} = \sqrt{\frac{238.051 + 6(18.998)}{235.044 + 6(18.998)}} = 1.0043$$

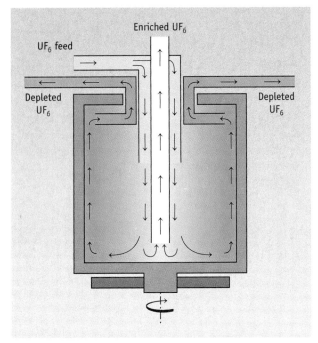

Oak Ridge National Lab

Figure 12.20   Isotope separation. Separation of uranium isotopes for use in atomic weaponry or in nuclear power plants was originally done by gas effusion. (These types of plants are still in use in the United States at Piketon, Ohio, and Paducah, Kentucky.) The more modern approach is to use a gas centrifuge (*left*). (*right*) UF_6 gas is injected into the centrifuge from a tube passing down through the center of a tall, spinning cylinder. The heavier $^{238}UF_6$ molecules experience more centrifugal force and move to the outer wall of the cylinder; the lighter $^{235}UF_6$ molecules stay closer to the center. A temperature difference inside the rotor causes the $^{235}UF_6$ molecules to move to the top of the cylinder. (See *The New York Times*, p. F1, March 23, 2004.)

we find that $^{235}UF_6$ will pass through a porous barrier 1.0043 times faster than $^{238}UF_6$. In other words, if we sample the gas that passes through the barrier, the fraction of $^{235}UF_6$ molecules will be larger. If the process is carried out again on the sample now higher in $^{235}UF_6$ concentration, the fraction of $^{235}UF_6$ would again increase in the effused sample, and the separation factor is now 1.0043×1.0043. If the cycle is repeated over and over again, the separation factor is 1.0043^n, where n is the number of enrichment cycles. To achieve a separation of about 99%, several hundred cycles are required (Figure 12.20)!

Deep Sea Diving

Diving with a self-contained underwater breathing apparatus (SCUBA) is exciting. If you want to dive much beyond about 60 ft (18 m), however, you need to take special precautions.

When you breathe air from a SCUBA tank (Figure 12.21), the pressure of the gas in your lungs is equal to the pressure exerted on your body. When you are at the surface, atmospheric pressure is about 1 atm and, because air has an oxygen concentration of 21%, the partial pressure of O_2 is about 0.21 atm. If you are at a depth of about 33 ft, the water pressure is 2 atm. Thus the oxygen partial pressure at this depth is double the surface partial pressure, or about 0.4 atm. Similarly, the partial pressure of N_2, which is about 0.8 atm at the surface, doubles to about 1.6 atm at a depth of 33 ft. The solubility of gases in water (and in blood) is directly proportional

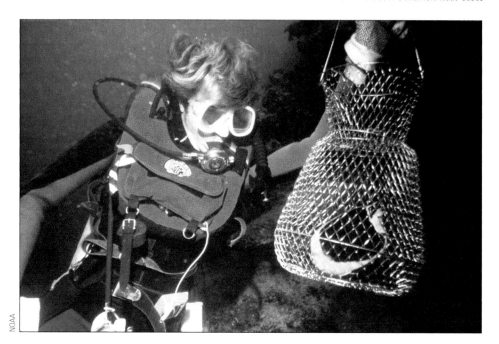

Figure 12.21 SCUBA diving. Ordinary recreational dives can be made with compressed air to depths of about 60 feet or so. With a gas mixture called Nitrox (which contains a maximum of 64% N_2), a person can stay at such depths for a longer period. To go even deeper, however, divers must breathe special gas mixtures such as Trimix. This breathing mixture consists of oxygen, helium, and nitrogen.

to pressure. Therefore, more oxygen and nitrogen dissolve in blood under these conditions, which creates a problem called "nitrogen narcosis."

Nitrogen narcosis, also called "rapture of the deep" or the "martini effect," results from the toxic effect on nerve conduction of N_2 dissolved in blood. Its effect is comparable to drinking a martini on an empty stomach or taking laughing gas (nitrous oxide, N_2O) at the dentist; it makes you slightly giddy. In severe cases, it can impair a diver's judgment and even cause a diver to take the regulator out of his or her mouth and hand it to a fish! Some people can go as deep as 130 ft with no problem, but others experience nitrogen narcosis at 80 ft.

Another problem with breathing air at depths beyond 100 ft or so is oxygen toxicity. Our bodies are regulated for a partial pressure of O_2 of 0.21 atm. At a depth of 130 ft, the partial pressure of O_2 is comparable to breathing 100% oxygen at sea level. These higher partial pressures can harm the lungs and cause central nervous system damage. Oxygen toxicity is the reason deep dives are done not with compressed air but rather with gas mixtures containing a much lower percentage of O_2—say, about 10%.

Because of the risk of nitrogen narcosis, divers going beyond about 130 ft, such as those who work for offshore oil drilling companies, use a mixture of oxygen and helium. This solves the nitrogen narcosis problem, but it introduces another side effect. If the diver has a voice link to the surface, the diver's speech sounds like Donald Duck! Speech is altered because the velocity of sound in helium is different from that in air, and the density of gas at several hundred feet is much higher than at the surface.

12.9—Nonideal Behavior: Real Gases

If you are working with a gas at approximately room temperature and a pressure of 1 atm or less, the ideal gas law is remarkably successful in relating the amount of gas and its pressure, volume, and temperature. At higher pressures or lower temperatures, however, deviations from the ideal gas law occur. The origin of these deviations is explained by the breakdown of the assumptions used when describing ideal

■ **Assumptions of the KMT—Revisited**
The assumptions of the kinetic-molecular theory were given on page 567.
1. Gases consist of particles (molecules or atoms), whose separation is much greater than the size of the particles themselves.
2. The particles of a gas are in continual, random, and rapid motion. As they move, they collide with one another and with the walls of their container, but they do so without loss of energy.
3. The average kinetic energy of gas particles is proportional to the gas temperature. All gases, regardless of their molecular mass, have the same average kinetic energy at the same temperature.

Table 12.2   Van der Waals Constants

Gas	a Values (atm · L²/mol²)	b Values (L/mol)
He	0.034	0.0237
Ar	1.34	0.0322
H_2	0.244	0.0266
N_2	1.39	0.0391
O_2	1.36	0.0318
CO_2	3.59	0.0427
Cl_2	6.49	0.0562
H_2O	5.46	0.0305

gases. Specifically, ideality assumes gas molecules have no volume and that no forces act between them.

At standard temperature and pressure (STP), the volume occupied by a single molecule is *very* small relative to its share of the total gas volume. A helium atom with a radius of 31 pm, for example, has roughly the same space to move about as a pea has inside a basketball. Now suppose the pressure is increased significantly, to 1000 atm. The volume available to each molecule is a sphere with a radius of only about 200 pm, which means the situation is now like that of a pea inside a sphere a bit larger than a ping-pong ball.

The kinetic-molecular theory and the ideal gas law are concerned with the volume available to the molecules to move about, not the volume of the molecules themselves. It is clear that the volume occupied by gas molecules is not negligible at higher pressures. For example, suppose you have a flask marked with a volume of 500 mL. This does not mean the space available to molecules is 500 mL. In reality, the available volume is less than 500 mL, especially at high gas pressures, because the molecules themselves occupy some of the volume.

Another assumption of the kinetic-molecular theory is that collisions between molecules are elastic—that is, that the atoms or molecules of the gas never stick to one another by some type of intermolecular force. This is also clearly not true. All gases can be liquefied, although some gases require a very low temperature to do so (see Figure 12.12). The only way that this phase change can happen is if there are forces between the molecules. When a molecule is about to strike the wall of its container, other molecules in its vicinity exert a slight attraction for the molecule and pull it away from the wall. As a result of the intermolecular forces, molecules strike the wall with less force than they would in the absence of intermolecular attractive forces. Thus, because collisions between molecules in a real gas and the wall are softer, the observed gas pressure is less than that predicted by the ideal gas law. This effect can be particularly pronounced when the temperature is low.

The Dutch physicist Johannes van der Waals (1837–1923) studied the breakdown of the ideal gas law equation and developed an equation to correct for the errors arising from nonideality. This equation is known as the **van der Waals equation**:

Observed pressure   Container V

$$\left(P + a\left[\frac{n}{V}\right]^2\right)(V - bn) = nRT \tag{12.11}$$

Correction for intermolecular forces   Correction for molecular volume

where a and b are experimentally determined constants (Table 12.2). Although Equation 12.11 might seem complicated at first glance, the terms in parentheses are those of the ideal gas law, each corrected for the effects discussed previously. The pressure correction term, $a(n/V)^2$, accounts for intermolecular forces. Owing to intermolecular forces the observed gas pressure is lower than the ideal pressure ($P_{observed} < P_{ideal}$, where P_{ideal} is calculated using the equation $PV = nRT$). Therefore, the term $a(n/V)^2$ is added to the observed pressure. The constant a typically has values in the range 0.01 to 10 atm · L²/mol². The actual volume available to the molecules is smaller than the volume of the container because the molecules themselves take up space. Therefore, we subtract an amount from the container volume ($= bn$) to take this factor into account. Here n is the number of moles of gas, and b is an experimental quantity that corrects for the molecular volume. Typical values of b range from 0.01 to 0.1 L/mol, roughly increasing with increasing molecular size.

Chemical Perspectives

The Earth's Atmosphere

Earth's atmosphere is a fascinating mixture of gases in more or less distinct layers with widely differing temperatures.

Up to the tropopause, there is a gradual decline in temperature (and pressure) with altitude. The temperature climbs again in the stratosphere due to the absorption of energy from the sun by stratospheric ozone, O_3.

Above the stratosphere, the pressure declines because fewer molecules are present. At still higher altitudes, we observe a dramatic increase in temperature in the thermosphere. This trend illustrates the difference between *temperature* and *heat*. The temperature of a gas reflects the *average* kinetic energy of the molecules of the gas, whereas the heat present in an object is the *total* kinetic energy of the molecules. In the thermosphere, the few molecules present have a very high temperature, but the heat content is exceedingly small because there are so few molecules.

Gases within the troposphere are well mixed by convection. Pollutants that are evolved on the earth's surface can rise into the stratosphere, but the stratosphere then acts as a "thermal lid" on the troposphere, preventing significant mixing of polluting gases into the stratosphere and beyond.

The pressure of the atmosphere declines with altitude; in conjunction with this trend, the partial pressure of O_2 declines. The figure shows why climbers have a hard time breathing on Mount Everest. At the mountain's peak, the altitude is 29,028 ft (8848 m), and the O_2 partial pressure is only 30% of the sea-level partial pressure. With proper training, a climber can reach the summit without supplemental oxygen. However, this same feat would not be possible if Mount Everest were located farther north. The earth's atmosphere thins toward the poles, so the O_2 partial pressure would be even lower if Mount Everest's summit were in North America, for example.

See G. N. Eby: *Environmental Geochemistry*, Belmont, CA, Thomson/Brooks/Cole, 2004.

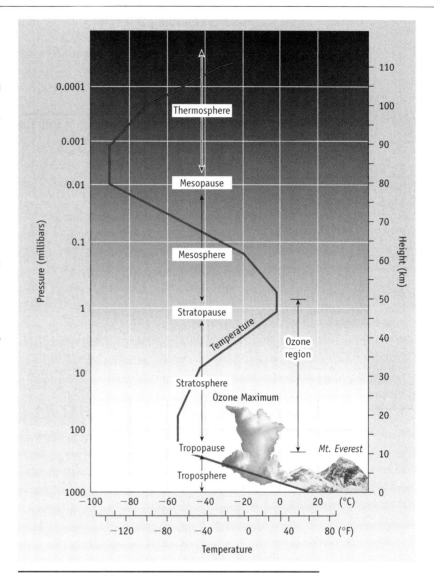

Average Composition of the Earth's Atmosphere to a Height of 25 km

Gas	Volume %	Source
N_2	78.08	biologic
O_2	20.95	biologic
Ar	0.93	radioactivity
Ne	0.0018	Earth's interior
He	0.0005	radioactivity
H_2O	0 to 4	evaporation
CO_2	0.036	biologic, industrial
CH_4	0.00017	biologic
N_2O	0.00003	biologic, industrial
O_3	0.000004	photochemical

As an example of the importance of these corrections, consider a sample of 8.00 mol of chlorine gas, Cl_2, in a 4.00-L tank at 27.0 °C. The ideal gas law would lead you to expect a pressure of 49.2 atm. A better estimate of the pressure, obtained from the van der Waals equation, is 29.5 atm, about 20 atm less than the ideal pressure!

Exercise 12.13—The van der Waals Equation

Using both the ideal gas law and the van der Waals equation, calculate the pressure expected for 10.0 mol of helium gas in a 1.00-L container at 25 °C.

GENERAL
Chemistry ⚛ Now™

See the General ChemistryNow CD-ROM or website to:
- Assess your understanding with homework questions keyed to each goal
- Check your readiness for an exam by taking the exam-prep quiz and exploring the resources in the personalized Learning Plan it provides

Chapter Goals Revisited

When you have finished studying this chapter, you should ask if you have met the chapter goals. In particular you should be able to

Understand the basis of the gas laws and know how to use those laws
a. Describe how pressure measurements are made and work with the units of pressure, especially atmospheres (atm) and millimeters of mercury (mm Hg) (Section 12.1). General ChemistryNow homework: Study Question(s) 1

b. Understand the origins of the gas laws (Boyle's law, Charles's law, and Avogadro's hypothesis) and know how to apply them (Section 12.2). General ChemistryNow homework: SQ(s) 6, 8, 10, 12

Use the ideal gas law
a. Understand the origin of the ideal gas law and know how to use the equation (Section 12.3). General ChemistryNow homework: SQ(s) 18, 22, 24, 65

b. Calculate the molar mass of a compound from a knowledge of the pressure of a known quantity of a gas in a given volume at a known temperature (Section 12.3). General ChemistryNow homework: SQ(s) 26, 30

Apply the gas laws to stoichiometric calculations (Section 12.4)
a. Stoichiometric calculations involving gases. General ChemistryNow homework: SQ(s) 32, 34, 71, 93

b. Use Dalton's law of partial pressures (Section 12.5). General ChemistryNow homework: SQ(s) 39, 40, 74

Understand kinetic-molecular theory as it is applied to gases, especially the distribution of molecular speeds (energies) (Section 12.6)
a. Apply the kinetic-molecular theory of gas behavior at the molecular level (Section 12.6). General ChemistryNow homework: SQ(s) 41, 45

b. Understand the phenomena of diffusion and effusion and know how to use Graham's law (Section 12.7). General ChemistryNow homework: SQ(s) 47

Recognize why real-world gases do not behave like ideal gases
a. Appreciate the fact that gases usually do not behave as ideal gases (Section 12.9). Deviations from ideal behavior are largest at high pressure and low temperature.

Key Equations

Equation 12.1 (page 551)
Boyle's law (where P is the gas pressure and V is its volume)

$$P_1V_1 = P_2V_2$$

Equation 12.2 (page 553)
Charles's law (where T is the temperature in kelvins)

$$\frac{V_1}{T_1} = \frac{V_2}{T_2}$$

Equation 12.3 (page 554)
General gas law (combined gas law)

$$\frac{P_1V_1}{T_1} = \frac{P_2V_2}{T_2}$$

Equation 12.4 (page 557)
Ideal gas law (where n is the amount of gas in moles and R is the universal gas constant, $0.082057 \, \text{L} \cdot \text{atm}/\text{K} \cdot \text{mol}$)

$$PV = nRT$$

Equation 12.5 (page 559)
Density of gases (where d is the gas density in grams per liter)

$$d = \frac{m}{V} = \frac{PM}{RT}$$

Equation 12.6 (page 564)
Dalton's law of partial pressures: The total pressure of a gas mixture is the sum of the partial pressures of the component gases (P_n)

$$P_{\text{total}} = P_1 + P_2 + P_3 + \cdots$$

Equation 12.7 (page 565)
The total pressure of a gas mixture is equal to the total number of moles of gases multiplied by (RT/V)

$$P_{\text{total}} = n_{\text{total}}\left(\frac{RT}{V}\right)$$

Equation 12.8 (page 565)
The pressure of a gas (A) in a mixture is the product of its mole fraction (X_A) and the total pressure of the mixture

$$P_A = X_A P_{\text{total}}$$

Equation 12.9 (page 569)
Maxwell's equation, which relates the rms speed $\sqrt{\overline{u^2}}$ to the molar mass of a gas (M) and its temperature (T) (where R = $8.314472 \, \text{J}/\text{K} \cdot \text{mol}$)

$$\sqrt{\overline{u^2}} = \sqrt{\frac{3RT}{M}}$$

Equation 12.10 (page 572)
Graham's law: The rate of effusion of a gas—the amount of material moving from one place to another in a given time—is inversely proportional to the square root of its molar mass

$$\frac{\text{Rate of effusion of gas 1}}{\text{Rate of effusion of gas 2}} = \sqrt{\frac{\text{molar mass of gas 2}}{\text{molar mass of gas 1}}}$$

Equation 12.11 (page 576)
The van der Waals equation: Relates pressure, volume, temperature, and amount of gas for a nonideal gas

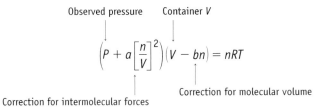

$$\left(P + a\left[\frac{n}{V}\right]^2\right)(V - bn) = nRT$$

Observed pressure   Container V

Correction for intermolecular forces   Correction for molecular volume

---

Study Questions

▲ denotes more challenging questions.

■ denotes questions available in the Homework and Goals section of the General ChemistryNow CD-ROM or website.

Blue numbered questions have answers in Appendix O and fully worked solutions in the *Student Solutions Manual*.

Structures of many of the compounds used in these questions are found on the General ChemistryNow CD-ROM or website in the Models folder.

GENERAL
Chemistry••Now™ Assess your understanding of this chapter's topics with additional quizzing and conceptual questions at **http://now.brookscole.com/kotz6e**

Practicing Skills

Pressure
(See Example 12.1 and the General ChemistryNow Screen 12.2.)

1. ■ The pressure of a gas is 440 mm Hg. Express this pressure in units of (a) atmospheres, (b) bars, and (c) kilopascals.

2. The average barometric pressure at an altitude of 10 km is 210 mm Hg. Express this pressure in atmospheres, bars, and kilopascals.

3. Indicate which represents the higher pressure in each of the following pairs:
   (a) 531 mm Hg or 0.754 bar
   (b) 534 mm Hg or 650 kPa
   (c) 1.34 bar or 934 kPa

4. Put the following in order of increasing pressure: 363 mm Hg, 363 kPa, 0.256 atm, and 0.523 bar.

Boyle's Law and Charles's Law
(See Examples 12.2 and 12.3 and the General ChemistryNow Screen 12.3.)

5. A sample of nitrogen gas has a pressure of 67.5 mm Hg in a 500.-mL flask. What is the pressure of this gas sample when it is transferred to a 125-mL flask at the same temperature?

6. ■ A sample of CO_2 gas has a pressure of 56.5 mm Hg in a 125-mL flask. The sample is transferred to a new flask, where it has a pressure of 62.3 mm Hg at the same temperature. What is the volume of the new flask?

7. You have 3.5 L of NO at a temperature of 22.0 °C. What volume would the NO occupy at 37 °C? (Assume the pressure is constant.)

8. ■ A 5.0-mL sample of CO_2 gas is enclosed in a gas-tight syringe (see Figure 12.4) at 22 °C. If the syringe is immersed in an ice bath (0 °C), what is the new gas volume, assuming that the pressure is held constant?

The General Gas Law
(See Example 12.4.)

9. You have 3.6 L of H_2 gas at 380 mm Hg and 25 °C. What is the pressure of this gas if it is transferred to a 5.0-L flask at 0.0 °C?

10. ■ You have a sample of CO_2 in a flask A with a volume of 25.0 mL. At 20.5 °C, the pressure of the gas is 436.5 mm Hg. To find the volume of another flask B, you move the CO_2 to that flask and find that its pressure is now 94.3 mm Hg at 24.5 °C. What is the volume of flask B?

11. You have a sample of gas in a flask with a volume of 250 mL. At 25.5 °C the pressure of the gas is 360 mm Hg. If you decrease the temperature to −5.0 °C, what is the gas pressure at the lower temperature?

12. ■ A sample of gas occupies 135 mL at 22.5 °C; the pressure is 165 mm Hg. What is the pressure of the gas sample when it is placed in a 252-mL flask at a temperature of 0.0 °C?

---

13. One of the cylinders of an automobile engine has a volume of 400. cm^3. The engine takes in air at a pressure of 1.00 atm and a temperature of 15 °C and compresses the air to a volume of 50.0 cm^3 at 77 °C. What is the final pressure of the gas in the cylinder? (The ratio of before and after volumes—in this case, 400:50 or 8:1—is called the compression ratio.)

14. A helium-filled balloon of the type used in long-distance flying contains 420,000 ft^3 (1.2×10^7 L) of helium. Suppose you fill the balloon with helium on the ground, where the pressure is 737 mm Hg and the temperature is 16.0 °C. When the balloon ascends to a height of 2 miles, where the pressure is only 600. mm Hg and the temperature is −33 °C, what volume is occupied by the helium gas? Assume the pressure inside the balloon matches the external pressure. Comment on the result.

Avogadro's Hypothesis
(See Example 12.5 and the General ChemistryNow Screen 12.3.)

15. Nitrogen monoxide reacts with oxygen to give nitrogen dioxide.

$$2\,NO(g) + O_2(g) \longrightarrow 2\,NO_2(g)$$

   (a) If you mix NO and O$_2$ in the correct stoichiometric ratio, and NO has a volume of 150 mL, what volume of O$_2$ is required (at the same pressure and temperature)?

   (b) After reaction is complete between 150 mL of NO and the stoichiometric volume of O$_2$, what is the volume of NO$_2$ (at the same pressure and temperature)?

16. Ethane, C$_2$H$_6$, burns in air according to the equation

$$2\,C_2H_6(g) + 7\,O_2(g) \longrightarrow 4\,CO_2(g) + 6\,H_2O(g)$$

   What volume of O$_2$ (L) is required for complete reaction with 5.2 L of C$_2$H$_6$? What volume of H$_2$O vapor (L) is produced? Assume all gases are measured at the same temperature and pressure.

Ideal Gaw Law
(See Example 12.6 and the General ChemistryNow Screen 12.4.)

17. A 1.25-g sample of CO$_2$ is contained in a 750.-mL flask at 22.5 °C. What is the pressure of the gas?

18. ■ A balloon holds 30.0 kg of helium. What is the volume of the balloon if the final pressure is 1.20 atm and the temperature is 22 °C?

19. A flask is first evacuated so that it contains no gas at all. Then, 2.2 g of CO$_2$ is introduced into the flask. On warming to 22 °C, the gas exerts a pressure of 318 mm Hg. What is the volume of the flask?

20. A steel cylinder holds 1.50 g of ethanol, C$_2$H$_5$OH. What is the pressure of the ethanol vapor if the cylinder has a volume of 251 cm^3 and the temperature is 250 °C? (Assume all of the ethanol is in the vapor phase at this temperature.)

21. A balloon for long-distance flying contains 1.2×10^7 L of helium. If the helium pressure is 737 mm Hg at 25 °C, what mass of helium (in grams) does the balloon contain? (See Study Question 14 and page 546.)

22. ■ What mass of helium, in grams, is required to fill a 5.0-L balloon to a pressure of 1.1 atm at 25 °C?

Gas Density
(See Examples 12.7 and 12.8 and the General ChemistryNow Screen 12.5.)

23. Forty miles above the earth's surface the temperature is 250 K and the pressure is only 0.20 mm Hg. What is the density of air (in grams per liter) at this altitude? (Assume the molar mass of air is 28.96 g/mol.)

24. ■ Diethyl ether, (C$_2$H$_5$)$_2$O, vaporizes easily at room temperature. If the vapor exerts a pressure of 233 mm Hg in a flask at 25 °C, what is the density of the vapor?

25. A gaseous organofluorine compound has a density of 0.355 g/L at 17 °C and 189 mm Hg. What is the molar mass of the compound?

26. ■ Chloroform is a common liquid used in the laboratory. It vaporizes readily. If the pressure of chloroform vapor in a flask is 195 mm Hg at 25.0 °C, and the density of the vapor is 1.25 g/L, what is the molar mass of chloroform?

Ideal Gas Laws and Determining Molar Mass
(See Examples 12.7 and 12.8 and the General ChemistryNow Screen 12.6.)

27. A 1.007-g sample of an unknown gas exerts a pressure of 715 mm Hg in a 452-mL container at 23 °C. What is the molar mass of the gas?

28. A 0.0125-g sample of a gas with an empirical formula of CHF$_2$ is placed in a 165-mL flask. It has a pressure of 13.7 mm Hg at 22.5 °C. What is the molecular formula of the compound?

29. A new boron hydride, B$_x$H$_y$, has been isolated. To find its molar mass, you measure the pressure of the gas in a known volume at a known temperature. The following experimental data are collected:

   Mass of gas = 12.5 mg
   Pressure of gas = 24.8 mm Hg
   Temperature = 25 °C
   Volume of flask = 125 mL

   Which formula corresponds to the calculated molar mass?
   (a) B$_2$H$_6$
   (b) B$_4$H$_{10}$
   (c) B$_5$H$_9$
   (d) B$_6$H$_{10}$
   (e) B$_{10}$H$_{14}$

30. ■ Acetaldehyde is a common liquid compound that vaporizes readily. Determine the molar mass of acetaldehyde from the following data:

   Sample mass = 0.107 g   Volume of gas = 125 mL
   Temperature = 0.0 °C   Pressure = 331 mm Hg

Gas Laws and Stoichiometry

(See Examples 12.9 and 12.10 and the General ChemistryNow Screen 12.7.)

31. Iron reacts with hydrochloric acid to produce iron(II) chloride and hydrogen gas:

$$Fe(s) + 2 HCl(aq) \longrightarrow FeCl_2(aq) + H_2(g)$$

The H_2 gas from the reaction of 2.2 g of iron with excess acid is collected in a 10.0-L flask at 25 °C. What is the pressure of the H_2 gas in this flask?

32. ■ Silane, SiH_4, reacts with O_2 to give silicon dioxide and water:

$$SiH_4(g) + 2 O_2(g) \longrightarrow SiO_2(s) + 2 H_2O(\ell)$$

A 5.20-L sample of SiH_4 gas at 356 mm Hg pressure and 25 °C is allowed to react with O_2 gas. What volume of O_2 gas, in liters, is required for complete reaction if the oxygen has a pressure of 425 mm Hg at 25 °C?

33. Sodium azide, the explosive compound in automobile air bags, decomposes according to the following equation:

$$2 NaN_3(s) \longrightarrow 2 Na(s) + 3 N_2(g)$$

What mass of sodium azide is required to provide the nitrogen needed to inflate a 75.0-L bag to a pressure of 1.3 atm at 25 °C?

34. ■ The hydrocarbon octane (C_8H_{18}) burns to give CO_2 and water vapor:

$$2 C_8H_{18}(g) + 25 O_2(g) \longrightarrow 16 CO_2(g) + 18 H_2O(g)$$

If a 0.095-g sample of octane burns completely in O_2, what will be the pressure of water vapor in a 4.75-L flask at 30.0 °C? If the O_2 gas needed for complete combustion was contained in a 4.75-L flask at 22 °C, what would its pressure be?

35. Hydrazine reacts with O_2 according to the following equation:

$$N_2H_4(g) + O_2(g) \longrightarrow N_2(g) + 2 H_2O(\ell)$$

Assume the O_2 needed for the reaction is in a 450-L tank at 23 °C. What must the oxygen pressure be in the tank to have enough oxygen to consume 1.00 kg of hydrazine completely?

36. A self-contained breathing apparatus uses canisters containing potassium superoxide. The superoxide consumes the CO_2 exhaled by a person and replaces it with oxygen.

$$4 KO_2(s) + 2 CO_2(g) \longrightarrow 2 K_2CO_3(s) + 3 O_2(g)$$

What mass of KO_2, in grams, is required to react with 8.90 L of CO_2 at 22.0 °C and 767 mm Hg?

Gas Mixtures and Dalton's Law

(See Example 12.11 and the General ChemistryNow Screen 12.8.)

37. What is the total pressure in atmospheres of a gas mixture that contains 1.0 g of H_2 and 8.0 g of Ar in a 3.0-L container at 27 °C? What are the partial pressures of the two gases?

38. A cylinder of compressed gas is labeled "Composition (mole %): 4.5% H_2S, 3.0% CO_2, balance N_2." The pressure gauge attached to the cylinder reads 46 atm. Calculate the partial pressure of each gas, in atmospheres, in the cylinder.

39. ■ A halothane–oxygen mixture ($C_2HBrClF_3 + O_2$) can be used as an anesthetic. A tank containing such a mixture has the following partial pressures: P (halothane) = 170 mm Hg and P (O_2) = 570 mm Hg.
    (a) What is the ratio of the number of moles of halothane to the number of moles of O_2?
    (b) If the tank contains 160 g of O_2, what mass of $C_2HBrClF_3$ is present?

40. ■ A collapsed balloon is filled with He to a volume of 12.5 L at a pressure of 1.00 atm. Oxygen, O_2 is then added so that the final volume of the balloon is 26 L with a total pressure of 1.00 atm. The temperature, which remains constant throughout, is 21.5 °C.
    (a) What mass of He does the balloon contain?
    (b) What is the final partial pressure of He in the balloon?
    (c) What is the partial pressure of O_2 in the balloon?
    (d) What is the mole fraction of each gas?

Kinetic-Molecular Theory

(See Section 12.6, Example 12.12, and the General ChemistryNow Screens 12.9–12.12.)

41. ■ You have two flasks of equal volume. Flask A contains H_2 at 0 °C and 1 atm pressure. Flask B contains CO_2 gas at 25 °C and 2 atm pressure. Compare these two gases with respect to each of the following:
    (a) average kinetic energy per molecule
    (b) average molecular velocity
    (c) number of molecules
    (d) mass of gas

42. Equal masses of gaseous N_2 and Ar are placed in separate flasks of equal volume at the same temperature. Tell whether each of the following statements is true or false. Briefly explain your answer in each case.
    (a) There are more molecules of N_2 present than atoms of Ar.
    (b) The pressure is greater in the Ar flask.
    (c) The Ar atoms have a greater average speed than the N_2 molecules.
    (d) The N_2 molecules collide more frequently with the walls of the flask than do the Ar atoms.

43. If the speed of an oxygen molecule is 4.28×10^4 cm/s at 25 °C, what is the speed of a CO_2 molecule at the same temperature?

44. Calculate the rms speed for CO molecules at 25 °C. What is the ratio of this speed to that of Ar atoms at the same temperature?

45. ■ Place the following gases in order of increasing average molecular speed at 25 °C: Ar, CH_4, N_2, CH_2F_2.

46. The reaction of SO_2 with Cl_2 gives dichlorine oxide, which is used to bleach wood pulp and to treat wastewater:

$$SO_2(g) + 2\ Cl_2(g) \longrightarrow OSCl_2(g) + Cl_2O(g)$$

All of the compounds involved in the reaction are gases. List them in order of increasing average speed.

Diffusion and Effusion

(See Example 12.13 and the General ChemistryNow Screen 12.12.)

47. ■ In each pair of gases below, tell which will effuse faster:
    (a) CO_2 or F_2
    (b) O_2 or N_2
    (c) C_2H_4 or C_2H_6
    (d) two chlorofluorocarbons: $CFCl_3$ or $C_2Cl_2F_4$

48. Argon gas is ten times denser than helium gas at the same temperature and pressure. Which gas is predicted to effuse faster? How much faster?

49. A gas whose molar mass you wish to know effuses through an opening at a rate one-third as fast as that of helium gas. What is the molar mass of the unknown gas?

50. ▲ A sample of uranium fluoride is found to effuse at the rate of 17.7 mg/h. Under comparable conditions, gaseous I_2 effuses at the rate of 15.0 mg/h. What is the molar mass of the uranium fluoride? (*Hint:* Rates must be converted to units of moles per time.)

Nonideal Gases

(See Section 12.9.)

51. In the text it is stated that the pressure of 8.00 mol of Cl_2 in a 4.00-L tank at 27.0 °C should be 29.5 atm if calculated using the van der Waals's equation. Verify this result and compare it with the pressure predicted by the ideal gas law.

52. You want to store 165 g of CO_2 gas in a 12.5-L tank at room temperature (25 °C). Calculate the pressure the gas would have using (a) the ideal gas law and (b) the van der Waals equation. (For CO_2, $a = 3.59$ atm · L^2/mol^2 and $b = 0.0427$ L/mol.)

General Questions

These questions are not designated as to type or location in the chapter. They may combine several concepts.

53. Complete the following table:

	atm	mm Hg	kPa	bar
Standard atmosphere	____	____	____	____
Partial pressure of N_2 in the atmosphere	____	593	____	____
Tank of compressed H_2	____	____	____	133
Atmospheric pressure at the top of Mount Everest	____	____	33.7	____

54. You want to fill a cylindrical tank with CO_2 gas at 865 mm Hg and 25 °C. The tank is 20.0 m long with a 10.0-cm radius. What mass of CO_2 (in grams) is required?

55. On combustion, 1.0 L of a gaseous compound of hydrogen, carbon, and nitrogen gives 2.0 L of CO_2, 3.5 L of H_2O vapor, and 0.50 L of N_2 at STP. What is the empirical formula of the compound?

56. To what temperature, in degrees Celsius, must a 25.5-mL sample of oxygen at 90 °C be cooled for its volume to decrease to 21.5 mL? Assume the pressure and mass of the gas are constant.

57. ▲ You have a sample of helium gas at −33 °C, and you want to increase the average speed of helium atoms by 10.0%. To what temperature should the gas be heated to accomplish this?

58. If 12.0 g of O_2 is required to inflate a balloon to a certain size at 27 °C, what mass of O_2 is required to inflate it to the same size (and pressure) at 5.0 °C?

59. You have two gas-filled balloons, one containing He and the other containing H_2. The H_2 balloon is twice the size of the He balloon. The pressure of gas in the H_2 balloon is 1 atm, and that in the He balloon is 2 atm. The H_2 balloon is outside in the snow (−5 °C), and the He balloon is inside a warm building (23 °C).
    (a) Which balloon contains the greater number of molecules?
    (b) Which balloon contains the greater mass of gas?

60. A bicycle tire has an internal volume of 1.52 L and contains 0.406 mol of air. The tire will burst if its internal pressure reaches 7.25 atm. To what temperature, in degrees Celsius, does the air in the tire need to be heated to cause a blowout?

61. The temperature of the atmosphere on Mars can be as high as 27 °C at the equator at noon, and the atmospheric pressure is about 8 mm Hg. If a spacecraft could collect 10. m^3 of this atmosphere, compress it to a small volume, and send it back to Earth, how many moles would the sample contain?

62. If you place 2.25 g of solid silicon in a 6.56-L flask that contains CH_3Cl with a pressure of 585 mm Hg at 25 °C, what mass of dimethyldichlorosilane, $(CH_3)_2SiCl_2(g)$, can be formed?

$$Si(s) + 2\ CH_3Cl(g) \longrightarrow (CH_3)_2SiCl_2(g)$$

What pressure of $(CH_3)_2SiCl_2(g)$ would you expect in this same flask at 95 °C on completion of the reaction? (Dimethyldichlorosilane is one starting material used to make silicones, polymeric substances used as lubricants, antistick agents, and water-proofing caulk.)

63. $Ni(CO)_4$ can be made by reacting finely divided nickel with gaseous CO. If you have CO in a 1.50-L flask at a pressure of 418 mm Hg at 25.0 °C, along with 0.450 g of Ni powder, what is the theoretical yield of $Ni(CO)_4$?

64. The gas B_2H_6 burns in air to give H_2O and B_2O_3.

$$B_2H_6(g) + 3\ O_2(g) \longrightarrow B_2O_3(s) + 3\ H_2O(g)$$

(a) Three gases are involved in this reaction. Place them in order of increasing molecular speed. (Assume all are at the same temperature.)

(b) A 3.26-L flask contains B_2H_6 at a pressure of 256 mm Hg and a temperature of 25 °C. Suppose O_2 gas is added to the flask until B_2H_6 and O_2 are in the correct stoichiometric ratio for the combustion reaction. At this point, what is the partial pressure of O_2?

65. ■ You have four gas samples:

1. 1.0 L of H_2 at STP

2. 1.0 L of Ar at STP

3. 1.0 L of H_2 at 27 °C and 760 mm Hg

4. 1.0 L of He at 0 °C and 900 mm Hg

(a) Which sample has the largest number of gas particles (atoms or molecules)?

(b) Which sample contains the smallest number of particles?

(c) Which sample represents the largest mass?

66. An automobile tire has a volume of 17 L. What mass of air is contained in the tire at 25 °C and a pressure of 3.2 atm? (Molar mass of air = 28.96 g/mol.)

67. Diborane, B_2H_6, reacts with oxygen to give boric oxide and water vapor.

$$B_2H_6(g) + 3\ O_2(g) \longrightarrow B_2O_3(s) + 3\ H_2O(g)$$

If you mix B_2H_6 and O_2 in the correct stoichiometric ratio, and if the total pressure of the mixture is 228 mm Hg, what are the partial pressures of B_2H_6 and O_2? If the temperature and volume do not change, what is the pressure of the water vapor?

68. Analysis of a gaseous chlorofluorocarbon, CCl_xF_y, shows that it contains 11.79% C and 69.57% Cl. In another experiment you find that 0.107 g of the compound fills a 458-mL flask at 25 °C with a pressure of 21.3 mm Hg. What is the molecular formula of the compound?

69. There are five compounds in the family of sulfur–fluorine compounds with the general formula S_xF_y. One of these compounds is 25.23% S. If you place 0.0955 g of the compound in a 89-mL flask at 45 °C, the pressure of the gas is 83.8 mm Hg. What is the molecular formula of S_xF_y?

70. A miniature volcano can be made in the laboratory with ammonium dichromate. When ignited, it decomposes in a fiery display.

$$(NH_4)_2Cr_2O_7(s) \longrightarrow N_2(g) + 4\ H_2O(g) + Cr_2O_3(s)$$

If 0.95 g of ammonium dichromate is used, and if the gases from this reaction are trapped in a 15.0-L flask at 23 °C, what is the total pressure of the gas in the flask? What are the partial pressures of N_2 and H_2O?

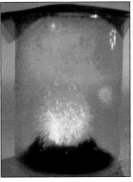

Charles D. Winters

Ammonium dichromate, $(NH_4)_2Cr_2O_7$, decomposes on heating to give nitrogen gas, water vapor, and the green solid, chromium(III) oxide.

71. ■ Iron carbonyl can be made by the direct reaction of iron metal and carbon monoxide.

$$Fe(s) + 5\ CO(g) \longrightarrow Fe(CO)_5(\ell)$$

What is the theoretical yield of $Fe(CO)_5$ if 3.52 g of iron is treated with CO gas having a pressure of 732 mm Hg in a 5.50-L flask at 23 °C?

72. You are given a solid mixture of $NaNO_2$ and $NaCl$ and are asked to analyze it for the amount of $NaNO_2$ present. To do so you allow the mixture to react with sulfamic acid, HSO_3NH_2, in water according to the equation

$$NaNO_2(aq) + HSO_3NH_2(aq) \longrightarrow$$
$$NaHSO_4(aq) + H_2O(\ell) + N_2(g)$$

What is the weight percentage of $NaNO_2$ in 1.232 g of the solid mixture if reaction with sulfamic acid produces 295 mL of N_2 gas with a pressure of 713 mm Hg at 21.0 °C?

73. The density of air 20 km above the earth's surface is 92 g/m³. The pressure of the atmosphere is 42 mm Hg and the temperature is −63 °C.

(a) What is the average molar mass of the atmosphere at this altitude?

(b) If the atmosphere at this altitude consists of only O_2 and N_2, what is the mole fraction of each gas?

74. ■ A 3.0-L bulb containing He at 145 mm Hg is connected by a valve to a 2.0-L bulb containing Ar at 355 mm Hg. (See the accompanying figure.) Calculate the partial pressure of each gas and the total pressure after the valve between the flasks is opened.

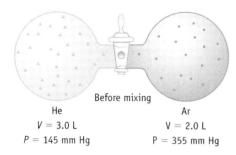

Before mixing

He
V = 3.0 L
P = 145 mm Hg

Ar
V = 2.0 L
P = 355 mm Hg

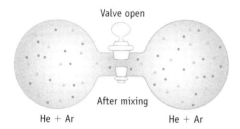

Valve open

After mixing

He + Ar   He + Ar

75. Phosphine gas, PH$_3$, is toxic when it reaches a concentration of 7×10^{-5} mg/L. To what pressure does this correspond at 25 °C?

76. A xenon fluoride can be prepared by heating a mixture of Xe and F$_2$ gases to a high temperature in a pressure-proof container. Assume that xenon gas was added to a 0.25-L container until its pressure reached 0.12 atm at 0.0 °C. Fluorine gas was then added until the total pressure reached 0.72 atm at 0.0 °C. After the reaction was complete, the xenon was consumed completely and the pressure of the F$_2$ remaining in the container was 0.36 atm at 0.0 °C. What is the empirical formula of the xenon fluoride?

77. Chlorine dioxide, ClO$_2$, reacts with fluorine to give a new gas that contains Cl, O, and F. In an experiment you find that 0.150 g of this new gas has a pressure of 17.2 mm Hg in a 1850-mL flask at 21 °C. What is the identity of the unknown gas?

78. A balloon at the circus is filled with helium gas to a gauge pressure of 22 mm Hg at 25 °C. The volume of the gas is 305 mL, and the barometric pressure is 755 mm Hg. What amount of helium is in the balloon? (Remember that gauge pressure = total pressure − barometric pressure. See page 550.)

79. Acetylene can be made by allowing calcium carbide to react with water:

$$CaC_2(s) + 2 H_2O(\ell) \longrightarrow C_2H_2(g) + Ca(OH)_2(s)$$

Suppose you react 2.65 g of CaC$_2$ with excess water. If you collect the acetylene and find that the gas has a volume of 795 mL at 25.2 °C with a pressure of 735.2 mm Hg, what is the percent yield of acetylene?

80. If you have a sample of water in a closed container, some of the water will evaporate until the pressure of the water vapor, at 25 °C, is 23.8 mm Hg. How many molecules of water per cubic centimeter exist in the vapor phase?

81. You are given 1.56 g of a mixture of KClO$_3$ and KCl. When heated, the KClO$_3$ decomposes to KCl and O$_2$,

$$2 KClO_3(s) \longrightarrow 2 KCl(s) + 3 O_2(g)$$

and 327 mL of O$_2$ with a pressure of 735 mm Hg is collected at 19 °C. What is the weight percentage of KClO$_3$ in the sample?

82. ▲ A study of climbers who reached the summit of Mount Everest without supplemental oxygen showed that the partial pressures of O$_2$ and CO$_2$ in their lungs were 35 mm Hg and 7.5 mm Hg, respectively. The barometric pressure at the summit was 253 mm Hg. Assume the lung gases are saturated with moisture at a body temperature of 37 °C [which means the partial pressure of water vapor in the lungs is P (H$_2$O) = 47.1 mm Hg]. If you assume the lung gases consists of only O$_2$, N$_2$, CO$_2$, and H$_2$O, what is the partial pressure of N$_2$?

83. Nitrogen monoxide reacts with oxygen to give nitrogen dioxide:

$$2 NO(g) + O_2(g) \longrightarrow 2 NO_2(g)$$

(a) Place the three gases in order of increasing rms speed at 298 K.
(b) If you mix NO and O$_2$ in the correct stoichiometric ratio, and NO has a partial pressure of 150 mm Hg, what is the partial pressure of O$_2$?
(c) After reaction between NO and O$_2$ is complete, what is the pressure of NO$_2$ if the NO originally had a pressure of 150 mm Hg and O$_2$ was added in the correct stoichiometric amount?

84. ▲ Ammonia gas is synthesized by combining hydrogen and nitrogen:

$$3 H_2(g) + N_2(g) \longrightarrow 2 NH_3(g)$$

(a) If you want to produce 562 g of NH$_3$, what volume of H$_2$ gas, at 56 °C and 745 mm Hg, is required?
(b) To produce 562 g of NH$_3$, what volume of air (the source of N$_2$) is required if the air is introduced at 29 °C and 745 mm Hg? (Assume the air sample has 78.1 mole % N$_2$.)

85. ▲ You have a 550-mL tank of gas with a pressure of 1.56 atm at 24 °C. You thought the gas was pure carbon monoxide gas, CO, but you later found it was contaminated by small quantities of gaseous CO$_2$ and O$_2$. Analysis shows that the tank pressure is 1.34 atm (at 24 °C) if the CO$_2$ is removed. Another experiment shows that 0.0870 g of O$_2$ can be removed chemically. What are the masses of CO and CO$_2$ in the tank, and what is the partial pressure of each of the three gases at 25 °C?

86. ▲ Methane is burned in a laboratory Bunsen burner to give CO_2 and water vapor. Methane gas is supplied to the burner at the rate of 5.0 L/min (at a temperature of 28 °C and a pressure of 773 mm Hg). At what rate must oxygen be supplied to the burner (at a pressure of 742 mm Hg and a temperature of 26 °C)?

87. ▲ Iron forms a series of compounds of the type $Fe_x(CO)_y$. In air they are oxidized to Fe_2O_3 and CO_2 gas. After heating a 0.142-g sample of $Fe_x(CO)_y$ in air, you isolate the CO_2 in a 1.50-L flask at 25 °C. The pressure of the gas is 44.9 mm Hg. What is the formula of $Fe_x(CO)_y$?

88. ▲ Group 2A metal carbonates are decomposed to the metal oxide and CO_2 on heating:

$$MCO_3(s) \longrightarrow MO(s) + CO_2(g)$$

You heat 0.158 g of a white, solid carbonate of a Group 2A metal (M) and find that the evolved CO_2 has a pressure of 69.8 mm Hg in a 285-mL flask at 25 °C. Identify M.

89. Silane, SiH_4, reacts with O_2 to give silicon dioxide and water vapor:

$$SiH_4(g) + 2 O_2(g) \longrightarrow SiO_2(s) + 2 H_2O(g)$$

If you mix SiH_4 with O_2 in the correct stoichiometric ratio, and if the total pressure of the mixture is 120 mm Hg, what are the partial pressures of SiH_4 and O_2? When the reactants have been completely consumed, what is the total pressure in the flask? (Assume T is constant.)

90. Chlorine trifluoride, ClF_3, is a valuable reagent because it can be used to convert metal oxides to metal fluorides:

$$6 NiO(s) + 4 ClF_3(g) \longrightarrow 6 NiF_2(s) + 2 Cl_2(g) + 3 O_2(g)$$

   (a) What mass of NiO will react with ClF_3 gas if the gas has a pressure of 250 mm Hg at 20 °C in a 2.5-L flask?

   (b) If the ClF_3 described in part (a) is completely consumed, what are the partial pressures of Cl_2 and of O_2 in the 2.5-L flask at 20 °C (in mm Hg)? What is the total pressure in the flask?

91. One way to synthesize diborane, B_2H_6, is the reaction

$$2 NaBH_4(s) + 2 H_3PO_4(aq) \longrightarrow$$
$$B_2H_6(g) + 2 NaH_2PO_4(aq) + 2 H_2(g)$$

   (a) If you have 0.136 g of $NaBH_4$ and excess H_3PO_4, and you collect the B_2H_6 in a 2.75 L flask at 25 °C, what is the pressure of the B_2H_6 in the flask?

   (b) A byproduct of the reaction is H_2 gas. If both B_2H_6 and H_2 gas come from this reaction, what is the *total* pressure in the 2.75-L flask (after reaction of 0.136 g of $NaBH_4$ with excess H_3PO_4) at 25 °C?

92. Calcium carbide reacts with water to produce acetylene and calcium hydroxide:

$$CaC_2(s) + 2 H_2O(\ell) \longrightarrow C_2H_2(g) + Ca(OH)_2(s)$$

Suppose you combine 13.0 g of CaC_2 with 4.65 g of water and collect the acetylene in a 4.66-L flask. What is the pressure of the acetylene at 23 °C?

93. ▲ You have 1.249 g of a mixture of $NaHCO_3$ and Na_2CO_3. You find that 12.0 mL of 1.50 M HCl is required to convert the sample completely to NaCl, H_2O, and CO_2.

$$NaHCO_3(aq) + HCl(aq) \longrightarrow$$
$$NaCl(aq) + H_2O(\ell) + CO_2(g)$$

$$Na_2CO_3(aq) + 2 HCl(aq) \longrightarrow$$
$$2 NaCl(aq) + H_2O(\ell) + CO_2(g)$$

What volume of CO_2 is evolved at 745 mm Hg and 25 °C?

94. ▲ A mixture of $NaHCO_3$ and Na_2CO_3 has a mass of 2.50 g. When treated with HCl(aq), 665 mL of CO_2 gas is liberated with a pressure of 735 mm Hg at 25 °C. What is the weight percent of $NaHCO_3$ and Na_2CO_3 in the mixture? (See Study Question 93 for the reactions that occur.)

95. ▲ Relative humidity is the ratio of the partial pressure of water in air at a given temperature to the vapor pressure of water at that temperature. Calculate the mass of water per liter of air under the following conditions.

   (a) at 20 °C and 45% relative humidity

   (b) at 0 °C and 95% relative humidity

   Under which circumstances is the mass of H_2O per liter greater? (See Appendix G for the vapor pressure of water.)

96. How much water vapor is present in a dormitory room when the relative humidity is 55% and the temperature is 23 °C? The dimensions of the room are 4.5 m^2 floor area and 3.5 m ceiling height. (See Study Question 95 for a definition of relative humidity and Appendix G for the vapor pressure of water.)

Summary and Conceptual Questions

The following questions may use concepts from the preceding chapters.

97. A 1.0-L flask contains 10.0 g each of O_2 and CO_2 at 25 °C.

   (a) Which gas has the greater partial pressure, O_2 or CO_2, or are they the same?

   (b) Which molecules have the greater average speed, or are they the same?

   (c) Which molecules have the greater average kinetic energy, or are they the same?

98. If equal masses of O_2 and N_2 are placed in separate containers of equal volume at the same temperature, which of the following statements is true? If false, tell why it is false.

   (a) The pressure in the flask containing N_2 is greater than that in the flask containing O_2.

   (b) There are more molecules in the flask containing O_2 than in the flask containing N_2.

99. You have two pressure-proof steel cylinders of equal volume, one containing 1.0 kg of CO and the other containing 1.0 kg of acetylene, C_2H_2.

   (a) In which cylinder is the pressure greater at 25 °C?

   (b) Which cylinder contains the greater number of molecules?

100. Two flasks, each with a volume of 1.00 L, contain O_2 gas with a pressure of 380 mm Hg. Flask A is at 25 °C, and flask B is at 0 °C. Which flask contains the greater number of O_2 molecules?

101. ▲ State whether each of the following samples of matter is a gas. If there is not enough information for you to decide, write "insufficient information."

    (a) A material is in a steel tank at 100 atm pressure. When the tank is opened to the atmosphere, the material suddenly expands, increasing its volume by 10%.

    (b) A 1.0-mL sample of material weighs 8.2 g.

    (c) The material is transparent and pale green in color.

    (d) One cubic meter of material contains as many molecules as 1.0 m^3 of air at the same temperature and pressure.

102. Each of the four tires of a car is filled with a different gas. Each tire has the same volume, and each is filled to the same pressure, 3.0 atm, at 25 °C. One tire contains 116 g of air, another tire has 80.7 g of neon, another tire has 16.0 g of helium, and the fourth tire has 160. g of an unknown gas.

    (a) Do all four tires contain the same number of gas molecules? If not, which one has the greatest number of molecules?

    (b) How many times heavier is a molecule of the unknown gas than an atom of helium?

    (c) In which tire do the molecules have the largest kinetic energy? The highest average speed?

103. The sodium azide required for automobile air bags is made by the reaction of sodium metal with dinitrogen oxide in liquid ammonia:

$$3\, N_2O(g) + 4\, Na(s) + NH_3(\ell) \longrightarrow$$
$$NaN_3(s) + 3\, NaOH(s) + 2\, N_2(g)$$

    (a) You have 65.0 g of sodium and a 35.0-L flask containing N_2O gas with a pressure of 2.12 atm at 23 °C. What is the theoretical yield (in grams) of NaN_3?

    (b) Draw a Lewis structure for the azide ion. Include all possible resonance structures. Which resonance structure is most likely?

    (c) What is the shape of the azide ion?

104. ▲ Chlorine gas (Cl_2) is used as a disinfectant in municipal water supplies, although chlorine dioxide (ClO_2) and ozone are becoming more widely used. ClO_2 is a better choice than Cl_2 in this application because it leads to fewer chlorinated byproducts, which are themselves pollutants.

    (a) How many valence electrons are in ClO_2?

    (b) The chlorite ion, ClO_2^-, is obtained by reducing ClO_2. Draw a possible electron dot structure for ClO_2^-. (Cl is the central atom.)

    (c) What is the hybridization of the central Cl atom in ClO_2^-? What is the shape of the ion?

    (d) Which species has the larger bond angle, O_3 or ClO_2^-? Explain briefly.

    (e) Chlorine dioxide, ClO_2, a yellow-green gas, can be made by the reaction of chlorine with sodium chlorite:

$$2\, NaClO_2(s) + Cl_2(g) \longrightarrow 2\, NaCl(s) + 2\, ClO_2(g)$$

    Assume you react 15.6 g of $NaClO_2$ with chlorine gas, which has a pressure of 1050 mm Hg in a 1.45-L flask at 22 °C. What mass of ClO_2 can be produced?

105. If the absolute temperature of a gas doubles, by how much does the average speed of the gaseous molecules increase? (*See General ChemistryNow Screen 12.9.*)

106. Screen 12.10 of the General ChemistryNow CD-ROM or website shows animations describing the following relationships on the molecular scale: *P* versus *n*, *P* versus *T*, and *P* versus *V*. Sketch a molecular-scale animation for the relationship between *n* and *V*.

Do you need a live tutor for homework problems?
Access vMentor at General ChemistryNow at
http://now.brookscole.com/kotz6e
for one-on-one tutoring from a chemistry expert

List of Appendices

$$\log 563 = 2.751$$
$$\log 125 = 2.097$$
$$\overline{\log xy = 4.848}$$
$$xy = 10^{4.848} = 10^4 \times 10^{0.848} = 7.05 \times 10^4$$

One number (x) can be divided by another (y) by subtraction of their logarithms:

$$\log \frac{x}{y} = \log x - \log y$$

For example, to divide 125 by 742,

$$\log 125 = 2.097$$
$$-\log 742 = \underline{2.870}$$
$$\log \frac{x}{y} = -0.773$$
$$\frac{x}{y} = 10^{-0.773} = 10^{0.227} \times 10^{-1} = 1.68 \times 10^{-1}$$

Similarly, powers and roots of numbers can be found using logarithms.

$$\log x^y = y(\log x)$$
$$\log \sqrt[y]{x} = \log x^{1/y} = \frac{1}{y} \log x$$

As an example, find the fourth power of 5.23. We first find the log of 5.23 and then multiply it by 4. The result, 2.874, is the log of the answer. Therefore, we find the antilog of 2.874:

$$(5.23)^4 = ?$$
$$\log (5.23)^4 = 4 \log 5.23 = 4(0.719) = 2.874$$
$$(5.23)^4 = 10^{2.874} = 748$$

As another example, find the fifth root of 1.89×10^{-9}:

$$\sqrt[5]{1.89 \times 10^{-9}} = (1.89 \times 10^{-9})^{1/5} = ?$$
$$\log (1.89 \times 10^{-9})^{1/5} = \frac{1}{5} \log (1.89 \times 10^{-9}) = \frac{1}{5}(-8.724) = -1.745$$

The answer is the antilog of -1.745:

$$(1.89 \times 10^{-9})^{1/5} = 10^{-1.745} = 1.80 \times 10^{-2}$$

A.2—Quadratic Equations

Algebraic equations of the form $ax^2 + bx + c = 0$ are called **quadratic equations.** The coefficients a, b, and c may be either positive or negative. The two roots of the equation may be found using the *quadratic formula:*

$$x = \frac{-b \pm \sqrt{b^2 - 4ac}}{2a}$$

As an example, solve the equation $5x^2 - 3x - 2 = 0$. Here $a = 5$, $b = -3$, and $c = -2$. Therefore,

$$x = \frac{3 \pm \sqrt{(-3)^2 - 4(5)(-2)}}{2(5)}$$

$$= \frac{3 \pm [2(5)/\sqrt{9 - (-40)}]}{10} = \frac{3 \pm \sqrt{49}}{10} = \frac{3 \pm 7}{10}$$

$$= 1 \text{ and } -0.4$$

How do you know which of the two roots is the correct answer? You have to decide in each case which root has physical significance. It is *usually* true in this course, however, that negative values are not significant.

When you have solved a quadratic expression, you should always check your values by substitution into the original equation. In the previous example, we find that $5(1)^2 - 3(1) - 2 = 0$ and that $5(-0.4)^2 - 3(-0.4) - 2 = 0$.

The most likely place you will encounter quadratic equations is in the chapters on chemical equilibria, particularly in Chapters 16 through 18. Here you will often be faced with solving an equation such as

$$1.8 \times 10^{-4} = \frac{x^2}{0.0010 - x}$$

This equation can certainly be solved using the quadratic equation (to give $x = 3.4 \times 10^{-4}$). You may find the *method of successive approximations* to be especially convenient, however. Here we begin by making a reasonable approximation of x. This approximate value is substituted into the original equation, which is then solved to give what is hoped to be a more correct value of x. This process is repeated until the answer converges on a particular value of x—that is, until the value of x derived from two successive approximations is the same.

Step 1:   First assume that x is so small that $(0.0010 - x) \approx 0.0010$. This means that

$$x^2 = 1.8 \times 10^{-4}(0.0010)$$
$$x = 4.2 \times 10^{-4} \text{ (to 2 significant figures)}$$

Step 2:   Substitute the value of x from Step 1 into the denominator of the original equation, and again solve for x:

$$x^2 = 1.8 \times 10^{-4}(0.0010 - 0.00042)$$
$$x = 3.2 \times 10^{-4}$$

Step 3:   Repeat Step 2 using the value of x found in that step:

$$x = \sqrt{1.8 \times 10^{-4}(0.0010 - 0.00032)} = 3.5 \times 10^{-4}$$

Step 4:   Continue repeating the calculation, using the value of x found in the previous step:

$$x = \sqrt{1.8 \times 10^{-4}(0.0010 - 0.00035)} = 3.4 \times 10^{-4}$$

Step 5: $\quad x = \sqrt{1.8 \times 10^{-4}(0.0010 - 0.00034)} = 3.4 \times 10^{-4}$

Here we find that iterations after the fourth step give the same value for x, indicating that we have arrived at a valid answer (and the same one obtained from the quadratic formula).

Here are several final thoughts on using the method of successive approximations. First, in some cases the method does not work. Successive steps may give answers that are random or that diverge from the correct value. In Chapters 16 through 18, you confront quadratic equations of the form $K = x^2/(C - x)$. The method of approximations works as long as $K < 4C$ (assuming one begins with $x = 0$ as the first guess, that is, $K \approx x^2/C$). This is always going to be true for weak acids and bases (the topic of Chapters 17 and 18), but it may *not* be the case for problems involving gas-phase equilibria (Chapter 16), where K can be quite large.

Second, values of K in the equation $K = x^2/(C - x)$ are usually known only to two significant figures. We are therefore justified in carrying out successive steps until two answers are the same to two significant figures.

Finally, we highly recommend this method of solving quadratic equations, especially those in Chapters 17 and 18. If your calculator has a memory function, successive approximations can be carried out easily and rapidly.

Appendix B*

Some Important Physical Concepts

B.1—Matter

The tendency to maintain a constant velocity is called inertia. Thus, unless acted on by an unbalanced force, a body at rest remains at rest, and a body in motion remains in motion with uniform velocity. Matter is anything that exhibits inertia; the quantity of matter is its mass.

B.2—Motion

Motion is the change of position or location in space. Objects can have the following classes of motion:

- Translation occurs when the center of mass of an object changes its location. Example: a car moving on the highway.
- Rotation occurs when each point of a moving object moves in a circle about an axis through the center of mass. Examples: a spinning top, a rotating molecule.
- Vibration is a periodic distortion of and then recovery of original shape. Examples: a struck tuning fork, a vibrating molecule.

B.3—Force and Weight

Force is that which changes the velocity of a body; it is defined as

$$\text{Force} = \text{mass} \times \text{acceleration}$$

The SI unit of force is the **newton,** N, whose dimensions are kilograms times meter per second squared ($kg \cdot m/s^2$). A newton is therefore the force needed to change the velocity of a mass of 1 kilogram by 1 meter per second in a time of 1 second.

Because the earth's gravity is not the same everywhere, the weight corresponding to a given mass is not a constant. At any given spot on earth gravity is constant, however, and therefore weight is proportional to mass. When a balance tells us that a given sample (the "unknown") has the same weight as another sample (the "weights," as given by a scale reading or by a total of counterweights), it also tells us that the two masses are equal. The balance is therefore a valid instrument for measuring the mass of an object independently of slight variations in the force of gravity.

---

*Adapted from F. Brescia, J. Arents, H. Meislich, et al.: *General Chemistry,* 5th ed. Philadelphia, Harcourt Brace, 1988.

B.4—Pressure[*]

Pressure is force per unit area. The SI unit, called the pascal, Pa, is

$$1 \text{ pascal} = \frac{1 \text{ newton}}{m^2} = \frac{1 \text{ kg} \cdot m/s^2}{m^2} = \frac{1 \text{ kg}}{m \cdot s^2}$$

The International System of Units also recognizes the bar, which is 10^5 Pa and which is close to standard atmospheric pressure (Table 1).

Table 1   Pressure Conversions

From	To	Multiply By
atmosphere	mm Hg	760 mm Hg/atm (exactly)
atmosphere	lb/in^2	14.6960 lb/(in^2 atm)
atmosphere	kPa	101.325 kPa/atm
bar	Pa	10^5 Pa/bar (exactly)
bar	lb/in^2	14.5038 lb/(in^2 bar)
mm Hg	torr	1 torr/mm Hg (exactly)

Chemists also express pressure in terms of the heights of liquid columns, especially water and mercury. This usage is not completely satisfactory, because the pressure exerted by a given column of a given liquid is not a constant but depends on the temperature (which influences the density of the liquid) and the location (which influences gravity). Such units are therefore not part of the SI, and their use is now discouraged. The older units are still used in books and journals, however, and chemists must be familiar with them.

The pressure of a liquid or a gas depends only on the depth (or height) and is exerted equally in all directions. At sea level, the pressure exerted by the earth's atmosphere supports a column of mercury about 0.76 m (76 cm, or 760 mm) high.

One **standard atmosphere** (atm) is the pressure exerted by exactly 76 cm of mercury at 0 °C (density, 13.5951 g/cm^3) and at standard gravity, 9.80665 m/s^2. The **bar** is equivalent to 0.9869 atm. One **torr** is the pressure exerted by exactly 1 mm of mercury at 0 °C and standard gravity.

B.5—Energy and Power

The SI unit of energy is the product of the units of force and distance, or kilograms times meter per second squared (kg $\cdot$ m/s^2) times meters ($\times$ m), which is kg $\cdot$ m^2/s^2; this unit is called the **joule,** J. The joule is thus the work done when a force of 1 newton acts through a distance of 1 meter.

Work may also be done by moving an electric charge in an electric field. When the charge being moved is 1 coulomb (C), and the potential difference between its initial and final positions is 1 volt (V), the work is 1 joule. Thus,

$$1 \text{ joule} = 1 \text{ coulomb volt (CV)}$$

Another unit of electric work that is not part of the International System of Units but is still in use is the **electron volt,** eV, which is the work required to move an electron against a potential difference of 1 volt. (It is also the kinetic energy acquired

[*]See Section 12.1.

by an electron when it is accelerated by a potential difference of 1 volt.) Because the charge on an electron is 1.602×10^{-19} C, we have

$$1 \text{ eV} = 1.602 \times 10^{-19} \text{ CV} \times \frac{1 \text{ J}}{1 \text{ CV}} = 1.602 \times 10^{-19} \text{ J}$$

If this value is multiplied by Avogadro's number, we obtain the energy involved in moving 1 mole of electron charges (1 faraday) in a field produced by a potential difference of 1 volt:

$$1 \frac{\text{eV}}{\text{particle}} = \frac{1.602 \times 10^{-19} \text{ J}}{\text{particle}} \times \frac{6.022 \times 10^{23} \text{ particles}}{\text{mol}} \cdot \frac{1 \text{ kJ}}{1000 \text{ J}} = 96.49 \text{ kJ/mol}$$

Power is the amount of energy delivered per unit time. The SI unit is the watt, W, which is 1 joule per second. One kilowatt, kW, is 1000 W. Watt hours and kilowatt hours are therefore units of energy (Table 2). For example, 1000 watts, or 1 kilowatt, is

$$1.0 \times 10^3 \text{ W} \times \frac{1 \text{ J}}{1 \text{ W} \cdot \text{s}} \cdot \frac{3.6 \times 10^3 \text{ s}}{1 \text{ h}} = 3.6 \times 10^6 \text{ J}$$

Table 2   Energy Conversions

From	To	Multiply By
calorie (cal)	joule	4.184 J/cal (exactly)
kilocalorie (kcal)	cal	10^3 cal/kcal (exactly)
kilocalorie	joule	4.184×10^3 J/kcal (exactly)
liter atmosphere (L·atm)	joule	101.325 J/L·atm
electron volt (eV)	joule	1.60218×10^{-19} J/eV
electron volt per particle	kilojoules per mole	96.485 kJ·particle/eV·mol
coulomb volt (CV)	joule	1 CV/J (exactly)
kilowatt hour (kWh)	kcal	860.4 kcal/kWh
kilowatt hour	joule	3.6×10^6 J/kWh (exactly)
British thermal unit (Btu)	calorie	252 cal/Btu

Appendix C

Abbreviations and Useful Conversion Factors

Table 3 Some Common Abbreviations and Standard Symbols

Term	Abbreviation	Term	Abbreviation
Activation energy	E_a	Face-centered cubic	fcc
Ampere	A	Faraday constant	F
Aqueous solution	aq	Gas constant	R
Atmosphere, unit of pressure	atm	Gibbs free energy	G
Atomic mass unit	u	Standard free energy	$G°$
Avogadro's constant	N_A	Standard free energy of formation	$\Delta G_f°$
Bar, unit of pressure	bar	Free energy change for reaction	$\Delta G_{rxn}°$
Body-centered cubic	bcc	Half-life	$t_{1/2}$
Bohr radius	a_0	Heat	q
Boiling point	bp	Hertz	Hz
Celsius temperature, °C	T	Hour	h
Charge number of an ion	z	Joule	J
Coulomb, electric charge	C	Kelvin	K
Curie, radioactivity	Ci	Kilocalorie	kcal
Cycles per second, hertz	Hz	Liquid	ℓ
Debye, unit of electric dipole	D	Logarithm, base 10	log
Electron	e^-	Logarithm, base e	ln
Electron volt	eV	Minute	min
Electronegativity	χ	Molar	M
Energy	E	Molar mass	M
Enthalpy	H	Mole	mol
Standard enthalpy	$H°$	Osmotic pressure	Π
Standard enthalpy of formation	$\Delta H_f°$	Planck's constant	h
Standard enthalpy of reaction	$\Delta H_{rxn}°$	Pound	lb
Entropy	S	Pressure	
Standard entropy	$S°$	Pascal, unit of pressure	Pa
Entropy change for reaction	$\Delta S_{rxn}°$	In atmospheres	atm
Equilibrium constant	K	In millimeters of mercury	mm Hg
Concentration basis	K_c	Proton number	Z
Pressure basis	K_p	Rate constant	k
Ionization weak acid	K_a	Simple cubic (unit cell)	sc
Ionization weak base	K_b	Standard temperature and pressure	STP
Solubility product	K_{sp}	Volt	V
Formation constant	K_{form}	Watt	W
Ethylenediamine	en	Wavelength	λ

C.1—Fundamental Units of the SI System

The metric system was begun by the French National Assembly in 1790 and has undergone many modifications. The International System of Units or *Système International* (SI), which represents an extension of the metric system, was adopted by the 11th General Conference of Weights and Measures in 1960. It is constructed from seven base units, each of which represents a particular physical quantity (Table 4).

Table 4   SI Fundamental Units

Physical Quantity	Name of Unit	Symbol
Length	meter	m
Mass	kilogram	kg
Time	second	s
Temperature	kelvin	K
Amount of substance	mole	mol
Electric current	ampere	A
Luminous intensity	candela	cd

The first five units listed in Table 4 are particularly useful in general chemistry and are defined as follows:

1. The *meter* was redefined in 1960 to be equal to 1,650,763.73 wavelengths of a certain line in the emission spectrum of krypton-86.
2. The *kilogram* represents the mass of a platinum–iridium block kept at the International Bureau of Weights and Measures at Sèvres, France.
3. The *second* was redefined in 1967 as the duration of 9,192,631,770 periods of a certain line in the microwave spectrum of cesium-133.
4. The *kelvin* is 1/273.15 of the temperature interval between absolute zero and the triple point of water.
5. The *mole* is the amount of substance that contains as many entities as there are atoms in exactly 0.012 kg of carbon-12 (12 g of ^{12}C atoms).

C.2—Prefixes Used with Traditional Metric Units and SI Units

Decimal fractions and multiples of metric and SI units are designated by using the prefixes listed in Table 5. Those most commonly used in general chemistry appear in italics.

C.3—Derived SI Units

In the International System of Units, all physical quantities are represented by appropriate combinations of the base units listed in Table 4. A list of the derived units frequently used in general chemistry is given in Table 6.

Table 5   Traditional Metric and SI Prefixes

Factor	Prefix	Symbol	Factor	Prefix	Symbol
10^{12}	tera	T	10^{-1}	*deci*	d
10^9	giga	G	10^{-2}	*centi*	c
10^6	mega	M	10^{-3}	*milli*	m
10^3	*kilo*	k	10^{-6}	micro	μ
10^2	hecto	h	10^{-9}	*nano*	n
10^1	deka	da	10^{-12}	*pico*	p
			10^{-15}	femto	f
			10^{-18}	atto	a

Table 6   Derived SI Units

Physical Quantity	Name of Unit	Symbol	Definition
Area	square meter	m^2	
Volume	cubic meter	m^3	
Density	kilogram per cubic meter	kg/m^3	
Force	newton	N	$kg \cdot m/s^2$
Pressure	pascal	Pa	N/m^2
Energy	joule	J	$kg \cdot m^2/s^2$
Electric charge	coulomb	C	$A \cdot s$
Electric potential difference	volt	V	$J/(A \cdot s)$

Table 7   Common Units of Mass and Weight

1 Pound = 453.39 Grams

1 kilogram = 1000 grams = 2.205 pounds

1 gram = 1000 milligrams

1 gram = 6.022×10^{23} atomic mass units

1 atomic mass unit = 1.6605×10^{-24} gram

1 short ton = 2000 pounds = 907.2 kilograms

1 long ton = 2240 pounds

1 metric tonne = 1000 kilograms = 2205 pounds

Table 8   Common Units of Length
1 inch = 2.54 centimeters (Exactly)
1 mile = 5280 feet = 1.609 kilometers
1 yard = 36 inches = 0.9144 meter
1 meter = 100 centimeters = 39.37 inches = 3.281 feet = 1.094 yards
1 kilometer = 1000 meters = 1094 yards = 0.6215 mile
1 Ångstrom = 1.0×10^{-8} centimeter = 0.10 nanometer = 100 picometers
$= 1.0 \times 10^{-10}$ meter = 3.937×10^{-9} inch

Table 9   Common Units of Volume
1 quart = 0.9463 liter
1 liter = 1.0567 quarts
1 liter = 1 cubic decimeter = 1000 cubic centimeters = 0.001 cubic meter
1 milliliter = 1 cubic centimeter = 0.001 liter = 1.056×10^{-3} quart
1 cubic foot = 28.316 liters = 29.924 quarts = 7.481 gallons

Appendix E

Naming Organic Compounds

It seems a daunting task—to devise a systematic procedure that gives each organic compound a unique name—but that is what has been done. A set of rules was developed to name organic compounds by the International Union of Pure and Applied Chemistry (IUPAC). The IUPAC nomenclature allows chemists to write a name for any compound based on its structure or to identify the formula and structure for a compound from its name. In this book, we have generally used the IUPAC nomenclature scheme when naming compounds.

In addition to the systematic names, many compounds have common names. The common names came into existence before the nomenclature rules were developed, and they have continued in use. For some compounds, these names are so well entrenched that they are used most of the time. One such compound is acetic acid, which is almost always referred to by that name and not by its systematic name, ethanoic acid.

The general procedure for systematic naming of organic compounds begins with the nomenclature for hydrocarbons. Other organic compounds are then named as derivatives of hydrocarbons. Nomenclature rules for simple organic compounds are given in the following section.

E.1—Hydrocarbons

Alkanes

The names of alkanes end in "-ane." When naming a specific alkane, the root of the name identifies the longest carbon chain in a compound. Specific substituent groups attached to this carbon chain are identified by name and position.

Alkanes with chains of from one to ten carbon atoms are given in Table 11.2. After the first four compounds, the names derive from Latin numbers—pentane, hexane, heptane, octane, nonane, decane—and this regular naming continues for higher alkanes. For substituted alkanes, the substituent groups on a hydrocarbon chain must be identified both by a name and by the position of substitution; this information precedes the root of the name. The position is indicated by a number that refers to the carbon atom to which the substituent is attached. (Numbering of the carbon atoms in a chain should begin at the end of the carbon chain that allows the substituent groups to have the lowest numbers.)

Names of hydrocarbon substituents are derived from the name of the hydrocarbon. The group $-CH_3$, derived by taking a hydrogen from methane, is called the methyl group; the C_2H_5 group is the ethyl group. The nomenclature scheme is easily extended to derivatives of hydrocarbons with other substituent groups such as $-Cl$ (chloro), $-NO_2$ (nitro), $-CN$ (cyano), $-D$ (deuterio), and so on (Table 13). If two or more of the same substituent groups occur, the prefixes "di-," "tri-," and "tetra-" are added. When different substituent groups are present, they are generally listed in alphabetical order.

Table 13   Names of Common Substituent Groups

Formula	Name	Formula	Name
$-CH_3$	methyl	$-D$	deuterio
$-C_2H_5$	ethyl	$-Cl$	chloro
$-CH_2CH_2CH_3$	1-propyl (*n*-propyl)	$-Br$	bromo
$-CH(CH_3)_2$	2-propyl (isopropyl)	$-F$	fluoro
$-CH=CH_2$	ethenyl (vinyl)	$-CN$	cyano
$-C_6H_5$	phenyl	$-NO_2$	nitro
$-OH$	hydroxo		
$-NH_2$	amino		

Example:

$$\begin{array}{cc} CH_3 & C_2H_5 \\ | & | \\ \end{array}$$
$$CH_3CH_2CHCH_2CHCH_2CH_3$$

Step	Information to include	Contribution to name
1.	An alkane	name will end in "-ane"
2.	Longest chain is 7 carbons	name as a *heptane*
3.	$-CH_3$ group at carbon 3	*3-methyl*
4.	$-C_2H_5$ group at carbon 5	*5-ethyl*

Name: 5-ethyl-3-methylheptane

Cycloalkanes are named based on the ring size and by adding the prefix "cyclo"; for example, the cycloalkane with a six-member ring of carbons is called cyclohexane.

Alkenes

Alkenes have names ending in "-ene." The name of an alkene must specify the length of the carbon chain and the position of the double bond (and when appropriate, the configuration, either *cis* or *trans*). As with alkanes, both identity and position of substituent groups must be given. The carbon chain is numbered from the end that gives the double bond the lowest number.

Compounds with two double bonds are called dienes and they are named similarly—specifying the positions of the double bonds and the name and position of any substituent groups.

For example, the compound $H_2C=C(CH_3)CH(CH_3)CH_2CH_3$ has a five-carbon chain with a double bond between carbon atoms 1 and 2 and methyl groups on carbon atoms 2 and 3. Its name using IUPAC nomenclature is **2,3-dimethyl-1-pentene.** The compound $CH_3CH=CHCCl_3$ with a *cis* configuration around the double bond is named **1,1,1-trichloro-*cis*-2-butene.** The compound $H_2C=C(Cl)CH=CH_2$ is **2-chloro-1,3-butadiene.**

Alkynes

The naming of alkynes is similar to the naming of alkenes, except that *cis–trans* isomerism isn't a factor. The ending "-yne" on a name identifies a compound as an alkyne.

Benzene Derivatives

The carbon atoms in the six-member ring are numbered 1 through 6, and the name and position of substituent groups are given. The two examples shown here are **1-ethyl-3-methylbenzene** and **1,4-diaminobenzene.**

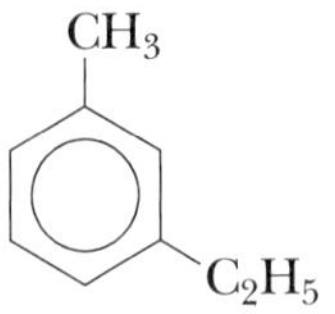

1-ethyl-3-methylbenzene 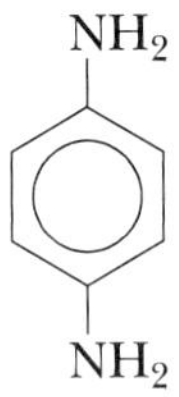   1,4-diaminobenzene

E.2—Derivatives of Hydrocarbons

The names for alcohols, aldehydes, ketones, and acids are based on the name of the hydrocarbon with an appropriate suffix to denote the class of compound, as follows:

- **Alcohols:** Substitute "-ol" for the final "-e" in the name of the hydrocarbon, and designate the position of the —OH group by the number of the carbon atom. For example, $CH_3CH_2CHOHCH_3$ is named as a derivative of the 4-carbon hydrocarbon butane. The —OH group is attached to the second carbon, so the name is 2-butanol.
- **Aldehydes:** Substitute "-al" for the final "-e" in the name of the hydrocarbon. The carbon atom of an aldehyde is, by definition, carbon-1 in the hydrocarbon chain. For example, the compound $CH_3CH(CH_3)CH_2CH_2CHO$ contains a 5-carbon chain with the aldehyde functional group being carbon-1 and the —CH_3 group at position 4; thus the name is **4-methylpentanal.**
- **Ketones:** Substitute "-one" for the final "-e" in the name of the hydrocarbon. The position of the ketone functional group (the carbonyl group) is indicated by the number of the carbon atom. For example, the compound $CH_3COCH_2CH(C_2H_5)CH_2CH_3$ has the carbonyl group at the 2 position and an ethyl group at the 4 position of a 6-carbon chain; its name is **4-ethyl-2-hexanone.**
- **Carboxylic acids (organic acids):** Substitute "-oic" for the final "-e" in the name of the hydrocarbon. The carbon atoms in the longest chain are counted beginning with the carboxylic carbon atom. For example, *trans*-$CH_3CH{=}CHCH_2CO_2H$ is named as a derivative of *trans*-3-pentene—that is, ***trans*-3-pentenoic acid.**

An **ester** is named as a derivative of the alcohol and acid from which it is made. The name of an ester is obtained by splitting the formula RCO_2R' into two parts, the RCO_2— portion and the —R' portion. The —R' portion comes from the alcohol and is identified by the hydrocarbon group name; derivatives of ethanol, for example, are called *ethyl* esters. The acid part of the compound is named by dropping the "-oic" ending for the acid and replacing it by "-oate." The compound $CH_3CH_2CO_2CH_3$ is named **methyl propanoate.**

Notice that an anion derived from a carboxylic acid by loss of the acidic proton is named the same way. Thus, $CH_3CH_2CO_2^-$ is the **propanoate anion,** and the sodium salt of this anion, $Na(CH_3CH_2CO_2)$, is **sodium propanoate.**

Appendix F

Values for the Ionization Energies and Electron Affinities of the Elements

1A (1)																	8 (18)
H 1312	2A (2)											3A (13)	4A (14)	5A (15)	6A (16)	7A (17)	He 2371
Li 520	Be 899											B 801	C 1086	N 1402	O 1314	F 1681	Ne 2081
Na 496	Mg 738	3B (3)	4B (4)	5B (5)	6B (6)	7B (7)	8B (8,9,10)		1B (11)	2B (12)		Al 578	Si 786	P 1012	S 1000	Cl 1251	Ar 1521
K 419	Ca 599	Sc 631	Ti 658	V 650	Cr 652	Mn 717	Fe 759	Co 758	Ni 757	Cu 745	Zn 906	Ga 579	Ge 762	As 947	Se 941	Br 1140	Kr 1351
Rb 403	Sr 550	Y 617	Zr 661	Nb 664	Mo 685	Tc 702	Ru 711	Rh 720	Pd 804	Ag 731	Cd 868	In 558	Sn 709	Sb 834	Te 869	I 1008	Xe 1170
Cs 377	Ba 503	La 538	Hf 681	Ta 761	W 770	Re 760	Os 840	Ir 880	Pt 870	Au 890	Hg 1007	Tl 589	Pb 715	Bi 703	Po 812	At 890	Rn 1037

Table 14   Electron Affinity Values for Some Elements (kJ/mol)*

H −72.77							
Li	Be	B	C	N	O	F	
−59.63	0[†]	−26.7	−121.85	0	−140.98	−328.0	
Na	Mg	Al	Si	P	S	Cl	
−52.87	0	−42.6	−133.6	−72.07	−200.41	−349.0	
K	Ca	Ga	Ge	As	Se	Br	
−48.39	0	−30	−120	−78	−194.97	−324.7	
Rb	Sr	In	Sn	Sb	Te	I	
−46.89	0	−30	−120	−103	−190.16	−295.16	
Cs	Ba	Tl	Pb	Bi	Po	At	
−45.51	0	−20	−35.1	−91.3	−180	−270	

*Data taken from H. Hotop and W. C. Lineberger: *Journal of Physical Chemistry, Reference Data,* Vol. 14, p. 731, 1985. (This paper also includes data for the transition metals.) Some values are known to more than two decimal places.

[†] Elements with an electron affinity of zero indicate that a stable anion A⁻ of the element does not exist in the gas phase.

Appendix G

Vapor Pressure of Water at Various Temperatures

Table 15   Vapor Pressure of Water at Various Temperatures

Temperature (°C)	Vapor Pressure (torr)	Temperature (°C)	Vapor Pressure (torr)	Temperature (°C)	Vapor Pressure (torr)	Temperature (°C)	Vapor Pressure (torr)
−10	2.1	21	18.7	51	97.2	81	369.7
−9	2.3	22	19.8	52	102.1	82	384.9
−8	2.5	23	21.1	53	107.2	83	400.6
−7	2.7	24	22.4	54	112.5	84	416.8
−6	2.9	25	23.8	55	118.0	85	433.6
−5	3.2	26	25.2	56	123.8	86	450.9
−4	3.4	27	26.7	57	129.8	87	468.7
−3	3.7	28	28.3	58	136.1	88	487.1
−2	4.0	29	30.0	59	142.6	89	506.1
−1	4.3	30	31.8	60	149.4	90	525.8
0	4.6	31	33.7	61	156.4	91	546.1
1	4.9	32	35.7	62	163.8	92	567.0
2	5.3	33	37.7	63	171.4	93	588.6
3	5.7	34	39.9	64	179.3	94	610.9
4	6.1	35	42.2	65	187.5	95	633.9
5	6.5	36	44.6	66	196.1	96	657.6
6	7.0	37	47.1	67	205.0	97	682.1
7	7.5	38	49.7	68	214.2	98	707.3
8	8.0	39	52.4	69	223.7	99	733.2
9	8.6	40	55.3	70	233.7	100	760.0
10	9.2	41	58.3	71	243.9	101	787.6
11	9.8	42	61.5	72	254.6	102	815.9
12	10.5	43	64.8	73	265.7	103	845.1
13	11.2	44	68.3	74	277.2	104	875.1
14	12.0	45	71.9	75	289.1	105	906.1
15	12.8	46	75.7	76	301.4	106	937.9
16	13.6	47	79.6	77	314.1	107	970.6
17	14.5	48	83.7	78	327.3	108	1004.4
18	15.5	49	88.0	79	341.0	109	1038.9
19	16.5	50	92.5	80	355.1	110	1074.6
20	17.5						

Appendix H

Ionization Constants for Weak Acids at 25 °C

Table 16   Ionization Constants for Weak Acids at 25 °C

Acid	Formula and Ionization Equation	K_a
Acetic	$CH_3CO_2H \rightleftharpoons H^+ + CH_3CO_2^-$	1.8×10^{-5}
Arsenic	$H_3AsO_4 \rightleftharpoons H^+ + H_2AsO_4^-$	$K_1 = 2.5 \times 10^{-4}$
	$H_2AsO_4^- \rightleftharpoons H^+ + HAsO_4^{2-}$	$K_2 = 5.6 \times 10^{-8}$
	$HAsO_4^{2-} \rightleftharpoons H^+ + AsO_4^{3-}$	$K_3 = 3.0 \times 10^{-13}$
Arsenous	$H_3AsO_3 \rightleftharpoons H^+ + H_2AsO_3^-$	$K_1 = 6.0 \times 10^{-10}$
	$H_2AsO_3^- \rightleftharpoons H^+ + HAsO_3^{2-}$	$K_2 = 3.0 \times 10^{-14}$
Benzoic	$C_6H_5CO_2H \rightleftharpoons H^+ + C_6H_5CO_2^-$	6.3×10^{-5}
Boric	$H_3BO_3 \rightleftharpoons H^+ + H_2BO_3^-$	$K_1 = 7.3 \times 10^{-10}$
	$H_2BO_3 \rightleftharpoons H^+ + HBO_3^{2-}$	$K_2 = 1.8 \times 10^{-13}$
	$HBO_3^{2-} \rightleftharpoons H^+ + BO_3^{3-}$	$K_3 = 1.6 \times 10^{-14}$
Carbonic	$H_2CO_3 \rightleftharpoons H^+ + HCO_3^-$	$K_1 = 4.2 \times 10^{-7}$
	$HCO_3^- \rightleftharpoons H^+ + CO_3^{2-}$	$K_2 = 4.8 \times 10^{-11}$
Citric	$H_3C_6H_5O_7 \rightleftharpoons H^+ + H_2C_6H_5O_7^-$	$K_1 = 7.4 \times 10^{-3}$
	$H_2C_6H_5O_7^- \rightleftharpoons H^+ + HC_6H_5O_7^{2-}$	$K_2 = 1.7 \times 10^{-5}$
	$HC_6H_5O_7^{2-} \rightleftharpoons H^+ + C_6H_5O_7^{3-}$	$K_3 = 4.0 \times 10^{-7}$
Cyanic	$HOCN \rightleftharpoons H^+ + OCN^-$	3.5×10^{-4}
Formic	$HCO_2H \rightleftharpoons H^+ + HCO_2^-$	1.8×10^{-4}
Hydrazoic	$HN_3 \rightleftharpoons H^+ + N_3^-$	1.9×10^{-5}
Hydrocyanic	$HCN \rightleftharpoons H^+ + CN^-$	4.0×10^{-10}
Hydrofluoric	$HF \rightleftharpoons H^+ + F^-$	7.2×10^{-4}
Hydrogen peroxide	$H_2O_2 \rightleftharpoons H^+ + HO_2^-$	2.4×10^{-12}
Hydrosulfuric	$H_2S \rightleftharpoons H^+ + HS^-$	$K_1 = 1 \times 10^{-7}$
	$HS^- \rightleftharpoons H^+ + S^{2-}$	$K_2 = 1 \times 10^{-19}$
Hypobromous	$HOBr \rightleftharpoons H^+ + OBr^-$	2.5×10^{-9}
Hypochlorous	$HOCl \rightleftharpoons H^+ + OCl^-$	3.5×10^{-8}
Nitrous	$HNO_2 \rightleftharpoons H^+ + NO_2^-$	4.5×10^{-4}
Oxalic	$H_2C_2O_4 \rightleftharpoons H^+ + HC_2O_4^-$	$K_1 = 5.9 \times 10^{-2}$
	$HC_2O_4^- \rightleftharpoons H^+ + C_2O_4^{2-}$	$K_2 = 6.4 \times 10^{-5}$
Phenol	$C_6H_5OH \rightleftharpoons H^+ + C_6H_5O^-$	1.3×10^{-10}

(continued)

Table 16 *(continued)*

Acid	Formula and Ionization Equation	K_a
Phosphoric	$H_3PO_4 \rightleftharpoons H^+ + H_2PO_4^-$	$K_1 = 7.5 \times 10^{-3}$
	$H_2PO_4^- \rightleftharpoons H^+ + HPO_4^{2-}$	$K_2 = 6.2 \times 10^{-8}$
	$HPO_4^{2-} \rightleftharpoons H^+ + PO_4^{3-}$	$K_3 = 3.6 \times 10^{-13}$
Phosphorous	$H_3PO_3 \rightleftharpoons H^+ + H_2PO_3^-$	$K_1 = 1.6 \times 10^{-2}$
	$H_2PO_3^- \rightleftharpoons H^+ + HPO_3^{2-}$	$K_2 = 7.0 \times 10^{-7}$
Selenic	$H_2SeO_4 \rightleftharpoons H^+ + HSeO_4^-$	$K_1 = $ very large
	$HSeO_4^- \rightleftharpoons H^+ + SeO_4^{2-}$	$K_2 = 1.2 \times 10^{-2}$
Selenous	$H_2SeO_3 \rightleftharpoons H^+ + HSeO_3^-$	$K_1 = 2.7 \times 10^{-3}$
	$HSeO_3^- \rightleftharpoons H^+ + SeO_3^{2-}$	$K_2 = 2.5 \times 10^{-7}$
Sulfuric	$H_2SO_4 \rightleftharpoons H^+ + HSO_4^-$	$K_1 = $ very large
	$HSO_4^- \rightleftharpoons H^+ + SO_4^{2-}$	$K_2 = 1.2 \times 10^{-2}$
Sulfurous	$H_2SO_3 \rightleftharpoons H^+ + HSO_3^-$	$K_1 = 1.2 \times 10^{-2}$
	$HSO_3^- \rightleftharpoons H^+ + SO_3^{2-}$	$K_2 = 6.2 \times 10^{-8}$
Tellurous	$H_2TeO_3 \rightleftharpoons H^+ + HTeO_3^-$	$K_1 = 2 \times 10^{-3}$
	$HTeO_3^- \rightleftharpoons H^+ + TeO_3^{2-}$	$K_2 = 1 \times 10^{-8}$

Appendix I

Ionization Constants for Weak Bases at 25 °C

Table 17 Ionization Constants for Weak Bases at 25 °C

Base	Formula and Ionization Equation	K_b
Ammonia	$NH_3 + H_2O \rightleftharpoons NH_4^+ + OH^-$	1.8×10^{-5}
Aniline	$C_6H_5NH_2 + H_2O \rightleftharpoons C_6H_5NH_3^+ + OH^-$	4.0×10^{-10}
Dimethylamine	$(CH_3)_2NH + H_2O \rightleftharpoons (CH_3)_2NH_2^+ + OH^-$	7.4×10^{-4}
Ethylenediamine	$H_2NCH_2CH_2NH_2 + H_2O \rightleftharpoons H_2NCH_2CH_2NH_3^+ + OH^-$	$K_1 = 8.5 \times 10^{-5}$
	$H_2NCH_2CH_2NH_3^+ + H_2O \rightleftharpoons H_3NCH_2CH_2NH_3^{2+} + OH^-$	$K_2 = 2.7 \times 10^{-8}$
Hydrazine	$N_2H_4 + H_2O \rightleftharpoons N_2H_5^+ + OH^-$	$K_1 = 8.5 \times 10^{-7}$
	$N_2H_5^+ + H_2O \rightleftharpoons N_2H_6^{2+} + OH^-$	$K_2 = 8.9 \times 10^{-16}$
Hydroxylamine	$NH_2OH + H_2O \rightleftharpoons NH_3OH^+ + OH^-$	6.6×10^{-9}
Methylamine	$CH_3NH_2 + H_2O \rightleftharpoons CH_3NH_3^+ + OH^-$	5.0×10^{-4}
Pyridine	$C_5H_5N + H_2O \rightleftharpoons C_5H_5NH^+ + OH^-$	1.5×10^{-9}
Trimethylamine	$(CH_3)_3N + H_2O \rightleftharpoons (CH_3)_3NH^+ + OH^-$	7.4×10^{-5}
Ethylamine	$C_2H_5NH_2 + H_2O \rightleftharpoons C_2H_5NH_3^+ + OH^-$	4.3×10^{-4}

Appendix J

Solubility Product Constants for Some Inorganic Compounds at 25 °C

Table 18A Solubility Product Constants (25 °C)

Cation	Compound	K_{sp}	Cation	Compound	K_{sp}
Ba^{2+}	*$BaCrO_4$	1.2×10^{-10}	Mg^{2+}	$MgCO_3$	6.8×10^{-6}
	$BaCO_3$	2.6×10^{-9}		MgF_2	5.2×10^{-11}
	BaF_2	1.8×10^{-7}		$Mg(OH)_2$	5.6×10^{-12}
	*$BaSO_4$	1.1×10^{-10}	Mn^{2+}	$MnCO_3$	2.3×10^{-11}
Ca^{2+}	$CaCO_3$ (calcite)	3.4×10^{-9}		*$Mn(OH)_2$	1.9×10^{-13}
	*CaF_2	5.3×10^{-11}			
	*$Ca(OH)_2$	5.5×10^{-5}	Hg_2^{2+}	*Hg_2Br_2	6.4×10^{-23}
	$CaSO_4$	4.9×10^{-5}		Hg_2Cl_2	1.4×10^{-18}
$Cu^{+,2+}$	$CuBr$	6.3×10^{-9}		*Hg_2I_2	2.9×10^{-29}
	CuI	1.3×10^{-12}		Hg_2SO_4	6.5×10^{-7}
	$Cu(OH)_2$	2.2×10^{-20}	Ni^{2+}	$NiCO_3$	1.4×10^{-7}
	$CuSCN$	1.8×10^{-13}		$Ni(OH)_2$	5.5×10^{-16}
Au^+	$AuCl$	2.0×10^{-13}	Ag^+	*$AgBr$	5.4×10^{-13}
				*$AgBrO_3$	5.4×10^{-5}
Fe^{2+}	$FeCO_3$	3.1×10^{-11}		$AgCH_3CO_2$	1.9×10^{-3}
	$Fe(OH)_2$	4.9×10^{-17}		$AgCN$	6.0×10^{-17}
				Ag_2CO_3	8.5×10^{-12}
Pb^{2+}	$PbBr_2$	6.6×10^{-6}		*$Ag_2C_2O_4$	5.4×10^{-12}
	$PbCO_3$	7.4×10^{-14}		*$AgCl$	1.8×10^{-10}
	$PbCl_2$	1.7×10^{-5}		Ag_2CrO_4	1.1×10^{-12}
	$PbCrO_4$	2.8×10^{-13}		*AgI	8.5×10^{-17}
	PbF_2	3.3×10^{-8}		$AgSCN$	1.0×10^{-12}
	PbI_2	9.8×10^{-9}		*Ag_2SO_4	1.2×10^{-5}
	$Pb(OH)_2$	1.4×10^{-15}			
	$PbSO_4$	2.5×10^{-8}			

(continued)

Table 18A *(continued)*

Cation	Compound	K_{sp}	Cation	Compound	K_{sp}
Sr^{2+}	$SrCO_3$	5.6×10^{-10}	Zn^{2+}	$Zn(OH)_2$	3×10^{-17}
	SrF_2	4.3×10^{-9}		$Zn(CN)_2$	8.0×10^{-12}
	$SrSO_4$	3.4×10^{-7}			
Tl^+	$TlBr$	3.7×10^{-6}			
	$TlCl$	1.9×10^{-4}			
	TlI	5.5×10^{-8}			

The values reported in this table were taken from J. A. Dean: *Lange's Handbook of Chemistry,* 15th Edition. New York, McGraw Hill Publishers, 1999. Values have been rounded off to two significant figures.

* Calculated solubility from these K_{sp} values will match experimental solubility for this compound within a factor of 2. Experimental values for solubilities are given in R. W. Clark and J. M. Bonicamp: *Journal of Chemical Education,* Vol. 75, p. 1182, 1998.

Table 18B K_{spa} Values* for Some Metal Sulfides (25 °C)

Substance	K_{spa}
HgS (red)	4×10^{-54}
HgS (black)	2×10^{-53}
Ag_2S	6×10^{-51}
CuS	6×10^{-37}
PbS	3×10^{-28}
CdS	8×10^{-28}
SnS	1×10^{-26}
FeS	6×10^{-19}

* The equilibrium constant value K_{spa} for metal sulfides refers to the equilibrium $MS(s) + H_2O(\ell) \rightleftharpoons M^{2+}(aq) + OH^-(aq) + HS^-(aq)$; see R. J. Myers, *Journal of Chemical Education,* Vol. 63, p. 687, 1986.

Appendix K

Formation Constants for Some Complex Ions in Aqueous Solution

Table 19   Formation Constants for Some Complex Ions in Aqueous Solution

Formation Equilibrium	K
$Ag^+ + 2\ Br^- \rightleftharpoons [AgBr_2]^-$	1.3×10^7
$Ag^+ + 2\ Cl^- \rightleftharpoons [AgCl_2]^-$	2.5×10^5
$Ag^+ + 2\ CN^- \rightleftharpoons [Ag(CN)_2]^-$	5.6×10^{18}
$Ag^+ + 2\ S_2O_3^{2-} \rightleftharpoons [Ag(S_2O_3)_2]^{3-}$	2.0×10^{13}
$Ag^+ + 2\ NH_3 \rightleftharpoons [Ag(NH_3)_2]^+$	1.6×10^7
$Al^{3+} + 6\ F^- \rightleftharpoons [AlF_6]^{3-}$	5.0×10^{23}
$Al^{3+} + 4\ OH^- \rightleftharpoons [Al(OH)_4]^-$	7.7×10^{33}
$Au^+ + 2\ CN^- \rightleftharpoons [Au(CN)_2]^-$	2.0×10^{38}
$Cd^{2+} + 4\ CN^- \rightleftharpoons [Cd(CN)_4]^{2-}$	1.3×10^{17}
$Cd^{2+} + 4\ Cl^- \rightleftharpoons [CdCl_4]^{2-}$	1.0×10^4
$Cd^{2+} + 4\ NH_3 \rightleftharpoons [Cd(NH_3)_4]^{2+}$	1.0×10^7
$Co^{2+} + 6\ NH_3 \rightleftharpoons [Co(NH_3)_6]^{2+}$	7.7×10^4
$Cu^+ + 2\ CN^- \rightleftharpoons [Cu(CN)_2]^-$	1.0×10^{16}
$Cu^+ + 2\ Cl^- \rightleftharpoons [CuCl_2]^-$	1.0×10^5
$Cu^{2+} + 4\ NH_3 \rightleftharpoons [Cu(NH_3)_4]^{2+}$	6.8×10^{12}
$Fe^{2+} + 6\ CN^- \rightleftharpoons [Fe(CN)_6]^{4-}$	7.7×10^{36}
$Hg^{2+} + 4\ Cl^- \rightleftharpoons [HgCl_4]^{2-}$	1.2×10^{15}
$Ni^{2+} + 4\ CN^- \rightleftharpoons [Ni(CN)_4]^{2-}$	1.0×10^{31}
$Ni^{2+} + 6\ NH_3 \rightleftharpoons [Ni(NH_3)_6]^{2+}$	5.6×10^8
$Zn^{2+} + 4\ OH^- \rightleftharpoons [Zn(OH)_4]^{2-}$	2.9×10^{15}
$Zn^{2+} + 4\ NH_3 \rightleftharpoons [Zn(NH_3)_4]^{2+}$	2.9×10^9

Appendix L

Selected Thermodynamic Values

Table 20   Selected Thermodynamic Values*

Species	ΔH_f° (298.15 K) (kJ/mol)	S° (298.15 K) (J/K · mol)	ΔG_f° (298.15 K) (kJ/mol)
Aluminum			
Al(s)	0	28.3	0
AlCl$_3$(s)	−705.63	109.29	−630.0
Al$_2$O$_3$(s)	−1675.7	50.92	−1582.3
Barium			
BaCl$_2$(s)	−858.6	123.68	−810.4
BaCO$_3$(s)	−1213	112.1	−1134.41
BaO(s)	−548.1	72.05	−520.38
BaSO$_4$(s)	−1473.2	132.2	−1362.2
Beryllium			
Be(s)	0	9.5	0
Be(OH)$_2$(s)	−902.5	51.9	−815.0
Boron			
BCl$_3$(g)	−402.96	290.17	−387.95
Bromine			
Br(g)	111.884	175.022	82.396
Br$_2$(ℓ)	0	152.2	0
Br$_2$(g)	30.91	245.47	3.12
BrF$_3$(g)	−255.60	292.53	−229.43
HBr(g)	−36.29	198.70	−53.45
Calcium			
Ca(s)	0	41.59	0
Ca(g)	178.2	158.884	144.3
Ca^{2+}(g)	1925.90	—	—
CaC$_2$(s)	−59.8	70.	−64.93
CaCO$_3$(s, calcite)	−1207.6	91.7	−1129.16
CaCl$_2$(s)	−795.8	104.6	−748.1
CaF$_2$(s)	−1219.6	68.87	−1167.3
CaH$_2$(s)	−186.2	42	−147.2
CaO(s)	−635.09	38.2	−603.42
CaS(s)	−482.4	56.5	−477.4
Ca(OH)$_2$(s)	−986.09	83.39	−898.43

(continued)

* Most thermodynamic data are taken from the NIST Webbook at **http://webbook.nist.gov**.

Table 20   *(continued)*

Species	ΔH_f° (298.15 K) (kJ/mol)	S° (298.15 K) (J/K · mol)	ΔG_f° (298.15 K) (kJ/mol)
Ca(OH)$_2$(aq)	−1002.82		−868.07
CaSO$_4$(s)	−1434.52	106.5	−1322.02
Carbon			
C(s, graphite)	0	5.6	0
C(s, diamond)	1.8	2.377	2.900
C(g)	716.67	158.1	671.2
CCl$_4$(ℓ)	−128.4	214.39	−57.63
CCl$_4$(g)	−95.98	309.65	−53.61
CHCl$_3$(ℓ)	−134.47	201.7	−73.66
CHCl$_3$(g)	−103.18	295.61	−70.4
CH$_4$(g, methane)	−74.87	186.26	−50.8
C$_2$H$_2$(g, ethyne)	226.73	200.94	209.20
C$_2$H$_4$(g, ethene)	52.47	219.36	68.35
C$_2$H$_6$(g, ethane)	−83.85	229.2	−31.89
C$_3$H$_8$(g, propane)	−104.7	270.3	−24.4
C$_6$H$_6$(ℓ, benzene)	48.95	173.26	124.21
CH$_3$OH(ℓ, methanol)	−238.4	127.19	−166.14
CH$_3$OH(g, methanol)	−201.0	239.7	−162.5
C$_2$H$_5$OH(ℓ, ethanol)	−277.0	160.7	−174.7
C$_2$H$_5$OH(g, ethanol)	−235.3	282.70	−168.49
CO(g)	−110.525	197.674	−137.168
CO$_2$(g)	−393.509	213.74	−394.359
CS$_2$(ℓ)	89.41	151	65.2
CS$_2$(g)	116.7	237.8	66.61
COCl$_2$(g)	−218.8	283.53	−204.6
Cesium			
Cs(s)	0	85.23	0
Cs$^+$(g)	457.964	—	—
CsCl(s)	−443.04	101.17	−414.53
Chlorine			
Cl(g)	121.3	165.19	105.3
Cl$^-$(g)	−233.13	—	—
Cl$_2$(g)	0	223.08	0
HCl(g)	−92.31	186.2	−95.09
HCl(aq)	−167.159	56.5	−131.26
Chromium			
Cr(s)	0	23.62	0
Cr$_2$O$_3$(s)	−1134.7	80.65	−1052.95
CrCl$_3$(s)	−556.5	123.0	−486.1

(continued)

Table 20   (continued)

Species	ΔH_f° (298.15 K) (kJ/mol)	S° (298.15 K) (J/K · mol)	ΔG_f° (298.15 K) (kJ/mol)
Copper			
Cu(s)	0	33.17	0
CuO(s)	−156.06	42.59	−128.3
$CuCl_2$(s)	−220.1	108.07	−175.7
$CuSO_4$(s)	−769.98	109.05	−660.75
Fluorine			
F_2(g)	0	202.8	0
F(g)	78.99	158.754	61.91
F^-(g)	−255.39	—	—
F^-(aq)	−332.63		−278.79
HF(g)	−273.3	173.779	−273.2
HF(aq)	−332.63	88.7	−278.79
Hydrogen			
H_2(g)	0	130.7	0
H(g)	217.965	114.713	203.247
H^+(g)	1536.202	—	—
$H_2O(\ell)$	−285.83	69.95	−237.15
H_2O(g)	−241.83	188.84	−228.59
$H_2O_2(\ell)$	−187.78	109.6	−120.35
Iodine			
I_2(s)	0	116.135	0
I_2(g)	62.438	260.69	19.327
I(g)	106.838	180.791	70.250
I^-(g)	−197	—	—
ICl(g)	17.51	247.56	−5.73
Iron			
Fe(s)	0	27.78	0
FeO(s)	−272	—	—
Fe_2O_3(s, hematite)	−825.5	87.40	−742.2
Fe_3O_4(s, magnetite)	−1118.4	146.4	−1015.4
$FeCl_2$(s)	−341.79	117.95	−302.30
$FeCl_3$(s)	−399.49	142.3	−344.00
FeS_2(s, pyrite)	−178.2	52.93	−166.9
$Fe(CO)_5(\ell)$	−774.0	338.1	−705.3
Lead			
Pb(s)	0	64.81	0
$PbCl_2$(s)	−359.41	136.0	−314.10
PbO(s, yellow)	−219	66.5	−196
PbO_2(s)	−277.4	68.6	−217.39
PbS(s)	−100.4	91.2	−98.7

(continued)

Table 20   *(continued)*

Species	ΔH_f° (298.15 K) (kJ/mol)	S° (298.15 K) (J/K · mol)	ΔG_f° (298.15 K) (kJ/mol)
Lithium			
Li(s)	0	29.12	0
Li$^+$(g)	685.783	—	—
LiOH(s)	−484.93	42.81	−438.96
LiOH(aq)	−508.48	2.80	−450.58
LiCl(s)	−408.701	59.33	−384.37
Magnesium			
Mg(s)	0	32.67	0
MgCl$_2$(s)	−641.62	89.62	−592.09
MgCO$_3$(s)	−1111.69	65.84	−1028.2
MgO(s)	−601.24	26.85	−568.93
Mg(OH)$_2$(s)	−924.54	63.18	−833.51
MgS(s)	−346.0	50.33	−341.8
Mercury			
Hg(ℓ)	0	76.02	0
HgCl$_2$(s)	−224.3	146.0	−178.6
HgO(s, red)	−90.83	70.29	−58.539
HgS(s, red)	−58.2	82.4	−50.6
Nickel			
Ni(s)	0	29.87	0
NiO(s)	−239.7	37.99	−211.7
NiCl$_2$(s)	−305.332	97.65	−259.032
Nitrogen			
N$_2$(g)	0	191.56	0
N(g)	472.704	153.298	455.563
NH$_3$(g)	−45.90	192.77	−16.37
N$_2$H$_4$(ℓ)	50.63	121.52	149.45
NH$_4$Cl(s)	−314.55	94.85	−203.08
NH$_4$Cl(aq)	−299.66	169.9	−210.57
NH$_4$NO$_3$(s)	−365.56	151.08	−183.84
NH$_4$NO$_3$(aq)	−339.87	259.8	−190.57
NO(g)	90.29	210.76	86.58
NO$_2$(g)	33.1	240.04	51.23
N$_2$O(g)	82.05	219.85	104.20
N$_2$O$_4$(g)	9.08	304.38	97.73
NOCl(g)	51.71	261.8	66.08
HNO$_3$(ℓ)	−174.10	155.60	−80.71
HNO$_3$(g)	−135.06	266.38	−74.72
HNO$_3$(aq)	−207.36	146.4	−111.25

(continued)

Table 20   *(continued)*

Species	ΔH_f° (298.15 K) (kJ/mol)	S° (298.15 K) (J/K · mol)	ΔG_f° (298.15 K) (kJ/mol)
Oxygen			
$O_2(g)$	0	205.07	0
$O(g)$	249.170	161.055	231.731
$O_3(g)$	142.67	238.92	163.2
Phosphorus			
$P_4(s,\ white)$	0	41.1	0
$P_4(s,\ red)$	−17.6	22.80	−12.1
$P(g)$	314.64	163.193	278.25
$PH_3(g)$	22.89	210.24	30.91
$PCl_3(g)$	−287.0	311.78	−267.8
$P_4O_{10}(s)$	−2984.0	228.86	−2697.7
$H_3PO_4(\ell)$	−1279.0	110.5	−1119.1
Potassium			
$K(s)$	0	64.63	0
$KCl(s)$	−436.68	82.56	−408.77
$KClO_3(s)$	−397.73	143.1	−296.25
$KI(s)$	−327.90	106.32	−324.892
$KOH(s)$	−424.72	78.9	−378.92
$KOH(aq)$	−482.37	91.6	−440.50
Silicon			
$Si(s)$	0	18.82	0
$SiBr_4(\ell)$	−457.3	277.8	−443.9
$SiC(s)$	−65.3	16.61	−62.8
$SiCl_4(g)$	−662.75	330.86	−622.76
$SiH_4(g)$	34.31	204.65	56.84
$SiF_4(g)$	−1614.94	282.49	−1572.65
$SiO_2(s,\ quartz)$	−910.86	41.46	−856.97
Silver			
$Ag(s)$	0	42.55	0
$Ag_2O(s)$	−31.1	121.3	−11.32
$AgCl(s)$	−127.01	96.25	−109.76
$AgNO_3(s)$	−124.39	140.92	−33.41
Sodium			
$Na(s)$	0	51.21	0
$Na(g)$	107.3	153.765	76.83
$Na^+(g)$	609.358	—	—
$NaBr(s)$	−361.02	86.82	−348.983
$NaCl(s)$	−411.12	72.11	−384.04
$NaCl(g)$	−181.42	229.79	−201.33
$NaCl(aq)$	−407.27	115.5	−393.133

(continued)

Table 20   (continued)

Species	ΔH_f° (298.15 K) (kJ/mol)	$S°$ (298.15 K) (J/K · mol)	ΔG_f° (298.15 K) (kJ/mol)
NaOH(s)	−425.93	64.46	−379.75
NaOH(aq)	−469.15	48.1	−418.09
Na_2CO_3(s)	−1130.77	134.79	−1048.08
Sulfur			
S(s, rhombic)	0	32.1	0
S(g)	278.98	167.83	236.51
S_2Cl_2(g)	−18.4	331.5	−31.8
SF_6(g)	−1209	291.82	−1105.3
H_2S(g)	−20.63	205.79	−33.56
SO_2(g)	−296.84	248.21	−300.13
SO_3(g)	−395.77	256.77	−371.04
$SOCl_2$(g)	−212.5	309.77	−198.3
H_2SO_4(ℓ)	−814	156.9	−689.96
H_2SO_4(aq)	−909.27	20.1	−744.53
Tin			
Sn(s, white)	0	51.08	0
Sn(s, gray)	−2.09	44.14	0.13
$SnCl_4$(ℓ)	−511.3	258.6	−440.15
$SnCl_4$(g)	−471.5	365.8	−432.31
SnO_2(s)	−577.63	49.04	−515.88
Titanium			
Ti(s)	0	30.72	0
$TiCl_4$(ℓ)	−804.2	252.34	−737.2
$TiCl_4$(g)	−763.16	354.84	−726.7
TiO_2(s)	−939.7	49.92	−884.5
Zinc			
Zn(s)	0	41.63	0
$ZnCl_2$(s)	−415.05	111.46	−369.398
ZnO(s)	−348.28	43.64	−318.30
ZnS(s, sphalerite)	−205.98	57.7	−201.29

Appendix M

Standard Reduction Potentials in Aqueous Solution at 25 °C

Table 21   Standard Reduction Potentials in Aqueous Solution at 25 °C

Acidic Solution	Standard Reduction Potential, $E°$ (volts)
$F_2(g) + 2\,e^- \longrightarrow 2\,F^-(aq)$	2.87
$Co^{3+}(aq) + e^- \longrightarrow Co^{2+}(aq)$	1.82
$Pb^{4+}(aq) + 2\,e^- \longrightarrow Pb^{2+}(aq)$	1.8
$H_2O_2(aq) + 2\,H^+(aq) + 2\,e^- \longrightarrow 2\,H_2O$	1.77
$NiO_2(s) + 4\,H^+(aq) + 2\,e^- \longrightarrow Ni^{2+}(aq) + 2\,H_2O$	1.7
$PbO_2(s) + SO_4^{2-}(aq) + 4\,H^+(aq) + 2\,e^- \longrightarrow PbSO_4(s) + 2\,H_2O$	1.685
$Au^+(aq) + e^- \longrightarrow Au(s)$	1.68
$2\,HClO(aq) + 2\,H^+(aq) + 2\,e^- \longrightarrow Cl_2(g) + 2\,H_2O$	1.63
$Ce^{4+}(aq) + e^- \longrightarrow Ce^{3+}(aq)$	1.61
$NaBiO_3(s) + 6\,H^+(aq) + 2\,e^- \longrightarrow Bi^{3+}(aq) + Na^+(aq) + 3\,H_2O$	≈ 1.6
$MnO_4^-(aq) + 8\,H^+(aq) + 5\,e^- \longrightarrow Mn^{2+}(aq) + 4\,H_2O$	1.51
$Au^{3+}(aq) + 3\,e^- \longrightarrow Au(s)$	1.50
$ClO_3^-(aq) + 6\,H^+(aq) + 5\,e^- \longrightarrow \frac{1}{2}\,Cl_2(g) + 3\,H_2O$	1.47
$BrO_3^-(aq) + 6\,H^+(aq) + 6\,e^- \longrightarrow Br^-(aq) + 3\,H_2O$	1.44
$Cl_2(g) + 2\,e^- \longrightarrow 2\,Cl^-(aq)$	1.36
$Cr_2O_7^{2-}(aq) + 14\,H^+(aq) + 6\,e^- \longrightarrow 2\,Cr^{3+}(aq) + 7\,H_2O$	1.33
$N_2H_5^+(aq) + 3\,H^+(aq) + 2\,e^- \longrightarrow 2\,NH_4^+(aq)$	1.24
$MnO_2(s) + 4\,H^+(aq) + 2\,e^- \longrightarrow Mn^{2+}(aq) + 2\,H_2O$	1.23
$O_2(g) + 4\,H^+(aq) + 4\,e^- \longrightarrow 2\,H_2O$	1.229
$Pt^{2+}(aq) + 2\,e^- \longrightarrow Pt(s)$	1.2
$IO_3^-(aq) + 6\,H^+(aq) + 5\,e^- \longrightarrow \frac{1}{2}\,I_2(aq) + 3\,H_2O$	1.195
$ClO_4^-(aq) + 2\,H^+(aq) + 2\,e^- \longrightarrow ClO_3^-(aq) + H_2O$	1.19
$Br_2(\ell) + 2\,e^- \longrightarrow 2\,Br^-(aq)$	1.08
$AuCl_4^-(aq) + 3\,e^- \longrightarrow Au(s) + 4\,Cl^-(aq)$	1.00
$Pd^{2+}(aq) + 2\,e^- \longrightarrow Pd(s)$	0.987
$NO_3^-(aq) + 4\,H^+(aq) + 3\,e^- \longrightarrow NO(g) + 2\,H_2O$	0.96
$NO_3^-(aq) + 3\,H^+(aq) + 2\,e^- \longrightarrow HNO_2(aq) + H_2O$	0.94
$2\,Hg^+(aq) + 2\,e^- \longrightarrow Hg_2^{2+}(aq)$	0.920
$Hg^{2+}(aq) + 2\,e^- \longrightarrow Hg(\ell)$	0.855
$Ag^+(aq) + e^- \longrightarrow Ag(s)$	0.7994
$Hg_2^{2+}(aq) + 2\,e^- \longrightarrow 2\,Hg(\ell)$	0.789
$Fe^{3+}(aq) + e^- \longrightarrow Fe^{2+}(aq)$	0.771

(continued)

Table 21 *(continued)*

Acidic Solution	Standard Reduction Potential, $E°$ (volts)
$SbCl_6^-(aq) + 2\,e^- \longrightarrow SbCl_4^-(aq) + 2\,Cl^-(aq)$	0.75
$[PtCl_4]^{2+}(aq) + 2\,e^- \longrightarrow Pt(s) + 4\,Cl^-(aq)$	0.73
$O_2(g) + 2\,H^+(aq) + 2\,e^- \longrightarrow H_2O_2(aq)$	0.682
$[PtCl_6]^{2-}(aq) + 2\,e^- \longrightarrow [PtCl_4]^{2-}(aq) + 2\,Cl^-(aq)$	0.68
$H_3AsO_4(aq) + 2\,H^+(aq) + 2\,e^- \longrightarrow H_3AsO_3(aq) + H_2O$	0.58
$I_2(s) + 2\,e^- \longrightarrow 2\,I^-(aq)$	0.535
$TeO_2(s) + 4\,H^+(aq) + 4\,e^- \longrightarrow Te(s) + 2\,H_2O$	0.529
$Cu^+(aq) + e^- \longrightarrow Cu(s)$	0.521
$[RhCl_6]^{3-}(aq) + 3\,e^- \longrightarrow Rh(s) + 6\,Cl^-(aq)$	0.44
$Cu^{2+}(aq) + 2\,e^- \longrightarrow Cu(s)$	0.337
$Hg_2Cl_2(s) + 2\,e^- \longrightarrow 2\,Hg(\ell) + 2\,Cl^-(aq)$	0.27
$AgCl(s) + e^- \longrightarrow Ag(s) + Cl^-(aq)$	0.222
$SO_4^{2-}(aq) + 4\,H^+(aq) + 2\,e^- \longrightarrow SO_2(g) + 2\,H_2O$	0.20
$SO_4^{2-}(aq) + 4\,H^+(aq) + 2\,e^- \longrightarrow H_2SO_3(aq) + H_2O$	0.17
$Cu^{2+}(aq) + e^- \longrightarrow Cu^+(aq)$	0.153
$Sn^{4+}(aq) + 2\,e^- \longrightarrow Sn^{2+}(aq)$	0.15
$S(s) + 2\,H^+ + 2\,e^- \longrightarrow H_2S(aq)$	0.14
$AgBr(s) + e^- \longrightarrow Ag(s) + Br^-(aq)$	0.0713
$2\,H^+(aq) + 2\,e^- \longrightarrow H_2(g)\,(\text{reference electrode})$	0.0000
$N_2O(g) + 6\,H^+(aq) + H_2O + 4\,e^- \longrightarrow 2\,NH_3OH^+(aq)$	−0.05
$Pb^{2+}(aq) + 2\,e^- \longrightarrow Pb(s)$	−0.126
$Sn^{2+}(aq) + 2\,e^- \longrightarrow Sn(s)$	−0.14
$AgI(s) + e^- \longrightarrow Ag(s) + I^-(aq)$	−0.15
$[SnF_6]^{2-}(aq) + 4\,e^- \longrightarrow Sn(s) + 6\,F^-(aq)$	−0.25
$Ni^{2+}(aq) + 2\,e^- \longrightarrow Ni(s)$	−0.25
$Co^{2+}(aq) + 2\,e^- \longrightarrow Co(s)$	−0.28
$Tl^+(aq) + e^- \longrightarrow Tl(s)$	−0.34
$PbSO_4(s) + 2\,e^- \longrightarrow Pb(s) + SO_4^{2-}(aq)$	−0.356
$Se(s) + 2\,H^+(aq) + 2\,e^- \longrightarrow H_2Se(aq)$	−0.40
$Cd^{2+}(aq) + 2\,e^- \longrightarrow Cd(s)$	−0.403
$Cr^{3+}(aq) + e^- \longrightarrow Cr^{2+}(aq)$	−0.41
$Fe^{2+}(aq) + 2\,e^- \longrightarrow Fe(s)$	−0.44
$2\,CO_2(g) + 2\,H^+(aq) + 2\,e^- \longrightarrow H_2C_2O_4(aq)$	−0.49
$Ga^{3+}(aq) + 3\,e^- \longrightarrow Ga(s)$	−0.53
$HgS(s) + 2\,H^+(aq) + 2\,e^- \longrightarrow Hg(\ell) + H_2S(g)$	−0.72
$Cr^{3+}(aq) + 3\,e^- \longrightarrow Cr(s)$	−0.74
$Zn^{2+}(aq) + 2\,e^- \longrightarrow Zn(s)$	−0.763
$Cr^{2+}(aq) + 2\,e^- \longrightarrow Cr(s)$	−0.91
$FeS(s) + 2\,e^- \longrightarrow Fe(s) + S^{2-}(aq)$	−1.01
$Mn^{2+}(aq) + 2\,e^- \longrightarrow Mn(s)$	−1.18
$V^{2+}(aq) + 2\,e^- \longrightarrow V(s)$	−1.18
$CdS(s) + 2\,e^- \longrightarrow Cd(s) + S^{2-}(aq)$	−1.21
$ZnS(s) + 2\,e^- \longrightarrow Zn(s) + S^{2-}(aq)$	−1.44
$Zr^{4+}(aq) + 4\,e^- \longrightarrow Zr(s)$	−1.53

(continued)

Table 21   *(continued)*

Acidic Solution	Standard Reduction Potential, $E°$ (volts)
$Al^{3+}(aq) + 3\ e^- \longrightarrow Al(s)$	-1.66
$Mg^{2+}(aq) + 2\ e^- \longrightarrow Mg(s)$	-2.37
$Na^+(aq) + e^- \longrightarrow Na(s)$	-2.714
$Ca^{2+}(aq) + 2\ e^- \longrightarrow Ca(s)$	-2.87
$Sr^{2+}(aq) + 2\ e^- \longrightarrow Sr(s)$	-2.89
$Ba^{2+}(aq) + 2\ e^- \longrightarrow Ba(s)$	-2.90
$Rb^+(aq) + e^- \longrightarrow Rb(s)$	-2.925
$K^+(aq) + e^- \longrightarrow K(s)$	-2.925
$Li^+(aq) + e^- \longrightarrow Li(s)$	-3.045

Basic Solution

	Standard Reduction Potential, $E°$ (volts)
$ClO^-(aq) + H_2O + 2\ e^- \longrightarrow Cl^-(aq) + 2\ OH^-(aq)$	0.89
$OOH^-(aq) + H_2O + 2\ e^- \longrightarrow 3\ OH^-(aq)$	0.88
$2\ NH_2OH(aq) + 2\ e^- \longrightarrow N_2H_4(aq) + 2\ OH^-(aq)$	0.74
$ClO_3^-(aq) + 3\ H_2O + 6\ e^- \longrightarrow Cl^-(aq) + 6\ OH^-(aq)$	0.62
$MnO_4^-(aq) + 2\ H_2O + 3\ e^- \longrightarrow MnO_2(s) + 4\ OH^-(aq)$	0.588
$MnO_4^-(aq) + e^- \longrightarrow MnO_4^{2-}(aq)$	0.564
$NiO_2(s) + 2\ H_2O + 2\ e^- \longrightarrow Ni(OH)_2(s) + 2\ OH^-(aq)$	0.49
$Ag_2CrO_4(s) + 2\ e^- \longrightarrow 2\ Ag(s) + CrO_4^{2-}(aq)$	0.446
$O_2(g) + 2\ H_2O + 4\ e^- \longrightarrow 4\ OH^-(aq)$	0.40
$ClO_4^-(aq) + H_2O + 2\ e^- \longrightarrow ClO_3^-(aq) + 2\ OH^-(aq)$	0.36
$Ag_2O(s) + H_2O + 2\ e^- \longrightarrow 2\ Ag(s) + 2\ OH^-(aq)$	0.34
$2\ NO_2^-(aq) + 3\ H_2O + 4\ e^- \longrightarrow N_2O(g) + 6\ OH^-(aq)$	0.15
$N_2H_4(aq) + 2\ H_2O + 2\ e^- \longrightarrow 2\ NH_3(aq) + 2\ OH^-(aq)$	0.10
$[Co(NH_3)_6]^{3+}(aq) + e^- \longrightarrow [Co(NH_3)_6]^{2+}(aq)$	0.10
$HgO(s) + H_2O + 2\ e^- \longrightarrow Hg(\ell) + 2\ OH^-(aq)$	0.0984
$O_2(g) + H_2O + 2\ e^- \longrightarrow OOH^-(aq) + OH^-(aq)$	0.076
$NO_3^-(aq) + H_2O + 2\ e^- \longrightarrow NO_2^-(aq) + 2\ OH^-(aq)$	0.01
$MnO_2(s) + 2\ H_2O + 2\ e^- \longrightarrow Mn(OH)_2(s) + 2\ OH^-(aq)$	-0.05
$CrO_4^{2-}(aq) + 4\ H_2O + 3\ e^- \longrightarrow Cr(OH)_3(s) + 5\ OH^-(aq)$	-0.12
$Cu(OH)_2(s) + 2\ e^- \longrightarrow Cu(s) + 2\ OH^-(aq)$	-0.36
$S(s) + 2\ e^- \longrightarrow S^{2-}(aq)$	-0.48
$Fe(OH)_3(s) + e^- \longrightarrow Fe(OH)_2(s) + OH^-(aq)$	-0.56
$2\ H_2O + 2\ e^- \longrightarrow H_2(g) + 2\ OH^-(aq)$	-0.8277
$2\ NO_3^-(aq) + 2\ H_2O + 2\ e^- \longrightarrow N_2O_4(g) + 4\ OH^-(aq)$	-0.85
$Fe(OH)_2(s) + 2\ e^- \longrightarrow Fe(s) + 2\ OH^-(aq)$	-0.877
$SO_4^{2-}(aq) + H_2O + 2\ e^- \longrightarrow SO_3^{2-}(aq) + 2\ OH^-(aq)$	-0.93
$N_2(g) + 4\ H_2O + 4\ e^- \longrightarrow N_2H_4(aq) + 4\ OH^-(aq)$	-1.15
$[Zn(OH)_4]^{2-}(aq) + 2\ e^- \longrightarrow Zn(s) + 4\ OH^-(aq)$	-1.22
$Zn(OH)_2(s) + 2\ e^- \longrightarrow Zn(s) + 2\ OH^-(aq)$	-1.245
$[Zn(CN)_4]^{2-}(aq) + 2\ e^- \longrightarrow Zn(s) + 4\ CN^-(aq)$	-1.26
$Cr(OH)_3(s) + 3\ e^- \longrightarrow Cr(s) + 3\ OH^-(aq)$	-1.30
$SiO_3^{2-}(aq) + 3\ H_2O + 4\ e^- \longrightarrow Si(s) + 6\ OH^-(aq)$	-1.70

Appendix N

Answers to Exercises

Chapter 1

1.1 Based on appearance, the bottle on the right appears to contain a homogeneous solution. The pile on the left appears to contain a heterogeneous mixture of at least two solid substances.

1.2 (a) Na = sodium; Cl = chlorine; Cr = chromium

    (b) Zinc = Zn; nickel = Ni; potassium = K

1.3 (a) Iron: lustrous solid, metallic, good conductor of heat and electricity, malleable, ductile, attracted to a magnet

    (b) Water: colorless liquid (at room temperature), melting point is 0 °C and boiling point is 100 °C, density ~ 1 g/cm^2

    (c) Table salt: solid, white crystals, soluble in water

    (d) Oxygen: colorless gas (at room temperature), low solubility in water

1.4 15.5 g (1 cm^3/1.18 × 10^{-3} g) = 1.31 × 10^4 cm^3

1.5 Density decreases by about 0.025 g/cm^3 for each 10 °C increase in temperature. The density at 30 °C is expected to be about 13.546 g/cm^3 − 0.025 g/cm^3 = 13.521 g/cm^3.

1.6 Chemical changes: the fuel in the campfire burns in air (combustion). Physical changes: water boils. Energy evolved in combustion is transferred to the water, to the water container, and to the surrounding air.

1.7 77 K − 273.15 K (1 °C/K) = −196 °C

1.8 Length: 25.3 cm (1 m/100 cm) = 0.253 m;
          25.3 cm (10 mm/1 cm) = 253 mm

    Width: 21.6 cm (1 m/100 cm) = 0.216 m;
         21.6 cm (10 mm/1 cm) = 216 mm

    Area = length × width = (25.3 cm)(21.6 cm) = 546 cm^2

    546 cm^2 (1 m/100 cm)^2 = 0.0546 m^2

1.9 Area of sheet = (2.50 cm)^2 = 6.25 cm^2

    Volume = 1.656 g (1 cm^3/21.45 g) = 0.07720 cm^3

    Thickness = volume/area = (0.07720 cm^3/6.25 cm^2)
                = 0.0124 cm

    0.0124 cm (10 mm/1 cm) = 0.124 mm

1.10 (a) 750 mL (1 L/1000 mL) = 0.75 L
         0.75 L (10 dL/L) = 7.5 dL

    (b) 2.0 qt = 0.50 gal
        0.50 gal (3.786 L/gal) = 1.9 L
        1.9 L (1 dm^3/1 L) = 1.9 dm^3

1.11 (a) Mass in kilograms = 5.59 g (1 kg/1000 g)
                     = 0.00559 kg

      Mass in milligrams = 5.59 g (10^3 mg/g)
                      = 5.59 × 10^{-3} mg

    (b) 0.02 μg/L (1 g/10^6 μg) = 2 × 10^{-8} g/L

1.12 Student A: average = −0.1 °C; average deviation = 0.2 °C; error = −0.1 °C. Student B: average = 273.16 K; average deviation = 0.02 K; error = +0.01 K. Student B's values are more accurate and more precise.

1.13 (a) 2.33 × 10^7 has three significant figures; 50.5 has three significant figures; 200 has one significant figure. (200. would express this number with three significant figures.)

    (b) The product of 10.26 and 0.063 is 0.65, a number with two significant figures. (10.26 has four significant figures, whereas 0.063 has two.)
    The sum of 10.26 and 0.063 is 10.32. The number 10.26 has only two numbers to the right of the decimal, so the sum must also have two numbers after the decimal.

    (c) x = 3.9 × 10^6. The difference between 110.7 and 64 is 47. Dividing 47 by 0.056 and 0.00216 gives an answer with two significant figures.

1.14 (a) 198 cm (1 m/100 cm) = 1.98 m;
        198 cm (1 ft/30.48 cm) = 6.50 ft

    (b) 2.33 × 10^7 m^2(1 km^2/10^6 m^2) = 23.3 km^2

    (c) 19,320 kg/m^3(10^3 g/1 kg)(1 m^3/10^6 cm^3) =
       19.32 g/cm^3

    (d) 9.0 × 10^3 pc(206,265 AU/1 pc)(1.496 × 10^8 km/
       1 AU) = 2.8 × 10^{17} km

1.15 Read from the graph, the mass of 50 beans is about 123 g.

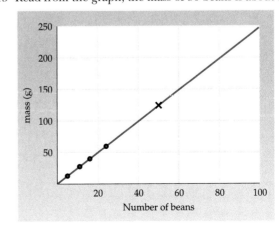

1.16 Change dimensions to centimeters: 7.6 m = 760 cm; 2.74 m = 274 cm; 0.13 mm = 0.013 cm.

Volume of paint = (760 cm)(274 cm)(0.013 cm)
= 2.7×10^3 cm^3

Volume (L) = (2.7×10^3 cm^3)(1 L/10^3 cm^3) = 2.7 L

Mass = (2.7×10^3 cm^3)(0.914 g/cm^3) = 2.5×10^3 g

Chapter 2

2.1 The ratio of the atom radius to the radius of the nucleus is 1×10^5 to 1. If the radius of the atom is 100 m, then the radius of the nucleus is 0.0010 m or 1.0 mm. The head of an ordinary pin has a diameter of about 1.0 mm.

2.2 (a) Mass number with 26 protons and 30 neutrons is 56
(b) 59.930788 u (1.661×10^{-24} g/u) = 9.955×10^{-23} g
(c) ^{64}Zn has 30 protons, 30 electrons, and (64 − 30) = 34 neutrons.

2.3 (a) Argon has an atomic number of 18. ^{36}Ar, ^{38}Ar, ^{40}Ar
(b) ^{69}Ga: 31 protons, 38 neutrons; ^{71}Ga: 31 protons, 40 neutrons; % abundance of ^{71}Ga = 39.9 %

2.4 (0.7577)(34.96885 u) + (0.2423)(36.96590 u) = 35.45 u

2.5 (a) 1.5 mol Si (28.1 g/mol) = 42 g Si
(b) 454 g S (1 mol S/32.07 g) = 14.2 mol S

14.2 mol S (6.022×10^{23} atoms/mol) = 8.53×10^{24} atoms S
(c) (32.07 g S/1 mol S) (1 mol S/6.022×10^{23} atoms) = 5.325×10^{-23} g/atom

2.6 2.6×10^{24} atoms (1 mol/6.022×10^{23} atoms) (197.0 g Au/1 mol) = 850 g Au

Volume = 850 g Au (1 cm^3/19.32 g) = 44 cm^3

Volume = 44 cm^3 = (thickness)(area) = (0.10 cm)(area)

Area = 440 cm^2

Length = width = (440 cm^2)$^{1/2}$ = 21 cm

2.7 There are eight elements in the third period. Sodium (Na), magnesium (Mg), and aluminum (Al) are metals. Silicon (Si) is a metalloid. Phosphorus (P), sulfur (S), chlorine (Cl), and argon (Ar) are nonmetals.

Chapter 3

3.1 The molecular formula for styrene is C_8H_8; the condensed formula, $C_6H_5CH{=}CH_2$, contains the same information and is more descriptive of its structure.

3.2 The molecular formula is $C_3H_7NO_2S$. You will often see its formula written as $HSCH_2CH(N^+H_3)CO_9^-$ to emphasize the molecule's structure.

3.3 (a) K^+ is formed if K loses one electron. K^+ has the same number of electrons as Ar.
(b) Se^{2-} is formed by adding two electrons to an atom of Se. It has the same number of electrons as Kr.

(c) Ba^{2+} is formed if Ba loses two electrons; Ba^{2+} has the same number of electrons as Xe.
(d) Cs^+ is formed if Cs loses one electron. It has the same number of electrons as Xe.

3.4 (a) (1) NaF: 1 Na^+ and 1 F^- ion. (2) $Cu(NO_3)_2$: 1 Cu^{2+} and 2 NO_3^- ions. (3) $NaCH_3CO_2$: 1 Na^+ and 1 $CH_3CO_2^-$ ion.
(b) $FeCl_2$, $FeCl_3$
(c) Na_2S, Na_3PO_4, BaS, $Ba_3(PO_4)_2$

3.5 (1) (a) NH_4NO_3; (b) $CoSO_4$; (c) $Ni(CN)_2$; (d) V_2O_3; (e) $Ba(CH_3CO_2)_2$; (f) $Ca(ClO)_2$
(2) (a) magnesium bromide; (b) lithium carbonate; (c) potassium hydrogen sulfite; (d) potassium permanganate; (e) ammonium sulfide; (f) copper(I) chloride and copper(II) chloride

3.6 The force of attraction between ions is proportional to the product of the ion charges (Coulomb's law). The force of attraction between Mg^{2+} and O^{2-} ions in MgO is approximately four times greater than the force of attraction between Na^+ and Cl^- ions in NaCl, so a much higher temperature is required to disrupt the orderly array of ions in crystalline MgO.

3.7 (1) (a) CO_2; (b) PI_3; (c) SCl_2; (d) BF_3; (e) O_2F_2; (f) XeO_3
(2) (a) dinitrogen tetrafluoride; (b) hydrogen bromide; (c) sulfur tetrafluoride; (d) boron trichloride; (e) tetraphosphorus decaoxide; (f) chlorine trifluoride

3.8 (a) Citric acid: 192.1 g/mol; magnesium carbonate: 84.3 g/mol
(b) 454 g citric acid (1 mol /192.1 g) = 2.36 mol citric acid
(c) 0.125 mol $MgCO_3$ (84.3 g/mol) = 10.5 g $MgCO_3$

3.9 (a) 1.00 mol $(NH_4)_2CO_3$ (molar mass 96.09 g/mol) has 28.0 g of N (29.2%), 8.06 g of H (8.39%), 12.0 g of C (12.5%), and 48.0 g of O (50.0%)
(b) 454 g C_8H_{18} (1 mol C_8H_{18}/114.2 g)(8 mol C/1 mol C_8H_{18})(12.01 g C/1 mol C) = 382 g C

3.10 (a) C_5H_4 (b) $C_2H_4O_2$

3.11 88.17 g C (1 mol C/12.011 g C) = 7.341 mol C

11.83 g H (1 mol H/1.008 g H) = 11.74 mol H

11.74 mol H/7.341 mol C = 1.6 mol H/1 mol C = (8/5)(mol H/1 mol C) = 8 mol H/5 mol C

The empirical formula is C_5H_8. The molar mass, 68.11 g/ mol, closely matches this formula, so C_5H_8 is also the molecular formula.

3.12 78.90 g C (1 mol C/12.011 g C) = 6.569 mol C

10.59 g H (1 mol H/1.008 g H) = 10.51 mol H

10.51 g O (1 mol O/16.00 g O) = 0.6569 mol O

10.51 mol H/0.6569 mol O = 16 mol H/1 mol O

6.569 mol C/0.6569 mol O = 10 mol C/1 mol O

The empirical formula is $C_{10}H_{16}O$.

3.13   0.586 g K (1 mol K/39.10 g K) = 0.0150 mol K

0.480 g O(1 mol O/16.00 g O) = 0.0300 mol O

The ratio of moles K to moles O atoms is 1 to 2; the empirical formula is KO_2.

3.14   Mass of water lost on heating is 0.235 g − 0.128 g = 0.107 g

0.107 g H_2O (1 mol H_2O/18.016 g H_2O) = 0.00594 mol H_2O

0.128 g $NiCl_2$ (1 mol $NiCl_2$/129.6 g $NiCl_2$) = 0.000988 mol $NiCl_2$

Mole ratio = 0.00594 mol H_2O/0.000988 mol $NiCl_2$ = 6.01

The formula for the hydrate is $NiCl_2 \cdot 6\ H_2O$.

Chapter 4

4.1   (a) Reactants: Al, aluminum, a solid, and Br_2, bromine, a liquid. Product: Al_2Br_6, dialuminum hexabromide, a solid

     (b) Stoichiometric coefficients: 2 for Al, 3 for Br_2, and 1 for Al_2Br_6

     (c) 8000 atoms of Al requires (3/2)8000 = 12,000 molecules of Br_2

4.2   (a) $2\ C_4H_{10}(g) + 13\ O_2(g) \longrightarrow 8\ CO_2(g) + 10\ H_2O(g)$

     (b) $2\ Pb(C_2H_5)_4(\ell) + 27\ O_2(g) \longrightarrow$
$$2\ PbO(s) + 16\ CO_2(g) + 20\ H_2O(\ell)$$

4.3   454 g C_3H_8 (1 mol C_3H_8/44.10 g C_3H_8) = 10.3 mol C_3H_8

10.3 mol C_3H_8 (5 mol O_2/1 mol C_3H_8) (32.00 g O_2/1 mol O_2) = 1650 g O_2

10.3 mol C_3H_8 (3 mol CO_2/1 mol C_3H_8) (44.01 g CO_2/1 mol CO_2) = 1360 g CO_2

10.3 mol C_3H_8 (4 mol H_2O/1 mol C_3H_8) (18.02 g H_2O/1 mol H_2O) = 742 g H_2O

4.4   Amount C = 125 g C (1 mol C/12.01 g C) = 10.4 mol C

Amount Cl_2 = 125 g Cl_2 (1 mol Cl_2/70.91 g Cl_2) = 1.76 mol Cl_2

Mol C/mol Cl_2 = 10.41/1.763 = 5.90

This is more than the 1 : 2 ratio required, so the limiting reactant is Cl_2.

Mass $TiCl_4$ = 1.763 mol Cl_2 (1 mol $TiCl_4$/2 mol Cl_2) (189.7 g $TiCl_4$/1 mol $TiCl_4$) = 167 g $TiCl_4$

4.5   (a) Amount Al = 50.0 g Al (1 mol Al/26.98 g Al) = 1.85 mol Al

Amount Fe_2O_3 = 50.0 g Fe_2O_3 (1 mol Fe_2O_3/159.7 g Fe_2O_3) = 0.313 mol Fe_2O_3

Mol Al/mol Fe_2O_3 = 1.853/0.3131 = 5.92

This is more than the 2 : 1 ratio required, so the limiting reactant is Fe_2O_3.

Mass Fe = 0.313 mol Fe_2O_3 (2 mol Fe/1 mol Fe_2O_3) (55.85 g Fe/1 mol Fe) = 35.0 g Fe

4.6   Theoretical yield = 125 g CH_3OH (1 mol CH_3OH/32.04 g CH_3OH) (2 mol H_2/1 mol CH_3OH) (2.016 g H_2/1 mol H_2) = 15.7 g H_2

Percent yield = (13.6 g/15.7 g) (100%) = 86.5%

4.7   0.143 g O_2 (1 mol O_2/32.00 g O_2) (3 mol TiO_2/3 mol O_2) (79.88 g TiO_2/1 mol TiO_2) = 0.357 g TiO_2

Percent TiO_2 in sample = (0.357 g/2.367 g) (100%) = 15.1%

4.8   1.612 g CO_2 (1 mol CO_2/44.01 g CO_2) (1 mol C/1 mol CO_2) = 0.03663 mol C

0.7425 g H_2O (1 mol H_2O/18.01 g H_2O) (2 mol H/1 mol H_2O) = 0.08243 mol H

0.08243 mol H/0.03663 mol = 2.250 H/1 C = 9 H/4 C

The empirical formula is C_4H_9, which has a molar mass of 57 g/mol. This is one half of the molar mass, so the molecular formula is $(C_4H_9)_2$ or C_8H_{18}.

4.9   0.240 g CO_2 (1 mol CO_2/44.01 g CO_2) (1 mol C/1 mol CO_2) (12.01 g C/1 mol C) = 0.06549 g C

0.0982 g H_2O (1 mol H_2O/18.02 g H_2O) (2 mol H/1 mol H_2O) (1.008 g H/1 mol H) = 0.01099 g H

Mass O (by difference) = 0.1342 g − 0.06549 g − 0.01099 g = 0.05772 g

Amount C = 0.06549 g (1 mol C/12.01 g C) = 0.00545 mol C

Amount H = 0.01099 g H (1 mol H/1.008 g H) = 0.01090 mol H

Amount O = 0.05772 g O (1 mol O/16.00 g O) = 0.00361 mol O

To find a whole-number ratio, divide each value by 0.00361; this gives 1.51 mol C : 3.02 mol H : 1 mol O. Multiply each value by 2 and round off to 3 mol C : 6 mol H : 2 mol O. The empirical formula is $C_3H_6O_2$; given the molar mass of 74.1, this is also the molecular formula.

Chapter 5

5.1   Epsom salt is an electrolyte and methanol is a nonelectrolyte.

5.2   (a) $LiNO_3$ is soluble and gives Li^+(aq) and NO_3^-(aq) ions.

     (b) $CaCl_2$ is soluble and gives Ca^{2+}(aq) and Cl^-(aq) ions.

     (c) CuO is not water-soluble.

     (d) $NaCH_3CO_2$ is soluble and gives Na^+(aq) and $CH_3CO_2^-$(aq) ions.

5.3   (a) $Na_2CO_3(aq) + CuCl_2(aq) \longrightarrow$
$$2\ NaCl(aq) + CuCO_3(s)$$

     (b) No reaction; no insoluble compound is produced.

     (c) $NiCl_2(aq) + 2\ KOH(aq) \longrightarrow$
$$Ni(OH)_2(s) + 2\ KCl(aq)$$

5.4   (a) $AlCl_3(aq) + Na_3PO_4(aq) \longrightarrow AlPO_4(s) + 3\ NaCl(aq)$
$$Al^{3+}(aq) + PO_4^{3-}(aq) \longrightarrow AlPO_4(s)$$

(b) $FeCl_3(aq) + 3\ KOH(aq) \longrightarrow$
$$Fe(OH)_3(s) + 3\ KCl(aq)$$
$$Fe^{3+}(aq) + 3\ OH^-(aq) \longrightarrow Fe(OH)_3(s)$$

(c) $Pb(NO_3)_2(aq) + 2\ KCl(aq) \longrightarrow$
$$PbCl_2(s) + 2\ KNO_3(aq)$$
$$Pb^{2+}(aq) + 2\ Cl^-(aq) \longrightarrow PbCl_2(s)$$

5.5 (a) $H^+(aq)$ and $NO_3^-(aq)$
(b) $Ba^{2+}(aq)$ and $2\ OH^-(aq)$

5.6 Metals form basic oxides; nonmetals form acidic oxides.
(a) SeO_2 is an acidic oxide; (b) MgO is a basic oxide; and (c) P_4O_{10} is an acidic oxide.

5.7 $Mg(OH)_2(s) + 2\ HCl(aq) \longrightarrow MgCl_2(aq) + 2\ H_2O(\ell)$

Net ionic equation: $Mg(OH)_2(s) + 2\ H^+(aq) \longrightarrow$
$$Mg^{2+}(aq) + 2\ H_2O(\ell)$$

5.8 (a) $BaCO_3(s) + 2\ HNO_3(aq) \longrightarrow$
$$Ba(NO_3)_2(aq) + CO_2(g) + H_2O(\ell)$$
Barium carbonate and nitric acid produce barium nitrate, carbon dioxide, and water.

(b) $(NH_4)_2SO_4(aq) + 2\ NaOH(aq) \longrightarrow$
$$2\ NH_3(g) + Na_2SO_4(aq) + 2\ H_2O(\ell)$$

5.9 (a) Gas-forming reaction: $CuCO_3(s) + H_2SO_4(aq) \longrightarrow$
$$CuSO_4(aq) + H_2O(\ell) + CO_2(g)$$
Net ionic equation: $CuCO_3(s) + 2\ H^+(aq) \longrightarrow$
$$Cu^{2+}(aq) + H_2O(\ell) + CO_2(g)$$

(b) Acid–base reaction: $Ba(OH)_2(s) + 2\ HNO_3(aq) \longrightarrow$
$$Ba(NO_3)_2(aq) + 2\ H_2O(\ell)$$
Net ionic equation: $Ba(OH)_2(s) + 2\ H^+(aq) \longrightarrow$
$$Ba^{2+}(aq) + 2\ H_2O(\ell)$$

(c) Precipitation reaction: $CuCl_2(aq) + (NH_4)_2S(aq) \longrightarrow$
$$CuS(s) + 2\ NH_4Cl(aq)$$
Net ionic equation: $Cu^{2+}(aq) + S^{2-}(aq) \longrightarrow CuS(s)$

5.10 (a) Fe in Fe_2O_3, +3; (b) S in H_2SO_4, +6;
(c) C in CO_3^{2-}, +4; (d) N in NO_2^+, +5

5.11 Dichromate ion is the oxidizing agent and is reduced. (Cr with a +6 oxidation number is reduced to Cr^{3+} with a +3 oxidation number). Ethanol is the reducing agent and is oxidized. (The C atoms in ethanol have an oxidation number of -2, whereas this number is 0 in acetic acid.)

5.12 (a) Acid–base reaction ($H^+ + OH^- \longrightarrow H_2O$).
(b) Oxidation–reduction reaction; Cu is oxidized (oxidation number changes from 0 in Cu to +2 in $CuCl_2$), and Cl_2 is reduced (oxidation number for each Cl changes from 0 to -1). Cu is the reducing agent and Cl_2 is the oxidizing agent.
(c) Gas-forming reaction (gaseous CO_2 is evolved).
(d) Oxidation–reduction reaction; S in $S_2O_3^{2-}$ is oxidized (oxidation number changes from +2 in $S_2O_3^{2-}$ to +2.5 in $S_4O_6^{2-}$), and I_2 is reduced (oxidation number of each I atom changes from 0 to -1). $S_2O_3^{2-}$ is the reducing agent and I_2 is the oxidizing agent.

5.13 26.3 g (1 mol $NaHCO_3$/84.01 g $NaHCO_3$) = 0.313 mol $NaHCO_3$

0.313 mol $NaHCO_3$/0.200 L = 1.57 M

Ion concentrations: $[Na^+] = [HCO_3^-] = 1.57$ M

5.14 First, determine the mass of $AgNO_3$ required.

Amount of $AgNO_3$ required = (0.0200 M)(0.250 L)
$$= 5.00 \times 10^{-3}\ mol$$

Mass of $AgNO_3$ = $(5.00 \times 10^{-3}\ mol)(169.9\ g/mol)$
$$= 0.850\ g\ AgNO_3$$

Weigh out 0.850 g $AgNO_3$. Then, dissolve it in a small amount of water in the volumetric flask. After the solid is dissolved, fill the flask to the mark.

5.15 $(0.15\ M)(0.0060\ L) = (0.0100\ L)(c_{dilute})$
$c_{dilute} = 0.090$ M

5.16 $(2.00\ M)(V_{conc}) = (1.00\ M)(0.250\ L)$; $V_{conc} = 0.125$ L

To prepare the solution, measure accurately 125 mL of 2.00 M NaOH into a 250-mL volumetric flask, and add water to give a total volume of 250 mL.

5.17 (a) pH = $-\log(2.6 \times 10^{-2})$ = 1.59
(b) $-\log[H^+]$ = 3.80; $[H^+] = 1.5 \times 10^{-4}$ M

5.18 HCl is the limiting reagent.

(0.350 mol HCl/1 L)(0.0750 L)(1 mol CO_2/2 mol HCl)
(44.01 g CO_2/1 mol CO_2) = 0.578 g CO_2

5.19 (0.953 mol NaOH/1 L)(0.02833 L NaOH) = 0.0270 mol NaOH

(0.0270 mol NaOH)(1 mol CH_3CO_2H /1 mol NaOH) = 0.0270 mol CH_3CO_2H

(0.0270 mol CH_3CO_2H)(60.05 g/mol) = 1.62 g CH_3CO_2H

0.0270 mol CH_3CO_2H/0.0250 L = 1.08 M

5.20 (0.100 mol HCl/1 L)(0.02967 L) = 0.00297 mol HCl

(0.00297 mol HCl)(1 mol NaOH/1 mol HCl) = 0.00297 mol NaOH

0.00297 mol NaOH/0.0250 L = 0.119 M NaOH

5.21 Mol acid = mol base = (0.323 mol/L)(0.03008 L) = 9.716×10^{-3} mol

Molar mass = 0.856 g acid/9.716×10^{-3} mol acid = 88.1 g/mol

5.22 (0.196 mol $Na_2S_2O_3$/1 L)(0.02030 L) = 0.00398 mol $Na_2S_2O_3$

(0.00398 mol $Na_2S_2O_3$)(1 mol I_2/2 mol $Na_2S_2O_3$) = 0.00199 mol I_2

0.00199 mol I_2 is in excess, and was not used in the reaction with ascorbic acid.

I_2 originally added = (0.0520 mol I_2/1 L)(0.05000 L) = 0.00260 mol I_2

I_2 used in reaction with ascorbic acid = 0.00260 mol $-$ 0.00199 mol = 6.1×10^{-4} mol I_2

$(6.1 \times 10^{-4}$ mol $I_2)(1$ mol $C_6H_8O_6$ /1 mol $I_2)(176.1$ g/ 1 mol) = 0.11 g $C_6H_8O_6$

Chapter 6

6.1 The chemical potential energy of a battery can be converted to work (to run a motor), heat (an electric space heater), and light (a light bulb).

6.2 (a) $(3800 \text{ calories})(4.184 \text{ J/calorie}) = 1.6 \times 10^4 \text{ J}$

     (b) $(250 \text{ Calories})(1000 \text{ calories/Calorie})(4.184 \text{ J/calorie})(1 \text{ kJ/1000 J}) = 1.0 \times 10^3 \text{ kJ}$

6.3 $C = 59.8 \text{ J}/[(25.0 \text{ g})(1.00 \text{ K})] = 2.39 \text{ J/g} \cdot \text{K}$

6.4 $(15.5 \text{ g})(C_{\text{metal}})(18.9 \text{ °C} - 100.0 \text{ °C}) + (55.5 \text{ g})(4.184 \text{ J/g} \cdot \text{K})(18.9 \text{ °C} - 16.5 \text{ °C}) = 0$

     $C_{\text{metal}} = 0.44 \text{ J/g} \cdot \text{K}$

6.5 $(400. \text{ g iron})(0.449 \text{ J/g} \cdot \text{K})(32.8 \text{ °C} - T_{\text{initial}}) + (1000. \text{ g})(4.184 \text{ J/g} \cdot \text{K})(32.8 \text{ °C} - 20.0 \text{ °C}) = 0$

     $T_{\text{initial}} = 331 \text{ °C}$

6.6 To heat methanol: $(25.0 \text{ g CH}_3\text{OH})(2.53 \text{ J/g} \cdot \text{K})(64.6 \text{ °C} - 25.0 \text{ °C}) = 2.50 \times 10^3 \text{ J}$

     To evaporate methanol:
$(25.0 \text{ g CH}_3\text{OH})(2.00 \times 10^3 \text{ J/g}) = 5.00 \times 10^4 \text{ J}$

     Total heat: $0.250 \times 10^4 \text{ J} + 5.00 \times 10^4 \text{ J} = 5.25 \times 10^4 \text{ J}$

6.7 Heat transferred from tea + heat to melt ice = 0

     $(250 \text{ g})(4.2 \text{ J/g} \cdot \text{K})(273.2 \text{ K} - 291.4 \text{ K}) + x \text{ g }(333 \text{ J/g}) = 0$

     $x = 57.4 \text{ g}$

     57 g of ice melts with heat supplied by cooling 250 g of tea from 18.2 °C (291.4 K) to 0 °C (273.2 K)

     Mass of ice remaining = mass of ice initially − mass of ice melted

     Mass of ice remaining = 75 g − 57 g = 18 g

6.8 (a) $(12.6 \text{ g H}_2\text{O})(1 \text{ mol/18.02 g})(285.8 \text{ kJ/mol}) = 2.00 \times 10^2 \text{ kJ}$

     (b) $\Delta H = (15.0 \text{ g})(1 \text{ mol/30.07 g})(-2857.3 \text{ kJ}/(2 \text{ mol}) = -713 \text{ kJ}$

6.9 Mass of final solution = 400. g; $\Delta T = 27.78 \text{ °C} - 25.10 \text{ °C} = 2.68 \text{ °C} = 2.68 \text{ K}$

     Amount of HCl used = amount of NaOH used = $CV = 0.0800 \text{ mol}$

     Heat transferred by acid–base reaction + heat absorbed to warm solution = 0

     $q_{\text{rxn}} + (4.20 \text{ J/g} \cdot \text{K})(400. \text{ g})(2.68 \text{ K}) = 0$

     $q_{\text{rxn}} = -4.50 \times 10^3 \text{ J}$

     This represents the heat evolved in the reaction of 0.0800 mol HCl.

     Heat per mole = $\Delta H_{\text{rxn}} = -4.50 \text{ kJ}/0.0800 \text{ mol HCl} = -56.3 \text{ kJ/mol of HCl}$

6.10 (a) Heat evolved in reaction + heat absorbed by H_2O + heat absorbed by bomb = 0

         $q_{\text{rxn}} + (1.50 \times 10^3 \text{ g})(4.20 \text{ J/g} \cdot \text{K})(27.32 \text{ °C} - 25.00 \text{ °C}) + (837 \text{ J/K})(27.32 \text{ K} - 25.00 \text{ K}) = 0$

$q_{\text{rxn}} = -16,600 \text{ J}$ (heat evolved in burning 1.0 g sucrose)

     (b) Heat per mole = $(-16.6 \text{ kJ/g})(342.2 \text{ g sucrose}/1 \text{ mol sucrose}) = -5650 \text{ kJ/mol}$

6.11 (a) $\text{C(graphite)} + O_2(g) \longrightarrow CO_2(g)$ $\Delta H_1 = -393.5 \text{ kJ}$

         $CO_2(g) \longrightarrow \text{C(diamond)} + O_2(g)$ $\Delta H_2 = +395.4 \text{ kJ}$

         Net: $\text{C(graphite)} \longrightarrow \text{C(diamond)}$

                 $\Delta H_{\text{net}} = \Delta H_1 + \Delta H_2 = +1.9 \text{ kJ}$

     (b)

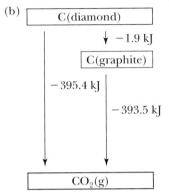

6.12 $\text{C}(s) + O_2(g) \longrightarrow CO_2(g)$ $\Delta H_1 = -393.5 \text{ kJ}$

     $2 [\text{S}(s) + O_2(g) \longrightarrow SO_2(g)]$ $\Delta H_2 = 2(-296.8)$

                             $= -593.6 \text{ kJ}$

     $CO_2(g) + 2 SO_2(g) \longrightarrow CS_2(g) + 3 O_2(g)$

                           $\Delta H_3 = +1103.9 \text{ kJ}$

     Net: $\text{C}(s) + 2 \text{ S}(s) \longrightarrow CS_2(g)$

         $\Delta H_{\text{net}} = \Delta H_1 + \Delta H_2 + \Delta H_3 = +116.8 \text{ kJ}$

6.13 Standard states: $Br_2(\ell)$, $Hg(\ell)$, $Na_2SO_4(s)$, $C_2H_5OH(\ell)$

6.14 $\text{Fe}(s) + \frac{3}{2} Cl_2(g) \longrightarrow FeCl_3(s)$

     $12 \text{ C}(s, \text{ graphite}) + 11 H_2(g) + \frac{11}{2} O_2(g) \longrightarrow$
                                      $C_{12}H_{22}O_{11}(s)$

6.15 $\Delta H^{\circ}_{\text{rxn}} = 6 \Delta H^{\circ}_f[CO_2(g)] + 3 \Delta H^{\circ}_f[H_2O(\ell)] - \{\Delta H^{\circ}_f[C_6H_6(\ell)] + \frac{15}{2} \Delta H^{\circ}_f[O_2(g)]\}$
         $= (6 \text{ mol})(-393.5 \text{ kJ/mol}) + (3 \text{ mol})(-285.8 \text{ kJ/mol}) - (1 \text{ mol})(+49.0 \text{ kJ/mol}) - 0$
         $= -3267.4 \text{ kJ}$

6.16 (a) $\Delta H^{\circ}_{\text{rxn}} = -2\Delta H^{\circ}_f[HBr(g)]$
            $= -(2 \text{ mol})(-36.29 \text{ kJ/mol}) = +72.58 \text{ kJ}$

     The reaction is reactant-favored.

     (b) $\Delta H^{\circ}_{\text{rxn}} = \Delta H^{\circ}_f[\text{C(graphite)}] - \Delta H^{\circ}_f[\text{C(diamond)}]$
            $= 0 - 1.8 \text{ kJ} = -1.8 \text{ kJ}$

     The reaction is product-favored.

Chapter 7

7.1 (a) 10 cm

     (b) 5 cm

     (c) 4 waves. There are 9 nodes, 7 in the middle and 2 at the ends.

7.2   (a) Highest frequency, violet; lowest frequency, red

(b) The FM radio frequency, 91.7 MHz, is lower than the frequency of a microwave oven, 2.45 GHz.

(c) The wavelength of x-rays is shorter than the wavelength of ultraviolet light.

7.3   Orange light: 6.25×10^2 nm = 6.25×10^{-7} m

$\nu = (2.998 \times 10^8 \text{ m/s})/6.25 \times 10^{-7} \text{ m} = 4.80 \times 10^{14} \text{ s}^{-1}$

$E = (6.626 \times 10^{-34} \text{ J} \cdot \text{s/photon})(4.80 \times 10^{14} \text{ s}^{-1})$
$(6.022 \times 10^{23} \text{ photons/mol})$

$= 1.92 \times 10^5 \text{ J/mol}$

Microwave: $E = (6.626 \times 10^{-34} \text{ J} \cdot \text{s/photon})(2.45 \times 10^9 \text{ s}^{-1})(6.022 \times 10^{23} \text{ photons/mol}) = 0.978 \text{ J/mol}$

$E(\text{orange light})/E(\text{microwave}) = 1.96 \times 10^5$; orange (625-nm) light is 196,000 times more energetic than 2.45-GHz microwaves.

7.4   (a) E (per atom) $= -Rhc/n^2 = (-2.179 \times 10^{-18})/(3^2)$ J/atom $= -2.421 \times 10^{-19}$ J/atom

(b) E (per mol) $= (-2.421 \times 10^{-19} \text{ J/atom})$
$(6.022 \times 10^{23} \text{ atoms/mol})(1 \text{ kJ}/10^3 \text{ J})$
$= -145.8 \text{ kJ/mol}$

7.5   The least energetic line is from the electron transition from $n = 2$ to $n = 1$.

$\Delta E = -Rhc[1/1^2 - 1/2^2]$
$= -(2.179 \times 10^{-18} \text{ J/ atom})(3/4)$
$= -1.634 \times 10^{-18} \text{ J/atom}$

$\nu = \Delta E/h = (-1.634 \times 10^{-18} \text{ J/atom})/(6.626 \times 10^{-34} \text{ J} \cdot \text{s})$
$= 2.466 \times 10^{15} \text{ s}^{-1}$

$\lambda = c/\nu = (2.998 \times 10^8 \text{ m/s}^{-1})/(2.466 \times 10^{15} \text{ s}^{-1})$
$= 1.216 \times 10^{-7}$ m (or 121.6 nm)

7.6   First, calculate the velocity of the neutron:

$v = [2E/m]^{1/2} = [2(6.21 \times 10^{-21} \text{ kg} \cdot \text{m}^2 \text{ s}^{-2})/(1.675 \times 10^{-27} \text{ kg})]^{1/2} = 2720 \text{ m s}^{-1}$

Use this value in the de Broglie equation:

$\lambda = h/mv = (6.626 \times 10^{-34} \text{ kg} \cdot \text{m}^2 \text{ s}^{-2})/(1.675 \times 10^{-31} \text{ kg}) (2720 \text{ m s}^{-1}) = 1.45 \times 10^{-6}$ m

7.7   (a) $\ell = 0$ or 1; (b) $m_\ell = -1$, 0, or +1, p subshell; (c) d subshell; (d) $\ell = 0$ and $m_\ell = 0$; (e) 3 orbitals in the p subshell; (f) 7 values of m_ℓ and 7 orbitals

7.8   (a)

Orbital	n	ℓ
6s	6	0
4p	4	1
5d	5	2
4f	4	3

(b) A $4p$ orbital has one nodal plane; a $6d$ orbital has two nodal planes.

Chapter 8

8.1   (a) $4s$ ($n + \ell = 4$) filled before $4p$ ($n + \ell = 5$)

(b) $6s$ ($n + \ell = 6$) filled before $5d$ ($n + \ell = 7$)

(c) $5s$ ($n + \ell = 5$) filled before $4f$ ($n + \ell = 7$)

8.2   (a) chlorine (Cl)

(b) $1s^2 2s^2 2p^6 3s^2 3p^3$

(c) Calcium has two valence electrons in the $4s$ subshell. Quantum numbers for these two electrons are $n = 4$, $\ell = 0$, $m_\ell = 0$, and $m_s = \pm 1/2$

8.3   Obtain the answers from Table 8.3.

8.4

All three ions are paramagnetic with three, two, and four unpaired electrons, respectively.

8.5   Increasing atomic radius: $C < Si < Al$

8.6   (a) H—O distance = 37 pm + 66 pm = 103 pm; H—S distance = 37 pm + 104 pm = 141 pm

(b) $r_{Br} = 114$ pm; Br—Cl bond length = $r_{Br} + r_{Cl} = 114$ pm + 99 pm = 213 pm

8.7   (a) Increasing atomic radius: $C < B < Al$

(b) Increasing ionization energy: $Al < B < C$

(c) Carbon is predicted to have the most negative electron affinity.

8.8   Decreasing ionic radius: $N^{3-} > O^{2-} > F^-$. In this series of isoelectronic ions, the size decreases with increased nuclear charge.

8.9 $MgCl_3$, if it existed, would contain one Mg^{3+} ion (and three Cl^- ions). The formation of Mg^{3+} is energetically unfavorable, with a huge input of energy being required to remove the third electron (a core electron).

Chapter 9

9.1   · Ba · Ba, Group 2A, has 2 valence electrons.

· Äs · As, Group 5A, has 5 valence electrons.

· B̈r : Br, Group 7A, has 7 valence electrons.

9.2 ΔH_f° for Na(g) = +107.3 kJ/mol

ΔH_f° for I(g) = +106.8 kJ/mol

ΔH° [for Na(g) $\longrightarrow$ Na$^+$(g) + e$^-$] = +496 kJ/mol

ΔH° [for I(g) + e$^-$ $\longrightarrow$ I$^-$(g)] = −295.2 kJ/mol

ΔH° [for Na$^+$(g) + I$^-$(g)] = −702 kJ/mol

The sum of these values = $\Delta H_f^\circ[\text{NaI}(s)] = -287$ kJ/mol; the literature value (from calorimetry) is -287.8 kJ/mol.

9.3
$$\left[\begin{array}{c} \text{H} \\ | \\ \text{H}-\text{N}-\text{H} \\ | \\ \text{H} \end{array}\right]^+ \quad :\text{C}{\equiv}\text{O}: \quad \left[:\text{N}{\equiv}\text{O}:\right]^+ \quad \left[\begin{array}{c} :\ddot{\text{O}}: \\ | \\ :\ddot{\text{O}}-\text{S}-\ddot{\text{O}}: \\ | \\ :\ddot{\text{O}}: \end{array}\right]^{2-}$$

9.4
$$\text{H}-\overset{\overset{\displaystyle\text{H}}{|}}{\underset{\underset{\displaystyle\text{H}}{|}}{\text{C}}}-\ddot{\text{O}}-\text{H} \qquad \text{H}-\overset{\overset{\displaystyle\text{H}}{|}}{\underset{\underset{\displaystyle\text{H}}{|}}{\text{N}}}-\ddot{\text{O}}-\text{H}$$
methanol   hydroxylamine

9.5
$$\left[\text{H}-\ddot{\text{O}}-\overset{\overset{\displaystyle:\ddot{\text{O}}:}{|}}{\underset{\underset{\displaystyle:\ddot{\text{O}}:}{|}}{\text{P}}}-\ddot{\text{O}}-\text{H}\right]^-$$

9.6 (a) The acetylide ion, C_2^{2-}, and the N_2 molecule have the same number of valence electrons (10) and identical electronic structures; that is, they are isoelectronic.

(b) Ozone, O_3, is isoelectronic with NO_2^-; hydroxide ion, OH^-, is isoelectronic with HF.

9.7 Resonance structures for the nitrate ion:

Lewis electron dot structure for nitric acid, HNO_3:

9.8 $\left[:\ddot{\text{F}}-\ddot{\text{Cl}}-\ddot{\text{F}}:\right]^+$ ClF_2^+, 2 bond pairs and 2 lone pairs.

$\left[:\ddot{\text{F}}-\ddot{\underset{..}{\text{Cl}}}-\ddot{\text{F}}:\right]^-$ ClF_2^-, 2 bond pairs and 3 lone pairs.

9.9 Tetrahedral geometry around carbon. The Cl—C—Cl bond angle will be close to $109.5°$.

9.10 For each species, the electron-pair geometry and the molecular shape are the same. BF_3: trigonal planar; BF_4^-: tetrahedral. Adding F^- to BF_3 adds an electron pair to the central atom and changes the shape.

9.11 The electron-pair geometry around the I atom is trigonal bipyramidal. The molecular geometry of the ion is linear.

9.12 (a) In PO_4^{3-}, there is tetrahedral electron-pair geometry. The molecular geometry is tetrahedral.

$$\left[\begin{array}{c} :\ddot{\text{O}}: \\ | \\ :\ddot{\text{O}}-\text{P}-\ddot{\text{O}}: \\ | \\ :\ddot{\text{O}}: \end{array}\right]^{3-}$$

(b) In SO_3^{2-}, there is tetrahedral electron-pair geometry. The molecular geometry is trigonal pyramidal.

$$\left[\begin{array}{c} :\ddot{\text{O}}-\text{S}-\ddot{\text{O}}: \\ | \\ :\ddot{\text{O}}: \end{array}\right]^{2-}$$

(c) In IF_5, there is octahedral electron-pair geometry. The molecular geometry is square pyramidal.

9.13 (a) CN^-: formal charge on C is -1; formal charge on N is 0.

(b) SO_3: formal charge on S is $+2$; formal charge on each O is $-\frac{2}{3}$.

9.14 (a) The H atom is positive in each case. H—F ($\Delta\chi = 1.8$) is more polar than H—I($\Delta\chi = 0.5$).

(b) B—F ($\Delta\chi = 2.0$) is more polar than B—C ($\Delta\chi = 0.5$). In B—F, F is the negative pole and B is the positive pole. In B—C, C is the negative pole and B is the positive pole.

(c) C—Si ($\Delta\chi = 0.6$) is more polar than C—S ($\Delta\chi = 0.1$). In C—Si, C is the negative pole and Si is the positive pole. In C—S, S is the negative pole and C the positive pole.

9.15
$$\overset{-1\ +1\ \ 0}{:\ddot{\text{O}}-\text{S}{=}\ddot{\text{O}}} \longleftrightarrow \overset{0\ \ +1\ -1}{\ddot{\text{O}}{=}\text{S}-\ddot{\text{O}}:}$$

The S—O bonds are polar, with the negative end being the O atom. (The O atom is more electronegative than the S atom.) Formal charges show that these bonds are, in fact, polar, with the O atom being the more negative atom.

9.16 (a) $BFCl_2$, polar, negative side is the F atom because F is the most electronegative atom in the molecule.

(b) NH_2Cl, polar, negative side is the Cl atom.

(c) SCl_2, polar, Cl atoms are on the negative side.

9.17 (a) C—N: bond order 1; C=N: bond order 2; C≡N: bond order 3. Bond length: C—N > C=N > C≡N

(b) $\left[:\overset{..}{\underset{..}{O}}\text{—}\overset{..}{N}\text{=}\overset{..}{\underset{..}{O}} \right]^{-} \longleftrightarrow \left[\overset{..}{\underset{..}{O}}\text{=}\overset{..}{N}\text{—}\overset{..}{\underset{..}{O}}: \right]^{-}$

The bond order in NO_2^- is 1.5. Therefore, the NO bond length (124 pm) should be between the length of a N—O single bond (136 pm) and a N=O double bond (115 pm).

9.18 $CH_4(g) + 2\,O_2(g) \longrightarrow CO_2(g) + 2\,H_2O(g)$

Break 4 C—H bonds and 2 O=O bonds:
(4 mol)(413 kJ/mol) + (2 mol)(498 kJ/mol) = 2648 kJ

Make 2 C=O bonds and 4 H—O bonds:
(2 mol)(745 kJ/mol) + (4 mol)(463 kJ/mol) = 3342 kJ

ΔH_{rxn}° = 2648 kJ − 3342 kJ = −694 kJ

Chapter 10

10.1 The oxygen atom in H_3O^+ is sp^3 hybridized. The three O—H bonds are formed by overlap of oxygen sp^3 and hydrogen 1s orbitals. The fourth sp^3 orbital contains a lone pair of electrons.

The carbon and nitrogen atoms in CH_3NH_2 are sp^3 hybridized. The C—H bonds arise from overlap of carbon sp^3 orbitals and hydrogen 1s orbitals. The bond between C and N is formed by overlap of sp^3 orbitals from these atoms. Overlap of nitrogen sp^3 and hydrogen 1s orbitals gives the two N—H bonds, and there is a lone pair in the remaining sp^3 orbital on nitrogen.

10.2 The Lewis structure and the electron-pair and molecular geometries are shown below. The Xe atom is sp^3d^2 hybridized. Lone pairs of electrons reside in two of these orbitals; the four others overlap with 2p orbitals on fluorine, forming sigma bonds.

electron dot structure   molecular geometry

10.3 (a) BH_4^-, tetrahedral electron-pair geometry, sp^3

(b) SF_5^-, octahedral electron-pair geometry, sp^3d^2

(c) SOF_4, trigonal-bipyramidal electron-pair geometry, sp^3d

(d) ClF_3, trigonal-bipyramidal electron-pair geometry, sp^3d

(e) BCl_3, trigonal-planar electron-pair geometry, sp^2

(f) XeO_6^{4-}, octahedral electron-pair geometry, sp^3d^2

10.4 The two CH_3 carbon atoms are sp^3 hybridized and the center carbon atom is sp^2 hybridized. For each of the carbon atoms in the methyl groups, three orbitals are used to form C—H bonds and the fourth is used to bond to the central carbon atom. Overlap of carbon and oxygen sp^2 orbitals gives the sigma bond. The pi bond arises by overlap of p orbitals on these elements.

10.5 A triple bond links the two nitrogen atoms, each of which also has one lone pair. Each nitrogen is sp hybridized. One sp orbital contains the lone pair; the other is used to form the sigma bond between the two atoms. Two pi bonds arise by overlap of p orbitals on the two atoms.

10.6 Bond angles: H—C—H = 109.5°, H—C—C = 109.5°, C—C—N = 180°. Carbon in the CH_3 group is sp^3 hybridized; the central C and the N are sp hybridized. The three C—H bonds form by overlap of an H 1s orbital with one of the sp^3 orbitals of the CH_3 group; the fourth sp^3 orbital overlaps with an sp orbital on the central C to form a sigma bond. The triple bond between C and N is a combination of a sigma bond (the sp orbital on C overlaps with the sp orbital on N) and two pi bonds (overlap of two sets of p orbitals on these elements). The remaining sp orbital on N contains a lone pair.

10.7 H_2^+: $(\sigma 1s)^1$ The ion has a bond order of $\frac{1}{2}$ and is expected to exist. A bond order of $\frac{1}{2}$ is predicted for He_2^+ and H_2^-, both of which are predicted to have electron configurations $(\sigma 1s)^2 (\sigma^*1s)^1$.

10.8 Li_2^- is predicted to have an electron configuration $(\sigma 1s)^2 (\sigma^*1s)^2 (\sigma 2s)^2 (\sigma^*2s)^1$ and a bond order of $\frac{1}{2}$, implying that the ion might exist.

10.9 O_2^+: [core electrons] $(\sigma 2s)^2 (\sigma^*2s)^2 (\pi 2p)^4 (\sigma 2p)^2 (\pi^*2p)^1$. The bond order is 2.5. The ion is paramagnetic with one unpaired electron.

Chapter 11

11.1 (a) Isomers of C_7H_{16}

$CH_3CH_2CH_2CH_2CH_2CH_2CH_3$ heptane

$$CH_3CH_2CH_2CH_2\overset{\displaystyle CH_3}{\overset{|}{C}}HCH_3 \qquad \text{2-methylhexane}$$

$$CH_3CH_2CH_2\overset{\displaystyle CH_3}{\overset{|}{C}}HCH_2CH_3 \qquad \text{3-methylhexane}$$

$$CH_3CH_2\overset{\displaystyle CH_3}{\overset{|}{C}}H\underset{\underset{\displaystyle CH_3}{|}}{C}HCH_3 \qquad \text{2,3-dimethylpentane}$$

$$CH_3CH_2CH_2\overset{\displaystyle CH_3}{\underset{\underset{\displaystyle CH_3}{|}}{\overset{|}{C}}}CH_3 \qquad \text{2,2-dimethylpentane}$$

$$CH_3CH_2\overset{\displaystyle CH_3}{\underset{\underset{\displaystyle CH_3}{|}}{\overset{|}{C}}}CH_2CH_3 \qquad \text{3,3-dimethylpentane}$$

$$CH_3\overset{\displaystyle CH_3}{\overset{|}{C}}HCH_2\underset{\underset{\displaystyle CH_3}{|}}{C}HCH_3 \qquad \text{2,4-dimethylpentane}$$

2-Ethylpentane is pictured on page 484.

$$H_3C \quad CH_3$$
$$CH_3C{-}CHCH_3 \qquad \text{2,2,3-trimethylbutane}$$
$$CH_3$$

(b) Two isomers, 3-methylhexane, and 2,3-dimethylpentane, are chiral.

11.2 The names accompany the structures in the answer to Exercise 11.1.

11.3 Isomers of C_6H_{12} in which the longest chain has six C atoms:

$$\underset{H}{\overset{H}{\diagdown}}C{=}C\underset{CH_2CH_2CH_2CH_3}{\overset{H}{\diagup}}$$

$$\underset{H_3C}{\overset{H}{\diagdown}}C{=}C\underset{CH_2CH_2CH_3}{\overset{H}{\diagup}}$$

$$\underset{H_3C}{\overset{H}{\diagdown}}C{=}C\underset{H}{\overset{CH_2CH_2CH_3}{\diagup}}$$

$$\underset{H_3CCH_2}{\overset{H}{\diagdown}}C{=}C\underset{CH_2CH_3}{\overset{H}{\diagup}}$$

$$\underset{H_3CCH_2}{\overset{H}{\diagdown}}C{=}C\underset{H}{\overset{CH_2CH_3}{\diagup}}$$

Names: 1-hexene, *cis*-2-hexene, *trans*-2-hexene, *cis*-3-hexene, *trans*-3-hexene. None of these isomers is chiral.

11.4 (a) $\underset{H}{\overset{H}{H{-}C{-}C}}{-}Br$ (b) $\underset{H}{\overset{Br\ Br}{H_3C{-}C{-}C}}{-}CH_3$

      bromoethane   2,3-dibromobutane

11.5 1,4-diaminobenzene

$$\begin{array}{c} NH_2 \\ \bighexagon \\ NH_2 \end{array}$$

11.6 $CH_3CH_2CH_2CH_2OH$ 1-butanol

$$\underset{CH_3CH_2CHCH_3}{\overset{OH}{|}} \qquad \text{2-butanol}$$

$$\underset{CH_3CHCH_2OH}{} \qquad \text{2-methyl-1-propanol}$$
$$CH_3$$

$$\underset{CH_3CCH_3}{\overset{OH}{|}} \qquad \text{2-methyl-2-propanol}$$
$$CH_3$$

11.7 (a) $CH_3CH_2CH_2\overset{O}{\overset{||}{C}}CH_3$ 2-pentanone

$CH_3CH_2\overset{O}{\overset{||}{C}}CH_2CH_3$ 3-pentanone

$CH_3CH_2CH_2CH_2\overset{O}{\overset{||}{C}}H$ pentanal

$\underset{CH_3}{CH_3CHCH_2}\overset{O}{\overset{||}{C}}H$ 3-methylbutanal

(b) $\underset{CH_3CHCH_2CH_2CH_3,}{\overset{OH}{|}}$ 2-pentanol

11.8 (a) 1-butanol gives butanal $CH_3CH_2CH_2\overset{O}{\overset{||}{C}}H$

(b) 2-butanol gives butanone $CH_3CH_2\overset{O}{\overset{||}{C}}CH_3$

(c) 2-methyl-1-propanol gives 2-methylpropanal

$$CH_3\overset{H}{\overset{|}{C}}{-}\overset{O}{\overset{||}{C}}H$$
$$CH_3$$

11.9 (a) $CH_3CH_2\overset{O}{\overset{||}{C}}OCH_3$ methyl propanoate

(b) $CH_3CH_2CH_2\overset{O}{\overset{||}{C}}OCH_2CH_2CH_2CH_3$
                              butyl butanoate

(c) $CH_3CH_2CH_2CH_2CH_2\overset{O}{\overset{||}{C}}OCH_2CH_3$
                              ethyl hexanoate

11.10 (a) Propyl acetate is formed from acetic acid and propanol:

$$CH_3\overset{O}{\overset{||}{C}}OH + CH_3CH_2CH_2OH$$

(b) 3-Methylpentyl benzoate is formed from benzoic acid and 3-methylpentanol:

(c) Ethyl salicylate is formed from salicylic acid and ethanol:

11.11 (a) $CH_3CH_2CH_2OH$: 1-propanol, has an alcohol ($—OH$) group

CH$_3$CO$_2$H: ethanoic acid (acetic acid), has a carboxylic acid ($—CO_2H$) group

$CH_3CH_2NH_2$: ethylamine, has an amino ($—NH_2$) group

(b) 1-propyl ethanoate (propyl acetate)

(c) Oxidation of this primary alcohol first gives propanal, CH_3CH_2CHO. Further oxidation gives propanoic acid, $CH_3CH_2CO_2H$.

(d) N-ethylacetamide, $CH_3CONHCH_2CH_3$

(e) The amine is protonated by hydrochloric acid, forming ethylammonium chloride, $[CH_3CH_2NH_3]Cl$.

11.12 The polymer is a polyester.

Chapter 12

12.1 0.83 bar (0.82 atm) > 75 kPa (0.74 atm) > 0.63 atm > 250 mm Hg (0.33 atm)

12.2 $P_1 = 55$ mm Hg and $V_1 = 125$ mL; $P_2 = 78$ mm Hg and $V_2 = ?$

$V_2 = V_1(P_1/P_2) = (125$ mL$)(55$ mm Hg$/78$ mm Hg$)$
$= 88$ mL

12.3 $V_1 = 45$ L and $T_1 = 298$ K; $V_2 = ?$ and $T_2 = 263$ K

$V_2 = V_1(T_2/T_1) = (45$ L$)(263$ K$/298$ K$) = 40.$ L

12.4 $V_2 = V_1\ (P_1/P_2)(T_2/T_1)$
$= (22$ L$)(150$ atm$/0.993$ atm$)(295$ K$/304$ K$)$
$= 3200$ L

At 5.0 L per balloon, there is sufficient He to fill 640 balloons.

12.5 44.8 L of O_2 is required; 44.8 L of $H_2O(g)$ and 22.4 L $CO_2(g)$ are produced.

12.6 $PV = nRT$

$(750/760$ atm$)(V) =$
$(1300$ mol$)(0.082057$ L $\cdot$ atm$/$ mol $\cdot$ K$)(296$ K$)$

$V = 3.2 \times 10^4$ L

12.7 $d = PM/RT$; $M = dRT/P$

$M = (5.02$ g/L$)(0.082057$ L $\cdot$ atm/mol $\cdot$ K$)(288.2$ K$)/$
$(745/760$ atm$) = 121$ g/mol

12.8 $PV = (m/M)RT$; $M = mRT/PV$

$M = (0.105$ g$)(0.082057$ L $\cdot$ atm/mol $\cdot$ K$)(296.2$ K$)/$
$[(561/760$ atm$)(0.125$ L$)] = 27.7$ g/mol

12.9 $n(H_2) = PV/RT = (542/760$ atm$)(355$ L$)/$
$(0.082057$ L $\cdot$ atm/mol $\cdot$ K$)(298.2$ K$)$

$n(H_2) = 10.3$ mol

$n(NH_3) = (10.3$ mol $H_2)(2$ mol $NH_3/3$ mol $H_2)$
$= 6.87$ mol NH_3

$PV = nRT$; $P(125$ L$)$
$= (6.87$ mol$)(0.082057$ L $\cdot$ atm$/$ mol $\cdot$ K$)(298.2$ K$)$

$P(NH_3) = 1.35$ atm

12.10 $P_{halothane}(5.00$ L$) =$
$(0.0760$ mol$)(0.082057$ L $\cdot$ atm/mol $\cdot$ K$)\ (298.2$ K$)$

$P_{halothane} = 0.372$ atm (or 283 mm Hg)

$P_{oxygen}(5.00$ L$) = (0.734$ mol$)(0.082057$ L $\cdot$ atm/mol $\cdot$ K$)$
$(298.2$ K$)$

$P_{oxygen} - 3.59$ atm (or 2730 mm Hg)

$P_{total} = P_{halothane} + P_{oxygen}$
$= 283$ mm Hg $+ 2730$ mm Hg $= 3010$ mm Hg

12.11 For He: Use Equation 12.9, with $M = 4.00 \times 10^{-3}$ kg/mol, $T = 298$ K, and $R = 8.314$ J/mol $\cdot$ K to calculate the rms speed of 1360 m/s. A similar calculation for N_2, with $M = 28.01 \times 10^{-3}$ kg/mol, gives an rms speed of 515 m/s.

12.12 The molar mass of CH_4 is 16.0 g/mol.

$$\frac{\text{Rate for CH}_4}{\text{Rate for unknown}} = \frac{n \text{ molecules}/1.50 \text{ min}}{n \text{ molecules}/4.73 \text{ min}} = \sqrt{\frac{M_{unknown}}{16.0}}$$

$M_{unknown} = 159$ g/mol

12.13 $P(1.00$ L$) =$
$(10.0$ mol$)(0.082057$ L $\cdot$ atm/mol $\cdot$ K$)\ (298$ K$)$

$P = 245$ atm (calculated by $PV = nRT$)

$P = 320$ atm (calculated by van der Waals equation)

Chapter 13

13.1 Because F^- is the smaller ion, water molecules can approach most closely and interact more strongly. Thus, F^- should have the more negative heat of hydration.

13.2 Water is a polar solvent, while hexane and CCl_4 are nonpolar. London dispersion forces are the primary forces of attraction between all pairs of dissimilar solvents. For mixtures of water with the other solvents, dipole–induced dipole forces will also be present. When mixed, the three liquids will form two separate layers—the first being water and the second consisting of a mixture of the two nonpolar liquids.

13.3
$$H_3C - \overset{\displaystyle H}{\underset{\displaystyle H - O}{O}}$$
$$\underset{\displaystyle CH_3}{}$$

Hydrogen bonding in methanol entails the attraction of the hydrogen atom bearing a partial positive charge (δ^+) on one molecule to the oxygen atom bearing a partial negative charge (δ^-) on a second molecule. The strong attractive force of hydrogen bonding will cause the boiling point and the heat of vaporization of methanol to be quite high.

13.4   (a) O_2: induced dipole–induced dipole forces only.

(b) CH_3OH: strong hydrogen bonding (dipole–dipole forces) as well as induced dipole–induced dipole forces.

(c) Forces between water molecules: strong hydrogen bonding and induced dipole–induced dipole forces. Between O_2 and H_2O: dipole–induced dipole forces and induced dipole–induced dipole forces.

Relative strengths: a < forces between O_2 and H_2O in c < b < forces between water molecules in c.

13.5 $(1.00 \times 10^3 \text{ g})(1 \text{ mol}/32.04 \text{ g})(35.2 \text{ kJ/mol}) = 1.10 \times 10^3 \text{ kJ}$

13.6   (a) At 40 °C, the vapor pressure of ethanol is about 120 mm Hg.

(b) The equilibrium vapor pressure of ethanol at 60 °C is about 320 mm Hg. At 60 °C and 600 mm Hg, ethanol is a liquid. If vapor is present, it will condense to a liquid.

13.7 $PV = nRT$; $P = 0.50 \text{ g } (1 \text{ mol}/18.02 \text{ g})(0.0821 \text{ L} \cdot \text{atm/mol} \cdot \text{K})(333 \text{ K})/5.0 \text{ L}$

$P = 0.15 \text{ atm (or 120 mm Hg)}$

The vapor pressure of water at 60 °C is 149.4 mm Hg (Appendix G). The calculated pressure is lower than this, so all the water (0.50 g) evaporates. If 2.0 g of water is used, the calculated pressure, 460 mm Hg, exceeds the vapor pressure. In this case, only part of the water will evaporate.

13.8   Use the Clausius-Clapeyron equation, with $P_1 = 57.0$ mm Hg, $T_1 = 250.4$ K, $P_2 = 534$ mm Hg, and $T_2 = 298.2$ K.

$\ln [P_2/P_1] = \Delta H_{vap}/R [1/T_1 - 1/T_2] = \Delta H_{vap}/R[(T_2 - T_1)/T_1 T_2]$

$\ln [534/57.0] = \Delta H_{vap}/(0.0083145 \text{ kJ/K} \cdot \text{mol})[47.8/(250.4)(298.2)]$

$\Delta H_{vap} = 29.1 \text{ kJ/mol}$

13.9   Glycerol is predicted to have a higher viscosity than ethanol. It is a larger molecule than ethanol, and there are higher forces of attraction between molecules because each molecule has three OH groups that hydrogen-bond to other molecules.

13.10 M_2X; In a face-centered cubic unit cell, there are four anions and eight tetrahedral holes in which to place metal ions. All of the tetrahedral holes are inside the unit cell, so the ratio of atoms in the unit cell is 2 : 1.

Chapter 14

14.1   10.0 g sucrose = 0.0292 mol; 250. g H_2O = 13.9 mol

$X = (0.0292 \text{ mol})/(0.0292 \text{ mol} + 13.9 \text{ mol}) = 0.00210$

$(0.0292 \text{ mol sucrose})/(0.250 \text{ kg solvent}) = 0.117 \text{ m}$

% sucrose = (10.0 g sucrose/260. g soln)($\times 100\%$) = 3.85%

14.2 $1.08 \times 10^4 \text{ ppm} \equiv 1.08 \times 10^4 \text{ mg NaCl per 1000 g soln}$

$(1.08 \times 10^4 \text{ mg Na}/1000 \text{ g soln})(1050 \text{ g soln}/1 \text{ L}) = 1.13 \times 10^4 \text{ mg Na/L} = 11.3 \text{ g Na/L}$

$(11.3 \text{ g Na/L})(58.44 \text{ g NaCl}/23.0 \text{ g Na}) = 28.7 \text{ g NaCl/L}$

14.3 $\Delta H^{\circ}_{soln} = \Delta H^{\circ}_f [\text{NaOH(aq)}] - \Delta H^{\circ}_f [\text{NaOH(s)}]$
$= -469.2 \text{ kJ/mol} - (-425.9 \text{ kJ/mol})$
$= -43.3 \text{ kJ/mol}$

14.4   Solubility (CO_2) = $(4.48 \times 10^{-5} \text{ M/mm Hg})(251 \text{ mm Hg})$
$= 1.1 \times 10^{-2} \text{ M}$

14.5   The solution contains sucrose [(10.0 g)(1 mol/342.3 g) = 0.0292 mol] in water [(225 g)(1 mol/18.02 g) = 12.5 mol].

$X_{water} = (12.5 \text{ mol } H_2O)/(12.5 \text{ mol} + 0.0292 \text{ mol}) = 0.998$

$P_{water} = 0.998(149.4 \text{ mm Hg}) = 149 \text{ mm Hg}$

14.6 $m = \Delta T_{bp}/K_{bp} = 1.0 \text{ °C}/(0.512 \text{ °C}/m) = 1.95 \text{ m}$
$(1.95 \text{ mol/kg})(0.125 \text{ kg})(62.02 \text{ g/mol}) = 15 \text{ g glycol}$

14.7   Concentration = (525 g)(1 mol/62.07 g)/(3.00 kg)
$= 2.82 \text{ m}$

$\Delta T_{fp} = K_{fp} \times m = (-1.86 \text{ °C}/m)(2.82 \text{ m}) = -5.24 \text{ °C}$

You will be protected only to about -5 °C and not to -25 °C.

14.8 $\Delta T_{bp} = 80.23 \text{ °C} - 80.10 \text{ °C} = 0.13 \text{ °C}$

$m = \Delta T_{bp}/K_{bp} = 0.13 \text{ °C}/(2.53 \text{ °C}/m) = 0.051 \text{ m}$
$(0.051 \text{ mol/kg})(0.099 \text{ kg}) = 0.0051 \text{ mol}$

Molar mass = 0.640 g/0.0051 mol = 130 g/mol

The formula $C_{10}H_8$ (molar mass = 128.2 g/mol) is the closest match to this value.

14.9   Concentration = (25.0 g NaCl)(1 mol/58.44 g)/(0.525 kg) = 0.815 m

$\Delta T_{fp} = K_{fp} \times m \times i = (-1.86 \text{ °C}/m)(0.815 \text{ m})(1.85)$
$= -2.80 \text{ °C}$

14.10 $M = \Pi/RT = [(1.86 \text{ mm Hg})(1 \text{ atm}/760 \text{ mm Hg})]/$
$[(0.08206 \text{ L} \cdot \text{atm/mol} \cdot \text{K})(298 \text{ K})]$
$= 1.00 \times 10^{-4} \text{ M}$

$(1.00 \times 10^{-4} \text{ mol/L})(0.100 \text{ L}) = 1.0 \times 10^{-5} \text{ mol}$

Molar mass $= 1.40 \text{ g}/1.00 \times 10^{-5} \text{ mol} = 1.4 \times 10^5 \text{ g/mol}$

Assuming the polymer is composed of CH_2 units, the polymer is about 10,000 units long.

Chapter 15

15.1 $-\frac{1}{2}(\Delta[\text{NOCl}]/\Delta t) = \frac{1}{2}(\Delta[\text{NO}]/\Delta t) = \Delta[\text{Cl}_2]/\Delta t$

15.2 For the first two hours:

$-\Delta[\text{sucrose}]/\Delta t = [(0.033 - 0.050) \text{ mol/L}]/(2.0 \text{ h})$
$= 0.0080 \text{ mol/L} \cdot \text{h}$

For the last two hours:

$\Delta[\text{sucrose}]/\Delta t = -[(0.010 - 0.015) \text{ mol/L}]/(2.0 \text{ h})$
$= 0.0025 \text{ mol/L} \cdot \text{h}$

Instantaneous rate at 4 h $= 0.0045 \text{ mol/L} \cdot \text{h}$

15.3 Compare experiments 1 and 2: Doubling $[\text{O}_2]$ causes the rate to double, so the rate is first order in $[\text{O}_2]$. Compare experiments 2 and 4: Doubling $[\text{NO}]$ causes the rate to increase by a factor of 4, so the rate is second order in $[\text{NO}]$. Thus, the rate law is

Rate $= k[\text{NO}]^2[\text{O}_2]$

Using the data in experiment 1 to determine k:

$0.028 \text{ mol/L} \cdot \text{s} = k[0.020 \text{ mol/L}]^2[0.010 \text{ mol/L}]$

$k = 7.0 \times 10^3 \text{ L}^2/\text{mol}^2 \cdot \text{s}$

15.4 Rate $= k[\text{Pt(NH}_3)_2\text{Cl}_2] = (0.090 \text{ h}^{-1})(0.020 \text{ mol/L})$
$= 0.0018 \text{ mol/L} \cdot \text{h}$

The rate of formation of Cl^- is the same value, 0.0018 mol/L $\cdot$ h.

15.5 $\ln([\text{sucrose}]/[\text{sucrose}]_o) = -kt$

$\ln([\text{sucrose}]/[0.010]) = -(0.21 \text{ h}^{-1})(5.0 \text{ h})$

$[\text{sucrose}] = 0.0035 \text{ mol/L}$

15.6 (a) The fraction remaining is $[\text{NO}_2]/[\text{NO}_2]_o$.

$\ln([\text{NO}_2]/[\text{NO}_2]_o) = -(3.6 \times 10^{-3} \text{ s}^{-1})(150 \text{ s})$

$[\text{NO}_2]/[\text{NO}_2]_o = 0.58$

(b) The fraction remaining after the reaction is 99% complete is 0.010.

$\ln(0.010) = -(3.6 \times 10^{-3} \text{ s}^{-1})(t)$

$t = 1300 \text{ s}$

15.7 $1/[\text{HI}] - 1/[\text{HI}]_o = kt$

$1/[\text{HI}] - 1/[0.010 \text{ M}] = (30. \text{ L/mol} \cdot \text{min})(12 \text{ min})$

$[\text{HI}] = 0.0022 \text{ M}$

15.8

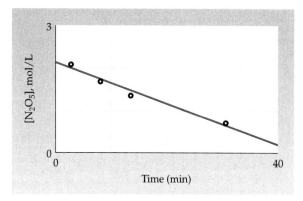

Concentration versus time

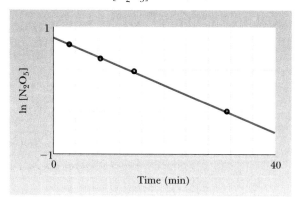

ln $[\text{N}_2\text{O}_5]$ versus time

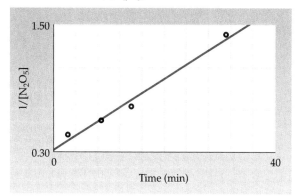

$1/[\text{N}_2\text{O}_5]$ versus time

The plot of $\ln[\text{N}_2\text{O}_5]$ versus time is linear, indicating that this is a first-order reaction. The rate constant is determined from the slope: $k = -\text{slope} = 0.038 \text{ min}^{-1}$.

15.9 (a) For ^{241}Am, $t_{1/2} = 0.693/k = 0.693/(0.0016 \text{ y}^{-1})$
$= 430 \text{ y}$

For ^{125}I, $t_{1/2} = 0.693/(0.011 \text{ d}^{-1}) = 63 \text{ d}$

(b) ^{125}I decays much faster.

(c) $\ln[(n)/(1.6 \times 10^{15} \text{ atoms})] = -(0.011 \text{ d}^{-1})(2.0 \text{ d})$

$n/1.6 \times 10^{15} \text{ atoms} = 0.978$; $n = 1.57 \times 10^{-15} \text{ atoms}$

Since the answer should have two significant figures, we should round this off to 1.6×10^{15} atoms. The approximately 2% that has decayed is not noticeable within the limits of accuracy of the data presented.

15.10 $\ln (k_2/k_1) = (-E_a/R)(1/T_2 - 1/T_1)$

$\ln [(1.00 \times 10^4)/(4.5 \times 10^3)] = -(E_a/8.315 \times 10^{-3}$
$\text{kJ/mol} \cdot \text{K})(1/283 \text{ K} - 1/274 \text{ K})$

$E_a = 57 \text{ kJ/mol}$

15.11 All three steps are bimolecular.

For step 3: Rate = $k[N_2O][H_2]$

When the three equations are added, N_2O_2 (a product in the first step and a reactant in the second step) and N_2O (a product in the second step and a reactant in the third step) cancel, leaving the net equation: $2 NO(g) + 2 H_2(g) \longrightarrow N_2(g) + 2 H_2O(g)$

15.12 (a) $2 NH_3(aq) + OCl^-(aq) \longrightarrow$
$N_2H_4(aq) + Cl^-(aq) + H_2O(\ell)$

(b) The second step is the rate-determining step.

(c) Rate = $k[NH_2Cl][NH_3]$

(d) NH_2Cl, $N_2H_5^+$, and OH^- are intermediates.

15.13 Overall reaction: $2 NO_2Cl(g) \longrightarrow 2 NO_2(g) + Cl_2(g)$

Rate = $k[NO_2Cl]^2/[NO_2]$

The presence of NO_2 causes the reaction rate to decrease.

Chapter 16

16.1 (a) $K = [PCl_3][Cl_2]/[PCl_5]$

(b) $K = [CO]^2/[CO_2]$

(c) $K = [Cu^{2+}][NH_3]^4/[Cu(NH_3)_4^{2+}]$

(d) $K = [H_3O^+][CH_3CO_2^-]/[CH_3CO_2H]$

16.2 (a) Both reactions are reactant-favored ($K \ll 1$).

(b) $[NH_3]$ in the second solution is greater. K for this reaction is larger, so the reactant, $Cd(NH_3)_4^{2+}$, dissociates to a greater extent.

16.3 (a) $Q = [2.18]/[0.97] = 2.3$. The system is not at equilibrium; $Q < K$. To reach equilibrium, [isobutane] will increase and [butane] will decrease.

(b) $Q = [2.60]/[0.75] = 3.5$. The system is not at equilibrium; $Q > K$. To reach equilibrium, [butane] will increase and [isobutane] will decrease.

16.4 $Q = [NO]^2/[N_2][O_2] = [4.2 \times 10^{-3}]^2/[0.50][0.25]$
$= 1.4 \times 10^{-4}$

$Q < K$, so the reaction is not at equilibrium. To reach equilibrium, [NO] will increase and $[N_2]$ and $[O_2]$ will decrease.

16.5 (a)

Equation	$C_6H_{10}I_2$ ⇌	C_6H_{10} +	I_2
Initial (M)	0.050	0	0
Change (M)	−0.035	+0.035	+0.035
Equilibrium (M)	0.015	0.035	0.035

(b) $K = (0.035)(0.035)/(0.015) = 0.082$

16.6

Equation	H_2 +	I_2 ⇌	2 HI
Initial (M)	6.00×10^{-3}	6.00×10^{-3}	0
Change (M)	−x	−x	+2x
Equilibrium (M)	0.00600 − x	0.00600 − x	+2x

$K_c = 33 = \dfrac{(2x)^2}{(0.00600 - x)^2}$

$x = 0.0045$ M, so $[H_2] = [I_2] = 0.0015$ M and $[HI] = 0.0090$ M.

16.7

Equation	C(s) + CO_2 (g) ⇌		2 CO(g)
Initial (M)	0.012		0
Change (M)	−x		+2x
Equilibrium (M)	0.012 − x		2x

$K_c = 0.021 = \dfrac{(2x)^2}{(0.012 - x)}$

$x = [CO_2] = 0.0057$ M and $2x = [CO] = 0.011$ M

16.8 (a) $K' = K^2 = (2.5 \times 10^{-29})^2 = 6.3 \times 10^{-58}$

(b) $K'' = 1/K^2 = 1/(6.3 \times 10^{-58}) = 1.6 \times 10^{57}$

16.9 Manipulate the equations and equilibrium constants as follows:

$\frac{1}{2} H_2(g) + \frac{1}{2} Br_2(g) \rightleftharpoons HBr(g)$
$\qquad\qquad\qquad K_1' = (K_1)^{1/2} = 8.9 \times 10^5$

$H(g) \rightleftharpoons \frac{1}{2} H_2(g) \qquad K_2' = 1/(K_2)^{1/2} = 1.4 \times 10^{20}$

$Br(g) \rightleftharpoons \frac{1}{2} Br_2(g) \qquad K_3' = 1/(K_3)^{1/2} = 2.1 \times 10^7$

Net: $H(g) + Br(g) \rightleftharpoons HBr(g)$
$\qquad\qquad\qquad K_{net} = K_1'K_2'K_3' = 2.6 \times 10^{33}$

16.10 (a) [NOCl] decreases with an increase in temperature

(b) $[SO_3]$ decreases with an increase in temperature

16.11

Equation	butane ⇌	isobutane
Initial (M)	0.20	0.50
After adding 2.0 M more isobutene	0.20	2.0 + 0.50
Change (M)	+x	−x
Equilibrium (M)	0.20 + x	2.50 − x

$K = \dfrac{[\text{isobutane}]}{[\text{butane}]} = \dfrac{(2.50 - x)}{(0.20 + x)} = 2.50$

Solving for x gives $x = 0.57$ M. Therefore, [isobutene] = 1.93 M and [butane] = 0.77 M.

16.12 (a) Adding H_2 shifts the equilibrium to the right, increasing $[NH_3]$. Adding NH_3 shifts the equilibrium to the left, increasing $[N_2]$ and $[H_2]$.

(b) An increase in volume shifts the equilibrium to the left.

Chapter 17

17.1 (a) $H_3PO_4(aq) + H_2O(\ell) \rightleftharpoons H_3O^+(aq) + H_2PO_4^-(aq)$

$H_2PO_4^-$ is amphiprotic.

(b) $CN^-(aq) + H_2O(\ell) \rightleftharpoons HCN(aq) + OH^-(aq)$

CN^- is a Brønsted base.

17.2 NO_3^- is the conjugate base of the acid HNO_3; NH_4^+ is the conjugate acid of the base NH_3.

17.3 $[H_3O^+] = 4.0 \times 10^{-3}$ M; $[OH^-] = K_w/[H_3O^+]$
$= 2.5 \times 10^{-12}$

17.4 (a) $pOH = -\log [0.0012] = 2.92$; $pH = 14.00 - pOH$
$= 11.08$

(b) $[H^+] = 4.8 \times 10^{-5}$ M; $[OH^-] = 2.1 \times 10^{-10}$ M

(c) $pOH = 14.00 - 10.46 = 3.54$; $[OH^-] = 2.9 \times 10^{-4}$ M. The solubility of $Sr(OH)_2$ is half of this value, or 1.4×10^{-4} M, because dissolving 1 mol of $Sr(OH)_2$ will give 2 mol of OH^- in solution.

17.5 Answer this question by comparing values of K_a and K_b from Table 17.3.

(a) H_2SO_4 is stronger than H_2SO_3.

(b) $C_6H_5CO_2H$ is a stronger acid than CH_3CO_2H.

(c) The conjugate base of boric acid, $B(OH)_4^-$, is a stronger base than the conjugate base of acetic acid, $CH_3CO_2^-$.

(d) Ammonia is a stronger base than acetate ion.

(e) The conjugate acid of acetate ion, CH_3CO_2H, is a stronger acid than the conjugate acid of ammonia, NH_4^+.

17.6 (a) $pH = 7$

$pH < 7$ (NH_4^+ is an acid)

$pH < 7$ $[Al(H_2O)_6]^{3+}$ is an acid

$pH > 7$ (HPO_4^{2-} is a stronger base than it is an acid)

17.7 (a) $pK_a = -\log [6.3 \times 10^{-5}] = 4.20$

(b) $ClCH_2CO_2H$ is stronger (a pK_a of 2.87 is less than a pK_a of 4.20)

(c) pK_a for NH_4^+, the conjugate acid of NH_3, is $-\log [5.6 \times 10^{-10}] = 9.26$. It is a weaker acid than acetic acid, for which $K_a = 1.8 \times 10^{-5}$.

17.8 K_b for the lactate ion $= K_w/K_a = 7.1 \times 10^{-11}$. It is a slightly stronger base than the formate, nitrite, and fluoride ions, and weaker than the benzoate ion.

17.9 (a) NH_4^+ is a stronger acid than HCO_3^-. CO_3^{2-}, the conjugate base of HCO_3^-, is a stronger base than NH_3, the conjugate base of NH_4^+.

(b) Reactant-favored; the reactants are the weaker acid and base.

(c) Reactant-favored; the reactants are the weaker acid and base.

17.10 $CH_3CO_2H(aq) + HSO_4^-(aq) \rightleftharpoons$
$CH_3CO_2^-(aq) + H_2SO_4(aq)$

The equilibrium favors the weaker acid and base, which in this equation are the reactants.

17.11 (a) The two compounds react and form a solution containing HCN and NaCl. The solution is acidic (HCN is an acid).

(b) $CH_3CO_2H(aq) + SO_3^{2-}(aq) \rightleftharpoons$
$HSO_3^-(aq) + CH_3CO_2^-(aq)$

The solution is acidic, because HSO_3^- is a stronger acid than $CH_3CO_2^-$ is a base.

17.12 From the pH we can calculate $[H_3O^+] = 1.9 \times 10^{-3}$ M. Also, $[butanoate^-] = [H_3O^+] = 1.9 \times 10^{-3}$ M. Use these values along with [butanoic acid] to calculate K_a.

$K_a = [1.9 \times 10^{-3}] [1.9 \times 10^{-3}]/(0.055 - 1.9 \times 10^{-3})$
$= 6.8 \times 10^{-5}$

17.13 $K_a = 1.8 \times 10^{-5} = [x][x]/(0.10 - x)$

$x = [H_3O^+] = [CH_3CO_2^-] = 1.3 \times 10^{-3}$ M; $[CH_3CO_2H]$
$= 0.099$ M; $pH = 2.89$

17.14 $K_a = 7.2 \times 10^{-4} = [x][x]/(0.015 - x)$

The x in the denominator cannot be dropped. This equation must be solved with the quadratic formula or by successive approximations.

$x = [H_3O^+] = [F^-] = 2.9 \times 10^{-3}$ M

$[HF] = 0.015 - 2.9 \times 10^{-3} = 0.012$ M

$pH = 2.54$

17.15 $OCl^-(aq) + H_2O(\ell) = HOCl(aq) + OH^-(aq)$

$K_b = 2.9 \times 10^{-7} = [x][x]/(0.015 - x)$

$x = [OH^-] = [HOCl] = 6.6 \times 10^{-5}$ M

$pOH = 4.18$; $pH = 9.82$

17.16 Equivalent amounts of acid and base were used. The solution will contain $CH_3CO_2^-$ and Na^+. Acetate ion hydrolyzes to a small extent, giving CH_3CO_2H and OH^-. We need to determine $[CH_3CO_2^-]$ and then solve a weak base equilibrium problem to determine $[OH^-]$.

Amount $CH_3CO_2^-$ = moles base = 0.12 M $\times$ 0.015 L
$= 1.8 \times 10^{-3}$ mol

Total volume = 0.030 L

so $[CH_3CO_2^-] = (1.8 \times 10^{-3}$ mol$)/0.030$ L $= 0.060$ M

$CH_3CO_2^-(aq) + H_2O(\ell) \rightleftharpoons CH_3CO_2H(aq) + OH^-(aq)$

$K_b = 5.6 \times 10^{-10} = [x][x]/(0.060 - x)$

$x = [OH^-] = [CH_3CO_2H] = 5.8 \times 10^{-6}$ M

$pOH = 5.24$; $pH = 8.76$

17.17 $H_2C_2O_4(aq) + H_2O(\ell) \rightleftharpoons H_3O^+(aq) + HC_2O_4^-(aq)$

$K_{a_1} = 5.9 \times 10^{-2} = [x][x]/(0.10 - x)$

The x in the denominator cannot be dropped. This equation must be solved with the quadratic formula or by successive approximations.

$x = [H_3O^+] = [HC_2O_4^-] = 5.3 \times 10^{-2}$ M

$pH = 1.28$

$[C_2O_4^{2-}] = K_{a_2} = 6.4 \times 10^{-5}$ M

17.18 (a) Lewis base (electron-pair donor)

(b) Lewis acid (electron-pair acceptor)

(c) Lewis base (electron-pair donor)

(d) Lewis base (electron-pair donor)

17.19 (a) H_2SeO_4

(b) $Fe(H_2O)_6^{3+}$

(c) HOCl

(d) Amphetamine is a primary amine and a (weak) base. It is both a Brønsted base and a Lewis base.

Chapter 18

18.1 pH of 0.30 M HCO_2H:

$K_a = [H_3O^+][HCO_2^-]/[HCO_2H]$

$1.8 \times 10^{-4} = [x][x]/[0.30 - x]$

$x = 7.3 \times 10^{-3}$ M; pH = 2.14

pH of 0.30 M formic acid + 0.10 M $NaHCO_2$

$K_a = [H_3O^+][HCO_2^-]/[HCO_2H]$

$1.8 \times 10^{-4} = [x][0.10 + x]/(0.30 - x)$

$x = 5.4 \times 10^{-4}$ M; pH = 3.27

18.2 NaOH: $(0.100 \text{ mol/L})(0.0300 \text{ L}) = 3.00 \times 10^{-3}$ mol

CH_3CO_2H: $(0.100 \text{ mol/L})(0.0450 \text{ L}) = 4.50 \times 10^{-3}$ mol

3.00×10^{-3} mol NaOH reacts with 3.00×10^{-3} mol CH_3CO_2H, forming 3.00×10^{-3} mol $CH_3CO_2^-$; 1.50×10^{-3} mol unreacted CH_3CO_2H remains in solution. The total volume is 75.0 mL. Use these values to calculate $[CH_3CO_2H]$ and $[CH_3CO_2^-]$, and use the concentrations in a weak acid equilibrium calculation to obtain $[H_3O^+]$ and pH.

$[CH_3CO_2H] = 1.5 \times 10^{-3} \text{ mol}/0.075 \text{ L} = 0.0200$ M

$[CH_3CO_2^-] = 3.0 \times 10^{-3} \text{ mol}/0.075 \text{ L} = 0.0400$ M

$K_a = [H_3O^+][CH_3CO_2^-]/[CH_3CO_2H]$

$1.8 \times 10^{-5} = [x][0.0400 + x]/(0.0200 - x)$

$x = [H_3O^+] = 9.0 \times 10^{-6}$ M; pH = 5.05

18.3 pH = pK_a + log {[base]/[acid]}

pH = $-$log (1.8×10^{-4}) + log {[0.70]/[0.50]}

pH = 3.74 + 0.15 = 3.89

18.4 $(15.0 \text{ g NaHCO}_3)(1 \text{ mol}/84.01 \text{ g}) = 0.179$ mol $NaHCO_3$, and $(18.0 \text{ g Na}_2\text{CO}_3)(1 \text{ mol}/106.0 \text{ g}) = 0.170$ mol Na_2CO_3

pH − pK_a + log {[base]/[acid]}

pH = $-$log (4.8×10^{-11}) + log {[0.170]/[0.179]}

pH = 10.32 − 0.02 = 10.30

18.5 pH = pK_a + log {[base]/[acid]}

$5.00 = -\log (1.8 \times 10^{-5}) + \log \{[\text{base}]/[\text{acid}]\}$

$5.00 = 4.74 + \log \{[\text{base}]/[\text{acid}]\}$

[base]/[acid] = 1.8

To prepare this buffer solution, the ratio [base]/[acid] must equal 1.8. For example, you can dissolve 1.8 mol (148 g) of $NaCH_3CO_2$ and 1.0 mol (60.05 g) of CH_3CO_2H in enough water to make 1.0 L of solution.

18.6 Initial pH (before adding acid):

pH = pK_a + log {[base]/[acid]}

$= -\log (1.8 \times 10^{-4}) + \log \{[0.70]/[0.50]\}$

$= 3.74 + 0.15 = 3.89$

After adding acid, the added HCl will react with the weak base (formate ion) and form more formic acid. The net effect is to change the ratio of [base]/[acid] in the buffer solution.

Initial amount $HCO_2H = 0.50$ M $\times 0.500$ L = 0.250 mol

Initial amount $HCO_2^- = 0.70$ M $\times 0.50$ L = 0.350 mol

Amount HCl added = 1.0 M $\times 0.010$ L = 0.010 mol

Amount HCO_2H after HCl addition = 0.250 mol + 0.010 mol = 0.26 mol

Initial amount HCO_2^- after HCl addition = 0.350 mol − 0.010 mol = 0.34 mol

pH = pK_a + log {[base]/[acid]}

pH = $-$log (1.8×10^{-4}) + log {[0.340]/[0.260]}

pH = 3.74 + 0.12 = 3.86

18.7 After addition of 25.0 mL base, half of the acid has been neutralized.

Initial amount HCl = 0.100 M $\times 0.0500$ L = 0.00500 mol

Amount NaOH added = 0.100 M $\times 0.0250$ L = 0.00250 mol

Amount HCl after reaction: $0.00500 - 0.00250 = 0.00250$ mol HCl

[HCl] after reaction = 0.00250 mol/0.0750 L = 0.0333 M

This is a strong acid and completely ionized, so $[H_3O^+] = 0.0333$ M and pH = 1.48.

After 50.50 mL base is added, a small excess of base is present in the 100.5 mL of solution. (Volume of excess base added is 0.50 mL = 5.0×10^{-4} L.)

Amount excess base = 0.100 M $\times 5.0 \times 10^{-4}$ L $= 5.0 \times 10^{-5}$ mol

$[OH^-] = 5.0 \times 10^{-5} \text{ mol}/0.1005 \text{ L} = 4.9 \times 10^{-4}$ M

pOH = $-$log (4.9×10^{-4}) = 3.31; pH = 14.00 − pOH = 10.69

18.8 35.0 mL base will partially neutralize the acid.

Initial amount $CH_3CO_2H = (0.100 \text{ M})(0.1000 \text{ L}) = 0.0100$ mol

Amount NaOH added = $(0.10 \text{ M})(0.035 \text{ L}) = 0.0035$ mol

Amount CH_3CO_2H after reaction = $0.0100 - 0.0035 = 0.0065$ mol

Amount $CH_3CO_2^-$ after reaction = 0.0035 mol

$[CH_3CO_2H]$ after reaction = 0.0065 mol/0.135 L = 0.0481 M

$[CH_3CO_2^-]$ after reaction $= 0.00350$ mol$/0.135$ L
$$= 0.0259 \text{ M}$$

$K_a = [H_3O^+][CH_3CO_2^-]/[CH_3CO_2H]$

$1.8 \times 10^{-5} = [x][0.0259 + x]/[0.0481 - x]$

$x = [H_3O^+] = 3.34 \times 10^{-5}$ M; pH $= 4.48$

18.9 75.0 mL acid will partially neutralize the base.

Initial amount $NH_3 = (0.100 \text{ M})(0.1000 \text{ L}) = 0.0100$ mol

Amount HCl added $= (0.100 \text{ M})(0.0750 \text{ L}) = 0.00750$ mol

Amount NH_3 after reaction $= 0.0100 - 0.00750$
$$= 0.0025 \text{ mol}$$

Amount NH_4^+ after reaction $= 0.00750$ mol

Solve using the Henderson-Hasselbach equation; use K_a for the weak acid NH_4^+:

pH $= pK_a + \log \{[\text{base}]/[\text{acid}]\}$

pH $= -\log (5.6 \times 10^{-10}) + \log \{[0.0025]/[0.00750]\}$

pH $= 9.25 - 0.48 = 8.77$

18.10 An indicator that changes color near the pH at the equivalence point is required. Possible indicators include methyl red, bromcresol green, and Eriochrome black T; all change color in the pH range of 5–6.

18.11 (a) $AgI \rightleftharpoons Ag^+ + I^-$
$$K_{sp} = [Ag^+][I^-]; K_{sp} = 8.5 \times 10^{-17}$$
  (b) $BaF_2 \rightleftharpoons Ba^{2+} + 2 F^-$
$$K_{sp} = [Ba^{2+}][F^-]^2; K_{sp} = 1.8 \times 10^{-7}$$
  (c) $Ag_2CO_3 \rightleftharpoons 2 Ag^+ + CO_3^{2-}$
$$K_{sp} = [Ag^+]^2[CO_3^{2-}]; K_{sp} = 8.5 \times 10^{-12}$$

18.12 $[Ba^{2+}] = 3.6 \times 10^{-3}$ M; $[F^-] = 7.2 \times 10^{-3}$ M

$K_{sp} = [Ba^{2+}][F^-]^2$

$K_{sp} = [3.6 \times 10^{-3}][7.2 \times 10^{-3}]^2 = 1.9 \times 10^{-7}$

18.13 $Ca(OH)_2 \rightleftharpoons Ca^{2+} + 2 OH^-$
$$K_{sp} = [Ca^{2+}][OH^-]^2; K_{sp} = 5.5 \times 10^{-5}$$

$5.5 \times 10^{-5} = [x][2x]^2$ (where $x =$ solubility in mol/L)

$x = 2.4 \times 10^{-2}$ mol/L

sol. in g/L $= (2.4 \times 10^{-2}$ mol/L$)(74.1$ g/mol$)$
$$= 1.8 \text{ g/L}$$

18.14 (a) AgCl
  (b) $Ca(OH)_2$
  (c) Because these compounds have different stoichiometries, the most soluble cannot be identified without doing a calculation. The solubility of $Ca(OH)_2$ is 2.4×10^{-2} M (from Exercise 18.13); $Ca(OH)_2$ is more soluble than $CaSO_4$, whose solubility is 7.0×10^{-3} M $\{K_{sp} = [Ca^{2+}][SO_4^{2-}]; 4.9 \times 10^{-5} = [x][x];$ $x = 7.0 \times 10^{-3}$ M$\}$.

18.15 (a) In pure water:
$$K_{sp} = [Ba^{2+}][SO_4^{2-}]; 1.1 \times 10^{-10} = [x][x];$$
$$x = 1.0 \times 10^{-5} \text{ M}$$
  (b) In 0.010 M $Ba(NO_3)_2$, which furnishes 0.010 M Ba^{2+} in solution:
$$K_{sp} = [Ba^{2+}][SO_4^{2-}]; 1.1 \times 10^{-10} = [0.010 + x][x];$$
$$x = 1.1 \times 10^{-8} \text{ M}$$

18.16 (a) In pure water:
$$K_{sp} = [Zn^{2+}][CN^-]^2; 8.0 \times 10^{-12} = [x][2x]^2 = 4x^3$$
Solubility $= x = 1.3 \times 10^{-4}$ M
  (b) In 0.10 M $Zn(NO_3)_2$, which furnishes 0.10 M Zn^{2+} in solution:
$$K_{sp} = [Zn^{2+}][CN^-]^2; 8.0 \times 10^{-12} = [0.10 + 2x][2x]^2$$
Solubility $= x = 4.5 \times 10^{-6}$ M

18.17 When $[Pb^{2+}] = 1.1 \times 10^{-3}$ M, $[I^-] = 2.2 \times 10^{-3}$ M.

$Q = [Pb^{2+}][I^-]^2 = [1.1 \times 10^{-3}][2.2 \times 10^{-3}]^2 = 5.3 \times 10^{-9}$

This value is less than K_{sp}, which means that the system has not yet reached equilibrium and more PbI_2 will dissolve.

18.18 $Q = [Sr^{2+}][SO_4^{2-}] = [2.5 \times 10^{-4}][2.5 \times 10^{-4}]$
$$= 6.3 \times 10^{-8}$$

This value is less than K_{sp}, which means that the system has not yet reached equilibrium. Precipitation will not occur.

18.19 $K_{sp} = [Pb^{2+}][I^-]^2$. Let x be the concentration of I^- required at equilibrium. If an amount greater than x is used, precipitation will occur.

$9.8 \times 10^{-9} = [0.050][x]^2$

$x = [I^-] = 4.4 \times 10^{-5}$ M

Let x be the concentration of Pb^{2+} in solution, in equilibrium with 0.0015 M I^-.

$9.8 \times 10^{-9} = [x][1.5 \times 10^{-3}]^2$

$x = [Pb^{2+}] = 4.4 \times 10^{-3}$ M

18.20 First determine the concentrations of Ag^+ and Cl^-; then calculate Q and see whether it is greater than or less than K_{sp}. Concentrations are calculated using the final volume, 105 mL, in the equation $C_{dil} \times V_{dil} = C_{conc} \times V_{conc}$.

$[Ag^+](0.105 \text{ L}) = (0.0010 \text{ mol/L})(0.100 \text{ L})$
$$[Ag^+] = 9.5 \times 10^{-4} \text{ M}$$

$[Cl^-](0.105 \text{ L}) = (0.025 \text{ M})(0.005 \text{ L})$
$$[Cl^-] = 1.2 \times 10^{-3} \text{ M}$$

$Q = [Ag^+][Cl^-] = [9.5 \times 10^{-4}][1.2 \times 10^{-3}] = 1.1 \times 10^{-6}$

$Q > K_{sp}$; precipitation occurs.

18.21 $Cu(OH)_2 \rightleftharpoons Cu^{2+} + 2 OH^-$ $\qquad K_{sp} = [Cu^{2+}][OH^-]^2$

$Cu^{2+} + 4 NH_3 \rightleftharpoons Cu(NH_3)_4^{2+}$
$$K_{form} = [Cu(NH_3)_4^{2+}]/[Cu^{2+}][NH_3]^4$$

Net: $Cu(OH)_2 + 4 NH_3 \rightleftharpoons Cu(NH_3)_4^{2+} + 2 OH^-$

$K_{net} = K_{sp} \times K_{form} = (2.2 \times 10^{-20})(6.8 \times 10^{12}) = 1.5 \times 10^{-7}$

18.22 (a) Add NaCl(aq). AgCl will precipitate; $CaCl_2$ is soluble.
  (b) Add Na_2S(aq) or NaOH(aq) to precipitate FeS or $Fe(OH)_2$; K_2S and KOH are soluble.

Chapter 19

19.1 (a) O_3; larger molecules generally have higher entropies than smaller molecules.

(b) $SnCl_4(g)$; gases have higher entropies than liquids.

19.2 (a) $\Delta S° = \sum S°\,(\text{products}) - \sum S°\,(\text{reactants})$

$\Delta S° = S°\,[NH_4Cl(aq)]) - S°\,[NH_4Cl(s)]$

$\Delta S° = (1\,\text{mol})(169.9\,\text{J/mol} \cdot \text{K}) - (1\,\text{mol})(94.85\,\text{J/mol} \cdot \text{K}) = 75.1\,\text{J/K}$

A gain in entropy for the formation of a mixture (solution) is expected.

(b) $\Delta S° = 2\,S°\,(NH_3) - [S°\,(N_2) + 3\,S°\,(H_2)]$

$\Delta S° = (2\,\text{mol})(192.77\,\text{J/mol} \cdot \text{K}) - [(1\,\text{mol})(191.56\,\text{J/mol} \cdot \text{K}) + (3\,\text{mol})(130.7\,\text{J/mol} \cdot \text{K})]$

$\Delta S° = -198.1\,\text{J/K}$

A decrease in entropy is expected because there is a decrease in the number of moles of gases.

19.3 (a) Type 2

(b) Type 3

(c) Type 1

(d) Type 2

19.4 $\Delta S°_{sys} = 2\,S°\,(HCl) - [S°\,(H_2) + S°\,(Cl_2)]$

$\Delta S°_{sys} = (2\,\text{mol})(186.2\,\text{J/mol} \cdot \text{K}) - [(1\,\text{mol})(130.7\,\text{J/mol} \cdot \text{K}) + (1\,\text{mol})(223.08\,\text{J/mol} \cdot \text{K})] = 18.6\,\text{J/K}$

$\Delta S°_{surr} = -\Delta H°_{sys}/T = -(-184{,}620\,\text{J}/298\,\text{K}) = 619.5\,\text{J/K}$

$\Delta S°_{univ} = \Delta S°_{sys} + \Delta S°_{surr} = 18.6\,\text{J/K} + 619.5\,\text{J/K} = 638.1\,\text{J/K}$

19.5 At 298 K, $\Delta S°_{surr} = -(467{,}900\,\text{J/K})/298\,\text{K} = -1570\,\text{J/K}$

$\Delta S°_{univ} = \Delta S°_{sys} + \Delta S°_{surr} = 560.7\,\text{J/K} - 1570\,\text{J/K}$
$= -1010\,\text{J/K}$

The negative sign indicates that the process is not spontaneous. At higher temperature, the value of $-\Delta H°_{sys}/T$ will be less negative. At a high enough temperature, matter dispersal will outweigh the energy dispersal in the system and the reaction will be spontaneous.

19.6 For the reaction $N_2(g) + 3\,H_2(g) \longrightarrow 2\,NH_3(g)$:

$\Delta H°_{rxn} = 2\,\Delta H°_f$ for $NH_3(g) = (2\,\text{mol})(-45.90\,\text{kJ/mol})$
$= -91.80\,\text{kJ}$

$\Delta S° = 2\,S°\,(NH_3) - [S°\,(N_2) + 3\,S°\,(H_2)]$

$\Delta S° = (2\,\text{mol})(192.77\,\text{J/mol} \cdot \text{K}) - [(1\,\text{mol})(191.56\,\text{J/mol} \cdot \text{K}) + (3\,\text{mol})(130.7\,\text{J/mol} \cdot \text{K})]$

$\Delta S° = -198.1\,\text{J/K}$

$\Delta G°_f = \Delta H°_f - T\Delta S° = -91.80\,\text{kJ} - (298\,\text{K})(-0.198\,\text{kJ/K})$

$\Delta G°_f = -32.8\,\text{kJ}$

19.7 $SO_2(g) + \tfrac{1}{2}\,O_2(g) \longrightarrow SO_3(g)$

$\Delta G° = \sum \Delta G°\,(\text{products}) - \sum \Delta G°\,(\text{reactants})$

$\Delta G° = \Delta G°\,[SO_3(g)] - \{\Delta G°\,[SO_2(g)] + \tfrac{1}{2}\,\Delta G°\,[O_2(g)]\}$

$\Delta G° = -371.04\,\text{kJ} - (-300.13\,\text{kJ} + 0) = -70.91\,\text{kJ}$

19.8 $HgO(s) \longrightarrow Hg(\ell) + \tfrac{1}{2}\,O_2(g)$; Determine the temperature at which $\Delta G°_f = 0$, in which case $\Delta H° - T\Delta S° = 0$. T is the unknown in this problem.

$\Delta H° = 90.83\,\text{kJ} = [-\Delta H°_f$ for $HgO(s)]$

$\Delta S° = S°\,[Hg(\ell)]) + \tfrac{1}{2}\,S°\,(O_2) - S°\,(HgO(s)$

$\Delta S° = (1\,\text{mol})(76.02\,\text{J/mol} \cdot \text{K}) + [(0.5\,\text{mol})(205.07\,\text{J/mol} \cdot \text{K}) - (1\,\text{mol})(70.29\,\text{J/mol} \cdot \text{K})] = 108.26\,\text{J/K}$

$\Delta H° - T\Delta S° = 90{,}830\,\text{J} - T(108.27\,\text{J/K}) = 0$

$T = 839\,\text{K (566 °C)}$

19.9 $C(s) + CO_2(g) \rightleftharpoons 2\,CO(g)$

$\Delta G°_{rxn} = 2\,\Delta G°_f\,(CO) - \Delta G°_f\,(CO_2)$

$\Delta G°_{rxn} = (2\,\text{mol})(-137.17\,\text{kJ/mol}) - (1\,\text{mol})(-394.36\,\text{kJ/mol})$

$\Delta G°_{rxn} = 120.02\,\text{kJ}$

$\Delta G°_{rxn} = -RT\ln K$

$120{,}020\,\text{J/mol} = -(8.3145\,\text{J/mol} \cdot \text{K})(298\,\text{K})(\ln K)$

$K = 8.94 \times 10^{-22}$

19.10 $\Delta G°_{rxn} = -RT\ln K$

$\Delta G°_{rxn} = -(8.3145\,\text{J/mol} \cdot \text{K})(298\,\text{K})\,[\ln\,(1.6 \times 10^7)]$

$\Delta G°_{rxn} = -41{,}100\,\text{J/mol}\,(-41.1\,\text{KJ/mol})$

Chapter 20

20.1 Oxidation half-reaction: $Al(s) \longrightarrow Al^{3+}(aq) + 3\,e^-$

Reduction half-reaction: $2\,H^+(aq) + 2\,e^- \longrightarrow H_2(g)$

Overall reaction: $2\,Al(s) + 6\,H^+(aq) \longrightarrow 2\,Al^{3+}(aq) + 3\,H_2(g)$

Al is the reducing agent and is oxidized; $H^+(aq)$ is the oxidizing agent and is reduced.

20.2 $2\,VO^{2+}(aq) + Zn(s) + 4\,H^+(aq) \longrightarrow Zn^{2+}(aq) + 2\,V^{3+}(aq) + 2\,H_2O(\ell)$

$2\,V^{3+}(aq) + Zn(s) \longrightarrow 2\,V^{2+}(aq) + Zn^{2+}(aq)$

20.3 Oxidation half-reaction: $Fe^{2+}(aq) \longrightarrow Fe^{3+}(aq) + e^-$

Reduction half-reaction:
$MnO_4^-(aq) + 8\,H^+(aq) + 5\,e^- \longrightarrow Mn^{2+}(aq) + 4\,H_2O(\ell)$

Overall reaction:
$MnO_4^-(aq) + 8\,H^+(aq) + 5\,Fe^{2+}(aq) \longrightarrow Mn^{2+}(aq) + 5\,Fe^{3+}(aq) + 4\,H_2O(\ell)$

20.4 (a) Oxidation half-reaction: $Al(s) + 3\,OH^-(aq) \longrightarrow Al(OH)_3(s) + 3\,e^-$

Reduction half-reaction: $S(s) + H_2O(\ell) + 2\,e^- \longrightarrow HS^-(aq) + OH^-(aq)$

Overall reaction:
$2\,Al(s) + 3\,S(s) + 3\,H_2O(\ell) + 3\,OH^-(aq) \longrightarrow 2\,Al(OH)_3(s) + 3\,HS^-(aq)$

(b) Aluminum is the reducing agent and is oxidized; sulfur is the oxidizing agent and is reduced.

20.5 Construct two half-cells, the first with a silver electrode and a solution containing $Ag^+(aq)$, and the second with a nickel electrode and a solution containing $Ni^{2+}(aq)$. Connect the two half-cells with a salt bridge. When the electrodes are connected through an external circuit, electrons will flow from the anode (the nickel electrode) to the cathode (the silver electrode). The overall cell reaction is $Ni(s) + 2\,Ag^+(aq) \longrightarrow Ni^{2+}(aq) + 2\,Ag(s)$. To maintain electrical neutrality in the two half-cells, negative ions will flow from the Ag|Ag⁺ half-cell to the Ni|Ni²⁺ half-cell, and positive ions will flow in the opposite direction.

20.6 Anode reaction: $Zn(s) \longrightarrow Zn^{2+}(aq) + 2\,e^-$

Cathode reaction: $2\,Ag^+(aq) + 2\,e^- \longrightarrow 2\,Ag(s)$

$E°_{cell} = E°_{cathode} - E°_{anode} = 0.80\,V - (-0.76\,V) = 1.56\,V$

20.7 Mg is easiest to oxidize, and Au is the most difficult. (See Table 20.1.)

20.8 Use the "northwest-southeast rule" or calculate the cell voltage to determine whether a reaction is product-favored. Reactions (a) and (c) are reactant-favored; reactions (b) and (d) are product-favored.

20.9 Overall reaction: $Fe(s) + 2\,H^+(aq) \longrightarrow$
$Fe^{2+}(aq) + H_2(g)$ $(E°_{cell} = 0.44\,V, n = 2)$

$E_{cell} = E°_{cell} - (0.0257/n)\,\ln\{[Fe^{2+}]P_{H_2}/[H^+]^2\}$

$= 0.44 - (0.0257/2)\,\ln\{[0.024]1.0/[0.056]^2\}$

$= 0.44\,V - 0.026\,V = 0.41\,V$

20.10 Overall reaction: $2\,Al(s) + 3\,Fe^{2+}(aq) \longrightarrow$
$2\,Al^{3+}(aq) + 3\,Fe(s)$
$(E°_{cell} = 1.22\,V, n = 6)$

$E_{cell} = E°_{cell} - (0.0257/n)\,\ln\{[Al^{3+}]^2/[Fe^{2+}]^3\}$

$= 1.22 - (0.0257/6)\,\ln\{[0.025]^2/[0.50]^3\}$

$= 1.22\,V - (-0.023)\,V = 1.24\,V$

20.11 $\Delta G° = -nFE° = -(2\,mol\,e^-)(96,500\,C/mol\,e^-)$
$(-0.76\,V)(1\,J/1\,C\cdot V)$

$= 146,680\,J\ (= 150\,kJ)$

The negative value of $E°$ and the positive value of $\Delta G°$ both indicate a reactant-favored reaction.

20.12 $E°_{cell} = E°_{cathode} - E°_{anode} = 0.80\,V - 0.855\,V = -0.055\,V$;
$n = 2$

$E° = (0.0257/n)\,\ln K$

$-0.055 = (0.0257/2)\,\ln K$

$K = 0.014$

20.13 Cathode: $Zn^{2+}(aq) + 2\,e^- \longrightarrow Zn(s)$
$E°_{cathode} = -0.76\,V$

Anode: $Zn(s) + 4\,CN^-(aq) \longrightarrow [Zn(CN)_4^{2-}] + 2\,e^-$
$E°_{anode} = -1.26\,V$

Overall: $Zn^{2+}(aq) + 4\,CN^-(aq) \longrightarrow [Zn(CN)_4^{2-}]$
$E°_{cell} = 0.50\,V$

$E° = (0.0257/n)\,\ln K$

$0.50 = (0.0257/2)\,\ln K$

$K = 7.9 \times 10^{16}$

20.14 Cathode: $2\,H_2O(\ell) + 2\,e^- \longrightarrow 2\,OH^-(aq) + H_2(g)$
$E°_{cathode} = -0.83\,V$

Anode: $4\,OH^-(aq) \longrightarrow O_2(g) + 2\,H_2O(\ell) + 4\,e^-$
$E°_{anode} = 0.40\,V$

Overall: $2\,H_2O(\ell) \longrightarrow 2\,H_2(g) + O_2(g)$
$E°_{cell} = E°_{cathode} - E°_{anode} = -0.83\,V - 0.40\,V = -1.23\,V$

This is the minimum voltage needed to cause this reaction to occur.

20.15 O_2 is formed at the anode, by the reaction $2\,H_2O(\ell) \longrightarrow 4\,H^+(aq) + O_2(g) + 4\,e^-$.

$(0.445\,A)(45\,min)(60\,s/min)(1\,C/1\,A\cdot s)$
$(1\,mol\,e^-/96,500\,C)(1\,mol\,O_2/4\,mol\,e^-)(32\,g\,O_2/1\,mol\,O_2) = 0.10\,g\,O_2$

20.16 The cathode reaction (electrolysis of molten NaCl) is $Na^+(\ell) + e^- \longrightarrow Na(\ell)$.

$(25 \times 10^3\,A)(60\,min)(60\,s/min)(1\,C/1\,A\cdot s)$
$(1\,mol\,e^-/96,500\,C)(1\,mol\,Na/mol\,e^-)(23\,g\,Na/1\,mol\,Na) = 21,450\,g\,Na = 21\,kg$

Chapter 21

21.1 (a) $2\,Na(s) + Br_2(\ell) \longrightarrow 2\,NaBr(s)$

(b) $Ca(s) + Se(s) \longrightarrow CaSe(s)$

(c) $4\,K(s) + O_2(g) \longrightarrow 2\,K_2O(s)$

K_2O is one of the possible products of this reaction. The primary product from the reaction of potassium and oxygen is KO_2, potassium superoxide.

(d) $2\,Al(s) + 3\,Cl_2(g) \longrightarrow 2\,AlCl_3(s)$

21.2 (a) H_2Te

(b) Na_3AsO_4

(c) $SeCl_6$

(d) $HBrO_4$

21.3 (a) NH_4^+ (ammonium ion)

(b) O_2^{2-} (peroxide ion)

(c) N_2H_4 (hydrazine)

(d) NF_3 (nitrogen trifluoride)

21.4 (a) ClO is an odd-electron molecule, with Cl having the unlikely oxidation number of +2.

(b) In Na_2Cl, chlorine would have the unlikely charge of 2− (to balance the two positive charges of the two Na^+ ions).

(c) This compound would require either the calcium ion to have the formula Ca^+ or the acetate ion to have the formula $CH_3CO_2^{2-}$. In all of its compounds, calcium occurs as the Ca^{2+} ion. The acetate ion, formed from acetic acid by loss of H^+, has a 1− charge.

(d) No octet structure for C_3H_7 can be drawn. This species has an odd number of electrons.

21.5 $CH_4(g) + H_2O(g) \longrightarrow 3\,H_2(g) + CO(g)$

Bonds broken: 4 C—H and 2 O—H (sum = 2578 kJ)

$4\,D(C—H) = 4(413\,kJ) = 1652\,kJ$

$2\,D(O—H) = 4(463\,kJ) = 926\,kJ$

Bonds formed: 3 H—H and 1 C≡O (sum = 2354 kJ)

$3\,D(H—H) = 3(436\,kJ) = 1308\,kJ$

$D(CO) = 1046\,kJ$

Estimated energy of reaction = 2578 kJ − 2354 kJ
$= +224\,kJ$

The triple-bond energy is small, relative to the energies of the three single bonds. Six bonds were broken and made, but the incremental energy to form the second and third bonds between C and O is not sufficient to overcome the energy required to break two single bonds.

21.6 Cathode reaction: $Na^+ + e^- \longrightarrow Na(\ell)$; 1 F, or 96,500 C, is required to form 1 mol of Na. There are (24 h) (60 min/h)(60 s/min) = 86,400 s in 1 day. 1000. kg = 1.000×10^6 g.
$(1.000 \times 10^6$ g Na)(1 mol Na/23.00 g Na)(96,500 C/mol Na)(1 A · s/1 C)(1/86,400 s) = 4.855×10^4 A

21.7 Some interesting topics: gemstones of the mineral beryl; uses of Be in the aerospace industry and in nuclear reactors; beryllium–copper alloys; severe health hazards when beryllium or its compounds get into the lungs.

21.8 (a) $Ga(OH)_3(s) + 3 H^+(aq) \longrightarrow Ga^{3+}(aq) + 3 H_2O(\ell)$
$Ga(OH)_3(s) + OH^-(aq) \longrightarrow Ga(OH)_4^-(aq)$
(b) $Ga^{3+}(aq)$ $(K_a = 1.2 \times 10^{-3})$ is stronger acid than $Al^{3+}(aq)$ $(K_a = 7.9 \times 10^{-6})$

21.9

21.10 (a) $: N \equiv N - \overset{..}{\underset{..}{O}} : \longleftrightarrow : \overset{..}{\underset{..}{N}} = N = \overset{..}{O} :$

The first resonance structure places the negative charge on oxygen; the second places it on the terminal nitrogen. Because oxygen is more electronegative and better able to accommodate the negative charge, the first structure is favored.
(b) $NH_4NO_3(s) \longrightarrow N_2O(g) + 2 H_2O(g)$
$\Delta H°$ (reaction) = $\Delta H_f°$ (products) − $\Delta H_f°$ (reactants)
$\Delta H°$ (reaction) = $\Delta H_f°$ (N_2O) + 2 $\Delta H_f°$ (H_2O) − $\Delta H_f°$ (NH_4NO_3)
= 82.05 kJ + 2(−241.83 kJ) − (−365.56 kJ) = −36.05 kJ

The reaction is exothermic.

21.11 First, calculate $\Delta G°$, $\Delta H°$, and $\Delta S°$ for this reaction, using data from Appendix L.
$\Delta G° = \Delta G_f°$ (products) − $\Delta G_f°$ (reactants)
$\Delta G° = 2 \Delta G_f°$ (ZnO) + 2 $\Delta G_f°$ (SO_2) − 2 $\Delta G_f°$ (ZnS) − 3 $\Delta G_f°(O_2)$
= 2(−318.30 kJ) + 2(−300.13 kJ) − 2(−201.29 kJ) − 0 = −834.28 kJ. The reaction is product-favored at 298 K.
$\Delta H° = 2 \Delta H_f°$ (ZnO) + 2 $\Delta H_f°$ (SO_2) − 2 $\Delta H_f°$ (ZnS) − 3 $\Delta H°_f (O_2)$
= 2(−348.28 kJ) + 2(−296.84 kJ) − [2(−205.98 kJ) + 0] = −878.28 kJ
$\Delta S° = 2 S°$ (ZnO) + 2 $S°$ (SO_2) − 2 $S°$ (ZnS) − 3 $S°$ (O_2)
= 2(43.64 J/K) + 2(248.21 J/K) − [2(57.7 J/K) + 3(205.07 J/K)] = −146.9 J/K

This reaction is enthalpy favored and entropy disfavored. The reaction will become less favored at higher temperatures. See Table 19.2, page 920.

21.12 For the reaction $HX + Ag \longrightarrow AgX + \frac{1}{2} H_2$:
$\Delta G° = \Delta G_f°$ (products) − $\Delta G_f°$ (reactants)
$\Delta G° = \Delta G_f°$ (AgX) − $\Delta G_f°$ (HX)
For HF: $\Delta G° = +79.4$ kJ; reactant-favored
For HCl: $\Delta G° = −14.67$ kJ; product-favored
For HBr: $\Delta G° = −43.45$ kJ; product-favored
For HI: $\Delta G° = −67.75$ kJ; product-favored

Chapter 22

22.1 (a) $Co(NH_3)_3Cl_3$
(b) (i) $K_3[Co(NO_2)_6]$: a complex of cobalt(III) with a coordination number of 6
(ii) $Mn(NH_3)_4Cl_2$: a complex of manganese(II) with a coordination number of 6

22.2 (a) hexaaquanickel(II) sulfate
(b) dicyanobis(ethylenediamine)chromium(III) chloride
(c) potassium amminetrichloroplatinate(II)
(d) potassium dichlorocuprate(I)

22.3 (a) These are geometric isomers (with the NH_3 ligands in *cis* and *trans* positions).
(b) Only a single structure is possible.
(c) Only a single structure is possible.
(d) This compound is chiral; there are two optical isomers.
(e) Only a single structure is possible.
(f) Two structural isomers are possible based on coordination of the NO_2^- ligand through oxygen or nitrogen.

22.4 (a) $[Ru(H_2O)_6]^{2+}$: An octahedral complex of ruthenium(II) (d^6). A low-spin complex has no unpaired electrons and is diamagnetic. A high-spin complex has four unpaired electrons and is paramagnetic.

high-spin Ru^{2+} low-spin Ru^{2+}

(b) $[Ni(NH_3)_6]^{2+}$: An octahedral complex of nickel(II) (d^8). Only one electron configuration is possible; it has two unpaired electrons and is paramagnetic.

Ni^{2+} ion (d^8)

Chapter 23

23.1 (a) $^{222}_{86}Rn \longrightarrow ^{218}_{84}Po + ^{4}_{2}\alpha$

(b) $^{218}_{84}Po \longrightarrow ^{218}_{85}At + ^{0}_{-1}\beta$

23.2 (a) E (per photon) $= h\nu = hc/\lambda$

$E = [(6.626 \times 10^{-34} \, J \cdot s/photon)(3.00 \times 10^{8} \, m/s)]/(2.0 \times 10^{-12} \, m)$

$E = 9.94 \times 10^{-14} \, J/photon$

E (per mole) $= (9.94 \times 10^{-14} \, J/photon)(6.022 \times 10^{23} \, photons/mol)$

E (per mole) $= 5.99 \times 10^{10} \, J/mol$

23.3 (a) Emission of six α particles leads to a decrease of 24 in the mass number and a decrease of 12 in the atomic number. Emission of four β particles increases the atomic number by 4, but doesn't affect the mass. The final product of this process has a mass number of $232 - 24 = 208$ and an atomic number of $90 - 12 + 4 = 82$, identifying it as $^{208}_{82}Pb$.

(b) Step 1: $^{232}_{90}Th \longrightarrow ^{228}_{88}Ra + ^{4}_{2}\alpha$

Step 2: $^{228}_{88}Ra \longrightarrow ^{228}_{89}Ac + ^{0}_{-1}\beta$

Step 3: $^{228}_{89}Ac \longrightarrow ^{228}_{90}Th + ^{0}_{-1}\beta$

23.4 (a) $^{0}_{+1}\beta$

(b) $^{41}_{19}K$

(c) $^{0}_{-1}\beta$

(d) $^{22}_{12}Mg$

23.5 (a) $^{32}_{14}Si \longrightarrow ^{32}_{15}P + ^{0}_{-1}\beta$

(b) $^{45}_{22}Ti \longrightarrow ^{45}_{21}Sc + ^{0}_{+1}\beta$ or $^{45}_{22}Ti + ^{0}_{-1}e \longrightarrow ^{45}_{21}Sc$

(c) $^{239}_{94}Pu \longrightarrow ^{235}_{92}U + ^{4}_{2}\alpha$

(d) $^{42}_{19}K \longrightarrow ^{42}_{20}Ca + ^{0}_{-1}\beta$

23.6 $\Delta m = 0.03438 \, g/mol$

$\Delta E = (3.438 \times 10^{-5} \, kg/mol)(2.998 \times 10^{8})^{2}$

$= 3.090 \times 10^{12} \, J/mol \, (= 3.090 \times 10^{9} \, kJ/mol)$

$E_b = 5.150 \times 10^{8} \, kJ/mol \, nucleons$

23.7 (a) 49.2 years is exactly 4 half-lives; quantity remaining $= 1.5 \, mg(1/2)^{4} = 0.094 \, mg$

(b) 3 half-lives, 36.9 years

(c) 1% is between 6 half-lives, 73.8 years (1/64 remains), and 7 half-lives, 86.1 years (1/128 remains)

23.8 $\ln([A]/[A_o]) = -kt$

$\ln([3.18 \times 10^{3}]/[3.35 \times 10^{3}]) = -k(2.00 \, d)$

$k = 0.0260 \, d^{-1}$

$t_{1/2} = 0.693/k = 0.693/(0.0260 \, d^{-1}) = 26.7 \, d$

23.9 $k = 0.693/t_{1/2} = 0.693/200. \, y = 3.47 \times 10^{-3} \, y^{-1}$

$\ln([A]/[A_o]) = -kt$

$\ln([3.00 \times 10^{3}]/[6.50 \times 10^{12}]) = -(3.47 \times 10^{-3} \, y^{-1})t$

$\ln(4.62 \times 10^{-10}) = -(3.47 \times 10^{-3} \, y^{-1})t$

$t = 6190 \, y$

23.10 $\ln([A]/[A_o]) = -kt$

$\ln([9.32]/[13.4]) = -(1.21 \times 10^{-4} \, y^{-1})t$

$t = 3.00 \times 10^{3} \, y$

23.11 $3000 \, dpm/x = 1200 \, dpm/60.0 \, mg$

$x = 150 \, mg$

Appendix O

Answers to Selected Study Questions

Chapter 1

1.1 (a) C, carbon (d) P, phosphorus
(b) K, potassium (e) Mg, magnesium
(c) Cl, chlorine (f) Ni, nickel

1.3 (a) Ba, barium (d) Pb, lead
(b) Ti, titanium (e) As, arsenic
(c) Cr, chromium (f) Zn, zinc

1.5 (a) Na (element) and NaCl (compound)
(b) Sugar (compound) and carbon (element)
(c) Gold (element) and gold chloride (compound)

1.7 (a) Physical property (d) Physical property
(b) Chemical property (e) Physical property
(c) Chemical property (f) Physical property

1.9 (a) Physical (colorless) and chemical (burns in air)
(b) Physical (shiny metal, orange liquid) and chemical (reacts with bromine)

1.11 555 g

1.13 2.79 cm^3 or 2.79 mL

1.15 Al, aluminum

1.17 298 K

1.19 (a) 289 K
(b) 97 °C
(c) 310 K

1.21 4.2195×10^4 m; 26.219 miles

1.23 5.3 cm^2; 5.3×10^{-4} m^2

1.25 250 cm^3; 0.25 L, 2.5×10^{-4} m^3; 0.25 dm^3

1.27 2.52×10^3 g

1.29 (a) Method A with all data included: average = 2.4 g/cm^3
Method B with all data included: average = 3.480 g/cm^3
For B the reading of 5.811 can be excluded because it is more than twice as large as all other readings. Using only the first three readings, you find average = 2.703 g/cm^3.
(b) Method A: error = 0.3 g/cm^3 or about 10%
Method B: error = 0.001 g/cm^3 or about 0.04%

(c) Before excluding a data point for B, method A gives more accurate and more precise answer. After excluding data for B, this method gives a more accurate and more precise result.

1.31 (a) Qualitative: blue-green color, solid physical state
Quantitative: density = 2.65 g/cm^3 and mass = 2.5 g
(b) Density, physical state, and color are intensive properties, whereas mass is an extensive property.
(c) Volume = 0.94 cm^3

1.33 (a) Al, Si, O
(b) All are solids. Aluminum metal (Al) and silicon (Si) chips are shiny, whereas the aquamarine crystal is a blue-gray color.

1.35 24.6 K, 27.1 K

1.37 0.197 nm; 197 pm

1.39 (a) 7.5×10^{-6} m; (b) 7.5×10^3 nm; (c) 7.5×10^6 pm

1.41 0.995 g Pt

1.43 50. mg procaine hydrochloride

1.45 The piece of brass has a volume of 18.0 cm^3, so the water volume increases by 18.0 mL. The final water volume is 68.0 mL.

1.47 The experimental density of the necklace is 19 g/cm^3, so the necklace is gold. At $380 per troy ounce it is worth about $820, so $300 is a bargain (and too good to be true).

1.49 The mixture is heterogeneous. Iron is magnetic, so a small magnet will attract the iron chips and remove them from the nonmagnetic sand.

1.51 The density of the plastic is less than that of CCl_4, so the plastic will float on the liquid CCl_4. Aluminum is more dense than CCl_4, so aluminum will sink when placed in CCl_4.

1.53 Comparing the two substances on the basis of three properties:

	Sucrose	NaCl
Density (g/cm³)	1.587	2.164
Melting temperature (°C)	160–186 (decomp)	800
Solubility (g in 100 mL water)	200	36

The melting (or decomposition) temperatures are significantly different, as are their water solubilities.

1.55   Your normal body temperature (about 98.6 °F) is 37 °C. As this is higher than gallium's melting point, the metal will melt in your hand.

1.57   HDPE will float in ethylene glycol, water, acetic acid, and glycerol.

1.59

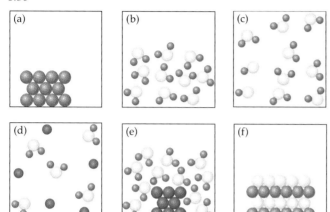

1.61   One could check for an odor, check the boiling or freezing point, or determine the density. If the density is approximately 1 g/cm³ at room temperature, the liquid could be water. If it boils at about 100 °C and freezes about 0 °C, that would be consistent with water. To check for the presence of salt, boil the liquid away. If a substance remains, it could be salt, but further testing would be required.

1.63

1.65   (a) Solid potassium metal reacts with liquid water to produce gaseous hydrogen and a homogeneous mixture (solution) of potassium hydroxide in liquid water.

(b) The reaction is a chemical change.

(c) The reactants are potassium and water. The products are hydrogen gas and a water (aqueous) solution of potassium hydroxide. Heat and light are also evolved.

(d) Among the qualitative observations are (i) the reaction is violent and (ii) heat and light (a purple flame) are produced.

1.67   (a) The water could be evaporated by heating the solution, leaving the salt behind.

(b) Use a magnet to attract the iron away from lead, which is not magnetic.

(c) Mixing the solids with water will dissolve only the sugar. Filtration would separate the solid sulfur from the sugar solution. Finally, the sugar could be separated from the water by evaporating the water.

1.69   As described on Screen 1.18 of the General ChemistryNow CD-ROM or website, using a reasonably powerful magnet will separate the iron from the cereal.

1.71   (a) The reactants are P_4 and Cl_2 and the product is PCl_3.

(b) The P_4 molecules are tetrahedra (four-sided polyhedra), and the chlorine molecules consist of two atoms. The PCl_3 molecule is a triangular pyramid.

Mathematics: Section 1.8

1.73   (a) 5.4×10^{-2}

(b) 5.462×10^3

(c) 7.92×10^{-4}

1.75   (a) 9.44×10^{-3}

(b) 5.69×10^3

(c) 1.19×10^1 (or 11.9)

1.77   (a) 3   (c) 5

(b) 3   (d) 4

1.79   0.122

1.81   Equation: $y = 248.4x + 0.0022$ (where y is the absorbance and x is the concentration). The slope is 248.4. The concentration is 2.548×10^{-3} when the absorbance is 0.635.

1.83   (a) $x = 0.21$ when $y = 4.0$

(b) $y = 5.6$ when $x = 0.30$

(c) Slope = 18 and intercept = 0.20

(d) When $x = 1.0$, $y = 18$

1.85 $C = 0.0823$

1.87 $T = 295$

1.89   Volume = 0.0854 cm³

1.91   Mass = 22 g

1.93   Correct conversion factor = 0.803 kg/L

Fuel in tank = 7682 L = 6170 kg

Additional fuel needed = 22,300 kg − 6170 kg = 16,130 kg (or 20,100 L)

1.95   Thickness of aluminum foil = 1.8×10^{-2} mm

1.97   Oil layer thickness = 2×10^{-7} cm. This is likely related to the "length" of the oil molecules.

1.99 $(0.546 \text{ g}) \left(\dfrac{1 \text{ cm}^3}{8.96 \text{ g}} \right) \left(\dfrac{1 \text{ L}}{1000 \text{ cm}^3} \right) = 6.09 \times 10^{-5} \text{ L}$

1.101 Area $= 9.6 \times 10^3 \text{ m}^2$

1.103 (a) Volume $= 65 \text{ m}^3$ or $6.5 \times 10^4 \text{ L}$

(b) Mass $= 78 \text{ kg}$ or 170 lb

1.105 (a) Calculated density $= 1.11 \text{ g/mL}$. The unknown is ethylene glycol.

(b) If the volume were 3.5 mL, the calculated density would be 1.1 g/mL. Although still indicating ethylene glycol, it is close enough to the density of acetic acid that you would be unsure of the answer. (Fortunately, acetic acid and ethylene glycol have very different odors and could be identified that way.)

1.107 (a,b) Density $= 1.25 \text{ g/L}$. To three significant figures, three gases (N_2, C_2H_4, and CO) have this density.

(c) Using the more accurate mass, the density is 1.249 g/L. This eliminates C_2H_4 from consideration, but N_2 and CO are still possible choices.

1.109. Mass of Hg in the tube $= 0.153 \text{ g}$

Volume of Hg in the tube $= 0.0113 \text{ cm}^3$

Using the formula for the volume of a cylinder, the radius of the cylinder of Hg in the tube is 0.0463 cm, which gives a diameter of 0.0927 cm.

Chapter 2

2.1 Atoms contain the following fundamental particles: protons (+1 charge), neutrons (zero charge), and electrons (−1 charge). Protons and neutrons are in the nucleus of an atom. Electrons are the least massive of the three particles.

2.3 The discovery of radioactivity showed that atoms must be divisible; that is, atoms must be composed of even smaller, subatomic particles.

2.5 Exercise 2.1 provides the relative sizes of the nuclear and atomic diameters, with the nuclear radius on the order of 0.001 pm and the atomic radius approximately 100 pm. If the nuclear diameter is 6 cm, then the atomic diameter is 600,000 cm (or 6 km).

2.7 Radon, Rn

2.9 (a) ^{27}Mg, mass number $= 12 + 15 = 27$

(b) ^{48}Ti, mass number $= 22 + 26 = 48$

(c) ^{62}Zn, mass number $= 30 + 32 = 62$

2.11 (a) $^{39}_{19}$K

(b) $^{84}_{36}$Kr

(c) $^{60}_{27}$Co

2.13

Element	^{24}Mg	^{119}Sn	^{232}Th
Electrons	12	50	90
Protons	12	50	90
Neutrons	12	69	142

2.15 ^{99}Tc has 43 protons, 43 electrons, and 56 neutrons.

2.17 $^{57}_{27}$Co, $^{58}_{27}$Co, $^{60}_{27}$Co

2.19 ^{205}Tl is more abundant (70.5%) than ^{203}Tl (29.5%). Its atomic weight is closer to 205 than to 203.

2.21 $(0.0750)(6.015121) + (0.9250)(7.016003) = 6.94$

2.23 (c), About 50%. Actual percent ^{107}Ag $= 51.839\%$.

2.25 ^{69}Ga, 60.12%; ^{71}Ga, 39.88%

2.27 (a) 68 g Al   (c) 0.60 g Ca

(b) 0.0698 g Fe   (d) 1.32×10^4 g Ne

2.29 (a) 1.9998 mol Cu   (c) 2.1×10^{-5} mol Am

(b) 0.0017 mol Li   (d) 0.250 mol Al

2.31 He has the smallest molar mass, and Fe has the largest molar mass. Therefore, 1.0 g of He has the largest number of atoms in these samples, and 1.0 g of Fe has the smallest number of atoms.

2.33 1.0552×10^{-22} g for 1 Cu atom

2.35 Five elements. Nitrogen (N) and phosphorus (P) are nonmetals, arsenic (As) and antimony (Sb) are metalloids, and bismuth (Bi) is a metal.

2.37 8 elements: periods 2 and 3. 18 elements: periods 4 and 5. 32 elements: period 6.

2.39 (a) Nonmetals: C, Cl

(b) Main group elements: C, Cl, Cs, Ca

(c) Lanthanides: Ce

(d) Transition elements: Cr, Co, Cd, Cu, Ce, Cf, and Cm

(e) Actinides: Cm, Cf

(f) Gases: Cl

2.41 Metals: Na, Ni, Np.
Metalloids: None in this list
Nonmetals: N, Ne

2.43 Metals: sodium, scandium, strontium, silver, and samarium

Main group elements: sodium, silicon, sulfur, selenium, strontium

Transition metals: scandium, silver (some chemists include the lanthanides, such as samarium, in the transition elements)

2.45

Symbol	^{58}Ni	^{33}S	^{20}Ne	^{55}Mn
Protons	28	16	10	25
Neutrons	30	17	10	30
Electrons	28	16	10	25
Name	Nickel	Sulfur	Neon	Manganese

2.47 Potassium has an atomic weight of 39.0983. This mass is close to the mass of the ^{39}K isotope. Therefore, the abundance of ^{41}K is low (6.73%).

2.49 (a) Mg   (c) Si
(b) H   (d) Fe
(e) F, Cl, and Br. Chlorine is more abundant.

2.51 (a), (b), and (c) are all possible. (d) is impossible because one atom of S has a mass of 5.325×10^{-23} g. Therefore, one mole of molecules consisting of eight S atoms cannot be less than the mass of one atom.

2.53 (a) Beryllium, magnesium, calcium, strontium, barium, radium
(b) Sodium, magnesium, aluminum, silicon, phosphorus, sulfur, chlorine, argon
(c) Carbon   (g) Krypton
(d) Sulfur   (h) Sulfur
(e) Iodine   (i) Germanium or arsenic
(f) Magnesium

2.55 (a) Three elements—Co, Ni, and Cu—have densities of about 9 g/cm^3.
(b) Boron in the second period and aluminum in the third period have the largest densities. Both are in Group 3A.
(c) Elements that have very low densities are all gases. These include hydrogen, helium, nitrogen, oxygen, fluorine, neon, chlorine, argon, and krypton.

2.57 (a) 0.5 mol Si
(b) 0.5 mol Na
(c) 10 atoms Fe

2.59 Boron; 1.4 mol or 8.4×10^{23} atoms

2.61 9.42×10^{-5} mol Kr; 5.67×10^{19} atoms Kr

2.63 40.2 g H$_2$ (b) < 103 g C (c) < 182 g Al (f) < 210 g Si (d) < 212 g Na (e) < 351 g Fe (a) < 650 g Cl$_2$(g)

2.65 (a) Average charge = 1.59×10^{-19} C; error is about 0.5%

(b)
Drop	Number of Electrons
1	1
2	7
3	6
4	10
5	4

2.67 K = 78 weight %; Na = 22 weight %

2.69 The drawing should have two protons and two neutrons in the nucleus and two electrons outside the nucleus. The electrons do not trace a particular path around the nucleus but rather exist as a "cloud" (as in Figure 2.1). The radius of a He atom is 128 pm. If the nucleus is only 1×10^{-5} as large as the atom (see Exercise 2.1),

then the nuclear radius is about 0.00128 pm. If you depict the nucleus as a penciled dot on paper (with a radius of about 0.1 mm), then the radius of the atom is 10 m!

2.71 Required data: density of iron, molar mass of iron, Avogadro's number

$$1.0 \text{ cm}^3 \left(\frac{7.87 \text{ g}}{1 \text{ cm}^3} \right) \left(\frac{1 \text{ mol}}{55.85 \text{ g}} \right) \left(\frac{6.02 \times 10^{23} \text{ atoms}}{1 \text{ mol}} \right) = 8.5 \times 10^{22} \text{ atoms Fe}$$

2.73 Barium would be more reactive than calcium, so a more vigorous evolution of hydrogen should occur. Reactivity increases on descending the periodic table, at least for Groups 1A and 2A.

2.75 Volume = 3.33×10^3 cm^3. Edge = 14.9 cm.

2.77 4.9×10^{24} atoms Na

2.79 1.0028×10^{23} atoms C. If the accuracy is ±0.0001 g, the maximum mass could be 2.0001 g, which also represents 1.0028×10^{23} atoms C.

2.81 See Screen 2.20 of the General ChemistryNow CD-ROM or website for a possible method.

Chapter 3

3.1 Sulfuric acid, H$_2$SO$_4$. The structure is not flat. Chemists describe the structure as a tetrahedron of O atoms around the S atom.

3.3 Pt(NH$_3$)$_2$Cl$_2$

structural formula

$$\text{H}_3\text{N} \diagdown \text{Pt} \diagup \text{NH}_3$$
$$\text{Cl} \diagup \quad \diagdown \text{Cl}$$

3.5 (a) Mg^{2+} (c) Ni^{2+}
(b) Zn^{2+} (d) Ga^{3+}

3.7 (a) Ba^{2+} (e) S^{2-}
(b) Ti^{4+} (f) ClO$_4^-$
(c) PO$_4^{3-}$ (g) Co^{2+}
(d) HCO$_3^-$ (h) SO$_4^{2-}$

3.9 K loses one electron per atom to form a K$^+$ ion. It has the same number of electrons as an Ar atom.

3.11 Ba^{2+} and Br$^-$ ions. Compound formula is BaBr$_2$.

3.13 (a) Two K$^+$ ions and one S^{2-} ion
(b) One Co^{2+} ion and one SO$_4^{2-}$ ion
(c) One K$^+$ ion and one MnO$_4^-$ ion
(d) Three NH$_4^+$ ions and one PO$_4^{3-}$ ion
(e) One Ca^{2+} ion and two ClO$^-$ ions

3.15 Co^{2+} gives CoO and Co^{3+} gives Co$_2$O$_3$.

3.17 (a) AlCl$_2$ should be AlCl$_3$ (based on one Al^{3+} ion and three Cl$^-$ ions).
(b) KF$_2$ should be KF (based on one K$^+$ ion and one F$^-$ ion).
(c) Ga$_2$O$_3$ is correct.
(d) MgS is correct.

3.19 (a) Potassium sulfide
      (b) Cobalt(II) sulfate
      (c) Ammonium phosphate
      (d) Calcium hypochlorite

3.21 (a) $(NH_4)_2CO_3$ (d) $AlPO_4$
      (b) CaI_2 (e) $AgCH_3CO_2$
      (c) $CuBr_2$

3.23 Compounds with Na^+: Na_2CO_3 and NaI
      Compounds with Ba^{2+}: $BaCO_3$ and BaI_2

3.25 The force of attraction is stronger in NaF than in NaI because the distance between ion centers is smaller in NaF (235 pm) than in NaI (322 pm).

3.27 (a) Nitrogen trifluoride
      (b) Hydrogen iodide
      (c) Boron triiodide
      (d) Phosphorus pentafluoride

3.29 (a) SCl_2 (c) $SiCl_4$
      (b) N_2O_5 (d) B_2O_3

3.31 (a) 159.7 g/mol
      (b) 117.2 g/mol
      (c) 176.1 g/mol

3.33 (a) 290.8 g/mol
      (b) 249.7 g/mol

3.35 (a) 1.53 g
      (b) 4.60 g
      (c) 4.60 g

3.37 60.9 mol CH_3CN

3.39 Amount of SO_3 = 12.5 mol
      Number of molecules = 7.52×10^{24} molecules
      Number of S atoms = 7.52×10^{24} atoms
      Number of O atoms = 2.26×10^{25} atoms

3.41 (a) 86.60% Pb and 13.40% S
      (b) 81.71% C and 18.29% H
      (c) 79.96% C, 9.394% H, and 10.65% O

3.43 86.60% lead. There is 8.66 g of Pb in 10.0 g of PbS.

3.45 66.46% copper in CuS. 15.0 g of CuS is needed to obtain 10.0 g of Cu.

3.47 $C_4H_6O_4$

3.49 (a) CH, 26.0 g/mol; C_2H_2
      (b) CHO, 116.1 g/mol; $C_4H_4O_4$
      (c) CH_2, 112.2 g/mol; C_8H_{16}

3.51 Empirical formula, CH; molecular formula, C_2H_2

3.53 Empirical formula, C_3H_4; molecular formula, C_9H_{12}

3.55 Empirical and molecular formulas, $C_8H_8O_3$

3.57 Formula is $MgSO_4 \cdot 7\ H_2O$

3.59 XeF_2

3.61 ZnI_2

3.63 $(NH_4)_2CO_3$, $(NH_4)_2SO_4$, $NiCO_3$, $NiSO_4$

3.65 A strontium atom has 38 electrons, but can lose 2 electrons to form a Sr^{2+} ion. (This characteristic is shared by all Group 2A metals.) The Sr^{2+} ion has 36 electrons, the same number as in Kr.

3.67 All of these compounds have one atom of some element plus three Cl atoms. The highest weight percent of chlorine will occur in the compound having the lightest central element. Here that is B, so BCl_3 should have the highest weight percent of Cl (90.77%).

3.69 All of these compounds have one atom of some element plus one O atom. The highest weight percent of oxygen will occur in the compound having the lightest central element. Here that is C, so CO should have the highest weight percent of O (57.12%).

3.71 Borate anion has the formula BO_3^{3-}.

3.73 3.0×10^{23} molecules represents 0.50 mol adenine. The molar mass of adenine ($C_5H_5N_5$) is 135.13 g/mol, so 0.50 mol adenine has a mass of 68 g. Thus, 0.50 mol has a larger mass than 40.0 g of the compound.

3.75 2×10^{21} molecules of water

3.77 245.8 g/mol. Mass percent: 25.86% Cu, 22.80% N, 5.742% H, 13.05% S, and 32.55% O. In 10.5 g of compound there are 2.72 g Cu and 0.770 g H_2O.

3.79 Empirical formula of malic acid: $C_4H_6O_5$

3.81 FeC_2O_4

3.83 (a) $C_{10}H_{15}NO$, molar mass = 165.23 g/mol
      (b) 72.69% C
      (c) 7.57×10^{-4} mol
      (d) 4.56×10^{20} molecules and 4.56×10^{21} C atoms

3.85 Ionic compounds
      (c) Li_2S, lithium sulfide
      (d) In_2O_3, indium oxide
      (g) CaF_2, calcium fluoride

3.87 (a) NaClO, ionic   (f) $(NH_4)_2SO_3$, ionic
      (b) BI_3 (g) KH_2PO_4, ionic
      (c) $Al(ClO_4)_3$, ionic   (h) S_2Cl_2
      (d) $Ca(CH_3CO_2)_2$, ionic (i) ClF_3
      (e) $KMnO_4$, ionic   (j) PF_3

3.89

Cation	Anion	Name	Formula
Li^+	ClO_4^-	Lithium perchlorate	$LiClO_4$
Al^{3+}	PO_4^{3-}	Aluminum phosphate	$AlPO_4$
Li^+	Br^-	Lithium bromide	$LiBr$
Ba^{2+}	NO_3^-	Barium nitrate	$Ba(NO_3)_2$
Al^{3+}	O^{2-}	Aluminum oxide	Al_2O_3
Fe^{3+}	CO_3^{2-}	Iron(III) carbonate	$Fe_2(CO_3)_3$

3.91 Empirical formula, C_5H_4; molecular formula, $C_{10}H_8$

3.93 Empirical and molecular formulas, $C_5H_{14}N_2$

3.95 $C_9H_7MnO_3$

3.97 1200 kg Cr_2O_3

3.99 Empirical formula, ICl_3; molecular formula, I_2Cl_6

3.101 7.35 kg iron

3.103 (d) Na_2MoO_4

3.105 5.52×10^{-4} mol $C_{21}H_{15}Bi_3O_{12}$; 0.346 g Bi

3.107 The unknown element is carbon.

3.109 $n = 19$

3.111 (a) 0.766 g Ni or 0.0130 mol

(b) NiF_2

(c) Nickel(II) fluoride

3.113 Answer (d) is correct. The other students apparently did not correctly calculate the number of moles of material in 100.0 g or they improperly calculated the ratio of those moles in determining their empirical formula.

3.115 (a) 7.10×10^{-4} mol U was used. The empirical formula is U_3O_8, and 2.37×10^{-4} mol U_3O_8 was obtained.

(b) ^{238}U is more abundant.

(c) The formula of the hydrated compound is $UO_2(NO_3)_2 \cdot 6\ H_2O$.

3.117 First, multiply the volume of the cube (27.0 cm^3) by the density of alum (1.757 g/cm^3). This gives the mass of alum in the cube. Next, multiply this mass by the weight percent of Al in alum (5.688% Al) to give the mass of Al in the cube. Finally, convert the mass of aluminum to the amount of aluminum (mol) and then use Avogadro's number to calculate the number of Al atoms in the crystal.

$$27.0\ \text{cm}^3 \left(\frac{1.757\ \text{g alum}}{1\ \text{cm}^3} \right) \left(\frac{5.688\ \text{g Al}}{100\ \text{g alum}} \right)$$
$$\times \left(\frac{1\ \text{mol Al}}{26.98\ \text{g Al}} \right) \left(\frac{6.022 \times 10^{23}\ \text{atoms Al}}{1\ \text{mole Al}} \right)$$

Chapter 4

4.1 $C_5H_{12}(\ell) + 8\ O_2(g) \longrightarrow 5\ CO_2(g) + 6\ H_2O(g)$

4.3 (a) $4\ Cr(s) + 3\ O_2(g) \longrightarrow 2\ Cr_2O_3(s)$

(b) $Cu_2S(s) + O_2(g) \longrightarrow 2\ Cu(s) + SO_2(g)$

(c) $C_6H_5CH_3(\ell) + 9\ O_2(g) \longrightarrow 4\ H_2O(\ell) + 7\ CO_2(g)$

4.5 (a) $Fe_2O_3(s) + 3\ Mg(s) \longrightarrow 3\ MgO(s) + 2\ Fe(s)$

Reactants = iron(III) oxide, magnesium

Products = magnesium oxide, iron

(b) $AlCl_3(s) + 3\ NaOH(aq) \longrightarrow$
$$Al(OH)_3(s) + 3\ NaCl(aq)$$

Reactants = aluminum chloride, sodium hydroxide

Products = aluminum hydroxide, sodium chloride

(c) $2\ NaNO_3(s) + H_2SO_4(\ell) \longrightarrow$
$$Na_2SO_4(s) + 2\ HNO_3(\ell)$$

Reactants = sodium nitrate, sulfuric acid

Products = sodium sulfate, nitric acid

(d) $NiCO_3(s) + 2\ HNO_3(aq) \longrightarrow$
$$Ni(NO_3)_2(aq) + CO_2(g) + H_2O(\ell)$$

Reactants = nickel(II) carbonate, nitric acid

Products = nickel(II) nitrate, water

4.7 4.5 mol O_2; 310 g Al_2O_3

4.9 22.7 g Br_2; 25.3 g Al_2Br_6

4.11 (a) $4\ Fe(s) + 3\ O_2(g) \longrightarrow 2\ Fe_2O_3(s)$

(b) 3.83 g Fe_2O_3

(c) 1.15 g O_2

4.13 (a) 242 g $CaCO_3$; (b) 329 g $CaSO_4$

4.15

Equation	2 PbS(s)	+ 3 O$_2$(g)	$\longrightarrow$	2 PbO(s)	+ 2 SO$_2$(g)
Initial (mol)	2.5	0		0	0
Change (mol)	−2.5	(−3/2)(2.5)		(+2/2)(2.5)	(2/2)(2.5)
Final (mol)	0	0		2.5	2.5

The amounts table shows that 2.5 mol PbS requires $3/2(2.5) = 3.75$ mol O_2 and produces 2.5 mol PbO and 2.5 mol SO_2.

4.17 (a) Balanced equation: $4\ Cr(s) + 3\ O_2(g) \longrightarrow$
$$2\ Cr_2O_3(s)$$

(b, c) 0.175 g Cr is equivalent to 0.00337 mol

Equation	4 Cr(s)	+ 3 O$_2$(g)	$\longrightarrow$	2 Cr$_2$O$_3$(s)
Initial (mol)	0.00337	0		0
Change (mol)	−0.00337	(−3/4)(0.00337)		(2/4)(0.00337)
Final (mol)	0	0		(2/4)(0.00337)

The reaction produces $(2/4)(0.00337\ \text{mol})$ Cr_2O_3 or 0.00169 mol. This is equivalent to 0.256 g. Mass of O_2 required = 0.081 g.

4.19   0.11 mol Na_2SO_4 and 0.62 mol C are mixed. Sodium sulfate is the limiting reactant. Therefore, 0.11 mol Na_2S is formed, or 8.2 g.

4.21 F_2 is the limiting reactant.

4.23   (a) CH_4 is the limiting reactant.
        (b) 375 g H_2
        (c) Excess H_2O = 1390 g

4.25   (a) $2\ C_6H_{14}(\ell) + 19\ O_2(g) \longrightarrow$
                                $12\ CO_2(g) + 14\ H_2O(g)$
        (b) O_2 is the limiting reactant. Products are 187 g of CO_2 and 89.2 g of H_2O.
        (c) 154 g of hexane remains

4.27   (332 g/407 g)(100%) = 81.6%

4.29   (a) 14.3 g $Cu(NH_3)_4SO_4$
        (b) 88.3% yield

4.31   91.9% hydrate

4.33   84.3% $CaCO_3$

4.35   1.467% Tl_2SO_4

4.37   Empirical formula, CH

4.39   Empirical formula, CH_2; molecular formula, C_5H_{10}

4.41   Empirical formula, CH_3O, and molecular formulas, $C_2H_6O_2$

4.43 $Ni(CO)_4$

4.45   (a) $CO_2(g) + 2\ NH_3(g) \longrightarrow NH_2CONH_2(s) + H_2O(\ell)$
        (b) $UO_2(s) + 4\ HF(aq) \longrightarrow UF_4(s) + 2\ H_2O(\ell)$
            $UF_4(s) + F_2(g) \longrightarrow UF_6(s)$
        (c) $TiO_2(s) + 2\ Cl_2(g) + 2\ C(s) \longrightarrow$
                                $TiCl_4(\ell) + 2\ CO(g)$
            $TiCl_4(\ell) + 2\ Mg(s) \longrightarrow Ti(s) + 2\ MgCl_2(s)$

4.47   (a) Products = $CO_2(g)$ and $H_2O(g)$
        (b) $2\ C_6H_6(\ell) + 15\ O_2(g) \longrightarrow 12\ CO_2(g) + 6\ H_2O(g)$
        (c) 49.28 g O_2
        (d) 65.32 g products (= sum of C_6H_6 mass and O_2 mass)

4.49   71.1 mg

4.51   (a) $2\ Fe(s) + 3\ Cl_2(g) \longrightarrow 2\ FeCl_3(s)$
        (b) 19.0 g Cl_2 required; 29.0 g $FeCl_3$ produced
        (c) 63.7% yield
        (d) 15.3 g $FeCl_3$

4.53   (a) Titanium(IV) chloride, water, titanium(IV) oxide, hydrogen chloride
        (b) 4.60 g H_2O
        (c) 10.2 TiO_2, 18.6 g HCl

4.55   8.33 g NaN_3

4.57   399 g Cu

4.59   The H : B ratio is 1.4 : 1.0. The empirical formula is B_5H_7.

4.61   Empirical formula, $C_{10}H_{20}O$

4.63   (a) $FeCl_2(aq) + Na_2S(aq) \longrightarrow FeS(s) + 2\ NaCl(aq)$
        (b) $FeCl_2$
        (c) 28 g FeS produced
        (d) 15 g Na_2S remains
        (e) 40. g Na_2S requires 65 g $FeCl_2$

4.65   The metal is most likely copper, Cu.

4.67 Ti_2O_3 (which could be a mixture of TiO and TiO_2)

4.69   11.48% 2,4-D

4.71   (a) 333 kg $Na_2S_2O_4$
        (b) 369 kg of commercial product

4.73   858 kg H_2SO_4

4.75   1.59 g C_4H_{10} (55.6%) and 1.27 g C_4H_8 (44.4%)

4.77   85.4% $NaHCO_3$

4.79   (a) In the reactions represented by the sloping portion of the graph, Fe is the limiting reactant. At the point at which the yield of product begins to be constant (at 2.0 g Fe), the reactants are present in stoichiometric amounts. That is, 10.6 g of product contains 2.0 g Fe and 8.6 g Br_2.
        (b) 2.0 g Fe = 0.036 mol Fe; 8.6 g Br_2 = 0.054 mol Br_2. The mole ratio is 1.5 mol Br_2 to 1.0 mol Fe.
        (c) The mole ratio is 1.5 mol Br_2/1.0 mol Fe = 3 Br/1 Fe. The empirical formula is $FeBr_3$.
        (d) $2\ Fe(s) + 3\ Br_2(\ell) \longrightarrow 2\ FeBr_3(s)$
        (e) Iron(III) bromide
        (f) Statement (i) is correct.

4.81   (a) 65.02% Pt, 9.34% N, and 23.63% Cl
        (b) 1.31 g NH_3 required and 11.6 g $Pt(NH_3)_2Cl_2$ produced

4.83   See General ChemistryNow CD-ROM or website Screen 4.8 for the calculations.

Chapter 5

5.1   Electrolytes are compounds whose solutions conduct electricity. Substances whose solutions are good electrical conductors are strong electrolytes (such as NaCl), whereas poor electrical conductors are weak electrolytes (such as acetic acid).

5.3   (a) $CuCl_2$
       (b) $AgNO_3$
       (c) All are water-soluble.

5.5   (a) K^+ and OH^- ions   (c) Li^+ and NO_3^- ions
       (b) K^+ and SO_4^{2-} ions   (d) NH_4^+ and SO_4^{2-} ions

5.7   (a) Soluble, Na^+ and CO_3^{2-} ions
       (b) Soluble, Cu^{2+} and SO_4^{2-} ions
       (c) Insoluble
       (d) Soluble, Ba^{2+} and Br^- ions

5.9 $CdCl_2(aq) + 2\,NaOH(aq) \longrightarrow$
$$Cd(OH)_2(s) + 2\,NaCl(aq)$$
$Cd^{2+}(aq) + 2\,OH^-(aq) \longrightarrow Cd(OH)_2(s)$

5.11   (a) $NiCl_2(aq) + (NH_4)_2S(aq) \longrightarrow$
$$NiS(s) + 2\,NH_4Cl(aq)$$
$Ni^{2+}(aq) + S^{2-}(aq) \longrightarrow NiS(s)$

(b) $3\,Mn(NO_3)_2(aq) + 2\,Na_3PO_4(aq) \longrightarrow$
$$Mn_3(PO_4)_2(s) + 6\,NaNO_3(aq)$$
$3\,Mn^{2+}(aq) + 2\,PO_4^{3-}(aq) \longrightarrow Mn_3(PO_4)_2(s)$

5.13 $HNO_3(aq) \longrightarrow H^+(aq) + NO_3^-(aq)$

5.15 $H_2C_2O_4(aq) \longrightarrow H^+(aq) + HC_2O_4^-(aq)$
$HC_2O_4^-(aq) \longrightarrow H^+(aq) + C_2O_4^{2-}(aq)$

5.17 $MgO(s) + H_2O(\ell) \longrightarrow Mg(OH)_2(s)$

5.19   (a) Acetic acid reacts with magnesium hydroxide to give magnesium acetate and water.
$2\,CH_3CO_2H(aq) + Mg(OH)_2(s) \longrightarrow$
$$Mg(CH_3CO_2)_2(aq) + 2\,H_2O(\ell)$$

(b) Perchloric acid reacts with ammonia to give ammonium perchlorate.
$HClO_4(aq) + NH_3(aq) \longrightarrow NH_4ClO_4(aq)$

5.21 $Ba(OH)_2(aq) + 2\,HNO_3(aq) \longrightarrow$
$$Ba(NO_3)_2(aq) + 2\,H_2O(\ell)$$

5.23   (a) $(NH_4)_2CO_3(aq) + Cu(NO_3)_2(aq) \longrightarrow$
$$CuCO_3(s) + 2\,NH_4NO_3(aq)$$
$CO_3^{2-}(aq) + Cu^{2+}(aq) \longrightarrow CuCO_3(s)$

(b) $Pb(OH)_2(s) + 2\,HCl(aq) \longrightarrow PbCl_2(s) + 2\,H_2O(\ell)$
$Pb(OH)_2(s) + 2\,H^+(aq) + 2\,Cl^-(aq) \longrightarrow$
$$PbCl_2(s) + 2\,H_2O(\ell)$$

(c) $BaCO_3(s) + 2\,HCl(aq) \longrightarrow$
$$BaCl_2(aq) + H_2O(\ell) + CO_2(g)$$
$BaCO_3(s) + 2\,H^+(aq) \longrightarrow$
$$Ba^{2+}(aq) + H_2O(\ell) + CO_2(g)$$

5.25   (a) $AgNO_3(aq) + KI(aq) \longrightarrow AgI(s) + KNO_3(aq)$
$Ag^+(aq) + I^-(aq) \longrightarrow AgI(s)$

(b) $Ba(OH)_2(aq) + 2\,HNO_3(aq) \longrightarrow$
$$Ba(NO_3)_2(aq) + 2\,H_2O(\ell)$$
$OH^-(aq) + H^+(aq) \longrightarrow H_2O(\ell)$

(c) $2\,Na_3PO_4(aq) + 3\,Ni(NO_3)_2(aq) \longrightarrow$
$$Ni_3(PO_4)_2(s) + 6\,NaNO_3(aq)$$
$2\,PO_4^{3-}(aq) + 3\,Ni^{2+}(aq) \longrightarrow Ni_3(PO_4)_2(s)$

5.27 $FeCO_3(s) + 2\,HNO_3(aq) \longrightarrow$
$$Fe(NO_3)_2(aq) + CO_2(g) + H_2O(\ell)$$
Iron(II) carbonate reacts with nitric acid to give iron(II) nitrate, carbon dioxide, and water.

5.29   (a) Acid–base
$Ba(OH)_2(aq) + 2\,HCl(aq) \longrightarrow BaCl_2(aq) + H_2O(\ell)$
(b) Gas-forming
$2\,HNO_3(aq) + CoCO_3(s) \longrightarrow$
$$Co(NO_3)_2(aq) + H_2O(\ell) + CO_2(g)$$

(c) Precipitation
$2\,Na_3PO_4(aq) + 3\,Cu(NO_3)_2(aq) \longrightarrow$
$$Cu_3(PO_4)_2(s) + 6\,NaNO_3(aq)$$

5.31   (a) Precipitation
$MnCl_2(aq) + Na_2S(aq) \longrightarrow MnS(s) + 2\,NaCl(aq)$
$Mn^{2+}(aq) + S^{2-}(aq) \longrightarrow MnS(s)$

(b) Precipitation
$K_2CO_3(aq) + ZnCl_2(aq) \longrightarrow ZnCO_3(s) + 2\,KCl(aq)$
$CO_3^{2-}(aq) + Zn^{2+}(aq) \longrightarrow ZnCO_3(s)$

5.33   (a) Precipitation of CuS
(b) Formation of water in an acid–base reaction

5.35   (a) Br = +5 and O = −2
(b) C = +3 each and O = −2
(c) F = −1
(d) Ca = +2 and H = −1
(e) H = +1, Si = +4, and O = −2
(f) H = +1, S = +6, and O = −2

5.37   (a) Oxidation–reduction
Zn is oxidized from 0 to +2, and N in NO_3^- is reduced from +5 to +4 in NO_2.
(b) Acid–base reaction
(c) Oxidation–reduction
Calcium is oxidized from 0 to +2 in $Ca(OH)_2$, and H is reduced from +1 in H_2O to 0 in H_2.

5.39   (a) O_2 is the oxidizing agent (as it always is), so C_2H_4 is the reducing agent. As is always the case in a combustion, O_2 oxidizes the other reactant (a C-containing compound), which is reduced.
(b) Si is oxidized from 0 in Si to +4 in $SiCl_4$. Cl_2 is reduced from 0 in Cl_2 to −1 in Cl^-.

5.41 $[Na_2CO_3] = 0.254$ M; $[Na^+] = 0.508$ M; $[CO_3^{2-}] = 0.254$ M

5.43 0.494 g $KMnO_4$

5.45 5.08×10^3 mL

5.47   (a) 0.50 M NH_4^+ and 0.25 M SO_4^{2-}
(b) 0.246 M Na^+ and 0.123 M CO_3^{2-}
(c) 0.056 M H^+ and 0.056 M NO_3^-

5.49   A mass of 1.06 g Na_2CO_3 is required. After weighing out this quantity of Na_2CO_3, transfer it to a 500.-mL volumetric flask. Rinse any solid from the neck of the flask while filling the flask with distilled water. Add water until the bottom of the meniscus of the water is at the top of the scribed mark on the neck of the flask.

5.51   0.0750 M

5.53   Method (a) is correct. Method (b) gives an acid concentration of 0.15 M.

5.55 $[H^+] = 10^{-pH} = 4.0 \times 10^{-4}$ M

5.57 HNO_3 is a strong acid, so $[H^+] = 0.0013$ M. pH = 2.89.

5.59

	pH	[H⁺]	Acidic/Basic
(a)	1.00	0.10 M	Acidic
(b)	10.50	3.2×10^{-11} M	Basic
(c)	4.89	1.3×10^{-5} M	Acidic
(d)	7.64	2.3×10^{-8} M	Basic

5.61   268 mL

5.63   210 g NaOH and 190 g Cl_2

5.65   174 mL $Na_2S_2O_3$

5.67   1500 mL $Pb(NO_3)_2$

5.69   44.6 mL

5.71   1.052 M HCl

5.73   104 g/mol

5.75   12.8% Fe

5.77   (a) NaBr, KBr, or other alkali metal bromides; Group 2A bromides; other metal bromides

     (b) $Al(OH)_3$ and transition metal hydroxides

     (c) Alkaline earth carbonates ($CaCO_3$) or transition metal carbonates ($NiCO_3$)

     (d) Metal nitrates are generally water-soluble [e.g., $NaNO_3$, $Ni(NO_3)_2$].

5.79   Water-soluble: $Cu(NO_3)_2$, $CuCl_2$. Water-insoluble: $CuCO_3$, $Cu_3(PO_4)_2$.

5.81   Spectator ion, NO_3^-. Acid–base reaction.

     $2\ H^+(aq) + Mg(OH)_2(s) \longrightarrow 2\ H_2O(\ell) + Mg^{2+}(aq)$

5.83   (a) Cl_2 is reduced (to Cl^-) and Br^- is oxidized (to Br_2).

     (b) Cl_2 is the oxidizing agent and Br^- is the reducing agent.

     (c) 0.678 g Cl_2

5.85   6 g each of NaCl and Na_2CO_3

5.87   The mass of Na_2CO_3 required is 11 g. Weigh out 11 g Na_2CO_3 and place it in the 500.0-mL flask. Add a small amount of distilled water and mix until the solute dissolves. Add water until the meniscus of the solution rests at the calibrated mark on the neck of the volumetric flask. Cap the flask and swirl to ensure complete mixing.

5.89   In 0.015 M HCl, $[H^+] = 0.015$ M. In a pH 1.2 solution, $[H^+] = 0.06$ M. The pH 1.2 solution has a higher hydrogen ion concentration.

5.91   (a) $MgCO_3(s) + 2\ H^+(aq) \longrightarrow$
            $CO_2(g) + Mg^{2+}(aq) + H_2O(\ell)$

     Chloride ion (Cl^-) is the spectator ion.

     (b) Gas-forming reaction

     (c) 0.15 g

5.93   (a) H_2O, NH_3, NH_4^+, and OH^- (and a trace of H^+)

     (b) H_2O, CH_3CO_2H, $CH_3CO_2^-$, and H^+ (and a trace of OH^-)

     (c) H_2O, Na^+, and OH^- (and a trace of H^+)

     (d) H_2O, H^+, and Br^- (and a trace of OH^-)

5.95   15.0 g $NaHCO_3$ requires 1190 mL 0.15 M acetic acid. Therefore, acetic acid is the limiting reactant. (Conversely, 125 mL 0.15 M acetic acid requires only 1.58 g $NaHCO_3$.) 1.54 g $NaCH_3CO_2$ produced.

5.97   3.13 g $Na_2S_2O_3$, 96.8%

5.99   (a) Water-soluble: $Cu(NO_3)_2$ [copper(II) nitrate] or $CuCl_2$ [copper(II) chloride].

       Water-insoluble: CuS [copper(II) sulfide] or $CuCO_3$ [copper(II) carbonate]

     (b) Water-soluble: $BaCl_2$ (barium chloride) or $Ba(NO_3)_2$ (barium nitrate)

       Water-insoluble: $BaSO_4$ (barium sulfate) or barium phosphate [$Ba_3(PO_4)_2$]

5.101   (a) $Pb(NO_3)_2(aq) + 2\ KOH(aq) \longrightarrow$
            $Pb(OH)_2(s) + 2\ KNO_3(aq)$

       $Pb^{2+}(aq) + 2\ OH^-(aq) \longrightarrow Pb(OH)_2(s)$

     (b) $Cu(NO_3)_2(aq) + Na_2CO_3(aq) \longrightarrow$
            $CuCO_3(s) + 2\ NaNO_3(aq)$

       $Cu^{2+}(aq) + CO_3^{2-}(aq) \longrightarrow CuCO_3(s)$

5.103   0.029 g $NaHCO_3$

5.105   0.00263 mol HCl reacted; 0.0004 mol HCl remains in 0.500 L; pH = 3.13

5.107   3.7 spoonfuls of $NaHCO_3$ is required.

5.109   1.56 g $CaCO_3$ required; 1.00 g $CaCO_3$ remains; 1.73 g $CaCl_2$ produced

5.111 $x = 6$; $Co(NH_3)_6Cl_3$

5.113   Volume of water in the pool = 7.6×10^4 L

5.115   (a) First reaction: oxidizing agent = Cu^{2+} and reducing agent = I^-

       Second reaction: oxidizing agent = I_3^- and reducing agent = $S_2O_3^{2-}$

     (b) 67.3% copper

5.117   (a) Au, gold, has been oxidized and is the reducing agent.

       O_2, oxygen, has been reduced and is the oxidizing agent.

     (b) 26 L NaCN solution

5.119   If both students base their calculations on the amount of HCl solution pipeted into the flask (20 mL), then the second student's result will be (e) the same as the first student's. However, if the HCl concentration is calculated using the diluted solution volume, student 1 will use a volume of 40 mL and student 2 will use a volume of 80 mL in the calculation. The second student's result will be (c) half that of the first student's.

5.121   100 mL of 0.10 M HCl contains 0.010 mol HCl. This requires 0.0050 mol Zn or 3.17 g for complete reaction. Thus, in flask 2 the reaction just uses all of the Zn and produces 0.0050 mol H_2 gas. In flask 1, containing 7.00 g Zn, some Zn remains after the HCl has been consumed; 0.005 mol H_2 gas is produced. In flask 3, there is insufficient Zn, so less hydrogen is produced.

5.123   (a) Several precipitation reactions are possible:

   i. $BaCl_2(aq) + H_2SO_4(aq) \longrightarrow$
   $BaSO_4(s) + 2\ HCl(aq)$

   ii. $BaCl_2(aq) + Na_2SO_4(aq) \longrightarrow$
   $BaSO_4(s) + 2\ NaCl(aq)$

   iii. $BaCO_3(aq) + H_2SO_4(aq) \longrightarrow$
   $BaSO_4(s) + CO_2(g) + H_2O(\ell)$

   iv. $Ba(OH)_2(aq) + H_2SO_4(aq) \longrightarrow$
   $BaSO_4(s) + 2\ H_2O(\ell)$

   (b) Gas-forming reaction: reaction (iii) in part (a) is also a gas-forming reaction.

5.125   150 mg/dL. The person is intoxicated.

Chapter 6

6.1   Mechanical energy is used to move the lever, which in turns moves gears. The device produces electrical energy and radiant energy.

6.3 5.0×10^6 J

6.5   170 kcal is equivalent to 710 kJ, considerably greater than 280 kJ.

6.7   0.140 J/g · K

6.9   2.44 kJ

6.11   32.8 °C

6.13   20.7 °C

6.15   47.8 °C

6.17   0.40 J/g · K

6.19   330 kJ

6.21   49.3 kJ

6.23   273 J

6.25 9.97×10^5 J

6.27   Reaction is exothermic because ΔH°_{rxn} is negative. The heat evolved is 2.38 kJ.

6.29 3.3×10^4 kJ

6.31 $\Delta H_{rxn} = -56$ kJ/mol

6.33   0.52 J/g · K

6.35 $\Delta H = +23$ kJ/mol

6.37   297 kJ/mol SO_2

6.39 3.09×10^3 kJ/mol

6.41   0.236 J/g · K

6.43   (a) $\Delta H^\circ_{rxn} = -126$ kJ

   (b)

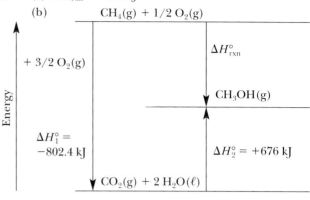

6.45 $\Delta H^\circ_{rxn} = +90.3$ kJ

6.47 $C(s) + 2\ H_2(g) + \frac{1}{2} O_2(g) \longrightarrow CH_3OH(\ell)$
   $\Delta H_f^\circ = -238.4$ kJ/mol

6.49   (a) $2\ Cr(s) + \frac{3}{2} O_2(g) \longrightarrow Cr_2O_3(s)$
   $\Delta H_f^\circ = -1134.7$ kJ/mol

   (b) 2.4 g is equivalent to 0.046 mol Cr. This will produce 26 kJ of heat energy.

6.51   (a) $\Delta H^\circ_{rxn} = -24$ kJ for 1.0 g phosphorus

   (b) $\Delta H^\circ_{rxn} = -18$ kJ for 0.2 mol NO

   (c) $\Delta H^\circ_{rxn} = -16.9$ kJ for the formation of 2.40 g NaCl(s)

   (d) $\Delta H^\circ_{rxn} = -1.8 \times 10^3$ kJ for the oxidation of 250 g iron

6.53   (a) $\Delta H^\circ_{rxn} = -906.2$ kJ (or -226.5 for 1.00 mol NH_3)

   (b) The heat evolved is 133 kJ for the oxidation of 10.0 g NH_3.

6.55   (a) $\Delta H^\circ_{rxn} = +80.8$ kJ

   (b)

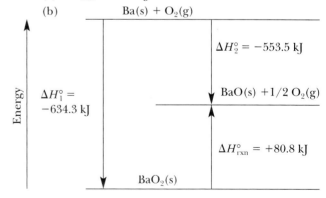

6.57 $\Delta H_f^\circ = +77.7$ kJ/mol for naphthalene

6.59   (a) $\Delta H°_{rxn} = -705.63$ kJ; reaction is expected to be product-favored

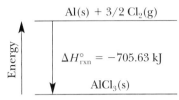

     (b) $\Delta H°_{rxn} = +90.83$ kJ; reaction is expected to be reactant-favored

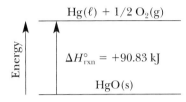

6.61   (a) Exothermic: a process in which heat is transferred from a system to its surroundings. (Heat is evolved in the combustion of methane.)

Endothermic: a process in which heat is transferred from the surroundings to the system. (Ice melting absorbs heat.)

     (b) System: the object or collection of objects being studied. (A chemical reaction—the system—taking place inside a calorimeter—the surroundings.)

Surroundings: everything outside the system that can exchange energy with the system. (The calorimeter and everything outside the calorimeter.)

     (c) Specific heat capacity: the quantity of heat required to raise the temperature of 1 gram of a substance by 1 kelvin. (The specific heat capacity of water is 4.184 J/g · K.)

     (d) State function: a quantity that is characterized by changes that do not depend on the path chosen to go from the initial state to the final state. (Enthalpy and internal energy.)

     (e) Standard state: the most stable form of a substance in the physical state that exists at a pressure of 1 bar and at a specified temperature. (The standard state of carbon at 25 °C is graphite.)

     (f) Enthalpy change, ΔH: the difference between the final and initial heat content of a substance at constant pressure. (The enthalpy change for melting ice at 0 °C is 6.00 kJ/mol.)

     (g) Standard enthalpy of formation: the enthalpy change for the formation of one mole of a compound in its standard state directly from the component elements in their standard states. ($\Delta H°_f$ for liquid water is -285.83 kJ/mol.)

6.63   (a) System: reaction between methane and oxygen.

Surroundings: the furnace and the rest of the universe.

Heat flows from the system to the surroundings.

     (b) System: water drops.

Surroundings: skin and the rest of the universe.

Heat flows from the surroundings to the system.

     (c) System: water

Surroundings: freezer and the rest of the universe.

Heat flows from the system to the surroundings.

     (d) System: reaction of aluminum and iron(III) oxide.

Surroundings: flask, laboratory bench, and rest of the universe.

Heat flows from system to the surroundings.

6.65 $\Delta E = q + w$. ΔE is the change in energy content, q is the heat transferred to or from the system, and w is the work transferred to or from the system.

6.67   Standard state of oxygen is gas, $O_2(g)$.
$O_2(g) \longrightarrow 2\,O(g)$,
$\Delta H°_{rxn} = +498.34$ kJ, endothermic
$\frac{3}{2}\,O_2(g) \longrightarrow O_3(g)$, $\Delta H°_{rxn} = +142.67$ kJ

6.69 $SnBr_2(s) + TiCl_2(s) \longrightarrow SnCl_2(s) + TiBr_2(s)$
$\Delta H°_{rxn} = -4.2$ kJ

$SnCl_2(s) + Cl_2(g) \longrightarrow SnCl_4(\ell)$ $\Delta H°_{rxn} = -195$ kJ
$TiCl_4(\ell) \longrightarrow TiCl_2(s) + Cl_2(g)$ $\Delta H°_{rxn} = +273$ kJ

$SnBr_2(s) + TiCl_4(\ell) \longrightarrow SnCl_4(\ell) + TiBr_2(s)$
$\Delta H°_{net} = +74$ kJ

6.71 $q_{water} = -8400$ kJ and $q_{ethanol} = -9800$ kJ. Ethanol sample gives up more heat than water sample.

6.73 $C_{Ag} = 0.24$ J/g · K

6.75   Mass of ice melted = 75.4 g

6.77   Final temperature = 278 K (4.8 °C)

6.79 $\Delta H_{rxn} = -69$ kJ/mol

6.81   36.0 kJ evolved per mol NH_4NO_3

6.83   (a) When summed, the following equations give the balanced equation for the formation of $B_2H_6(g)$ from the elements.

$2\,B(s) + \frac{3}{2}\,O_2(g) \longrightarrow B_2O_3(s)$ $\Delta H°_{rxn} = -1271.9$ kJ
$3\,H_2(g) + \frac{3}{2}\,O_2(g) \longrightarrow 3\,H_2O(g)$ $\Delta H°_{rxn} = -725.4$ kJ
$B_2O_3(s) + 3\,H_2O(g) \longrightarrow B_2H_6(g) + 3\,O_2(g)$
$\Delta H°_{rxn} = +2032.9$ kJ

$2\,B(s) + 3\,H_2(g) \longrightarrow B_2H_6(g)$ $\Delta H°_{rxn} = +35.6$ kJ

     (b) The enthalpy of formation of $B_2H_6(g)$ is +35.6 kJ/mol.

     (c)

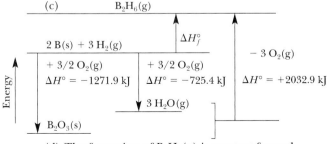

     (d) The formation of $B_2H_6(g)$ is reactant-favored.

6.85 The standard enthalpy change, ΔH°_{rxn}, is -352.88 kJ. The quantity of magnesium needed is 0.43 g.

6.87 (a) $\Delta H^\circ_{rxn} = +131.31$ kJ

(b) Reactant-favored

(c) 1.0932×10^7 kJ

6.89 Assuming $CO_2(g)$ and $H_2O(\ell)$ are the products of combustion:

ΔH°_{rxn} for isooctane is -5461.3 kJ/mol or -47.81 kJ per gram

ΔH°_{rxn} for liquid methanol is -726.77 kJ/mol or -22.682 kJ per gram

6.91 (a) Adding the equations as they are given in the question results in the desired equation for the formation of $SrCO_3(s)$. The calculated $\Delta H^\circ_{rxn} = -1220.$ kJ/mol.

(b) $Sr(s) + 1/2\,O_2(g) + C(graphite) + O_2(g)$

6.93 $\Delta H^\circ_{rxn} = -305.3$ kJ

6.95 3.28×10^4 kJ from 1.00 kg C. 1.00 kg C produces 83.3 mol each of C and H_2 gas. These produce a total heat of 4.37×10^4 kJ. Although the water gas from 1.00 kg of C produces more heat than 1.00 kg C, some carbon must be burned to provide the heat for the water gas reaction in the first place (Question 94).

6.97 Yes, the first law of thermodynamics is a version of the general principle of the conservation of energy applied specifically to a system.

6.99 (a) Product-favored

(b) Reactant-favored

6.101 (a) The temperature of the cooler object increases, and the motions of its particles are faster. The temperature of the warmer object decreases, and its particles move more slowly.

(b) The two objects have the same temperature.

6.103 The enthalpy change for each of the three reactions below is known or can be measured by calorimetry. The three equations sum to give the enthalpy of formation of $CaSO_4(s)$.

$$Ca(s) + \tfrac{1}{2}\,O_2(g) \longrightarrow CaO(s)$$
$$\Delta H^\circ_{rxn} = \Delta H^\circ_f = -635.09 \text{ kJ}$$
$$\tfrac{1}{8}\,S_8(s) + \tfrac{3}{2}\,O_2(g) \longrightarrow SO_3(g)$$
$$\Delta H^\circ_{rxn} = \Delta H^\circ_f = -395.77 \text{ kJ}$$
$$CaO(s) + SO_3(g) \longrightarrow CaSO_4(s)$$
$$\Delta H^\circ_{rxn} = -402.7 \text{ kJ}$$
$$\overline{\phantom{Ca(s) + \tfrac{1}{8}\,S_8(s) + \tfrac{3}{2}\,O_2(g) \longrightarrow CaSO_4(s)}}$$
$$Ca(s) + \tfrac{1}{8}\,S_8(s) + \tfrac{3}{2}\,O_2(g) \longrightarrow CaSO_4(s)$$

$$\Delta H^\circ_{rxn} = \Delta H^\circ_f = -1433.6 \text{ kJ}$$

6.105

Metal	Molar Heat Capacity (J/mol · K)
Al	24.2
Fe	25.1
Cu	24.5
Au	25.4

All the metals have a molar heat capacity of 24.8 J/mol · K plus or minus 0.6 J/mol · K. Therefore, assuming the molar heat capacity of Ag is 24.8 J/mol · K, its specific heat capacity is 0.230 J/g · K. This is very close to the experimental value of 0.236 J/g · K.

6.107 120 g CH_4 required

6.109 1.6×10^{11} kJ released to the surroundings. This is equivalent to 3.8×10^4 tons of dynamite.

Chapter 7

7.1 (a) Microwaves

(b) Red light

(c) Infrared

7.3 (a) Green light has a higher frequency than amber light.

(b) 5.04×10^{14} s^{-1}

7.5 Frequency = 6.0×10^{14} s^{-1}; energy per photon = 4.0×10^{-19} J; energy per mol photons = 2.4×10^5 J

7.7 302 kJ/mol photons

7.9 In order of increasing energy: FM station < microwaves < yellow light < x-rays

7.11 Light with a wavelength as long as 600 nm would be sufficient. This is in the visible region.

7.13 (a) The light of shortest wavelength has a wavelength of 253.652 nm.

(b) Frequency = 1.18190×10^{15} s^{-1}. Energy per photon = 7.83139×10^{-19} J/photon.

(c) The lines at 404 and 436 nm are in the visible region of the spectrum.

7.15 The color is violet. $n_{initial} = 6$ and $n_{final} = 2$.

7.17 (a) 10 lines possible

(b) Highest frequency (highest energy), $n = 5$ to $n = 1$

(c) Longest wavelength (lowest energy), $n = 5$ to $n = 4$

7.19 (a) $n = 3$ to $n = 2$

(b) $n = 4$ to $n = 1$

7.21 Wavelength $= 102.6$ nm and frequency $= 2.923 \times 10^{15}$ s^{-1}. Light with these properties is in the ultraviolet region.

7.23 Wavelength $= 0.29$ nm

7.25 The wavelength is 2.2×10^{-25} nm. (Calculated from $\lambda = h/m \cdot v$, where m is the ball's mass in kilograms and v is the velocity.) To have a wavelength of 5.6×10^{-3} nm, the ball would have to travel at 1.2×10^{-21} m/s.

7.27 (a) $n = 4$, $\ell = 0, 1, 2, 3$

(b) When $\ell = 2$, $m_\ell = -2, -1, 0, 1, 2$

(c) For a $4s$ orbital, $n = 4$, $\ell = 0$, and $m_\ell = 0$

(d) For a $4f$ orbital, $n = 4$, $\ell = 3$, and $m_\ell = -3, -2, -1, 0, 1, 2, 3$

7.29 Set 1: $n = 4$, $\ell = 1$, and $m_\ell = -1$

Set 2: $n = 4$, $\ell = 1$, and $m_\ell = 0$

Set 3: $n = 4$, $\ell = 1$, and $m_\ell = +1$

7.31 4 subshells. (The number of subshells in a shell is always equal to n.)

7.33 (a) ℓ must have a value no greater than $n - 1$.

(b) m_ℓ can only equal 0 in this case.

(c) m_ℓ can only equal 0 in this case.

7.35 (a) None. The quantum number set is not possible. Here m_ℓ can only equal zero.

(b) 3 orbitals

(c) 11 orbitals

(d) 1 orbital

7.37 $2d$ and $3f$ orbitals cannot exist. The $n = 2$ shell consists only of s and p subshells. The $n = 3$ shell consists only of s, p, and d subshells.

7.39 (a) For $2p$: $n = 2$, $\ell = 1$, and $m_\ell = -1, 0$, or $+1$

(b) For $3d$: $n = 3$, $\ell = 2$, and $m_\ell = -2, -1, 0, +1$, or $+2$

(c) For $4f$: $n = 4$, $\ell = 3$, and $m_\ell = -3, -2, -1, 0, +1, +2$, or $+3$

7.41 $4d$

7.43 Considering only angular nodes (the planes that pass through the nucleus):

(a) $2s$ has zero nodal surfaces.

(b) $5d$ has two nodal surfaces.

(c) $5f$ has three nodal surfaces.

7.45 (a) Correct

(b) Incorrect; the intensity of a light beam is independent of frequency and is related to the number of photons of light with a certain energy

(c) Correct

7.47 Considering only angular nodes (the planes that pass through the nucleus):

s orbital	Zero nodal surface
p orbitals	One nodal surface or plane passing through the nucleus
d orbitals	Two nodal surfaces or planes passing through the nucleus
f orbitals	Three nodal surfaces or planes passing through the nucleus

7.49

ℓ value	Orbital Type
3	f
0	s
1	p
2	d

7.51 Considering only angular nodes (the planes that pass through the nucleus):

Orbital Type	Number of Orbitals in a Given Subshell	Number of Surfaces
s	1	0
p	3	1
d	5	2
f	7	3

7.53 (a) Green light

(b) Red light has a wavelength of 680 nm, and green light has a wavelength of 500 nm.

(c) Green light has a higher frequency than red light.

7.55 (a) Wavelength $= 0.35$ m

(b) Energy $= 0.34$ J/mol

(c) Blue light (with $\lambda = 420$ nm) has an energy of 280 kJ/mol photons.

(d) Blue light has an energy (per mole of photons) that is 840,000 times greater than a mole of photons from a cell phone.

7.57 The ionization energy for He$^+$ is 5248 kJ/mol. This is four times the ionization energy for the H atom.

7.59 $1s < 2s = 2p < 3s = 3p = 3d < 4s$

In the H atom, orbitals in the same shell (e.g., $2s$ and $2p$) have the same energy.

7.61 Frequency $= 2.836 \times 10^{20}$ s^{-1} and wavelength $= 1.057 \times 10^{-12}$ m

7.63 260 s or 4.3 min

7.65   (a) size

(b) ℓ

(c) more

(d) 7 (when $\ell = 3$ these are f orbitals)

(e) one orbital

(f) (left to right) d, s, and p

(g) $\ell = 0, 1, 2, 3, 4$ (or 5 orbitals)

(h) 16 orbitals (1 s, 3 p, 5 d, and 7 f) ($= n^2$)

7.67   An electron orbiting the nucleus could occupy only certain orbits or energy levels in which it is stable. An electron in an atom will remain in its lowest energy level unless disturbed.

7.69   (c)

7.71   An experiment can be done showing that the electron can behave as a particle, and another experiment can be done showing that it has wave properties. (However, no single experiment shows both properties of the electron.) The modern view of atomic structure is based on the wave properties of the electron.

7.73   (a) and (b)

7.75   Radiation with a wavelength of 93.8 nm is sufficient to raise the electron to the $n = 6$ quantum level (see Figure 7.12). There should be 15 emission lines involving transitions from $n = 6$ to lower energy levels. (There are 5 lines for transitions from $n = 6$ to lower levels, 4 lines for $n = 5$ to lower levels, 3 for $n = 4$ to lower levels, 2 lines for $n = 3$ to lower levels, and 1 line for n = 2 to n = 1.) Wavelengths for many of the lines are given in Figure 7.12. For example, there will be an emission involving an electron moving from $n = 6$ to $n = 2$ with a wavelength of 410.2 nm.

7.77   De Broglie's equation (7.6) states that the wavelength of an object is given by h/mv. Planck's constant, h, has a very small value. A golf ball is a relatively massive object (compared with an electron). Its velocity, while small compared with an electron, is measurable. The quotient h/mv, the wavelength, will be exceedingly small—so small, in fact, that it cannot be measured.

7.79   The pickle glows because it was made by soaking a cucumber in brine, a concentrated solution of NaCl. The sodium atoms in the pickle are excited by the electric current and release energy as yellow light as they return to the ground state. Excited sodium atoms are the source of the yellow light you see in fireworks and in certain kinds of street lighting.

Chapter 8

8.1   (a) Phosphorus: $1s^2 2s^2 2p^6 3s^2 3p^3$

$1s$ $2s$ $2p$ $3s$ $3p$

The element is in the third period in Group 5A. Therefore, it has five electrons in the third shell.

(b) Chlorine: $1s^2 2s^2 2p^6 3s^2 3p^5$

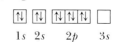
$1s$ $2s$ $2p$ $3s$ $3p$

The element is in the third period and in Group 7A. Therefore, it has seven electrons in the third shell.

8.3   (a) Chromium: $1s^2 2s^2 2p^6 3s^2 3p^6 3d^5 4s^1$

(b) Iron: $1s^2 2s^2 2p^6 3s^2 3p^6 3d^6 4s^2$

8.5   (a) Arsenic: $1s^2 2s^2 2p^6 3s^2 3p^6 3d^{10} 4s^2 4p^3$

[Ar]$3d^{10} 4s^2 4p^3$

(b) Krypton: $1s^2 2s^2 2p^6 3s^2 3p^6 3d^{10} 4s^2 4p^6 =$ [Kr]

8.7   (a) Tantalum: This is the third element in the transition series in the sixth period. Therefore, it has a core equivalent to Xe plus 2 $6s$ electrons, 14 $4f$ electrons, and 3 $5d$ electrons: [Xe]$4f^{14} 5d^3 6s^2$.

(b) Platinum: This is the eighth element in the transition series in the sixth period. Therefore, it has a core equivalent to Xe plus 2 $6s$ electrons, 14 $4f$ electrons, and 8 $5d$ electrons: [Xe]$4f^{14} 5d^8 6s^2$. The actual configuration (Table 8.3) is [Xe]$4f^{14} 5d^9 6s^1$.

8.9   Americium: [Rn]$5f^7 7s^2$ (see Table 8.3)

8.11   (a) Mg^{2+} ion

$1s$ $2s$ $2p$ $3s$

(b) K$^+$ ion

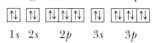
$1s$ $2s$ $2p$ $3s$ $3p$

(c) Cl$^-$ ion (Note that both Cl$^-$ and K$^+$ have the same configuration; both are equivalent to Ar.)

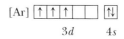
$1s$ $2s$ $2p$ $3s$ $3p$

(d) O^{2-} ion

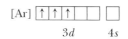
$1s$ $2s$ $2p$

8.13   (a) V (paramagnetic; three unpaired electrons)

[Ar] □□□ □ □
      $3d$ $4s$

(b) V^{2+} ion (paramagnetic, three unpaired electrons)

[Ar] □□□ □ □
      $3d$ $4s$

(c) V^{5+} ion. This ion has an electron configuration equivalent to argon. It is diamagnetic, with no unpaired electrons.

8.15   (a) Manganese

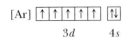
         3d   4s

(b) Manganese(II) ion, Mn^{2+}

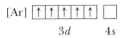
         3d   4s

(c) The 2+ ion is paramagnetic to the extent of five unpaired electrons.

(d) 5

8.17   (a) The spin quantum number cannot be 0. The set is correct if $m_s = \pm 1/2$.

(b) m_ℓ cannot be larger than ℓ. The set is correct if $m_\ell = -1, 0,$ or $+1$.

(c) ℓ can be no larger than $n - 1$. The set is correct if $\ell = 1$ or 2.

8.19   (a) 14

(b) 2

(c) 0 (because ℓ cannot equal n)

8.21   Magnesium: $1s^2 2s^2 2p^6 3s^2$

1s 2s   2p   3s

Quantum numbers for the two electrons in the 3s orbital:

$n = 3, \ell = 0, m_\ell = 0, m_s = +1/2$

$n = 3, \ell = 0, m_\ell = 0, m_s = -1/2$

8.23   Gallium: $1s^2 2s^2 2p^6 3s^2 3p^6 3d^{10} 4s^2 4p^1$

[Ar] ⬆⬇⬆⬇⬆⬇⬆⬇⬆⬇ ⬆⬇ ⬆ ☐☐
         3d   4s   4p

Quantum numbers for the 4p electron:

$n = 4, \ell = 1, m_\ell = 1, m_s = +1/2$

8.25   Increasing size: C < B < Al < Na < K

8.27   (a) Cl^-

(b) Al

(c) In

8.29   (c)

8.31   (a) Largest radius, Na

(b) Most negative electron affinity: O

(c) Ionization energy: Na < Mg < P < O

8.33   (a) Increasing ionization energy: S < O < F. S is less than O because the IE decreases down a group. F is greater than O because IE generally increases across a period.

(b) Largest IE: O. See part (a).

(c) Most negative electron affinity: Cl. Electron affinity becomes increasingly more negative across the periodic table and on ascending a group.

(d) Largest I radius, O^{2-}

8.35   (a) Drawing (a) is a ferromagnetic solid, (b) is a diamagnetic solid, and (c) is a paramagnetic solid.

(b) Substance (a) would be most strongly attracted to a magnet, whereas (b) would be least strongly attracted.

8.37   Uranium configuration: $[Rn]5f^3 6d^1 7s^2$

         5f   6d   7s

Uranium(IV ion, U^{4+}): $[Rn]5f^2$

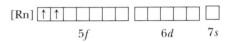
         5f   6d   7s

Both U and U^{4+} are paramagnetic.

8.39   (a) Atomic number = 20

(b) Total number of s electrons = 8

(c) Total number of p electrons = 12

(d) Total number of d electrons = 0

(e) The element is Ca, calcium, a metal.

8.41   (b) The maximum value of ℓ is $(n - 1)$.

8.43   (a) Neodymium, Nd: $[Xe]4f^4 6s^2$ (Table 8.3)

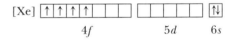
         4f   5d   6s

Iron, Fe: $[Ar]3d^6 4s^2$

[Ar] ⬆⬇⬆⬆⬆⬆ ⬆⬇
         3d   4s

Boron, B: $1s^2 2s^2 2p^1$

⬆⬇ ⬆⬇ ⬆☐☐
1s 2s   2p

(b) All three elements have unpaired electrons and so should be paramagnetic.

(c) Neodymium(III) ion, Nd^{3+}: $[Xe]4f^3$

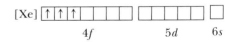
         4f   5d   6s

Iron(III) ion, Fe^{3+}: $[Ar]3d^5$

[Ar] ⬆⬆⬆⬆⬆ ☐
         3d   4s

Both neodymium(III) and iron(III) have unpaired electrons and are paramagnetic.

8.45   K < Ca < Si < P

8.47   (a) metal   (c) B

(b) B   (d) A

8.49 In^{4+}: Indium has three outer shell electrons, so it is unlikely to form a 4+ ion.

Fe^{6+}: Although iron has eight electrons in its $3d$ and $4s$ orbitals, so ions with a 6+ charge are highly unlikely. The ionization energy is too large.

Sn^{5+}: Tin has four outer shell electrons, so it is unlikely to form a 5+ ion.

8.51   (a) Se   (d) N
(b) Br$^-$ (e) N^{3-}
(c) Na

8.53   (a) Na   (c) Na < Al < B < C
(b) C

8.55   (a) Cobalt   (c) 4 unpaired electrons
(b) Paramagnetic

8.57   Li has 3 electrons ($1s^2 2s^1$) and Li$^+$ has only two electrons ($1s^2$). The ion is smaller than the atom because there are only two electrons to be held by three protons in the ion. Also, an electron in a larger orbital has been removed. Fluorine atoms have 9 electrons and 9 protons ($1s^2 2s^2 2p^5$). The anion, F$^-$, has one additional electron, which means that 10 electrons must be held by only 9 protons, and the ion is larger than the atom.

8.59   K ($[Ar]4s^1$) $\longrightarrow$ K$^+$ ($[Ar]$)   IE = 419 kJ/mol
K$^+$ ($[Ar]$) $\longrightarrow$ K^{2+} ($[Ne]3s^2 3p^5$)   IE = 3051 kJ/mol

The second electron must be removed from a positive ion and is a core electron, whereas the first electron is removed from a neutral atom.

8.61   (a) In going from one element to the next across the period, the effective nuclear charge increases slightly and the attraction between the nucleus and the electrons increases. (See the General ChemistryNow CD-ROM or website Screen 8.9.)

(b) The size of fourth-period transition elements, for example, is a reflection of the size of the $4s$ orbital. As d electrons are added across the series, protons are added to the nucleus. Adding protons should lead to a decreased atom size, but the effects of the protons are balanced by $3d$ electrons, and the atom size is changed little.

8.63   Arguments for a compound composed of Mg^{2+} and O^{2-}:

(a) Chemical experience suggests that all Group 2A elements form 2+ cations, and that oxygen is typically the O^{2-} ion in its compounds.

(b) Other alkaline earth elements form oxides such as BeO, CaO, and BaO.

A possible experiment is to measure the melting point of the compound. An ionic compound such as NaF (with ions having 1+ and 1− charges) melts at 990 °C, whereas a compound analogous to MgO, CaO, melts at a much higher temperature (2580 °C).

8.65   (a) The effective nuclear charge increases, causing the valence orbital energies to become more negative on moving across the period.

(b) As the valence orbital energies become more negative, it is increasingly difficult to remove an electron from the atom, and the IE increases. Toward the end of the period, the orbital energies have become so negative that removing an electron requires significant energy. Instead, the effective nuclear charge has reached the point that it is energetically more favorable for the atom to gain an electron.

(c) Valence orbital energies:
Li (-530.7 kJ) < Be (-897.3 kJ) > B (-800.8 kJ) < C (-1032 kJ)

It is more difficult to remove an electron from Be than from either Li or B. The energy is more negative for C than for B, so it is more difficult to remove an electron from C than from B.

8.67   The size declines across this series of elements while the mass increases. Thus, the mass per volume—the density—increases.

8.69   (a) Element 113: $[Rn]5f^{14}6d^{10}7s^2 7p^1$
Element 115: $[Rn]5f^{14}6d^{10}7s^2 7p^3$

(b) Element 113 is in Group 3A (with elements such as boron and aluminum), and element 115 is in Group 5A (with elements such as nitrogen and phosphorus).

(c) Americium ($Z = 95$) + argon ($Z = 18$) = element 113

8.71   (a) Sulfur electron configuration

↑↓	↑↓	↑↓ ↑↓ ↑↓	↑↓	↑↓ ↑ ↑
$1s$	$2s$	$2p$	$3s$	$3p$

(b) $n = 3$, $\ell = 1$, $m_\ell = 1$, $m_s = +1/2$

(c) S has the smallest ionization energy and O has the smallest radius.

(d) S is smaller than the S^{2-} ion.

(e) 584 g SCl$_2$

(f) 10.0 g SCl$_2$ is the limiting reactant, and 11.6 g SOCl$_2$ can be produced.

(g) ΔH_f° [SCl$_2$(g)] = -17.6 kJ/mol

8.73

Atom Distance	Calculated (pm)	Measured (pm)
B—F	154	130
P—F	6	178
C—H	114	109
C—O	143	150

With the exception of B—F, the agreement is quite good.

Chapter 9

9.1   (a) Group 6A, 6 valence electrons

(b) Group 3A, 3 valence electrons

(c) Group 1A, 1 valence electron

(d) Group 2A, 2 valence electrons

(e) Group 7A, 7 valence electrons

(f) Group 6A, 6 valence electrons

9.3   Group 3A, 3 bonds

Group 4A, 4 bonds

Group 5A, 3 bonds (for a neutral compound)

Group 6A, 2 bonds (for a neutral compound)

Group 7A, 1 bond (for a neutral compound)

9.5   Most negative, MgS; least negative, KI

9.7   Increasing lattice energy: RbI < LiI < LiF < CaO

9.9   As the ion–ion distance decreases, the force of attraction between ions increases. This should make the lattice more stable, and more energy should be required to melt the compound.

9.11   (a) NF_3, 26 valence electrons

:F—N—F:
|
:F:

(b) ClO_3^-, 26 valence electrons

[:O—Cl—O:]⁻
|
:O:

(c) HOBr, 14 valence electrons

H—O—Br:

(d) SO_3^{2-}, 26 valence electrons

[:O—S—O:]²⁻
|
:O:

9.13   (a) $CHClF_2$, 26 valence electrons

H
|
:Cl—C—F:
|
:F:

(b) CH_3CO_2H, 24 valence electrons

H :O:
| ||
H—C—C—O—H
|
H

(c) CH_3CN, 16 valence electrons

H
|
H—C—C≡N:
|
H

(d) H_2CCCH_2, 16 valence electrons

H   H
| |
H—C=C=C—H

9.15   (a) SO_2, 18 valence electrons

:O—S=O ⟷ O=S—O:

(b) NO_2^-, 18 valence electrons

[:O—N=O]⁻ ⟷ [O=N—O:]⁻

(c) SCN^-, 16 valence electrons

[S=C=N]⁻ ⟷ [:S≡C—N:]⁻ ⟷ [:S—C≡N:]⁻

9.17   (a) BrF_3, 28 valence electrons   (b) I_3^-, 22 valence electrons

:F:
|
:Br—F:
|
:F:

[:I: I: :I:]⁻

(c) XeO_2F_2, 34 valence electrons   (d) XeF_3^+, 28 valence electrons

:F:
|
:O—Xe—O:
|
:F:

:F:
|
:Xe—F:
|
:F:

9.19   (a) Electron-pair geometry around N is tetrahedral. Molecular geometry is trigonal pyramidal.

:Cl—N—H
|
H

(b) Electron-pair geometry around O is tetrahedral. Molecular geometry is bent.

:Cl—O—Cl:

(c) Electron-pair geometry around C is linear. Molecular geometry is linear.

[S=C=N]⁻

(d) Electron-pair geometry around O is tetrahedral. Molecular geometry is bent.

H—O—F:

9.21   (a) Electron-pair geometry around C is linear. Molecular geometry is linear.

O=C=O

(b) Electron-pair geometry around N is trigonal planar. Molecular geometry is bent.

$$\left[\ddot{\text{O}}-\text{N}=\ddot{\text{O}}\right]^{-}$$

(c) Electron-pair geometry around O is trigonal planar. Molecular geometry is bent.

$$\ddot{\text{O}}=\ddot{\text{O}}-\ddot{\text{O}}:$$

(d) Electron-pair geometry around Cl atom is tetrahedral. Molecular geometry is bent.

$$\left[:\ddot{\text{O}}-\ddot{\text{Cl}}-\ddot{\text{O}}:\right]^{-}$$

All have two atoms attached to the central atom. As the bond and lone pairs vary, the molecular geometries vary from linear to bent.

9.23 (a) Electron-pair geometry around Cl is trigonal bipyramidal. Molecular geometry is linear.

$$\left[:\ddot{\text{F}}-\overset{..}{\text{Cl}}-\ddot{\text{F}}:\right]^{-}$$

(b) Electron-pair geometry around Cl is trigonal bipyramidal. Molecular geometry is T-shaped.

$$:\ddot{\text{F}}-\overset{|}{\underset{\underset{:\ddot{\text{F}}:}{|}}{\text{Cl}}}-\ddot{\text{F}}:$$

(c) Electron-pair geometry around Cl is octahedral. Molecular geometry is square planar.

$$\left[\begin{array}{c}:\ddot{\text{F}}:\\ :\ddot{\text{F}}-\overset{|}{\underset{|}{\text{Cl}}}-\ddot{\text{F}}:\\ :\ddot{\text{F}}:\end{array}\right]^{-}$$

(d) Electron-pair geometry around Cl is octahedral. Molecular geometry is square pyramidal.

$$\ddot{\text{F}}_{\text{in}}\overset{:\ddot{\text{F}}:}{\underset{\ddot{\text{F}}}{\overset{|}{\text{Cl}}}}{}^{\text{F}:}$$

9.25 (a) Ideal O—S—O angle = 120°

(b) 120°

(c) 120°

(d) H—C—H = 109° and C—C—N angle = 180°

9.27 1 = 120°; 2 = 109°; 3 = 120°; 4 = 109°; 5 = 109°

9.29 (a) N = 0; H = 0

(b) P = +1; O = −1

(c) B = −1; H = 0

(d) All are zero.

9.31 (a) N = +1; O = 0.

(b) The central N is 0. The singly bonded O atom is −1, and the doubly bonded O atom is 0.

$$\left[:\ddot{\text{O}}-\text{N}=\ddot{\text{O}}\right]^{-} \longleftrightarrow \left[\ddot{\text{O}}=\text{N}-\ddot{\text{O}}:\right]^{-}$$

(c) N and F are both 0.

(d) The central N atom is +1, one of the O atoms is −1, and the other two O atoms are both 0.

$$\text{H}-\overset{0}{\ddot{\text{O}}}-\overset{+1}{\text{N}}=\overset{0}{\ddot{\text{O}}}$$
$$\underset{\underset{-1}{:\ddot{\text{O}}:}}{|}$$

9.33 (a) $\overset{\longrightarrow}{\underset{+\delta\;\;-\delta}{\text{C}-\text{O}}}$ $\overset{\longrightarrow}{\underset{+\delta\;\;-\delta}{\text{C}-\text{N}}}$ (c) $\overset{\longrightarrow}{\underset{+\delta\;\;-\delta}{\text{B}-\text{O}}}$ $\overset{\longrightarrow}{\underset{+\delta\;\;-\delta}{\text{B}-\text{S}}}$

CO is planar   BO is more polar

(b) $\overset{\longrightarrow}{\underset{+\delta\;\;-\delta}{\text{P}-\text{Cl}}}$ $\overset{\longrightarrow}{\underset{+\delta\;\;-\delta}{\text{P}-\text{Br}}}$ (d) $\overset{\longrightarrow}{\underset{+\delta\;\;-\delta}{\text{B}-\text{F}}}$ $\overset{\longrightarrow}{\underset{+\delta\;\;-\delta}{\text{B}-\text{I}}}$

PCl is more polar   BF is more polar

9.35 (a) CH and CO bonds are polar.

(b) The CO bond is most polar, and O is the most negative atom.

9.37 (a) OH⁻: The formal charge on O is −1 and on H is 0.

(b) BH₄⁻: Even though the formal charge on B is −1 and on H is 0, H is slightly more electronegative than B. The four H atoms are therefore more likely to bear the −1 charge of the ion. The BH bonds are polar, with the H atom being the negative end.

(c) The CH and CO bonds are all polar (but the C—C bond is not). The negative charge in the CO bonds lies on the O atoms.

9.39 Structure C is most reasonable. The charges are as small as possible and the negative charge resides on the more electronegative atom.

$$\underset{\text{A}}{:\overset{-2}{\text{N}}-\overset{+1}{\text{N}}\equiv\overset{+1}{\text{O}}:} \longleftrightarrow \underset{\text{B}}{\overset{-1}{\text{N}}=\overset{+1}{\text{N}}=\overset{0}{\ddot{\text{O}}}} \longleftrightarrow \underset{\text{C}}{:\overset{0}{\text{N}}\equiv\overset{+1}{\text{N}}-\overset{-1}{\ddot{\text{O}}}:}$$

9.41 If an H⁺ ion were to attack NO₂⁻, it would attach to an O atom.

$$\left[\overset{-1}{:\ddot{\text{O}}}-\overset{0}{\text{N}}=\overset{0}{\ddot{\text{O}}}\right]^{-} \longleftrightarrow \left[\overset{0}{\ddot{\text{O}}}=\overset{0}{\text{N}}-\overset{-1}{\ddot{\text{O}}}:\right]^{-}$$

9.43 (i) The most polar bonds are in H₂O (because O and H have the largest difference in electronegativity).

(ii) Not polar: CO₂ and CCl₄

(iii) F

9.45 (a) Not polar; linear molecule

(b) HBF₂, polar trigonal-planar molecule with F atoms at the negative end of the dipole and the H atom at the positive end.

(c) CH₃Cl, polar tetrahedral molecule. The Cl atom is the negative end of the dipole, and the three H atoms are on the positive side of the molecule.

(d) SO₃, nonpolar trigonal-planar molecule

9.47 (a) C—H bonds, bond order is 1; 1 C=O bond, bond order is 2

(b) 3 S—O single bonds, bond order is 1

(c) 2 nitrogen-oxygen double bonds, bond order is 2

(d) 1 N=O double bond, bond order is 2; 1 N—Cl bond, bond order is 1

9.49 (a) B—Cl (c) P—O

(b) C—O (d) C=O

9.51 NO bond orders: 2 in NO_2^+, 1.5 in NO_2^-; 1.33 in NO_3^-, The NO bond is longest in NO_3^- and shortest in NO_2^+.

9.53 The CO bond in carbon monoxide is a triple bond, so it is both shorter and stronger than the CO double bond in H_2CO.

9.55 $\Delta H°_{rxn} = -126$ kJ

9.57 O—F bond dissociation energy = 192 kJ/mol

9.59

Element	Valence Electrons
Li	1
Ti	4
Zn	2
Si	4
Cl	7

9.61 Ionic: KI and MgS

Covalent: CS_2 and P_4O_{10}

9.63 Group 2A elements form 2+ ions (such as Ca^{2+}), and Group 7A elements form 1− ions when combined with a metal. Therefore, only $CaCl_2$ is a reasonable formula.

9.65 SeF_4, BrF_4^-, XeF_4

9.67 The C—H bonds in C_2H_2 have a bond order of 1, whereas the carbon–carbon bond has an order of 3. In phosgene the C—Cl bonds are single bonds, whereas the C=O bond is a double bond with an order of 2.

9.69 NO bond order in NO_3^- is 1.33.

9.71 To estimate the enthalpy change we need energies for the following bonds: O=O, H—H, and H—O.

Energy to break bonds = 498 kJ (for O=O) + 2 × 436 kJ (for H—H) = +1370 kJ.

Energy evolved when bonds are made = 4 × 463 kJ (for O—H) = −1852 kJ

Total energy = −482 kJ

9.73 All the molecules in the series have 16 valence electrons and all are linear.

(a)

(b)

(c)

9.75 The N—O bonds in NO_2^- have a bond order of 1.5, while in NO_2^+ the bond order is 2. The shorter bonds (110 pm) are the NO bonds with the higher bond order (in NO_2^+), whereas the longer bonds (124 pm) in NO_2^- have a lower bond order.

9.77 The F—Cl—F bond angle in ClF_2^+, which has a tetrahedral electron-pair geometry, is approximately 109°.

The ClF_2^- ion has a trigonal-bipyramidal electron-pair geometry, with F atoms in the axial positions and the lone pairs in the equatorial positions. Therefore, the F—Cl—F angle is 180°.

9.79 An H^+ ion will attach to an O atom of SO_3^{2-} and not to the S atom. Each O atom has a formal charge of −1, whereas the S atom has a formal charge of +1.

9.81 (a) Calculation from bond energies: $\Delta H°_{rxn} = -509$ kJ/mol CH_3OH

(b) Calculation from thermochemical data: $\Delta H°_{rxn} = -676$ kJ/mol CH_3OH

9.83 (a)

(b) The third resonance structure is the most reasonable because the negative formal charge is on the most electronegative atom.

(c) Carbon, the least electronegative element in the ion, has a negative formal charge. In addition, all three resonance structures have an unfavorable charge distribution.

9.85

(a) XeF_2 has three lone pairs around the Xe atom. The electron-pair geometry is trigonal bipyramidal. Because lone pairs require more space than bond pairs, it is better to place the lone pairs in the equator of the bipyramid, where the angles between them are 120°.

(b) Like XeF_2, ClF_3 has a trigonal-bipyramidal electron-pair geometry, but only two lone pairs around the Cl. These are better placed in the equatorial plane, where the angle between them is 120°.

9.87 (a) Angle 1 = 109°; angle 2 = 120°; angle 3 = 109°; angle 4 = 109°; angle 5 = 109°

(b) O—H bonds

9.89 $\Delta H_{rxn} = +146 \text{ kJ} = 2 (D_{C-N}) + D_{C=O} - [D_{N-N} + D_{C=O}]$

9.91 (a) Two C—H bonds and one O=O bond are broken and two O—C bonds and two H—O bonds are made in the reaction. $\Delta H_{rxn} = -318 \text{ kJ}$.

(b) Acetone is polar.

(c) The O—H hydrogen atoms are the most positive in dihydroxyacetone.

9.93 (a) The C=C bond is stronger than the C—C bond.

(b) The C—C single bond is longer than the C=C double bond.

(c) Ethylene is nonpolar, whereas acrolein is polar.

(d) The reaction is exothermic ($\Delta H_{rxn} = -45 \text{ kJ}$).

9.95 $\Delta H_{rxn} = -211 \text{ kJ}$

9.97 (a) Angle 1 = 109°; angle 2 = 120°; angle 3 = 120°; angle 4 = 109°; angle 5 = 109°

(b) The OH bonds are the most polar bonds in the molecule.

9.99 The molecule can have a pyramidal structure if there are three bond pairs and one lone pair at the corners of a tetrahedron (e.g., NH_3). The bond angles are likely to be slightly less than 109°. The molecule can have a bent structure with two lone pairs and two bond pairs. The bond angle is likely to be less than 109° (e.g., H_2O).

9.101 (a) Odd-electron molecules: BrO (13 electrons) and OH (7 electrons)

(b) $Br_2(g) \longrightarrow 2 Br(g)$ $\Delta H_{rxn} = +193 \text{ kJ}$

$2 Br(g) + O_2(g) \longrightarrow 2 BrO(g)$ $\Delta H_{rxn} = +96 \text{ kJ}$

$BrO(g) + H_2O(g) \longrightarrow HOBr(g) + OH(g)$
                    $\Delta H_{rxn} = 0 \text{ kJ}$

(c) $\Delta H [HOBr(g)] = -101 \text{ kJ/mol}$

(d) The reactions in part (b) are endothermic (or thermal-neutral), and the heat of formation in part (c) is exothermic.

9.103 Lattice energy depends directly on ion charges and inversely on the distance between ions. The sizes of the Cl^-, Br^-, and I^- ions fall in a relatively narrow range (181, 196, and 220 pm, respectively), and the ion sizes change by only 15–24 pm from one ion to the next. Therefore, their lattice energies are expected to decrease in a narrow range. The F^- ion (133 pm), is only 74% as large as the Cl^- ion, so the lattice energy of NaF is much more negative.

9.105 (a) BF_3 is not a polar molecule, and replacing one of two F atoms with an H atom (HBF_2 and H_2BF) gives polar molecules.

(b) $BeCl_2$ is not polar, whereas replacing a Cl atom with a Br atom, gives a polar molecule (BeClBr).

Chapter 10

10.1 The electron-pair and molecular geometries of $CHCl_3$ are both tetrahedral. An sp^3 hybrid orbital on the C atom overlaps a p orbital on a Cl atom to form a sigma bond. A C—H sigma bond is formed by a C atom orbital overlapping an H atom $1s$ orbital.

$$\begin{array}{c} \quad :\ddot{C}l: \\ \quad | \\ H-\underset{\displaystyle |}{C}-\ddot{C}l: \\ \quad :\ddot{C}l: \end{array}$$

10.3 (a) sp^2 (c) sp^3

(b) sp (d) sp^2

10.5 (a) C, sp^3; O, sp^3

(b) CH_3, sp^3; middle C, sp^2; CH_2, sp^2

(c) CH_2, sp^3; CO_2H, sp^2; N, sp^3

10.7 (a) Electron-pair geometry is octahedral. Molecular geometry is octahedral. Si: sp^3d^2.

(b) Electron-pair geometry is trigonal bipyramidal. Molecular geometry is seesaw. Se: sp^3d.

(c) Electron-pair geometry is trigonal bipyramidal. Molecular geometry is linear. I: sp^3d.

(d) Electron-pair geometry is octahedral. Molecular geometry is square planar. Xe: sp^3d^2

$$\begin{array}{c} :\ddot{F}_{\text{\tiny{II}}} \quad | \quad \ddot{F}: \\ \quad \diagdown Xe \diagup \\ :F \quad \diagup \;| \;\diagdown \quad \ddot{F}: \end{array}$$

10.9 There are 32 valence electrons in both HPO_2F_2 and its anion. Both have a tetrahedral molecular geometry, so the P atom in both is sp^3 hybridized.

$$\begin{array}{cc} \quad :\ddot{O}: & \qquad\quad :\ddot{O}: \\ \quad || & \qquad\quad || \\ H-\ddot{O}-P\cdots\cdots\ddot{F}: & \quad \left[:\ddot{O}-P\cdots\cdots\ddot{F}: \right]^- \\ \qquad | & \qquad\qquad | \\ \qquad :\ddot{F}: & \qquad\qquad :\ddot{F}: \end{array}$$

10.11 The C atom is sp^2 hybridized. Two of the sp^2 hybrid orbitals are used to form C—Cl sigma bonds, and the third is used to form the C—O sigma bond. The p orbital not used in the C atom hybrid orbitals is used to form the CO pi bond.

10.13

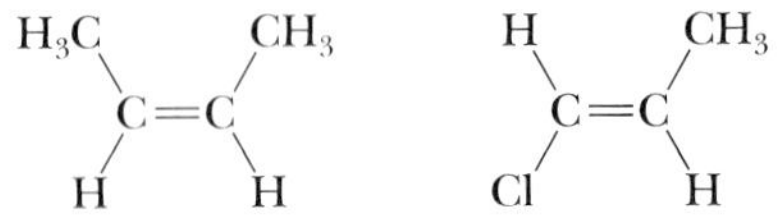

cis isomer   *trans* isomer

10.15 H_2^+ ion: $(\sigma_{1s})^1$. Bond order = 0.5. The bond in H_2^+ is weaker than in H_2 (bond order = 1).

10.17 MO diagram for C_2^{2-} ion:

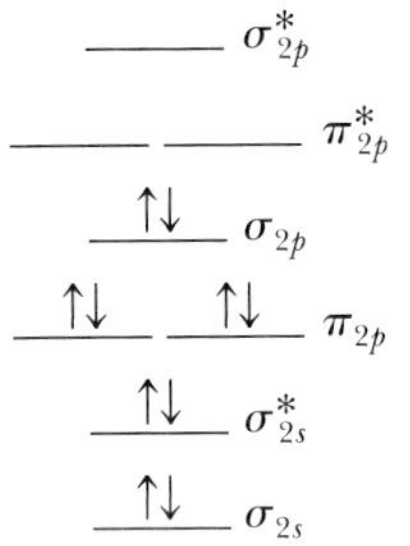

The ion has 10 valence electrons (isoelectronic with N_2). There is one net sigma bond and two net pi bonds, for a bond order of 3. The bond order increases by 1 on going from C_2 to C_2^{2-}. The ion is not paramagnetic.

10.19 (a) CO has 10 valence electrons.
  $[\text{core}](\sigma_{2s})^2(\sigma_{2s}^*)^2(\pi_{2p})^4(\sigma_{2p})^2$
  (b) HOMO, σ_{2p}
  (c) Diamagnetic
  (d) 1 σ bond and 2 π bonds; bond order is 3

10.21

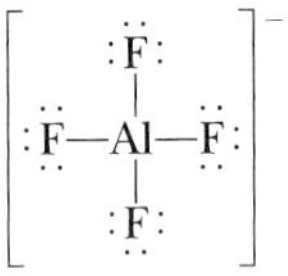

The electron-pair and molecular geometries are both tetrahedral. The Al atom is sp^3 hybridized, and so the Al—F bonds are formed by overlap of an Al sp^3 orbital with a p orbital on the F atom.

10.23

Molecule/Ion	O—S—O Angle	Hybrid Orbitals
SO_2	120°	sp^2
SO_3	120°	sp^2
SO_3^{2-}	109°	sp^3
SO_4^{2-}	109°	sp^3

10.25

The electron-pair geometry is trigonal planar. The molecular geometry is bent or angular. The O—N—O

angle will be about 120°, the average N—O bond order is 3/2, and the N atom is sp^2 hybridized.

10.27 The resonance structures of N_2O, with formal charges, are shown here.

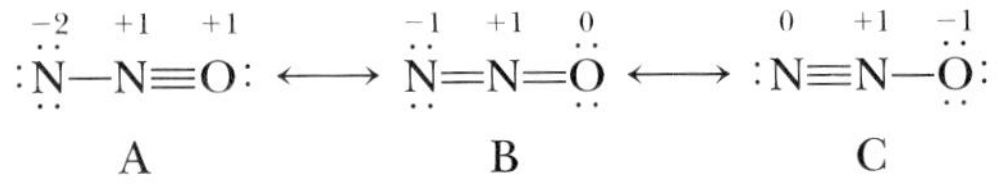

The central N atom is sp hybridized in all structures.

10.29 (a) All three have the formula C_2H_4O. They are usually referred to as structural isomers.
  (b) Ethylene oxide: both C atoms are sp^3 hybridized, and the bond angles in the ring are only 60° (which makes this a relatively unstable molecule).
  Acetaldehyde: The CH_3 carbon atom has sp^3 hybridization (bond angles of 109°), and the other C atom is sp^2 hybridized (bond angles of 120°).
  Vinyl alcohol: Both C atoms are sp^2 hybridized, and the bond angles are 120°.
  (c) H—C—H angles in ethylene oxide and acetaldehyde = about 109°. Angle in vinyl alcohol = 120°.
  (d) All are polar.
  (e) Acetaldehyde has the strongest CO bond, and vinyl alcohol has the strongest C—C bond.

10.31 (a) CH_3 carbon atom: sp^3
  C=N carbon atom: sp^2
  N atom: sp^2
  (b) C—N—O bond angle = 120°

10.33 (a) C(1) = sp^2; O(2) = sp^3; N(3) = sp^3; C(4) = sp^3; P(5) = sp^3
  (b) Angle A = 120°; angle B = 109°; C = 109°; angle D = 109°
  (c) The P—O and O—H bonds are most polar ($\Delta\chi = 1.4$).

10.35 (a) The geometry about the boron atom is trigonal planar in BF_3, but tetrahedral in $H_3N—BF_3$.
  (b) Boron is sp^2 hybridized in BF_3 but sp^3 hybridized in $H_3N—BF_3$.
  (c) Yes

10.37 (a) Then C=O bond is most polar.
  (b) 18 sigma bonds and 5 pi bonds
  (c)

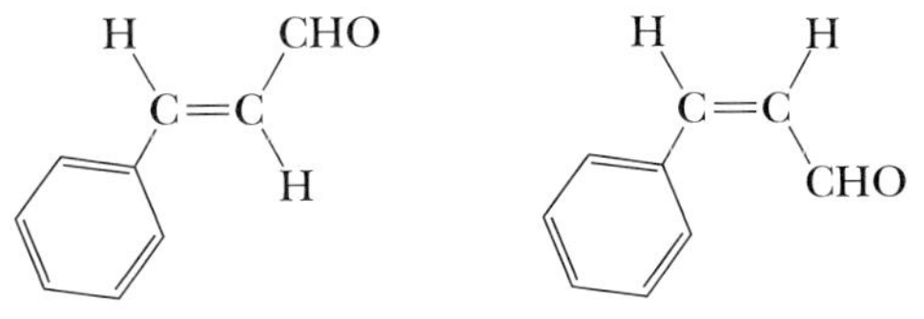

trans isomer   *cis* isomer

  (d) All C atoms are sp^2 hybridized.
  (e) All bond angles are 120°.

10.39   (a) The Sb in SbF_5 is sp^3d hybridized, whereas it is sp^3d^2 hybridized in SbF_6^-.

(b) The molecular geometry of the H_2F^+ ion is bent or angular, and the F atom is sp^3 hybridized.

10.41   (a) The peroxide ion has a bond order of 1.

(b) [core electrons] $(\sigma_{2s})^2(\sigma_{2s}^*)^2(\pi_{2p})^4(\sigma_{2p})^2(\pi_{2p}^*)^4$

This configuration also leads to a bond order of 1.

(c) Both theories lead to a diamagnetic ion with a bond order of 1.

10.43   See Table 10.1 on page 463 for the paramagnetism of diatomic molecules.

(a) Paramagnetic diatomic molecules: B_2 and O_2

(b) Bond order of 1: Li_2, B_2, F_2

(c) Bond order of 2: C_2 and O_2

(d) Highest bond order: N_2

10.45   CN has 9 valence electrons.

[core electrons] $(\sigma_{2s})^2(\sigma_{2s}^*)^2(\pi_{2p})^4(\sigma_{2p})^1$

(a) HOMO, σ_{2p}

(b, c) Bond order = 2.5 (0.5 σ bond and 2 π bonds)

(d) Paramagnetic

10.47   (a) All C atoms are sp^3 hybridized.

(b) About 109°

(c) Polar

(d) The six-membered ring cannot be planar owing to the tetrahedral C atoms of the ring. The bond angles are all 109°.

10.49   (a) The keto and enol forms are not resonance structures because both electron pairs and atoms have been rearranged.

(b) In the enol form, the terminal —CH_3 carbon atoms are sp^3 hybridized and the central C atoms are sp^2 hybridized. In the keto form, the terminal —CH_3 carbon atoms and the central C atom are sp^3 hybridized and the two C=O carbon atoms are sp^2 hybridized.

(c) Enol form: The —CH_3 groups have tetrahedral electron-pair and molecular geometries. The other three C atoms all have trigonal-planar electron-pair and molecular geometries.

Keto form: The —CH_3 groups and the central C atom have tetrahedral electron-pair and molecular geometries. The other two C atoms have trigonal-planar electron-pair and molecular geometries.

(d)

(e) *Cis-trans* isomerism is possible in the enol form.

10.51   A C atom may form, at most, four hybrid orbitals (sp^3). The minimum number is two—for example, the sp hybrid orbitals used by carbon in CO.

10.53   (a) C, sp^2; N, sp^3

(b) The amide or peptide link has two resonance structures (shown here with formal charges on the O and N atoms). Structure B is less favorable owing to the separation of charges.

(c) The fact that the amide link is planar indicates that structure B has some importance.

10.55   MO theory is better to use when explaining or understanding the effect of adding energy to molecules. A molecule can absorb energy and an electron can thus be promoted to a higher level. Using MO theory one can see how this can occur. Additionally, MO allows us to understand how a molecule can be paramagnetic.

10.57   (a) The number of hybrid orbitals equals the number of atomic orbitals used in their creation.

(b) No

(c) The hybrid orbitals have an energy that is the weighted average of their constituent atomic orbitals.

(d,e) The hybrid orbital shapes are the same but the hybrids lie along different axes (or in different planes).

10.59   (a) The C atom in the center of the molecule is sp hybridized, so two unhybridized p orbitals remain. These could be the p_x and p_y orbitals, for example. Because these p orbitals lie at 90° angles to each other, they form pi bonds to the end CH_2 groups that are in planes that lie at 90° angles to each other.

(b) The C atoms in benzene are all sp^2 hybridized. These hybrid orbitals all lie in the same plane.

(c) C atom 1 = sp^3; C atoms 2 and 3 = sp^2

Chapter 11

11.1   Heptane

11.3 $C_{14}H_{30}$ is an alkane and C_5H_{10} could be a cycloalkane.

11.5   2,3-dimethylbutane

11.7   (a) 2,3-Dimethylhexane

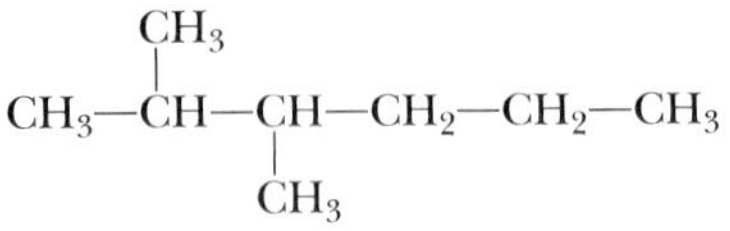

$$CH_3-\underset{\underset{CH_3}{|}}{\overset{\overset{CH_3}{|}}{CH}}-CH-CH_2-CH_2-CH_3$$

(b) 2,3-Dimethyloctane

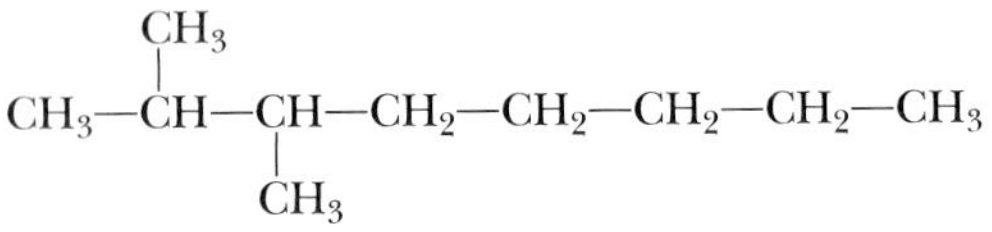

$$CH_3-\underset{\underset{CH_3}{|}}{\overset{\overset{CH_3}{|}}{CH}}-CH-CH_2-CH_2-CH_2-CH_2-CH_3$$

(c) 3-Ethylheptane

$$CH_3-CH_2-\underset{\underset{}{|}}{\overset{\overset{CH_2CH_3}{|}}{CH}}-CH_2-CH_2-CH_2-CH_3$$

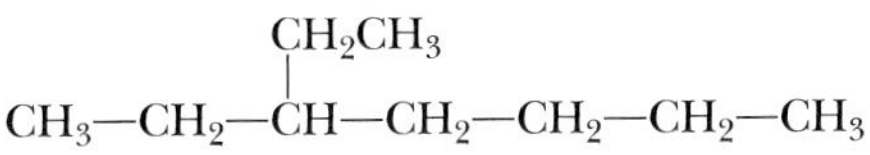

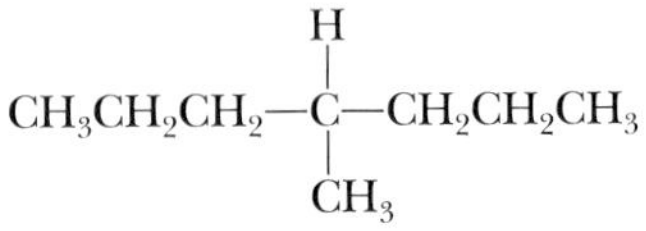

(d) 3-Ethyl-2-methylhexane

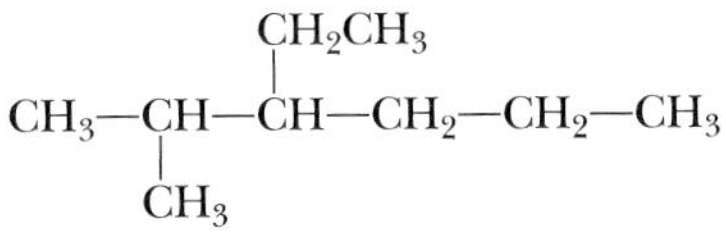

$$CH_3-\underset{\underset{CH_3}{|}}{\overset{}{CH}}-\underset{}{\overset{\overset{CH_2CH_3}{|}}{CH}}-CH_2-CH_2-CH_3$$

11.9

$$H_3C-\underset{\underset{CH_3}{|}}{\overset{\overset{H}{|}}{C}}-CH_2CH_2CH_2CH_2CH_3 \quad \text{2-methylheptane}$$

$$CH_3CH_2CH_2-\underset{\underset{CH_3}{|}}{\overset{\overset{H}{|}}{C}}-CH_2CH_2CH_3 \quad \text{4-methylheptane}$$

$$CH_3CH_2-\underset{\underset{CH_3}{|}}{\overset{\overset{H}{|}}{C^*}}-CH_2CH_2CH_2CH_3 \quad \begin{array}{l}\text{3-methylheptane. The C}\\\text{atom with an asterisk is}\\\text{chiral.}\end{array}$$

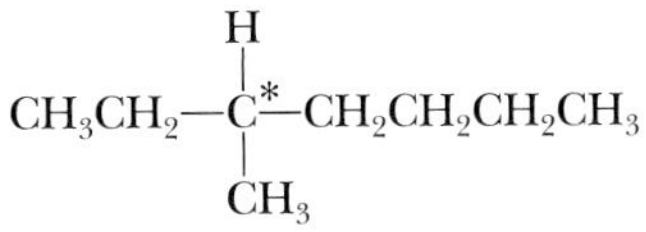

11.11

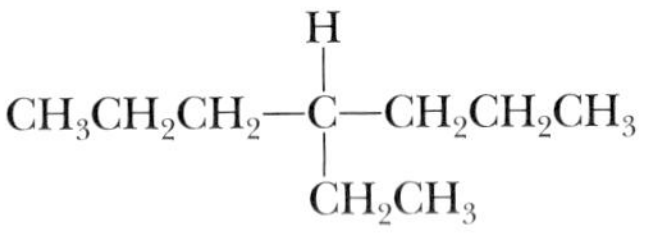

$$CH_3CH_2CH_2-\underset{\underset{CH_2CH_3}{|}}{\overset{\overset{H}{|}}{C}}-CH_2CH_2CH_3 \quad \begin{array}{l}\text{4-ethylheptane. The}\\\text{compound is not chiral.}\end{array}$$

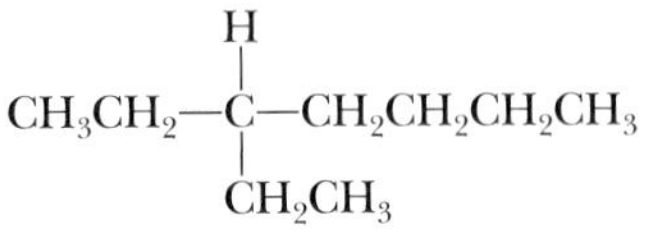

$$CH_3CH_2-\underset{\underset{CH_2CH_3}{|}}{\overset{\overset{H}{|}}{C}}-CH_2CH_2CH_2CH_3 \quad \text{3-ethylheptane. Not chiral.}$$

11.13 C_4H_{10}, butane: a low-molecular weight-fuel gas at room temperature and pressure. Slightly soluble in water.

$C_{12}H_{26}$, dodecane: a colorless liquid at room temperature. Expected to be insoluble in water but quite soluble in nonpolar solvents.

11.15

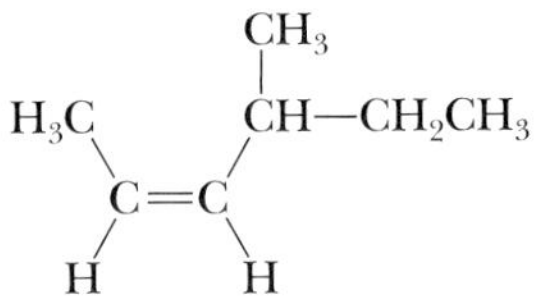

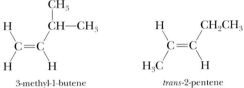

cis-4-methyl-2-hexene

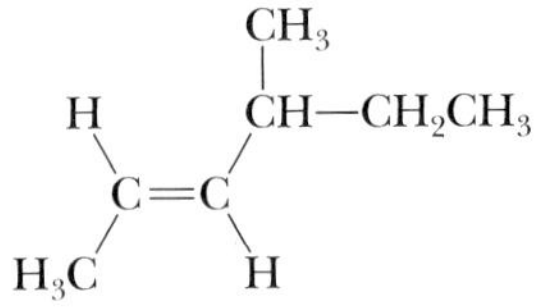

trans-4-methyl-2-hexene

11.17   (a)

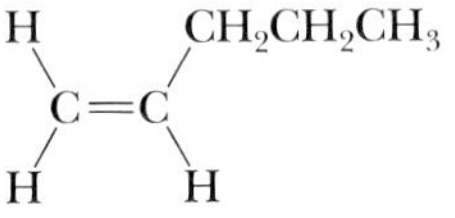

1-pentene 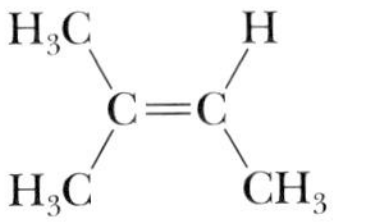   2-methyl-2-butene

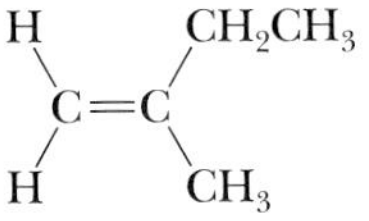

2-methyl-1-butene 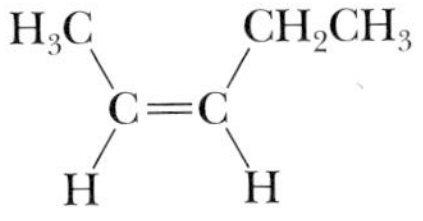   *cis*-2-pentene

3-methyl-1-butene   *trans*-2-pentene

(b)

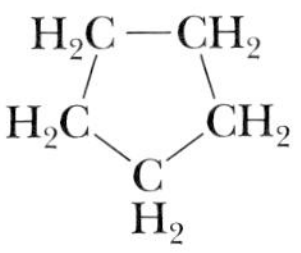

cyclopentane

11.19   (a) 1,2-Dibromopropane, $CH_3CHBrCH_2Br$

(b) Pentane, C_5H_{12}

11.21   1-Butene, $CH_3CH_2CH{=}CH_2$

11.23   Four isomers are possible.

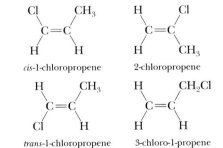

cis-1-chloropropene   2-chloropropene

trans-1-chloropropene   3-chloro-1-propene

11.25

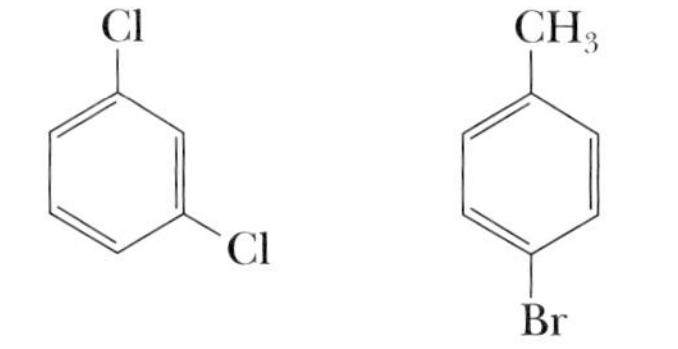

m-dichlorobenzene   *o*-bromotoluene

11.27

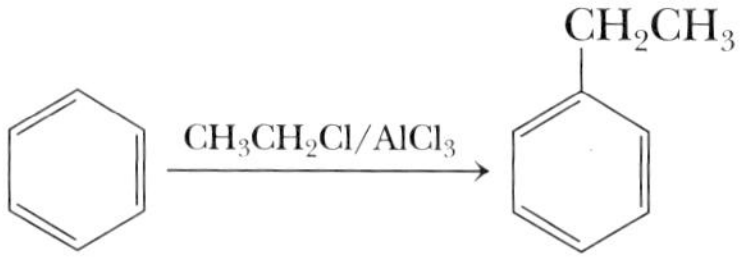

$$CH_3CH_2Cl/AlCl_3$$

ethylbenzene

11.29

$$CH_3Cl/AlCl_3$$

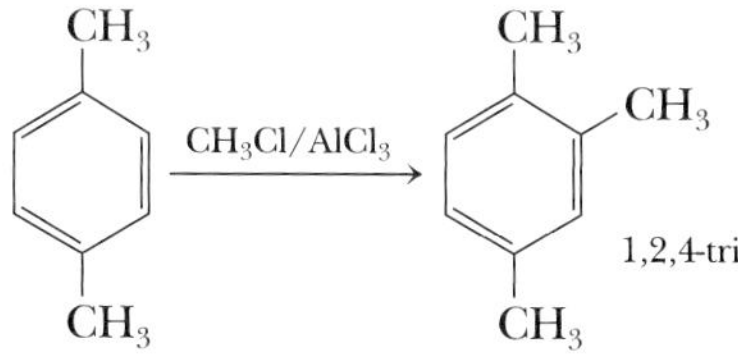

1,2,4-trimethylbenzene

11.31 (a) 1-Propanol, primary
       (b) 1-Butanol, primary
       (c) 2-Methyl-2-propanol, tertiary
       (d) 2-Methyl-2-butanol, tertiary

11.33 (a) Ethylamine, $CH_3CH_2NH_2$
       (b) Dipropylamine, $(CH_3CH_2CH_2)_2NH$

$$CH_3CH_2CH_2-\underset{\underset{H}{|}}{N}-CH_2CH_2CH_3$$

       (c) Butyldimethylamine

$$CH_3CH_2CH_2CH_2-\underset{\underset{CH_3}{|}}{N}-CH_3$$

       (d) triethylamine

$$CH_3CH_2-\underset{\underset{CH_2CH_3}{|}}{N}-CH_2CH_3$$

11.35 (a) 1-butanol, $CH_3CH_2CH_2CH_2OH$
       (b) 2-butanol

$$CH_3CH_2-\underset{\underset{H}{|}}{\overset{\overset{OH}{|}}{C}}-CH_3$$

       (c) 2-methyl-1-propanol

$$CH_3-\underset{\underset{CH_3}{|}}{\overset{\overset{H}{|}}{C}}-CH_2OH$$

(d) 2-methyl-2-propanol

$$CH_3-\underset{\underset{CH_3}{|}}{\overset{\overset{OH}{|}}{C}}-CH_3$$

11.37 (a) $C_6H_5NH_2(aq) + HCl(aq) \longrightarrow (C_6H_5NH_3)Cl(aq)$
       (b) $(CH_3)_3N(aq) + H_2SO_4(aq) \longrightarrow$
                                $[(CH_3)_3NH]HSO_4(aq)$

11.39 (a) $CH_3-\overset{\overset{O}{\|}}{C}-CH_2CH_2CH_3$

       (b) $H-\overset{\overset{O}{\|}}{C}-CH_2CH_2CH_2CH_2CH_3$

       (c) $CH_3CH_2CH_2CH_2-\overset{\overset{O}{\|}}{C}-OH$

11.41 (a) Acid, 3-methylpentanoic acid
       (b) Ester, methyl propanoate
       (c) Ester, butyl acetate (or butyl ethanoate)
       (d) Acid, *p*-bromobenzoic acid

11.43 (a) Pentanoic acid (see Question 39c)
       (b) 1-Pentanol, $CH_3CH_2CH_2CH_2CH_2OH$
       (c) 2-Octanol

$$H_3C-\underset{\underset{H}{|}}{\overset{\overset{OH}{|}}{C}}-CH_2CH_2CH_2CH_2CH_2CH_3$$

       (d) No reaction. A ketone is not oxidized by $KMnO_4$.

11.45 Step 1: Oxidize 1-propanol to propanoic acid.

$$CH_3CH_2-\underset{\underset{H}{|}}{\overset{\overset{H}{|}}{C}}-OH \xrightarrow{\text{oxidizing agent}} CH_3CH_2-\overset{\overset{O}{\|}}{C}-OH$$

Step 2: Combine propanoic acid and 1-propanol.

$$CH_3CH_2-\overset{\overset{O}{\|}}{C}-OH + CH_3CH_2-\underset{\underset{H}{|}}{\overset{\overset{H}{|}}{C}}-OH \xrightarrow[-H_2O]{}$$

$$CH_3CH_2-\overset{\overset{O}{\|}}{C}-O-CH_2CH_2CH_3$$

11.47 Sodium acetate, $NaCH_3CO_2$, and 1-butanol,
       $CH_3CH_2CH_2CH_2OH$

structure	name	bonds
H—C—H (methane with 4 H)	methane	four single bonds
formaldehyde structure	formaldehyde	one double bond and two single bonds
allene structure	allene	two double bonds
H—C≡C—H	acetylene	one single bond and one triple bond

11.93 (a) Cross-linking makes the material very rigid and inflexible.

(b) The OH groups give the polymer a high affinity for water.

(c) Hydrogen bonding allows the chains to form coils and sheets with high tensile strength.

11.95 (a) Ethane heat of combustion = -47.51 kJ/g

Ethanol heat of combustion = -26.82 kJ/g

(b) The heat obtained from the combustion of ethanol is less negative than for ethane, so partially oxidizing ethane to form ethanol decreases the amount of energy per mole available from the combustion of the substance.

11.97 2-Propanol will react with an oxidizing agent such as $KMnO_4$ (to give the ketone), whereas methyl ethyl ether ($CH_3OC_2H_5$) will not react. In addition, the alcohol should be more soluble in water than the ether.

11.99 (a) Empirical formula, CHO

(b) Molecular formula, $C_4H_4O_4$

(c) HO—C—C=C—C—OH (with two C=O groups and H H)

(d) All four C atoms are sp^2 hybridized.

(e) 120°

11.101 (a) reaction of propadiene + HBr → 2-bromopropane

2-bromopropane

(b) reaction + H_2O →

(c) reaction + H_2O →

Adding H_2O to 2-methyl-2-butene gives the same product as in part (b).

11.103 (a) In a substitution reaction, one atom or group of atoms is substituted for another. In an elimination reaction, a small molecule is removed or eliminated from a larger molecule.

(b) The elimination reaction produces an alkene, whereas the hydrogenation reaction has an alkene as a reactant. Both involve a small molecule, either H_2 or H_2O, being added to or eliminated from an organic molecule.

11.105 (a) Double bonds. See page 517.

(b) Termination occurs when the chain reaches 14 atoms. It could have been terminated earlier than this, or it could have continued to grow.

(c) The termination step

(d) Addition

Chapter 12

12.1 (a) 0.58 atm

(b) 0.59 bar

(c) 59 kPa

12.3 (a) 0.754 bar

(b) 650 kPa

(c) 934 kPa

12.5 2.70×10^2 mm Hg

12.7 3.7 L

12.9 250 mm Hg

12.11 3.2×10^2 mm Hg

12.13 9.72 atm

12.15 (a) 75 mL O_2

(b) 150 mL NO_2

12.17 0.919 atm

12.19 $V = 2.9$ L

12.21 1.9×10^6 g He

12.23 3.7×10^{-4} g/L

12.25 34.0 g/mol

12.27 57.5 g/mol

12.29 Molar mass = 74.9 g/mol; B_6H_{10}

12.31 0.039 mol H_2; 0.096 atm; 73 mm Hg

12.33 170 g NaN_3

12.35 1.7 atm O_2

12.37 4.1 atm H_2; 1.6 atm Ar; total pressure = 5.7 atm

12.39 (a) 0.30 mol halothane/1 mol O_2

(b) 3.0×10^2 g halothane

12.41 (a) They all have the same kinetic energy.

(b) The average speed of the H_2 molecules is greater than the average speed of the CO_2 molecules.

(c) The number of CO_2 molecules is greater than the number of H_2 molecules [$n(CO_2) = 1.8n(H_2)$].

(d) The mass of CO_2 is greater than the mass of H_2.

12.43 Average speed of CO_2 molecule = 3.65×10^4 cm/s

12.45 Average speed (and molar mass) increases in the order $CH_2F_2 < Ar < N_2 < CH_4$.

12.47 (a) F_2 (38 g/mol) effuses faster than CO_2 (44 g/mol)

(b) N_2 (28 g/mol) effuses faster than O_2 (32 g/mol)

(c) C_2H_4 (28.1 g/mol) effuses faster than C_2H_6 (30.1 g/mol)

(d) $CFCl_3$ (137 g/mol) effuses faster than $C_2Cl_2F_4$ (171 g/mol)

12.49 36 g/mol

12.51 P from the van der Waals equation = 29.5 atm

P from the ideal gas law = 49.3 atm

12.53 (a) Standard atmosphere: 1 atm; 760 mm Hg; 101.325 kPa; 1.013 bar.

(b) N_2 partial pressure: 0.780 atm; 593 mm Hg; 79.1 kPa; 0.791 bar

(c) H_2 pressure: 131 atm; 9.98×10^4 mm Hg; 1.33×10^4 kPa; 133 bar

(d) Air: 0.333 atm; 253 mm Hg; 33.7 kPa; 0.337 bar

12.55 C_2H_7N

12.57 $T = 290.$ K or 17 °C

12.59 (a) There are more molecules of H_2 than atoms of He.

(b) The mass of He is greater than the mass of H_2.

12.61 4 mol

12.63 Ni is the limiting reactant; 1.31 g $Ni(CO)_4$

12.65 (a, b) Sample 4 (He) has the largest number of molecules and sample 3 (H_2 at 27 °C and 760 mm Hg) has the fewest number of molecules.

(c) Sample 2 (Ar)

12.67 $P_{total} = 228$ mm Hg $= P(B_2H_6) + P(O_2)$

Stoichiometry requires that there be three times as many moles of O_2 as B_2H_6, so $P(O_2) = 3 P(B_2H_6)$. Therefore, 228 mm Hg $= 4 P(B_2H_6)$ and $P(B_2H_6) = 57$ mm Hg. This means $P(O_2) = 171$ mm Hg. The water vapor pressure is the same as O_2 pressure, or 171 mm Hg.

12.69 S_2F_{10}

12.71 8.54 g $Fe(CO)_5$

12.73 (a) 28.7 g/mol $\cong$ 29 g/mol

(b) X of $O_2 = 0.18$ and X of $N_2 = 0.82$

12.75 $P = 5 \times 10^{-8}$ atm

12.77 Molar mass = 86.4 g/mol. The gas is probably ClO_2F.

12.79 Yield = 76.0%

12.81 Weight percent $KClO_3$ = 69.1%

12.83 (a) $NO_2 < O_2 < NO$

(b) $P(O_2) = 75$ mm Hg

(c) $P(NO_2) = 150$ mm Hg

12.85 The mixture contains 0.22 g CO_2 and 0.77 g CO.

$P(CO_2) = 0.22$ atm; $P(O_2) = 0.12$ atm; $P(CO) = 1.22$ atm

12.87 Formula of the iron compound: $Fe(CO)_5$

12.89 $P(SiH_4) = 40.$ mm Hg and $P(O_2) = 80.$ mm Hg

Pressure after reaction = 80. mm Hg

12.91 (a) $P(B_2H_6) = 0.0160$ atm

(b) $P(H_2) = 0.0320$ atm, so $P_{total} = 0.0480$ atm

12.93 Amount of $Na_2CO_3 = 0.00424$ mol

Amount of $NaHCO_3 = 0.00951$ mol

Amount of CO_2 produced = 0.0138 mol

Volume of CO_2 produced = 0.343 L

12.95 At 20 °C, there is 7.8×10^{-3} g H_2O/L. At 0 °C, there is 4.6×10^{-3} g H_2O/L.

12.97 (a) 10.0 g of O_2 represents more molecules than 10.0 g of CO_2. Therefore, O_2 has the greater partial pressure.

(b) The average speed of the O_2 molecules is greater than the average speed of the CO_2 molecules.

(c) The gases are at the same temperature and so have the same average kinetic energy.

12.99 (a) $P(C_2H_2) > P(CO)$

(b) There are more molecules in the C_2H_2 container than in the CO container.

12.101 (a) Not a gas. A gas would expand to an infinite volume.

(b) Not a gas. A density of 8.2 g/mL is typical of a solid.

(c) Insufficient information

(d) Gas

12.103 (a) 46.0 g NaN_3

(b)

$$\left[\ddot{N}\!=\!N\!=\!\ddot{N}\right]^{-} \longleftrightarrow \left[:\!\ddot{N}\!-\!N\!\equiv\!N:\right]^{-} \longleftrightarrow \left[:\!N\!\equiv\!N\!-\!\ddot{N}:\right]^{-}$$
$$_{-1 \quad +1 \quad -1} _{-2 \quad +1 \quad 0} _{0 \quad +1 \quad -2}$$

(c) The N_3^- ion is linear.

12.105 The speed of gas molecules is related to the square root of the absolute temperature, so a doubling of the temperature will lead to an increase of about $(2)^{1/2}$ or 1.4.

Chapter 13

13.1 (a) Dipole–dipole interactions (and hydrogen bonds)

(b) Induced dipole–induced dipole forces

(c) Dipole–dipole interactions (and hydrogen bonds)

13.3   (a) Induced dipole–induced dipole forces

       (b) Induced dipole–induced dipole forces

       (c) Dipole–dipole forces

       (d) Dipole–dipole forces (and hydrogen bonding)

13.5   The predicted order of increasing strength is Ne < CH_4 < CO < CCl_4. In this case, prediction does not quite agree with reality. The boiling points are Ne ($-246\ °C$) < CO ($-192\ °C$) < CH_4 ($-162\ °C$) < CCl_4 ($77\ °C$).

13.7   (c) HF; (d) acetic acid; (f) CH_3OH

13.9   (a) LiCl. The Li^+ ion is smaller than Cs^+ (Figure 8.15), which makes the ion-ion forces of attraction stronger in LiCl.

       (b) $Mg(NO_3)_2$. The Mg^{2+} ion is smaller than the Na^+ ion (Figure 8.15), and the magnesium ion has a 2+ charge (as opposed to 1+ for sodium). Both of these effects lead to stronger ion–ion forces of attraction in magnesium nitrate.

       (c) $NiCl_2$. The nickel(II) ion has a larger charge than Rb^+ and is considerably smaller. Both effects mean that there are stronger ion–ion forces of attraction in nickel(II) chloride.

13.11 $q = +90.1$ kJ

13.13   (a) Water vapor pressure is about 150 mm Hg at 60 °C. (Appendix G gives a value of 149.4 mm Hg at 60 °C.)

       (b) 600 mm Hg at about 93 °C

       (c) At 70 °C, ethanol has a vapor pressure of about 520 mm Hg, whereas that of water is about 225 mm Hg.

13.15   At 30 °C the vapor pressure of ether is about 590 mm Hg. (This pressure requires 0.23 g of ether in the vapor phase at the given conditions, so there is sufficient ether in the flask.) At 0 °C the vapor pressure is about 160 mm Hg, so some ether condenses when the temperature declines.

13.17   (a) O_2 ($-183\ °C$) (bp of $N_2 = -196\ °C$)

       (b) SO_2 ($-10\ °C$) (CO_2 sublimes at $-78\ °C$)

       (c) HF ($+19.7\ °C$) (HI, $-35.6\ °C$)

       (d) GeH_4 ($-90.0\ °C$) (SiH_4, $-111.8\ °C$)

13.19   (a) CS_2, about 620 mm Hg; CH_3NO_2, about 80 mm Hg

       (b) CS_2, induced dipole–induced dipole forces; CH_3NO_2, dipole–dipole forces

       (c) CS_2, about 46 °C; CH_3NO_2, about 100 °C

       (d) About 39 °C

       (e) About 34 °C

13.21   (a) 80.1 °C

       (b) At about 48 °C the liquid has a vapor pressure of 250 mm Hg. The vapor pressure is 650 mm Hg at 75 °C.

       (c) 33.5 kJ/mol (from slope of plot)

13.23   Two possible unit cells are illustrated here. The simplest formula is AB_8.

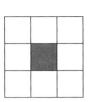 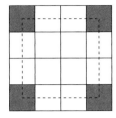

13.25 Ca^{2+} ions at 8 corners = 1 net Ca^{2+} ion

       O^{2-} ions in 6 faces = 3 net O^{2-} ions

       Ti^{4+} ion in center of unit cell = 1 net Ti^{4+} ion

       Formula = $CaTiO_3$

13.27   (a) There are eight O^{2-} ions at the corners and one in the center for a net of two O^{2-} ions per unit cell. There are four Cu ions in the interior in tetrahedral holes. The ratio of ions is Cu_2O.

       (b) The oxidation number of copper must be +1.

13.29   (a) 8 C atoms per unit cell. There are 8 corners (= 1 net C atom), 6 faces (= 3 net C atoms), and 4 internal C atoms.

       (b) Face-centered cubic (fcc) with C atoms in the tetrahedral holes.

13.31 q (for fusion) = -1.97 kJ; q (for melting) = $+1.97$ kJ

13.33   (a) The density of liquid CO_2 is less than that of solid CO_2.

       (b) CO_2 is a gas at 5 atm and 0 °C.

       (c) Critical temperature = 31 °C, so CO_2 cannot be liquefied at 45 °C.

13.35 q (to heat the liquid) = 9.4×10^2 kJ

       q (to vaporize NH_3) = 1.6×10^4 kJ

       q (to heat the vapor) = 8.8×10^2 kJ

       q_{total} = 1.8×10^4 kJ

13.37   Yes. The critical temperature (416 K, 143 °C) is well above room temperature.

13.39   Ar < CO_2 < CH_3OH

13.41 O_2 phase diagram. (i) Note the slight positive slope of the solid–liquid equilibrium line. It indicates that the density of solid O_2 is greater than that of liquid O_2. (ii) Using the diagram here, the vapor pressure of O_2 at 77 K is between 150 mm Hg and 200 mm Hg.

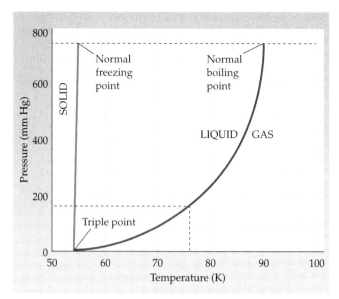

13.43 Less ethanol is available (17 mol) than would be required to completely fill the room with vapor (60. mol), so all of the ethanol will evaporate.

13.45 Li^+ ions are smaller than Cs^+ ions (78 pm and 165 pm, respectively; see Figure 8.15). Thus, there will be a stronger attractive force between Li^+ ion and water molecules than between Cs^+ ions and water molecules.

13.47 (a) 350 mm Hg
       (b) Ethanol (lower vapor pressure at every temperature)
       (c) 84 °C
       (d) CS_2, 46 °C; C_2H_5OH, 78 °C; C_7H_{16}, 99 °C
       (e) CS_2, gas; C_2H_5OH, gas; C_7H_{16}, liquid

13.49 Radius of silver = 145 pm

13.51 Ca^{2+}: there are 8 corner Ca^{2+} ions and 1 internal Ca^{2+} ion or a total of 2 Ca^{2+} ions. C atoms: there are 8 C atoms on edges. At 1/4 per atom, there are 2 within the unit cell. There are 2 more C atoms internal to the cell. Thus, there is a total of 4 C atoms per unit cell. The formula is CaC_2.

13.53 Molar enthalpy of vaporization increases with increasing intermolecular forces: C_2H_6 (14.69 kJ/mol; induced dipole) < HCl (16.15 kJ/mol; dipole) < CH_3OH (35.21 kJ/mol, hydrogen bonds) (The molar enthalpies of vaporization here are given at the boiling point of the liquid.)

13.55 1.356×10^{-8} cm (literature value is 1.357×10^{-8} cm)

13.57
Assumed Unit Cell	Calculated Density (g/cm³)
Simple cubic	4.60
Body-centered cubic	5.97
Face-centered cubic	6.52

The calculated density for a body-centered cubic unit cell is closest to the experimental value. In fact, vanadium has a body-centered cubic unit cell.

13.59 Mass of 1 CaF_2 unit calculated from crystal data = 1.2963×10^{-22} g. Divide molar mass of CaF_2 (78.077 g/mol) by mass of 1 CaF_2 to obtain Avogadro's number. Calculated value = 6.0230×10^{23} CaF_2/mol.

13.61 Diagram A leads to a surface coverage of 78.5%. Diagram B leads to 90.7% coverage.

13.63 (a) 70.3 °C
       (b)

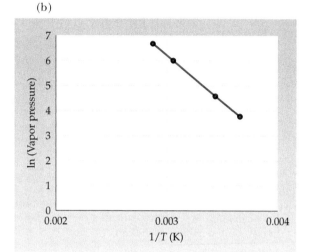

Using the equation for the straight line in the plot

$\ln P = -3885 \, (1/T) + 17.949$

we calculate that $T = 312.6$ K (39.5 °C) when $P = 250$ mm Hg. When $P = 650$ mm Hg, $T = 338.7$ K (65.5 °C).

       (c) Calculated ΔH_{vap} = 32.3 kJ/mol

13.65 Acetone and water can interact by hydrogen bonding.

13.67 Glycol's viscosity will be greater than ethanol's owing to the greater hydrogen-bonding capacity of glycol.

13.69 (a) Water has two OH bonds and two lone pairs, whereas the O atom of ethanol has only one OH bond (and two lone pairs). More extensive Hydrogen bonding is likely for water.
       (b) Water and ethanol interact extensively through hydrogen bonding, so the volume is expected to be slightly smaller than the sum of the two volumes.

13.71 No. NaCl has a 1 : 1 ratio of cations and anions in the unit cell, whereas the unit cell of $CaCl_2$ must have a 1 : 2 ratio of cations to anions.

13.73 Two pieces of evidence for $H_2O(\ell)$ having considerable intermolecular attractive forces:

(a) Based on the boiling points of the Group 6A hydrides (Figure 13.8), the boiling point of water should be approximately −80 °C. The actual boiling point of 100 °C reflects the significant hydrogen bonding that occurs.

(b) Liquid water has a specific heat capacity that is higher than almost any other liquid. This reflects the fact that a relatively larger amount of energy is necessary to overcome intermolecular forces and raise the temperature of the liquid.

13.75 (a) HI, hydrogen iodide

(b) The large iodine atom in HI leads to a significant polarizability for the molecule and thus to a large dispersion force.

(c) The dipole moment of HCl (1.07 D, Table 9.8) is larger than for HI (0.38 D).

(d) HI. See part (b).

13.77 When the can is inverted in cold water, the water vapor pressure in the can, which was approximately 760 mm Hg, drops rapidly—say, to 9 mm Hg at 10 °C. This creates a partial vacuum in the can, and the can is crushed because of the difference in pressure inside the can and the pressure of the atmosphere pressing down on the outside of the can.

13.79 (a) About −27 °C

(b) Pressure is about 6.5 atm.

(c) As the more energetic molecules leave the liquid phase, enter the gas phase, and escape from the tank, only lower-energy molecules remain. These have a lower temperature. The tank is thereby cooled, and water vapor can condense on the tank's surface. As the temperature drops, the vapor pressure of the remaining liquid drops and the flow of gas out of the tank slows.

(d) Cool the tank in dry ice (to −78 °C). The vapor pressure of the liquid is less than one atmosphere, so the tank can be opened safely and the liquid poured out.

13.81 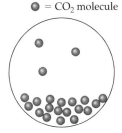 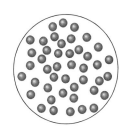 ● = CO_2 molecule

Separate liquid and vapor phases in equilibrium

Supercritical CO_2. Distinct liquid and vapor phases not visible. Molecules are closer together than in vapor phase.

13.83 The Zn^{2+} ions are in a face-centered cubic arrangement and the S^{2-} ions fill half of the tetrahedral holes. There are four Zn^{2+} ions and four S^{2-} ions per unit cell, a 1:1 ratio that matches the compound formula.

13.85 (a) The Ca^{2+} ions are in a face-centered arrangement and the F^- ions fill all of the tetrahedral holes.

(b) There are four Ca^{2+} ions and eight F^- ions per unit cell, a 1:2 ratio that matches the compound formula.

(c) The CaF_2 and ZnS structures both have a face-centered cubic arrangement of cations. The ZnS structure has anions in only one-half of the tetrahedral holes, whereas F^- ions fill all of the tetrahedral holes in CaF_2.

13.87 (a) Structure of aspartame:

(b) There are three C=O groups that are highly polar and can interact with H atoms of water. In addition, there are two NH groups and one —OH group that can hydrogen bond.

Chapter 14

14.1 (a) Concentration $(m) = 0.0434\ m$

(b) Mole fraction of acid = 0.000781

(c) Weight percent of acid = 0.509%

14.3 NaI: 0.15 m; 2.2%; $X = 2.7 \times 10^{-3}$

CH_3CH_2OH: 1.1 m; 5.0%; $X = 0.020$

$C_{12}H_{22}O_{11}$: 0.15 m; 4.9%; $X = 2.7 \times 10^{-3}$

14.5 2.65 g Na_2CO_3; $X(Na_2CO_3) = 3.59 \times 10^{-3}$

14.7 220 g glycol; 5.7 m

14.9 16.2 m; 37.1%

14.11 Molality = 2.6×10^{-5} m (assuming that 1 kg of sea water is equivalent to 1 kg of solvent)

14.13 (b) and (c)

14.15 $\Delta H^\circ_{solution}$ for LiCl = −36.9 kJ/mol. This is an exothermic heat of solution, as compared with the very slightly endothermic value for NaCl.

14.17 Above about 40 °C the solubility increases with temperature; therefore, add more NaCl and raise the temperature.

14.19 See the discussion and data on page 593.

(a) The heat of hydration of LiF is more negative than that for RbF because the Li^+ ion is much smaller than the Rb^+ ion.

(b) The heat of hydration for $Ca(NO_3)_2$ is larger than that for KNO_3 owing to the $+2$ charge on the Ca^{2+} ion (and its smaller size).

(c) The heat of hydration is greater for $CuBr_2$ than for $CsBr$ because Cu^{2+} has a larger charge than Cs^+, and the Cu^{2+} ion is smaller than the Cs^+ ion.

14.21 2×10^{-3} g O_2

14.23 1130 mm Hg or 1.49 atm

14.25 35.0 mm Hg

14.27 $X(H_2O) = 0.869$; 16.7 mol glycol; 1040 g glycol

14.29 Calculated boiling point = 84.2 °C

14.31 $\Delta T_{bp} = 0.808$ °C; solution boiling point = 62.51 °C

14.33 Molality = 0.16 m; 0.0081 mol solute; 1.4 g solute

14.35 Molality = 8.60 m; 28.4%

14.37 Molality = 0.195 m; $\Delta T_{fp} = -0.362$ °C

14.39 Molar mass = 360 g/mol; $C_{20}H_{16}Fe_2$

14.41 Molar mass = 150 g/mol

14.43 Molar mass = 170 g/mol

14.45 Molar mass = 130 g/mol

14.47 Freezing point = -24.6 °C

14.49 0.080 m $CaCl_2 < 0.10$ m $NaCl < 0.040$ m $Na_2SO_4 < 0.10$ sugar

14.51 (a) $\Delta T_{fp} = -0.348$ °C; fp = -0.348 °C

(b) $\Delta T_{bp} = +0.0959$ °C; bp = 100.0959 °C

(c) $\Pi = 4.58$ atm

The osmotic pressure is large and can be measured with a small experimental error.

14.53 Molar mass = 6.0×10^3 g/mol

14.55 (a) $BaCl_2(aq) + Na_2SO_4(aq) \longrightarrow BaSO_4(s) + 2\ NaCl(aq)$

(b) Initially the $BaSO_4$ particles form a colloidal suspension.

(c) Over time the particles of $BaSO_4(s)$ grow and precipitate.

14.57 Li_2SO_4 should have a more negative heat of hydration than Cs_2SO_4 because the Li^+ ion is smaller than the Cs^+ ion.

14.59 (a) Increase in vapor pressure of water

0.20 m $Na_2SO_4 < 0.50$ m sugar < 0.20 m $KBr < 0.35$ m ethylene glycol

(b) Increase in boiling point

0.35 m ethylene glycol < 0.20 m $KBr < 0.50$ m sugar < 0.20 m Na_2SO_4

14.61 (a) 0.456 mol DMG and 11.4 mol ethanol; $X(DMG) = 0.0385$

(b) 0.869 m

(c) VP ethanol over the solution at 78.4 °C = 730.7 mm Hg

(d) bp = 79.5 °C

14.63 For ammonia: 23 m; 28%; $X(NH_3) = 0.29$

14.65 0.592 g Na_2SO_4

14.67 (a) 0.20 m KBr   (b) 0.10 m Na_2CO_3

14.69 Freezing point = -11 °C

14.71 4.0×10^2 g/mol

14.73 4.93×10^{-4} mol/L; 1.38×10^{-2} g/L

14.75 (a) Molar mass = 4.9×10^4 g/mol

(b) $\Delta T_{fp} = -3.8 \times 10^{-4}$ °C

14.77 Molar mass in benzene = 1.20×10^2 g/mol; molar mass in water = 62.4 g/mol. The actual molar mass of acetic acid is 60.1 g/mol. In benzene the molecules of acetic acid form "dimers." That is, two molecules form a single unit through hydrogen bonding. See Figure 13.9 on page 601.

14.79 $\Delta H°_{solution} [Li_2SO_4] = -28.0$ kJ/mol

$\Delta H°_{solution} [LiCl] = -36.9$ kJ/mol

$\Delta H°_{solution} [K_2SO_4] = +23.7$ kJ/mol

$\Delta H°_{solution} [KCl] = +17.2$ kJ/mol

Both lithium compounds have exothermic heats of solution, whereas both potassium compounds have endothermic values. Consistent with this is the fact that lithium salts (LiCl) are often more water-soluble than potassium salts (KCl) (see Figure 14.11).

14.81 X(benzene in solution) = 0.67 and X(toluene in solution) = 0.33

$P_{total} = P_{toluene} + P_{benzene} = 7.3$ mm Hg + 50. mm Hg
$= 57$ mm Hg

$$X(\text{toluene in vapor}) = \frac{7.3 \text{ mm Hg}}{57 \text{ mm Hg}} = 0.13$$

$$X(\text{benzene in vapor}) = \frac{50. \text{ mm Hg}}{57 \text{ mm Hg}} = 0.87$$

14.83 The calculated molality at the freezing point of benzene is 0.47 m, whereas it is 0.99 m at the boiling point. A higher molality at the higher temperature indicates more molecules are dissolved. Therefore, assuming benzoic acid forms dimers like acetic acid (Figure 13.9), dimer formation is more prevalent at the lower temperature. In this process two molecules become one entity, lowering the number of separate species in solution and lowering the molality.

14.85 $i = 1.7$. That is, there is 1.7 mol of ions in solution per mole of compound.

14.87 (a) Calculate the number of moles of ions in 10^6 g H_2O: 550. mol Cl^-; 470. mol Na^+; 53.1 mol Mg^{2+}; 9.42 mol SO_4^{2-}; 10.3 mol Ca^{2+}; 9.72 mol K^+; 0.84 mol Br^-. Total moles of ions = 1.103×10^3 per 10^6 g water. This gives ΔT_{fp} of $-2.05\ °C$.

(b) $\Pi = 27.0$ atm. This means that a minimum pressure of 27 atm would have to be used in a reverse osmosis device.

14.89 (a) $i = 2.06$

(b) There are approximately two particles in solution, so $H^+ + HSO_4^{2+}$ best represents H_2SO_4 in aqueous solution.

14.91 (a) Molar mass = 97.6 g/mol; empirical formula, $BF_2^°$, and molecular formula, B_2F_4

(b)

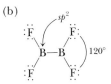

14.93 Colligative properties depend on the number of ions or molecules in solution. Each mole of $CaCl_2$ provides 1.5 times as many ions as each mole of NaCl.

14.95 At 0 °C some solid NaCl remains in the beaker or flask, and Na^+ and Cl^- ions are in solution. As some Na^+ and Cl^- ions are removed from the surface of the solid NaCl, enter the solution, and are hydrated by water, other Na^+ and Cl^- ions move to the solid surface.

14.97 Benzene is a nonpolar solvent. Thus, ionic substances such as $NaNO_3$ and NH_4Cl will certainly not dissolve. However, naphthalene is also nonpolar and resembles benzene in its structure; it should dissolve very well. (A chemical handbook gives a solubility of 33 g naphthalene per 100 g benzene.) Diethyl ether is weakly polar and will also be miscible to some extent with benzene.

14.99 The C—C and C—H bonds in hydrocarbons are nonpolar or weakly polar and tend to make such dispersions hydrophobic (water-hating). The C—O and O—H bonds in starch present opportunities for hydrogen bonding with water. Hence, starch is expected to be more hydrophilic.

14.101 [NaCl] = 1.0 M and [KNO_3] = 0.88 M. The KNO_3 solution has a higher solvent concentration, so solvent will flow from the KNO_3 solution to the NaCl solution.

Chapter 15

15.1 (a) $-\dfrac{1}{2}\dfrac{\Delta[O_3]}{\Delta t} = \dfrac{1}{3}\dfrac{\Delta[O_2]}{\Delta t}$

(b) $-\dfrac{1}{2}\dfrac{\Delta[HOF]}{\Delta t} = \dfrac{1}{2}\dfrac{\Delta[HF]}{\Delta t} = \dfrac{\Delta[O_2]}{\Delta t}$

15.3 $\dfrac{1}{3}\dfrac{\Delta[O_2]}{\Delta t} = -\dfrac{1}{2}\dfrac{\Delta[O_3]}{\Delta t}$ or $\dfrac{\Delta[O_2]}{\Delta t} = -\dfrac{2}{3}\dfrac{\Delta[O_2]}{\Delta t}$

so $\Delta[O_3]/\Delta t = -1.0 \times 10^{-3}$ mol/L · s.

15.5 (a) The graph of [B] (product concentration) versus time shows [B] increasing from zero. The line is curved, indicating the rate changes with time; thus that the rate depends on concentration. Rates for the four 10-s intervals are as follows: 0–10 s, 0.0326 mol/L · s; from 10–20 s, 0.0246 mol/L · s; 20–30 s, 0.0178 mol/L · s; 30–40 s, 0.0140 mol/L · s.

(b) $-\dfrac{\Delta[A]}{\Delta t} = \dfrac{1}{2}\dfrac{\Delta[B]}{\Delta t}$ throughout the reaction

In the interval 10–20 s, $\dfrac{\Delta[A]}{\Delta t} = -0.0123\ \dfrac{mol}{L \cdot s}$

(c) Instantaneous rate when [B] = 0.750 mol/L

$= \dfrac{\Delta[B]}{\Delta t} = 0.0163\ \dfrac{mol}{L \cdot s}$

15.7 The reaction is second order in A, first order in B, and third order overall.

15.9 (a) Rate = $k[NO_2][O_3]$

(b) If [NO_2] is tripled, the rate triples.

(c) If [O_3] is halved, the rate is halved.

15.11 (a) The reaction is second order in [NO] and first order in [O_2].

(b) $\dfrac{-\Delta[NO]}{\Delta t} = k[NO]^2[O_2]$

(c) $k = 13$ L^2/mol^2 · s

(d) $\dfrac{-\Delta[NO]}{\Delta t} = -1.4 \times 10^{-5}$ mol/L · s

(e) When $-\Delta[NO]/\Delta t = 1.0 \times 10^{-4}$ mol/L · s, $\Delta[O_2]/\Delta t = 5.0 \times 10^{-5}$ mol/L · s and $\Delta[NO_2]/\Delta t = 1.0 \times 10^{-4}$ mol/L · s.

15.13 (a) Rate = $k[NO]^2[O_2]$

(b) $k = 50.$ L^2/mol^2 · h

(c) Rate = 8.4×10^{-9} mol/L · h

15.15 (a) Rate = $k[CO]^2[O_2]$

(b) Third order overall; 1st order in O_2 and 2nd order in CO.

(c) $k = 5$ L^2/mol^2 · min

15.17 $k = 0.0392$ h^{-1}

15.19 5.0×10^2 min

15.21 105 min

15.23 (a) 153 min

(b) 1790 min

15.25 (a) $t_{1/2} = 1400$ s   (b) 4600 s

15.27 4.48×10^{-3} mol (0.260 g) azomethane remains; 0.0300 mol N_2 formed

15.29 Fraction of ^{64}Cu remaining = 0.030

15.31 72 s represents two half-lives, so $t_{1/2} = 36$ s.

15.33 (a) A graph of ln[sucrose] versus time produces a straight line, indicating that the reaction is first order in [sucrose].

(b) $-\Delta[\text{sucrose}]/\Delta t = +k[\text{sucrose}]$; $k = 3.7 \times 10^{-3}$ min^{-1}

(c) At 175 min, [sucrose] = 0.167 M

15.35   The straight line obtained in a graph of $\ln[N_2O]$ versus time indicates a first-order reaction.

$k = (-\text{ slope}) = 0.0127$ min^{-1}

The rate when $[N_2O] = 0.035$ mol/L is 4.4×10^{-4} mol/L · min.

15.37   The graph of $1/[NO_2]$ versus time gives a straight line, indicating the reaction is second order with respect to $[NO_2]$ (see Table 15.1 on page 619). The slope of the line is k, so $k = 1.1$ L/mol · s.

15.39 $-\Delta[C_2F_4]/\Delta t = k[C_2F_4]^2 = (0.04 \text{ L/mol} \cdot \text{s})[C_2F_4]^2$

15.41   Activation energy = 102 kJ/mol

15.43 $k = 0.3$ s^{-1}

15.45

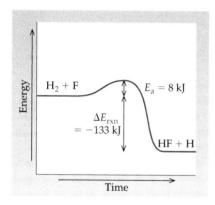

15.47   (a) Rate = $k[NO_3][NO]$

(b) Rate = $k[Cl][H_2]$

(c) Rate = $k[(CH_3)_3CBr]$

15.49   (a) The Second step   (b) Rate = $k[O_3][O]$

15.51   (a) The substances OI$^-$ and HOI cancel out to give the equation for the overall reaction.

(b) Steps 1 and 2 are bimolecular, whereas step 3 is termolecular.

(c) Rate = $k[H_2O_2][I^-]$

(d) OI$^-$ and HOI are intermediates. They are produced and consumed during the reaction and do not appear in the equation for the overall reaction.

15.53 NO_2 is a reactant in the first step and a product in the second step. CO is a reactant in the second step. NO_3 is an intermediate, and CO_2 is a product. NO is a product.

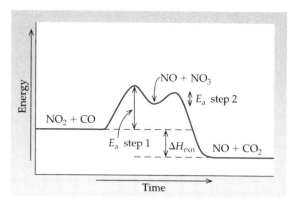

15.55   The reaction rate will double.

15.57   After measuring pH as a function of time, one could then calculate pOH and then [OH$^-$]. Finally, a plot of $1/[OH^-]$ versus time would give a straight line with a slope equal to k.

15.59   (a) Rate = $k[NH_3]$

(b) $k = 0.050$ s^{-1}

(c) Half life = 14 s

15.61   (a) Rate = $k[CO_2]$

(b) $k = 0.028$ s^{-1}

(c) 24 s

15.63   (a) A plot of $1/[C_2F_4]$ versus time indicates the reaction is second order with respect to $[C_2F_4]$. The rate law is Rate = $k[C_2F_4]^2$.

(b) The rate constant (= slope of the line) is about 0.045 L/mol · s. (The graph does not allow a very accurate calculation.)

(c) Using $k = 0.045$ L/mol · s, the concentration after 600 s is 0.03 M (to 1 significant figure).

(d) Time = 2000 s (using k from part a).

15.65   (a) A plot of $1/[NH_4NCO]$ versus time is linear, so the reaction is second order with respect to NH_4NCO.

(b) Slope = $k = 0.0109$ L/mol · min.

(c) $t_{1/2} = 200.$ min

(d) $[NH_4NCO] = 0.0997$ mol/L

15.67   Mechanism 2

15.69 $k = 0.037$ h^{-1} and $t_{1/2} = 19$ h

15.71   (a) After 125 min, 0.251 g remains. After 145, 0.144 g remains.

(b) Time = 43.9 min

(c) Fraction remaining = 0.016

15.73   (a) $2\,NO(g) + Br_2(g) \longrightarrow 2\,BrNO(g)$

(b) Mechanism 1 is termolecular.

Mechanism 2 has two bimolecular steps.

Mechanism 3 has two bimolecular steps.

(c) Br_2NO is the intermediate in mechanism 2 and N_2O_2 is the intermediate in mechanism 3.

(d) Assuming step 1 in each mechanism is the slow step, the rate equations will all differ. Mechanism 1 would be second order in NO and first order in Br_2. Mechanism 2 would be first order in both NO and Br_2. Mechanism 3 would be second order in NO and zero order in Br_2.

15.75 The rate equation for the slow step is Rate = $k[O_3][O]$. The equilibrium constant, K, for step 1 is $K = [O_2][O]/[O_3]$. Solving this for $[O]$, we have $[O] = K[O_3]/[O_2]$. Substituting the expression for $[O]$ into the rate equation we find

Rate = $k[O_3]\{K[O_3]/[O_2]\} = kK[O_3]^2/[O_2]$

15.77 The slope of the $\ln k$ versus $1/T$ plot is -6370. From slope = $-E_a/R$, we derive $E_a = 53.0$ kJ/mol.

15.79 Estimated time at 90 °C = 4.76 min

15.81 After 30 min (one half-life), $P_{HOF} = 50.0$ mm Hg and $P_{total} = 125.0$ mm Hg. After 45 min, $P_{HOF} = 35.4$ mm Hg and $P_{total} = 132$ mm Hg.

15.83 (a) The slow step is unimolecular and the fast step is bimolecular.

(b) Rate = $k[Ni(CO)_4]$. Yes, this agrees with the mechanism proposed.

(c) $[Ni(CO)_3L]$ after 5.0 min = 0.023 M

15.85 The finely divided rhodium metal will have a significantly greater surface area than the small block of metal. This leads to a large increase in the number of reaction sites and vastly increases the reaction rate.

15.87 (a) False. The reaction may occur in a single step but this does not have to be true.

(b) True

(c) False. Raising the temperature increases the value of k.

(d) False. Temperature has no effect on the value of E_a.

(e) False. If the concentrations of both reactants are doubled, the rate will increase by a factor of 4.

(f) True

15.89 (a) True

(b) True

(c) False. As a reaction proceeds, the reactant concentration decreases and the rate decreases.

(d) False. It is possible to have a one-step mechanism for a third-order reaction if the slow, rate-determining step is termolecular.

15.91 (a) Decrease   (d) No change

(b) Increase   (e) No change

(c) No change   (f) No change

15.93 (a) There are three mechanistic steps.

(b) The overall reaction is exothermic.

15.95 (a) The average rate is calculated over a period of time, whereas the instantaneous rate is the rate of reaction at some instant in time.

(b) The reaction rate decreases with time as the dye concentration decreases.

(c) See part (b).

15.97 (a) Molecules must collide with enough energy to overcome the activation energy, and they must be in the correct orientation.

(b) In animation 2 the molecules are moving faster, so they are at a higher temperature.

(c) Less sensitive. The O_3 must collide with NO in the correct orientation for a reaction to occur. The O_3 and N_2 collisions do not depend to the same extent on orientation because N_2 is a symmetrical, diatomic molecule.

15.99 (a) I^- is regenerated during the second step in the mechanism.

(b) The activation energy is smaller for the catalyzed reaction.

Chapter 16

16.1 (a) $K = \dfrac{[H_2O]^2[O_2]}{[H_2O_2]^2}$

(b) $K = \dfrac{[CO_2]}{[CO][O_2]^{1/2}}$

(c) $K = \dfrac{[CO_2]^2}{[CO_2]}$

(d) $K = \dfrac{[CO_2]}{[CO]}$

16.3 $Q = (2.0 \times 10^{-8})^2/(0.020) = 2.0 \times 10^{-14}$

$Q < K$ so the reaction proceeds to the right.

16.5 $Q = 1.0 \times 10^3$, so $Q > K$ and the reaction is not at equilibrium. It proceeds to the left to convert products to reactants.

16.7 $K = 1.2$

16.9 (a) $K = 0.025$

(b) $K = 0.025$

(c) The amount of solid does not affect the equilibrium.

16.11 (a) $[COCl_2] = 0.00308$ M; $[CO] = 0.00712$ M

(b) $K = 144$

16.13 $[isobutane] = 0.024$ M; $[butane] = 0.010$ M

16.15 $[I_2] = 6.14 \times 10^{-3}$ M; $[I] = 4.79 \times 10^{-3}$ M

16.17 $[COBr_2] = 0.107$ M; $[CO] = [Br_2] = 0.143$ M

57.1% of the $COBr_2$ has decomposed.

16.19 (b)

16.21 (e), $K_2 = 1/(K_1)^2$

16.23 $K = 13.7$

16.25   (a) Equilibrium shifts to the right

(b) Equilibrium shifts to the left

(c) Equilibrium shifts to the right

(d) Equilibrium shifts to the left

16.27   Equilibrium concentrations are the same under both circumstances: [butane] = 1.1 M and [isobutane] = 2.9 M

16.29 $K = 3.9 \times 10^{-4}$

16.31   For decomposition of $COCl_2$, $K = 1/(K$ for $COCl_2$ formation$) = 1/(6.5 \times 10^{11}) = 1.5 \times 10^{-12}$

16.33 $K = 4$

16.35 Q is less than K, so the system shifts to form more isobutane.

At equilibrium, [butane] = 0.86 M and [isobutane] = 2.14 M.

16.37   The second equation has been reversed and multiplied by 2.

(c) $K_2 = 1/K_1^2$

16.39   (a) No change   (d) Shifts right

(b) Shifts left   (e) Shifts right

(c) No change

16.41   (a) The equilibrium will shift to the left on adding more Cl_2.

(b) K is calculated (from the quantities of reactants and products at equilibrium) to be 0.0470. After Cl_2 is added, the concentrations are: $[PCl_5]$ = 0.0199 M, $[PCl_3]$ = 0.0231 M, and $[Cl_2]$ = 0.0403 M.

16.43 $K_p = 0.215$

16.45 $P_{total} = 1.21$ atm

16.47 $[NH_3] = 0.67$ M; $[N_2] = 0.57$ M; $[H_2] = 1.7$ M; $P_{total} = 180$ atm

16.49   (a) $[NH_3] = [H_2S] = 0.013$ M

(b) $[NH_3] = 0.027$ M and $[H_2S] = 0.0067$ M

16.51   (a) Fraction dissociated = 0.15

(b) Fraction dissociated = 0.189. If the pressure decreases, the equilibrium shifts to the right, increasing the fraction of N_2O_4 dissociated.

16.53   (a) The flask containing $(H_3N)B(CH_3)_3$ will have the largest partial pressure of $B(CH_3)_3$.

(b) $P[B(CH_3)_3] = P(NH_3) = 2.1$ and $P[(H_3N)B(CH_3)_3] = 1.0$ atm

$P_{total} = 5.2$ atm

Percent dissociation = 69%

16.55 $P(CO) = 0.0010$ atm

16.57 1.7×10^{18} O atoms

16.59   Glycerin concentration should be 1.7 M

16.61   (a) $K_p = 0.20$

(b) When initial $[N_2O_4] = 1.00$ atm, the equilibrium pressures are $[N_2O_4] = 0.80$ atm and $[NO_2] = 0.40$ atm. When initial $[N_2O_4] = 0.10$ atm, the equilibrium pressures are $[N_2O_4] = 0.050$ atm and $[NO_2] = 0.10$ atm. The percent dissociation is now 50.%. This is in accord with Le Chatelier's principle: If the initial pressure of the reactant decreases, the equilibrium shifts to the right, increasing the fraction of the reactant dissociated. See also Question 16.51.

16.63   (a) False. The magnitude of K is always dependent on temperature.

(b) True

(c) False. The equilibrium constant for a reaction is the reciprocal of the value of K for its reverse.

(d) True

(e) False. $\Delta n = 1$ so $K_p = K_c(RT)$

16.65   (a) Product-favored, $K \gg 1$

(b) Reactant-favored, $K \ll 1$

(c) Product-favored, $K \gg 1$

16.67   (a) $K_p = K_c = 56$. Because 2 mol of reactants gives 2 mol of product, Δn does not change and $K_p = K_c$ (see page 264).

(b,c) Initial $P(H_2) = P(I_2) = 2.6$ atm and $P_{total} = 5.2$ atm

At equilibrium, $P(H_2) = P(I_2) = 0.54$ atm and $P(HI) = 4.1$ atm. Therefore, $P_{total} = 5.2$ atm. The initial total pressure and the equilibrium total pressure are the same owing to the reaction stoichiometry.

16.69   The reaction is endothermic. Adding heat shifts an equilibrium in the endothermic direction.

16.71   An elementary chemical step can occur both in the forward and reverse directions. Solid lead chloride forms when solutions containing lead ions and chloride ions are mixed, and a solution containing lead ions and chloride ions forms when pure lead chloride is placed in water and heated.

Chapter 17

17.1   (a) CN^-, cyanide ion

(b) SO_4^{2-}, sulfate ion

(c) F^-, fluoride ion

17.3   (a) $H_3O^+(aq) + NO_3^-(aq)$; $H_3O^+(aq)$ is the conjugate acid of H_2O, and $NO_3^-(aq)$ is the conjugate base of HNO_3.

(b) $H_3O^+(aq) + SO_4^{2-}(aq)$; $H_3O^+(aq)$ is the conjugate acid of H_2O, and $SO_4^{2-}(aq)$ is the conjugate base of HSO_4^-.

(c) $H_2O + HF$; H_2O is the conjugate base of H_3O^+, and HF is the conjugate acid of F^-.

17.5   Brønsted acid: $HC_2O_4^-(aq) + H_2O(\ell) \rightleftharpoons$
$$H_3O^+(aq) + C_2O_4^{2-}(aq)$$

Brønsted base: $HC_2O_4^-(aq) + H_2O(\ell) \rightleftharpoons$
$$H_2C_2O_4(aq) + OH^-(aq)$$

17.7

	Acid (A)	Base (B)	Conjugate Base of A	Conjugate Acid of B
(a)	HCO_2H	H_2O	HCO_2^-	H_3O^+
(b)	H_2S	NH_3	HS^-	NH_4^+
(c)	HSO_4^-	OH^-	SO_4^{2-}	H_2O

17.9 $[H_3O^+] = 1.8 \times 10^{-4}$ M; acidic

17.11   HCl is a strong acid, so $[H_3O^+]$ = concentration of the acid. $[H_3O^+] = 0.0075$ M and $[OH^-] = 1.3 \times 10^{-12}$ M. pH = 2.12.

17.13 $Ba(OH)_2$ is a strong base, so $[OH^-] = 2 \times$ concentration of the base.

$[OH^-] = 3.0 \times 10^{-3}$ M; pOH = 2.52; and pH = 11.48

17.15   (a) The strongest acid is HCO_2H (largest K_a) and the weakest acid is C_6H_5OH (smallest K_a).

(b) The strongest acid (HCO_2H) has the weakest conjugate base.

(c) The weakest acid (C_6H_5OH) has the strongest conjugate base.

17.17   (c) HClO, the weakest acid in this list (Table 17.3), has the strongest conjugate base.

17.19 $CO_3^{2-}(aq) + H_2O(\ell) \longrightarrow HCO_3^-(aq) + OH^-(aq)$

17.21   Highest pH, Na_2S; lowest pH, $AlCl_3$ (which gives the weak acid $[Al(H_2O)_6]^{3+}$ in solution)

17.23 $pK_a = 4.19$

17.25 $K_a = 3.0 \times 10^{-10}$

17.27   2-Chlorobenzoic acid has the smaller pK_a value.

17.29 $K_b = 7.4 \times 10^{-12}$

17.31 $K_b = 6.3 \times 10^{-5}$

17.33 $CH_3CO_2H(aq) + HCO_3^-(aq) \rightleftharpoons$
$$CH_3CO_2^-(aq) + H_2CO_3(aq)$$

Equilibrium lies predominantly to the right because CH_3CO_2H is a stronger acid than H_2CO_3.

17.35   (a) Left; NH_3 and HBr are the stronger base and acid, respectively.

(b) Left; PO_4^{3-} and CH_3CO_2H are the stronger base and acid, respectively.

(c) Right; $Fe(H_2O)_6^{3+}$ and HCO_3^- are the stronger acid and base, respectively.

17.37   (a) $OH^-(aq) + HPO_4^{2-}(aq) \rightleftharpoons H_2O(\ell) + PO_4^{3-}(aq)$

(b) OH^- is a stronger base than PO_4^{3-}, so the equilibrium will lie to the right. (The predominant species in solution is PO_4^{3-}, so the solution is likely

to be basic because PO_4^{3-} is the conjugate base of a weak acid.)

17.39   (a) $CH_3CO_2H(aq) + HPO_4^{2-}(aq) \rightleftharpoons$
$$CH_3CO_2^-(aq) + H_2PO_4^-(aq)$$

(b) CH_3CO_2H is a stronger acid than $H_2PO_4^-$, so the equilibrium will lie to the right.

17.41   (a) 2.1×10^{-3} M; (b) $K_a = 3.5 \times 10^{-4}$

17.43 $K_b = 6.6 \times 10^{-9}$

17.45   (a) $[H_3O^+] = 1.6 \times 10^{-4}$ M

(b) Moderately weak; $K_a = 1.1 \times 10^{-5}$

17.47 $[CH_3CO_2^-] = [H_3O^+] = 1.9 \times 10^{-3}$ M and $[CH_3CO_2H] = 0.20$ M

17.49 $[H_3O^+] = [CN^-] = 3.2 \times 10^{-6}$ M; $[HCN] = 0.025$ M; pH = 5.50

17.51 $[NH_4^+] = [OH^-] = 1.64 \times 10^{-3}$ M; $[NH_3] = 0.15$ M; pH = 11.22

17.53 $[OH^-] = 0.0102$ M; pH = 12.01; pOH = 1.99

17.55   pH = 3.25

17.57 $[H_3O^+] = 1.1 \times 10^{-5}$ M; pH = 4.98

17.59 $[HCN] = [OH^-] = 3.3 \times 10^{-3}$ M; $[H_3O^+] = 3.0 \times 10^{-12}$ M; $[Na^+] = 0.441$ M

17.61 $[H_3O^+] = 1.5 \times 10^{-9}$ M; pH = 8.81

17.63   (a) The reaction produces acetate ion, the conjugate base of acetic acid. The solution is weakly basic. pH is greater than 7.

(b) The reaction produces NH_4^+, the conjugate acid of NH_3. The solution is weakly acidic. pH is less than 7.

(c) The reaction mixes equal molar amounts of strong base and strong acid. The solution will be neutral. pH will be 7.

17.65   (a) pH = 1.17; (b) $[SO_3^{2-}] = 6.2 \times 10^{-8}$ M

17.67   (a) $[OH^-] = [N_2H_5^+] = 9.2 \times 10^{-8}$ M; $[N_2H_6^{2+}] = 8.9 \times 10^{-16}$ M

(b) pH = 9.96

17.69   (a) Lewis base

(b) Lewis acid

(c) Lewis base (owing to lone pair of electrons on the N atom)

17.71   CO is a Lewis base in its reactions with transition metal atoms. It donates a lone pair of electrons on the C atom.

17.73   HOCN should be a stronger acid than HCN because the H atom in HOCN is attached to a highly electronegative O atom. This induces a positive charge on the H atom, making it more readily removed by an interaction with water.

17.75   The S atom is surrounded by four highly electronegative O atoms. The inductive effect of these atoms induces a positive charge on the H atom, making it susceptible to removal by water.

17.77 pH = 2.671

17.79 The weaker acid (smaller K_a) will have the higher pH in solution. Thus, the pH of a benzoic acid solution is higher than that of 4-chlorobenzoic acid.

17.81 $H_2S(aq) + CH_3CO_2^-(aq) \rightleftharpoons$

$$CH_3CO_2H(aq) + HS^-(aq)$$

The equilibrium lies to the left and favors the reactants.

17.83 $[X^-] = [H_3O^+] = 3.0 \times 10^{-3}$ M; $[HX] = 0.007$ M; pH = 2.52

17.85 $K_a = 1.4 \times 10^{-5}$; $pK_a = 4.86$

17.87 pH = 5.84

17.89 (a) Ethylamine is a stronger base than ethanolamine.
(b) For ethylamine, the pH of the solution is 11.82.

17.91 The K_b for pyridine (Appendix I) is 1.5×10^{-9}. Therefore, K_a for the conjugate acid, the pyridinium ion, is $K_a = K_w/K_b = 6.7 \times 10^{-6}$. The pH of the pyridinium hydrochloride solution is 3.39.

17.93 Acidic: $NaHSO_4$, NH_4Br, $FeCl_3$

Neutral: $KClO_4$, $NaNO_3$, $LiBr$

Basic: Na_2CO_3, $(NH_4)_2S$, Na_2HPO_4

17.95 $K_{net} = K_{a_1} \times K_{a_2} = 3.8 \times 10^{-6}$

17.97 For the reaction $HCO_2H(aq) + OH^-(aq) \longrightarrow H_2O(\ell) + HCO_2^-(aq)$, $K_{net} = K_a$ (for HCO_2H) $\times [1/K_w] = 1.8 \times 10^{10}$

17.99 To double the percent ionization, you must dilute 100 mL of solution to 400 mL.

17.101 pH = 7.97

17.103 $HCl < NH_4Cl < NaCl < NaCH_3CO_2 < KOH$

17.105 $[H_3O^+] = [(1.12 \times 10^{-3}) \times (3.91 \times 10^{-6})]^{1/2} = 6.62 \times 10^{-5}$ M

pH = 4.180

17.107 Water can both accept a proton (a Brønsted base) and donate a lone pair (a Lewis base). Water can also donate a proton (Brønsted acid), but it cannot accept a pair of electrons (and act as a Lewis acid).

17.109 Measure the pH of the 0.1 M solutions of the three bases. The solution containing the strongest base will have the highest pH. The solution having the weakest base will have the lowest pH.

17.111 Species in solution listed in order of decreasing concentration:

$H_2O > HCN$ (0.10 M) $> H_3O^+$ and
CN^- (6.3×10^{-6} M $> OH^-$ (1.6×10^{-9} M)

17.113 Mixing the NaOH and acetic acid solutions gives 60.0 mL of 0.075 M $NaCH_3CO_2$ (sodium acetate). This solution is weakly basic and has water, sodium ion, acetate

ions, acetic acid, and hydrogen and hydroxide ions. In order of decreasing concentration, these are

$H_2O > Na^+$ and $CH_3CO_2^-$ (0.075 M) $> OH^-$ and CH_3CO_2H (6.5×10^{-6} M) $> H_3O^+$ (1.5×10^{-9} M)

17.115 (a) $HClO_4 + H_2SO_4 \rightleftharpoons ClO_4^- + H_3SO_4^+$

(b) The O atoms on sulfuric acid have lone pairs of electrons that can be used to bind to an H^+ ion.

17.117 The possible cation–anion combinations are NaCl (neutral), NaOH (basic), NH_4Cl (acidic), NH_4OH (basic), HCl (acidic), and H_2O (neutral).

$A = H^+$ solution; $B = NH_4^+$ solution; $C = Na^+$ solution; $Y = Cl^-$ solution; $Z = OH^-$ solution

17.119 (a) Add the three equations.

$NH_4^+(aq) + H_2O(\ell) \longrightarrow NH_3(aq) + H_3O^+(aq)$
$$K_1 = K_w/K_b$$

$CN^-(aq) + H_2O(\ell) \longrightarrow HCN(aq) + OH^-(aq)$
$$K_2 = K_w/K_a$$

$H_3O^+(aq) + OH^-(aq) \longrightarrow 2 H_2O(\ell) \quad K_3 = 1/K_w$

$NH_4^+(aq) + CN^-(aq) \longrightarrow NH_3(aq) + HCN(aq)$

$K_{net} = K_1K_2K_3 = K_w/K_aK_b$

(b) Substitute expressions for K_w, K_a, and K_b into the equation.

$$[H_3O^+] = \sqrt{\frac{K_wK_a}{K_b}}$$

$$\sqrt{\frac{K_wK_a}{K_b}} = \sqrt{\frac{[H_3O^-][OH^-]\left(\frac{[H_3O^+][NH_3]}{[NH_4^+]}\right)}{\frac{[OH^-][HCN]}{[CN^-]}}}$$

In a solution of NH$_4$CN, we have $[NH_4^+] = [CN^-]$ and $[NH_3] = [HCN]$. When these and $[OH^-]$, are canceled from the expression, we see it is equal to $[H_3O^+]$.

(c) pH = 9.33

Chapter 18

18.1 (a) Decrease pH; (b) increase pH; (c) no change in pH

18.3 pH = 9.25

18.5 pH = 4.38

18.7 pH = 9.12; pH of buffer is lower than the pH of the original solution of NH$_3$(pH = 11.17).

18.9 4.7 g

18.11 pH = 4.92

18.13 (a) pH = 3.59; (b) $[HCO_2H]/[HCO_2^-] = 0.45$

18.15   (b), $NH_3 + NH_4Cl$

18.17   The buffer must have a ratio of 0.51 mol NaH_2PO_4 to 1 mol Na_2HPO_4. For example, dissolve 0.51 mol NaH_2PO_4 (61 g) and 1.0 mol Na_2HPO_4 (140 g) in some amount of water.

18.19   (a) pH = 4.95; (b) pH = 5.05

18.21   (a) pH = 9.55; (b) pH = 9.50

18.23   (a) Original pH = 5.62

       (b) $[Na^+] = 0.0323$ M, $[OH^-] = 1.5 \times 10^{-3}$ M, $[H_3O^+] = 6.5 \times 10^{-12}$ M, and $[C_6H_5O^-] = 0.0308$ M

       (c) pH = 11.19

18.25   (a) Original NH_3 concentration = 0.0154 M

       (b) At the equivalence point $[H_3O^+] = 1.9 \times 10^{-6}$ M, $[OH^-] = 5.3 \times 10^{-9}$ M, $[NH_4^+] = 6.25 \times 10^{-3}$ M.

       (c) pH at equivalence point = 5.73

18.27   The titration curve begins at pH = 13.00 and drops slowly as HCl is added. Just before the equivalence point (when 30.0 mL of acid has been added), the curve falls steeply. The pH at the equivalence point is exactly 7. Just after the equivalence point, the curve flattens again and begins to approach the final pH of just over 1.0. The total volume at the equivalence point is 60.0 mL.

18.29   (a) Starting pH = 11.12

       (b) pH at equivalence point = 5.28

       (c) pH at midpoint (half-neutralization point) = 9.25

       (d) Methyl red, bromcresol green

       (e)

Acid (mL)	Added pH
5.00	9.85
15.0	9.08
20.0	8.65
22.0	8.39
30.0	2.04

18.31   See Figure 18.10 on page 872.

       (a) Thymol blue or bromphenol blue

       (b) Phenolphthalein

       (c) Methyl red; thymol blue

18.33   (a) Silver chloride, AgCl; lead chloride, $PbCl_2$

       (b) Zinc carbonate, $ZnCO_3$; zinc sulfide, ZnS

       (c) Iron(II) carbonate, $FeCO_3$; iron(II) oxalate, FeC_2O_4

18.35   (a) and (b) are soluble, (c) and (d) are insoluble.

18.37   (a) $AgCN(s) \longrightarrow Ag^+(aq) + CN^-(aq)$, $K_{sp} = [Ag^+][CN^-]$

       (b) $NiCO_3(s) \longrightarrow Ni^{2+}(aq) + CO_3^{2-}(aq)$, $K_{sp} = [Ni^{2+}][CO_3^{2-}]$

       (c) $AuBr_3(s) \longrightarrow Au^{3+}(aq) + 3\,Br^-(aq)$, $K_{sp} = [Au^{3+}][Br^-]^3$

18.39 $K_{sp} = (1.9 \times 10^{-3})^2 = 3.6 \times 10^{-6}$

18.41 $K_{sp} = 4.37 \times 10^{-9}$

18.43 $K_{sp} = 1.4 \times 10^{-15}$

18.45   (a) 9.2×10^{-9} M; (b) 2.2×10^{-6} g/L

18.47   (a) 2.4×10^{-4} M; (b) 0.018 g/L

18.49   Only 2.1×10^{-4} g dissolves.

18.51   (a) $PbCl_2$; (b) FeS; (c) $Fe(OH)_2$

18.53   Solubility in pure water = 1.0×10^{-6} mol/L; solubility in 0.010 M SCN^- = 1.0×10^{-10} mol/L

18.55   (a) Solubility in pure water = 2.2×10^{-6} mg/mL

       (b) Solubility in 0.020 M $AgNO_3$ = 1.0×10^{-12} mg/mL

18.57   (a) PbS

       (b) Ag_2CO_3

       (c) $Al(OH)_3$

18.59 $Q < K_{sp}$, so no precipitate forms.

18.61 $Q > K_{sp}$; $Zn(OH)_2$ will precipitate.

18.63 $[OH^-]$ must exceed 1.0×10^{-5} M.

18.65 $AuCl(s) \rightleftharpoons Au^+(aq) + Cl^-(aq)$

       $Au^+(aq) + 2\,CN^-(aq) \rightleftharpoons Ag(CN)_2^-(aq)$

       Net: $AuCl(s) + 2\,CN^-(aq) \rightleftharpoons Au(CN)_2^-(aq) + Cl^-(aq)$

       $K_{net} = K_{sp} \times K_f = 4.0 \times 10^{25}$

18.67   (a) Add H_2SO_4, precipitating $BaSO_4$ and leaving $Na^+(aq)$ in solution.

       (b) Add HCl or another source of chloride ion. $PbCl_2$ will precipitate, but $NiCl_2$ is water-soluble.

18.69   (a) $NaBr(aq) + AgNO_3(aq) \longrightarrow NaNO_3(aq) + AgBr(s)$

       (b) $2\,KCl(aq) + Pb(NO_3)_2(aq) \longrightarrow 2\,KNO_3(aq) + PbCl_2(s)$

18.71 $Q > K_{sp}$, so $BaSO_4$ precipitates.

18.73 $[H_3O^+] = 1.9 \times 10^{-10}$ M; pH = 9.73

18.75 $BaCO_3 < Ag_2CO_3 < Na_2CO_3$

18.77   Original pH = 8.62; dilution will not affect the pH.

18.79   (a) pH = 2.81

       (b) pH at equivalence point = 8.72

       (c) pH at the midpoint = pK_a = 4.62

       (d) Phenolphthalein

       (e) After 10.0 mL, pH = 4.39.

          After 20.0 mL, pH = 5.07.

          After 30.0 mL, pH = 11.84.

       (f) A plot of pH versus volume of NaOH added would begin at a pH of 2.81, rise slightly to the midpoint at pH = 4.62, and then begin to rise more steeply as the equivalence point is approached (when the volume of NaOH added is 27.0 mL). The pH rises vertically through the equivalence point, and then begins to level off above a pH of about 11.0.

18.81   110 mL NaOH

18.83   The K_b value for ethylamine (4.27×10^{-4}) is found in Appendix I.

(a) pH = 11.89

(b) Midpoint pH = 10.63

(c) pH = 10.15

(d) pH = 5.93 at the equivalence point

(e) pH = 2.13

(g) Alizarin or bromcresol purple (see Figure 18.10)

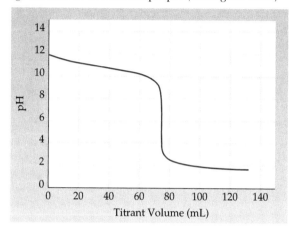

18.85   (a) 0.100 M acetic acid has a pH of 2.87. Adding sodium acetate slowly raises the pH.

(b) Adding $NaNO_3$ to 0.100 M HNO_3 has no effect on the pH.

(c) In part (a), adding the conjugate base of a weak acid creates a buffer solution. In part (b), HNO_3 is a strong acid, and its conjugate base (NO_3^-) is so weak that the base has no effect on the complete ionization of the acid.

18.87   (a) HPO_4^{2-}

(b) 32 g of Na_2HPO_4

(c) Add 10. g of the base, Na_3PO_4, to raise the pH.

18.89   (a) $BaSO_4$ will precipitate first.

(b) $[Ba^{2+}] = 1.8 \times 10^{-7}$ M

18.91 $K = 2.1 \times 10^6$; yes, AgI forms

18.93   (a) $[F^-] = 1.3 \times 10^{-3}$ M; (b) $[Ca^{2+}] = 2.9 \times 10^{-5}$ M

18.95   (a) $PbSO_4$ will precipitate first.

(b) $[Pb^{2+}] = 5.1 \times 10^{-6}$ M

18.97 $Cu(OH)_2$ will dissolve in a non-oxidizing acid such as HCl, whereas CuS will not.

18.99   The strong base (OH^-) is consumed completely in a reaction with the weak acid present in the buffer. In an acetic acid/sodium acetate buffer, the acetic acid reacts with OH^- to produce more of the conjugate base, the acetate ion.

$$OH^-(aq) + CH_3CO_2H(aq) \longrightarrow CH_3CO_2^-(aq) + H_2O(\ell)$$

18.101 When Ag_3PO_4 dissolves slightly, it produces a small concentration of the phosphate ion, PO_4^{3-}. This ion is a strong base and hydrolyzes to HPO_4^{2-}. As this reaction removes the PO_4^{3-} ion from equilibrium with Ag_3PO_4, the equilibrium shifts to the right, producing more PO_4^{3-} and Ag^+ ions. Thus, Ag_3PO_4 dissolves to a greater extent than might be calculated from a K_{sp} value (unless the K_{sp} value was actually determined experimentally).

18.103 (a) Base is added to increase the pH. The added base reacts with acetic acid to form more acetate ions in the mixture. Thus, the fraction of acid declines and the fraction of conjugate base rises (i.e., the ratio $[CH_3CO_2H]/[CH_3CO_2^-]$ decreases) as the pH rises.

(b) At pH = 4, acid predominates (85% acid and 15% acetate ions). At pH = 6, acetate ions predominate (95% acetate ions and 5% acid).

(c) At the point the lines cross, $[CH_3CO_2H] = [CH_3CO_2^-]$. At this point pH = pK_a, so pK_a for acetic acid is 4.75.

18.105 (a) C—C—C angle, 120°; O—C=O, 120°; C—O—H, 109°; C—C—H, 120°

(b) Both the ring C atoms and the C in CO_2H are sp^2 hybridized.

(c) $K_a = 1 \times 10^{-3}$

(d) 10%

(e) pH at half-way point = pK_a = 3.0; pH at equivalence point = 7.3

18.107 $PbCl_2(s) \longrightarrow Pb^{2+}(aq) + 2 Cl^-(aq)$

Adding Cl^- ions to the test tube shifts the equilibrium to the left, forming more $PbCl_2$ and decreasing the concentration of Pb^{2+} ions.

18.109 Decreasing the pH by 1.0 is equivalent to decreasing $[OH^-]$ by a factor of 10. More $Ca(OH)_2$ will dissolve to replace OH^- removed from solution. The hydroxide ion concentration in the equilibrium constant expression for $Ca(OH)_2$ is squared, so decreasing $[OH^-]$ by a factor of 10 results in a solubility increase of 10^2 (i.e., 100).

Chapter 19

19.1   (a) Disorder increases as CO_2 goes from the solid phase to the gas phase.

(b) Liquid water at 50 °C

(c) Ruby

(d) One mole of N_2 at 1 bar

19.3   (a) $CH_3OH(g)$; (b) HBr; (c) $NH_4Cl(aq)$; (d) $HNO_3(g)$

19.5   (a) $\Delta S° = +12.7$ J/K. Disorder increases.

(b) $\Delta S° = -102.55$ J/K. Significant decrease in disorder.

(c) $\Delta S° = +93.2$ J/K. Disorder increases.

(d) $\Delta S° = -129.7$ J/K. The solution is more ordered (with H^+ forming H_3O^+ and hydrogen bonding occurring) than HCl in the gaseous state.

19.7 $\Delta S° = -174.1$ J/K

19.9   (a) $\Delta S° = +9.3$ J/K; (b) $\Delta S° = -293.97$ J/K

19.11   (a) $\Delta S° = -507.3$ J/K; (b) $\Delta S° = +313.25$ J/K

19.13 $\Delta S°_{sys} = -134.18$ J/K; $\Delta H°_{sys} = -662.75$ kJ; $\Delta S°_{surr} = +2222.9$ J/K; $\Delta S°_{univ} = +2088.7$ J/K

19.15 $\Delta S°_{sys} = +163.3$ J/K; $\Delta H°_{sys} = +285.83$ kJ; $\Delta S°_{surr} = -958.68$ J/K; $\Delta S°_{univ} = -795.4$ J/K

The reaction is not spontaneous, because the overall entropy change in the universe is negative. The reaction is disfavored by energy dispersal.

19.17   (a) Type 2. The reaction is enthalpy-favored but entropy-disfavored. It is more favorable at low temperatures.

(b) Type 4. This endothermic reaction is not favored by the enthalpy change nor is it favored by the entropy change. It is not spontaneous under any conditions.

19.19   (a) $\Delta S°_{sys} = +174.75$ J/K; $\Delta H°_{surr} = +116.94$ kJ; $\Delta S°_{surr} = -392.4$ J/K

(b) $\Delta S°_{univ} = -217.67$ J/K. The reaction is not spontaneous at 298 K.

(c) As the temperature increases, $\Delta S°_{surr}$ becomes less important, so $\Delta S°_{univ}$ will become positive at a sufficiently high temperature.

19.21   (a) $\Delta H°_{rxn} = -438$ kJ; $\Delta S°_{rxn} = -201.7$ J/K; $\Delta G°_{rxn} = -378$ kJ

The reaction is product-favored and is enthalpy-driven.

(b) $\Delta H°_{rxn} = -86.61$ kJ; $\Delta S°_{rxn} = -79.4$ J/K; $\Delta G°_{rxn} = -62.9$ kJ

The reaction is product-favored. The enthalpy change favors the reaction.

19.23   (a) $\Delta H°_{rxn} = +116.7$ kJ; $\Delta S°_{rxn} = +168.0$ J/K; $\Delta G°_f = +66.6$ kJ/mol

(b) $\Delta H°_{rxn} = -425.93$ kJ; $\Delta S°_{rxn} = -154.6$ J/K; $\Delta G°_f = -379.82$ kJ/mol

(c) $\Delta H°_{rxn} = +17.51$ kJ; $\Delta S°_{rxn} = +77.95$ J/K; $\Delta G°_f = -5.73$ kJ/mol

19.25   (a) $\Delta G°_{rxn} = -817.54$ kJ; spontaneous

(b) $\Delta G°_{rxn} = +256.6$ kJ; not spontaneous

(c) $\Delta G°_{rxn} = -1101.14$ kJ; spontaneous

19.27 $\Delta G°_f[BaCO_3(s)] = -1134.4$ kJ/mol

19.29   (a) $\Delta H°_{rxn} = +66.2$ kJ; $\Delta S°_{rxn} = -121.62$ J/K; $\Delta G°_{rxn} = +102.5$ kJ

Both the enthalpy and the entropy changes indicate the reaction is not spontaneous. There is no temperature to which it will become spontaneous. This is a case like that in the right panel in Figure 19.12 and is a Type 4 reaction (Table 19.2).

(b) $\Delta H°_{rxn} = -221.05$ kJ; $\Delta S°_{rxn} = +179.1$ J/K; $\Delta G°_{rxn} = -283.99$ kJ

The reaction is favored by both enthalpy and entropy and is product-favored at all temperatures. This is a case like that in the left panel in Figure 19.12 and is a Type 1 reaction.

(c) $\Delta H°_{rxn} = -179.0$ kJ; $\Delta S°_{rxn} = -160.2$ J/K; $\Delta G°_{rxn} = -131.4$ kJ

The reaction is favored by the enthalpy change but disfavored by the entropy change. The reaction becomes less product-favored as the temperature increases; it is a case like the upper line in the middle panel of Figure 19.12.

(d) $\Delta H°_{rxn} = +822.2$ kJ; $\Delta S°_{rxn} = +181.28$ J/K; $\Delta G°_{rxn} = +768.08$ kJ

The reaction is not favored by the enthalpy change but favored by the entropy change. The reaction becomes more product-favored as the temperature increases; it is a case like the lower line in the middle panel of Figure 19.12.

19.31 $\Delta H°_{rxn} = +337.2$ kJ and $\Delta S°_{rxn} = +161.5$ J/K. When $\Delta G°_{rxn} = 0$, $T = 2088$ K.

19.33 $K = 6.8 \times 10^{-16}$. Note that K is very small and that $\Delta G°$ is positive. Both indicate a reactant-favored process.

19.35 $\Delta G°_{rxn} = -100.24$ kJ and $K_p = 3.64 \times 10^{17}$. Both the free energy change and K indicate a product-favored process.

19.37   Reaction 1: $\Delta S°_1 = -80.7$ J/K

Reaction 2: $\Delta S°_2 = -161.60$ J/K

Reaction 3: $\Delta S°_3 = -242.3$ J/K

$\Delta S°_1 + \Delta S°_2 = \Delta S°_3$

19.39 $\Delta H°_{rxn} = -1428.66$ kJ; $\Delta S°_{rxn} = +47.1$ J/K; $\Delta S°_{univ} = +4840$ J/K

Combustion reactions are spontaneous, and this is confirmed by the sign of $\Delta S°_{univ}$.

19.41   The reaction occurs spontaneously and is product-favored. Therefore, $\Delta S°_{univ}$ is positive and $\Delta G°_{rxn}$ is negative. The reaction is likely to be exothermic, so $\Delta H°_{rxn}$ is negative, and $\Delta S°_{surr}$ is positive. $\Delta S°_{sys}$ is expected to be negative because two moles of gas form one mole of solid. The calculated values are as follows:

$\Delta S^\circ_{sys} = -284.2$ J/K

$\Delta H^\circ_{rxn} = -176.34$ kJ

$\Delta S^\circ_{surr} = +591.45$ J/K

$\Delta S^\circ_{univ} = +307.3$ J/K

$\Delta G^\circ_{rxn} = -91.64$ kJ

$K_p = 1.13 \times 10^{16}$

19.43 $K_p = 1.3 \times 10^{29}$ at 298 K ($\Delta G^\circ = -166.1$ kJ). The reaction is already extremely product-favored at 298 K. A higher temperature, however, would make the reaction less product-favored because ΔS°_{rxn} has a negative value (-242.3 J/K).

19.45   At the boiling point, $\Delta G^\circ = 0 = \Delta H^\circ - T\Delta S^\circ$.

Here $\Delta S^\circ = \Delta H^\circ / T = 112$ J/K · mol at 351.15 K.

19.47   For $C_2H_5OH(\ell) \longrightarrow C_2H_5OH(g)$, $\Delta S^\circ = +122.0$ J/K and $\Delta H^\circ = +41.7$ kJ.

$T = \Delta H^\circ / \Delta S^\circ = 341.8$ K or 68.7 °C.

19.49 ΔS°_{rxn} is $+137.2$ J/K. A positive entropy change means that raising the temperature will increase the product favorability of the reaction (because $T\Delta S^\circ$ will become more negative).

19.51   (a) The reaction is endothermic and reactant-favored. Predicted results: $\Delta H^\circ_{rxn} > 0$, $\Delta S^\circ_{surr} < 0$, $\Delta S^\circ_{univ} < 0$, and $\Delta G^\circ_{sys} > 0$. ΔS°_{sys} is > 0 because 1 mol of gas and 2 mol of liquid are produced from 2 mol of solid.

Calculated results: $\Delta H^\circ_{rxn} = +181.66$ kJ, $\Delta S^\circ_{sys} = +216.53$ J/K, $\Delta G^\circ_{sys} = +117$ kJ.

(b) $K_p = 3.1 \times 10^{-21}$ Reaction is reactant-favored.

19.53   The reaction is exothermic, so ΔH°_{rxn} should be negative. Also, a gas and an aqueous solution are formed, so ΔS°_{rxn} should be positive. The calculated values are

$\Delta H^\circ_{rxn} = -183.32$ kJ (with a negative sign as expected)

$\Delta S^\circ_{rxn} = -7.7$ J/K

The entropy change is slightly negative, not positive as predicted. The reason for this is the negative entropy change upon dissolving NaOH. Apparently the OH$^-$ ions in water hydrogen-bond with water molecules and lead to a slight ordering of the system relative to pure water.

19.55 $\Delta H^\circ_{rxn} = +126.03$ kJ; $\Delta S^\circ_{rxn} = +78.2$ J/K; and $\Delta G^\circ_{rxn} = +103$ kJ.

The reaction is not predicted to be spontaneous under standard conditions.

19.57 $\Delta G^\circ_{rxn} = 6.98$ kJ/mol

19.59 $\Delta G^\circ_{rxn} = -98.9$ kJ

The reaction is spontaneous under standard conditions and is enthalpy-driven

19.61   (a) $\Delta G^\circ_{rxn} = +141.82$ kJ, so the reaction is not spontaneous.

(b) $\Delta H^\circ_{rxn} = +197.86$ kJ; $\Delta S^\circ_{rxn} = +187.95$ J/K

$T = \Delta H^\circ_{rxn} / \Delta S^\circ_{rxn} = 1052.7$ K or 779.6 °C

(c) ΔG°_{rxn} at 1500 °C (1773 K) $= -135.4$ kJ

K_p at 1500 °C $= 1 \times 10^4$

19.63 $\Delta S^\circ_{rxn} = -459.0$ J/K; $\Delta H^\circ_{rxn} = -793$ kJ; $\Delta G^\circ_{rxn} = -657$ kJ

The reaction is spontaneous and enthalpy-driven.

19.65   (a) ΔG° at 80.0 °C $= +0.14$ kJ

ΔG° at 110.0 °C $= -0.12$ kJ

Rhombic sulfur is more stable than monoclinic sulfur at 80 °C, but the reverse is true at 110 °C.

(b) $T = 370$ K or about 96 °C

19.67   (a) $\Delta S^\circ_{rxn} = +24.89$ J/K; $\Delta H^\circ_{rxn} = +180.58$ kJ; $\Delta G^\circ_{rxn} = +173.16$ kJ

$K_p = 4.62 \times 10^{-31}$, so the reaction is reactant-favored at 298 K.

(b) At 700 °C (973 K), $\Delta G^\circ_{rxn} = +156.6$ kJ

$K_p = 4 \times 10^{-9}$, so the reaction is still reactant-favored at 700 °C, but less so than at 298 K.

(c) $P(NO) = 6 \times 10^{-5}$; $P(O_2) = P(N_2) = 1$ bar

19.69 $\Delta G^\circ_f[HI(g)] = -10.9$ kJ/mol

19.71 $K_p = P_{Hg(g)}$ at any temperature.

$K_p = 1$ at 620.3 K or 347.2 °C when $P_{Hg(g)} = 1.000$ bar.

T when $P_{Hg(g)} = (1/760)$ bar is 393.3 K or 125.2 °C.

19.73   (a) Reactant-favored (mercury is a liquid under standard conditions)

(b) Product-favored (water vapor will condense to a liquid)

(c) Reactant-favored (a continuous supply of energy is required)

(d) Product-favored (carbon will burn)

(e) Product-favored (salt will dissolve)

(f) Reactant-favored (calcium carbonate is insoluble)

19.75   (a) True

(b) False. Whether an exothermic system is spontaneous also depends on the entropy change for the system.

(c) False. Reactions with $+\Delta H^\circ_{rxn}$ and $+\Delta S^\circ_{rxn}$ are spontaneous at higher temperatures.

(d) True

19.77   Dissolving a solid such as NaCl in water is a spontaneous process. Thus, $\Delta G^\circ < 0$. If $\Delta H^\circ = 0$, then the only way the free energy change can be negative is if ΔS° is positive. Generally the entropy change is the driving force in forming a solution.

19.79 $2\,C_2H_6(g) + 7\,O_2(g) \longrightarrow 4\,CO_2(g) + 6\,H_2O(g)$

(a) Not only is this an exothermic combustion reaction, but there is also an increase in the number of molecules from reactants to products. Therefore, we would predict an increase in $\Delta S°$ for both the system and the surroundings and thus for the universe as well.

(b) The exothermic reaction has $\Delta H°_{rxn} < 0$. Combined with a positive $\Delta S°_{sys}$, the value of $\Delta G°_{rxn}$ is negative.

(c) The value of K_p is likely to be much greater than 1. Further, because $\Delta S°_{sys}$ is positive, the value of K_p will be even larger at a higher temperature. (See the left panel of Figure 19.12.)

19.81 In solid NaCl, the particles are fixed in a solid lattice. When the solid is dissolved, the particles (Na^+ and Cl^- ions) are dispersed throughout the solution.

19.83 Iodine dissolves readily, so the process is spontaneous and $\Delta G°$ must be less than zero. Because $\Delta H° = 0$, the process is entropy-driven.

19.85 (a) $\Delta H°_{rxn} = -352.88$ kJ and $\Delta S°_{rxn} = +21.31$ J/K. Therefore, at 298 K, $\Delta G°_{rxn} = -359.23$ kJ.

(b) 4.84 g of Mg is required.

19.87 (a) $N_2H_4(\ell) + O_2(g) \longrightarrow 2 H_2O(\ell) + N_2(g)$
O_2 is the oxidizing agent and N_2H_4 is the reducing agent.

(b) $\Delta H°_{rxn} = -622.29$ kJ and $\Delta S°_{rxn} = +4.87$ J/K. Therefore, at 298 K, $\Delta G°_{rxn} = -623.77$ kJ.

(c) 0.0027 K

(d) 7.5 mol O_2

(e) 4.8×10^3 g solution

(f) 7.5 mol $N_2(g)$ occupies 170 L at 273 K and 1.0 atm of pressure.

19.89 (a) $\Delta S°_{sys} = -60.49$ J/K

(b) No

(c) Yes

(d) Yes

(e) Yes. $\Delta S°_{surr}$ always changes with T.

(f) The reaction is spontaneous at 400 K but not at 700 K.

19.91 (a) $\Delta G°$ decreases as temperature increases.

(b) No, the reaction is always spontaneous.

(c) The spontaneity of a reaction is dependent on temperature because $\Delta G°_{rxn}$ is determined in part by the $T\Delta S°$ term.

Chapter 20

20.1 (a) $Cr(s) \longrightarrow Cr^{3+}(aq) + 3 e^-$
Cr is a reducing agent; this is an oxidation reaction.

(b) $AsH_3(g) \longrightarrow As(s) + 3 H^+(aq) + 3 e^-$
AsH_3 is a reducing agent; this is an oxidation reaction.

(c) $VO_3^-(aq) + 6 H^+(aq) + 3 e^- \longrightarrow V^{2+}(aq) + 3 H_2O(\ell)$
$VO_3^-(aq)$ is an oxidizing agent; this is a reduction reaction.

(d) $2 Ag(s) + 2 OH^-(aq) \longrightarrow Ag_2O(s) + H_2O(\ell) + 2e^-$
Silver is a reducing agent; this is an oxidation reaction.

20.3 (a) $Ag(s) \longrightarrow Ag^+(aq) + e^-$
$e^- + NO_3^-(aq) + 2 H^+(aq) \longrightarrow NO_2(g) + H_2O(\ell)$
$\overline{Ag(s) + NO_3^-(aq) + 2 H^+(aq) \longrightarrow}$
$Ag^+(aq) + NO_2(g) + H_2O(\ell)$

(b) $2[MnO_4^-(aq) + 8 H^+(aq) + 5 e^- \longrightarrow Mn^{2+}(aq) + 4 H_2O(\ell)]$
$5[HSO_3^-(aq) + H_2O(\ell) \longrightarrow SO_4^{2-}(aq) + 3 H^+(aq) + 2 e^-]$
$\overline{2 MnO_4^-(aq) + H^+(aq) + 5 HSO_3^-(aq) \longrightarrow}$
$2 Mn^{2+}(aq) + 3 H_2O(\ell) + 5 SO_4^{2-}(aq)$

(c) $4[Zn(s) \longrightarrow Zn^{2+}(aq) + 2 e^-]$
$2 NO_3^-(aq) + 10 H^+(aq) + 8 e^- \longrightarrow N_2O(g) + 5 H_2O(\ell)$
$\overline{4 Zn(s) + 2 NO_3^-(aq) + 10 H^+(aq) \longrightarrow}$
$4 Zn^{2+}(aq) + N_2O(g) + 5 H_2O(\ell)$

(d) $Cr(s) \longrightarrow Cr^{3+}(aq) + 3 e^-$
$3 e^- + NO_3^-(aq) + 4 H^+(aq) \longrightarrow NO(g) + 2 H_2O(\ell)$
$\overline{Cr(s) + NO_3^-(aq) + 4 H^+(aq) \longrightarrow}$
$Cr^{3+}(aq) + NO(g) + 2 H_2O(\ell)$

20.5 (a) $2[Al(s) + 4 OH^-(aq) \longrightarrow Al(OH)_4^-(aq) + 3 e^-]$
$3[2 H_2O(\ell) + 2 e^- \longrightarrow H_2(g) + 2 OH^-(aq)]$
$\overline{2 Al(s) + 2 OH^-(aq) + 6 H_2O(\ell) \longrightarrow}$
$2 Al(OH)_4^-(aq) + 3 H_2(g)$

(b) $2[CrO_4^{2-}(aq) + 4 H_2O(\ell) + 3 e^- \longrightarrow Cr(OH)_3(s) + 5 OH^-(aq)]$
$3[SO_3^{2-}(aq) + 2 OH^-(aq) \longrightarrow SO_4^{2-}(aq) + H_2O(\ell) + 2 e^-]$
$\overline{2 CrO_4^{2-}(aq) + 3 SO_3^{2-}(aq) + 5 H_2O(\ell) \longrightarrow}$
$2 Cr(OH)_3(s) + 3 SO_4^{2-}(aq) + 4 OH^-(aq)$

(c) $Zn(s) + 4 OH^-(aq) \longrightarrow Zn(OH)_4^{2-}(aq) + 2 e^-$
$Cu(OH)_2(s) + 2 e^- \longrightarrow Cu(s) + 2 OH^-(aq)$
$\overline{Zn(s) + 2 OH^-(aq) + Cu(OH)_2(s) \longrightarrow}$
$Zn(OH)_4^{2-}(aq) + Cu(s)$

(d) $3[HS^-(aq) + OH^-(aq) \longrightarrow S(s) + H_2O(\ell) + 2 e^-]$
$ClO_3^-(aq) + 3 H_2O(\ell) + 6 e^- \longrightarrow Cl^-(aq) + 6 OH^-(aq)$
$\overline{3 HS^-(aq) + ClO_3^-(aq) \longrightarrow}$
$3 S(s) + Cl^-(aq) + 3 OH^-(aq)$

20.7 Electrons flow from the Cr electrode to the Fe electrode. Negative ions move via the salt bridge from the Fe/Fe^{2+} half-cell to the Cr/Cr^{3+} half-cell (and positive ions move in the opposite direction).
Anode (oxidation): $Cr(s) \longrightarrow Cr^{3+}(aq) + 3 e^-$
Cathode (reduction): $Fe^{2+}(aq) + 2 e^- \longrightarrow Fe(s)$

20.9   (a) Oxidation: $Fe(s) \longrightarrow Fe^{2+}(aq) + 2\,e^-$

Reduction: $O_2(g) + 4\,H^+(aq) + 4\,e^- \longrightarrow 2\,H_2O(\ell)$

Overall: $2\,Fe(s) + O_2(g) + 4\,H^+(aq) \longrightarrow$
$2\,Fe^{2+}(aq) + 2\,H_2O(\ell)$

(b) Anode, oxidation: $Fe(s) \longrightarrow Fe^{2+}(aq) + 2\,e^-$

Cathode, reduction:
$O_2(g) + 4\,H^+(aq) + 4\,e^- \longrightarrow 2\,H_2O(\ell)$

(c) Electrons flow from the negative anode (Fe) to the positive cathode (site of the O_2 half-reaction). Negative ions move through the salt bridge from the cathode compartment in which the O_2 reduction occurs to the anode compartment in which Fe oxidation occurs (and positive ions move in the opposite direction).

20.11   (a) All are primary batteries, not rechargeable.

(b) Dry cells and alkaline batteries have Zn anodes. Ni-Cd batteries have a cadmium anode.

(c) Dry cells have an acidic environment, whereas the environment is alkaline for alkaline and Ni-Cd cells.

20.13   (a) $E^\circ_{cell} = -1.298$ V; not product-favored

(b) $E^\circ_{cell} = -0.51$ V; not product-favored

(c) $E^\circ_{cell} = -1.023$ V; not product-favored

(d) $E^\circ_{cell} = +0.029$ V; product-favored

20.15   (a) $Sn^{2+}(aq) + 2\,Ag(s) \longrightarrow Sn(s) + 2\,Ag^+(aq)$

$E^\circ_{cell} = -0.94$ V; not product-favored

(b) $3\,Sn^{4+}(aq) + 2\,Al(s) \longrightarrow 3\,Sn^{2+}(aq) + 2\,Al^{3+}(aq)$

$E^\circ_{cell} = +1.81$ V; product-favored

(c) $ClO_3^-(aq) + 6\,Ce^{3+}(aq) + 6\,H^+(aq) \longrightarrow$
$Cl^-(aq) + 6\,Ce^{4+}(aq) + 3\,H_2O(\ell)$

$E^\circ_{cell} = -0.99$ V; not product-favored

(d) $3\,Cu(s) + 2\,NO_3^-(aq) + 8\,H^+(aq) \longrightarrow$
$3\,Cu^{2+}(aq) + 2\,NO(g) + 4\,H_2O(\ell)$

$E^\circ_{cell} = +0.62$ V; product-favored

20.17   (a) Al

(b) Zn and Al

(c) $Fe^{2+}(aq) + Sn(s) \longrightarrow Fe(s) + Sn^{2+}(aq)$; reactant-favored

(d) $Zn^{2+}(aq) + Sn(s) \longrightarrow Zn(s) + Sn^{2+}(aq)$; not product-favored

20.19   Best reducing agent, $Zn(s)$

20.21 Ag^+

20.23   See Example 20.5, page 971.

(a) F_2, most readily reduced

(b) F_2 and Cl_2

20.25 $E^\circ_{cell} = +0.3923$ V. When $[Zn(OH)_4{}^{2-}] = [OH^-] = 0.025$ M and $P(H_2) = 1.0$ bar, $E_{cell} = 0.345$ V.

20.27 $E^\circ_{cell} = +1.563$ V and $E_{cell} = +1.58$ V.

20.29   When $E^\circ_{cell} = +1.563$ V, $E_{cell} = 1.48$ V, $n = 2$, and $[Zn^{2+}] = 1.0$ M, the concentration of $Ag^+ = 0.040$ M.

20.31   (a) $\Delta G^\circ = -45.5$ kJ; $K = 9 \times 10^7$

(b) $\Delta G^\circ = +110$ kJ; $K = 4 \times 10^{-19}$

20.33 E°_{cell} for $AgBr(s) \longrightarrow Ag^+(aq) + Br^-(aq)$ is -0.7281.

$K_{sp} = 4.9 \times 10^{-13}$

20.35 $K_{formation} = 2 \times 10^{25}$

20.37   (a) $Fe^{2+}(aq) + 2\,e^- \longrightarrow Fe(s)$

$2[Fe^{2+}(aq) \longrightarrow Fe^{3+}(aq) + e^-]$

$3\,Fe^{2+}(aq) \longrightarrow Fe(s) + 2\,Fe^{3+}(aq)$

(b) $E^\circ_{cell} = -1.21$ V; not product-favored

(c) $K = 1 \times 10^{-41}$

20.39   See Figure 20.18.

20.41 O_2 from the oxidation of water is more likely than F_2. See Example 20.11.

20.43   See Example 20.11.

(a) Cathode: $2\,H_2O(\ell) + 2\,e^- \longrightarrow H_2(g) + 2\,OH^-(aq)$

(b) Anode: $2\,Br^-(aq) \longrightarrow Br_2(\ell) + 2\,e^-$

20.45   Mass of Ni = 0.0334 g

20.47   Time = 2300 s or 38 min

20.49   Time = 250 h

20.51   (a) $UO_2^+(aq) + 4\,H^+(aq) + e^- \longrightarrow$
$U^{4+}(aq) + 2\,H_2O(\ell)$

(b) $ClO_3^-(aq) + 6\,H^+(aq) + 6\,e^- \longrightarrow$
$Cl^-(aq) + 3\,H_2O(\ell)$

(c) $N_2H_4(aq) + 4\,OH^-(aq) \longrightarrow$
$N_2(g) + 4\,H_2O(\ell) + 4\,e^-$

(d) $ClO^-(aq) + H_2O(\ell) + 2\,e^- \longrightarrow$
$Cl^-(aq) + 2\,OH^-(aq)$

20.53 (a,c) The electrode at the right is a magnesium anode. (Magnesium metal supplies electrons and is oxidized to Mg^{2+} ions.) Electrons pass through the wire to the silver cathode, where Ag^+ ions are reduced to silver metal. Nitrate ions move via the salt bridge from the $AgNO_3$ solution to the $Mg(NO_3)_2$ solution (and Na^+ ions move in the opposite direction).

(b) Anode: $Mg(s) \longrightarrow Mg^{2+}(aq) + 2\,e^-$

Cathode: $Ag^+(aq) + e^- \longrightarrow Ag(s)$

Net reaction: $Mg(s) + 2\,Ag^+(aq) \longrightarrow$
$Mg^{2+}(aq) + 2\,Ag(s)$

20.55   (a) For 1.7 V:

Use chromium as the anode to reduce $Ag^+(aq)$ to $Ag(s)$ at the cathode. The cell potential is $+1.71$ V.

(b) For 0.5 V:

Use copper as the anode to reduce silver ions to silver metal at the cathode. The cell potential is $+0.46$ V.

Use silver as the anode to reduce chlorine to chloride ions. The cell potential would be +0.56 V. (In practice, this setup is not likely to work well because the product would be insoluble silver chloride.)

20.57 (a) $Zn^{2+}(aq)$ (c) $Zn(s)$

       (b) $Au^+(aq)$ (d) $Au(s)$

       (e) Yes, $Sn(s)$ will reduce Cu^{2+} (as well as Ag^+ and Au^+).

       (f) No, $Ag(s)$ can only reduce $Au^+(aq)$.

       (g) See part (e).

       (h) $Ag^+(aq)$ can oxidize Cu, Sn, Co, and Zn.

20.59 (a) The cathode is the site of reduction, so the half-reaction must be $2 H^+(aq) + 2 e^- \longrightarrow H_2(g)$. This is the case with the following half-reactions: $Cr^{3+}(aq)|Cr(s)$, $Fe^{2+}(aq)|Fe(s)$, and $Mg^{2+}(aq)|Mg(s)$.

       (b) Choosing from the half-cells in part (a), the reaction of $Mg(s)$ and $H^+(aq)$ would produce the most positive potential (2.37 V), and the reaction of H_2 with Cu^{2+} would produce the least positive potential (+0.337 V).

20.61 8.1×10^5 g Al

20.63 (a) $E°_{anode} = -0.268$ V

       (b) $K_{sp} = 2 \times 10^{-5}$

20.65 $\Delta G° = -562$ kJ

20.67 6700 kWh

20.69 $Ru(NO_3)_2$

20.71 9.5×10^6 g Cl_2 per day

20.73 0.054 g Au

20.75 (a) $2[Ag^+(aq) + e^+ \longrightarrow Ag(s)]$

$C_6H_5CHO(aq) + H_2O(\ell) \longrightarrow$
$C_6H_5CO_2H(aq) + 2 H^+(aq) + 2 e^-$
_____
$2Ag^+(aq) + C_6H_5CHO(aq) + H_2O(\ell) \longrightarrow$
$C_6H_5CO_2H(aq) + 2 H^+(aq) + 2 Ag(s)$

       (b) $3[CH_3CH_2OH(aq) + H_2O(\ell) \longrightarrow$
$CH_3CO_2H(aq) + 4 H^+(aq) + 4 e^-]$

$2[Cr_2O_7^{2-}(aq) + 14 H^+(aq) + 6 e^- \longrightarrow$
$2 Cr^{3+}(aq) + 7 H_2O(\ell)]$
_____
$3 CH_3CH_2OH(aq) + 2 Cr_2O_7^{2-}(aq) + 16 H^+(aq) \longrightarrow$
$3 CH_3CO_2H(aq) + 4 Cr^{3+}(aq) + 11 H_2O(\ell)$

20.77 (a) 0.974 kJ/g

       (b) 0.60 kJ/g

       (c) The silver-zinc battery produces more energy per gram of reactants.

20.79 (a) 92 g HF required; 230 g CF_3SO_2F and 9.3 g H_2 isolated

       (b) H_2 is produced at the cathode.

       (c) 48 kWh

20.81 (a)

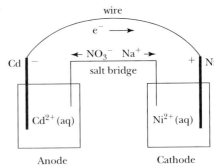

       (b) Anode: $Cd(s) \longrightarrow Cd^{2+}(aq) + 2 e^-$

         Cathode: $Ni^{2+}(aq) + 2 e^- \longrightarrow Ni(s)$

         Net: $Cd(s) + Ni^{2+}(aq) \longrightarrow Cd^{2+}(aq) + Ni(s)$

       (c) The anode is negative and the cathode is positive.

       (d) $E°_{cell} = E°_{cathode} - E°_{anode} = (-0.25$ V$) - (-0.40$ V$) = +0.15$ V

       (e) Electrons flow from anode (Cd) to cathode (Ni).

       (f) Na^+ ions move from the anode compartment to the cathode compartment. Anions move in the opposite direction.

       (g) $K = 1 \times 10^5$

       (h) $E_{cell} = 0.21$ V

       (i) 480 h

20.83 I^- is the strongest reducing agent of the three halide ions. Iodide ion reduces Cu^{2+} to Cu^+, forming insoluble $CuI(s)$.

$2 Cu^{2+}(aq) + 4 I^-(aq) \longrightarrow 2 CuI(s) + I_2(aq)$

20.85 290 h

20.87 (a) Au^{3+}, Br_2, Hg^{2+}, Ag^+, Hg_2^{2+}

       (b) Al, Mg, Li

Chapter 21

21.1 $4 Li(s) + O_2(g) \longrightarrow 2 Li_2O(s)$

$Li_2O(s) + H_2O(\ell) \longrightarrow 2 LiOH(aq)$

$2 Ca(s) + O_2(g) \longrightarrow 2 CaO(s)$

$CaO(s) + H_2O(\ell) \longrightarrow Ca(OH)_2(s)$

21.3 These are the elements of Group 3A: boron, B; aluminum, Al; gallium, Ga; indium, In; and thallium, Tl.

21.5 $2 Na(s) + Cl_2(g) \longrightarrow 2 NaCl(s)$

The reaction is exothermic and the product is ionic. See Figure 1.7.

21.7 The product, NaCl, is a colorless solid and is soluble in water. Other alkali metal chlorides have similar properties.

21.9 Calcium will not exist in the earth's crust because the metal reacts with water.

21.11   Increasing basicity: $CO_2 < SiO_2 < SnO_2$

21.13   (a) $2\ Na(s) + Br_2(\ell) \longrightarrow 2\ NaBr(s)$

(b) $2\ Mg(s) + O_2(g) \longrightarrow 2\ MgO(s)$

(c) $2\ Al(s) + 3\ F_2(g) \longrightarrow 2\ AlF_3(s)$

(d) $C(s) + O_2(g) \longrightarrow CO_2(g)$

21.15 $2\ H_2(g) + O_2(g) \longrightarrow 2\ H_2O(g)$

$H_2(g) + Cl_2(g) \longrightarrow 2\ HCl(g)$

$3\ H_2(g) + N_2(g) \longrightarrow 2\ NH_3(g)$

21.17 $CH_4(g) + H_2O(g) \longrightarrow CO(g) + 3\ H_2(g)$

$\Delta H^\circ_{rxn} = +20.62$ kJ; $\Delta S^\circ_{sys} = +214.7$ J/K; $\Delta G^\circ_{sys} = -43.4$ kJ (at 298.15 K).

21.19   Step 1: $2\ SO_2(g) + 4\ H_2O(\ell) + 2\ I_2(s) \longrightarrow$
$2\ H_2SO_4(\ell) + 4\ HI(g)$

Step 2: $2\ H_2SO_4(\ell) \longrightarrow 2\ H_2O(\ell) + 2\ SO_2(g) + O_2(g)$

Step 3: $4\ HI(g) \longrightarrow 2\ H_2(g) + 2\ I_2(g)$

Net: $2\ H_2O(\ell) \longrightarrow 2\ H_2(g) + O_2(g)$

21.21 $Na(s) + F_2(g) \longrightarrow 2\ NaF(s)$

$Na(s) + Cl_2(g) \longrightarrow 2\ NaCl(s)$

$Na(s) + Br_2(\ell) \longrightarrow 2\ NaBr(s)$

$Na(s) + I_2(s) \longrightarrow 2\ NaI(s)$

The alkali metal halides are white, crystalline solids. They have high melting and boiling points, and are soluble in water.

21.23 $2\ Cl^-(aq) + 2\ H_2O(\ell) \longrightarrow$
$Cl_2(g) + H_2(g) + 2\ OH^-(aq)$

If this were the only process used to produce chlorine, the mass of Cl_2 reported for industrial production would be 0.88 times the mass of NaOH produced (2 mol NaCl, 117 g, would yield 2 mol NaOH, 80 g, and 1 mol Cl_2, 70 g). The amounts quoted indicate a Cl_2-to-NaOH mass ratio 0.96. Chlorine is presumably also prepared by other routes than this one.

21.25 $2\ Mg(s) + O_2(g) \longrightarrow 2\ MgO(s)$

$3\ Mg(s) + N_2(g) \longrightarrow Mg_3N_2(s)$

21.27 $CaCO_3$ is used in agriculture to neutralize acidic soil, to prepare CaO for use in mortar, and in steel production.

$CaCO_3(s) + H_2O(\ell) + CO_2(g) \longrightarrow$
$Ca^{2+}(aq) + 2\ HCO_3^-(aq)$

21.29 1.4×10^6 g SO_2

21.31

$B_3O_6^{3-}$

$B_2O_5^{4-}$

21.33   (a) $2\ B_5H_9(g) + 12\ O_2(g) \longrightarrow 5\ B_2O_3(s) + 9\ H_2O(g)$

(b) Heat of combustion of $B_5H_9 = -4341.2$ kJ/mol. This is more than double the heat of combustion of B_2H_6.

(c) Heat of combustion of $C_2H_6(g)$ [to give $CO_2(g)$ and $H_2O(g)$] $= -1428.7$ kJ/mol. C_2H_6 produces 47.5 kJ/g whereas diborane produces much more (73.7 kJ/g).

21.35 $2\ Al(s) + 6\ HCl(aq) \longrightarrow$
$2\ Al^{3+}(aq) + 6\ Cl^-(aq) + 3\ H_2(g)$

$2\ Al(s) + 3\ Cl_2(g) \longrightarrow 2\ AlCl_3(s)$

$4\ Al(s) + 3\ O_2(g) \longrightarrow 2\ Al_2O_3(s)$

21.37 $2\ Al(s) + 2\ OH^-(aq) + 6\ H_2O(\ell) \longrightarrow$
$2\ Al(OH)_4^-(aq) + 3\ H_2(g)$

Volume of H_2 obtained from 13.2 g Al = 18.4 L

21.39 $Al_2O_3(s) + 3\ H_2SO_4(aq) \longrightarrow Al_2(SO_4)_3(s) + 3\ H_2O(\ell)$

Mass of H_2SO_4 required = 860 g and mass of Al_2O_3 required = 298 g

21.41

The ion has tetrahedral geometry. Aluminum is sp^3 hybridized.

21.43 SiO_2 is a network solid, with tetrahedral silicon atoms covalently bonded to four oxygens in an infinite array; CO_2 consists of individual molecules, with oxygen atoms double-bonded to carbon. Melting SiO_2 requires breaking very stable Si—O bonds. Weak intermolecular forces of attraction between CO_2 molecules result in this substance being a gas at ambient conditions.

21.45   (a) $2\ CH_3Cl(g) + Si(s) \longrightarrow (CH_3)_2SiCl_2(\ell)$

(b) 0.823 atm

(c) 12.2 g

21.47   Consider the general decomposition reaction:

$N_xO_y \longrightarrow {}^x/_2\ N_2 + {}^y/_2\ O_2$

The value of ΔG° can be obtained for all N_xO_y molecules because $\Delta G^\circ_{rxn} = -\Delta G^\circ_f$. These data show that the decomposition reaction is product-favored for all of the nitrogen oxides. All are unstable with respect to decomposition to the elements.

Compound	$-\Delta G_f^\circ$ (kJ/mol)
NO(g)	−86.58
NO_2	−51.23
N_2O	−104.20
N_2O_4	−97.73

21.49 $\Delta H_{rxn}^\circ = -114.4$ kJ; exothermic $\Delta G_{rxn}^\circ = -70.7$ kJ, product-favored

21.51 (a) $N_2H_4(aq) + O_2(g) \longrightarrow N_2(g) + 2\,H_2O(\ell)$
(b) 1.32×10^3 g

21.53 $5\,N_2H_5^+(aq) + 4\,IO_3^-(aq) \longrightarrow$
$\qquad\qquad 5\,N_2(g) + 2\,I_2(aq) + H^+(aq) + 12\,H_2O(\ell)$

$E_{net}^\circ = 1.43$ V

21.55 (a) Oxidation number = +3
(b) Diphosphonic acid ($H_4P_2O_5$) should be a diprotic acid (losing the two H atoms attached to O atoms).

$$\begin{array}{c}
\ddot{\text{O}}: \qquad \ddot{\text{O}}: \\
| \qquad\quad | \\
\text{H}-\text{P}-\ddot{\text{O}}-\text{P}-\text{H} \\
| \qquad\quad | \\
\text{H}-\ddot{\text{O}}: \quad :\ddot{\text{O}}-\text{H}
\end{array}$$

21.57 (a) 3.5×10^3 kg SO_2
(b) 4.1×10^3 kg $Ca(OH)_2$

21.59
$$\left[:\ddot{\text{S}}-\ddot{\text{S}}: \right]^{2-}$$

disulfide ion

21.61 $E_{cell}^\circ = E_{cathode}^\circ - E_{anode}^\circ = +1.44\text{ V} - (+1.51\text{ V}) = -0.07$ V

The reaction is not product-favored under standard conditions.

21.63 $Cl_2(aq) + 2\,Br^-(aq) \longrightarrow 2\,Cl^-(aq) + Br_2(aq)$

Cl_2 is the oxidizing agent, Br^- is the reducing agent; $E_{cell}^\circ = 0.28$ V.

21.65 The reaction consumes 4.32×10^8 C to produce 8.51×10^4 g F_2.

21.67

Element	Appearance	State
Na, Mg, Al	Silvery metal	Solids
Si	black, shiny metalloid	Solid
P	White, red, and black allotropes; nonmetal	Solid
S	Yellow nonmetal	Solid
Cl	Pale green nonmetal	Gas
Ar	Colorless nonmetal	Gas

21.69 (a) $2\,Na(s) + Cl_2(g) \longrightarrow 2\,NaCl(s)$
$Mg(s) + Cl_2(g) \longrightarrow MgCl_2(s)$
$2\,Al(s) + 3\,Cl_2(g) \longrightarrow 2\,AlCl_3(s)$
$Si(s) + 2\,Cl_2(g) \longrightarrow SiCl_4(\ell)$
$P_4(s) + 10\,Cl_2(g) \longrightarrow 4\,PCl_5(s)$ (excess Cl_2)
$S_8(s) + 8\,Cl_2(g) \longrightarrow 8\,SCl_2(s)$

(b) NaCl and $MgCl_2$ are ionic; the other products are covalent.
(c) $SiCl_4$ is tetrahedral; PCl_5 is trigonal bipyramidal.

$$\begin{array}{cc}
:\ddot{\text{Cl}}: & :\ddot{\text{Cl}}: \\
| & | \quad \ddot{\text{Cl}}: \\
\text{Si} \cdots \ddot{\text{Cl}}: & :\ddot{\text{Cl}}-\text{P} \\
| & | \quad \ddot{\text{Cl}}: \\
:\ddot{\text{Cl}} \quad \ddot{\text{Cl}}: & \ddot{\text{Cl}}
\end{array}$$

21.71 (a) $2\,KClO_3(s) \longrightarrow 2\,KCl(s) + 3\,O_2(g)$
(b) $2\,H_2S(g) + 3\,O_2(g) \longrightarrow 2\,H_2O(g) + 2\,SO_2(g)$
(c) $2\,Na(s) + O_2(g) \longrightarrow Na_2O_2(s)$
(d) $P_4(s) + 3\,KOH(aq) + 3\,H_2O(\ell) \longrightarrow$
$\qquad\qquad PH_3(g) + 3\,KH_2PO_4(aq)$
(e) $NH_4NO_3(s) \longrightarrow N_2O(g) + 2\,H_2O(g)$
(f) $2\,In(s) + 3\,Br_2(\ell) \longrightarrow 2\,InBr_3(s)$
(g) $SnCl_4(\ell) + 2\,H_2O(\ell) \longrightarrow SnO_2(s) + 4\,HCl(aq)$

21.73 (a)
$$\begin{array}{c}
:\ddot{\text{Cl}}: :\ddot{\text{Cl}}: \\
| \quad\quad | \\
:\ddot{\text{Cl}}-\text{B}-\text{B}-\ddot{\text{Cl}}:
\end{array}$$

(b) Each B atom is surrounded in a trigonal-planar arrangement by another B atom and two Cl atoms. Each B atom is sp^2 hybridized.

21.75 (a) $2\,BaO(s) + O_2(g) \longrightarrow 2\,BaO_2(s)$
(b) $3\,BaO_2(s) + 2\,Fe(s) \longrightarrow 3\,BaO(s) + Fe_2O_3(s)$

21.77 Cathode: $Li^+(\ell) + e^+ \longrightarrow Li(\ell)$

Anode: $2\,H^-(\ell) \longrightarrow H_2(g) + 2\,e^-$

Formation of H_2 at the anode is evidence for the presence of H^-.

21.79 Mg: $\Delta G_{rxn}^\circ = +64.9$ kJ

Ca: $\Delta G_{rxn}^\circ = +131.40$ kJ

Ba: $\Delta G_{rxn}^\circ = +219.4$ kJ

Relative tendency to decompose: $MgCO_3 > CaCO_3 > BaCO_3$

21.81 (a) ΔG_f° should be more negative than $(-95.1\text{ kJ}) \times n$
(b) Ba, Pb, Ti

21.83 O—F bond energy = 190 kJ/mol

21.85 (a) N_2O_4 is the oxidizing agent (N is reduced from +4 to 0 in N_2), and $H_2NN(CH_3)_2$ is the reducing agent.
(b) 1.3×10^4 kg N_2O_4 is required. Product masses: 5.7×10^3 kg N_2; 4.9×10^3 kg H_2O; 6.0×10^3 kg CO_2.

21.87 (a) The NO bond with a length of 114.2 pm is a double bond. The other two NO bonds (with a length of 121 pm) have a bond order of 1.5 (as there are two resonance structures involving these bonds).

(b) $K = 1.90$ (c) $\Delta H_f^\circ = 82.9$ kJ/mol

21.89 $\Delta H_{rxn}^\circ = -257.78$ kJ. This reaction is entropy-disfavored, however, with $\Delta S_{rxn}^\circ = -963$ J/K because of the decrease in the number of moles of gases. Combining these values gives $\Delta G_{rxn}^\circ = +29.19$ kJ, indicating that under standard conditions at 298 K the reaction is not spontaneous. (The reaction has a favorable ΔG_{rxn}° at temperatures less than 268 K, indicating that further research on this system might be worthwhile. Note that at that temperature water is a solid.)

21.91 3.5 kWh

21.93 The flask contains a fixed number of moles of gas at the given pressure and temperature. One could burn the mixture because only the H_2 will combust; the argon is untouched. Cooling the gases from combustion would remove water (the combustion product of H_2) and leave only Ar in the gas phase. Measuring its pressure in a calibrated volume at a known temperature would allow one to calculate the amount of Ar that was in the original mixture.

21.95 Generally a sodium fire can be extinguished by smothering it with sand. The worst choice is to use water (which reacts violently with sodium to give H_2 gas and NaOH).

21.97 Nitrogen is a relatively unreactive gas, so it will not participate in any reaction typical of hydrogen or oxygen. The most obvious property of H_2 is that it burns, so attempting to burn a small sample of the gas would immediately confirm or deny the presence of H_2. If O_2 is present, it can be detected by allowing it to react as an oxidizing agent. There are many reactions known with low-valent metals, especially transition metal ions in solution, that can be detected by color changes.

21.99 The reducing ability of the Group 3A metals declines considerably on descending the group, with the largest drop occurring on going from Al to Ga. The reducing ability of gallium and indium are similar, but another large change is observed on going to thallium. In fact, thallium is most stable in the +1 oxidation state. This same tendency for elements to be more stable with lower oxidation numbers is seen in Groups 4A (Ge and Pb) and 5A (Bi).

21.101 A through E, in order: $BaCO_3$, BaO, $CaCO_3$, $BaCl_2$, $BaSO_4$

Chapter 22

22.1 (a) Cr^{3+}: $[Ar]3d^3$, paramagnetic
(b) V^{2+}: $[Ar]3d^3$, paramagnetic
(c) Ni^{2+}: $[Ar]3d^8$, paramagnetic
(d) Cu^+: $[Ar]3d^{10}$, diamagnetic

22.3 (a) Fe^{3+}: $[Ar]3d^5$, isoelectronic with Mn^{2+}
(b) Zn^{2+}: $[Ar]3d^{10}$, isoelectronic with Cu^+
(c) Fe^{2+}: $[Ar]3d^6$, isoelectronic with Co^{3+}
(d) Cr^{3+}: $[Ar]3d^3$, isoelectronic with V^{2+}

22.5 (a) $Cr_2O_3(s) + 2\,Al(s) \longrightarrow Al_2O_3(s) + 2\,Cr(s)$
(b) $TiCl_4(\ell) + 2\,Mg(s) \longrightarrow Ti(s) + 2\,MgCl_2(s)$
(c) $2\,[Ag(CN)_2]^-(aq) + Zn(s) \longrightarrow$
$2\,Ag(s) + [Zn(CN)_4]^{2-}(aq)$
(d) $3\,Mn_3O_4(s) + 8\,Al(s) \longrightarrow 4\,Al_2O_3(s) + 9\,Mn(s)$

22.7 Monodentate: CH_3NH_2, CH_3CN, N_3^-, Br^-
Bidentate: en, phen (see Figure 22.14)

22.9 (a) Mn^{2+} (c) Co^{3+}
(b) Co^{3+} (d) Cr^{2+}

22.11 $[Ni(en)(NH_3)_3(H_2O)]^{2+}$

22.13 (a) $Ni(en)_2Cl_2$ (en = $H_2NCH_2CH_2NH_2$)
(b) $K_2[PtCl_4]$
(c) $K[Cu(CN)_2]$
(d) $[Fe(NH_3)_4(H_2O)_2]^{2+}$

22.15 (a) Diaquabis(oxalato)nickelate(II) ion
(b) Dibromobis(ethylenediamine)cobalt(III) ion
(c) Amminechlorobis(ethylenediamine)cobalt(III) ion
(d) Diammineoxalatoplatinum(II)

22.17 (a) $[Fe(H_2O)_5OH]^{2+}$
(b) Potassium tetracyanonickelate(II)
(c) Potassium diaquabis(oxalato)chromate(III)
(d) $(NH_4)_2[PtCl_4]$

22.19 (a)

(b)

(c)

(d) $\left[\begin{array}{c} \\ N\cdots Co\cdots Cl \\ Cl\quad Cl \\ N\quad Cl\quad Cl \end{array}\right]^{-}$ Only one structure possible. (N—N is the bidentate ethylenediamine ligand.)

22.21 For a discussion of chirality, see Chapter 11, pages 477–480).

(a) Fe^{2+} is a chiral center.

(b) Co^{3+} is not a chiral center.

(c) Neither of the two possible isomers is chiral.

(d) No. Square-planar complexes are never chiral.

22.23 (a) $[Mn(CN)_6]^{4-}$: d^5, low-spin complex is paramagnetic.

$\uparrow\downarrow \quad \uparrow\downarrow \quad \uparrow$

(b) $[Co(NH_3)_6]^{3+}$: d^6, low-spin complex is diamagnetic.

$\uparrow\downarrow \quad \uparrow\downarrow \quad \uparrow\downarrow$

(c) $[Fe(H_2O)_6]^{3+}$: d^5, low-spin complex is paramagnetic (1 unpaired electron; same as part a).

(d) $[Cr(en)_3]^{2+}$: d^4, complex is paramagnetic (2 unpaired electrons).

$\uparrow\downarrow \quad \uparrow \quad \uparrow$

22.25 (a) Fe^{2+}, d^6, paramagnetic, 4 unpaired electrons

(b) Co^{2+}, d^7, paramagnetic, 3 unpaired electrons

(c) Mn^{2+}, d^5, paramagnetic, 5 unpaired electrons

(d) Zn^{2+}, d^{10}, diamagnetic, 0 unpaired electrons

22.27 (a) 6; (b) octahedral; (c) +2; (d) 4 unpaired electrons (high spin); (e) paramagnetic

22.29 When $Co_2(SO_4)_3$ dissolves in water, it forms $[Co(H_2O)_6]^{3+}$. Addition of fluoride converts this to $[CoF_6]^{3-}$. The hexaaquo complex is low spin (diamagnetic, no unpaired electrons), and the fluoride complex is high spin (paramagnetic, 4 unpaired electrons.) Notice that fluoride is a weaker field ligand than water.

22.31 The light absorbed is in the blue region of the spectrum (page 1097). Therefore, the light transmitted—which is the color of the solution—is yellow.

22.33 Determine the magnetic properties of the complex. Square-planar Ni^{2+} (d^8) complexes are diamagnetic, whereas tetrahedral complexes are paramagnetic.

22.35 Fe^{2+} has a d^6 configuration. Low-spin octahedral complexes are diamagnetic, whereas high-spin complexes have four unpaired electrons and are paramagnetic.

22.37 Square-planar complexes most often arise from d^8 transition metal ions. Therefore, it is likely that $[Ni(CN)_4]^{2-}$ (Ni^{2+}) and $[Pt(CN)_4]^{2-}$ (Pt^{2+}) are square planar.

22.39 Two geometric isomers are possible.

22.41 Absorbing at 425 nm means the complex is absorbing light in the blue-violet end of the spectrum. Therefore, red and green light are transmitted, and the complex appears yellow (see Figure 22.27).

22.43 (a) Mn^{2+}; (b) 6; (c) octahedral; (d) 5; (e) paramagnetic; (f) cis and trans isomers exist.

22.45 Name: tetraamminedichlorocobalt(III) ion

$\left[\begin{array}{c} NH_3 \\ H_3N\cdots Co\cdots Cl \\ H_3N\quad Cl \\ NH_3 \end{array}\right]^{+}$ $\left[\begin{array}{c} NH_3 \\ Cl\cdots Co\cdots NH_3 \\ H_3N\quad Cl \\ NH_3 \end{array}\right]^{+}$

   cis   trans

22.47 $[Co(en)_2(H_2O)Cl]^{2+}$

22.49 (a)

mer   fac

(b)

trans chlorides   cis chlorides

(c) (three structures shown) $^{3+}$

22.51

H_2O and NH_3 cis, chiral

H_2O cis and NH_3 trans, not chiral

H_2O trans and NH_3 cis, not chiral

22.53   In $[Mn(H_2O)_6]^{2+}$ and $[Mn(CN)_6]^{4-}$, Mn has an oxidation number of $+2$ (Mn is a d^5 ion).

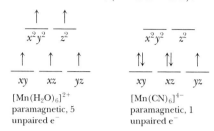

[$Mn(H_2O)_6$]$^{2+}$
paramagnetic, 5
unpaired e$^-$

[$Mn(CN)_6$]$^{4-}$
paramagnetic, 1
unpaired e$^-$

This shows that Δ_o for CN^- is greater than for H_2O.

22.55   A, dark violet isomer: $[Co(NH_3)_5Br]SO_4$

B, violet-red isomer: $[Co(NH_3)_5(SO_4)]Br$

$[Co(NH_3)_5Br]SO_4(aq) + BaCl_2(aq) \longrightarrow$
$[Co(NH_3)_5Br]Cl_2(aq) + BaSO_4(s)$

22.57   (a) The light absorbed is in the orange region of the spectrum (page 1098). Therefore, the light transmitted (the color of the solution) is blue or cyan.

(b) Using the cobalt(III) complexes in Table 22.3 as a guide, we might place CO_3^{2-} between F^- and the oxalato ion, $C_2O_4^{2-}$.

(c) Δ_o is small, so the complex should be high spin and paramagnetic.

22.59

N   O = $H_2N\!-\!CH_2\!-\!CO_2^-$

enantiometric pair

enantiometric pair

enantiometric pair

22.61   Volume of sulfuric acid required = 1.10 L; mass of TiO_2 obtained = 526 g

22.63   (a) There is 5.41×10^{-4} mol of $UO_2(NO_3)_2$, and this provides 5.41×10^{-4} mol or U^{n+} ions on reduction by Zn. The 5.41×10^{-4} mol U^{n+} requires $2.16 \times$

10^{-4} mol MnO_4^- to reach the equivalence point. This is a ratio of 5 mol of U^{n+} ions to 2 mol MnO_4^- ions. The 2 mol MnO_4^- ions requires 10 mol of e^- (to go to Mn^{2+} ions), so 5 mol of U^{n+} ions provide 10 mol e^- (on going to UO_2^{2+} ions, with a uranium oxidation number of $+6$). This means the U^{n+} ion must be U^{4+}.

(b) $Zn(s) \longrightarrow Zn^{2+}(aq) + 2\,e^-$

$UO_2^{2+}(aq) + 4\,H^+(aq) + 2\,e^- \longrightarrow$
$U^{4+}(aq) + 2\,H_2O(\ell)$

$\overline{UO_2^{2+}(aq) + 4\,H^+(aq) + Zn(s) \longrightarrow}$
$U^{4+}(aq) + 2\,H_2O(\ell) + Zn^{2+}(aq)$

(c) $5[U^{4+}(aq) + 2\,H_2O(\ell) \longrightarrow$
$UO_2^{2+}(aq) + 4\,H^+(aq) + 2\,e^-]$

$2[MnO_4^-(aq) + 8\,H^+(aq) + 5\,e^- \longrightarrow$
$Mn^{2+}(aq) + 4\,H_2O(\ell)]$

$\overline{5\,U^{4+}(aq) + 2\,MnO_4^-(aq) + 2\,H_2O(\ell) \longrightarrow}$
$5\,UO_2^{2+}(aq) + 4\,H^+(aq) + 2\,Mn^{2+}(aq)$

22.65

Ion	$K_{formation}$ (ammine complexes)
Co^{2+}	7.7×10^4
Ni^{2+}	5.6×10^8
Cu^{2+}	6.8×10^{12}
Zn^{2+}	2.9×10^9

The data for these hexammine complexes do indeed, verify the Irving-Williams series. In the book *Chemistry of the Elements* (N. N. Greenwood and A. Earnshaw: 2nd edition, p. 908, Oxford, England, Butterworth-Heinemann, 1997), it is stated: "the stabilities of corresponding complexes of the bivalent ions of the first transition series, irrespective of the particular ligand involved, usually vary in the Irving-Williams order, . . . , which is the reverse of the order for the cation radii. These observations are consistent with the view that, at least for metals in oxidation states $+2$ and $+3$, the coordinate bond is largely electrostatic. This was a major factor in the acceptance of the crystal field theory."

Chapter 23

23.12   (a) $^{56}_{28}Ni$; (b) $^{1}_{0}n$; (c) $^{32}_{15}P$; (d) $^{97}_{43}Tc$; (e) $^{0}_{-1}\beta$;
(f) $^{0}_{1}e$ (positron)

23.14   (a) $^{0}_{-1}\beta$; (b) $^{87}_{37}Rb$; (c) $^{4}_{2}\alpha$; (d) $^{226}_{88}Ra$; (e) $^{0}_{-1}\beta$; (f) $^{24}_{11}Na$

23.16 $^{235}_{92}U \longrightarrow {}^{231}_{90}Th + {}^{4}_{2}\alpha$

$^{231}_{90}Th \longrightarrow {}^{231}_{91}Pa + {}^{0}_{-1}\beta$

$^{231}_{91}Pa \longrightarrow {}^{227}_{89}Ac + {}^{4}_{2}\alpha$

$^{227}_{89}Ac \longrightarrow {}^{227}_{90}Th + {}^{0}_{-1}\beta$

$^{227}_{90}Th \longrightarrow {}^{223}_{88}Ra + {}^{4}_{2}\alpha$

$^{223}_{88}Ra \longrightarrow {}^{219}_{86}Rn + {}^{4}_{2}\alpha$

$^{219}_{86}Rn \longrightarrow {}^{215}_{84}Po + {}^{4}_{2}\alpha$

$^{215}_{84}Po \longrightarrow {}^{211}_{82}Pb + {}^{4}_{2}\alpha$

$^{211}_{82}Pb \longrightarrow {}^{211}_{83}Bi + {}^{0}_{-1}\beta$

$^{211}_{83}Bi \longrightarrow {}^{211}_{84}Po + {}^{0}_{-1}\beta$

$^{211}_{84}Po \longrightarrow {}^{207}_{82}Pb + {}^{4}_{2}\alpha$

23.18 (a) $^{198}_{79}Au \longrightarrow {}^{198}_{80}Hg + {}^{0}_{-1}\beta$

(b) $^{222}_{86}Rn \longrightarrow {}^{218}_{84}Po + {}^{4}_{2}\alpha$

(c) $^{137}_{55}Cs \longrightarrow {}^{137}_{56}Ba + {}^{0}_{-1}\beta$

(d) $^{110}_{49}In \longrightarrow {}^{110}_{48}Cd + {}^{0}_{1}e$

23.20 (a) $^{80}_{35}Br$ has a high neutron/proton ratio of 45/35: Beta decay will allow the ratio to decrease: $^{80}_{35}Br \longrightarrow {}^{80}_{36}Kr + {}^{0}_{-1}\beta$. Some ^{80m}Br decays by gamma emission.

(b) Alpha decay is likely: $^{240}_{98}Cf \longrightarrow {}^{236}_{96}Cm + {}^{4}_{2}\alpha$

(c) Cobalt-61 has a high n:p ratio so beta decay is likely: $^{61}_{27}Co \longrightarrow {}^{61}_{28}Kr + {}^{0}_{-1}\beta$

(d) Carbon-11 has only 5 neutrons, so K-capture or positron emission may occur:

$^{11}_{6}C + {}^{0}_{-1}e \longrightarrow {}^{11}_{5}B$

$^{11}_{6}C \longrightarrow {}^{11}_{5}B + {}^{0}_{1}e$

23.22 Generally beta decay will occur when the n/p ratio is high, whereas positron emission will occur when the n/p ratio is low.

(a) Beta decay: $^{20}_{9}F \longrightarrow {}^{20}_{10}Ne + {}^{0}_{-1}\beta$

$^{3}_{1}H \longrightarrow {}^{3}_{2}He + {}^{0}_{-1}\beta$

(b) Positron emission:

$^{22}_{11}Na \longrightarrow {}^{22}_{10}Ne + {}^{0}_{1}\beta$

23.24 Binding energy per nucleon for $^{11}B = 6.70 \times 10^{8}$ kJ

Binding energy per nucleon for $^{10}B = 6.26 \times 10^{8}$ kJ

23.26 8.256×10^{8} kJ/nucleon

23.28 7.700×10^{8} kJ/nucleon

23.30 $0.781 \ \mu g$

23.32 (a) $^{131}_{53}I \longrightarrow {}^{131}_{54}Xe + {}^{0}_{-1}\beta$

(b) $0.075 \ \mu g$

23.34 9.5×10^{-4} mg

23.36 (a) $^{222}_{86}Rn \longrightarrow {}^{218}_{84}Po + {}^{4}_{2}\alpha$

(b) Time = 8.87 d

23.38 About 2700 years old

23.40 (a) 15.8 yr; (b) 88%

23.42 If $t_{1/2} = 14.28$ d, then $k = 4.854 \times 10^{-2}$ d^{-1}. If the original disintegration rate is 3.2×10^{6} dpm, then (from the integrated first-order rate equation), the rate after 365 d is 0.065 dpm. The plot will resemble Figure 23.5.

23.44 Plot ln (activity) versus time. The slope of the plot is $-k$, the rate constant for decay. Here $k = 0.0050$ d^{-1}, so $t_{1/2} = 140$ d.

23.46 $^{239}_{94}Pu + + {}^{4}_{2}\alpha \longrightarrow {}^{240}_{95}Am + {}^{1}_{1}H + 2 {}^{1}_{0}n$

23.48 $^{48}_{20}Ca + {}^{242}_{94}Pu \longrightarrow {}^{287}_{114}Uuq + 3 {}^{1}_{0}n$

23.50 (a) $^{115}_{48}Cd$; (b) $^{7}_{4}Be$; (c) $^{4}_{2}\alpha$; (d) $^{63}_{29}Cu$

23.52 $^{10}_{5}B + {}^{1}_{0}n \longrightarrow {}^{7}_{3}Li + {}^{4}_{2}\alpha$

23.54 Time = 4.4×10^{10} yr

23.56 Time = 1.9×10^{9} yr

23.58 (a) $^{238}_{92}U + {}^{1}_{0}n \longrightarrow {}^{239}_{92}U$

(b) $^{239}_{92}U \longrightarrow {}^{239}_{93}Np + {}^{0}_{-1}\beta$

(c) $^{239}_{93}Np \longrightarrow {}^{239}_{94}Pu + {}^{0}_{-1}\beta$

(d) $^{239}_{94}Pu + {}^{1}_{0}n \longrightarrow 2 {}^{1}_{0}n +$ energy + other nuclei

23.60 Energy obtained from 1.000 lb (452.6 g) of $^{235}U = 4.05 \times 10^{+10}$ kJ

Mass of coal required = 1.6×10^{3} ton (or about 3 million lb of coal)

23.62 130 mL

23.64 27 tagged fish out of 5250 fish caught represents 0.51% of the fish in the lake. Therefore, 1000 fish put into the lake represents 0.51% of the fish in the lake, or 0.51% of 190,000 fish.

23.66 (a) The mass decreases by 4 units (with an $^{4}_{2}\alpha$ emission), or is unchanged (with a $^{0}_{-1}\beta$ emission) so the only masses possible are 4 units apart.

(b) ^{232}Th series, $m = 4n$; ^{235}U series, $m = 4n + 3$

(c) ^{226}Ra and ^{210}Bi, $4n + 2$ series; ^{215}At, $4n + 3$ series; ^{228}Th, $4n$ series

(d) Each series is headed by a long-lived isotope (on the order of 10^{9} years, the age of the earth). The $4n + 1$ series is missing because there is no long-lived isotope in this series. Over geologic time, all of the members of this series have decayed completely.

23.68 (a) The ^{231}Pa isotope belongs to the ^{235}U decay series (see Question 23.66b).

(b) $^{235}_{92}U \longrightarrow {}^{231}_{90}Th + {}^{4}_{2}\alpha$

$^{231}_{90}Th \longrightarrow {}^{231}_{91}Pa + {}^{0}_{-1}\beta$

(c) Pa-231 is present to the extent of 1 part per million. Therefore, 1 million g of pitchblende needs to be used to obtain 1 g of Pa-231.

(d) $^{231}_{91}Pa \longrightarrow {}^{227}_{89}Ac + {}^{4}_{2}\alpha$

Appendix P

Answers to Selected Interchapter Study Questions

The Chemistry of Fuels and Energy Sources

1.

100. g of Reactant	Mass H_2 Produced (g)
CH_4	37.7 g
CH_2 (petroleum)	28.7 g
C (coal)	16.8 g

3. 70. pounds of coal produces 1.1×10^6 kJ of heat energy per day.

5. 7.0 gal of fuel oil produces 9.5×10^5 kJ of heat energy. This is about 14% less than the heat produced by 70. pounds of coal.

7. (a) Energy per gram of isooctane = 47.7 kJ/g
   (b) Energy per liter of isooctane = 3.28×10^4 kJ/L

9. 940 kW-h per year is equivalent to 3.4×10^6 kJ per year

11. Methanol provides 23 kJ per gram or 1.8×10^4 kJ/L. At 100% efficiency the cell produces 5.0 kW-h of energy.

13. Energy striking the parking lot = 4.2×10^8 kJ/day

15. The formula of hydrogen-saturated palladium is $Pd_{1.35}H$ or $PdH_{0.74}$.

17. Heat content of gasoline = 35.5 kJ/mL or 1.34×10^5 kJ/gallon.

   Energy consumed per mile = 2.43×10^3 kJ

The Chemistry of Life: Biochemistry

1. (a)

$$
\begin{array}{ccc}
 & H & H & :O: \\
 & | & | & \| \\
H-N-C-C-\ddot{O}-H \\
 & | \\
 & H-C-CH_3 \\
 & | \\
 & CH_3
\end{array}
$$

   (b)

$$
\begin{array}{ccc}
 & H & H & :O: \\
 & | & +| & \| \\
H-N-C-C-\ddot{O}: ^- \\
 & | \\
 & H \\
 & | \\
 & H-C-CH_3 \\
 & | \\
 & CH_3
\end{array}
$$

   (c) The zwitterionic form will be the predominant form at physiological pH.

3.

$$
\begin{array}{c}
H-\ddot{N}-C-C-\ddot{N}-C-C-\ddot{O}-H \\
\\
H-\ddot{N}-C-C-\ddot{N}-C-C-\ddot{O}-H
\end{array}
$$

5.

7. The quaternary structure would tell us how the two subunits are arranged with respect to each other.

9.

11. They proposed A–T base pairs and C–G base pairs. There are two hydrogen bonds in an A–T pair and three in a C–G pair.

13. (a) 5′-GAATCGCGT-3′

    (b) 5′-GAAUCGCGU-3′

    (c) 5′-UUC-3′, 5′-CGA-3′, and 5′-ACG-3′

    (d) glutamic acid, serine, and arginine

15. (a) In transcription, a strand of RNA complementary to the segment of DNA is constructed.

    (b) In translation, an amino acid sequence is constructed based on the information in an mRNA sequence.

17. (a) False

    (b) True

    (c) True

    (d) True

19. (a) $6\,CO_2\,(g) + 6\,H_2O\,(\ell) \longrightarrow C_6H_{12}O_6\,(s) + 6\,O_2\,(g)$

$$\Delta H^\circ_{rxn} = \sum \Delta H^\circ_f(\text{products}) - \sum \Delta H^\circ_f(\text{reactants})$$

$$= \left[1\text{ mole} \cdot \Delta H^\circ_f(C_6H_{12}O_6(s)) + 6\text{ moles} \cdot \Delta H^\circ_f(O_2(g)) \right]$$

$$- \left[6\text{ moles} \cdot \Delta H^\circ_f(CO_2(g)) + 6\text{ moles} \cdot \Delta H^\circ_f(H_2O(l)) \right]$$

$$= \left[1\text{ mole} \left(-1273\frac{kJ}{mole} \right) + 6\text{ moles} \left(0\frac{kJ}{mole} \right) \right]$$

$$- \left[6\text{ mole} \left(-393.509\frac{kJ}{mole} \right) + 6\text{ moles} \left(-285.83\frac{kJ}{mole} \right) \right]$$

$$= +2803\text{ kJ}$$

(b) $\dfrac{2803\text{ kJ}}{\text{mole glucose}} \times \dfrac{1\text{ mole glucose}}{6.022 \times 10^{23}\text{ molecules glucose}} \times \dfrac{1000\text{ J}}{1\text{ kJ}} = 4.655 \times 10^{-18}\text{ J/molecules}$

(c) $650\text{ nm} \times \dfrac{1\text{m}}{1 \times 10^9\text{ nm}} = 6.50 \times 10^{-7}\text{ m}$

$$E = \frac{hc}{\lambda}$$

$$= \frac{(6.626 \times 10^{-34}\text{ J} \cdot \text{s})(300 \times 10^8\text{ m/s})}{6.50 \times 10^{-7}\text{ m}}$$

$$= 3.06 \times 10^{-19}\text{ J}$$

(d) The amount of energy per photon is less than the amount of required per molecule of glucose, therefore multiple photons must be absorbed.

The Chemistry of Modern Materials

1. The GaAs band gap is 140 kJ/mol. Using the equations $E = h\nu$ and $\lambda \cdot \nu = c$, we calculate a wavelength corresponding to this energy of 854 nm. This is in the infrared portion of the spectrum.

3. The amount of light falling on a single solar cell is 0.0925 W/cell. Using the conversion factor 1 W = 1 J/s, the energy is 5.55 J/(min · cell). At 25% efficiency, this is 1.39 J/min for each cell.

5. In the photo, there are approximately 3 gears across for the width of the spider mite. Because the spider mite is about 0.4 mm wide, then each of these gears measures approximately 0.1 to 0.13 mm in diameter, or 100 to 130 μm in diameter. A typical red blood cell is about 6 to 8 μm in diameter. This means that the gears are about 12 to 18 times bigger than a red blood cell.

7. Aerogel has a density of 2.3 g/cm^3. The mass of 1.0 cm^3 of aerogel is

$$(2.3\text{ g/cm}^3 \times 0.01) + (1.29 \times 10^{-3}\text{ g/cm}^3) = 0.024\text{ g}$$

The Chemistry of the Environment

1. $K = K_{sp}$ for $Ca(OH)_2 \times [1/K_{sp}$ for $Mg(OH)_2] = (5.5 \times 10^{-5})(1/5.6 \times 10^{-12}) = 9.8 \times 10^6$

3. 3.7×10^5 g CaO required

5. The following reactions were discussed in the text:

   (1) $HClO(aq) + NH_3(aq) \longrightarrow NH_2Cl(aq) + H_2O(\ell)$

   (2) $HClO(aq) + NH_2Cl(aq) \longrightarrow NHCl_2(aq) + H_2O(\ell)$

   (3) $HClO(aq) + NHCl_2(aq) \longrightarrow NCl_3(aq) + H_2O(\ell)$

   (4) $H_2SO_4(aq) + 2 NH_3(aq) \longrightarrow (NH_4)_2SO_4(aq)$

   (5) $HNO_3(aq) + NH_3(aq) \longrightarrow NH_4NO_3(aq)$

   All five reactions are acid/base reactions. $HClO(aq)$, $H_2SO_4(aq)$ and $HNO_3(aq)$ are all acids and $NH_3(aq)$, $NH_2Cl(aq)$, and $NHCl_2(aq)$ are Brønsted bases. Note that the products of all five reactions are salts and water. Notice that $HClO(aq)$ and $HNO_3(aq)$ can also be oxidizing agents, and $NH_3(aq)$, $NH_2Cl(aq)$, and $NHCl_2(aq)$ can be reducing agents. In Reaction 1, the oxidation number of N increases from -3 to -1, and Cl changes from $+1$ to -1, so NH_3 is oxidized, and HClO is reduced. In Reaction 2, NH_2Cl is oxidized (N changes from -1 to $+1$), and HClO is reduced (Cl again changes from $+1$ to -1). Finally, in Reaction 3, $NHCl_2$ is oxidized (N changes from -1 to $+3$) and HClO is reduced. The oxidation numbers of H, O, S, and N do not change in Reactions 4 and 5.

7. The answer to this question depends on where you live. The air quality of various locations within New York City may be found at **http://www.dec.state.ny.us/website/dar/bts/airmon/aqipage2.htm**. Current concentrations of sulfur dioxide, carbon monoxide, formaldehyde, and ozone are reported at various locations within New York City.

9. PM_{10} is the sum of the concentration of all particles larger than or equal to 10 micrometers in diameter.

   $PM_{10} = 0.012 + 0.012 + 0.009 + 0.001 + 0.001 = 0.035 \ \mu g/m^3$

   $PM_{2.5} = 0.065 \ \mu g/m^3$

Index/Glossary

Italicized page numbers indicate pages containing illustrations, and those followed by "t" indicate tables. Glossary terms are printed in blue.

Abba, Mohammed Bah, 232
abbreviations, A-10
absolute temperature scale. *See* Kelvin temperature scale.
absolute zero The lowest possible temperature, equivalent to −273.15 °C, used as the zero point of the Kelvin scale, 27, 552
zero entropy at, 912
absorption spectrum A plot of the intensity of light absorbed by a sample as a function of the wavelength of the light, 1100
abundance(s), of elements in Earth's crust, 82t, 87t, 1014
of essential elements in human body, 88t
of isotopes, 69
acceptor level, in semiconductor, 646
accuracy The agreement between the measured quantity and the accepted value, 32
acetaldehyde, 504t
structure of, 469
acetaminophen, structure of, 511
acetate ion, buffer solution of, 855t
acetic acid, 505t
as weak electrolyte, 178
buffer solution of, 855t
density of, 52
dimerization of, 793
formation of, 273
hydrogen bonding in, 601
ionization of, 799
equilibrium in, 760
orbital hybridization in, 452
production of, 502
quantitative analysis of, 159
reaction with calcium carbonate, *192*, 193

reaction with ethanol, 771
reaction with sodium bicarbonate, 815
structure of, 186, 477, 502
titration with sodium hydroxide, 864
acetic anhydride, 157
acetone, 504t
hydrogenation of, 423
structure of, *169, 453,* 502
acetonitrile, isomerization of, 750
structure of, 429, 432, *454*
acetylacetonate ion, as ligand, 1082
acetylacetone, enol and keto forms, 472
structure of, 429
acetylcholinesterase, 732t
acetylene, orbital hybridization in, 453
production of, 585
structure of, 477
N-acetylglucosamine (NAG), 535
N-acetylmuramic acid (NAM), 535
acetylsalicylic acid. *See* aspirin.
acid(s) A substance that, when dissolved in pure water, increases the concentration of hydrogen ions, 185. *See also* Brønsted acid(s), Lewis acid(s).
bases and, 796–847. *See also* acid–base reaction(s).
Brønsted definition, 799
carboxylic. *See* carboxylic acid(s).
common, 187t
Lewis definition, 828–832
molecular structure of, 832–837
properties of, *186*
reaction with bases, 191–195

strengths of, 807–809
direction of reaction and, 814
strong. *See* strong acid.
weak. *See* weak acid.
acid–base indicator(s), 870–872
acid–base pairs, conjugate, 802, 803t
acid–base reaction(s) An exchange reaction between an acid and a base producing a salt and water, 191–196
characteristics of, 817t
equivalence point of, 217, 862
pH after, 824–826
titration using, 216, 862–872
acid ionization constant (K_a) The equilibrium constant for the ionization of an acid in aqueous solution, 807, 808t
determining, 866
relation to conjugate base ionization constant, 813
values of, A-21t
acidic oxide(s) An oxide of a nonmetal that acts as an acid, 190
acidic solution A solution in which the concentration of hydronium ions is greater than the concentration of hydroxide ion, 804
acidosis, 861
acrolein, formation of, 433
structure of, 430, 470
acrylonitrile, structure of, 100, 428
actinide(s) The series of elements between actinium and rutherfordium in the periodic table, 87, 349
activation energy (E_a) The minimum amount of energy that must be absorbed by a

system to cause it to react, 724
experimental determination, 727–729
reduction by catalyst, 730
active site, in enzyme, 535
activity (A) A measure of the rate of nuclear decay, the number of disintegrations observed in a sample per unit time, 1123
actual yield The measured amount of product obtained from a chemical reaction, 157
addition polymer(s) A synthetic organic polymer formed by directly joining monomer units, 513–517
production from ethylene derivatives, 514t
addition reaction(s), of alkenes and alkynes, 490
adduct, acid–base, 829
adenine, hydrogen bonding to thymine, 537–538, 603
structure of, 136
adenosine 5'-triphosphate (ATP), 541
adhesive force A force of attraction between molecules of two different substances, 614
adhesives, 654
adipoyl chloride, 518
adrenaline, 833
aerobic fermentation, 497
aerogel, 652
aerosol, 687t
air, components of, 565t, 577t, 1004
density of, 560
environmental concerns, 1004–1007
fractionation of, 1052
air bags, 548, 556
alanine, 532

chlorine oxide, in chlorine catalytic cycle, 753

chlorine trifluoride, reaction with nickel(II) oxide, 586

chlorobenzene, structure of, 493

chlorocarbons, densities of, 57

chlorofluorocarbons (CFCs), 1008

chloroform, boiling point elevation and freezing point depression constants of, 676t

chloromethane, 273
  as refrigerant, 1008
  critical point of, 637
  molecular polarity of, 416, *417*

chlorophyll, 543, 1030

cholesterol, 3

chymotrypsin, 732t

cinnabar, 4, *5*, 1052

cinnamaldehyde, structure of, 470, 503

cisplatin, atomic distances in, 50
  discovery of, 7
  isomers of, 1089
  preparation of, 172, 230
  rate of substitution reaction, 709
  structure of, 132

cis-trans isomers, 454–455, 477, 729
  in coordination compounds, 1089

citric acid, 505t
  reaction with sodium hydrogen carbonate, 195
  structure of, *809*

Clapeyron, Émile, 612

Clark, L. C. Jr., 943

Clausius, Rudolf, 612

Clausius-Clapeyron equation, 612

clay(s), 651, 1041

cleavage, of crystalline solids, 626

cleavage reaction, enzyme-catalyzed, 535–536

coagulation, of colloids, 687, 1000

coal, air pollution from, 1006
  combustion of, 285
  impurities in, 285

coal gas, 278

coal tar, aromatic compounds from, 492t

cobalt, colors of complexes of, 1101t
  in alnico V, 1079

cobalt-60, gamma rays from, 329

cobalt blue, 128

cobalt(II) choride hexahydrate, 128

Cockcroft, J. D., 1128

codon A three-nucleotide sequence in mRNA that corresponds to a particular amino acid in protein synthesis, 539

coefficient(s), stoichiometric, 144, 762

cofactors, enzyme, 541

coffee, decaffeination with supercritical carbon dioxide, 614

coffee-cup calorimeter, 257

cohesive force A force of attraction between molecules of a single substance, 614

coke, in iron production, 1077
  water gas from, 1021

cold pack, 274

coliform bacteria, 999

collagen, 654

colligative properties The properties of a solution that depend only on the number of solute particles per solvent molecule and not on the nature of the solute or solvent, 659, 672–685
  of solutions of ionic compounds, 679

collision theory A theory of reaction rates that assumes that molecules must collide in order to react, 722–732

colloid(s) A state of matter intermediate between a solution and a suspension, in which solute particles are large enough to scatter light but too small to settle out, 686–690
  surface charge on, 1000
  types of, 687t

color(s), acid–base indicators, *872*

coordination compounds, 1097–1102

fireworks, 294

flowers, 848

light-emitting diodes, 648

"neon" signs, 331

transition metal compounds, 1071

visible light, 299, 1097

combined available chlorine, 1003

combined gas law. *See* general gas law.

combustion analysis, determining empirical formula by, 162

combustion calorimeter, 259

combustion reaction The reaction of a compound with molecular oxygen to form products in which all elements are combined with oxygen, 146
  of fossil fuels, 285

common ion effect The limiting of acid (or base) ionization caused by addition of its conjugate base (or conjugate acid), 850–853
  solubility and, 879–882

common logarithms, A-2

common names, 485
  of binary compounds, 115

compact disc player, light energy in, 304

complementary strands, in DNA, 538

completion, reaction going to, 766

complex(es), 829. *See also* coordination compound(s).
  formation constants of, 888, A-26t
  solubility and, 887–890

composition diagram(s), 900

compound(s) Matter that is composed of two or more kinds of atoms chemically combined in definite proportions, 18
  binary, naming, 115
  coordination. *See* coordination compound(s).
  determining formulas of, 121–128
  hydrated, 128, 595, 1080
  intermetallic, 645
  ionic, 103–114
  molecular, 114–116
  naming, 111
  nonexistent, 381

specific heat capacity of, 243t

standard enthalpy of formation of, 265–269, 266t

compressibility The change in volume with change in pressure, 550

computers, molecular modeling with, 102

concentration(s) The amount of solute dissolved in a given amount of solution, 206
  effect on equilibrium of changing, 783
  graph of, determining reaction rate from, 702
  in collision theory, 723
  in equilibrium constant expressions, 760
  known, preparation of, 209–211
  of ions in solution, 205–211
  partial pressures as, 763–764
  rate of change, 700–704
  reaction rate and, 705–712
  units of, 659–662

conch, shell structure, 653

condensation The movement of molecules from the gas to the liquid phase, 607

condensation polymer(s) A synthetic organic polymer formed by combining monomer units in such a way that a small molecule, usually water, is split out, 513, 517–519
  silicone, 1042

condensed formula A variation of a molecular formula that shows groups of atoms, 99, 478

condition(s), standard. *See* standard state.

conduction band, 646

conductor(s), band theory of, 643

conformations of cyclohexane, 487

conjugate acid–base pair(s) A pair of compounds or ions that differ by the presence of one H^+ unit, 802, 803t
  in buffer solutions, 854

conservation of, 237, 249, 252

density, in batteries vs. gasoline, 962t

direction of transfer, 238

dispersal of, 906

forms of, *235*, 236

internal, 252

ion pair formation, 378

ionization. *See* ionization energy.

kinetic. *See* kinetic energy.

lattice, 379–381

law of conservation of, 237, 249, 252

levels in hydrogen atom, 308

mass equivalence of, 1120

potential. *See* potential energy.

quantization of, 307, 316, 909

sign conventions for, 243

sources of, 282–293

state changes and, 246–249

temperature and, 237

units of, 240, A-8

energy level diagram, 262

enthalpy change (Δ*H*) Heat energy transferred at constant pressure, 253, 905

as state function, 254

for chemical reactions, 254–257

enthalpy of fusion (Δ*H*_{fusion}) The energy required to convert one mole of a substance from a solid to a liquid, 627, 628t

enthalpy of hydration, 592

enthalpy of solution (Δ*H*_{soln}) The amount of heat involved in the process of solution formation, 666–669

enthalpy of vaporization The quantity of heat required to convert 1 mol of a liquid to a gas at constant temperature, 246, 606

intermolecular forces and, 594

values of, A-15t

entropy (*S*) A measure of the disorder of a system, 906

molecular structure and, 913

origin of life and, 931

phase and, 912

second law of thermodynamics and, 917

solution process and, 663

standard molar, 912, 913t

time and, 932

entropy change (Δ*S*), equation for, 912

of reaction, 915

of universe, system, and surroundings, 917

environment, chemistry of, 998–1011

enzyme(s) A biological catalyst, 535, 698–699

theory of 732

enzyme cofactors, 541

ephedra, 836

ephedrine, structure of, 136, 836

epinephrine, 833

structure of, 434

Epsom salt, formula of, 135

equation(s), activation energy, 727–729

activity of nuclear decay, 1123

Arrhenius, 727

Bohr, 308

Boltzmann, 911

bond order, 419

Boyle's law, 551

buffer solution pH, 856

Celsius-Kelvin scale conversion, 28

Charles's law, 553

chemical reactions, 24, 142

Clausius-Clapeyron, 612

Coulomb's law, 112

Dalton's law, 564

de Broglie, 313

dilution, 210

Einstein's, 1120

enthalpy change of reaction, 256

entropy change, 912

of reaction, 915

equilibrium constant and standard free energy change, 928

equilibrium constant expression, 762

first law of thermodynamics, 251

formal charge, 405

free energy change at nonequilibrium conditions, 928

general gas law, 554

Gibbs free energy, 921

Graham's law, 572

half-life, 719

heat and temperature change, 242

Henderson-Hasselbalch, 856

Henry's law, 669

Hess's law, 261

ideal gas law, 557

integrated rate, 712–722

ionization constant, for acids and bases, 807

for water, 804

kinetic energy, 568

Maxwell's, 569

Nernst, 975

net ionic, 183

nuclear reactions, 1111

osmotic pressure, 683

pH, 805

Planck's, 300–305

pressure–volume work, 253

quadratic, 774–775, A-4

Raoult's law, 673

rate, 701, 707

reaction quotient, 928

Rydberg, 307

Schrödinger, 316, 909

second law of thermodynamics, 917

speed of a wave, 297

standard free energy change of reaction, 924

standard potential, 965

van der Waals, 576

equatorial position, in cyclohexane structure, 487

in trigonal-bipyramidal molecular geometry, 401

equilibrium A condition in which the forward and reverse reaction rates in a physical or chemical system are equal, 756–795

dynamic, 760

factors affecting, 781–786

in osmosis, 683

in reaction mechanism, 739

Le Chatelier's principle and, 671, 781–786

reversibility and, 914

solution process as, 662

successive, 888

thermal, 238

weak acid ionization, 798

equilibrium constant (*K*) The constant in the equilibrium constant expression, 760–766

calculating from initial concentrations and pH, 818

calculating from standard potential, 979

calculations with, 772–777

concentration vs. partial pressure, 763–764

determining, 770–772

meaning of, 765

product-favored versus reactant-favored reactions, 765

relation to reaction quotient, 767

relation to standard molar free energy change, 928–932

simplifying assumption in, 774, 820, A-5

values of, 765t

weak acid and base (*K*_a and *K*_b), 806–813

equilibrium constant expression A mathematical expression that relates the concentrations of the reactants and products at equilibrium at a particular temperature to a numerical constant, 762

gases and, 763–764

reverse reaction, 778

stoichiometric multipliers and, 777–780

equilibrium vapor pressure The pressure of the vapor of a substance at equilibrium in contact with its liquid or solid phase in a sealed container, 609–612

in phase diagram, 630

equivalence point The point in a titration at which one reactant has been exactly consumed by addition of the other reactant, 217

of acid–base reaction, 862

error The difference between the measured quantity and the accepted value, 34

ester(s) Any of a class of organic compounds structurally related to carboxylic acids, but in which the hydrogen atom of the carboxyl group is replaced by a hydrocarbon group, 502, 507–509

hydrolysis of, 508

naming of, A-18

PHYSICAL AND CHEMICAL CONSTANTS

Avogadro's number	$N = 6.02214155 \times 10^{23}/mol$
Electronic charge	$e = 1.60217653 \times 10^{-19}$ C
Faraday's constant	$F = 9.6485338 \times 10^{4}$ C/mol electrons
Gas constant	$R = 8.314472$ J/K · mol
	$= 0.082057$ L · atm/K · mol

π	$\pi = 3.1415926536$
Planck's constant	$h = 6.6260693 \times 10^{-34}$ J · sec
Speed of light (in a vacuum)	$c = 2.99792458 \times 10^{8}$ m/sec

USEFUL CONVERSION FACTORS AND RELATIONSHIPS

Length
SI unit: Meter (m)

1 kilometer = 1000 meters
$\qquad$ = 0.62137 mile
1 meter = 100 centimeters
1 centimeter = 10 millimeters
1 nanometer = 1.00×10^{-9} meter
1 picometer = 1.00×10^{-12} meter
$\qquad$ 1 inch = 2.54 centimeter (exactly)
1 Ångstrom = 1.00×10^{-10} meter

Mass
SI unit: Kilogram (kg)

1 kilogram = 1000 grams
$\qquad$ 1 gram = 1000 milligrams
$\qquad$ 1 pound = 453.59237 grams = 16 ounces
$\qquad$ 1 ton = 2000 pounds

Volume
SI unit: Cubic meter (m^3)

1 liter (L) = 1.00×10^{-3} m^3
$\qquad$ = 1000 cm^3
$\qquad$ = 1.056710 quarts
1 gallon = 4.00 quarts

Energy
SI unit: Joule (J)

1 joule = 1 kg · m^2/s^2
$\qquad$ = 0.23901 calorie
$\qquad$ = 1 C $\times$ 1 V
1 calorie = 4.184 joules

Pressure
SI unit: Pascal (Pa)

1 pascal = 1 N/m^2
$\qquad$ = 1 kg/m · s^2
1 atmosphere = 101.325 kilopascals
$\qquad$ = 760 mm Hg = 760 torr
$\qquad$ = 14.70 lb/in^2
$\qquad$ = 1.01325 bar
1 bar = 10^5 Pa (exactly)

Temperature
SI unit: kelvin (K)

0 K = −273.15 °C
K = °C + 273.15°C
? °C = (5 °C/9 °F)(°F − 32 °F)
? °F = (9 °F/5 °C)(°C) + 32 °F

LOCATION OF USEFUL TABLES AND FIGURES